ARCTIC SYSTEMS

NATO CONFERENCE SERIES

I   Ecology
II   Systems Science
III   Human Factors
IV   Marine Sciences
V   Air—Sea Interactions

II SYSTEMS SCIENCE

ARCTIC SYSTEMS

Edited by

P. J. Amaria and A. A. Bruneau
Memorial University of Newfoundland
St. John's, Newfoundland, Canada

and

P. A. Lapp
Philip A. Lapp Ltd.
Toronto, Ontario, Canada

Published in coordination with NATO Scientific Affairs Division
PLENUM PRESS · NEW YORK AND LONDON

Library of Congress Cataloging in Publication Data

Main entry under title:

Arctic systems.

(NATO conference series: II, Systems science; v. 2)
"Proceedings of a conference . . . held at the Memorial University of Newfoundland
in St. John's, Newfoundland, Canada, August 18-22, 1975, sponsored by the Nato
Special Program Panel on Systems Science."
Includes index.
1. Arctic regions—Congresses. 2. Engineering—Cold weather conditions—Congresses.
3. Systems engineering—Congresses. I. Amaria, P. J. II. Bruneau, Angus A. III. Lapp,
Philip A. IV. Nato Special Program Panel on Systems Science. V. Series.
G600.5.A73   919.8   77-3871
ISBN 978-1-4684-0801-0

ISBN 978-1-4684-0801-0   ISBN 978-1-4684-0799-0 (eBook)
DOI 10.1007/978-1-4684-0799-0

Proceedings of a conference on Arctic Systems held at the Memorial University of
Newfoundland in St. John's, Newfoundland, Canada, August 18—22, 1975, sponsored
by the NATO Special Program Panel on Systems Science

© 1977 Plenum Press, New York

Softcover reprint of the hardcover 1st edition 1977

A Division of Plenum Publishing Corporation
227 West 17th Street, New York, N.Y. 10011

Preface

For the purpose of publication of these Proceedings, the original conference programme has been rearranged to provide a more logical sequence of presentation. The beginning sections give the inaugural speech and the six keynote addresses which were delivered at the opening plenary session. Following these are the working papers, published more or less in the same sequence in which they were presented in the original programme. The order of presentation does not necessarily emphasise the importance of any one aspect of the Arctic Systems over others.

The final reports of the six working groups and their conclusions and recommendations are edited in such a manner as to present them in a standardised format for easy comprehension. The editors accept responsibility for any distortion inadvertently introduced in the summarising and editing processes.

Later sections of the Proceedings give a background to the Conference organization and deliberations, and an independent critique of the meeting.

The directors and those who attended the Conference were conscious of the debt of gratitude owed by them to the Conference chairmen, rapporteurs, authors of working papers, and many individuals for their contributions to the success of the meeting. We wish to thank them and it is a pleasure to record their names in these Proceedings.

Inaugural Speaker
>   Dr. J. Rennie Whitehead, Canada

Banquet Guest Speaker
>   Honourable Mr. T. Alex Hickman, Canada

Keynote Addresses
>   Mr. C. Bornemann, Denmark
>   Dr. A.E. Collin, Canada
>   Dr. R.E. Francois, U.S.A.
>   Mr. W. Ganong, Canada
>   Mr. L. Nitzki, Federal Republic of Germany
>   Mr. M. Turner, Canada

Chairmen
     Professor D. Clough, Canada
     Dr. R. E. Francois, U.S.A.
     Brigadier General K. Greenaway, Canada
     Dr. J. Keys, Canada
     Dr. N. Orvik, Canada
     Mr. A. E. Pallister, Canada
     Mr. E. Reimers, Norway
     Dr. H. Schroeder-Lanz, Germany
     Captain R. White, U.S.A.

Rapporteurs
     Dr. D. Bajzak, Canada
     Dr. D. Dunsiger, Canada
     Mr. J. English, Canada
     Mr. D. Grenville, Canada
     Dr. R. Peters, Canada
     Mr. H. Snyder, Canada

Editorial Assistance
     Mr. B. LeDrew, Centre for Cold Ocean Resources Engineering,
        Memorial University of Newfoundland, St. John's, Canada

Conference Social Hostesses
     Dr. Roda Amaria
     Mrs. Daphne Bajzak
     Mrs. Jean Bruneau
     Mrs. Jane Dunsiger
     Mrs. Ruth English
     Mrs. Riva Mackee
     Mrs. Pauline Newbury
     Mrs. Elizabeth Peters
     Mrs. Adelle Snyder
     Mrs. Grace Ward

Technical Assistants
  Audio Visual
     Mr. Tony Howell
     Mr. Neil Riggs
     Mr. Bob McIsaac
     Dr. T. Swami

  Educational Television (ETV)
     Mr. Andy Wells

  Tape Recording
     Mr. Rick Harris, Eastern Audio, St. John's

  Graphic Art Designs
     Mrs. Hanny Muggeridge
     Mrs. Gita Reddy
     Mrs. Jane Dunsiger

Thanks are also due to the NATO Special Program Panel on Systems Science and the NATO Science Committee who were the sponsors of the Conference, and to Mobil Oil Ltd. of Canada; Churchill Falls (Labrador) Corporation Ltd.; Iron Ore Company of Canada; the Ministry of State for Science and Technology, Government of Canada; and Memorial University of Newfoundland, who provided supplemental financial assistance towards the organization of the Conference.

We also gratefully acknowledge the assistance and cooperation of many members of the administrative, technical, planning and works, and security departments of the Memorial University of Newfoundland who provided effective staffing during the Conference, and of our secretary, Mrs. Margaret Rose, who not only assisted us during the planning stages but also typed the final manuscript of the Proceedings.

Contents

WORKING PAPERS

CONTENTS

WORKING SESSION CHAIRMEN'S REPORTS

INAUGURAL ADDRESS

NEED FOR INTERNATIONAL COOPERATION IN THE ARCTIC

J. Rennie Whitehead

Canadian Member, NATO Science Committee

Ottawa, Ontario, Canada

Ladies and Gentlemen,
    It is my pleasant duty to welcome you on behalf of NATO and
the NATO Science Committee, to this Conference on Arctic Systems:
Pleasant in several ways: First because it brings me to
Newfoundland with time perhaps to see a little of this beautiful
province. My only previous visit was a stopover of 3/4 hour at
Botwood between a 17 hour 20 minute flight from Shannon to Botwood
and an 11 hour flight from Botwood to Baltimore in a Boeing Clipper
Flying Boat in 1944. Secondly it is pleasant because it brings me
to Memorial University which is graciously acting as host for the
Conference. Thirdly, as the Canadian Member of the NATO Science
Committee for the past five years, it enables me to conform to a
policy of that Committee which I helped to formulate - that at
least one member of the plenary Committee should be present at
each of the major functions it sponsors, and if possible give some
account of the work of the Committee. Finally it gives me the
opportunity as a non-expert to make my small introductory
contribution to the substance of the Conference itself.
    The NATO Science Committee which sponsors this Conference,
has been financing a wide range of scientific activities for
seventeen years, since its first meeting in 1958. It was created
in recognition of the need to improve and extend cooperation
between the member nations in non-military fields. All its
activities are published and have been throughout its life.
Moreover, as you see today it is the policy not to limit attendance
at advanced study institutes and conferences such as this to
scientists from member countries and it gives me particular
pleasure today to welcome participants from several non-NATO
countries including Japan and the Soviet Union.

I thought I might say a few words about the objectives and nature of the NATO Science Committee and the scope of its activities.

With regard to objectives I can do no better than quote from an excellent publication "NATO and Science" which I recommend you to seek out and read.

"Since the dawn of history man has displayed both vast curiosity and innate ingenuity. When these two basic human characteristics are combined successfully, the result is what we call progress. Much of modern man's progress is measured by developments in science and technology.

In the past years authorities, the public, and the scientists have expressed growing concern over the unexpected problems facing modern, western societies and this has influenced science policies in the direction of research for immediate application. Science should obviously contribute to solving today's problems and to lessening their detrimental effects. But long-term needs are unlikely to be served only by narrow applied research: our dynamic society will continuously create new situations in which today's problems will be succeeded by quite different and unknown problems. Basic research, by its nature, deals with the unknown. Its fruits cannot be determined until the seed has been planted, nurtured and allowed to flower.

Thus, to seek relevance to immediate, specific problems in contrast to the basic search for new knowledge does not solve our problems of the future. A delicate balance must be maintained between the application of present knowledge and the creation of new knowledge for the unknown needs of the future. The Alliance is not satisfied by encouragement of applied research only, nor by consideration of just today's problems, however pressing they might seem.

The Science Committee believes that support of long-term research is an important investment for the future, and one which must be planned systematically. Active international cooperation ensures that any significant advance in basic science in any member country will, in a usefully short time, be shared by a sufficient number of scientists in all other countries.

In a world which demonstrates a growing tendency toward rigid economic justification of efforts in science and technology, and even an animosity toward continuing basic search for knowledge, the Science Committee is a staunch supporter of the long-range value of pure research guided by individual ingenuity and curiosity."

The Committee, which is a plenary committee of NATO, chaired by the Assistant Secretary-General for Scientific Affairs, "is composed of national representatives highly qualified to speak authoritatively on science policy in the name of their governments" but also, in practice to exert their judgements as individual scientists. The Committee has been fortunate over the years to

have members of the quality of "Ned" Steacie, late President of
NRC, Louis Néel of France and Isidore Rabi of the U.S., both Nobel
prize-winners who still serve, Eduard Pestel of F.R.G., - well
known for his joint authorship with Mihajlo Mesurovic of the book
Mankind at the Turning Point and Professor Ozdas of Turkey, who was
a member for many years and is now chairman in his capacity as
Assistant Secretary-General of NATO.

The Committee which meets three times a year - twice in
Brussels and once as guest of a Member Country-finances several
major programmes, including Science Fellowships, Advanced Study
Institutes, Conferences, Research Grants, Visiting Professorships,
Senior Research Fellowships and a number of special scientific
programmes.

The Science Fellowships, financed by the Committee, are
administered by a national agency under the general supervision of
the Science Committee and a Sub-Committee on Science Fellowships.
In Canada the agency is the National Research Council which, of
course also administers the National Fellowships scheme.

The main purpose of the programme is to stimulate countries
to enlarge the exchange of post-graduate and post-doctoral students
of the pure and applied sciences and thus to increase the mutual
scientific strength. Some $2.5 million are distributed each year
under this scheme, or a total of about $40 million since the
Committee was formed. The number of fellowships is about 600 a
year, and 90% are given to scientists under 35 years of age.

A portion of the funds does however go to encourage the
exchange of senior scientists and a small standing group of the
NATO Science Committee, on which I have served in recent years,
deals with the applications for Visiting Professorships, Senior
Fellowships and Science Lectureships for which funds are not
available under other NATO Science programmes.

The Advanced Study Institutes are generally summer schools at
which a carefully defined subject is presented in a systematic and
coherently-structured programme. The subject is treated in
considerable depth by the best available experts for the benefit
of other scientists who have already specialised in the same field
or possess an advanced general background. About 50 Advanced
Study Institutes are supported each year spread over several
countries and many disciplines. For instance the 1975 ASI's alone
cover such widely diverse subjects as gravitational waves, molecular
and cell biology and cardiovascular flow, and are held in locations
in England, France, Italy, Denmark, Germany, Belgium, Scotland,
Sicily, Canada, U.S.A., Norway and Corsica, to mention just a few.
It is our policy to keep the number of participants low, confined
to those who can contribute or benefit directly. However,
qualified lecturers and students are accepted from all over the
world.

Conferences, such as this one, differ from Advanced Study
Institutes. They are held in areas which the Science Committee

feels are in special need of development and are normally limited
to one or two a year. Conferences are regarded as working meetings
with all participants playing an active, contributory role. They
serve to identify particularly fruitful areas for research and to
make recommendations both to those who have a responsibility for
selecting and supporting research programmes and to the Science
Committee itself as a guide to the allocation of human and material
support.

The NATO Science Committee also has a programme of research
grants to stimulate, encourage and facilitate scientific research
carried out in collaboration by scientists working in different
member countries and thus to promote the flow of ideas and methods
across national boundaries. Almost all fields of science are
eligible for the grants, with emphasis on fundamental aspects
rather than application and on "small" science rather than "big"
science, because of the limitation of funds. The Grants are
administered by a special Panel of the Science Committee which
consists of scientists in a number of different fields. The
grants are awarded with due regard not only to scientific excellence
but timeliness, novelty and enthusiasm. They are grants in aid of
specific projects which are generally largely supported by
national funds but where international collaboration entails costs
which are not so easily met from these sources.

Finally, on the subject of the sponsoring Committee, it has
Special Programme Panels which change from time to time, but which
have covered, over the years, subjects such as Oceanography,
Meteorology, Human Factors, Operational Research, Radiometeorology,
Eco-Sciences and Air-Sea Interaction. These Advisory Panels, of
which there are currently some six or seven, generally have six to
eight members with three year terms and membership rotates among
the member countries. The panels are involved in stimulating
activities in their specialist areas, making recommendations and
planning cooperative research. They are often the source of
initiative for an Advanced Study Institute or Conference. For
instance, this Conference on Arctic Systems involved the Special
Programme Panel on Systems Sciences, with which George Lindsey of
DND and John Gratwick of CN and our Chairman Don Clough have been
associated.

Arctic Systems was a subject of which the importance was
readily recognised by the NATO Science Committee when this
Conference was first proposed. The use of the word "systems" in
this context, emphasises the international nature of the problem,
because the systems of the Arctic impinge on a number of bordering
countries and have implications for most countries of the world.

The Arctic is one of the planet's great natural systems and
the major phenomena associated with it all transcend national
boundaries. There are continuous geological formations across
the Arctic basin which form the relatively static component of the

system. But it is a dynamic, even turbulent system. The ice and current are in constant motion - what is off Siberia this year is off Greenland the next. The wildlife, such as whales, polar bears and seals, is wide-ranging and does not recognise frontiers.

Looked at from the South, as it usually is, the Arctic is, to most people in the countries which border it, a vague area somewhere to the North. But if you look at it from the top, you see it as an enormous ocean which raises unique problems which are largely characteristic of "it" rather than of any particular part of its perimeter. Moreover, this great area is the great international weather-kitchen in which are brewed many of the constituents of the global climate. Large variations of global climate occur over the years. They often correlate with variations in the extent of ice in the Arctic (and, of course, the Antarctic). Effects of these variations can be seen in alteration of sea levels at our shores, changes in the location and yield of commercial fisheries, and in the yield of inland crops such as grain and timber. The feasibility of the northward extension of agriculture and forestry depends on the ability to predict these variations.

But the basic causes of climatic variation are not well understood. Research on a large scale in the polar region as a whole is fundamental to an understanding of the role of the Arctic as a weather-kitchen.

The Arctic is also a home for a great number of natural species of bird, animal and fish. Fishing and hunting have been a way of life there since time immemorial. But it is only relatively recently that the relevant governments have introduced conservation measures. And it requires intensive international effort to make these measures consistent and to keep them so.

As man increasingly exploits the resources of the North to his economic benefit, even greater international effort is required to conserve the environment from harmful changes which result from his activities.

Oil spills associated with the transportation of oil out of the North are a major hazard and there is controversy over the long-term atmospheric effects of high-flying jets, of which many cross the Polar regions every day. Indeed both for sea and air transport the shortest routes between some of the highly-populated areas of the world, lie across the Arctic, and a considerable increase of the use of these routes for commercial purposes can be expected as more and more of them become technically possible. I am pleased to see that a significant part of the programme of the Conference will be devoted to this subject. The evaluation of alternative methods of transporting resources from the North, including pipelines, surface transport, submarines, seems to me to be particularly important and I see you have made provision for it.

The extension of telecommunications in the Arctic has already had an effect on social conditions, which will, no doubt, increase with time. I recall a deep awareness of this during the early

discussions on whether or not we should have a domestic
communications satellite. Today the effects can already be seen
and the technique is now available to use wisely in ways which
benefit rather than degrade the cultural environment.

    It seemed to me when I toured the Canadian Arctic in 1966 with
the then Minister of Northern Affairs, the late Arthur Laing - it
seemed to me, as an instant expert, that far too little thought
had been given to the design of northern communities - or to
individual houses for that matter. The layout of Inuvik looked
suspiciously like that of any small southern Canadian town. I am
sure other countries have given thought to the design of efficient
low-energy dwellings and of their arrangement into a logical
workable agreeable community characteristic of the northern
environment. The social and cultural implications of housing and
community systems technology are great. And they are common to
all peoples of the North because in the history of the Arctic not
only animals have migrated. I recall on the Northern tour to
which I referred, I found myself seated next to the Japanese
Ambassador in the DC-4 aircraft in which we flew for a total of 48
hours around the Western Arctic. A Canadian Eskimo stewardess
joined us in Churchill. The Japanese Ambassador expressed surprise
to me that a Canadian Airline would recruit a stewardess from
Hokkaido - he was quite certain she was from the Northern Islands
of Japan, nor would he be corrected until he had been assured by
the girl herself that she was Canadian born of Canadian Eskimo
ancestry and really couldn't speak Japanese!

    People must have built-in instincts or characteristics which
make them feel at home in a certain latitude and which encourage
migration along parallels of latitude. I was born on latitude 54°N
and I still feel more "at home" in the mid-fifties than in the mid-
forties where I am condemned to live. Or is this merely a prejudice
against living in Ottawa?

    Beyond all the practical problems of the Arctic is its
perennial challenge to men with courage, vision and scientific
curiosity. All countries with a northern perspective have common
aspirations for the North and people who recognise this challenge.
Rapid progress in the solution of most of the problems which will
preoccupy you here in St. John's, cannot be envisaged on a nation
to nation basis, but can only be foreseen as a result of concerted
international effort. You in this Conference, from many countries,
have the vision to look at the total dynamic system of the Arctic -
at the movement of national and man-made phenomena, machines and
messages and their economic social and environmental implications.
You have the techniques - at least many of them - to develop
options for action which governments might take and to forecast in
many cases the benefits which might accrue and the consequences of
those actions on the total Arctic system. Yes, out of this work
you may be able to develop realistic policy options. But this
effort may be to no avail unless the options are taken up on a

concerted international basis.

The problems of creating and achieving acceptance of policies relating to the Arctic system are serious and urgent. The real limits are neither scientific nor technological, but rather the familiar problems associated with international collaboration, which are more of a political and managerial nature. As David Judd[1] wrote to me before I came here: "if we are to cope with, understand and predict the behaviour and nature of these arctic phenomena which affect us all, we must look not only to our technology, but to our man-made systems of government, regulation and consultation which are becoming less and less adequate to meet human needs."

Fortunately there is a fine tradition of international cooperation between scientists in the Arctic. We should do everything we can, domestically and internationally to improve the formal and informal mechanisms for international cooperation in other sectors until it reaches the level already achieved by scientists.

On behalf of the NATO Science Committee and the NATO Secretariat, and on my own behalf, I should like to extend good wishes for an enjoyable and productive Conference. I should like to thank the General Conference Chairman Don Clough, the Directors Angus Bruneau and Phil Lapp and the Administrative Director Professor Amaria, for giving me this opportunity to welcome you.

1. An official of the Ministry of State for Science and Technology, Ottawa, Ontario, Canada.

ARCTIC ENVIRONMENTAL MANAGEMENT SYSTEMS

A. E. Collin

Environment Canada

Ottawa, Ontario, Canada

There is a line in the scientific literature of the Arctic which seems to be particularly appropriate to the theme of this meeting; it was written in 1928 by Dr. Diamond Jenness, the pioneer Canadian anthropologist who ended his work, the "People of the Twilight," by asking:
"Were we the harbingers of a brighter dawn, or only messengers of ill-omen, portending disaster?"
This quotation seems relevant now more than ever before, as exploration and exploitation of oil and gas forces us to seek a balance between the needs of the northerners on the one hand, and the resource requirements of Canada on the other. Clearly, we are at a critical time, for within the next few years we will have to make decisions that will influence the pattern of northern development for the foreseeable future.

The first part of the 1973 National Oceans Policy calls for Canada to "achieve world-recognized excellence in operations on and below ice-covered waters within five years." The geographic areas involved, of direct interest to this Conference, include the Beaufort Sea, the Arctic Archipelago, Foxe Basin, Baffin Bay and Davis Strait, the Labrador Shelf, and the waters of our eastern seaboard. The most important activity in these areas, with the possible exception of the Labrador Shelf, is, or will be, oil and gas exploration and exploitation. Towards the orderly execution of this major resource development several federal departments are increasing their level of operations in northern Canada. They include: Energy, Mines and Resources (EMR); Environment (DOE); Transport (MOT); Indian and Northern Affairs (DINA); National Research Council (NRC); National Defence (DND); and Communications (DOC).

The Department of Energy, Mines and Resources is conducting
remote sensing, geological and mineral assessment, and research
throughout the North. The Canada Centre for Remote Sensing
provides data and imagery on Canadian earth resources and a number
of environmental features. The Polar Continental Shelf Project
(PCSP) has been studying the shelf north of the arctic mainland
and the sea floor throughout the islands during the last fifteen
years, and, of course, the Geological Survey of Canada has been
very active in the north for much of its long history.

The Department of the Environment is also active in the north.
The Atmospheric Environment Service is responsible for data and
research on sea ice and meteorology and the Fisheries and Marine
Service is responsible for all the hydrographic charting and
oceanographic research undertaken by the government in these
regions.

Within the Department of Indian and Northern Affairs,
resource-oriented and cultural programs coexist. The Indian and
Eskimo Affairs Program is a people-directed activity, while the
Northern Economic Development Branch's Oil and Mineral Division
controls permits for the exploration and exploitation of all
minerals north of 60°.

The Department of National Defence performs non-military
functions such as search and rescue flights, surveillance of arctic
land and water, and flights in support of programs and requirements
for other departments, as well as its military tasks.

The National Research Council deals with three main areas of
northern involvement - the division of Building Research, Space
Research Facilities, and the Ship and Marine Dynamics Laboratory,
all of which have a direct interest in arctic technology.

Three groups within the Ministry of Transport are heavily
involved in arctic development: the Arctic Transportation Agency
(ATA), the Canadian Marine Transportation Administration (CMTA),
and the Transportation Development Agency (TDA). The Arctic
Transportation Agency controls transport infrastructure development
in the Northwest Territories. It is a coordinating body, bringing
together industry and the territorial government. The Canadian
Marine Transportation Administration coordinates movement of cargo
and personnel to and from the north. To that end, it provides
icebreaker support and supervised design of new icebreakers. New
aids to navigation for arctic shipping are also provided by the
Canadian Marine Transportation Administration.

The great distances between settlements and between the arctic
and southern Canada place communications high on the list of
priorities for northern development. From its establishment in
1969, the Department of Communications has been active in the
orderly development and operation of telecommunications in Canada.
Among its most effective and dramatic contributions has been the
launching and use of the telecommunications satellites ANIK I,
II, and III. Thus, one measure of this success may be the fact
that by March 1974 twenty-seven settlements in the remote parts

of Canada were receiving colour TV pictures. Small, transportable earth receiving stations were successfully tested by oil companies in the north during 1974. Future plans call for increased service, better communications for remote settlements and exploration parties, as well as further satellite applications.

A notable example of recent government research activities in the north is the Beaufort Sea Project - a multi-agency, multi-disciplinary program funded in part by the oil industry for assessing environmental parameters and effects in advance of major developmental commitments. The program is the first industry-government joint enterprise in this direction. Government coordinates the program, maintains cost control, and has final authority over work specifications and contract awards. On the industry side, an Industry Project Manager and individual study coordinators work with government personnel in planning and monitoring the programs. Above all, agreement is sought and maintained on the aims, objects, and scope of the project.

The objectives of this work are to achieve a level of information sufficient for a knowledgeable decision to be made on applications for offshore exploratory drilling; to define seasonal and geographical sensitivities of the area so as to minimize impact of drilling activities; and to enable the design of a weather/sea-ice prediction system in support of the operation of drill rigs and drill ships.

The critical decision-making period of the next five years holds, as well, for oil exploration in other areas of the north - particularly Baffin Bay and Lancaster Sound.

Thus, one may look upon the Beaufort Sea Project as an important step toward the direct application of a short-term, complex research undertaking to ocean resource development decisions. This project comes to an end this year and a number of critical decisions concerning exploratory drilling in the Beaufort Sea will have to be made early in 1976.

Mineral exploitation is continuing to develop in arctic Canada. Estimates point toward a reserve of seven billion tons of ore, the majority of which is iron and lead-zinc ore. However, only 3% of that total has so far been earmarked for exploitation - at Strathcona Sound, Baffin Island, and on Little Cornwallis Island.

As a measure of government response to arctic shipping priorities, two R-class icebreakers are now being constructed, in order to provide better support for arctic resupply. Further, a $35 million ice-strengthened cargo vessel is to be developed as part of the government's Transportation Policy; perhaps a forerunner of a series of vessels capable of operating for longer periods and able to reach more-distant ports when the "nominal" shipping season closes for conventional vessels.

Although industry is financing much of the Beaufort Sea Project and has provided financial and logistic support for other government operations, a large part of our northern activities is

totally government funded and managed. Without Ottawa's presence,
most scientific, technical and socio-economic programs would never
become reality. One example is the work conducted by the Geological
Survey of Canada. It is clear that were it not for Government
support most of these programs would never happen; local surveys
for petroleum and mineral exploitation purposes would occur, but
this would still leave much of our north relatively unknown.

One wonders, then, at the potential combined effects of
northern development.

The northern marine ecosystem is not likely to tolerate man-
induced changes graciously. One reason for this is the capacity
of arctic waters to disperse pollution. Within the Archipelago,
depths are not great and the circulation is such that exchanges
with the adjacent oceans are small compared to oceanic circulations.
Also, temperatures are near freezing, thus biological levels are
not high. Another factor on the physical side is the presence of
ice at sea which serves to restrict one's ability to do something
about pollution in the ocean while at the same time limiting the
exchange which can take place between the ocean and the atmosphere
at the sea surface. However, the presence of sea ice is not
likely to present overwhelming problems for all time. In fact,
there are considerable possibilities for sea ice to function in
useful ways - ice "islands" for offshore drill rig emplacement,
offshore vessel docking locations, and ocean air strips are obvious
examples.

The geography of the Arctic Ocean leads to the realization
that a major environmental event is likely to be felt throughout
the arctic region within a relatively short time. For example,
nuclear fallout generated by the Soviet Union during the 1960's
was detected in the boreal zone within the next few years.
Similarly, in the event of a serious oil spill on the arctic coast,
the effects would likely be felt by other circumpolar countries
within a short time, as the coastal currents would tend to be the
dominant factor in the spread of the oil. We are therefore faced
with important international considerations when dealing with
environmental issues in the Arctic Ocean.

One of the crucial problems before us, therefore, is that of
oil in northern waters. Most available information confirms the
fear that oil spill detection, containment, and cleanup will be a
very difficult problem to solve. Dispersal in ice-covered or ice-
infested waters is bound to be ineffective. Burning may remove
some surface oil. However, as a result of some of the Beaufort Sea
work we now know something of the behaviour of oil under ice and
it may be that this is not the insurmountable problem that it once
was thought to be, since the ice may in fact serve to confine the
spill. On the other hand, major oil spills on arctic shorelines
remains a very serious problem.

In the quest for a better understanding of the physical nature
of the Arctic, Canada is a participant in two major international

expeditions focusing on the arctic environment. The POLEX (Polar Experiment) will be managed by scientists from the United States, Canada, the USSR, and Norway. It began in 1964 through informal discussions concerning research on energy exchange between atmosphere and ocean in high latitudes. The program is to run concurrently with the Global Atmospheric Research Program and is to commence in 1977. However, the final stage of Canadian participation in the POLEX program is not yet determined.

Since 1970, the U.S. and Canada have been involved in a joint study of the pressure characteristics of sea ice in the Arctic Ocean. This project, AIDJEX (Arctic Ice Dynamics Joint Experiment), is scheduled for completion in April, 1976. Among the major programs of this project are arctic meteorology, physical oceanography, remote sensing, radiation, and geomagnetism.

In order for us to come to grips with a comprehensive approach to northern management systems, the demands of the natural environment must be balanced against the realities of economic and cultural development. To move quickly and flexibly when faced with sudden requirements, even in the absence of a fully detailed long-term policy directive, must remain a goal of our research programs. The present system of government in the north - whether situated in Ottawa, Whitehorse, Yellowknife, or Frobisher Bay - is forced to deal equitably with these problems of people, resources, and environment. To assist in solving this breadth of interests, two major interdepartmental committees serve to provide guidelines for management of the north.

The Advisory Committee on Northern Development (ACND) was established to advise the federal government on policy relating to civilian and military undertakings in northern Canada, and to provide federal government program coordination in the area. Subcommittees deal with specifics in transportation, science and technology, communications, and employment of native northerners.

The Task Force on Northern Oil Development was formed to advise the federal government on matters of national or regional importance, as they were affected by northern oil exploration. Studies and recommendations are made regarding oil and gas exploration, pipeline, marketing, and shipping. The subcommittee on environmental-social problems has been dealing with anticipated questions concerning environmental and sociological aspects of a Mackenzie Valley pipeline program, as well as those arising from construction of other pipeline routes.

The interrelationships among the various departments represented on these Task Forces and Committees are not particularly strong, and certainly do not represent the final word in Canadian policy. Rather, these bodies provide the opportunity for information exchange and senior level discussion.

Three general areas dominate governmental interest in the north: social aspects, environmental concerns and resource potential. The major objective of the social interest is to

provide a higher standard of living and a broadened field of
opportunity for native northerners. A variety of specific subjects
now receives much attention: health care, education, community
development, native land claims, political/social/administrative
development, transportation and communications. Guidelines for
the attainment of the goals of the above program areas stress a
number of requisites - assurances of participation in developmental
activities by native people, including prior consultation;
assurances that federally funded research projects in the north
are in full support of federal objectives generally - the
participation of Canadian universities and industries in federally
sponsored northern research; provision for adequate program review
and evaluation, both before and after completion; an adequate
system for handling research data; and the necessity of Canadian
control over all multilateral scientific programs carried out
wholly or partially in the Canadian north.

The environmental and scientific concerns of government stem
from the desire to expand northern development concurrently with
conservation and protection of the ecosystem. Therefore, a large
part of the scientific activities conducted by government and by
industry are undertaken with a view towards the design of
regulations which are likely to be required for environmental
protection. In this manner, subsequent developmental activities
will have been planned with environmental legislation and guidelines in mind.

Federal responsibilities are, for the most part, easily
recognized by looking at government activities; however,
inconsistencies between departments do exist. Search and rescue,
for example, is carried out both by National Defence and by
Transport. Another example is the jurisdictional cutoff at 60°N
for minerals and hydrocarbons - that is, south of that line, Energy,
Mines and Resources has sole responsibility for licensing, while
Indian and Northern Affairs is the responsible agency north of 60°N.
Within the responsibilities delegated to each department, outward
communication is not always easy. Behind all of the above is the
knowledge that political balances must be maintained. On one side
are government pronouncements regarding the need to protect the
environment and the duty to provide an intelligent framework for
northern development. On the other are the political necessities
of government operations and the pressure for resource potential.

Coming together, then, within the National Management
Situation are the goals and time limitations placed upon the
elected officials, and the scientific, technical, and socially
oriented bureaucracy of the federal civil service. The federal
Cabinet has a four-year timeframe within which to guide plans into
operating programs. However, the federal management system cannot
always confine itself to this timeframe. It must attempt to design
a management structure for the long term, while at the same time
responding to those immediate issues which are cast within the

shorter political interval. Thus, many federal programs are short-
term in nature while in fact serving as components of a much
larger longer-term approach. Development of policy, then, must
occur in this milieu of political expediency, longer-term
international or global considerations and economic reality.

The Department of Indian and Northern Affairs, for example,
must contend with environmental pressure groups, native people's
groups, other departments in the federal structure, and pressures
initiated from within itself and the keen interests of private
industry. In addition, opposition political parties criticize
almost every major activity. Is it any wonder, then, that
departments dealing with northern policy become technically
"paranoid" and are thus deliberative to the point of frustration
in accepting policy positions.

A major consideration is the dissemination of information -
new information and ideas within the management process. There is
also a need to question values and policy dimensions which we now
hold. Also, it seems clear that if management is to be effective,
decisions should be made through positive statements and not by a
silence - a silence of resignation or action by default.

I have already noted that federal departments are subject to
any number of stresses, all pulling more or less in the same
direction toward various objectives. Therefore, a program of
resource management is bound to be enmeshed in political, legal,
social, and economic concerns voiced from many quarters. All that
is in addition to the original dilemma of managing the resource's
exploration, exploitation, and utilization.

One such geographic area exhibiting signs of becoming a
resource management problem is the Labrador Shelf area. Extensive
fishing grounds exist in the Hamilton Bank region concurrent with
promising deposits of oil and natural gas.

Two factors adversely affect optimum resource management
possibilities for the area. The first factor, one which now
exists, is our inability to undertake extended offshore fishery
operations in ice-covered waters. The second obstacle is not yet
a reality, but could ultimately lead to a degradation of the
Labrador Shelf area as a viable fishing ground. Until now,
activity has been sparse; however, the potential for commercially
exploitable quantities of oil and gas is high.

The management dilemma is aggravated by the potential conflict
of two economically advantageous, politically sensitive and
socially rewarding resources - oil and gas versus fish. A careful
balance must be struck so as to accommodate both interests.
Examples of balanced cooperation between the North Sea oil industry
and local fishing interests point out the feasibility of management
schemes for the two resources. Until the full measure of
scientific knowledge of the effect of this development on fish
stocks is uncovered, no one is able to state conclusively that oil
and fish do not mix.

Similar problems do or will exist in other areas of the
north. Assuming development of a Beaufort Sea oilfield, how is
the oil brought down? What part is played by the native
northerner? And what form of stable management can be put in
place to ensure the long-term balanced exploitation of this
resource within the other major elements of the equation - the
technical capability, the need to husband our resources and the
regional acceptability?

In developing new technologies and methods of governmental
organization and resource management, society must accept trade-
offs. To quote Alec Hemstock (in M.J. Dunbar, Environment and
Good Sense, 1971):

"The petroleum industry's prime responsibility is to supply
Canadians with the energy chemicals they need, at lowest possible
cost ... and least possible disturbance to other life factors ...
disturbing the environment ... often cannot be measured in
dollars, and Canadians ... will not accept dirty rivers, scarred
landscapes, or polluted air. Yet if we want to go on using
packaged goods, and growing foods or producing minerals ... we
must fac at least some disturbance ... some alteration....
Canadians must decide what their priorities are and what balance
they really want between nature and technology.... What we must
do is stop unnecessary damage, and develop methods and procedures
that will prevent most serious accidents, and ... lessen the
impact of any that do occur."

Hemstock cautions that trade-offs must be acknowledged and
designed into our management systems. However, contrary to this
wise approach, it would seem that technology is leading our
decisions, and that we adapt most readily to changes brought about
through new methods and engineering rather than through a process
of conscious examination of social priorities.

If there were no technology for moving ore from the arctic
islands, would the federal government have made a decision to
research and develop the capability in an effort to initiate this
development? If the technology for drilling from artificial islands
in the Beaufort Sea were not presented as an accepted fact by the
oil industry, would the federal government have sponsored the
development of such technology? The point is that in regard to
resource development in the north it looks like many decisions are
being urged upon government as a reaction to the appearance of a
new technology and that, in some cases, the trade-off is not
technical potential against economic benefit, but technical
potential against environmental concept. I believe the challenging
point here is that although we understand the principle of what we
are trying to do within our management scheme, we have not yet
decided on how we measure our environmental concerns - thus we
find it very difficult to equate one to the other.

I would suggest that one obvious solution to this question is
the recognition that, from a management point of view, the
environmental element must be identified and costed in the

calculation of economic benefit. In these examples, the decision
is made with the national interest responding to the technological
and industrial potential. It seems clear, therefore, that in the
case of development in the north management systems have evolved
primarily as a result of economic requirements rather than as an
element of policy or government initiative.

Over the next five to ten years, options for achieving arctic
operational excellence and resource management will be developed
and presented to succeeding federal governments. Most of these
options will originate within the bureaucracy. Industry,
university groups, and pressure groups will make major contributions
and inject a large part of the initiative - nevertheless, the
questions will continue to come before Government for decision.

The eventual decision has to accommodate resource economics,
political tolerances, and environmental acceptability. That is,
how much environmental degradation can be permitted in order to
allow resource extraction? Second, is the extraction necessary at
all? Can we in fact get along with what we have without tapping
new, more costly sources of energy and minerals? A management
decision must also take into account the emerging question of
resource conservation and the balance which must be struck between
environmental hazard and increasing pressure upon known resources.

The overall policy, then, appears as the refinement of
management decisions. Currently, our compendium of decisions is
small, with the result that in many instances the policy reveals
obvious gaps or reaction by default.

There is no doubt that for some time to come arctic developers
will have to be flexible in the design of their goals and
objectives, quick to make decisions, and adept in responding to
situations which may diverge widely in the political sense. In
that vein, new ground rules for funding research may have to
emerge, particularly when the research is a cornerstone of
industrial activity. The statement that industry has been the
recipient of much unpaid-for government-conducted, government-
funded research is met by industry's answer: "the eventual product
benefits all Canadians, and therefore it is up to the government
to assist us." It follows that several questions arise as
important elements of management:

1. Who pays for northern research, much of which
    ultimately applies more directly to industrial
    development than to anything else ?

2. Where should the charges for navigation aids,
    oilspill research and development, icebreaker
    support, and hydrographic charting go? In short,
    who pays for user services?

3. On-going environmental monitoring following
    development is a necessity. Once again, who pays?

>   Is it government, through the taxpayer's dollar,
>   or does industry rightly share in the cost?

The ultimate management system for the north could do well to face those questions.

In fulfilling the need for overall coordination of government activities, some means must be found for the design and long-term support of northern research programs. An extensive research and survey activity is essential to northern management, and, to be effective, should be planned and run within a timeframe consistent with that of the industrial and cultural developments.

Even under the best of circumstances, management problems regarding highly technical resource development, which fall under a federal structure, bring to the fore complex economic, social and political variables within an effective time interval which is unusually short. When dealing with these problems, the most serious concerns seem to return to the engineering potential available within the allowable time limits of government reaction.

We seem to be faced with a situation in which technology is driving the management system rather than making a conscious effort to cast our plans within a timeframe whereby technology can be influenced in response to the long-term course of national interests.

ARCTIC TRAFFIC MANAGEMENT, NAVIGATION AND COMMUNICATION SYSTEM

M. A. Turner

Canadian Ministry of Transport

Ottawa, Ontario, Canada

INTRODUCTION

Geographically the Arctic, simplistically speaking, is composed of the Arctic Ocean together with the northern-most territorial reaches of various countries. The methods of transportation are sometimes quite different from one northern reach to another, essentially because of geographical differences. For example, Canada and the Soviet Union in the North are two quite different countries when we consider transportation. Whereas in the Soviet Union most of the North is part of the main land mass with river and overland routes connecting north and south, the Canadian North is composed mainly of islands. Because seawater, in one form or another, divides much of the North from the mainland, air and marine methods of transportation are prime modes of travel in the Canadian North. Beyond the Mackenzie River Delta and Hudson Bay areas, there is no such thing as overland travel from the South.

Because this is the geographical environment in which we in Canada have gained most of our experience, much of what is now offered will relate to the Canadian North. There we have a direct interest in about 35,000 airline route km and 22,000 sea route km. Overland travel, apart from local traffic is minimal.

Before discussing traffic management and the navigational and communications facilities such traffic in the North might need, let us briefly consider what such traffic is today and what it is likely to be in the future.

SURFACE TRANSPORTATION

North-South overland transportation is limited today and probably will not increase much in the future. Railways from the South stop at Churchill on Hudson Bay, although some thought has been given to extension of the rail system along the Mackenzie Valley into the Western Arctic and this might occur years hence. However, it is unlikely that in the foreseeable future railways will play a major role in the development of the Arctic, north of about 60°.

Road transportation today contributes little to the Arctic from a north-south point of view. Some studies are underway relative to the feasibility of a Mackenzie Valley Highway. However, there are many problems, not only associated with the actual road construction in heavy frost regions and in permafrost, but also associated with the economic and reliability aspects of the numerous ferry crossings which would be required.

Local surface transportation is mainly via tracked vehicles or other novel vehicles replacing the dog teams of yesteryear. Skidoos have become a way of life. Air cushion vehicles (ACVs) serve a purpose when they can be made cost effective for the mission at hand.

AIR TRANSPORTATION

Transportation by air offers the quickest year-round method available for moving people and cargo into, out of or between points within the Arctic today. Aircraft from multi jet to single engine play a valuable role in the North. Facilities are such that most areas of consequence are served by some sort of an airfield, although many are gravel tracks with few, if any, aids.

Arctic airports have all been classified into A, B or C categories, according to various criteria such as traffic demand and the size and degree of isolation of the community being served. The eleven "A" airports have been operated by the Canadian Ministry of Transport for some time, and gradually are being upgraded to southern standards. B and C categories meet community type standards and are being upgraded as required to meet these requirements.

Like the ACV, the prime question in respect to cargo transport is whether or not air transportation is cost-effective for the purpose intended.

Turning to air traffic control, today most aircraft movements are monitored as a matter of course by both civilian and military surveillance systems. If an aircraft encounters trouble, help can usually be brought to bear on short notice.

Of course long range aircraft flying high over the Arctic from Canada to Europe and the like are not aware of the environment which lies below them. All being well, their flights over the high

Figure 1   Skidoos used for Surface Transport

Figure 2   Air Cushion Vehicles (ACVs)

Figure 3   Air Transportation -- Aircraft

Figure 4   Air Transportation -- Helicopters

North are usually no different from similar flights over the more
temperate climes.

From the point of view of Feeder Aircraft operations in the
Arctic, local navigation aids and communications are all that most
require, although some improved traffic control would become
necessary if high density areas develop.

Local air traffic such as that flown by the "Northern Bush
Pilot", who is well known for his method of flying "by the seat of
his pants", wants little help from others. Generally speaking, he
does not want to be told where and when he may or may not go. All
he wants is a reliable weather forecast and basic navaids and
communications. Therefore, air traffic management of most local
air traffic would not be necessarily welcome.

Generally speaking, and taking a realistic view as to
resources available for developing Arctic air traffic management
systems, facilities in existence or being planned are adequate to
meet current traffic levels.

SEA TRANSPORTATION

Where size, weight and cost are predominant factors, shipping
by sea becomes viable. Despite limited shipping seasons and severe
environmental conditions, the bulk of the cargo moving in and out
of the Arctic today is by ship. Such movement is likely to continue.

Because of the severe environmental conditions, limited dock
facilities exist. Where vessels cannot go alongside, cargo is
moved principally between ship and shore by lighter and helicopter.
Liquid cargo is frequently moved by flexible pipeline.

As for transitting the Arctic by sea, the master of a ship
would tell you that all he needs is a dependable radar, reliable
weather forecasts and ice timely information and some lateral
communications, bridge to bridge and ship to shore. North-South
communications mean less to him than they do to the ships' owners,
agents and dispatchers.

OTHER MODES

We know that transportation in the North, whether it be by
land, sea, or air, is not easy and thus special solutions evolve
from particular requirements. For example, development of an oil
and gas industry in the Arctic make construction of pipelines to
the South cost-effective in some areas. Some construction is
planned in the western Arctic and studies are underway to bring
gas by pipeline under ice and water in the eastern Arctic to the
mainland and on South.

These latter activities do not need "traffic management" per
se, so need not be considered as directly bearing on the subject.
However, the success or failure of plans to build and operate

Figure 5   Sea Transportation -- Ice Strengthened Ships

pipelines from the Arctic will have a bearing on the amount of ship traffic we can expect in the future, if oil and gas is moved south in quantity. Currently, as studies progress, the pendulum appears to have swung in favour of pipelines for transporting oil and gas out of the North. Various factors could change on short notice though, and transport by sea could regain favour.

RECENT TRAFFIC TRENDS

Now let us consider the volumes of these types of traffic.

Statistics indicate a rise in air traffic annually since the 1950s, although volumes each year latterly have not increased as rapidly as forecast a few years ago.

Ship traffic in fact peaked out two or three years ago and there appears to have been a modest reduction in the actual number of cargo tons transported through the Arctic recently. The volume of overland traffic in the North has not changed very much in recent years.

Increased development of the North's natural resources together with changing world conditions could alter these present trends.

Some may ask then, why would we need a Traffic Management System in the Canadian Arctic? Aircraft have or will have adequate control facilities to meet their needs. Ships operating there are mainly limited to the summer months, and their masters are content with their present rather independent arrangements. Surface transport is minimal.

It has been suggested that the answer lies in the fact that the North has something very important to offer - namely natural resources. As we become fully aware of these resources, as their need in the South grows, and as they become more readily available, perhaps traffic - people and things - in and out of the North will increase. Perhaps even the often talked about, yet to materialize, year-round Arctic ship navigation might come about.

The question of sovereignty also arises, which involves the need to have a better knowledge and control of who goes into the Canadian North and what he does there. This leads us to the important aspect of Arctic pollution; pollution by accident or by neglect, and the resultant cleanup operations required.

Thus there may be a need for traffic control, not necessarily based on volumes of traffic so much as on the requirement to know who is doing what, where, and also the requirement to be able to react when necessary without delay.

AIR TRAFFIC CONTROL

There is no need to dwell on air traffic control because it has been with us in the South for many years. Its principles are well known.

VESSEL TRAFFIC MANAGEMENT (VTM)

Ship traffic control, or Vessel Traffic Management as we in Canada call it, is a development of the last decade and does bear mentioning as it or a modification of it could be deployed in the Arctic.

As early as 1966, the first formal system was devised for the St. Lawrence River with the objective of environmental protection through safe, expeditious and orderly traffic flow. This scheme saw the early beginnings of the integrated system which today is aimed at reducing those hazards from shipping which have the potentials of "ARROW" and "TORRIE CANYON" oil spill incidents.

The system provides initially for early contact with ships approaching Canadian waters to ensure their efficiency status in regard to manning, equipment and navigational information. Subsequently, it can regulate and monitor each ship's progress through the area by radar surveillance and VHF communications; and finally, it can immediately alert and call into action responsible authorities if an emergency situation should arise.

We now operate or are implementing nine VTM systems of varying complexities on the East and West coasts of Canada.

Objectives of a Vessel Traffic Management System have been described as follows:

1.   To enhance the safe and expeditious movement of maritime transportation by:
   (a) establishing an acceptable system for controlling vessel traffic;
   (b) providing real time information to ships on traffic, navigational dangers, weather and other pertinent advice on a routine basis and on request;
   (c) monitoring traffic movements to the degree required of a particular locality;
   (d) ensuring, through pre-clearance procedures and subsequent surveillance, vessel compliance with appropriate regulations, procedures and practices;
   (e) providing position fixing information on request;
   (f) utilizing standard procedures to avoid confusing the ship;
   (g) scheduling marine traffic movements; and
   (h) responding to emergency situations as circumstances demand.

2.   To provide a means of communication and co-ordination between the various activities vital to the ship and to its safe movement, by eliminating multiple sources the ship must contact for pilots, docking instructions and other information.

3.   To alert and if necessary in the initial stages to co-ordinate responsible authorities when an infraction of regulations is detected.

4.   To operate Vessel Traffic Management Systems and procedures which are compatible with those of other countries by:

(a) acquainting other countries with Canada's approach to
    Vessel Traffic Management;
(b) developing comprehensive and internationally acceptable
    methods of Vessel Traffic Management;
(c) developing with other countries standardized VTM
    procedures and regulations; and
(d) pressing other countries to accept these internationally
    agreed procedures and regulations.

Obviously one country cannot go it alone in developing VTM procedures.

There are four levels of ship control in Canada's VTM System. They are:

Level 1 - Ship to Ship Information System. At designated calling-in-points, vessels broadcast on specific frequencies information on their position and their intentions. There is no participating shore station in the Level 1 System.

The main feature of this system is that it provides, on a party line basis, information to shipping in the area on the intention and movements of other shipping and thus reduces the risk of collision situations developing.

This system can be used in open water areas where few hazards exist and can be made mandatory by regulation. However, its effectiveness is dependent upon the decision of ship Masters to participate. Enforcement is difficult.

Level 2 - Shore to Ship Advisory System. Vessels are required to obtain clearance from the VTM shore station prior to entering the area designated as a traffic management area. This clearance is related to regulations currently in effect concerning the capability of a vessel to navigate safely without pollution risk while in Canadian waters.

Ship-to-shore station reports provide position, courses, speed and other data at designated calling-in-points.

Ship-to-ship information is passed as per the Level 1 system.

VTM shore station monitors and records communications between participating vessels and the shore station.

VTM shore station broadcasts navigation information, traffic information and responds to ship originated requests for information.

This system includes the basic conditions of a Level-1 system in that there is ship-to-ship party line information available to shipping. In addition, the VTM shore station will regulate marine traffic to the extent of issuing clearances related to information provided by the vessel. Furthermore, the shore station will be broadcasting on a regular basis, information related to marine traffic and other information considered essential to the mariner.

This type of vessel traffic management is suitable for operation in coastal waters (and in North America, - the Great Lakes) where there is a large volume of traffic moving over a wide area along designated routes, and where the main interests on the part of Canadian authorities would be to identify deficient vessels.

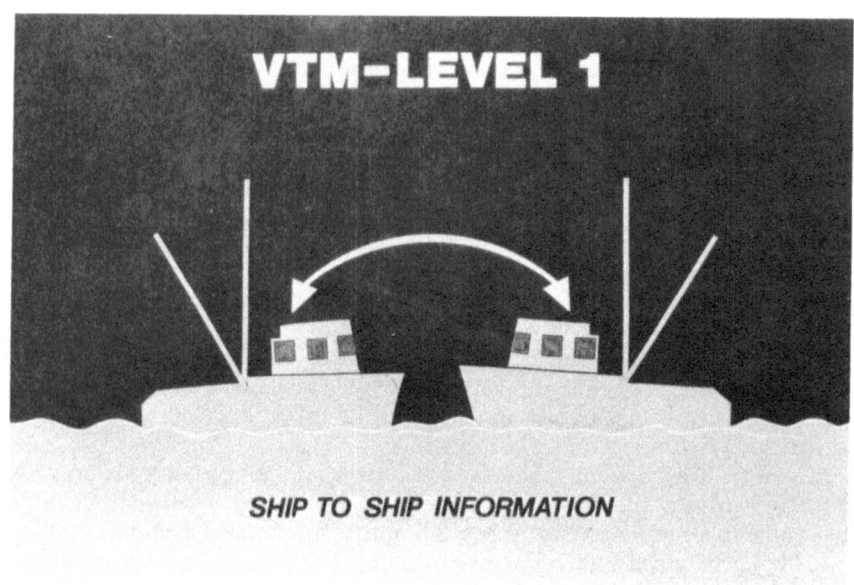

Figure 6   VTM – Level 1, Ship to Ship Information System

Figure 7   VTM – Level 2, Shore to Ship Advisory System

It permits the regulating of traffic flowing to and from the more congested areas.

Level 3 - Shore to Ship Regulating System. The Level-3 system provides:

    (a) Ship-to-ship information;
    (b) Ship-to-shore station position reporting;
    (c) Shore station monitoring and recording of all communications;
    (d) Shore station broadcasting navigation information and traffic summaries;
    (e) Shore station responding to shipboard requests for information;
    (f) Vessels required to obtain clearances from a VTM shore station prior to entering the area designated as a Traffic Management area. This clearance is related to regulations presently in effect as in Level 2;
    (g) Clearance required from the VTM shore station to proceed from one sector to another and for leaving berths and anchorages;
    (h) Updating of the clearance is required if a major change in status of shipboard equipment or an amendment of the ship's program occurs;
    (i) Shore record maintained of vessels in system.

Traffic management is enforced to the extent of scheduling vessel movements through the VTM area by instructing vessels to anchor, to hold their present position, to leave an area or to proceed.

This system incorporates the features of Levels 1 and 2 systems, where a ship-to-ship open line type situation applies. In addition, it requires close scheduling (based on information from shipping) of the marine traffic in the area. Marine traffic can be closely regulated, and complete information regarding the on-board status of ship's equipment and crew would be assessed by the shore centre.

Level 4 - Shore to Ship Control System. A Level 4 system incorporates the main features of Levels 1, 2 and 3 systems. Additionally the Level 4 system provides:

    (a) Shore station radar surveillance of area under traffic management and capability of video recording and computer logging of vessel movements and general information;
    (b) Shore station plot maintained of vessels in system;
    (c) Traffic management of the extent of instructing vessels to anchor, to hold their present position, to leave an area, or to proceed. In addition, in situations where the shore station identifies potential collision or grounding situations, *the shore station to direct the master of a vessel to take avoiding action.*

In a Level 4 system radar surveillance will be maintained over the entire VTM area to allow positive identification of targets and the actual plotting would be based on real time

Figure 8   VTM – Level 3, Shore to Ship Regulatory System

Figure 9   VTM – Level 4, Shore to Ship Control System

observed data. With radar surveillance of clearly identified
targets, it will be possible eventually to introduce a system of
traffic regulations which will include direction from a shore
station to a vessel requiring that vessel to take positive and
direct action in circumstances of an especially hazardous nature.
This capability, as a specific function of a Level 4 system, will
be assigned only to those VTM control centres having fully trained
and qualified traffic regulators.

Traditionally the master has ultimate authority over the
movements of his ship. Thus this provision may be unacceptable to
some. However, even the most traditional master would probably be
reluctant to countermand avoiding action directed from the VTM
centre ashore, knowing that the sophisticated equipment ashore can
provide a fully trained and experienced VTM traffic regulator (who
has local knowledge) with a better picture of the total situation.

The Level 4 system would have its major applications in the
approaches to high density ports, in locations where major marine
hazards exist, where there is a high density of pollutant carrying
cargo ships and in general in those locations where it is felt the
maximum degree of regulation is necessary.

Theoretically, any of these Systems could be applied to the
Canadian Arctic today.

The Level 4 system is the one which offers full control over
ship traffic, but is a system with its remotely situated radars
relaying their displays back to the VTM Shore Station or Centre
practicable in the North? Such a system covering the Arctic or
even the focal areas of the Arctic would be astronomically
expensive. Is the Level 1 system the only one appropriate to the
North? Are other methods more appropriate to the North available?

Before attempting to suggest answers to these questions, let
us consider first what traffic is up against and what it has
available to aid it in transitting the Arctic. Then we will be in
a better position to assess the requirements, if any, for traffic
management.

NAVIGATION

To start with, "The Pilot of Arctic Canada" in Volume 1 says,
and I quote:
"Navigation in the waters of the Arctic regions, is beset with
many problems not encountered in lower latitudes. These problems
are created through various factors such as weather conditions, ice,
charts, and lack of navigational aids.
Weather conditions, such as fog, ice and low cloud ceiling create
a problem for the mariner, to determine his accurate position.
These three adverse conditions are most prevalent during the months
when ice permits navigation, therefore celestial observations to
determine position, cannot be depended upon.
The present Arctic charts are largely based on aerial photography,
although a program is underway to update these charts from

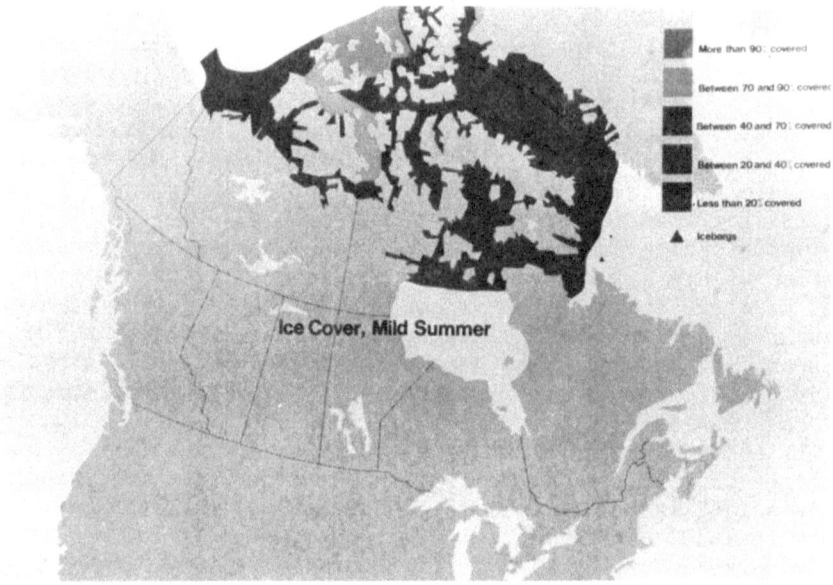

Figure 10   Ice Cover Over Canadian Arctic - Mild Summer

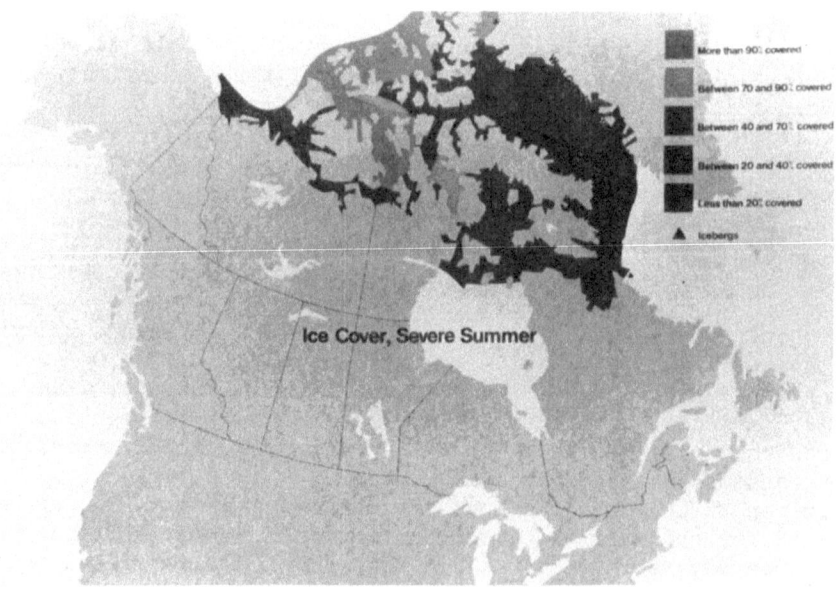

Figure 11   Ice Cover Over Canadian Arctic - Severe Summer

controlled topographic and geodetic surveys, which now extend
throughout the Arctic. There are, however, still some charts
where discrepancies of some magnitude exist in the charted
positions of islands relative to the adjacent coastline, also some
discrepancies in distances between coastlines forming channels and
sounds.
Some charts are lacking in topographic detail such as prominent
hills, mountains, glaciers etc. These features are essential for
the navigator to fix or identify his position by means of radar or
visually.
Soundings are shown on some charts, with the exceptions where
harbours and landing places have been systematically sounded, are
for the most part, a collation of individual ship track soundings,
while accurate as to depths, may be misleading in their charted
positions.
Aids to navigation, such as buoys, lights, fog signals and marine
radio beacons, are almost non-existent, except in such well
travelled routes as Hudson Strait, Hudson Bay, in the eastern
Arctic and in certain areas off the mainland coast in the western
Arctic.
The difficulties so far mentioned are not insurmountable if the
navigator accepts the fact that unconventional methods are
required to meet unconventional situations. It may be impossible
to fix a ship's position geographically, but fixing relative to
land masses, is usually possible and the fact the land mass itself
is inaccurately fixed, is immaterial."
     The topography of much of the eastern Arctic lends itself to
the master who said that all he needed was dependable radar. High
definition type radar sets are essentially all that is needed to
navigate much of the eastern Arctic with its abrupt coastline
rising high out of the ocean. However, because there are also
important low lying land masses in the Arctic, particularly in the
western Arctic, where positive radar response cannot be certain,
many 30 foot three legged aluminum towers with 36 inch radar
reflectors were erected in the 1950's and 1960's. These continue
to give yeoman service.
     To enhance radar signals in low lying areas, Radar transponder
beacons (RACONS) transmitting signals only when triggered by
marine radar in the band 9300-9500 MHz (X-band) are also in use in
parts of the Arctic. When activated by a ship's radar, they
transmit a coded signal which appears on the ship's radar screen,
and which tied in with his navigational chart, provides the
navigator with positive land mass identification ashore, a much
more positive one than can be obtained with the simple radar
reflectors. Radar transponder beacons have to be used sparingly,
though, so as not to clutter and confuse the radar scope. In all
there are 22 radar beacons currently operating in the Arctic.
     Additionally day beacons have been erected to provide a visual
aid on land masses which require prominent marking. These beacons
are usually diamond shaped and painted fluorescent orange.

Because daylight prevails around the clock for much of the summer resupply navigation season in the Arctic, shore navigation lights are not needed except perhaps at the beginning and end of the navigation season. Thus limited light installations only have been made.

A number of areas are marked with unlighted spar buoys. However, these buoys are frequently cast adrift or moved during the winter months and so are not necessarily reliable navigation aids from year to year, without confirmation. In fact, if dragged out of position by ice and the like, they become a menace to navigation rather than a navaid.

Consideration by some has been given to the installation of buoys which can be purposely sunk below the level of ice to produce an underwater signal. This signal would then be detected by passing ships for navigating purposes.

Aeronautical and aeromarine MF omni-directional radio beacons have been installed throughout many areas of the North. Coastal refraction and intervening land masses frequently make bearings from aeronautical radio beacons unreliable for marine use. However, the aeromarine beacons have been located in such a manner that they provide good service to both marine and air navigators.

In the western Arctic, the Mackenzie River System has developed beyond the rest of the Arctic, to the point where a modern and reliable system of beacons, buoys and lights is now operative.

Limited use of standard marine radio beacons as homing devices to lead ships through channels between land masses separated by broad expanses of navigable water has also been considered for critical areas and would be installed if traffic later warrants.

Of course, in the more open areas, navigation by celestial observations is also possible to the navigator but because of frequent inclement conditions in the Arctic, this traditional system is not always available. Therefore other methods are required.

Off the Canadian east and west coasts various systems including Loran-A and Decca are currently in use, and others such as Loran-C and Omega are under consideration. A decision regarding similar navigation systems off our north coast and between the Arctic islands lying within has not yet been made and probably won't be made until a better indication of our Arctic requirements is available.

LORAN-A, a 2 MHz pulse-matching hyperbolic system with an over-water daytime range of about 1200 km, has given long and reliable service to both air and marine interests in Canada. However, it has now been over taken by advancing technology and consideration is being given to closing down the present east and west coast Loran-A systems in the early 1980's. The United States used to operate a Loran-A chain in the Baffin Bay area, but this has recently been closed down.

DECCA NAVIGATOR has given and is continuing to give yeoman service. A hyperbolic system also, it operates continuous wave at four dedicated frequencies between 84 and 128 KHz, and gives a daytime range of about 450 km with an accuracy of between 50 and 500 metres. Installed in some of the confluence areas off the Canadian East Coast, it is very well thought of by many mariners. While it is very accurate by today's standards, its range is such that many stations would be needed if it were to blanket the Arctic. However, it could be considered as a possibility if limited confluence coverage only was required.

LORAN-C operates in the same frequency range as Decca Navigator, but transmits specially shaped bursts of carrier frequency energy instead of a constant amplitude carrier wave, which allows the receiver to discriminate against sky waves. Ground wave coverage is not degraded at night, nor is it normally affected by ionospheric disturbances. Accuracy varies from 90 m to 1 km with a range of about 1100 km overland and 1800 km over water. The effects of permafrost and sea-ice on the performance of Loran-C, particularly in respect to reductions in range, are not fully understood and some further study is required.

However, Loran-C is considered to be a front running contender in the Arctic, should such a system be required. Figure 12 gives a suggested Canadian Arctic coverage, utilizing 10 Loran-C stations, tying in with Canadian and U.S. Loran-C stations on the East Coast and Alaska.

The U.S. has chosen Loran-C as its prime coastal and confluence system. The U.S.S.R.'s use of Loran-C, although its coverage does not extend into the Canadian Arctic to any useable degree, also makes it a desirable system from an international standardization point of view, if ever marine Arctic transport were to operate between the two regions. From an air point of view, there are advantages although some technical co-ordination between Canada and the U.S.S.R. would be necessary.

OMEGA is a VLF hyperbolic system which has potential for the Arctic navigator once developed and in full operation. Employing 8 stations scattered throughout the world, this system uses 3 frequencies, 10.2, 11.33 and 13.6 KHz, time shared over a 10 second interval. Omega, unlike Loran-C depends on skywave and the ionosphere for propagation. Accuracy is a factor of predicting the ionospheric height and its stability during irregular solar activity. Ionospheric "Day effect" can be predicted but sudden ionospheric disturbances together with polar cap absorption cause unpredictable anomalies of the Omega pattern in the Arctic. Thus Differential Omega, which can monitor and "correct" Omega reception at a known point as close to the service area as possible becomes desirable if not essential for both marine and air use.

It has been estimated that 39 Differential stations would be needed to cover the entire Canadian coastal areas, of which 20 would be required for Arctic navigation north of 60°.

With the development of a shipborne receiver at about $5,000 (the cost of a medium priced car in Canada) capable of utilizing both Loran-C and Omega, a policy of Omega for deep sea and of

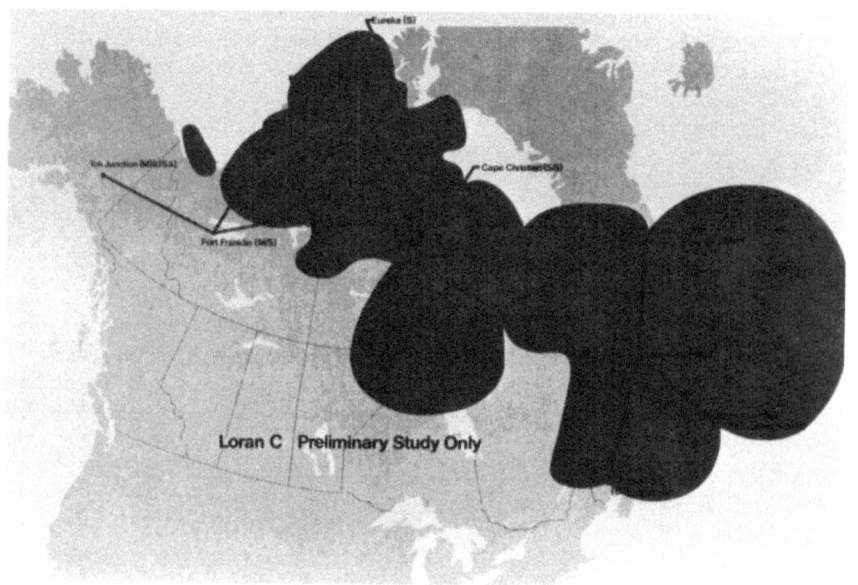

Figure 12   Proposed LORAN - C   Coverage

Figure 13   Ship's Navigation System

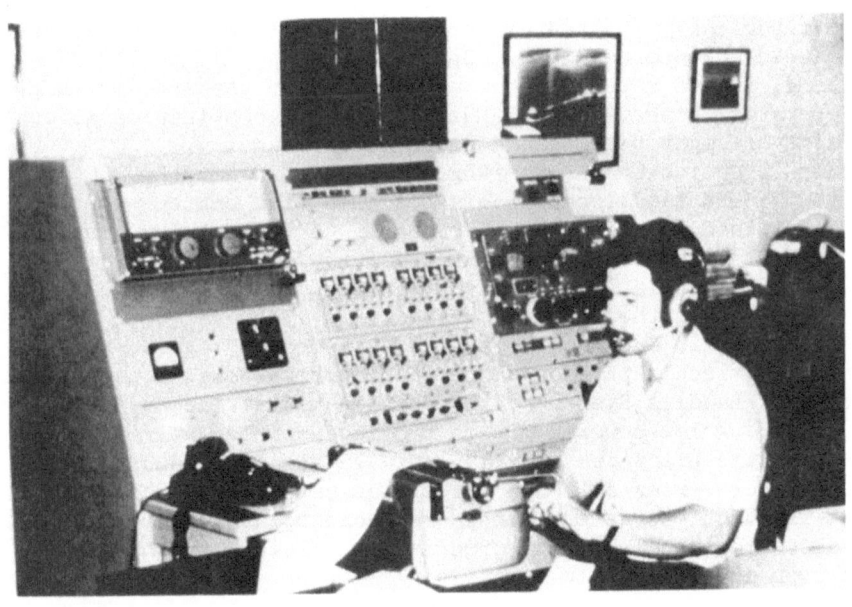

Figure 14   Shore Navigation Control System

Loran-C for coastal navigation has some merit. Air navigation, with its more sophisticated requirements, brings an estimated installation price tag of $20,000 plus, for its combined Omega/ Loran-C receivers.

Before leaving navaids, some developments in navigation satellites are worthy of consideration for the Arctic.

The U.S. TRANSIT System was originally developed for the U.S. Military but now is available commercially, although at $30,000 per receiver, computer and antenna, it is a little rich for the average commercial navigator. Twenty-two satellites in polar orbit have been flown since 1960, with six being operative at this time. Typical "single-pass fix" accuracy for a moving ship is between 150 and 300 metres but for locating oil rigs and the like, fixes within about 15 metre accuracy are possible.

NAVSTAR is the acronym for "Navigating System using Time and Range". Originally a military development, it too is being made available for commercial operation. The first of 24 satellites is scheduled for launch in 1977. The NAVSTAR Global Positioning System (GPS) is designed to provide accurate three dimensional fixes to within 9 m of true position for aircraft, ships and land vehicles. Velocity and a precise time reference are also available. Commercial Receiver costs are not yet known.

In reviewing navaids available for the shipborne navigator, we have included the air navigator in some of our comments. To complete the Arctic air picture we should touch on:

(a) VHF Omni Range/Distance Measuring Equipment (VOR/DME) facilities which are or will be installed on major and other strategically located airports throughout the Arctic as both navaid and approach aid. Operating in the 112 to 118 MHz band, VOR is usually combined with DME in the Arctic. DME operates in the 962-1213 MHz band. This combination gives line of sight bearing and distance coverage.

(b) Non-Directional Beacons (NDBs) transmitting in the 200 to 535 KHz band from which a position line can be determined, are located extensively throughout the area. Typical 1 KW NDBs have an Arctic range of about 300 km.

(c) VHF/DF installations are located or planned at most major airfields in the North.

Further, in addition to the usual approach lights, runway lights and visual approach slope indicator systems, conventional Instrument Landing Systems (ILS) are located at larger airports in the North. A low powered NDB is provided at the outer ILS marker to assist the pilot in acquiring the ILS signal. Unfortunately in some Northern areas although precipitation is relatively low, winds frequently create blowing snow conditions which, while the sky above may be clear, cut ground level visibility to zero and thus even ILS won't help.

COMMUNICATIONS

To review further our background knowledge upon which to consider traffic management requirements, let us now turn from navigational aids to communications.

Marine and aeradio communications in the North are fairly conventional at this time, and pretty well what one would expect, although there is one exception.

Ten Marine Coast Radio Stations are strategically located to cover the Canadian Arctic, providing conventional ship to shore facilities, weather forecasts, ice information and navigational warnings. SSB techniques are used exclusively in the MF and HF bands. Long-range morse and voice radio services are provided at some key stations. Within the Arctic, the highly variable ionospheric conditions lead to extensive use of ground wave propagation.

Marine radio telephone on MF/HF and VHF/FM is available. Medium range 4 MHz voice is utilized in the Eastern Arctic to provide increased coverage beyond the 2 MHz range. Radiotelegraph A-1 morse code is used extensively. HF Radio facsimile broadcasts covering the Arctic are operated from Edmonton and Halifax giving marine weather and ice advisory and forecast information.

In the past communications problem areas frequently encountered by ships operating in the Arctic have included:
(a) HF radio frequency interference caused by congestion;
(b) Poor propagation conditions created by the less stable ionospheric conditions of the Arctic;
(c) HF communications blackouts or polar cap absorption (PCA) sometimes lasting up to 10 days;
(d) Delays sometimes in handling telegraphic traffic caused by the high density of traffic handled by coast stations awaiting clearance south;
(e) Inability of ships and shore stations sometimes to re-establish contact, due to the above conditions, when initial contact is lost.

Facsimile, duplex radio telephone and a ship to shore Telex Service in the North, augmenting that soon to be provided by our long range Marine Coast Radio Stations at Halifax and Vancouver.

Pay radiotelephone and television broadcasting receiving facilities are added niceties which should be considered if ships and their crews are to remain north for extended periods.

Like the marine side, air traffic control and aeradio communications in the Arctic are fairly conventional. Similarly, in the past the unreliability of North-South point to point communications has been a problem. Aeradio stations are located throughout the North with additional stations being planned at about ten more locations over the next five years. HF single sideband and VHF/AM are used extensively. Some VHF air traffic control installations are controlled from the south (Edmonton).

ANIK satellites are the one exception to our present fairly
conventional communications package in the North. ANIK or
"Brother" as the name signifies in the INUIT language, brings to
much of the North communications similar to that of Southern
Canada - telephones, telegraph, entertainment radio and television.
For better or for worse, the North and South are now said to be
linked together as never before.

Satellites offer considerable potential in the Arctic. From
the communications point of view they can provide reliable North-
South communications across the ionospherically turbulent polar
cap regions. The ANIK satellites have excelled in breaking through
this barrier which hitherto gave conventional HF radio communications
to and from the North about 60 per cent reliability - and the ANIK
derived circuits are being extended peripherally within the Arctic
by means of conventional terrestrial MF and HF links.

For example, a ship in the high Arctic would pass its message
to an Arctic coast station. Then it used to be that the coast
station would try, frequently unsuccessfully for prolonged periods,
to clear the message southwards via HF radio morse. Latterly radio
teletype was used giving some improvement in reliability. Now the
Arctic coast station can relay its message via one of the ANIK
ground terminals located in the Arctic, such as Resolute, through
an ANIK satellite southwards to another ground terminal near
Toronto or Victoria, and then on to say Halifax, with near 100
per cent reliability. Conventional short distance lateral
communications in the Arctic generally are reasonably reliable,
although ANIK is assisting here too, but it is the North-South
barrier which heretofore has created most of the problems mentioned.
By accessing the satellite we can hurdle these atmospheric
anomalies which plague conventional HF radio transmissions.

Thus with the advent of satellites the words "isolated" and
"remote" have lost their meaning to much of the Canadian North.
In fact, no longer need any community in Canada be out of touch
with any other community in Canada, regardless of size or
location. The drilling rigs in the Beaufort Sea can be in instant
telephone or teletype communication with their head offices in
Edmonton, and a mine in Strathcona Sound can instantly communicate
with its head office in Toronto. The ice-breaker clearing Lancaster
Sound can instantly communicate with the Fleet Superintendent in
Dartmouth, N.S. Today, the only real limitation is the rate of
expansion of providing satellite terminals.

Canadian communication satellites to date have been placed in
synchronous or geo-stationary orbit. As you are aware this means
that being about 35,888 km above the equator, the satellites have
a northern horizon of about 78°; thus surface traffic beyond this
horizon distance is unable to access the geo-stationary orbiting
satellite. Polar orbiting (non-synchronous) satellites would be
needed for communications north of about the 78th parallel.

Additional Canadian communications satellites are planned. We
are also monitoring international developments such as INMARSAT,
MARISAT and AEROSAT.

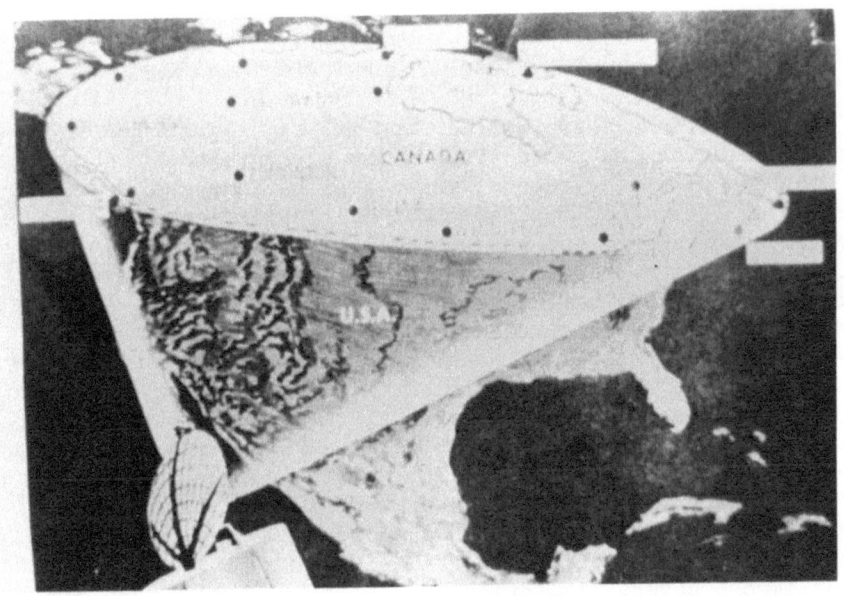

Figure 15   Satellite for Communications in the Arctic

A NEED FOR TRAFFIC MANAGEMENT?

Having touched on the navigation and communications systems available to the mariner and airman traversing the Arctic, let us return to the questions concerning the need for traffic management in the North. Is it needed?

If traffic patterns continue at about the same levels, it would appear that current and planned facilities are adequate to meet demands. Navigational facilities are satisfactory. With the advent of ANIK, north-south communications problems have been rectified. Therefore, the situation would appear to be more than adequate. But is it? The key seems to lie in what the traffic of the future is likely to be.

TRAFFIC OF THE FUTURE

If large tankers are to roam the Arctic, a multitude of new factors including the protection of the ecology, the need to know the location of these vessels, their destinations and estimated times of arrival, the improvement of oil spill clean-up, the improvement of search and rescue facilities and better control and utilization of ice-breaker resources will be needed. Are we suggesting improved traffic management?

The likelihood of hydrofoil craft operating in other than ice-free waters is negligible as the foils would be susceptible to ice damage but air cushion vehicles (ACVs) are already operating in the western Arctic and as previously mentioned, they offer considerable possibilities operating in marine, amphibian and terrestrial modes as transport, for search and rescue purposes and for ice-breaking. For the Voyageur type ACV, present operating ranges of about 270 km at 50 knots and a cost of about $800.00 Cdn. per operating hour (vehicle amortization, insurance, fuel and crew wages) impose severe limitations on their use today. But such operations have already been cost effective for replenishment activities in the western Arctic. Properly managed their potential is considerable. As an all-weather vehicle operating in whiteouts, low sun dazzle periods, fog and during the long winter nights of the Arctic, a good radar together with HF radiotelephone, HF Marine Radio, VHF-A/M (aeronautical) and VHF-F/M (marine) facilities are essential. But if ACVs become prevalent throughout the Arctic, will we not need to know where they are, what they are doing, when they will arrive and where? Is this traffic management?

And semi-submersibles and submersibles, what of them? Much has been said of the role of the submarine in the north, particularly in bringing out the Arctic's mineral wealth. Submersible and semi-submersible tankers and towed tankers are a possibility of the future. The same factors which applied to ships transporting oil also could apply to semi-submersibles and submersibles, but perhaps even more so as some would say the possibilities for their disaster is greater than for the surface vessel. If such operations become viable in the North, will they for the same reasons not warrant better systems of traffic management?

The future possibilities of unconventional vessels as well as the mammoth tankers operating in the Arctic, have been mentioned but what about conventional ships with their varying degrees of ice strengthened hulls? Present ship traffic is governed by Arctic Shipping Pollution Prevention Regulations which lay down shipping safety control zones whereby each class of ship may only enter particular zones at a given time of the year, depending on the categorized ice strengthened capability of its hull. For example, a vessel in category 14 (a ship with little or no ice strengthened capability) is allowed to enter zone 13 (Lancaster Sound area) between Aubust 15 and September 20 only and at no time of the year is it permitted to enter zones 1 through 6 (Gulf of Boothia, and generally those areas north and west of Resolute Bay). But can these regulations be enforced? Conventional ship traffic is likely to continue into the Arctic for many years. Is control of these vessels likely to pose a problem in the future? - a problem sufficient to warrant implementation of traffic management systems?

What sort of people will the traffic of the north serve in the future - Exploration and Mining engineers, and tradesmen, scientists, settlers, entrepreneurs, real estate salesmen, tourists as well as an enlightened native population?

... and what sort of communities? People living in southern style cities and towns, or communities continuing to exist as best they can, developing for the most part in make shift fashion with few creature comforts other than telephone and television? - or people living in sophisticated sub-surface habitats, underground and underwater, and above the ground in modern city-like conditions under astrodomes?

It is probably safe to assume that based on today's conventional technology the volume of future activities in the North, be they air, land or sea oriented, as we see them today, do not warrant conventional traffic management at, for example, the level 4, which we have described for ships. There is some argument for ship traffic management at level 1 as the resources needed to implement this low level traffic management system are minimal by comparison.

ARCTIC TRAFFIC CONTROL

But why *ship* traffic management or *air* traffic control? As we are talking of the future, let us think in the future.

Idealistically one system encompassing all requirements is more effective than many systems specializing in individual

operating modes.   Then, first of all, would it be better to forget
air traffic control and *vessel* traffic management?   Would Traffic
Control encompassing all activities on land, at sea and in the air
be better?   And secondly, would one overall area system be
preferable to many small regional systems?   Would *Arctic* Traffic
Control, encompassing the whole Canadian Arctic be better?   And,
where would we like to locate this Arctic Traffic Control Centre?
Need it be located in the environment in which it has control -
where the action is?   Some will say yes, - you must get the feel
of the environment to understand its problems.   If so, let us put
our Arctic Traffic Control Centre in Inuvik or Frobisher.   Both
have certain advantages.   Or let us compromise and put our Control
Centre at Resolute Bay, midway between east and west.

But what of those who say despite all the advanced technology
in the world, you are having to deal with people and you won't get
good people to stay in the North.   Or why do you need the Arctic
Control Centre in the North?   Why not locate it farther south
where conditions are more amenable to living year round, Whitehorse
or Yellowknife, or Edmonton or St. John's, or why not even in a
white ivory tower in Canada's capital Ottawa.

But the skeptics will say, *so far* from the scene of operations...   Located in the South or even in Resolute, how can you
cover the *entire* Arctic with air traffic control and with Level 4
type traffic management of ships without many radars and many
microwave links and the like?   The answer lies in - Satellite
technology, not only the satellites for the communicator and the
navigator but also earth resources type satellites of the present
decade and other satellite systems which are planned for the not
too distant future, which can keep track of all activities, land,
sea and air over large areas.

Yes, the answer to our Arctic operational problems, whether
its volume at sea, in the air and on the ground increases or not,
could lie in the utilization of a few satellites feeding a
computer inventory system displaying what, where and when
information.   In fact it is suggested that if it is necessary to
control the Arctic from an environmental point of view, from a
sovereignty point of view, from any other point of view, then an
Arctic Traffic Control Centre keeping track of what's going on in
the North is essential.   Such a centre, located wherever, is
possible and feasible with the technology of the not too distant
future, if we want to take the effort.

The essential elements of a traffic management system where
real time full control is required are:
(a)   Reliable communications between the Traffic Control Centre
      and the vehicle;
(b)   Real time information providing location (including height
      for aircraft), course and speed;
(c)   Real time and forecast meteorological information including
      ice reports.
Satellites can provide these elements.

Before concluding, let us briefly consider some of the satellites developments which could provide this information to an Arctic Traffic Control Centre.

We have said that by the present ANIK type satellites, we can have responsive reliable communications into the Arctic north to 78°, out of the Arctic and within the Arctic. Other satellites including a proposed Canadian Government Interdepartmental UHF Satellite will further augment ANIK. (As indicated non-synchronous satellites would be required for operations north of about 78°.)

Earth Resources Technology Satellites (ERTS) currently in being could provide information to an Arctic Traffic Control Centre. Future satellites associated with this program such as the United States SEASAT System could further augment the system and provide essential information. Let us consider SEASAT for a moment as it is a very interesting development and its type would have a lot to offer.

The SEASAT System, being developed as a NASA applications program, is scheduled to be launched by Space Shuttle starting in 1979 and is expected to provide a major step forward in developing and demonstrating an ability to chart the oceans of the world, locating ice fields and polynya, ocean temperatures, current and wave heights and wind characteristics. This is some of the information which our Arctic Traffic Control would need to have.

The SEASAT program will eventually consist of a number of satellites covering the majority of the globe. It is expected that the fully operational program, consisting of several multi sensor space craft together with ground data handling stations will be providing scientific and commercial information by 1985. Through SEASAT, a good insight into the various global and regional climatic trends and variations can be determined thus providing the potential for accurate long term weather forecasts. This information would be particularly beneficial in the routing of ships, port operations, off-shore exploration and ice navigation.

SEASAT-A, the first satellite of the SEASAT series scheduled to be launched in 3 years time, will have an all-weather capability in that its sensors both active and passive, unlike Earth Resources Technology Satellites (ERTS) of the past, will have a night time capability and will not be restricted to a sun-synchronous orbit.

SEASAT-A is scheduled to carry 5 types of sensors to monitor the earth's surface:

- a *compressed pulse radar altimeter* will have two functions, to insure the altitude of the space craft with an accuracy of ±10 cm and be capable of providing information on wave height from .5 to 20 metres.
- a *synthetic aperture imaging radar* is expected to produce high resolution pictures of ice fields, leads and open water as well as ship targets and the like. Because of power limitations, the radar will function only on real

time when it is over an ERTS compatible ground station.
We would need to locate two (?) such stations in the Arctic.
- a *microwave wind scatterometer*, looking out over two 450 km
  wide tracks will provide wind speed and wind direction.
- a *5 Frequency Scanning microwave radiometer* will provide
  all weather and temperature measurements as well as
  information to be used in constructing global ice images.
- a *scanning visible infra red radiometer* will provide images
  of visible and thermal infra red emissions from the oceans
  to amplify and augment information from the other satellite
  instruments.

We in Canada are particularly interested in the SEASAT program
although the initial satellite, because it does not provide complete
coverage of our Arctic and other coastal waters, falls short of
Canadian requirements.

Another new satellite is the promising new aircraft/satellite
picture navaid ICEWARN under development by the US (NASA) which
will give improved ice information for use by an Arctic Traffic
Centre.

In conclusion, satellites of various breeds could do much in
support of an Arctic Traffic Control Centre. But what will the
North of the future be like, - mines, drilling rigs and pipelines
abounding? - the underwater habitats and astro-dome cities we
mentioned? Conventional and fancy surface, marine and air vehicles
galore? Is a Canadian Arctic Traffic Control Centre located
somewhere in Canada with ANIKs, SEASATs and other communications
and resources applications' type satellites providing real time
information, - locating aircraft, ships and land vehicles,
directing their movement and tending their needs and the needs of
the environment socially viable and needed in the Arctic of the
future? ... and if so, should this centre control only activities
in the Canadian Arctic? In the long term, assuming the question
of international sovereignty and related sensitivities can be
overcome, is an International Traffic Control Centre located
somewhere controlling international traffic anywhere in the entire
Arctic archipelago socially viable and needed?

BIBLIOGRAPHY

1. Polhemus Navigation Sciences inc. "Radio Aids to Navigation for the U.S. Coastal Confluence Region". Report prepared for the United States Coast Guard, 1972.

2. Digital Method Ltd., - "Study to Evaluate Candidate Navigation Systems for Use in Canada North". Report prepared for Ministry of Transport, January, 1975.

3. Eaton, R.M., Canadian Hydrographic Service, "Loran C Compared with Other Navigation Aids in Meeting Future Canadian Needs", January, 1975.

4. United States Department of Transportation, "National Plan for Navigation", Annex, July, 1974.

5. Lapp, Philip A., "The Identification of Requirements Critical to Operations on and Below Ice-Covered Waters", September, 1974.

6. 1973-74 Government Activities in the North - Advisory Committee of Northern Development - "North of 60°".

7. Canadian Coast Guard - National Policy for Marine Search and Rescue - April, 1975.

8. Arctic Resources By Sea - Northern Associates (Holdings) Ltd. - November, 1973.

9. Efficient H.F. Channel Utilization - Channel Evaluation and Calling (CHEC) - 8 December, 1971, E.E. Stevens and E.D. Ducharme, Communications Research Centre, Ottawa.

10. Statistical Review of Ice Navigation in Eastern Canadian Waters, November 1974, - Canadian Coast Guard.

11. Radio Navigation in North America by J.M. Beukers, Vol. 22, Number 1, Spring, 1975 - Navigation Journal.

12. Navigation Aids Program for Northern Canada by D.E. Evans.

13. Commercial Applications Satellites - Flight International dated 19 December 1974.

14. "Marine Navigational Aids and Terminal Facilities" by J.N. Ballinger.

ACKNOWLEDGEMENTS

Thanks go to the many people within the Canadian Ministry of Transport who have discussed and contributed material used in this paper, also to those within the National Film Board and Indian and Northern Affairs Department who provided specific background information on the Arctic. A special thanks goes to the small team of associates in the Canadian Marine Transportation Administration of the Ministry of Transport who helped prepare the original outline of the paper and fill in some of the blanks, who helped edit and smooth out the final version and to the under-standing secretaries who took the dictation and who were kind enough to type and retype the various drafts.

Apologies go out to the aeronautically inclined readers of this paper. Speaking as a simple sailor, it is possible, in my haste to meet a deadline and at the same time in my attempts to ensure that the major areas of maritime involvement are sufficiently explored, that I have given the air aspects of the subject less than adequate coverage.

ARCTIC TRANSPORTATION SYSTEMS

L. Nitzki

A. G. Weser Shipyard

Bremen, Germany

BACKGROUND

The unexpected rise of oil prices draws the attention to alternative energies (nuclear power, solar energy, coal, oil tars and oil sands), to alternative production areas in politically stable regions (like the North Sea and the Arctic), and to alternative technologies (like coal liquefication, deep sea drilling). Within a short range of time public opinion in industrialized and developing countries became conscious of the very meaning of energy costs for global political development.

Besides oil and gas resources, large quantities of valuable mineral ores have become known in the last years - iron ore with more than 50% of iron, copper, and zinc, most of the deposits being suited for surface mining.

The countries bordering on the Arctic, and others with interests in natural resources and countries which could contribute their technology regard the development of the far North no longer as one of several options but as an urgent necessity.

While a finite number of completely integrated arctic systems will be examined in depth, this keynote address may serve as a general introduction into the main problems of arctic transportation systems.

TRANSPORTATION SYSTEMS

Air Transportation

At the moment air transport is the main system being used, but on a long run of low economy, for the transportation of bulk cargo and heavy machines and equipment. Future importance: Passenger transport and transportation of important or valuable spare parts.

Land Transportation

The land transport system is used to very little extent and at the moment is only of regional importance. Street cars and railway systems are expensive to construct and are not considered to be economically viable transport systems.

Pipelines

Pipeline is considered as an important alternative to the transportation of oil and gas by ships. It would act as a supplementary system. Main problems: Permafrost, earthquakes, and high costs in the Canadian archipelago.

Surface Effect Vehicles

Surface effect vehicles are used only for short distances and in extremely shallow waters.

Water Surface Transportation

Convoy technique of icebreaker and ice strengthened ships is necessary for the Northwest Passage. In deep waters the ship's mass should be mobilized for icebreaking assistance, so large ships with capabilities are most promising. A stepwise development from the existing ice strengthened ships to Arctic Ice Class is necessary for areas with increasing average ice thicknesses. Projects for Ungava Bay in the immediate future are known; projects for Baffin Island (Milne Inlet and Strathcona Sound) are to follow later, and little Cornwallis Island or Pt. Barrow could be the final aim for naval transportation systems for the Arctic, using ships of the Canadian Arctic Ice Class 8 to 10.

DESIGN PRINCIPLES AND OPERATING CONDITIONS FOR ICEBREAKING
BULK CARRIERS INCLUDING TANKERS AND OBOS

For the midbody of larger bulk carriers, trapezoidal <u>water lines</u> are recommended having the larger breadth at the forward shoulder. Generally small <u>breadth</u> reduces the ice resistance. However, for a certain required displacement and an unchanged block coefficient, the remaining dimensions should be increased. The <u>draught</u> selected should be as large as navigational conditions allow. An increased draught means imporved properties when operating in ice because, among other things, the smaller length and breadth reduces the friction surfaces. At the same time it ensures that a larger amount of the broken ice is pressed aside below the lateral ice and not below the bottom of the ship. Besides that, the top edge of the propeller is lower, ensuring that the propellers are more protected from touching ice.

In general, the midship section of icebreakers ranges between circular and oval. In the case of icebreaking bulk carriers, a <u>midship section</u> having a sharp bilge in favour of the cargo hold volume will have no adverse effect on icebreaking. It is necessary that the angle of flare of the ice belt is kept sufficiently large so that the ice pressure will not be excessive.

The resistance in ice strongly depends on the shape of the <u>forebody</u>.

The critical point of the forebody is the <u>forward shoulder</u>. If the broken channel is only slightly larger than the maximum breadth of the vessel and if an upright ice sheet is jammed there, then very large horizontal forces will result and wedge in the vessel. Such difficulties occurred with the "Manhattan". For this reason the flare of the forebody frames must be maintained through to the largest breadth of the vessel and the transition into the midbody should be an edge or a very short arch. The design of the forebody being as described above, an upright sheet will be pressed downwards at the shoulders because all longitudinal sections have the same angle as the stem. The increase of resistance when sailing in free water can be accepted.

By giving the forebody a <u>curved stem</u> the entrance angle of the latter and the flare can be adapted to the relative ice condition and ice thickness by trimming the vessel. The smallest icebreaking energy is needed if the force of inertia of the ship's mass, which can be influenced by the respective speed of the vessel and by the flare of the stem and the frames respectively, just attains the necessary icebreaking force.

The installation of <u>heeling tanks</u> is also advisable, in order to produce a rolling movement around the longitudinal axis if the vessel has been entrapped by the ice, thereby changing the static friction on the hull to the much lesser sliding friction.

Special Design and Operating Conditions

1. Lowest outside temperature - 60°C
2. Relative maximum wind velocity 26 m/sec.
3. Maximum ice thickness still allowing continuous sailing
    3.6m
4. Optimum lines of fore and after bodies for voyage
    through ice
5. Midship section with flared sides
6. At least single compartment vessel, or type "A" or type
    "B" - 60% according to freeboard regulations
7. Ice reinforcements conforming to the conditions to be
    expected for operation in ice
8. No filled cargo oil, fuel oil or dirty oil tanks along
    the shell and in the double bottom when sailing through
    heavy ice
9. When sailing through ice the ballast draft approximately
    corresponds to the freeboard draft
10. Heeling system for an average vertical shell movement of
    1 cm/sec. up to a heeling angle of approx. 4°
11. Clean ballast tanks, no changeable tanks
12. Navigation bridge forward
13. Deck machinery, etc. must be installed below deck
14. Since under extreme weather conditions a stay on the
    exposed deck is impossible, relative precautions must be
    taken
15. Two independent emergency spaces for the crew must be
    provided in the fore and after bodies allowing a stay of
    at least 10 days
16. Lifesaving devices (boats) must be accessible and
    operable without stay on the exposed deck and meet
    extreme weather conditions
17. Fire fighting must be possible without water and without
    stay on the free deck
18. Loading of the vessel must be possible without stay on
    deck
19. At least two independent main propulsion units must be
    installed
20. The ballasting system must be such that when loading the
    vessel the freeboard draught can be maintained
21. A helicopter landing area and hangar must be provided
22. All essential navigation instruments must be doubled
23. All submerged fittings and connections must be repairable
    from inside
24. Existing rules for operation in ice must be observed.

HULL OUTFIT AND EQUIPMENT MEETING ARCTIC CONDITIONS

Considering the requirements for Arctic proof equipment of
EOS vessels you nearly always find solutions that can be realized;
however, the technical equipment and knowledge involved are often
very high.

Where possible, items of equipment should be arranged below
deck where they are not excessively exposed to climate influences.

Official regulations have been observed as far as they can be
coordinated with the special requirements for voyages in the Arctic.

Mooring and Loading by Means of a Loading Tower

Based on the assumption that in frozen waters docking a
vessel alongside a quay wall would be very difficult, EOS vessels
in the Arctic are therefore moored to the loading tower by
the fore body, the former being firmly grounded and fitted with
the necessary loading equipment. The handling over of the mooring
hawser from the loading tower to the vessel is done by a trolley
which is not required for mooring the vessel. Heeling and trim
changes can be readily permitted and re-sitting of the mooring
hawser is not required.

Anchor and Mooring Equipments

The capstan, anchor windlass and mooring winches are arranged
in heated spaces below deck. The mooring hawsers and lines are
led outside through shell ports. Lateral mooring shall be done
only if the ice conditions allow this and if the temperatures are
such that working on deck is possible.

The use of ice anchors has not been intended because the
efficiency of these anchors is very doubtful and because their
transportation to and fixing at appropriate places of the ice cover
includes a risk for the crew on account of the number and size
required.

Life-Saving Equipment Suitable for Temperatures Down to -60°C

The EOS vessels will be equipped with plastic lifeboats of
closed construction. The boats will be stowed within the superstructure. Control of the launching device and opening of the
shell ports is operated from the closed boat after embarkation.
The shell opening is surrounded by a strip of ice-repellent plastic
which is to facilitate opening of the port. The temperature in the
boat is to be maintained at +15°C for a period of 10 days.
Considering the great risk of icing, the range of stability must
be given the utmost care.

Life capsules can be an alternative to the lifeboats. Owing
to their spherical shape these capsules have a relatively small
surface and thus a nearly optimum shape for the prevention of
icing and heat radiation. As an additional life-saving means
rafts will be provided which can be easily handled and transported.
On the ice they can also be used as tents.

Heated Windows

All windows on board will be heated in order to obtain a
comfortable living room temperature. As regards the wheelhouse
windows, however, it is of prime importance that icing is
prevented and a perfect view maintained. The wheelhouse window
is designed so that the outer covering glass is flush with the
window's frame. So, already when the boundary layer melts due to
its natural weight the ice can slip off over an ice-repellent
plastic layer.

Insulation of Living Rooms

Special attention is attached to the water vapour
impermeability for the insulation of the outer bulkheads at the
superstructure.

Fire Fighting Equipment

The actual problem is to select a suitable extinguishing
agent.

Conservation by Painting

The conservation by painting of the individual areas (bridge,
aerials, ice belt, etc.) is to meet, if possible, two functions,
i.e. the protection against corrosion and the reduction of ice
adhesion. There are also plastics which meet these requirements.

Furniture and Fitting of Living Spaces

When designing the accommodation spaces it should be
considered that in the Arctic the crew is exposed to extraordinary
physical and psychical burdens. Therefore, the colours for the
living rooms and their illumination must be planned and selected
very carefully.

Engine Plant

The following requirements are mandatory for the machinery
of icebreaking bulk-carriers suitable for year-round services in
the Arctic:
highest possible reliability
high availability
intensibility against shock loads
protection against freezing and glaciation
high propulsive output
astern power
satisfactory propeller efficiency with all speeds from full ahead
    to slow astern
increasing torque at decreasing speed
rapid power changes and reversing manoeuvres
adequate first and operation costs
adequate space requirement
reduces crew
suitable for full load as well as for very small partial load
    during continuous operation

A valuation analysis of 8 selected engine units, when 14
characteristic features were classified and valuated according to
their importance for the entire system, led to the following
results:
Well suited are:
  - Steam turbine units with c.p. propellers transmitting
    the power either mechanically by means of gearings or
    electrically by means of three-phase systems.
Of nearly the same suitability are:
  - Steam turbines with three-phase transmission and
    solid propellers.
  - Industrial gas turbines with three-phase transmission
    and c.p. propellers (this plant still has good
    prospects of an improved analysis as soon as more
    service experiences at sea have been gathered).
  - Medium-speed diesel engines with d.c. transmission
    and solid propellers as far as the output does not
    exceed approx. 50,000 - 60,000 SHP.
Less suitable units are:
  - Geared steam turbines with solid propellers (for
    trade routes requiring reduced manoeuvrability and
    shock resistance this plant is well suited).
  - Industrial gas turbines with gearing transmission
    and c.p. propellers (this arrangement does not
    comprise the advantages of electric power transmission
    so that the disadvantages of the industrial gas
    turbine, in particular the excessive purchase cost,
    fully apply).

- Aviation gas turbines with each power transmission
  examined (the disadvantages of the currently
  available aviation gas turbines such as maintenance
  at short intervals and high demands to POL's are so
  aggravating that even an excellent power transmission
  cannot compensate and that for the time being the
  installation of such units in icebreaking bulk-carriers
  cannot be recommended).

Provided these references are observed, the propulsion performances
at a wide ship types program can be realized with conventional,
widely approved arrangements.

NAUTICAL AND METEOROLOGICAL ENVIRONMENT

Simultaneously with the creation of year-round shipping the
following prerequisites must be provided:
- construction of Arctic-proof accommodation facilities
  and public supply centres to provide for life in the
  exacting Arctic surroundings. Using a few log-huts,
  and industry working all the year through can be
  scarcely established;
- careful selection and training of Arctic-proof and
  Arctic-experienced crews for the special ships to be
  built;
- upgrading of waterways by sufficiently lighting and
  beaconing the ship channels, installation of radio
  beacons and radar reflectors on the coasts, year-round
  ice pilot and icebreaker services, as well as
  continuous telephone weather and ice consultations,
  and the like;
- laying out discharging and loading sites.

In specialist circles the knowledge of the complex and rough
Arctic environment necessary for the year-round opening of the
Northwest Passage is still being stated as incomplete. This
specially applies to the winter months. In Canada the systematic
exploration of the Arctic was initiated in 1951.

The large-scale exploration of that area was possible later
on with the planned service of planes and helicopters by means of
which meteorological and oceanographic observations are carried
out today.

The characteristic features of the US American and Canadian
Arctic are:
- cold winters
- cool summers
- little precipitation
- frequent fog
- nearly always covered sky
- high atmospheric humidity
- heavy formation of ice

- little glacierization
- permafrost
- normally slow tidal current in the archipelago and along
  the north coasts of Alaska and Canada
- very slow ocean current
- very low storm frequency
- long winter night
- long summer day
- rocky, disrupted, dead landscape.

Sufficient Depths of Water in Wide Districts of the Arctic

There are only few soundings shown on official sea charts.
In certain Arctic districts soundings were made at large distances
only so that it is possible that between the soundings there are
shallows which may present risks, in particular to vessels having
a large draft.

The whole Arctic archipelago has only few shipping lanes
which cannot be navigated by large vessels on account of
insufficient depths of water, reefs and reefy small islands. For
most of the raw material deposits it will be possible to find
suitable landing places for such vessels. At any rate, the depth
of water conditions in the Canadian Arctic are such that large
vessels will find there a wide field of operation.

Ice Cover

Nature and shape of an ice cover depend on the conditions of
its origin, size, distribution of temperature and salt, strength,
on the elastic and plastic properties as well as on pressure
conditions.

If we distinguish the shape of an ice cover according to the
kind of its origin, then two characteristic types can be stated:
- an ice cover which has undisturbedly been formed by a
  thermal process only, i.e. by cooling the surface of
  the water, has an even surface and everywhere is of
  similar thickness,
- or an ice cover the outer shape of which is due to
  dynamic influence, that means movements and pressures,
  resulting in a very uneven ice cover. By piling up
  such as ridges and hummocks, it can take strange
  and chaotic shapes.

Pressed ice formations originate by ice sheets which often
pile on top of or below each other.

Thickness of Ice in the Northwest Passage up to 5m

Not only due to the seasons but also dependent upon the
severity of the winter, the thickness of the ice place exclusively
generated by cooling is different from year to year. Acoustic
measurements of U.S. planes and submarines uniformly demonstrated
that the thicknesses of the Arctic ice range between a few
centimeters and up to and beyond 5m.

When melting and breaking up, the ice conditions in the
passages of the archipelago are not improved simultaneously and to
the same extent because the ice disappears rather by driving away
caused by the wind than by melting on the spot. If the ice of a
passage is removed by the wind another route can then be blocked
by ice sheets and filled with ice. Hence it follows that in all
areas of the wide Arctic space in general, optimum or even rather
similar ice conditions can never exist.

The opening of the Northwest Passage for shipping and the
stay of people in this region the whole year through require an
adequate knowledge of the Arctic environment, at least along the
shipping routes. This knowledge is a prerequisite for the
construction of Arctic-proof vessels, for shipowners' and economic
reflections, and for the nautical handling of the ships. Man must
learn to comprehend the Arctic and adapt himself to its conditions
if he will start work in that region with the prospect of success.
This especially applies to the navigator who together with his
ship is directly faced with the hard Arctic environment.

ARCTIC OPERATIONAL INFORMATION AND FORECASTING SYSTEMS

W. F. Ganong

Atmospheric Environment Service

Downsview, Ontario, Canada

INTRODUCTION

To many, the popular concept of the Arctic is one of barren wasteland with a harsh winter climate. Figure 1 is an example of this cold climate as it produces ice on the foredeck of a Canadian icebreaker. Although this harsh climate, where man may be loath to venture outdoors, prevails for the long winter, there is another side to the Arctic. For about two brief summer months, climatic conditions approach those of the temperate zones. For this short summer growing season with its long days, it is possible to grow fine examples of a number of our common vegetables, and, incidentally, some of the world's largest and most ferocious mosquitoes and black flies.

This climate of the Arctic produces sea ice which covers most of the marine Arctic areas in the wintertime. During the summers, particularly in the areas close to large land masses, the ice melts and breaks up to the extent that normal shipping is possible in some regions, whereas only icebreaker-supported shipping is possible through much larger areas. Winter and summer ice cover is delineated in Figure 2 by W. E. Markham. This sea ice, in some cases, can act as a floating bridge, but basically it is an obstacle to shipping. Figure 3 is an example of tough going in Jones Sound in August, 1972. However, it can be readily shown that with the availability of accurate and detailed ice information, and its intelligent use, movement of ice-strengthened ships in many Arctic areas is possible and will be a key factor in future resource extraction from Northern areas.

My purpose here is to trace briefly the development of weather and ice services in these generally sparsely occupied areas, and to provide an overview of the present state of these services. In so

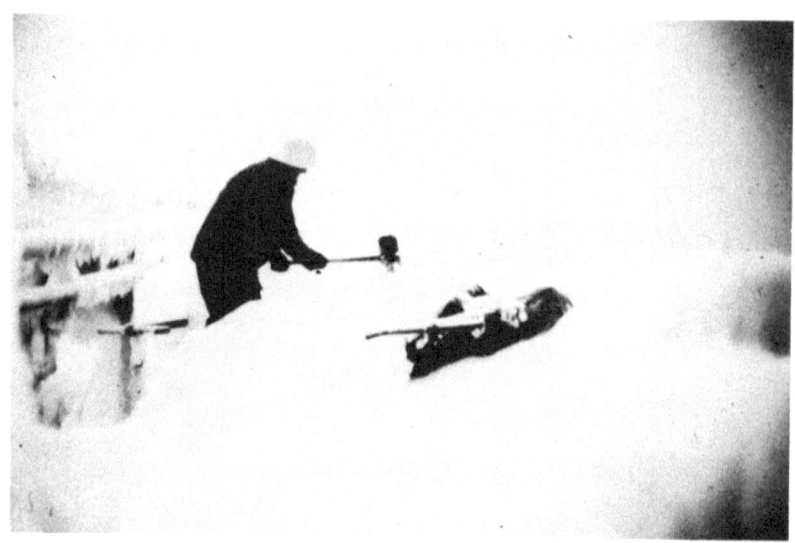

Figure 1   Seaman Removing Frozen Sea Spray
from Foredeck of an Ice Breaker

doing, a systems approach is applied to both weather and ice
services, each representing data-acquisition, data-transmittal,
data-processing, and information distribution.

In looking at total systems, three points of caution are
brought to your attention:
Firstly, the final output really can only be as good as the weakest
link or component in the system. To bring this into practical
perspective, it is not cost effective to spend additional sums on
data-processing if the data-acquisition system is not up to
providing basic data in sufficient quantity and quality to meet
the requirements and capability of the present data-processor.
Furthermore, the most effective system is of no use if its output
cannot get to the people who need it. For example, the best
weather forecast ever made by a meteorologist is only of academic
value if it does not reach, in a timely manner, the yachtsman,
farmer or pilot who needs to make a decision based on weather.
Secondly, the system must provide a useful and needed service, and
not exist for itself alone. Too frequently, experts have their
interests too wrapped up in the details of such things as computers,
satellites, fancy office or laboratory machinery to show any real
interest in what the total output does for society. Except perhaps
for pure research, our economy cannot support systems which are
introverted.
Thirdly, there is one other point to make - costs of doing things
in the Arctic are high - very high indeed, as compared to costs in
temperate zones.

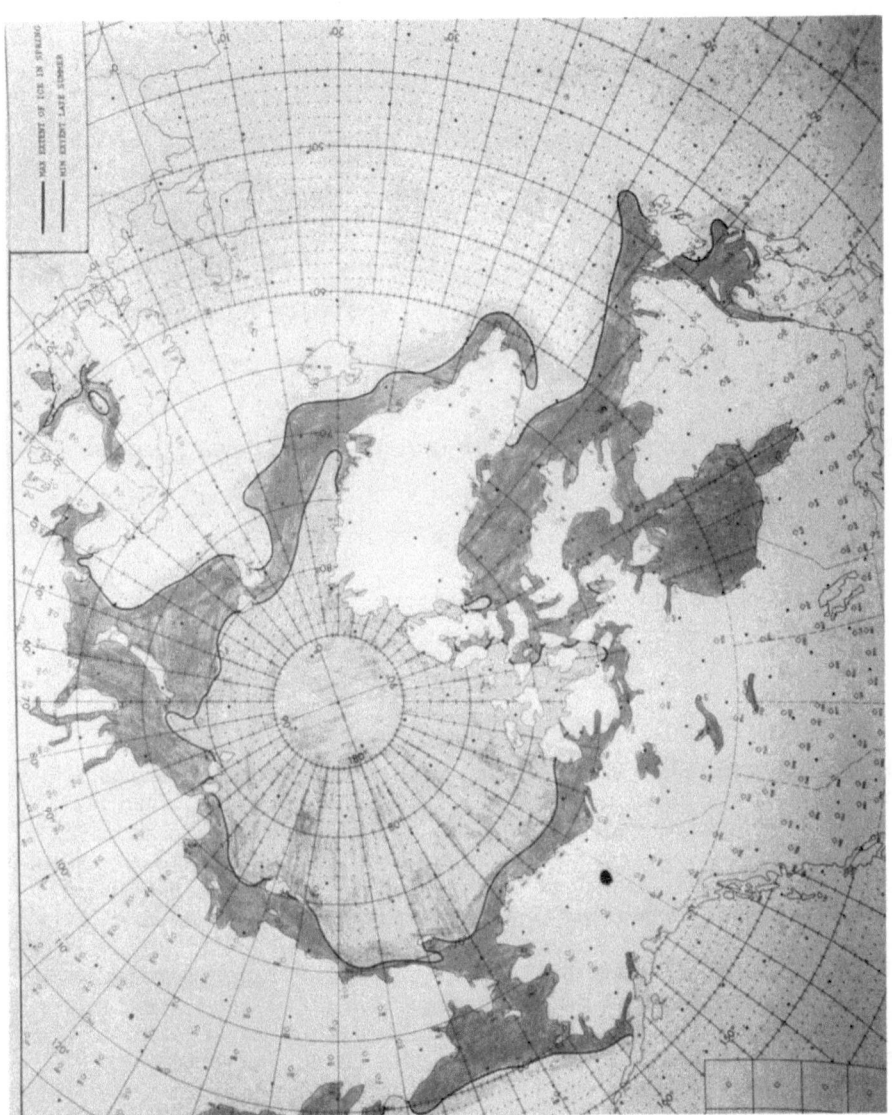

Figure 2 Winter and Summer Ice Cover over Arctic

Figure 3. CCGS d'Iberville stuck in ice in Jones Sound, August,
           1972.

Now turning to <u>Arctic Weather Systems</u>, or Arctic Operational
Information and Forecasting Systems.

Background

Prior to World War II, weather information became available
as settlements, mining, and bush flying moved northward. During
World War II, aviation requirements injected a tremendous stimulus
to meteorology. As part of this expansion, trans-Atlantic aviation
took on a new dimension, and to support this, new weather-observing
sites were added in Greenland and in north-eastern Canada and
Labrador.
     In the late 1940's and early 1950's, meteorological require-
ments in support of longer range aircraft, and the need for longer
range forecasts, resulted in the establishment of five joint
Canada-U.S. Arctic weather stations in the Canadian Arctic
Archipelago. As it turned out subsequently, these fitted well
into the global approach to meteorology. However, there are large
areas in the North, primarily the Arctic Ocean, from which reports
virtually are not available except for the occasional site manned
on a more-or-less temporary basis.
     Weather knows no national boundaries, and during the past
century meteorology has developed on an international basis
through a number of bilateral and multilateral agreements, and in

recent years, through the World Meteorological Organization
(W.M.O.), an agency of the United Nations.

Data Acquisition

Traditional methods of weather observing have been extended
into the Arctic by the establishment of weather observing
stations. These are expensive to build and expensive to man and
resupply. Basically, the atmosphere is sampled at regular
intervals from fixed points, and, as such, the data can input
nicely into numerical analysis and prediction methods. There are
two types of such stations:

(a) Surface weather stations which report atmospheric pressure
    tendency, temperature, dew point, wind direction and speed,
    sky condition, and precipitation (if any) and any weather
    phenomena that may be occurring.
(b) Upper air stations which report winds, pressure altitudes,
    temperatures and humidities as the ascending balloon and
    airborne instrumentation rise. The goal for these ascents
    is of the order of 100,000 feet, but these heights are not
    yet routinely achieved by many stations. Rockets are
    sometimes used to obtain upper air information, but because
    of costs, these are usually limited to special projects.

Figure 4 shows the distribution of surface and upper air
stations in Arctic areas. Obviously, the coverage is far from
ideal. One possible method of expanding the local surface data
input is by use of automatic weather stations. Another system,
the satellite, is proving itself as the new powerful tool capable
of providing a vast input of data on weather. It is a rapidly
developing field and the present VHRR systems offer real benefits
in delineating cloud cover and temperatures; also in deducing
temperature and wind profiles. Figure 5 is an example of visual
VHRR imagery showing a weather system on the right-hand side.
Although basic costs are high, so much data is available that it
becomes a cost effective system. For Arctic areas, earth-orbiting
satellites only are available. Geo-stationary satellites are very
effective in low latitudes, but for reasons of basic physics,
including Newton's Law of Gravitation, they are not available in
Arctic areas!

Data Transmission and Information Distribution

Weather observations are taken on a synoptic basis, that is,
on a near simultaneous basis around the world. Observed data is
collected regionally and transferred to national teletype systems
which, in turn, are tied in to a W.M.O. international meteorological
teletype network. Many Arctic stations have a direct connection
to national teletype systems, as do virtually all stations in

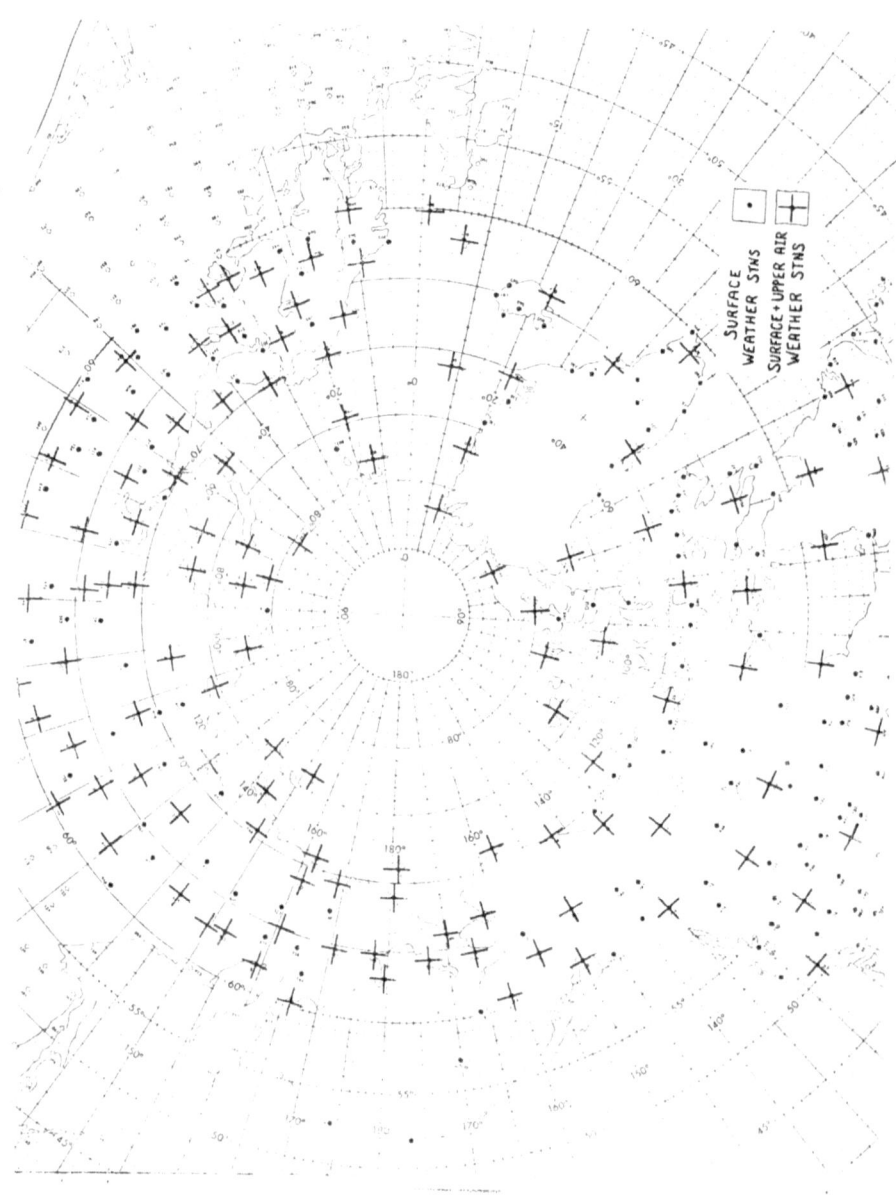

Figure 4   Surface and Upper Air Stations in Arctic areas

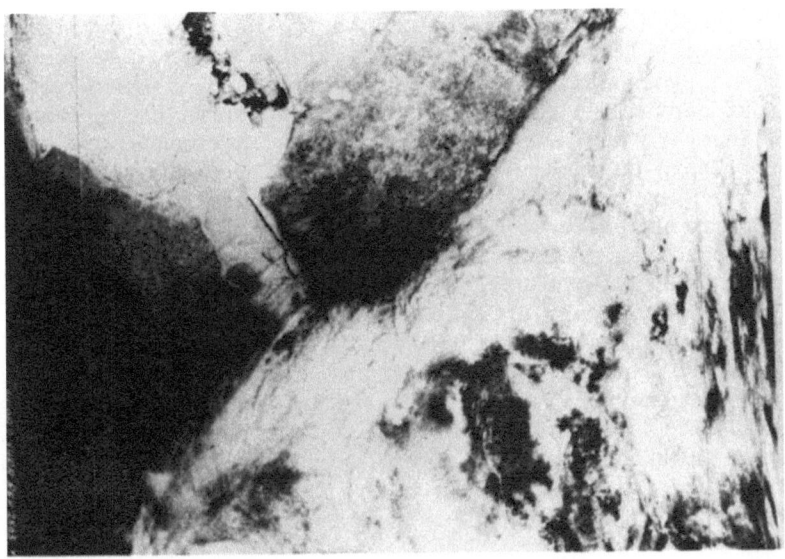

Figure 5.   VHRR image of weather system over Eastern Canada.

continental North America.   Observational data from isolated posts
such as those in the Canadian Archipelago are transmitted to
collecting points by various radio means, and more recently, some
of these reports come in via communications satellite.   As
observations are already in numerical brevity codes, these methods
suffice, but can be tedious.   Arctic weather observations are
transmitted to the forecast offices, and Canadian surface and
upper air observations are normally available at Canadian forecast
offices within 15-20 minutes from the time they are ready for
transmission.   International distribution to the W.M.O. Washington
Communications Centre takes place simultaneously.   This general
framework for communications systems has been in existence for
some years and is working effectively.   Technological progress has
brought the time delays very close to a minimum.
        Satellite data can be obtained from two sources:
(a) The main readout station where the data is usually processed
    and the processed data is transmitted to those facilities
    requiring it; and
(b) The direct readout station where the user has to make
    provision for initial processing of the data.
In either case, fairly sophisticated photo quality facsimile, or a
transmission system based on digitization of the data, is necessary.
Provision of this satellite data to the user meteorological offices,
even though it is a complex problem in communications technology,
is an essential link in the system.

Distribution of the end product to the user is always a problem in meteorology. It is fairly straightforward to provide weather analysis and prognostic charts and forecasts and even consultative services at central points such as major airports, or in large cities. However, how does one provide this information to a wide variety of users on a more-or-less continually available basis, particularly when the monitoring of changing weather conditions is also a continuous process? In populated areas, local TV and radio stations provide an outlet whereby the weather information is available on a regular basis. In the north, such information facilities exist in only a few places and there seems to be no mechanism for getting the information to small groups, many of which are to some degree mobile. And it is just these groups that are the most vulnerable to rapidly deteriorating Arctic weather conditions. This is a major deficiency in the whole system. This information distribution problem is somewhat similar to that in the distribution of ice information.

Data-Processing (Forecasting)

In the 1930's, weather forecasting was very subjective in nature and was performed for the general public and marine interests for the whole of Canada from one location. The system is now much more comprehensive, has much greater data input and much more detailed output, and has introduced a very large measure of objective numerical methods.

In many centres, broad-scale analyses and map prognoses are accomplished by high-speed and high-capacity computers. In Canada, a CDC Cyber 76 is used in the Canadian Meteorological Centre, Montreal. These broad-scale analyses and prognoses are passed by a facsimile network to six regional weather offices scattered strategically across the country. At these locations, the weather prognoses are refined, and forecasts prepared. In the case of the Canadian Arctic, the regional forecast office is in Edmonton. These forecasts are still primarily made by the skilled forecaster; however, the day is not too far distant when at least part of this function will be performed by a computer. In fact, the role of the meteorologist is becoming increasingly one of developing computer programs and monitoring the outputs, which is perhaps as it should be.

What are the areas where we need development of the forecasting system? In the last 35 or 40 years, demands in the temperate zone have brought the broad-scale weather forecasting system a long way forward. The three main features that are needed to improve the forecast output are:

(a) a better observational coverage, and as a corollary, continuing development of expertise in integrating satellite data into the processing system;

(b) more-refined models, particularly as related to higher latitudes,

and expansion to meso-scale output;
(c) development of better Arctic meso-scale forecasting. In this
latter respect, it may be interesting to note that in support
of oil-drilling operations in the Beaufort Sea near the mouth
of the Mackenzie River, a meso-scale weather and ice forecasting
service is being developed as a special project. Considerable
R & D effort has been allocated to the development of meso-
scale forecasting in an area where there is virtually no
observation to the north and north-west of the drilling site.
I will refer to this project later.

Data-Processing (Climatology)

Surface and upper air observations are repetitive observations
at regular intervals, and as such, lend themselves to machine data-
processing so that calculations of norms and departures from norms
are fairly straightforward. Production of climatological atlases
can be based on principles of drawing isopleths of the various
characteristics (temperature, precipitation, wind, etc.) using the
observing points as a data base. Because of local effects due to
topography, considerable professional skill is used in this
function. These methods can be used for climatic atlases in the
Arctic, a major problem being the scarcity of regular reporting
stations as a data base. An associated problem is that there is a
certain amount of bias in the observations from Arctic points in
that they practically all come from sheltered coastal points. A
more promising approach in the present time frame is the development of a system to use satellite data in building up a climatological base, particularly in the Arctic and open ocean areas, and
this approach is now being investigated. In the meantime, the
best we have is interpolation by the specialist of the presently
available broad-scale information.

Ice Information Systems

This shortened title includes data-acquisition, data-
transmission, processing (forecasting), processing (climatological)
and information distribution.

History. Exploration of the North American Arctic began in
the 17th and 18th centuries with the many voyages of the early
North American explorers searching for a Northwest Passage to Asia.
Most turned back after being unable to penetrate very far into the
ice pack. Others, less fortunate, lost their lives in this venture.
None succeeded in traversing the Arctic until the early 1940's when
RCMP vessel St. Roch did so, taking some two years to make the
transit West to East. The Royal Canadian Navy's icebreaker HMCS
Labrador (now a vessel of the Canadian Coast Guard) made the
transit East to West during the summer of 1954.

In Canada, the first organized ice data-gathering began in
the late 1920's and early 1930's over Hudson Bay and the Gulf of
St. Lawrence, primarily as a requirement of shipping interests.
In the late 1940's and early 1950's, joint Canadian-U.S. defence
interests necessitated the movement of large quantities of material
by sea lift to the Northern extremities of the continent proper,
and this requirement brought the potential of mass Arctic sea-lift
into focus. It also indicated the requirement for surveillance of
ice conditions as an integral part of this type of operation. For
this operation, ice reconnaissance was provided by the USN. As
activities continued to develop in Canada's Arctic, Canada moved
into this function first with aircraft of the RCAF (Figure 6),
later moving primarily to chartered civilian aircraft, and is now
using two Electra aircraft (Figure 7) in which the "eyeball" method
of observing ice is supplemented by an array of remote sensors.
On the data-processing side, an Ice Forecasting Centre was
established in 1958.

Figure 6.   A Canadian Forces Lancaster preparing for recco
            take-off, Resolute, late 1950's.

Figure 7. Electra ice reconnaissance aircraft on ramp at Gander
           airport. The aircraft are operated by Nordair under
           charter to DOE.

Nature of the Ice Problem

     Although first-year ice, formed during one season constitutes
a barrier to shipping, multi-year ice, which has eliminated most
of its brine content, is tougher than first-year ice and is even
more difficult for ships to penetrate, even with icebreaker
support. Figure 8 shows a sealing vessel beset in ice; Figure 9
shows the damage that ice can cause, and Figure 10 a convoy through
ice. The North American Arctic contains both first-year and
multi-year ice. The islands of the Arctic Archipelago cause
irregular distribution of the ice in the melt season. The degree
of melt, and consequently the degree of clearing, varies from
year to year as indicated in Figure 11 by W. E. Markham. In the
open coastal areas, such as along the north coast of Alaska, the
ice generally melts in the summer, leaving a period during which
navigation can usually take place near the coast with the Arctic
ice pack remaining some distance to the North. Similar conditions
prevail north of the USSR. Further south, over Hudson Bay and the
Gulf of St. Lawrence, first-year ice only is encountered. The
same applies to the Baltic Sea. Ocean currents cause some
anomalies, perhaps the two most striking being (a) the movement of
heavy ice and icebergs down the Labrador coast where they become
a menace to shipping as far south as Newfoundland waters; and

Figure 8.   A sealer caught in ice pressure off the coast of
            Newfoundland.

Figure 9.   Example of ice damage to marine vessels.

Figure 10. Shipping convoy passing through Norwegian Bay
            enroute to Eureka.

(b) the Gulf Stream and prevailing winds that keep the Norwegian
coast virtually ice-free. Figure 12 shows Arctic ice melting
under the midnight sun. Figure 13 shows one of the more picturesque
icebergs off the Labrador coast.

Data-Acquisition

Apart from local observations from ships and shore locations,
the present prime method of obtaining ice data is by aerial ice
reconnaissance. The present Canadian vehicle is the Electra
aircraft (Figure 7) equipped with visual observing facilities,
ground-mapping radar, a camera array, an airborne radiation
thermometer, a thermal mapper or infra red line scanner, and a
laser profilometer. With these, an experienced ice observer can
chart detailed ice coverage, including amount present, distribution,
age, floe size, estimated thickness, and extent of ridging and
puddling, all of which are required by the ship operating in ice.
As part of the ice observation, it is fundamental that the
navigation or positioning be of high calibre. In Canada, we find
that dual INS and dual Omega give the required accuracy of aircraft navigation even in high latitudes.

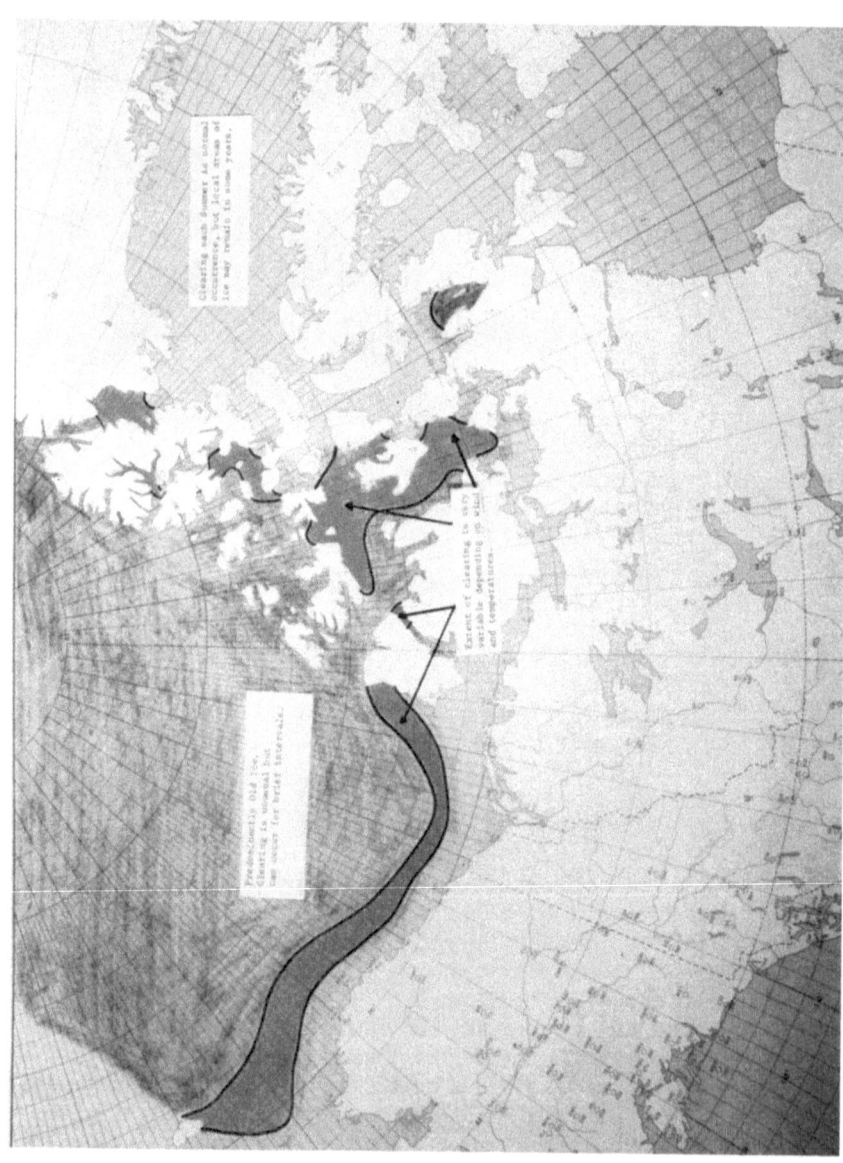

Figure 11   Irregular Distribution of Ice in the melt
season – Arctic Archipelago

Figure 12. Heavy puddling on old ice in the Arctic Islands.
Eureka Sound, 1973.

Figure 13. Sunset framed by an eroded iceberg. Labrador Coast.

A powerful new tool coming into use is the satellite. When
cloud and fog are not present, it provides excellent information
on ice distribution. Figure 14 is an example of VHRR imagery,
and Figure 15 of ERTS (or LANDSAT) imagery. With refinement and
with development of microwave-sensing, it will probably, in the
mid- or late 1980's, achieve the capability to collect virtually
all the data on ice that is now acquired by aircraft. Sideways-
Looking Airborne Radar (SLAR) is another powerful tool which has
been used by the military and which several U.S. agencies are
applying to the acquisition of ice data. Figure 16 is an example
of SLAR imagery over ice. In Canada, we hope to add this system
to the ice recco aircraft in late 1977. For purposes of ice
reconnaissance, we are looking at resolution of about 60 feet out
to a range of 30 miles which can be achieved by synthetic or a
coherent system. It should be recognized that this is an
expensive system, but because of the wide swath of coverage
possible in both good weather and bad, it will cut data acquisition
costs from approximately 40¢ per square mile to an estimated 20¢
per square mile.

The main deficiency in aerial ice reconnaissance data
acquisition is a method to measure thickness of sea ice from the
aircraft. R & D is going on in Canada in this area and it looks
as if we may have a suitable operating system in a few years.

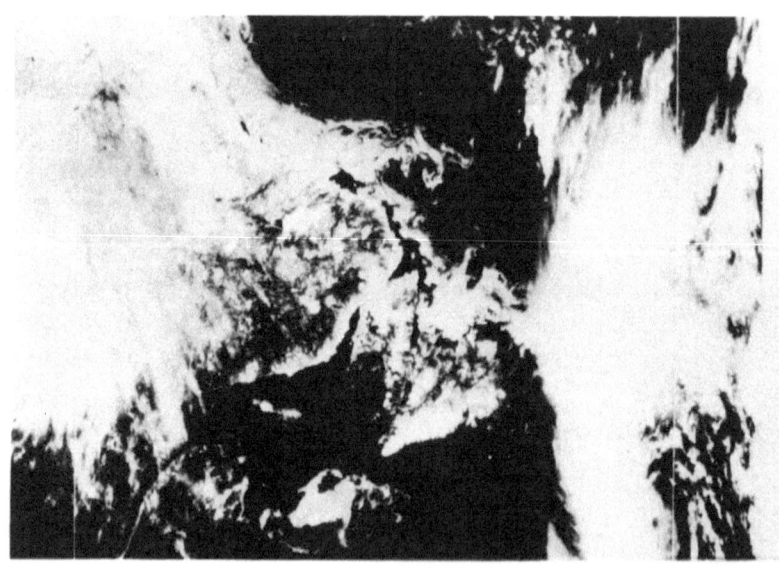

Figure 14.   A VHRR image of the East Coast showing ice drift
             along Labrador.   Spring time.

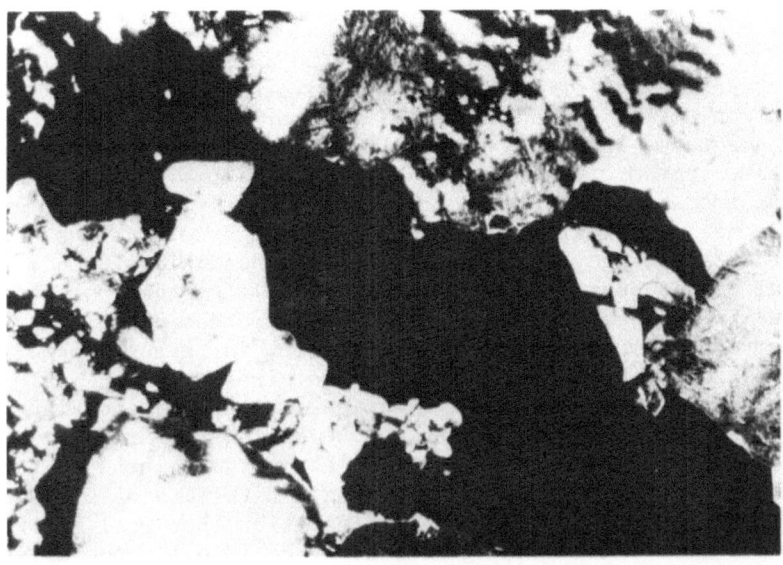

Figure 15. An ERTS image of Barrow Strait in mid summer. Note ice breaking out of Wellington Channel.

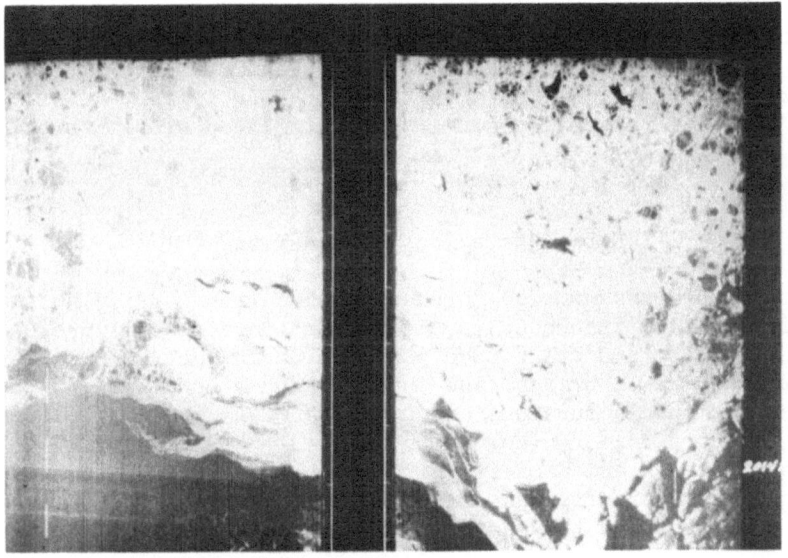

Figure 16. SLAR image of ice in Smith Sound, March 1973. Note ridges apparent in ice floes.

Data-Transmission and Information Distribution

In this area, it seems appropriate to address at one time the initial transmission of acquired data and also the information distribution to the ultimate user of the service. In some ways, this is adequate, in others it is the weakest link in the ice information system. Usually, ice is found in complex patterns which are frequently difficult to describe properly by words or a brevity code for a small area. Add to this characteristics of leads, thickness, age, pressure and ridging conditions, and a pictorial method becomes essential. Facsimile systems now in use can transmit ice chart data which has been processed either by the ice observer or forecaster. It is excellent for passing observed ice charts for tactical purposes from the aircraft to the ship it is supporting. Figure 17 shows a typical ice chart, and Figure 18 an airborne facsimile transmitter. However, as the area of data coverage becomes larger, there must be a compromise between total area of coverage and the detail provided. Satellite data and SLAR imagery will provide such detailed information that real effort must be put into communicating this detailed intelligence directly to the data-processing facility. Photo-facsimile systems have been used in some cases; however, the most promising systems, at least for SLAR imagery, appear to be the digitization of the data and transmitting it in digitized form.

Passing detailed near real-time ice information to the user is the weakest link in the present system, and will probably continue to be for some time to come, primarily because the high priority real-time user is the ship operating in ice. Although ground equipment is now expensive, communications satellites show some promise. The direct transmission of SLAR imagery to ships from a U.S. ice recco aircraft is being followed with interest in Canada.

Data-Processing (Forecasting)

Up to the present, Ice Forecasting has been carried out primarily by hand processes involving considerations of present ice distribution and its comparison with norms; ocean temperatures and currents; past, present and future weather conditions with particular emphasis on winds and heat transfer from the atmosphere to earth's surface, or vice versa.

The inter-relation of ice with weather is indeed strong at all times. The significant difference is that ice is usually slower changing than the weather, and consequently the time-frame of action for ice is roughly an order of magnitude slower.

Weather forecasting is going through the change to becoming highly automated, and numerical methods have been used for some time for analysis and prognosis of meteorological charts.

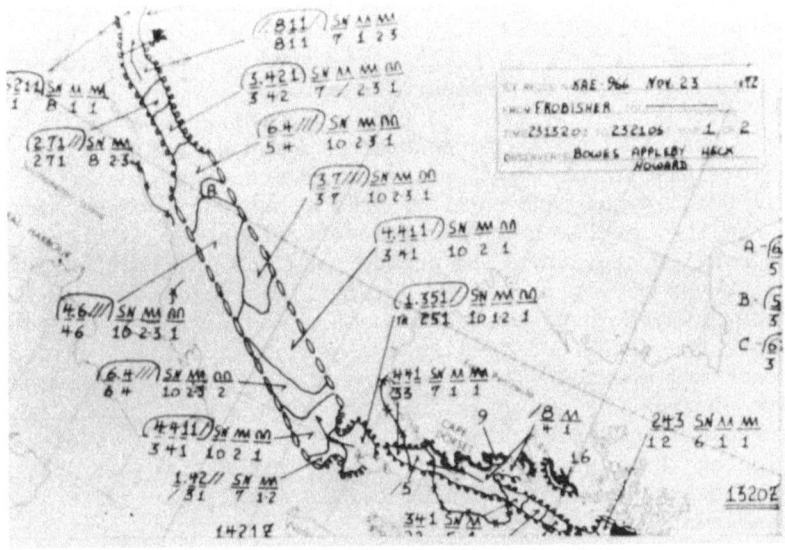

Figure 17. A typical ice chart prepared on board the aircraft.

Figure 18. Muirhead airborne facsimile transmitter used in
tactical support of shipping.

Forecasting of detailed local weather phenomena is still the
prerogative of the professional forecaster, but his (or her)
function too one day may be largely replaced by the machine as
current high levels of R & D bear fruit.

Earlier, I have referred to the interesting development of
meso-scale weather and ice forecasting to support oil drilling
planned for 1977 in the Beaufort Sea near the coast in the
Mackenzie Delta area. Here drilling is proposed in the summer
season when the ice has melted and moves off shore. A high
quality detailed and localized weather and ice information service
can provide the margin of safety for drilling to continue without
risking catastrophe due to a sudden storm and/or the ice suddenly
moving into the area, and causing a major oil spill. Traditional
ice-forecasting methods will continue, but detailed local wind
forecasts produced by numerical methods will make available ice
drift forecasts (based on wind) to augment the regular system.
In developing this system, wind forecasts (and hence ice drift) will
be updated every hour. Hopefully, before long, this system can be
proven and adapted elsewhere in order that such detail can be
provided anywhere it may be required.

Data-Processing (Climatological)

In addition to the availability of real-time data, there is a
growing need for information on ice condition norms and departures
from norms. These relate to development of the North, in particular
ship design, harbour development, pipeline crossings, etc.
Concentration of ice by types can be published in atlases.
However, there is significance to occurrence, orientation and size
of pressure ridges and characteristics which should also be
processed. The problem is to find a storage and retrieval system
capable of taking data from the various sources including sensors
and storing it in such a way that it can be retrieved as required.
This can only be done effectively by computer methods. Thus, we
seem to be moving to a mix of standard atlases for broad-scale
data portrayal and a live retrieval system by computer for more
detailed information.

In summary, the main areas requiring improvement to weather
and ice systems are:
1. Communication of: (a) the output data to individual users;
                     (b) high resolution data to central process-
                         ing or forecast offices.
2. Improvement of Data Input from Data-Barren Areas By:
                     (a) development of greater capability to use
                         satellite data;
                     (b) for weather - greater use of automatic
                         stations; for ice - further development of
                         the use of SLAR to provide an all-weather,
                         day/night data-acquisition capability.

3. <u>Development and Refinement of Computer Processing Methods to</u>
   <u>Provide</u>: (a) improved broad-scale processing models;
            (b) meso-scale analyses and forecasts.

You will note in the above the strong commonality between weather
and ice services where only in Para. 2(b) was it necessary to
separate the recommendations.

The state of the art in Arctic weather and ice information
services has been fairly broadly covered in the foregoing. As you
can see, much needs to be done to support human activity in these
areas, particularly as regards fields of surface and air transportation - and most of the hurdles to be overcome are scientific
and technical in nature. This is the challenge to you as
representing the research community.

ARCTIC UNDERWATER OPERATIONAL SYSTEMS

Robert E. Francois

University of Washington

Seattle, Washington, U.S.A.

INTRODUCTION

Among the topics to be addressed in this workshop is that of
Arctic Underwater Operational Systems. In the workshop announcement such things as life support systems, remote manipulator
systems, and underwater navigation and sensing systems, as well as
overall undersea system management or integration, were suggested
as sub-topics to be dealt with under this subject. In this paper,
I shall address the broader issues and problems associated with
the successful employment of such systems in the arctic undersea
environment and hopefully indicate areas in which some additional
research and development would be fruitful in extending existing
technologies to arctic undersea problems in general, as well as
delineate areas of unique knowledge gaps which hinder progress in
this area. I will conclude this paper with some examples from my
own experience which illustrate a method of approaching undersea
operational problems in the Arctic.

The Arctic has been defined, for the purposes of this meeting,
as those land areas where permafrost is found and ocean regions
where sea ice is encountered in the northern hemisphere. Now as we
all know, the reason for the new interest in the Arctic is its
known and still-to-be-discovered reserves of hydrocarbon and
mineral resources. Hence, the pressing set of arctic undersea
problems is related to the exploration and exploitation of these
resources. The things that make this difficult are the subject of
this workshop. Note that although exploration and exploitation are
members of a common set, the methods, technologies, and problems
of each are not necessarily the same. Oil exploration may require
extensive seismic surveying and test well drilling where, in
principle, one may select favorable times and make use of the most

mild annual environment so that conventional systems can be employed. On the other hand, this does little toward building the technological base for ultimate exploitation of the resource.

For the sea-related development of the hydrocarbon resource, we are concerned with operations that must take place in a water environment that is comparable to the depth limits of, say, the North Sea oil field, of a water temperature within only a few degrees of the latter -- but below the freezing temperature of fresh water, and of a relatively stable sea surface -- but composed of thick hard sea ice. We are concerned with the necessity of working from that sea ice surface and making use of it when we can while devising methods of coping with that interface when the transition from sea bottom to surface or shore is necessary.

Consider, for a moment, the problem of laying a pipeline on the sea bottom between two northern islands in the Canadian archipelago. Only general bathymetric data for the site is initially available and detailed information must be developed as part of the project. We can expect annual air temperatures to range between +10° and -50° Celsius, and precipitation (mostly snow) of about 25-30 cm per year. Winds typically may average 10 to 20 knots. The channel between the islands may never be ice free, even at the peak of the melt season in the first weeks of August. The channel will be accessible to 9,000-ton icebreakers for perhaps a month during favorable years. The ice will form to thicknesses of 1-1/2 to 2 meters over the season. Some pressure ridging can be expected. Typical pressure ridge heights may run from 1 to 3 meters, implying a corresponding keep depth of 3 to 25 meters. Many pieces of old pack ice or bergy bits may be trapped within the annual ice. Extended transit over the ice may not be possible. Some movement of the ice may occur at any time, depending upon wind, proximity of the adjacent land masses and movement of the arctic pack outside the archipelago. We assume that the use of divers and submersibles would be desirable to accomplish specific portions of the installation problem. Let us examine, then, the problems related to these technologies in the Arctic.

ARCTIC DIVING

Arctic diving has been accomplished successfully for many years, mostly by scientists in support of their research activities and by icebreaker personnel who inspect the ship hull and screws for damage at times of opportunity while on arctic patrol. The equipment modifications necessitated by the environment were largely found by trial and error and passed onto fellow divers. Recently an excellent publication on this subject has been prepared by Wallace Jenkins of the U.S. Naval Coastal Systems Laboratory.*

---

* "A Guide to Polar Diving," Wallace T. Jenkins, Naval Coastal System Laboratory, Panama City, Florida 32401.

Contributions from arctic investigators and government activities
were assembled by Jenkins and then subjected to an extensive
review. It is truly a "must" reference for anyone who wishes to
engage in arctic diving. I do not wish to infer that the problems
of arctic diving have been satisfactorily addressed by equipment
designers, however. To be sure, the modern air-inflatable dry
suit has moderated the hypothermia problem, but only for short dive
periods. The arctic diver will be able to well utilize and
appreciate the continued developments in conventional and
saturation diving related to personal comfort. The basic problem
with arctic diving is that the sea water temperature is or can be
below the freezing point of fresh water. Fresh water, that is
water vapor in the breath of the diver, may condense and freeze
within the breathing gas circuit, resulting in blockage. The
presence of water vapor in the breathing gas supply causes similar
problems.

For example, consider a conventional SCUBA system. The single-
hose regulator system is a source of freeze-up problems involving
the sea water itself. In this system, the regulator spring is
immersed in sea water. The regulator body temperature may drop
below the freezing temperature of the water vapor because of
expansion of the compressed gas. Ice forms within the mechanism
and blocks the air delivery system open, which further bootstraps
the freezing process. The modification to this system which
prevents the freeze-up of the sea water-exposed regulator spring
is to enclose the spring chamber within an antifreeze-filled rubber
cup. Jenkins wisely cautions against use of an unprotected single
hose regulator whenever the water temperature is below 38°F (3°C).

Two-hose regulator systems use a different mechanism arrangement. The first stage regulator spring derives its ambient pressure
reference from the delivery side of the breathing air circuit,
rather than from the water, so that the same kind of freezing problem
cannot occur. This mechanism is, however, somewhat more sensitive
to diver breath condensation-freezing problems than the single-hose
regulator system.

The necessity for use of moisture free air in arctic diving
cannot be overstressed. Fortunately, during the cold winter months,
the atmospheric air is a ready source of highly dried air. The
diving compressor should utilize this source of air, even though
it means extending the air intake outside of the building in which
the air compressor is operating or operating the air compressor
itself outside.

The seriousness of air supply blockage or free venting is
well understood for diving operations in general. However, in
arctic diving, dives that take place under the ice, all problems
are much more dangerous. The greatest risk, in my view, arises
when the diver allows his equipment and air supply to become overly
cooled prior to entering the water. The suits are uncomfortably
warm under room temperature conditioning and the diver, seeking
to obtain relief, exits from his dressing place and waits in the

cold prior to water entry. Freeze-up problems are very difficult
to avoid when this is allowed. For this reason, a well planned
dive, with quick water entry following dressing and a heated shelter
around the diving hole, is essential for safety purposes.

So far I have referred to arctic diving from the surface of a
sea-ice platform. How does saturation diving fit into such a
scenario? Well, this, I think, is one of the real challenges that
the environment poses. Normally saturation diving is accomplished
from a rather heavy support vessel, necessitated by the crane
requirements for handling the various chambers and deck decompression equipment. Can such systems be safely employed and be
useful in under-ice diving operations? What modification of
existing systems will be required to allow the sea ice platform to
substitute for the ship operations base? Obviously, the thickness
and stability of the ice cover are governing inputs. So is the
satisfactory resolution of logistic support problems. We hope to
address some of these areas in detail during the working sessions
of this meeting.

Before I leave the subject of diver life support, I should
mention some safety concerns of my own. I believe each diver
should have navigational aids which will allow him to return to
his recovery point -- I believe the diver tender should have precise
knowledge of the arctic diver's position at all time and that there
should be a communication link between the two. The required
sophistication of these aids depends upon the local environment --
visibility, ice cover conditions, distance and depth of the dive
from the diving platform, etc. Some of the elements are commercially
available. Other elements require development into a suitable
diver-oriented package, but utilize technology presently available
in other areas. For example, at our Laboratory, we have under
development a diver tracking/communication system which utilizes
a commercially available diver telephone system and a 3-dimensional
acoustic tracking system that uses pulsed signals from a diver-
carried transducer to establish his position. One of the problems
involves the location of the transducer on the diver to minimize
shadowing effects without interference with his working ability.
The importance of this type of aid for both safety and diver
operations management is slowly being recognized.

WORKING SUBMERSIBLES AND THE ARCTIC

A very recent summary of manned submersible and habitat
equipment and capabilities has been prepared by Joseph Vadus of

---

* J.R. Vadus, "International Review of Manned Submersibles and
Habitats," U.S. Dept. of Commerce, National Oceanographic and
Atmospheric Administration.

the Manned Undersea Science and Technology section of NOAA*. In this publication, a world-wide review and summary of manned submersibles including information on operator/owners, location, characteristics and utilization are presented. Of special interest to me was an appendix which presents a similar summary of unmanned submersibles. The free-swimming, untethered vehicles of our Laboratory, and, particularly, the Unmanned Arctic Research Submersible System (UARS), are types which I will later mention in some detail.

In the introduction to this paper, remote manipulator systems were identified as an item for discussion. In recent years the manned submersibles of the world have considerably expanded their working capabilities. In the early 60's, the small manned submersible's utility was largely restricted to those missions requiring human observation capability and photography under direct operator control. They satisfied the need for such necessary tasks as deep water sewer outfall inspection, pipeline and cable inspection, research related to the geology of the sea bottom, etc. Over the years, the manned submersible has greatly progressed from the passive observational role as more viable articulated arms and kindred systems have been developed and employed to do useful mechanical operations at the sea bottom. Such developments could be just as useful in the arctic regime as, say, in the North Sea. Consequently, we see no unique demand upon manipulator systems associated with this type of vehicle that would be related to use in the arctic environment per se. The major problem we see is in the use of small manned submersibles under the arctic ice canopy.

Consider for a moment the primary safety system of present manned submersibles. (Here I am referring to working vessels and not military mission submarines.) Safety or self-rescue must be achieved by the submersible in distress by reaching the sea surface. Almost invariably, this is accomplished by weight dropping or, in extremis, by releasing the manned compartment from the remainder of the vehicle. In arctic operations under sea ice cover, this is a non-viable approach. The reserve buoyancy of these submersibles is just not sufficient to rupture typical ice thicknesses.

In his book on ice breaking, MacDonald* presents an excellent summary of pertinent ice load bearing capacity (based on U.S. Army Cold Regions Research Engineering Laboratory research) for cargo handling upon sea ice. One example related to a 8-ton forklift which can safely operate on 20 inches (52 cm) of sea ice when the average air temperature has been 10°F (-12°C) for three days. To a first approximation, a buoyant force of 8 tons could be sustained by the 20-inch ice sheet as well. The typical manned submersible weight in air is about 12 tons (Pisces, Sea Link, etc.)

---

* MacDonald, "Polar Operations, U.S. Naval Institute, 1969"

and the maximum buoyant force is obviously only a small fraction of this. Ice cover can grow at very high rates. During this past March, I was engaged in acoustic research some 100 miles north of Pt. Barrow, Alaska. I measured the thickness of a wide refrozen lead at 40 cm (16 inches) after 10 days. These latitude and prevailing weather conditions are quite typical of those that apply throughout the southern portion of the Canadian Archipelago. For these reasons, dependence upon favorable conditions, that is, open water or thin ice, for ultimate manned submersible safety is precarious at best, at least during the winter season.

So we have a dilemma. During the winter season, the ice canopy can probably be used as a working platform for the manned submersible, but during the summer season, we can't depend upon the ice so we must support the submersible from a ship. But the ship cannot typically reach the site except for a brief period during late summer. It then appears that the period of usefulness for the small manned submersible in the Arctic is, at present, extremely limited.

What can be done to improve this situation? For one thing, a dependable rescue system that could cope with the interface would be extremely helpful. Such a system would undoubtedly require standardization of equipment on arctic manned submersibles. Acoustic location rescue gear would require suitable sources to be carried by under-ice submersibles. This could require automatic deployment of tethered acoustic pingers so that an acoustic path between the location gear and the "surfaced" submersible would be available, regardless of pressure ridge keels which could otherwise block the acoustic path. It could mean that a standard extreme positive buoyancy condition must be prescribed so that submersibles could be maneuvered to a safe recovery position. One recovery system would involve precisely locating the submersible, cutting a recovery hole through the ice and sending a diver to attach a line to a standard attachment point on the submersible. A weight slid down the line would allow the submersible to be maneuvered directly under the recovery hole. This type of system was developed for use with our UARS unmanned submersible which I will touch on later.

My personal feeling is that as long as the government entities concerned have a responsibility to provide emergency rescue service, that they have, in turn, the duty to require conformance with rescue equipment standardization on submersibles that may require that service. Certainly international cooperation in this matter is highly desirable.

UNMANNED SUBMERSIBLES

Now what about unmanned submersibles -- what is their possible role in development of the arctic resources? Lets look at the characteristics of these vehicles. Generally, they fall into two classes, tethered and untethered. During the 50's our Laboratory developed precision undersea 3-dimensional tracking ranges for the U.S. Navy. These ranges were used in the development and testing of sophisticated homing torpedos. Negative buoyant units were recovered from 600 feet of water at the Naval Torpedo Station, Keyport, Washington, using the known position on the bottom (from the 3-D range) and the similarly monitored position of a television camera and grabber. Jacobson Bros. of Seattle devised the latter system. It involved a television camera, recovery clamp and several guide lines, and anchored pulleys which allowed the recovery device to be brought to the known coordinates of a bottomed unit. Eventually, this system was expanded to one known as J-Star which has a lift capability of 12 tons in water depths to 3,000 feet. Somewhat similar units, SORD I and II, were later developed by the Naval Torpedo Station. These units can operate in water depths to 6,500 feet. Probably the best known cable-controlled devices belong to the CURV family (Figure 1). These devices, developed and operated by the Naval Undersea Center, San Diego, utilize a single cable for information transfer, control, power and lift. The instrument package framework is equipped with motor-driven thrusters which allow it to be maneuvered under surface operator control. Television and sonar sensors provide operator information. A variety of manipulators are attachable to the system depending upon the recovery mission.

The preceding tethered submersibles do not depend upon movement of the surface support vessel to gain limited horizontal mobility. An example of a submersible with large horizontal mobility is the MIZAR FISH which is a cable-towed body associated with the oceanographic research vessel MIZAR, operated by the Naval Research Laboratory. The FISH is a stabilized open frame which is equipped with a variety of cameras, magnetometers, and other sensors. It has been successfully employed in ocean bottom search operations as well as for geologic research in the central oceans.

A unique undersea submersible, the RUM (Remote Underwater Manipulator), was developed by the Marine Physical Laboratory of Scripps Institute of Oceanography some 15 years ago and has been operated by them since. It is a tracked vehicle, derived from a military personnel carrier or light tank, which is launched from the shore line and is guided and powered by a cable which it unreels from its back. It is used to inspect (television) and service (manipulator) submarine cables on the ocean floor within range of its cable supply. This approach could be a viable one for dealing with nearshore petroleum field service problems in the Arctic.

Figure 1   Unmanned Submersible - CURV III

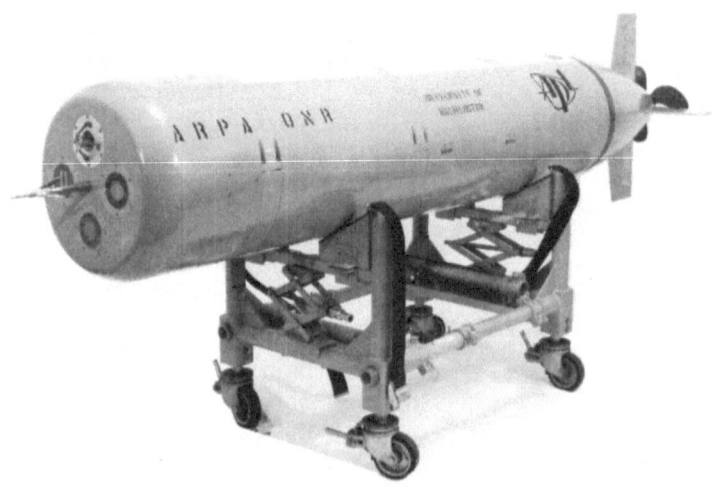

Figure 2   Unmanned Arctic Research Submersible System (UARS) -
University of Washington, Seattle

The next class of submersibles to be considered are unmanned, untethered, free-swimming vehicles. The first vehicle of this type was developed at our Laboratory in 1960. It was designed to support ocean research relating to long-range acoustic transmission. A considerable portion of the acoustic path under study occurs at deep depths so the system, known as SPURV, was designed to operate at depths to 3,500 meters and for distances of 30 miles in order to assess the horizontal distribution of the phenomena under study. The SPURV units operate at a typical speed of 5 knots. They are acoustically tracked with respect to their mother vessel, which maintains station more or less over the SPURV during operation. The acoustic tracking system is basically an inverted arrangement of the 3-D tracking system of our Laboratory. Several types of two-way acoustic communication links have been developed and used with this system.

In 1970, we began development of an arctic research version of this system. It would be fruitful, I think, to describe this system in some detail since its development illustrates most of the complexity facing arctic under-sea systems.

THE UNMANNED ARCTIC RESEARCH SUBMERSIBLE (UARS) SYSTEM*

As part of an Advanced Research Projects Agency-Office of Naval Research sponsored arctic technology program at the University of Washington, an Unmanned Arctic Research Submersible System was developed and successfully employed. This system, known as UARS, comprises two major elements -- the submersible which serves as a mobile instrument carrier, and an acoustic tracking, command, and recovery system. The first year of the program was devoted to design, with test hardware limited to breadboard assemblies, while during the second year the system was fabricated, tested, and its performance demonstrated in an arctic under-ice experimental program.

The UARS is a compact, torpedo-like vehicle which weighs 900 pounds in air and has a length of approximately 10 feet and a diameter of 19 inches (Figures 2 and 3). The hull is fabricated from aluminum forgings and filament-wound fiberglass to give a 1500-foot operating depth capability. Its tri-axial control system is designed to accommodate speeds as low as 3 knots; the speed of the unit operated in the Arctic was 3.7 knots. Higher speeds can be obtained by motor substitution. The main batteries (silver-zinc) supply sufficient energy for run times in excess of 10 hours with an equal propulsion and instrumentation power load.

---

* R.E. Francois, "The Unmanned Arctic Research Submersible System," Marine Technology Society Journal, Vol. 7, No. 1, 1973.

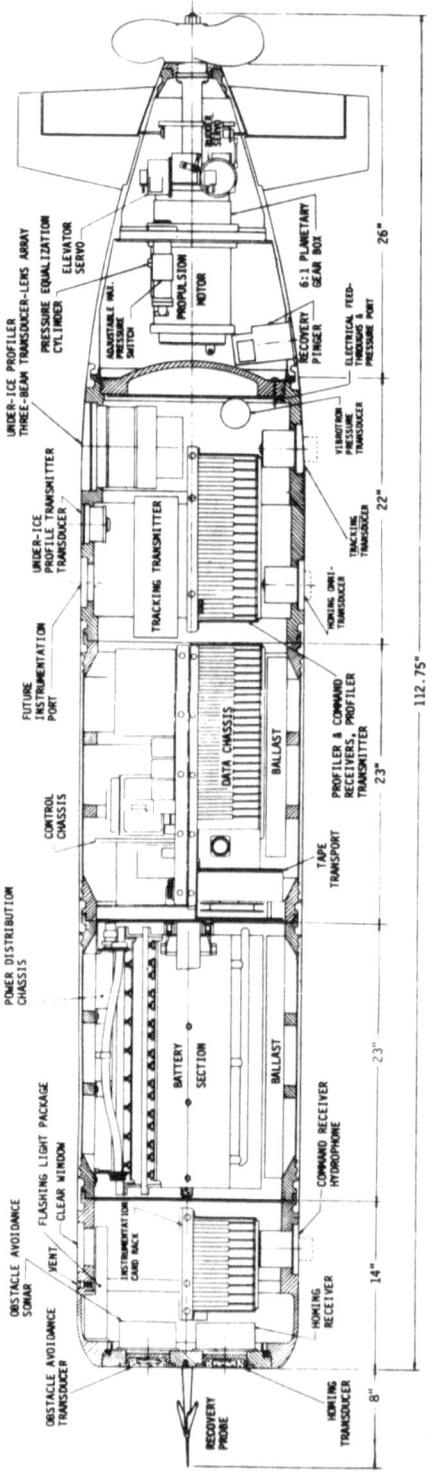

Figure 3 Unmanned Arctic Research Submersible System (UARS) —
University of Washington, Seattle

A reserve battery is carried for emergency purposes. At a speed
of 3 knots, the vehicle is operated at 10 pounds positive buoyancy.

In addition to acoustic instrumentation for measuring
physical phenomena, the vehicle carried several other acoustic
systems which are used for communications (both to and from UARS),
tracking, homing, collision avoidance, and emergency recovery.
The obstacle avoidance sonar is necessary because of potential
pressure ridge keel projections to the desired operating depth of
the UARS. Included in the fixed instrumentation suite of UARS is
an ice profiler system with which the elevation of the under-ice
surface is measured and digitally recorded five times a second,
on each of three separate, narrow, upward-looking beams, to provide
an overall elevation accuracy of 0.3 feet. Fine grain bathymetry
can be accomplished by inverting this profiler system.

The vehicle is launched in a horizontal attitude after being
lowered through a 4 x 12 foot hole in the ice. The motor is started
before release from a special launch rack; the UARS rises about 2
feet before full depth control is achieved and begins to dive or
climb to its preset initial depth. The vehicle controls its depth
to within 1/4 foot and within 1/2 degree maximum peak-to-peak
excursions in pitch. Roll is controlled during straight runs
within the same values although slightly larger excursions are
evident during homing maneuvers. All control functions are
recorded, along with measurement data, on a special 9-track digital
tape recorder which stores in excess of 1,000 binary bits per
second.

The position of the UARS is known at all times from
information derived from an acoustic tracking system (Figure 4).
The principal elements of this system are: a projector aboard the
UARS which transmits a unique pulse code, an array of four or more
hydrophone/decoder/RF-telemetering buoys arranged in a pattern
within the experiment or survey area, two base line acoustic
transducers within the experiment area which survey the location
of the tracking hydrophones and provide a coordinate reference
axis, and the timing units, data processors and computer system
which provide the real-time interpretation of acoustic information
and position calculations. The real-time position, along with
data measured by UARS, is derived from a ten-bit phase coded
acoustic pulse transmitted every second from the vehicle, and
provides the information for intelligent remote control of the
UARS during the experiment runs. The hydrophones are designed as
free-floating buoys, but are usually frozen in place. At the
power levels and frequencies used, the acoustic tracking has an
effective range in excess of 8,000 feet under typical arctic spring
ambient noise conditions. The command/communication and tracking
systems use the same acoustic frequency, 50 kHz. Up to 16 command
functions are presently available for controlling UARS.

The UARS is recovered by ensnaring it in a net. The capture
net contains an acoustic beacon which the UARS homing system is
commanded to seek at the appropriate time. In the event that

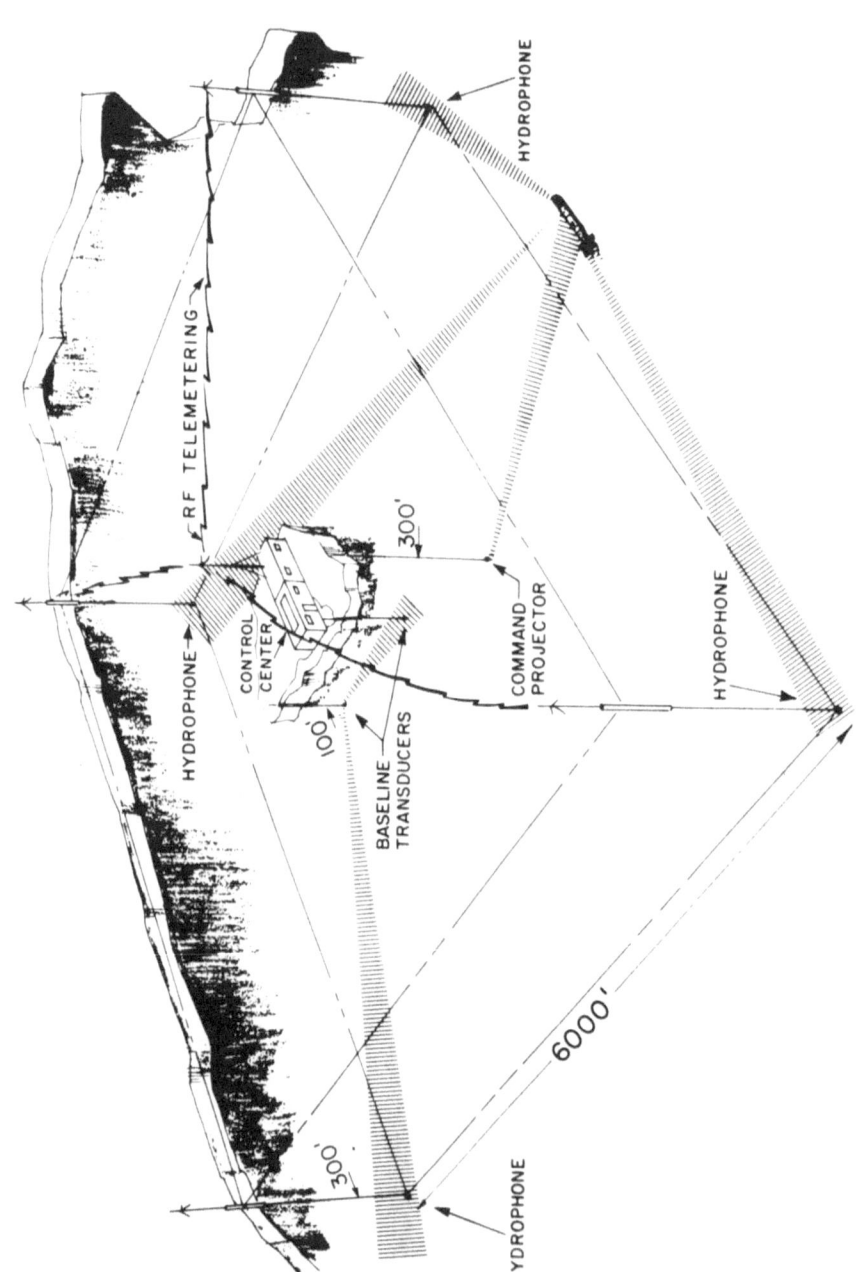

Figure 4 Acoustic Tracking System for UARS operation under water

command communication with the UARS is lost for a preset period of
time, the homing system will automatically activate and UARS will
begin a search for the beacon. At present power levels, the
effective range of the homing system is two to three miles,
depending upon noise background conditions. Internally programmed
logic, an inertial and depth-sensing guidance system, and the
command/tracking receivers provide retrieval redundancy. In the
event of massive power interruption or other catastrophic failure,
a further retrieval capability is provided. The submersible has
positive buoyancy and will rise to the undersurface of the ice and
automatically lower an acoustic beacon to aid an over-the-ice
search party. The system includes a beacon location device and
appropriate tools for emergency recovery of the vehicle once it is
located.

The UARS has about 150 pounds of reserve buoyancy which can
be devoted to instrumentation over and above that already installed.
There are several ports in the body which allow oceanographic,
optical, and acoustic instrumentation to be mounted without major
effort. After test runs; internally recorded data and externally
measured position data are merged on one tape so that spatial-
temporal correlation of observed phenomena can be accomplished.

The complete system configuration was first achieved in late
February, 1972, when tests of the UARS began in Lake Washington.
The usual interaction problems of complex systems were largely
resolved during the 20-run test program. This testing program was
concluded in late March and the entire system deployed to Fletcher's
Ice Island (T-3) for full scale arctic testing. At this time, the
ice island was at approximately 85°N lat, 85°W long, about 150
miles north of Ellesmere Island. The final development testing
progressed in an orderly manner, beginning with tethered operations
in the vicinity of the hydrohole and proceeding to untethered
vehicle runs at full design distances.

The vehicle's first free run in the under-ice environment was
on 3 May, 1972. This run lasted slightly over one hour. Almost
all acoustic systems indicated some problem areas related to the
arctic under-ice environment as compared with the Lake Washington
tests. The analyses and subsequent experimentation led to
resolution of these problems. The final run of the 1972 series
was made on 9 May (Figure 5). The vehicle was operated within a
radius of approximately one-half mile from the launch hydrohole
and was acoustically commanded to follow a rosette-type run
pattern. The run distance was in excess of 17 miles, and the run
duration was greater than 4 hours. The acoustic tracking and
communication from the vehicle was excellent throughout the run.
Over 200,000 position-correlated profile measurements of the under-
ice surface elevations were made. Pressure ridge keels to depths
exceeding 25 meters were observed. A small section of the ice
profile, combined in the upper surface observations, is shown in
Figure 6.

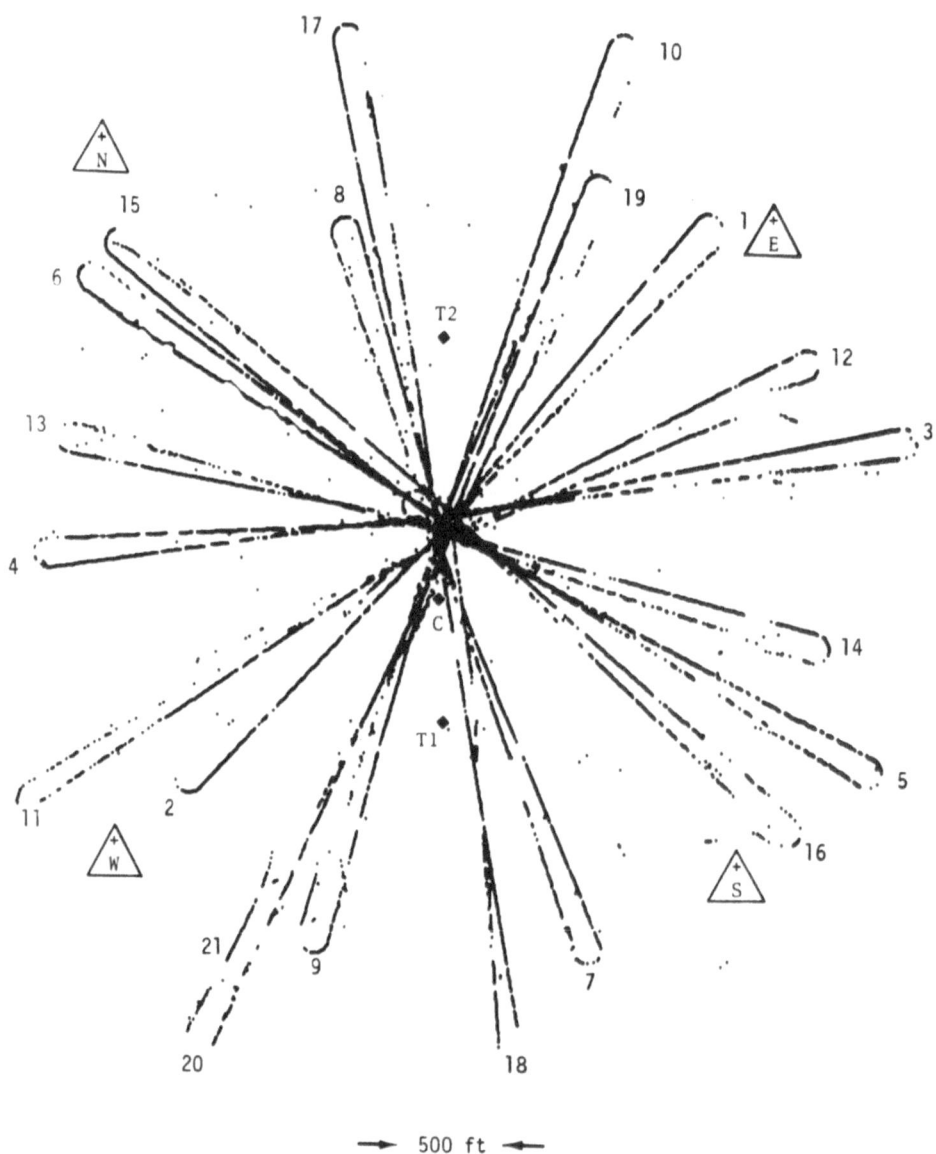

Figure 5   UARS Acoustically Commanded to Follow
a Rosette-Type Run Pattern

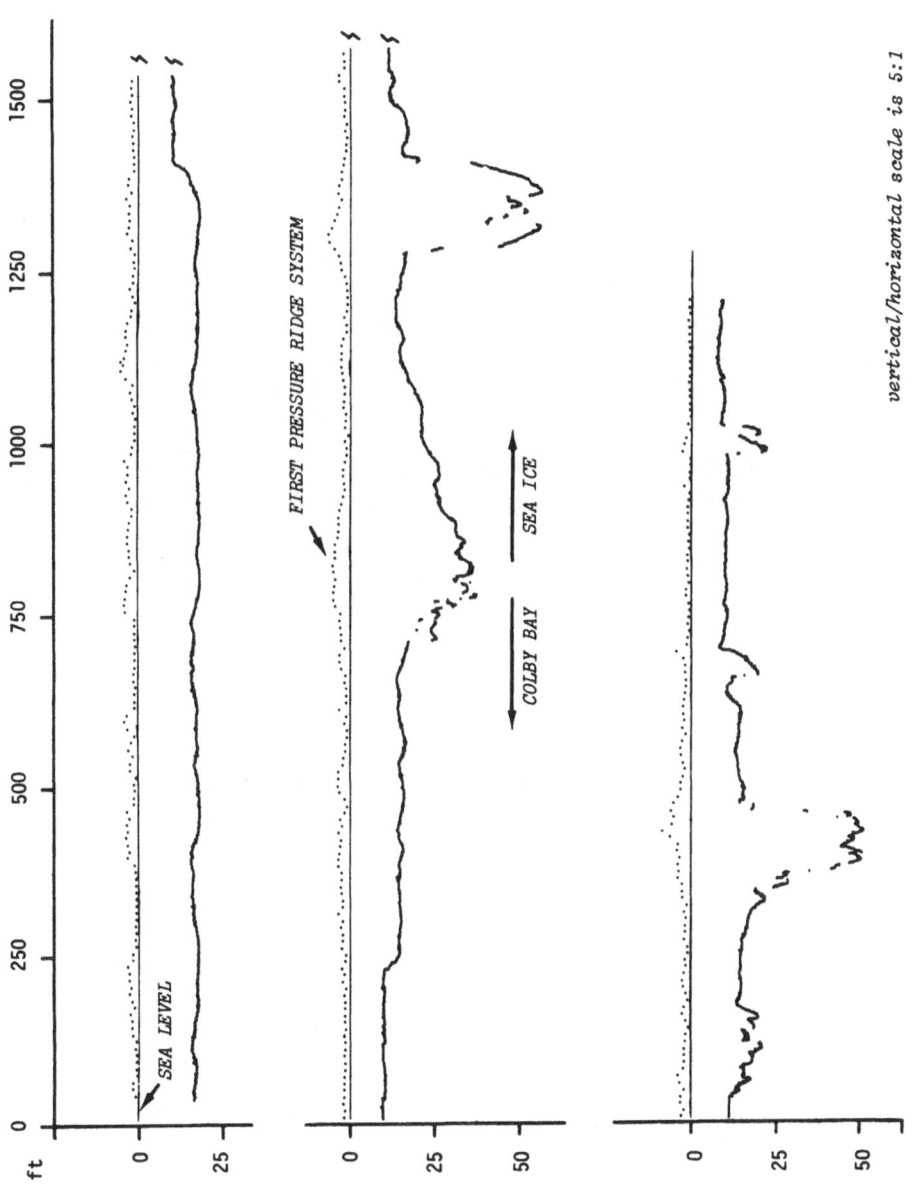

Figure 6 Position-correlated Ice Profile, Combined in the Upper Surface Observations

The normal vehicle recovery system was successful and there was no need to employ the emergency recovery system. At the conclusion of the field tests all frozen-in tracking instrumentation was recovered from the ice, using the thermal coring technique developed in this program. The launch/recovery hydrohole (4' x 12' x 28' deep) was also made using this device, and I'm going to say more about this later on.

The UARS technology is applicable to a wide variety of operations which require under-ice mobility. These include acoustic measurement programs, bathymetric surveying, messenger line laying, etc., as well as basic research of an oceanographic character.

GAINING ACCESS TO THE SEA THROUGH THE ICE INTERFACE

One of the major problems with working in the ice covered seas is that of gaining access to the sea from the ice canopy. Two past methods have been used by scientific researchers to make working hydroholes in sea ice. One method involves a quarrying approach wherein a chain saw is used to make vertical cuts in the ice to form multiple squares. Wedges are used to split out ice blocks, one layer at a time. The final layer is removed by chipping away with a chisel mounted on the end of a suitable length of pipe. (Sometimes explosives are used for the final layer.) The other method involves the use of explosives. In thick ice, this requires a significant quantity of explosive and results in a crater-like opening, filled with ice fragments of various sizes, which must be removed before a diver can safely submerge. Use of explosives in the vicinity of an under-ice submersible is highly questionable.

As part of the UARS system program, we developed a sure method of making large diameter holes in level sea ice of any thickness. The system uses thermal energy in the form of heated water delivered to the ice in a controlled manner to cut a groove of the desired shape. A delivery manifold, either circular or rectangular, with downward-facing orifices is designed to distribute water uniformly along the manifold or cutting head. Melting of the ice is caused almost entirely by convective heat transfer. A similar suction manifold, mounted directly above the delivery manifold, returns the mixed melt and delivery water to the heat source for reheating after first discarding the excess water. A dry groove is desired under winter conditions so that refreezing of the groove is impossible because of the absence of water. When penetration is complete, sea water floods the melted groove and the core is left floating in the water. If coring is done around an instrument, it is then free to be removed. If a hole in the ice is desired, rather than the ice core sample, the technique used to dispose of the core is to push it down through the hole where it will come to rest under the adjacent ice. This requires only 1/4 the energy and 1/8 the maximum force that lifting the core would entail.

Figure 7   Device to cut Sea Ice - 23 kw (80,000 Btu/hr).

Figure 8   Ice Hole 71 cm (28 inch) Diameter, 5.5m (18 feet) thick.

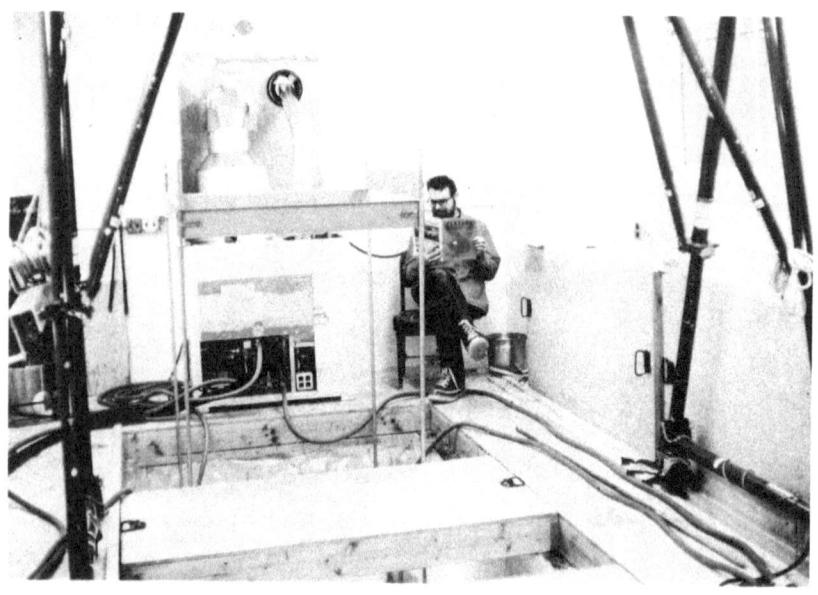

Figure 9   Device to cut Sea Ice - 64 kw (220,000 Btu/hr).

Figure 10   Ice Hole 1.22m (4 feet) wide by 3.66m (12 feet) long
and 8.54m (28 feet) thick

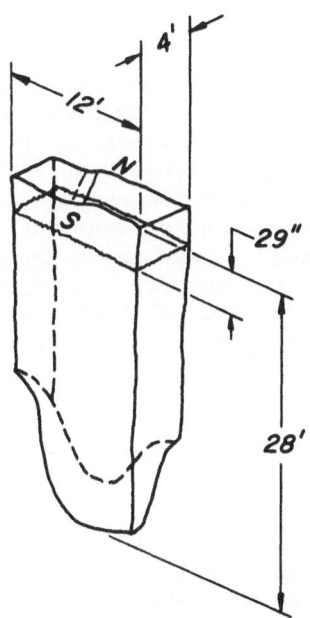

Figure 11 Profile of an Ice Block cut by
64 kw (220,000 Btu/hr) device –
300 miles from North Pole

This disposal system can be effectively used whenever the core
length-to-diameter ratio exceeds about 2. For very large hole
areas, the ice is cut into several columns which are lifted out of
the water with a ski-equipped frame hoist. Sometimes long columns
are broken into lengths that can be more easily handled.

Several sizes of this device have been built and successfully
employed. The first operational unit, 23 kW (80 000 Btu/hr)
output, made 71 cm (28 inch) diameter holes in Greenland Sea ice
up to 5.5 m (18 feet) thick at a 1.5 m (5 feet) per hour rate
(Figures 7 and 8). Another unit, 64 kW (220 000 Btu/hr) (Figures
9 and 10), made an ice hydrohole 1.22 m (4 feet) wide by 3.66 m
(12 feet) long in sea ice 8.54 m (28 feet) thick some 300 miles
from the North Pole (Figure 11). Large holes can be made in almost
any desired shape depending upon suitable cutter head designs and
multiple melts. The maximum melt depth is limited only by thermal
drill auxiliary equipment such as length of cutting hoses, weight
of cutter head and guide pipes (or vertical support mechanism),
and available fuel supply. The largest unit thus far built has an
output of 95 kW. It consists of two packages, a heat exchanger
(238 kg) and a support box which includes fuel, generator, hoses,
cutting head, etc. (225 kg). These units are helicopter or sled
transportable. (For complete description of this system see
R.E. Francois and J.G. Harrison, "A Thermal Drill for Making Large
Holes in Sea Ice," OCEAN 75 Record, IEEE Publication No. 75CHO
995-1 OCC.)

UNDERWATER NAVIGATION AND RELATED SENSING SYSTEMS

We have presented some of the background problems and some
applicable approaches to their solution for the employment of
conventional undersea technology to the unique problems of the
Arctic. Now we turn to the problems of undersea navigation and
other sensing systems.

In working in the arctic undersea environment with any type
of submersible, the navigation/location problem can be approached
in ways that are perhaps somewhat different from techniques that
would be employed in the open ocean or non-ice covered seas. When
one works on the sea bottom with divers or manned submersibles in
the ice covered sea, two separate navigation problems must be
resolved. The first problem is getting to the site of the work on
the bottom; the second is returning from the work site to a
specific point of egress. The first problem requires that a
reference system be established on the sea bottom and that the
reference system then be related to geographical space. This
involves a combination of, in general, acoustic transponders on
the sea bottom (minimum of 2 units, preferably 3) and an optical
or RF positioning system on the ice surface which can tie back to
geographical reference points on land. The elements of a workable
system exist, although there are some unique arctic system problems

related to the acoustic velocity structure, water depth, and ice
canopy drift. The generally positive sound velocity gradients
that obtain in the Arctic cause upward refraction of sound rays,
so that the direct acoustic path horizontal range is limited by
the bottom grazing ray. Velocity structure within the Canadian
Archipelago is not well known throughout the year. However, if
one assumes a typical Arctic Ocean gradient of +0.06 m/sec/m depth
(down to 300 meters) and a 300-meter water depth, the maximum direct
path horizontal range is 3800 meters (2700 m in 150-m depth).
Acoustic linking over greater distances requires use of a sound
transmission path involving reflection(s) from the ice undersurface. Because of the pressure ridge keel structure, the
reflection problem is highly complex. For example, we have derived
the slope statistics of the under-ice surface from direct
measurements made with the UARS system. Over a 1.5-km traverse
which includes several pressure ridge keel observations, the
standard deviation of the slope, over a 6-m sample interval
(corresponding to a 4-msec pulse length), was 16°. This implies
that large signal fluctuations will occur in multipath propagation
as indeed we have measured in our acoustic research work.

Now a range of 3800 meters may not appear particularly
alarming -- except when the return navigation problem for manned
submersibles is considered. The ice movement during the mission
must be considered and an immediate recall order issued when the
critical range is being approached. Normal communication with
submersibles is accomplished with some form of UQC -- a single
band communication system utilizing the 9-11 kHz band. In a
shallow water situation, there can be serious voice intelligibility
problems due to the multipath-wave guide effects, particularly
under ice. This is due to the roughness of the under-ice surface
and the fact that it, unlike the sea surface, is fixed so the
simple averaging or integration approach to signal enhancement is
not applicable.

We can summarize the acoustic problems briefly. It is not to
be expected that acoustic techniques which work satisfactorily in
the open ocean, with the tender vessel essentially over the
submersible, will automatically work in the arctic environment
where different geometries, velocity gradients, and surface
conditions obtain. On the other hand, we see no reason that
acoustic techniques cannot be reliably employed if the physical
limitations, the operational philosophy and procedures, and the
system integration are approached in a rational manner. This
requires, however, information on the environment that, to a large
extent, does not presently exist. In any case, acoustic systems
for use in supporting arctic under-ice operations should be tested
in the environment to establish the adequacy of the approach prior
to full deployment of the total undersea system.

While acoustic systems dominate medium-and long-range undersea
sensing requirements, at short range optical sensors play an
important role in submersible and diving operations. In contrast

with most areas of the world, the light transmission properties of
the arctic sea water, at least under ice cover, are excellent.
This is due to the blocking of solar energy transmission into the
sea water, which inhibits the photosynthesis process and the related
biological chain development which is important to the scattering
process, both for optical and acoustic energy. We have clearly
observed our UARS submersible operating at 80 meters depth, viewing
it from the control hydrohut when it passed through a wide-angle
light beam (100 watt light bulb equivalent) at its depth. So there
can be some advantages associated with the arctic sea medium.

SOME FINAL THOUGHTS

I have found that one of the most useful techniques in the
design of arctic systems is the preparation of a scenario of the
actual operation in outline form, from the very beginning of the
field operation through withdrawal from the field. Generally,
there is a mission objective that can be clearly stated. Then all
other functions are in support of that mission objective. As the
outline is more completely filled in, areas of uncertainty will
arise. How can a support objective, in the past accomplished
from a ship in more temperate waters, be carried out from the sea-
ice platform? Can the system be packaged in modules compatible
with specific freighter aircraft or helicopters? How do we avoid
cold-soaking equipment which must be lowered into the sea? How do
we prevent freeze-up of drain ports of submersible equipments?
What are the specific unknowns with respect to the environment that
apply to the various support systems design? These unknowns need
early identification so that workable engineering design criteria
can be established. In a sense, the outline scenario provides the
"straw men" needed to scope the project and its problems. Often,
we find that the auxiliary systems associated with a project
present much more formidable problems than does the central
objective.

The importance of identifying the unknowns and, concomittantly,
the risks in a given approach to an operational objective cannot
be overstressed. In the Arctic, we must not resort to brute force
approaches against nature. Rather we must seek ways to accommodate
our objectives with natural forces and to apply the latter to our
needs. Consequently, the better understanding that we acquire of
this region of the world, the greater will be the possibility of
harnessing this environment to better serve humanity's needs.

The problems in developing viable arctic undersea operational
systems to accomplish specified objectives are largely associated
with the water-sea ice interface. Successful accomplishment of
major undersea engineering projects in the Arctic depends upon how
well we can blend imagination and knowledge of that environment in
a complementary manner.

ARCTIC SOCIAL AND CULTURAL SYSTEMS

Claus Bornemann

Ministry for Greenland

Copenhagen, Denmark

Thank you for the great honour which I consider it to be to deliver this key-note paper at the opening of our Conference. I must admit, though, that the topic of "Arctic Social and Cultural Systems", on which the organizers of the Conference have asked me to speak, is a title that has given me considerable difficulties. I have given much thought to the concept of it, and feel that it is a topic of considerable universality, embracing geography, population and problematic issues. I have decided, therefore, to speak solely on the arctic minorities and their ways of life. This I have done, because the Conference Program has been concentrated to such a large extent on scientific and technological issues. I feel it important from the very beginning to point out to my audience that for thousands of years the territories of land which we are to deal with during the coming week, have been inhabited by peoples with other cultural traits and social habits than known to us from our own industrial countries.

My knowledge of these territories was gained first and foremost from my twenty-five years of employment in the Greenland Service and on several trips in Greenland, where I have been living for nine years. Moreover, I have been on a long trip to Alaska to study the Eskimo population there and have spent some time with Lapp friends in the Norwegian Finmark, apart from also having paid a number of visits to Canada, where I have had the opportunity of several useful conversations with natives and others who were familiar with arctic conditions. Though I have thus been in touch with other arctic territories, I am able to speak, with a certain degree of expert knowledge, on Greenland affairs exclusively.

Let me add also that on this occasion, I speak on my own behalf exclusively, and what I say may not always express the views held by my Ministry in the same respects. I have considered

it my duty on the basis of own experience and knowledge to present
a contribution that will give rise to a debate within the working
group set up to deal with cultural and social questions during the
next few days.

I may perhaps put rather too much emphasis on certain problems
by including too many quotations from natives and experts who tend
especially to support the views held by the minorities. This I
have chosen to do because it so seldom happens that such views are
being expressed at conferences of a technical nature, and also
because I want to point out to this audience that there is a
decisive tendency towards a political awakening extending from the
Alaskan Indians and Eskimos in the west to the Scandinavian Lapps
in the east, an awakening which will necessarily make its impact
felt on the discussions at this Conference. If my speech becomes
too one-sided, I trust that balance can be achieved in the working
group.

The population groups involved are but few and scattered.
Alaska has about 65,000 natives: Eskimos, Indians and Aleuts.
The Inuits (Eskimos) in Canada number 17,500, and apart therefrom
the Yukon and N.W.T. have about 10,000 status Indians and 15,000
metis and non-status Indians. Greenland has a population of some
42,000 Greenlanders and 7,000 Danish inhabitants. The 40,000
Lapps in Northern Scandinavia comprise 20,000 in Norway, 7-8,000
in Sweden and 2,000 in Finland. The arctic minorities in Soviet
Russia comprise 140,000 people distributed on twenty different
tribes including a few thousand Eskimos and Lapps. One thing
these circumpolar groups have in common is that they have been
living in these territories for thousands of years while the non-
natives have been there for a few centuries only. I do not wish
to take any part in the present tendency to romanticize the social
conditions of the so-called primitive peoples. I want to emphasize,
however, that these minorities have managed to survive in some of
the most inhospitable regions of the Earth for thousands of years,
a fact which proves that their social structure has been both
tough and flexible, and that they have been able to adapt their
culture to the rigors of Nature.

One thing that the arctic population groups have had in common
is that until a comparatively recent date they have existed almost
exclusively as hunters, trappers, fishermen and nomadic herdsmen
of reindeer. They lived in a state of nature. Indeed, you might
almost say that they became an integral part of their natural
environments. To control the land and its resources of game and
fish is to them not only an economic problem - it constitutes an
essential part of their culture.

The meeting between the arctic population groups and the
white peoples coming from the south did, in most regions, not take
place until a few centuries ago. Only the Lapps in northern
Scandinavia have had a thousand year old relationship with the
Europeans, and already in the ninth century some of them had to
pay tax to a Norwegian king. The colonization of the territories

occupied by the Lapps did not start properly until the seventeenth century. Greenland and Alaska had no permanent relationship with the dominant nations until the eighteenth century. The Danish/ Norwegian pastor, Hans Egede, came to Greenland in 1721 and started the Danish colonization there, while the Russians arrived in Alaska in 1741. Frobisher established contact with the Eskimos in northern Canada in 1670, but many years were to pass before non-natives became permanently settled in the present Canadian arctic regions, and it is not more than one generation ago that the most isolated Eskimo groups came into close contact with the civilization of the white man.

The meeting between the newcomers from the outside world and the arctic peoples became disastrous to the aborigines, with the exception, perhaps, of the Greenlanders as the Danish Government closed Greenland to outsiders in order to protect the native population. In Canada and Alaska alike, there was a heavy decline in the native population caused by epidemic disease, exploitation and disruption of the cultural pattern. When the Russians settled in Alaska in 1799 this region had a native population of 75,000. This figure had been reduced to 35,000 by the year 1867 when the U.S.A. bought Alaska at a price of U.S. $7.2 million.

The declining trend continued right up until the present time, and there was no marked recovery until after World War II when a social welfare policy was introduced. Thanks to an active effort within the field of public health and the establishment of an improved social welfare service, a regular population explosion, on the other hand, set in during the 1950's and 60's. In certain regions the growth rate did, in fact, reach a record high of as much as 4 per cent per year. This heavy population growth continued everywhere with the exception of Greenland where the implementation of an active family-planning program during the past five years has served to moderate the growth rate considerably.

Also in the present situation, we find a number of common features when comparing the territories in northern Scandinavia, Greenland, northern Canada and Alaska. Up until the Russian Revolution the Russian arctic territories were just as badly neglected and exploited as were the other arctic territories, but as will be known, they have undergone a special development since World War I. My information on the U.S.S.R. is, however, fairly limited, and in the following I will, therefore, concentrate on arctic population groups under the sovereignty of so-called western nations.

The first contact between the arctic populations and the white man took the form of meetings with whale-hunters, trappers, members of the clergy and subsequently also of the administrative authorities as for example the R.C.M.P. in northern Canada. A small number of white people only were involved, and the efforts made at administration were few and weak.

World War II brought a radical change in the situation of the arctic populations. The arctic territories became of military

importance, and money could be earned by working on the military
installations. After the war the governments have taken an ever
increasing interest in the arctic territories, and I feel that the
reasons for this have been two-fold. One was the wish to develop
the mineral and petroleum resources discovered by the geologists
up north; the other was a growing appreciation among the people
and politicians of the dominant nations that it would no longer do
to neglect the arctic population groups and leave them in a state
of social misery. Important was also the debate on the developing
countries and the Third World, and it is a characteristic feature
that in the 1960's the following slogan was used in Denmark:
Greenland is Denmark's developing country.

As a result of the mentioned situation, the arctic territories
witnessed an inrush of administrative personnel, business people,
scientists and military personnel. New communities were formed,
mostly of an urban character, and the newcomers brought their own
customs with them from the industrialized communities. The
indigenous populations, who had so far been living in small,
scattered groups, gradually drifted into the new urban communities
in ever larger numbers, which gave rise to a pronounced social
stratification, with the upper strata comprising people from
outside. Among them may especially be mentioned civil servants
sent out by their central governments. These officials were
provided with well-equipped official residences and assured a good
income to compensate them for the privations which they claimed
were inflicted upon them during their arctic stay. As to the
majority of outsiders their stay is of a temporary nature; very
few settle permanently in the respective territory. Therefore,
the arctic towns carry the stamp of something artificial. At the
very bottom of the urban communities are the natives, who because
of a lack of occupational training and unfamiliarity with the
industrial culture find it difficult to meet the competition.
Notwithstanding the vigorous efforts made by the various government
institutions, the economic development has so far, in most places,
left the natives in the lurch. Not wanting to take a too gloomy
view of the situation, I may add, though, that an ever increasing
number of natives, who have the necessary educational background,
manage to make a place for themselves in the modern urban
communities, with a greater degree of security than previously,
and special mention should be made of the fact that in Greenland
a rather large circle of well-educated Greenlanders form a sort of
bourgeois middle class with a standard of living considerably
higher than reached by the fishermen, hunters and unskilled
workers. Even in Greenland, however, the average income is only
between one-third and one-half of the average income for the rest
of Denmark. In Norway, the Lapps form the lowest income group of
all Norwegians, and as regards Alaska, reference may be made to
reports from reputed economists on the poverty of the Eskimos and
Indians, whose cultures have been disrupted by the inroads of

industrialized communities. From the publication "Alaska Natives and the Land", which appeared in 1968, the well known economist, George W. Rogers, quotes as follows:

"A great contrast exists today between the high income, moderate standard of living, and existence of reasonable opportunity of most Alaskans and the appallingly low income and standard of living and the virtual absence of opportunity for most Eskimos, Indians and Aleuts of Alaska. About four-fifths of the more than one-quarter million people of Alaska are not Alaska Natives. Most of them, living in or near urban places, lead lives very much like those of other Americans. They are, by and large, regularly employed. Most families earn more than $10,000 per year ... the median educational level is more than 12 years.

The other one-fifth ... live in widely scattered settlements ... are unemployed or only seasonally employed ... live in poverty ... in small, dilapidated or sub-standard houses under unsanitary conditions ... are more often victims of disease, and their life span is much shorter than that of other Alaskans ... They are not only undereducated for the modern world, but they are living where adequate education or training cannot be obtained, where there are few jobs, where little or no economic growth is taking place, and where little growth is forecast."

Reports from Canada seem to indicate that the native population there are up against the same difficulties as just described.

It is characteristic that until recently the fate of the arctic minorities everywhere was decided by central governments far away from the homes of these people. The Canadian sociologist, Frank G. Vallee, has characterized the administrative situation as follows: "Most decisions of vital importance to these minorities are made by non-natives."

In their efforts to improve the conditions of the natives, politicians and administrative executives from the south have to a wide extent copied the systems of the dominant communities. Here it should be pointed out that this procedure has not been altogether harmful, and many persons will probably maintain that it has been the only possible approach. In several arctic areas a considerable material improvement may be ascertained. This applies for example to Greenland. There has been an essential improvement of public health, a considerable lengthening of average life expectancy, a large housing construction program has been implemented, the standard of living has been raised, and there has been an enormous expansion of educational facilities. Concurrently herewith, however, developments have created serious difficulties of a social, cultural and psychological nature.

Before going further into these difficulties, please let me mention that, naturally, differences do exist among the arctic peoples. They are of different origin, they have developed different social structures, the Indians for instance lived in tribes, while the Eskimos until quite recently, except for those in Canada, have been living in small family groups.

The pronounced linguistic diffusion impedes a common cultural policy, and finally, there have also been regional differences in the manner in which the individual groups have been exploiting the natural resources, either as hunters or trappers, fishermen or herdsmen of reindeer.

The differences have also become apparent in the manner in which the nation States have dealt with their arctic territories after World War II. In the following I will describe the development that has taken place in the fields of culture and education, social conditions and exploitation of natural resources, and finally, I will try to outline the political developments, feeling that they will be of special interest at this Conference.

When dealing with culture and education in arctic territories, it must be realized that the societies found by the first whalers, fur trappers, traders, and missionaries on their way north were not without a culture of their own. In an article on "The Changing Canadian Eskimos" the well known administrator from the Department of Indian Affairs and Northern Development in Ottawa, Mr. A. Stevenson, has described the Eskimo culture as follows:

"They had no written literature, or writing of any sort, but they had a strong cultural heritage in the story, the folk song, the drum dance. They had ethics, taboos and philosophies of social relations, which were practical and effective for the good of all. In some respects, their morality shocked the newcomers from the South, but it was a morality born of the Arctic environment and not Europe's. In time, it was we who changed their philosophy and brought many tensions and contradictions to these hardy, resourceful people. They retained their sense of humour even in the extreme adversity that they faced daily. One of their philosophical or fatalistic phrases, passed down by generations till today, is Ayungamut: It cannot be helped." This information furnished by Mr. Stevenson may be supplemented by referring to the pronounced feeling of solidarity among members of a family to be found in the small societies and their strong traditionalism. The old and skilled hunters and trappers were held in the highest esteem and the catch was divided according to certain fixed rules. It was a creative people as may be seen for example from their beautifully ornamented tools. Ordinarily frictions were avoided if at all possible. They displayed mutual tolerance and respected the rights of the individual. Direct confrontations were evaded. A well known characteristic factor is also the very liberal manner in which the Eskimos brought up their children.

Many of these traits have been subject to disintegration in the encounter with the industrial culture and money economy. Catches are no longer being shared and the old and the skilled hunters are not held in such high esteem as previously. Now, the situation is dominated by those who have a good education and a knowledge of languages. And this brings us to educational policy.

To make the native competitive and give him equality, efforts
have after World War II been concentrated strongly on education
everywhere in the arctic territories. Numerous schools have been
built and educational programs have been implemented. There has
been a great expansion of vocational training with a view to
improving the conditions of the natives. Though the goals have
been the same, the means adopted have differed, in particular when
considering the use of the local language in the schools.

Right up until recent years a distinctive line could be
traced between the educational policy followed in Alaska, arctic
Canada and Norway and the guidelines followed in Greenland, in the
Lapp territories in Sweden and in the arctic territories of the
U.S.S.R. In his treatise on Alaska, Diamond Jenness tells of the
"Guidelines" which in 1890 formed the basis of the educational
system in Alaska. He writes: "... the education to be provided
for the Natives of Alaska should fit them for the social and
industrial life of the white population of the United States and
promote their not-too-distant assimilation .... The children
shall be taught in the English language, reading, writing,
arithmetic, geography, oral history, physiology, and temperance
hygiene. No textbooks printed in a foreign language shall be
allowed. Special efforts shall be put forth to train the pupils
in the use of the English language."

Correspondingly, a Norwegian Education Act from the 90's
stipulated that the language to be used in teaching the children
in the Lapp schools must be Norwegian, and the teachers were
forbidden the use of Lappish. This policy gave rise to great
dissatisfaction among the mountain Lapps and right up until the
present time the Lapps have entertained a deeprooted distrust of
the educational system. Until quite recently, the teaching in the
schools in the arctic territories of Canada has, likewise, been
conducted in English.

Gradually, however, the native reaction against the suppression
of the native language brought results, and in Alaska as well as
northern Canada and Finmarken in Norway changes in the educational
policy were introduced in the 1960's.

The Greenland educational system, which is based on a century-
old tradition, had until the 1950's protected Greenlandic and
Greenland culture. In the mid-50's, there was a change in approach
in this respect, however, when educationalists found that it might
make the teaching more effective if the children were being taught
in the Danish language from when they started school. The
Greenland parents were given the option of choosing whether the
teaching of the children in Greenlandic should be postponed until
grade 3, and every year a very large number of parents elected to
have their children taught in Danish in the hope thereby to ensure
that they were provided with a good education. This, however, also
gave rise to a reaction, and since the late 60's there has been an
ever increasing tendency to revert to the use of Greenlandic in
the elementary school.

In Sweden great care has been exercized in connection with
the cultural policy pursued in respect of the Swedish Lapps. The
first Lapp school was established in 1632, and during the past
centuries the Lappish language has been used in the teaching,
which has also comprised Lapp culture.

After World War I, the arctic minorities in Russia were
absorbed in the Russian Socialist Revolution. The primitive ways
of life were abandoned, trade and industry were modernized, and the
territories experienced an economic growth. In agreement with
Lenin's nationalistic policy, this development was carried out on
a national basis, meaning that consideration was paid to the
language and culture of the minorities. During the first few
years the children were taught in their native language, and it
was only as the children grew older that Russian was introduced
on the curriculum. The schools also helped to keep the cultural
pattern of the minorities alive such as their literature, music,
dance, etc. During the past fifty years a considerable number of
Russian and native teachers have received training in Leningrad in
the language and culture of the arctic minorities.

The difference in cultural policy when comparing the various
arctic territories has thus been quite considerable, but from the
mid-60's there has been an ever increasing tendency everywhere to
pay consideration to the local languages. In 1969 a conference of
wide scope was held in Montreal on the subject of "Education in
the North". The conference was attended by delegates from all
circumpolar countries, and particular interest was paid to the
Russian educational and cultural policy. The idea of teaching in
the mother tongue, at least in the first few grades, was given
strong support by teachers and sociologists as well as by delegates
representing the arctic minorities. On the part of the teachers
it was pointed out that the best teaching/learning situation was
provided by using the children's own native language when they
first started school, and it was claimed that, in actual fact,
school starters could suffer a shock if being required to start by
using a European language. In recent years there has been a
considerable increase in the interest taken in the native language
and today efforts are being made everywhere in the arctic territories
to support the position of the native language in the schools.
This development may, for one thing, be explained by an incipient
understanding on the part of the teachers of the importance of the
native language as affecting the development of the child's
personality but also by a constantly growing political pressure
exercized by the arctic minorities, who, in addition to putting up
a fight for their mother tongue, also has criticized the school
for providing an education which is not sufficiently interrelated
with the requirements of the local community.

But although there is a will to reform the school, the
difficulties to overcome are great. Everywhere, there is a shortage of teachers in command of the local language, whether it be
Indian, Eskimo or Lappish. Moreover, most schools are faced with

a dual task. They are to educate the children for life in the traditional society, but in view of the population explosion as a result of which there is not room for everybody in the traditional lines of occupation, the schools must at the same time provide the children with an education that will fit them for the modern society. At the Montreal conference the UNESCO expert, J.C. Cairns made the following statement:

"The ideal is an education balanced enough so the student can choose between (1) remaining a functioning member of his own group, and (2) fitting satisfactorily into the dominant society if he so wishes. Thus, he needs an expanded, not a restricted, education, covering the heritage and values of his group plus the knowledge required for living in the dominant society. Clearly, the same criteria apply to all forms of education - whether in school or out, whether for children, young persons or adults."

It is extremely difficult, however, to achieve the balance mentioned by Cairns, and even in the U.S.S.R. where the cultural policy adopted has been based on nationalistic views, the minorities apparently find it hard to hold their own when faced with the dominant society. In his comments on the Montreal conference, Professor Norman A. Chance emphasizes the good results achieved by the U.S.S.R. in the development of the arctic areas but continues:

"However, as more southern Russians move North to participate in the region's economic development, and National Minorities are more exposed to the Russian language at an early age, parents are expressing greater interest in seeing their children learn Russian as their first language and their Native tongue as the second. If this trend continues in the future, the disappearance of Native languages may complete the process of "cultural replacement" referred to in Professor Vallee's paper. While this process may be economically advantageous, the cultural loss is irreplaceable."

It is not only within the cultural and educational sectors that differences in the development policy followed by the territories become apparent. This is also the case when speaking of the employment policy, where the central governments, until recent years, have tried to solve the problems in vitally different ways.

In Greenland for example there has for many years been a construction program for fishery plants, and the population has been given support in connection with the purchase of fishing vessels. This has been done irrespective of immediate profitability, with a view to creating employment for the Greenland population. Today, Greenland has a considerable fleet of fishing vessels and a large-scale fishing industry, which is mainly being operated by the government. A large proportion of the processing plants show a permanent loss, which is being covered out of public funds, and it is thus a question of a masked form of social policy. Different in Alaska, where the government in good agreement with American traditions has not actually gone into trade or industry.

In Alaska efforts have, as mentioned, been concentrated on a rapid integration of the population by means of intensified occupational training and besides, social assistance has been provided for those who have failed to cope with conditions in an industrial society. This has involved increasingly larger social expenditures of a much larger magnitude in Alaska than in Greenland where better opportunities for employment have been provided thanks to the government operated concerns. Following the Montreal conference on Education, George W. Rogers has made an interesting analysis of the situation in Greenland and Alaska, respectively. He indicates the very different forms of development policy when comparing the two territories while at the same time pointing out the remarkable fact that at the present time there seems to be a switch in policy. In Alaska minor industries are being established to a steadily increasing extent with a view to cutting down on social expenditures, while in Greenland, the loss suffered on the fishing industry is increasing so fast that further expansion is being moderated which brings up the need for establishment of more and more social welfare institutions.

The employment situation is serious everywhere in the arctic territories associated with the western industrial countries. In many places the labour market cannot absorb the population bulge each year as the young people leave school, and this situation is aggravated because a considerable proportion of the school-leavers are without sufficient occupational training to fill any existing vacancies. It is a fact that the Indians, Eskimos and Greenlanders on an average rank below the rest of the population of their respective countries when it comes to employment and income. Few hold executive positions or jobs requiring an academic education and most key positions in the modern arctic communities are held by non-natives, which, of course, is a source of irritation among the minorities. The governments are fully aware of this situation. They try, through their education and labour market policy to provide more job opportunities, inter alia by committing private companies and contractors operating in arctic territories to employ the greatest possible number of natives.

On the background of the increasingly difficult employment situation, great hopes attach to the exploitation of the arctic resources. It is being hoped that it will be possible to create local jobs, and that the revenues from mineral and petroleum concessions will flow into the local exchequers, to be used for the financing of new enterprises. This optimistic view is, however, not shared by everybody, and many natives are becoming increasingly concerned about the larger areas that constantly are being occupied by the multinational companies who hold concessions on exploitation of petroleum or minerals.

This brings us to the most serious problem today in the arctic territories, i.e. the right to the land. During the past ten years the lodging of claims to land has been spreading like

wild fire on the part of the arctic minorities. They want to have
control of the land themselves because they greatly distrust the
modern way of exploiting the natural resources. They fear that
damage may be done to the vulnerable arctic environment, and doubt
that the local native population eventually will derive any great
benefits from the production of minerals and petroleum. During
the development stage they will have to put up with the heavy
inrush of experts and workers, which will disturb the peace in the
small communities and threaten the traditional ways of life. And
at the end of the development stage the mines, road systems and
hydro-electric power stations will not provide the local
population with any large number of jobs. Many feel that the
economic developments will be of benefit to the remote society,
but that the local society will be left with the disadvantages
such as the risk of pollution, social disturbance, inflation, etc.
The hunters and fishermen are no longer able to move about as
freely as previously, and the claims to the land are, therefore,
being voiced with increasing force.

I have already mentioned that it is not only a matter of
economic rights but that the very control of the land is of great
cultural importance to these minorities. This same view is
supported by Tomas Cramér, legal adviser to the Swedish Lapps, who
states that from a cultural point of view, the land base is
absolutely necessary to the minorities. He gives this as the
reason for taking legal action and states that in many cases the
minority cultures seem to be better adjusted to life in territories
with a scarcity of natural resources as they are based on renewal
of these resources whereas the industrial culture seems to permit
ruthless exploitation.

In their fight to obtain control of the land, the natives use
political means as well as court action. In this fight it is a
question of two different types of claims: treaty rights and
aboriginal rights, the former of which are based on treaties or
other commitments in writing and the latter on the fact that the
natives were the first to occupy the land in question. In respect
of this problem Cramér says that the governments are more inclined
to honour treaty rights on the basis of commitments in writing
undertaken by the respective government or its successors, whereas
claims based on aboriginal occupation of the land will not be
honoured if rights have not been specifically acknowledged.

In Alaska, the claims to land conflict reached a climax in
the late 1960's. Shortly after Alaska's admission to statehood,
the government in Washington entered into negotiations with the
government of the new State for transfer of the land which
previously had been owned by the federal government. That made
the native population go into action. In 1966 "The Alaska
Federation of Natives" was formed in direct consequence of the
fact that the natives considered their land threatened. For the
first time, the Eskimos, Indians and Aleuts stood side by side in
political cooperation, and as they jointly represented 30 per cent

of the Alaskan electorate, it soon became apparent that they had
political power. A claim to the land submitted to Washington
resulted in the so-called "land freeze" which suspended any
transfer of land to the state government of Alaska pending a
consideration of the claim lodged by the natives.

After negotiations over a period of four years, President
Nixon was able, on December 18, 1971, to sign "The Alaska Native
Claims Settlement Act". Expressed very briefly, this Act provides
that the 55,000 natives of Alaska will receive U.S. $962.5 mill.
and full title to 40 mill. acres of land, to be selected by the
approximately 205 native village corporations and twelve or
thirteen native regional corporations to be set up.

This is the largest single Native claims settlement in the
history of the United States, and it will have far-reaching
consequences for the political and economic future of Alaska.

It is no wonder that the implementation of the above Act gave
rise to large administrative problems. In a report to "The Arctic
Peoples' Conference" in Copenhagen in November 1973 the president
of the Alaska Federation of Natives, Mr. William L. Hensley,
called attention to the difficulties encountered by the natives
during their negotiations with the Department of the Interior in
connection with the implementation of the Act. The conclusion
drawn by Mr. Hensley is quoted in its entirety as follows:

"On the whole, the implementation of the Act is proceeding as
well as one could expect, particularly given the problems we
mentioned above with the Department of the Interior. If the
Department of the Interior would keep not only within the letter,
but also within the spirit of the Act, many of our problems would
be eliminated.

We would like to emphasize that this is probably the most
complicated Indian Land Settlement that the United States government has ever developed. It is forcing a whole new set of concepts
on a people that are not all that familiar with western society.
Concepts such as corporations, shareholders, dividends, proxies,
custodianship, and many others are completely unfamiliar to Alaska
Natives living in rural or even urban Alaska. The challenges faced
by the regional and village corporations are great, and unless we
have cooperation from the government agencies who will be
interpreting the Act, the settlement will ultimately fail. We
cannot continue having to fight tooth and nail for what is already
ours with the Department of the Interior and other federal agencies.
This at present, is our biggest problem."

In the opinion of experts, this Act can be construed to
acknowledge the concept of "aboriginal rights", and it is beyond
doubt, at least, that it has had great psychological effect among
the politically aware members of the arctic minorities.

In Canada the government has, until recently, not felt able
to acknowledge the concept of "aboriginal rights". In a White
Paper on Indian Policy, published by the government in June 1969,
it is stated that: "aboriginal claims to land ... are so general

and undefined that it is not realistic to think of them as specific claims capable of remedy...".

Shortly afterwards the Committee of the Privy Council stated "that the assertion of grievances based upon aboriginal title is so general and undefined that it cannot be settled except by a policy to enable Indians to participate fully as members of the Canadian community ..."

Later on, the Canadian government modified its attitude, and in August 1973 it sent out a declaration which - quote Professor Cumming - "implicitly recognized rights on the part of native peoples with respect to lands traditionally used and occupied, and left open the question of what a negotiated settlement in respect of the problem of competing land uses (in particular, due to exploration and development in the North) might be."

The natives in Canada have in recent years through court actions - and often with success - tried to stop large projects if they felt that they violated their title to land. Freshly remembered is, in particular, the case of the large James Bay Project. Incidentally, it is characteristic of the open attitude of the Canadian government that it grants economic support of native organizations which through legal action or by other means try to secure their claims to land. The question of claims to land in arctic Canada is in rapid development, and the native claims are now being investigated and negotiated with a view to clarification of this complicated problem.

In Greenland the discussion of title to land has just started. All land in Greenland belongs to the public. This has last been provided for in the Mining Act of 1969, which states that all mineral resources in Greenland are the property of the government. Gradually, as mining operations are being started and negotiations are carried on for concessions on petroleum, many Greenlanders have started to ponder over the actual meaning of public ownership of land.

In his report from 1973 on "Policies on Northern Development", Professor Ørvik writes in the Introduction that "The North is becoming politizised and the political problems there are likely to loom larger in the years to come." Anybody who has followed developments will agree with him that there has been a striking political awakening everywhere in the arctic territories. Since the mid-Sixties a considerable number of native organizations have been formed with political aims. I have already mentioned the "Alaska Federation of Natives". In Canada the "Indian Brotherhood of N.W.T." was formed in 1969 and "Inuit Tapirisat of Canada" in 1971. In September 1972 a federation of all native organizations in northern Canada and Alaska was formed under the name of the "Federation of Natives North of Sixty". In 1972 the northern Lapps got a Lapp Institute, intended to support the Lapp population and improve their economic, legal, cultural, and social conditions. Since 1967 the Greenland Provincial Council has elected its own chairman, and since the last election all the members of this

Council have been Greenlanders. Within the next few years the
Danish government will, on a large scale, delegate duties to the
Greenland municipal councils as for instance in the fields of
administration, technology and education. Effective January 1,
1975 a system of direct taxation has been introduced in Greenland,
and the tax revenues will be paid into the municipal exchequers.
A Government Commission whose members are all Greenlanders
submitted a report on the "introduction of Home Rule in Greenland"
at the beginning of this year. It is expected that in the course
of 4 years a number of important administrative tasks can be
delegated to the Greenland Provincial Council.

        One thing, which all minorities have in common, is that their
claims primarily are concerned with control of land and protection
of language and culture, as I have already described in this paper.
I should like to underline that naturally, there is not complete
consensus of opinion within the different groups on the subject of
political development. In Greenland for example, a clear division
is becoming noticeable between a modern population group and a
group with more radical aims. It is obvious, though, that the
groups with the radical program are those who advance, and
especially they are being supported by young college and university
students and natives with a higher education.

        The question of exploitation of natural resources is, I
believe, the most serious political issue. On one hand, we have
the minorities' claims to control of the resources, and on the
other hand, there is the rapidly growing need for energy and
mineral raw materials within the southern communities. The point
of view of the dominant society has been described as follows by
an executive from the Department of Indian Affairs in Ottawa:

        "The press for the needs of Canada, North America and the
world cannot be held up awaiting the needs of a small minority".

        In reply hereto it may probably be maintained that the
minorities have never rejected exploitation of the arctic
resources, all they want is co-influence. They want to be able to
control developments in order to safeguard the environment,
protect against social disturbance and get their fair share of
revenues.

        Occasionally, it has been maintained that when the large
companies take an interest in the arctic resources, this may, for
one thing, be due to the fact that the territory is considered
politically stable. A more thorough analysis of the present
situation may probably set a question mark against this point of
view. Admittedly, it is a question of small population groups
who have so far been without any great political influence. At
the present time, however, there are examples to show that if
driven to despair, even small groups may be able to create a great
deal of disturbance. Moreover, these small groups are now joining
forces for collective action. In November 1973 the Inuits,
Indians, Greenlanders and Lapps met in Copenhagen at the "Arctic
Peoples' Conference". At this Conference a number of joint
declarations were agreed upon and it was decided to continue the

cooperation. This is now being extended globally with the
planning of an "International Conference of Indigenous People",
to be held at Port Alberni, B.C., Canada in September 1975. At
this Conference Eskimos, Lapps and Greenlanders will meet natives
of Australia, Maories from New Zealand and Indians from the U.S.A.
and South America.

Apart from the question of the political power of the
minorities, the fact remains that by virtue of their status as
democratic states and members of the U.N., our countries are under
a moral obligation to meet the minorities with understanding.
This will place heavy demands on governments and central
administrations if a peaceful development in the arctic territories
is to be achieved within the next few years. A display of
flexibility, openness and understanding will be required in our
relations with the minorities, who are making themselves heard
more forcibly than ever before.

ENVIRONMENTAL PROTECTION AND QUALITY ENHANCEMENT IN AN ARCTIC REGION

PART I - ENVIRONMENTAL GUIDELINES AND UTILITIES DELIVERY

Daniel W. Smith

Environment Canada

Edmonton, Alberta, Canada

INTRODUCTION

Environmental protection in Arctic regions requires an understanding of the regional constraints, guidelines to be followed for water quality and waste management, treatment systems available and systemization techniques for handling total programs efficiently. It also requires a philosophy which recognizes the value of the natural environment. These topics will be discussed as they pertain to utilities integration and planning for services. It should be noted that two essential parts of the utilities package, the power and heating elements, are not discussed.

GENERAL CONSTRAINTS

Many regional constraints to development activities in the Arctic are well known. They include permafrost, temperature, sensitivity of the vegetation and the active layer to disruption, the long flushing time and interconnection of the lakes, and transportation problems.

The nature of problems with permafrost are completely dependent on the type of material present. Dry, coarse sands and gravels and rock can be dealth with easily. However, moist soils, which are more common, are a problem due to the potential of melting the ice and degrading of the soil structure and strength. In all designs careful heat balances must be made to prevent unnecessary stress on the environment.

The sensitivity of vegetation to disruption can be seen in nearly all parts of the north. Plants are easily damaged by vehicles and can be stimulated by small amounts of nutrients. Damage to the plants can cause decay of permafrost, resulting in irreparable scars. Likewise, lakes and streams can be seriously altered by small discharges of waste material. A discharge made up of carbonaceous material will cause dissolved oxygen reductions. During periods of ice cover, naturally occurring respiration reduces dissolved oxygen to critical levels in many streams (Gordon, 1970 and Shallock, 1975). The addition of oxygen demanding material through man's intervention can result in destruction of the aerobic populations before spring breakup and the resulting aeration. Similar oxygen depletion can occur when the addition of nutrients in waste discharges causes excessive algal growths. The movement of water between lakes and streams in Arctic regions is not well known. However, the potential for interconnection and the resultant movement of contaminants from one lake to another is a very real problem which warrants careful examination at each site.

Transportation plays an important role in environmental protection. Some modes are more damaging than others. Materials moved may be toxic, or otherwise dangerous. Accidents can and do happen and alternative measures need to be planned. This type of planning requires knowledge not only of location, time of year and time of accident, but also knowledge of the kinetics of dissolution and toxicity of the material concerned. Transportation constraints seriously affect equipment needs and backup requirements for utilities systems. Similarly, they affect response techniques devised for effectively countering threats to the environment.

A very serious constraint is the availability and requirement for highly trained operators to make the systems work properly. Regardless of how well designed and constructed utility facilities are, they will not operate effectively unless they are maintained and operated properly. As well, there is no substitute for a trained and resourceful person when responding to an emergency.

OBJECTIVES

The objective of the industrial projects in the north should be to obtain valuable materials in an environmentally sound fashion, at minimum cost. This normally means the creation of work camps which are safe, sanitary, easy to operate, and as unobtrusive as possible. It also means conducting field operations in such a way as to protect flora and fauna of the area. This discussion will concentrate on the types of existing environmental guidelines and regulations and on the utilities systems needed for work camps and permanent installations.

Water Supply

Water supply objectives for utilities systems must be determined for both quantity and quality. Factors affecting quantity
requirements will be discussed later. Quality requirements for
drinking water have been promulgated for virtually all parts of
the north. In Canada, the Department of National Health and Welfare (1969) presents standards which show the basic requirements
common to most of the standards which have been established (see
Apprendix I). Generally, they specify physical, bacteriological,
chemical and radioactive characteristics to be met before the water
is delivered to the public. Note that not only are the acceptable
and maximum allowable levels specified, but also the quality objective is given.

Wastewater Treatment

Wastewater treatment guidelines are being developed for northern regions. The Environmental Protection Service has released
interim guidelines for wastewater disposal practices in northern
communities (Water Pollution Control Directorate, 1974). These
guidelines state that "dumping of wastewater without appropriate
treatment shall be eliminated". The guidelines require treatment
adequate to protect public health and to maintain favorable conditions in receiving waters be provided. The requirements for
treatment vary with location and the community condition. The
important point is that new communities should be designed and
constructed in such a way as to eliminate pollution problems.

The second set of guidelines of particular interest are those
for federal establishments in Canada (Water Pollution Control
Directorate, 1973). These guidelines are based on effluent quality.
Limits for key parameters are presented in Table I.

These guidelines were established to set an example of concern for the sensitive ecological webs of waterways. The limits
require carefully designed and properly operated treatment systems.

A third document which has been released in Canada is the
Proposed Water and Sanitation Policy for the Northwest Territories
(Department of Local Government, 1973). This policy defines the
objectives of the Territorial Government with respect to cleanup
and management of community sanitation. This document has been
accepted in principle by the federal Treasury Board.

Solid Waste

Solid waste management codes of good practice have been promulgated for federal establishments (Environmental Protection

TABLE I

WASTEWATER EFFLUENT GUIDELINES FOR
FEDERAL ESTABLISHMENTS IN CANADA

Five-day Biochemical Oxygen Demand (BOD) - 20 mg/1 (1 = litre)

Suspended Solids - 20 mg/1

Fecal Coliform - 400/100 ml (MF Method, after disinfection)

Chlorine Residual - 0.5 mg/1 minimum after 30 min. contact time

pH - 6 to 9

Phenols - 20 mg/1

Oils and Greases - 15 mg/1

Temperature - not to alter the ambient water temperature by more
              than 1°C at perimeter of mixing zone

Service). Although the codes are applied only to federal establishments, they present solid waste management techniques designed
to protect the health and welfare of all land and aquatic life.
Management techniques applicable to northern Canadian communities
have received careful examination. Heinke, 1973; Underwood
McLellan, 1973; Stanley, 1973; and Associated Engineering Services,
1973 felt that communites in the north have a serious solid waste
problem, and that detailed study of actual systems in the north
was in order.

Air Quality

Air quality should be considered when planning industrial
activities in the north. In Canada, national ambient air quality
objectives have been established and can be expected to prevail
in Arctic regions (see Appendix II). These objectives have been
supplemented with national guidelines for the asphalt paving industry (Department of the Environment, 1975a), the cement industry
(Department of the Environment, 1974a), and metallurgical coke

manufacturing (Department of the Environment, 1975b). National regulations for leadfree gasoline (Clean Air Act, 1973), leaded gasoline (Clean Air Act, 1974), and secondary lead smelters (Department of the Environment, 1974b) have also been promulgated. Although these documents have limited application in the north at present, as activity increases the need for awareness and their applicability will increase. Currently, Arctic mining emissions guidelines are being prepared and will be issued within a year. Guidelines or regulations can be expected for virtually all industrial waste gas discharges in the future.

Air quality control goes beyond consideration of outdoor conditions. When considering encapsulated living conditions, air temperature, humidity, toxic or odorous gases and smoke must be considered. Carbon monoxide has been shown to be a very serious problem in some northern communities (Holty, 1973; State of Alaska 1972; Gilmore and Hanna, 1973; Joy, Tilsworth and Williams, 1975), including the interior of homes (Arctic Health Research Center, 1972).

WATER SUPPLY CONSIDERATIONS

The type of treatment required for potable water in Arctic regions varies with the season and the source. In planning temporary or permanent installations, factors to be considered in choosing a site must include the quantity and quality of water available. Failure to do so may greatly increase the treatment cost and endanger major operational functions.

Quantity

At many northern locations the quantity of water available for meeting year-round demands is minimal. Precipitation in the North American Arctic averages from 10 to 30 cm per year. Mean annual lake evaporation is between 10 and 20 cm per year (Department of Transport, 1970). Many of the ponds and lakes are less than 2 meters deep and may freeze to the bottom or provide very poor quality water during the winter. Flow in all but the largest of the spring fed rivers stops during the winter. In areas where unfrozen water is available in streams, fish may overwinter. For example, Arctic char or grayling stay in deeper areas of streams during the winter, and these will require protection.

Development of an adequate raw water supply may require special diking and snow fencing to trap and store water. Creation of such water storage areas can create problems with permafrost. Water stored in contact with soil will allow leaching processes to occur resulting in deteriorated water quality (Smith and Justice, 1975).

The cost of treatment and the problem of short supply often
requires that facilities be designed with water conservation techniques in mind. Low water use toilets can save substantial amounts
of water (Grainge and Slupsky, 1973). Providing saunas and fine
spray showers which are timed can reduce the total water use. Low
water use washers and wash water recycle can be used to reduce the
total water demand. Elimination of garbage grinders has also been
considered. Total water recycle is not yet socially accepted.

The last point concerning water quantity is the fact that in
many Arctic communities the local people do not have safe water
nor running water. The cost and effort involved in obtaining
adequate safe water is great and it is important that industrial
developments set positive and realistic examples with respect to
water management.

Quality

Arctic water quality varies drastically with time and source.
Both lakes and rivers go through somewhat predictable cycles.
Water quality is generally best in nonglacial streams during the
late summer. As winter proceeds, pure water is captured as ice
with the minerals and organics concentrated in the unfrozen portion. With spring, the rapid melting and runoff creates highly
turbid water. In glacial streams turbidity remains until cooler
temperatures stop melting of the glacier.

Several surveys have been conducted in Arctic regions to
develop an understanding of the quality of surface waters. Kalff
(1968) conducted a survey of 58 Arctic Alaskan lakes, rivers and
ponds, and a number of northwestern Canadian Arctic waters. The
majority of the grab samples were collected in the summer of 1964.
The color of the sampled water ranged from 15 to 30 pt. color units.
He also found that the iron ranged from 0.02 to 0.18 mg/1 in lakes
and 0.9 to 2.1 mg/1 in two ponds on the Alaska north slope. Lichens
and Johnson (1968) found 0.03 to 0.33 mg/1 of iron in 12 Arctic
slope lakes. Other values such as from the water supply reservoir
at Kotzebue, Alaska show iron concentrations as high as 6.7 mg/1
(Hargesheimer, 1975).

An extensive review of iron and color problems in permafrost
regions water was presented by Smith (1973). Great amounts of
data have been collected in the Alaskan Arctic areas as a result
of the trans Alaska pipeline construction activity. When this
information is published, a much better understanding of quality
and acceptable treatment alternatives will be possible.

Because of the long survival times in cold water, pathogens
are a serious matter of concern in northern regions. In a survey
of the Tanana River, Alaska, Gordon (1972) showed that indicator

organisms such as fecal coliform survive very well in cold, ice-covered rivers. Although populations in the north are low, the survival of microorganisms is higher than in warmer regions. Reports such as this point out the need for good disinfection of all water supplies and the need to disinfect wastewater discharges.

Taste and odor problems are dependent upon the source, with the most frequent problems originating in ponds. Turbidity, on the other hand, could be a problem in pond water, but is more closely associated with spring runoff and glacier fed streams.

Trace metals have been problems in localized areas. Analyses should be made before extensive use of the water for human consumption is planned.

The physical parameter temperature must be dealt with before any treatment processes are applied. For most of the year water temperatures are at or below 4 degrees C. For a short time in the summer the surface temperature of some ponds may reach or exceed 20 degrees C. The problem with the low temperature water is the slower reaction rates for chemical treatment processes and the susceptibility to freezing.

Treatment

Many water treatment techniques have been tried in the Arctic. Systems providing a range of treatment from most simple disinfection techniques to very complicated chemical coagulation and precipitation systems have been considered. Alter (1969c) reviewed much of the earlier work done on Arctic water treatment and distribution. In order to provide the reader with background information on past water treatment studies, a brief summary is included in Appendix III.

WASTEWATER CONTROL

Wastewater management in Arctic areas is very costly. Many of the costs and problems of present systems are the result of:

- improper choice of site for the facility or the waste water treatment system.

- the addition of treatment systems once the facility has already been constructed.

- failure to integrate the wastewater systems into the overall design.

- use of complicated equipment and systems which cannot be easily operated and maintained in isolated areas.

- the use of systems requiring transport of large amounts of chemicals and disposal of large amounts of sludge.

To summarize past efforts in the Arctic wastewater managements, some characteristics and treatment alternatives will be presented.

Quantity

In Arctic regions wastewater quantities are directly related to the water supply available, types of hardware installed (Grainge et al, 1973d) and the population fluctuations. Quality of the treated water will also affect usage. Very little reliable information has been published on water use in work camps. Water meters were installed at several locations in a Prudhoe Bay facility to help in determining the quantity of wastewater generated by various users. This data is shown in Table II.

Data reported by Edwards and Fahlman (1974) on water use at a washcar facility showed a monthly average of 100 lpcd. Other information on per capita flow is included in Table III.

Wastewater Characteristics

The biochemical oxygen demand (BOD), suspended solids (SS) and pathogen content of the wastewater from northern communities and work camps are very important parameters to be considered in designing and operating wastewater treatment plants. The flow rate and temperature may also affect performance of the system. Table III gives characteristics of wastewater generated in northern communities and work camps. Generally, BOD and SS concentrations exceeding 1,000 mg/l and flow rates near 190 litres per capita per day (50 U.S. gallons per capita per day) can be expected.

Wastewater Treatment

As with water treatment, many systems have been tested in the north. Alter (1969a) summarized much of the early work on wastewater collection and disposal. Recently a number of new treatment techniques have been examined. Much of the interest has been stimulated by the more stringent effluent guidelines. Lagoons have been one of the most popular methods of treatment in the North American Arctic. Increased activity, greater variability in the waste characteristics and flows, and new guidelines have resulted in the use of a variety of treatment methods including extended aeration, rotating biological contactors and physical

TABLE II

WATER USE IN AN ARCTIC WORK CAMP

	U.S. Gal/Capital/Day	Litres/Cap/Day
Drilling Pad		
Toilets and urinals	7.2	27.2
Kitchen	6.9	26.2
Showers	4.8	18.2
Wash basins	6.5	24.6
Laundry	6.0	22.7
Pusher and Camp Representative	1.8	6.8
Other	1.7	6.4
Total	34.9	132.1
Base Camp		
Kitchen	12.9	48.8
Wash Trailer No. 1	4.2	15.9
No. 2	4.8	18.2
No. 3	16.4	62.1
Kitchen cold water	2.2	8.3
hot water	0.6	2.3
Lease supervisor and medic	4.9	18.6
Lab and contractor	4.0	15.1
Sewage Plant	4.0	15.1
Total	54.0	204.4

TABLE III

CHARACTERISTICS OF WASTEWATER GENERATED IN THE ARCTIC

SOURCE TYPE	FLOW LITRES/ CAPITA/DAY	U.S. GALS./ CAPITA/ DAY	FIVE-DAY BOD			SUSPENDED SOLIDS		
			MG/L	GMS/CAPITA /DAY	POUNDS/ CAPITA/DAY	MG/L	GMS/CAPITA/ DAY	POUNDS/ CAPITA/ DAY
Common Values (Clark, Viessman and Hammer, 1971)	379	100	203	77	0.17	239	90.7	0.20
Construction Camp (Grainge et al., 1973)	189	50	360	68	0.15			
Alyeska Camps Design Estimate	246-265	65-70						
Observed	178-318	47-84						
Chandalar Camp 19 May 74 to to 16 Jan 75			216-1230					
Pump Station One 6 Dec. 74 to 28 Apr. 75			243-1920 (541)					
Old Man Camp 13 Sept. 74 (Zemansky, 1975)			615			416		

TABLE III (Continued)

CHARACTERISTICS OF WASTEWATER GENERATED IN THE ARCTIC

SOURCE TYPE	FLOW		FIVE-DAY BOD			SUSPENDED SOLIDS		
	LITRES/CAPITA/DAY	U.S. GALS./CAPITA/DAY	MG/L	GMS/CAPITA/DAY	POUNDS/CAPITA/DAY	MG/L	GMS/CAPITA/DAY	POUNDS/CAPITA/DAY
Yellowknife Correctional Camp (grab sample) 20 inmates (Grainge et al., 1973a)			516			523		
Washcar, Ft. Simpson (Edwards and Fahlman, 1974)	100	26.4						
Artificial Islands Mackenzie Bog (Heuchert, 1974a)	108	28.5	1900	230	0.51	1100	150	0.33

chemical treatment units. Appendix IV presents summary of data which has been published on wastewater treatment systems in the Arctic and subarctic.

SOLID WASTE MANAGEMENT

Solid waste management techniques have seen few major changes in the last twenty years. Refinements have included changes in the on-site storage and collection procedure (Heuchert, 1974b). Mechanized collection equipment often is of little value in northern regions and therefore has seen limited application. Since collection procedures are of a conventional type, this discussion will be limited to a brief review of quantities generated and disposal techniques. It is realized that in most cases, the most expensive part of the management program is collection and transportation.

Quantity

In remote locations the quantity of solid waste generated can be determined by several techniques such as daily estimates, examination of all shipments to and from the site noting the weight, volume and amount of waste to be expected, and daily weighing and volume measurements. The latter technique yields the most reliable and useful information.

The amount of solid waste generated in a remote area depends on economic conditions, transportation methods, and time of year. Smith and Straughn (1971) found that 2.3 kilograms (5.0 pounds) per capita per day were generated at a remote air force site. Pipeline construction camps in Alaska were reported to produce 2.7 kg. (6.0 pounds) per capita per day (Grundwaldt et al, 1975). Other estimates have been as high as 3.6 kg. (8.0 pounds) per capita per day. In all cases implementation of good plans for management of solid waste is a critical factor in maintaining pride in community and construction activities.

Disposal Techniques

Five general techniques for solid waste disposal have been considered for Arctic regions. Much of the past work was detailed by Alter (1969b). Disposal techniques include open burning, landfill, sanitary landfill, grinding and spreading, and incineration. These disposal techniques have been discussed by Grainge et al, (1973). Backhaul is another alternative which requires consideration in some cases. Sanitary landfills and incineration have been the most popular techniques for work camps. Many communities which are still using burning and landfills are slowly upgrading their techniques (Heinke, 1974b).

AIR QUALITY CONTROL

Encapsulated living quarters require proper planning and operation of air handling and heat exchange systems. Temperature and humidity control are two very important parameters. Carbon monoxide, smoke and other gases require careful monitoring and control.

Stack emissions from power generators, space heaters, and sludge and refuse burners must meet air emissions requirements. Conventional pollution control systems are used in most cases. Due to the prevalence of lichens and their importance as a food base, as well as their extreme sensitivity to SO_2 (Schofield and Hamilton 1970) it is possible that especially stringent requirements may be set for stack emissions in the Arctic. Stable inversions, which appear to be quite frequent (Eddy 1974) would reinforce the need for strict criteria limiting SO_2 emissions. Major design considerations include avoiding wet scrubbers, considering size and weight of combustion units, and the related transportation costs. Refuse and sludge incinerators are being used successfully at construction camps in Alaska. Operational and cost data have not been published to date.

INTEGRATED SYSTEMS

The use of integrated systems for environmental quality control has been discussed and practised to varying degrees for many years. The earliest systems combined water treatment and distribution operation with community power generation. Since most Arctic water supplies require treatment and are near 4 degrees Celsius most of the year, this was a logical combination. By utilizing cooling water from a power plant, the problem of low chemical reaction rates is reduced, waste heat is used and the chance of service lines freezing is reduced. If properly managed, such a system can result in a warmer influent to the waste water treatment plant and therefore more rapid treatment.

In the late 1960s and early 1970s integrated systems truly started to develop. The first of the community related efforts were the facilities at Emmonak and Wainwright, Alaska. These were constructed as part of the Alaska Village Demonstration Project (AVDP) of the Arctic Environmental Research Laboratory, U.S. Environmental Protection Agency (Puchtler and Reid, 1975). These facilities include heating, potable water treatment, graywater treatment, black water treatment, solid waste disposal, vehicular distribution and pickup of water and wastewater, laundry and personal hygiene facilities. The various pieces of equipment were assembled in a number of modules outside of Alaska and shipped to the communities. The Wainwright facility which was the first to be built, operated well until destroyed by fire in 1973. Some

very valuable information was collected. A replacement unit has
been completely redesigned and is currently under construction.
One of the most important changes is that construction of the
facility will take place on site.

The Emmonak facility actually came into full operation in
March 1975. Although little information has been published on the
operation of these facilities a few key points have surfaced.

1.   The facilities must be protected against the possibility of
      fire. A fire protection system and fire retardant materials
      should be utilized in construction.

2.   Water and wastewater systems should be designed for conditions
      present in each location. Generalizations lead to excessive-
      ly complicated systems.

3.   On-site construction appears to be more acceptable for the
      small community situation.

4.   Both the capital and operating cost of such systems are
      extremely high.

5.   Equipment utilized should be kept as simple as possible in
      order to minimize operational complication in breakdowns.

Shortly after the start of AVDP the State of Alaska, Department of Environmental Conservation started a program called the
Village Safe Water Projects (VSWP). Three projects similar to,
but smaller than the AVDP were constructed. Problems of cost and
operation with these facilities have resulted in a re-evaluation
of systems used to meet the project objectives of safe water
supply and proper waste management. Early in 1975 the State commissioned the design of three more units which hopefully will meet
these objectives in the most efficient manner possible. Reversion
to biological treatment systems properly designed for the waste
characteristics is being considered as a possible alternative.
Subsurface disposal of treated wastewater will be used if feasible.
Water supplied will be filtered and treated with activated carbon
if needed and then disinfected before distribution.

Work camps constructed on the North Slope of Alaska and along
the route of the Trans-Alaska pipeline are among the first attempts
at building integrated living facilities in the Arctic. A total
of eleven road and pipeline camps and five pump station camps have
been built north of the Yukon River (Zemansky, 1975). In addition,
a number of facilities have been built at Prudhoe Bay, Alaska.
These facilities made use of coagulation and sedimentation for
water treatment. Because the quality of the raw water is con-

stantly changing, regular adjustment of chemical doses based on jar tests is required.

The wastewater treatment systems generally include a coagulation and sedimentation step followed by sand filtration and carbon absorption (Floyd, 1974). The treated wastewater is chlorinated before discharge. This combination of treatment when operated properly produces a good effluent. However the number of camps which reported poor effluent points out the great need for well trained operators. In Alaska many of the camps have storage lagoons for holding poor quality effluent or feeding subsurface discharges. Sludge from plants is incinerated on site after thickening. This operation as well as all those preceding it is very expensive. Hopefully the mass of data collected for these sites will be analyzed and reported in the literature.

Data has been reported on utilities systems using extended aeration for wastewater treatment in common with small systems elsewhere. The major problems seen in all cases were:

1.   Organic overloading.

2.   Hydraulic overloading.

3.   Major fluctuation in camp population.

The third is particularly severe since a work force may change from five to fifty people in a week.

As more information is gained on the work schedules and design deficiencies in existing plants, better designs will evolve. Tilsworth (1973) has stated that biological systems would work properly if designed to meet the conditions. Heuchert (1974) recently reported on two units in Mackenzie Bay which were seriously overloaded and had a number of design problems. All this information must be documented and evaluated for future designs.

The rotating biological contactor has been investigated in two camp situations in the Canadian Arctic (Forgie et al, 1974). The unit designed for a five-man camp showed very good BOD reduction under a variety of loading conditions (see Appendix IV). The second unit was designed to treat 55,000 litres per day (15,000 USGPD) for approximately 300 men. This unit also had a good effluent quality under most loading conditions.

Another unitized system which has been studied is the washcar incinerator combination (Edwards and Fahlman, 1974). This approach combines the sanitary facilities housed in one trailer

with an incinerator in another. All wastewater generated is burned.
The important disadvantage is the large amount of fuel consumed in
destroying the wastes.

Problems with Integrated Utilities

The major problems with integrated utility systems are similar to those seen with any new approach to solving a problem.
Costs for engineering design, construction and transportation,
have been high. Inefficiencies and failures of equipment have been
common. Many of the errors will be eliminated as more information
is collected and made available to systems designing.

Data collection is difficult for most camps. Needed monitoring equipment is not normally included in the facilities since
it will not be of value in reaching short term goals. Many types
of tests, such as BOD and COD, cannot be performed on site. Time
delays between sample collection and analysis can lead to serious
errors. Governmental backup for monitoring important environmental
parameters is very valuable.

The lack of trained operating personnel has been a major factor in the repeated failure in the operation of utility systems.
The necessity of having competent operators cannot be overstressed.
Training programs should be developed early in the life of a system
and maintained on a year-round basis to reduce the number of
failures due to operating staff errors.

An environmentally responsible attitude is needed in the north.
Most of the land and water has not been used or visited by men.
In many cases future industrial and community growth will develop
near initial workcamps and transportation corridors. Many times
the lakes of an area are interconnected by visible or hidden
channels. These factors refute the old philosophy "polluting one
or two bodies of water does not matter". Patterns for environmental
protection for many years into the future are being set today.

SUMMARY

1. Guidelines and regulations for environmental quality control
    in Arctic regions are being prepared now. In general, regu-
    lations and guidelines from warmer climates have been applied
    to the Arctic. The future will see these standards refined
    to meet specific environmental conditions of Arctic regions.

2. Complete integration of utilities has not been achieved.
    Future efforts will see the entire system designed around
    the power-heating units. Total energy will be the objective.

3. Although management of solid waste is still a problem, refinements to conventional management techniques may be the only change in the foreseeable future.

4. The control of air quality is imperative in encapsulated living and utilities systems. Stack emissions may be strictly controlled due to the importance and sensitivities of lichens in the Arctic.

5. Provision for trained operating personnel is perhaps the most important factor to be considered in providing acceptable waste treatment for Northern communities and camps.

6. Back-up systems must be planned into each facility.

7. A feedback procedure must be established so that each new facility is an improvement over the last.

RESEARCH AND DEVELOPMENT REQUIREMENTS

1. Research and development is required on totally integrated utilities systems.

2. On-site evaluation studies of various wastewater treatment methods and systems are required to demonstrate their cost effectiveness and applicability to Arctic and subarctic regions.

3. The effects of discharging treated effluents to Arctic Lakes and streams during periods of ice cover require clarification.

4. Research on methods of delivering utility services to individual users requires further work. Community planning, population and geography considerations are essential to system selection.

5. The development of high quality design manuals specifically formulated for arctic applications is urgently required. Provision must be made for updating such manuals as new information becomes available.

REFERENCES

Alter, A.J. (1969a). Sewerage and sewage disposal in cold regions.
     Science and Engineering Monograph III-C5b, Cold Regions Research
     and Engineering Laboratory, Hanover, New Hampshire.

Alter, A.J. (1969b). Solid waste management in cold regions.
     Department of Health and Welfare, State of Alaska, Juneau.

Alter, A.J. (1969c). Water supply in cold regions. Science and
     Engineering Monograph III-C5a, Cold Regions Research and Engin-
     eering Laboratory, Hanover, New Hampshire.

Anonymous (1965). Report on operation of oxidation ditch sewage
     treatment plant, Glenwood, Minnesota. Department of Health,
     Division of Environmental Health, Section on Water Pollution
     Control, Minnesota.

Arctic Health Research Center (1972). Personal Communication.
     Fairbanks, Alaska.

Associated Engineering Services Limited (1973). Solid waste man-
     agement in the Canadian north. Solid Waste Management Division.
     Environmental Protection Service, Department of the Environment.

Balmér, P. (1970). Biological and chemical waste treatment exper-
     iments in far northern Sweden. In: Murphy, R.S. and Nyquist, D.,
     Symposium on Water Pollution Control in Cold Climates. U.S.
     Environmental Protection Agency, Washington, D.C.

Barmgartner, D.J. (1963). Color removal from surface water by
     carbon filter. AAL TDR 6Z-37. Arctic Aeromedical Laboratory,
     Fort Wainwright, Fairbanks, Alaska.

Barmgartner, D.J. (1964). Color removal from surface waters by
     hypochlorite. AAL TDR 63-38. Arctic Aeromedical Laboratory,
     Fort Wainwright, Fairbanks, Alaska.

Benson, B.E. (1966). Treatment for high iron content in remote
     Alaskan water supplies. Journal American Water Works Association,
     58, 10, 1356.

Buzzell, T.D., Reed, S.C., and Wilbur, P.F. (1974). Low tempera-
     ture extended aeration through the use of a floating tube
     settler and wood stave tankage. In: Davis, E., Symposium on
     wastewater treatment in cold climates. Report EPS 3-WP-74-3.
     Water Pollution Control Directorate, Ottawa.

Chambers, C.W. (1974). An overview of the problems of disinfection.
   In: Davis, E., Symposium on wastewater treatment in cold climates.
   Report EPS 3-WP-74-3. Water Pollution Control Directorate,
   Ottawa.

Christianson, C.D. and Smith, D.W. (1974). Diffusion systems for
   cold climate lagoons. In: Davis, E., Symposium on Wastewater
   Treatment in Cold Climates, Report EPS 3-WP-74-3. Environmental
   Protection Service, Ottawa.

Clark, J.W., Viessman, W.,Jr. and Hammer, M.J. (1971). Water
   supply and pollution control. International Textbook Company,
   Scranton, Pennsylvania.

Clean Air Act (1971). Assented to 23rd June, 1971.

Clean Air Act (1973). Lead-free gasoline regulations. The Canada
   Gazette, Vol. 107, No. 21, 73-663, November 14, 1973.

Clean Air Act (1974). Leaded gasoline regulations. The Canada
   Gazette, Part II, Vol. 108, No. 15, 74-459. August 8, 1974.
   Amendment, Vol. 109, No. 5, 75-104, March 12, 1975.

Cleasby, J.L. (1975). Wastewater filtration through coarse, sin-
   gle media, unstratified, granular filters. Paper presented at
   Region X wastewater technology transfer design seminar,
   Anchorage, Alaska.

Coutts, H.J. and Christianson, C.D. (1974). Extended aeration
   sewage treatment in cold climates. Arctic Environmental Research
   Laboratory, College, Alaska.

Damron, F.J. (1973). Water/Wastewater evaluation for an Arctic
   Alaskan industrial camp. Thesis. University of Alaska, Fairbanks.

Dawson, R.N. (1967). Lagoon sewage treatment in the Mackenzie
   District, Northwest Territories. M.S. No. 67-30. Division of
   Public Health Engineering, Department of National Health and
   Welfare, Edmonton, Alberta.

Dawson, R.N., Grainge, J.W., Shaw, J.W., and Greenwood, J.K. (1973).
   Sewage lagoon systems. In: J.W. Grainge, R. Edwards, K.R.
   Heuchert, and J.W. Shaw (1973). Management of Waste from Arctic
   and sub-Arctic Work Camps. Environmental - Social Committee,
   Northern Pipelines Task Force on Northern Oil Development.
   Government of Canada, Ottawa.

Department of Local Government (1973). Proposed water and sanitation policy for communities in the Northwest Territories. Government of the Northwest Territories, Yellowknife.

Department of National Health and Welfare (1969). Canadian drinking water standards and objectives. Ottawa.

Department of the Environment (1974a). Cement industry national emission guidelines. The Canada Gazette, Part I, 3868. October 12, 1974.

Department of the Environment (1974b). Proposed national emission standard regulations for secondary lead smelters. The Canada Gazette, Part I, 4563, December 7, 1974.

Department of the Environment (1975a). Asphalt paving industry national emission guidelines. The Canada Gazette. Part I, 1284, April 5, 1975.

Department of the Environment (1975b). Metallurgical coke manufacturing industry national emission guidelines. The Canada Gazette, Part I, 2219, May 31, 1975.

Department of Transport (1970). Atlas of climatic maps. Meteorological Branch, Ottawa, Ontario, Canada.

Edwards, A.C. and Fahlman, R. (1974). Work camps sewage disposal washcar - incineration complex. In: Arctic Waste Disposal, Environmental-Social Committee, Northern Pipelines, Task Force on Northern Oil Development, Government of Canada.

Eddy, W. (1974). Environmental cause/effect phenomena relating to technological development in the Canadian Arctic. National Research Council of Canada, Ottawa.

Environmental Protection Service. Codes of good practice on dump closing or conversion to sanitary landfill at federal establishments. Department of Environment, Ottawa, Ontario.

Fetner, R.H. and Ingols, R.S. (1959). Bactericidal activity of ozone and chlorine against Escherichia coli at $1^{\circ}C$. In: Ozone Chemistry and Technology, No. 21, American Chemical Society, New York.

Fahlman, R. and Edwards, R. (1974). Effects of land sewage disposal on sub-arctic vegetation. In: Arctic Waste Disposal, Environmental-Social Committee, Northern Pipelines, Task Force on Northern Oil Development, Government of Canada.

Floyd, P. (1974). The north slope center: how was it built? The Northern Engineer, 6, 3, 22.

Forgie, D. and Christensen, V. (1974). Work camp wastewater treatment, rotating biological contactor and physical-chemical treatment. Environment Canada, Environmental Protection Service, Edmonton, Alberta.

Gilmore, T.M. and Hanna, T.R. (1973). Regional monitoring of ambient air carbon monoxide in Fairbanks, Alaska. Department of Environmental Conservation, Juneau, 13 pp.

Given, P.W. and Chambers, H.G. (1975). Workcamp sewage disposal, Washcar incinerator complex, Ft. Simpson, N.W.T., Pre-publication copy of ALUR report on Arctic waste disposal.

Gordon, R.C. (1970). Depletion of oxygen by microorganisms in Alaskan rivers at low temperatures. Proceeding of the Water Pollution Control in Cold Climates Symposium, Institute of Water Resources, University of Alaska, Fairbanks, p. 71.

Gordon, R.C. (1972). Winter survival of fecal indicator bacteria in a sub-arctic Alaskan river. Arctic Environmental Research Laboratory, Fairbanks, Alaska.

Gordon, R.C., Davenport, C.V., and Reid, B.H. (1974). Chlorine disinfection of treated wastewater. In: Davis, E., Symposium on wastewater treatment in cold climates. Report EPS-3-WP-74-3 Water Pollution Control Directorate, Ottawa.

Grainge, J.W., Edwards, R., Heuchert, K.R. and Shaw, J.W. (1973a). Management of waste from arctic and subarctic work camps. Environmental-Social Committee, Northern Pipelines, Task Force on Northern Oil Development. Government of Canada, Ottawa.

Grainge, J.W., Shaw, J.W. and Greenwood, J.K. (1973b). Factors affecting the design and operation of activated sludge plants for wastewater treatment from northern work camps. In: J.W. Grainge, R. Edwards, K. Heuchert, and J.W. Shaw (1973). Management of Waste from Arctic and Sub-Arctic Work Camps. Environmental-Social Committee, Northern Pipelines, Task Force on Northern Oil Development, Government of Canada, Ottawa.

Grainge, J.W., Shaw, J.W., and Greenwood, J.K. (1973c). Physical-chemical treatment of wastewater from northern work camps. In: J.W. Grainge, R. Edwards, K.R. Heuchert, and J.W. Shaw, Management of Waste from Arctic and Sub-Arctic Work Camps. Environmental-Social Committee, Northern Pipelines, Task Force on Northern Oil Development, Government of Canada, Ottawa.

Grainge, J.W. and Slupsky, J.W. (1973d). Toilet units for northern
communities and work camps. In: J.W. Grainge, R. Edwards, K.R.
Heuchert and J.W. Shaw, Management of Waste from Arctic and Sub-
Arctic Work Camps. Environmental-Social Committee, Northern
Pipelines, Task Force on Northern Oil Development, Government of
Canada, Ottawa.

Grube, G.A. and Murphy, R.S. (1969). Oxidation ditch works well in
sub-arctic climates. Water and Sewage Works, 116, 7.

Grundwaldt, J.J., Tilsworth, T., and Clark, S.E. (1975). Solid
waste disposal in Alaska. In: Smith, D.W. and Tilsworth, T.,
Environmental Standards for Northern Regions, A Symposium,
Institute of Water Resources, University of Alaska, Fairbanks.

Hargesheimer, J. (1975). Ozone treatment of Arctic waters for iron
and colour removal. Master of Science Thesis in Environmental
Quality Engineering, University of Alaska, Fairbanks.

Harker, H., III (1973). Evaluation of batch freeze desalination
as a means of providing potable water for Arctic coastal villages.
Thesis. University of Alaska, Fairbanks.

Heinke, G.W. (1973). Solid waste disposal in communities of the
Northwest Territories. Solid Waste Management Division, Environmental Protection Service, Department of Environment, Ottawa.

Heinke, G.W. (1974b). Report on municipal services in communities
of the Northwest Territories. Department of Indian and Northern
Affairs, Ottawa.

Heuchert, K.R. (1974a). Evaluation of extended aeration package
sewage treatment plants on the Imperial Oil Limited artificial
islands IMMERK and ADGO-F28. In: Arctic Waste Disposal, Environmental-Social Committee, Northern Pipelines, Task Force on
Northern Oil Development, Government of Canada.

Heuchert, K.R. (1974b). Refuse sack collection system study. In:
Arctic Waste Disposal, Environmental-Social Committee, Northern
Pipelines, Task Force on Northern Oil Development, Government of
Canada.

Holty, J.G. (1973). Air quality in a subarctic community. Arctic
Vol. 26, No. 4, 292.

Joy, R.W., Tilsworth, T. and Williams, D.D. (1975). Carbon Monoxide exposure and human health. Institute of Water Resources,
University of Alaska, Fairbanks, 100 pp.

Kalff, J. (1968). Some physical and chemical characteristics of
  Arctic fresh waters in Alaska and Northwestern Canada.
  J. Fisheries Research Board of Canada, 25, 12, 2575.

Kim, S.W. (1971). The effectiveness of a contact filter for the
  removal of iron from ground water. Institute of Water Resources,
  University of Alaska, Fairbanks.

Kinman, R.N. (1974). Ozone disinfection of wastewaters at low
  temperatures. In: Davis, E., Symposium on wastewater treatment
  in cold climates, Report EPS 3-WP-74-3. Water Pollution Control
  Directorate, Ottawa.

Likens, G.E. and Johnson, P.L. (1968). A Limnological reconnaissance in interior Alaska. Cold Regions Research and Engineering
  Laboratory, Hanover, New Hampshire.

Middlebrooks, E.J. and Marshall, G.R. (1975). Stabilization pond
  upgrading with intermittent sand filters. In: Smith, D.W. and
  Tilsworth, T. Environmental Standards for Northern Regions. A
  Symposium, Institute of Water Resources, University of Alaska,
  Fairbanks.

Morrison, S.M. and Martin, K.L. (1974). Lime disinfection of sewage
  bacteria at low temperature. In: Davis, E., Symposium on wastewater treatment in cold climates. Report EPS 3-WP-74-3. Water
  Pollution Control Directorate, Ottawa.

Murphy, R.S. and Rananathan, K.R. (1974). Bio-processes of the
  oxidation ditch in a sub-arctic climate. In: Davis, E., Symposium on wastewater treatment in cold climates. Report EPS 3-WP-
  74-3. Water Pollution Control Directorate, Ottawa.

Patton, H.R., Johnson, P.R. and Behlke, C. (1967). Saline conversion and ice structure from artificially grown sea ice. Report
  No. 4, Arctic Environmental Engineering Laboratory and Institute
  of Water Resources, University of Alaska, Fairbanks.

Puchtler, B. and Reid, B. (1975). The Alaska village demonstration
  project. Submitted to The Northern Engineer, 7, 2.

Reid, L.C., Jr. (1969). Color removal from Arctic surface waters,
  Kotzebue Air Force Station. Environmental Engineering Section,
  Arctic Health Research Center, Fairbanks, Alaska.

Reid, L.C., and Potworowski, H.S. (1973). Ozone treats arctic
  waters. Arctic Health Research Center, Fairbanks, Alaska, 74 pp.

Schofield, E. and Hamilton, W.L. (1970). Probable damage to tundra biota by sulfur dioxide destruction of lichens. Biol. Conserv. 2: 278-80.

Shallock, E.W. (1975). Winter dissolved oxygen patterns in Arctic and subarctic rivers. In: Environmental Standards for Northern Regions: A Symposium, Institute of Water Resources, University of Alaska, Fairbanks.

Smith, D.W. (1973). Low temperature iron and color reduction using ozone; a literature review. In: L.C. Reid, Jr. and H.S. Potworowski, Ozone Treats Arctic Waters, Arctic Health Research Center, Fairbanks, Alaska.

Smith, D.W. (1975). Land disposal of secondary lagoon effluents (pilot project). Institute of Water Resources, University of Alaska, Fairbanks.

Smith, D.W. and Justice, S.R. (1975). Effects of reservoir clearing on water quality in the arctic and subarctic. Institute of Water Resources, University of Alaska, Fairbanks.

Smith, D.W. and Straughn, R.O. (1971). Refuse incineration at Murphy Dome Air Force Station, Arctic Health Research Center, Fairbanks, Alaska.

Stanley Associates Engineering Limited (1973). Solid waste management in the Canadian north. Solid Waste Management Division, Environmental Protection Service, Department of the Environment.

State of Alaska (1972). Air quality control plan, Volume 1, plan. Department of Environmental Conservation, Juneau, 143 pp.

Suhr, L.G. and Boyle, J.D. (1975). Region X-Technology transfer design seminar. Anchorage, Alaska (paper).

Tilsworth, T. (1974). Organic and color removal from water supplies by synthetic resinous adsorbants. IWR Report-50, Institute of Water Resources, University of Alaska, Fairbanks.

Tilsworth, T. and Damron, F.J. (1974). Industrial wastewater treatment in Arctic Alaska. In: Water-1974, American Institute of Chemical Engineers Symposium Series, New York.

Underwood McLellan and Associates Limited (1973). Solid waste management in the Canadian north. Solid Waste Management Division, Environmental Protection Service, Department of the Environment.

Water Pollution Control Directorate (1973). Guidelines for effluent quality and wastewater treatment at federal establishments,
Environmental Protection Service, Ottawa.

Water Pollution Control Directorate (1974). Interim Guidelines
for Wastewater Disposal in Northern Canadian Communities.
Environmental Protection Service, Department of the Environment,
Ottawa, Ontario.

Zemansky, G.M. (1975). Wastewater and Alyeska-north of the Yukon
River. Submitted to The Northern Engineer, 7, 2.

APPENDIX 1

DRINKING WATER STANDARDS

PARAMETER	OBJECTIVE	ACCEPTABLE LIMIT	MAXIMUM PERMISSIBLE LIMIT
PHYSICAL QUALITY			
Colour – TCU[3]	<5	15	
Odour – T.O.N.[4]	0	4	
Taste	Inoffensive	Inoffensive	
Turbidity – JTU[5]	<1	5	
Temperature – °C[6]	<10	15	
pH – Units[6]	–	6.5 – 8.3	
TOTAL COLIFORM STANDARDS (MF METHOD)[7]			
(a) No coliforms		At least 95% of the samples in any consecutive 30-day period should be "negative" for total coliform organisms.	At least 90% of the samples in any consecutive 30-day period should be "negative" for total coliform organisms.

APPENDIX 1 (Cont'd)

DRINKING WATER STANDARDS

PARAMETER	OBJECTIVE	ACCEPTABLE LIMIT	MAXIMUM PERMISSIBLE LIMIT
(b) No coliforms	None of the samples "positive" for total coliform organisms should have an MF Count greater than 4 per 200 ml or 10 per 500 ml portions.	None of the samples "positive" for total coliform organisms should have an MF Count greater than 6 per 200 ml or 15 per 500 ml portions.	
TOXIC CHEMICALS			
Arsenic as As	Not Detectable[8]	0.01	0.05
Barium as Ba	Not Detectable	<1.0	1.0
Boron as B	–	<5.0	5.0
Cadmium as Cd	Not detectable	<0.01	0.01
Chromium as Cr^{+6}	Not detectable	<0.05	0.05
Cyanide as CN	Not Detectable	0.01	0.20
Lead as Pb	Not Detectable	<0.05	0.05
Nitrate + Nitrite as N	<10.0	<10.0	10.0
Selenium as Se	Not Detectable	<0.01	0.01

APPENDIX 1 (Cont'd)

DRINKING WATER STANDARDS

PARAMETER	OBJECTIVE	ACCEPTABLE LIMIT	MAXIMUM PERMISSIBLE LIMIT
TOXIC CHEMICALS (Cont'd)			
Silver	-	-	0.05
OTHER CHEMICALS mg/1			
Alkalinity	(See Section 8.321)		
Ammonia as N	0.1	0.5	
Calcium as Ca	<75	200	
Chloride as Cl⁻	<250	250	
Copper as Cu	<0.01	1.0	
Corrosion & Incrustation	(See Section 8.323)		
Iron (dissolved) as Fe	<0.05	0.3	
Magnesium as Mg	<50	150	
Manganese as Mn	<0.01	0.05	
Methylene Blue Active Substances	<0.2	0.5	
Phenolic Substances as Phenol	Not Detectable	0.002	

APPENDIX 1 (Cont'd)

DRINKING WATER STANDARDS

PARAMETER	OBJECTIVE	ACCEPTANCE LIMIT	MAXIMUM PERMISSIBLE LIMIT
OTHER CHEMICALS mg/1 (Cont'd)			
Phosphates as PO_4 (inorganic)[4]	<0.2	0.2	
Total Dissolved Solids	<500	1 000	
Total Hardness as $CaCO_3$	<120	180	
Organics as CCE CAE	<0.05	0.2	
Sulphate as $SO_4^=$	<250	500	
Sulphide as H_2S	Not Detectable	0.05^{10}	
Uranyl Ion as $UO_2^=$	<1.0	5.0	
Zinc as Zn	<1.0	5.0	
BIOCIDES			
Aldrin		Not Detectable[11]	0.017
Chlordane		Not Detectable	0.003
DDT		Not Detectable	0.042
Dieldrin		Not Detectable	0.017

APPENDIX 1 (Cont'd)

DRINKING WATER STANDARDS

PARAMETER	OBJECTIVE	ACCEPTABLE LIMIT	MAXIMUM PERMISSIBLE LIMIT
BIOCIDES (Cont'd)			
Endrin		Not Detectable	0.001
Heptachlor		Not Detectable	0.018
Heptachlor Epoxide		Not Detectable	0.018
Lindane		Not Detectable	0.056
Methoxychlor		Not Detectable	0.035
Organic Phosphates + Carbamates12		Not Detectable	0.100
Toxaphene		Not Detectable	0.005
Herbicides, (e.g.: 2,4-D, 2,4,5-T, 2,4,5-TP)		Not Detectable	0.100
RADIOACTIVITY15	1/10 of the ICRP13 (MPC)$_\omega$ for 168-hour week.	1/3 of the ICRP (MPC)$_\omega$ for 168-hour week.	The ICRP (MPC)$_\omega$ for 168-hour week.

FOOTNOTES TO APPENDIX I

1 – Section numbers refer to the Canadian Drinking Water Standards and Objectives – 1968.

2 – To be examined according to the latest edition of Standard Methods for the Examination of Water and Wastewater (American Public Health Association, American Water Works Association, and Water Pollution Control Federation), or other acceptable methods as approved by the control agency.

3 – True Colour Unit, Platinum–Cobalt Scale.

4 – Threshold Odour Number.

5 – Jackson Turbidity Unit.

6 – Has significance in controlling corrosion and scaling tendency of the water (See Section 8.15).

7 – Where less than 10 samples are analyzed in any consecutive 30-day period, no more than one sample should be "positive" for total coliform organisms. When more than one "positive" sample is encountered, a special series of samples should be collected from the same location and analyzed for total coliform organisms.

8 – Not detectable by the method described in the latest edition of "Standard Methods" (APHA, AWWA, & WPCF), or by any other acceptable method approved by the control agency.

9 – Total of carbon chloroform and carbon alcohol extractibles.

10 – Based on taste and odour considerations. Concentration greater than 0.05 mg/l may be objected to by the majority.

FOOTNOTES TO APPENDIX I (CONT'D)

11 – Not detectable by an acceptable method of analysis as approved by the control agency.

12 – Expressed as parathion equivalents in cholinesterase inhibition.

13 – ICRP – International Commission on Radiological Protection.

14 – $(MPC)_\omega$ Maximum Permissible Concentration in Water.

15 – Water supplies may be considered as meeting the objective if analysis shows that the gross radioactivity is less than 10 picocuries per litre. Radioactivities above this limit require further investigation to determine the nature of the activity and the desirability of systematic surveillance or other action.

APPENDIX II

AIR QUALITY OBJECTIVES

AIR CONTAMINANTS	CONCENTRATIONS		RANGE OF QUALITY
1. Sulphur dioxide	(a) 0 to 30 micrograms per cubic metre arithmetic mean		Desirable
	(b) 0 to 150 micrograms per cubic metre average concentration over a 24-hour period		
	(c) 0 to 450 micrograms per cubic metre average concentration over a one hour period		
2. Sulphur dioxide	(a) 30 to 60 micrograms per cubic metre annual arithmetic mean		Acceptable
	(b) 150 to 300 micrograms per cubic metre average concentration over a 24 hour period		
	(c) 450 to 900 micrograms per cubic metre average concentration over a one hour period		
3. Suspended particulate matter	0 to 60 micrograms per cubic metre annual geometric mean		Desirable

APPENDIX II

AIR QUALITY OBJECTIVES

AIR CONTAMINANTS	CONCENTRATIONS	RANGE OF QUALITY
4. Supended particulate matter	(a) 60 to 70 micrograms per cubic metre annual. geometric mean	Acceptable
	(b) 0 to 120 micrograms per cubic metre average concentration over a 24 hour period	
5. Carbon monoxide	(a) 0 to 6 milligrams per cubic metre average concentration over an 8 hour period	Desirable
	(b) 0 to 15 milligrams per cubic metre average concentration over a one hour period	
6. Carbon monxide	(a) 6 to 15 milligrams per cubic metre average concentration over an 8 hour period	Acceptable
	(b) 15 to 35 milligrams per cubic metre average concentration over a one hour period	
7. Oxidants (ozone)	(a) 0 to 30 micrograms per cubic metre average concentration over a 24 hour period	Desirable

APPENDIX II

AIR QUALITY OBJECTIVES

AIR CONTAMINANTS	CONCENTRATIONS	RANGE OF QUALITY
8. Oxidants (ozone)	(b) 0 to 100 micrograms per cubic metre average concentration over a one hour period	
	(a) 0 to 30 micrograms per cubic metre annual arithmetic mean	
	(b) 30 to 50 micrograms per cubic metre average concentration over a 24 hour period	Acceptable
	(c) 100 to 160 micrograms per cubic metre average concentration over a one hour period	
9. Nitrogen Dioxide	0 to 60 micrograms per cubic metre annual arithmetic mean	Desirable
10. Nitrogen Dioxide	0 to 100 micrograms per cubic metre annual arithmetic mean	
	0 to 200 micrograms per cubic metre average concentration over a 24 hour period	Acceptable

APPENDIX II

AIR QUALITY OBJECTIVES

AIR CONTAMINANTS	CONCENTRATIONS	RANGE OF QUALITY
	0 to 400 micrograms per cubic metre average concentration over a one hour period	

APPENDIX III

SUMMARY OF WATER TREATMENT TECHNIQUES

A wide variety of water treatment techniques have been tested
in the Arctic. Information on much of the early work has been
compiled by Alter (1969). Freeze separation is probably the most
widely used method of water purification utilized by permanent
inhabitants. This process has been investigated as a potentially
low cost method of desalination (Harker, 1972 and Patton, et al,
1967). Much more development work must be done before controlled
systems can be used in Arctic communities. It appears that an ice
harvest technique for summer water supplies may be feasible in
some communities.

Disinfection is widely practised, although not universal, with
chlorine being the most popular disinfectant. Gaseous chlorine has
not been used to any extent. However, because of the transportation
problem hypochlorite additions from a premixed solution, powder or
tablets are the normal methods of application. Iodine has been
considered for some applications. However, it has not received
significant use. Ozone was studied as a disinfectant for Arctic
waters (Fetner and Ingols, 1959). To date no permanent ozone
installations have been planned for the North American Arctic.

Many variations and modifications of warm climate technologies
for water treatment have been tested in the north. Distillation
and reverse osmosis have both found applications in areas of
brackish water. In Alaska, the Bureau of Indian Affairs (BIA) has
installed three reverse osmosis units, and the Environmental
Health Section of the Public Health Service (PHS) installed one
unit. The BIA units have shown good consistent performance, although maintenance problems have occurred. The PHS unit has not
performed well. Distillation was used in the community of Kotzebue, but the operating cost and problems were excessive.

Water quality considerations noted earlier have led to consideration of a number of oxidation and coagulation processes.
Early work was directed at lime and alum coagulation followed by
a sedimentation step (Reid, 1969, Benson, 1966). Various poly-
electrolytes have also been used to increase removal of organics
and iron. The principal objective is to coagulate the color and
turbidity causing material. The most serious problem is the operation and maintenance problems which develop with chemical preparation and feed equipment. These problems have led to the investigation of many other approaches.

APPENDIX III (continued)

Resinous absorbants were examined at the University of Alaska for color removal. Both cationic and anionic resins were used. Although the long term removal efficiency was better for the anionic than for the cationic units, both performed poorly compared to the activated carbon control (Tilsworth, 1973).

The use of activated carbon, because of its highly efficient adsorption capabilities has been widespread in arctic water supply treatment. Tilsworth, as noted above, and Baumgartner (1973) have documented its effectiveness in treating northern water supplies. Some of the work camps constructed in Alaska use activated carbon as a final treatment step before chlorination.

Several approaches to the use of oxidants have been studied. Chlorine was the first oxidant considered in work by Baumgartner (1964). Results were mixed due to the need for large doses and resulting taste and odour problems.

Air is used to oxidize iron at the University of Alaska water treatment plant with potassium permanganate being added to complete the oxidation step. The use of permanganate is popular where the iron is not complexed with organics.

Greensand units are also used for iron removal. These units receive a very low dose of potassium permanganate during normal removal operation and a high dose during backwashing.

In 1973, work on the use of ozone to oxidize organics and iron was reported by Reid and Potworowski (1973). More recent work by Hargesheimer (1975) has yielded valuable energy and cost information on the use of ozone to treat colored, iron rich waters. Ozone requirements reached 59 mg per liter of water treated.

The application of the right combination of the above processes to meet quality criteria requires adequate knowledge of quality variations and treatment systems. Often systems are planned to handle all the possibile qualities of water. Such systems result in extremely high capital and operating costs.

Many of the processes noted above require suspended particulate removal before disinfection and distribution. Sedimentation tanks have been used to a very limited degree, because of their space requirements and problems with thermal gradients. The requirements for space can be handled with adequate capital expenditures. The thermal gradient problem is more complex. Temperature variation in a sedimentation tank results in short circuiting and

APPENDIX III (continued)

poor quality effluent. Such temperature differences can result from poor heating or air circulation, or variations in the temperature of influent. As a result of these problems tube and plate sedimentation systems have been used.

Filtration and disinfection are often the only treatment used. Considering the lack of trained operating personnel simple automated systems are the most desirable. Pressure sand and diatomaceous earth filters are two popular approaches. When conventional systems are utilized they are normally placed in heated buildings. Gravity units have been used in many of the package treatment plants. Some work has been done on media for iron floc removal (Kim, 1971), but little other information is available on Arctic applications of gravity filtration units.

APPENDIX IV

The following table presents a summary of some of the wastewater treatment systems which have been used in the Arctic and documented in the literature. Alter (1969a) summarized some of the earlier experiences with treatment systems. That information is not included in the table.

Grainge et al (1973a) reviewed the experience with extended aeration plants in northern Canada and found very poor performance being the rule rather than the exception. It was concluded that due to operational problems extended aeration systems were not meeting the objectives of avoiding any significant public health risk and public nuisance, and the prevention of significant environmental degradation.

Grainge et al (1973b) also reviewed physical chemical treatment experience in the north. Where high and consistant efficiency was needed it was stated that PC treatment was a reasonable alternative. Costs of equipment and operation were serious constraints as was the need for well trained operators.

APPENDIX IV

PERFORMANCE OF WASTEWATER TREATMENT SYSTEMS IN COLD REGIONS

Treatment Type	Subclassifications	Treatment Efficiency[1,2]			Comments	References
		BOD$_5$	SS	Other Parameters		
Lagoons	Short Retention[3]	41 (75)	69		Hay River – Summer, two cells	Dawson et al. 1973
		43 (75)	81		Hay River – Winter, two cells	Dawson et al. 1973
		30 (84)	15		Pine Point – Summer, two cells	Dawson et al. 1973
		42 (84)	82		Pine Point – Winter, two cells	Dawson et al. 1973
	Long Retention[4]	85 (207)	91	Total Coliforms 99.9	Inuvik – Summer	Dawson et al. 1973
		45 (207)	90	99.8	Inuvik – Winter	Dawson et al. 1973
		74 (135)			Yellowknife – Summer	Dawson, 1967

APPENDIX IV (CONT'D)

Treatment Type	Subclassifications	Treatment Efficiency[1, 2]				Comments	References
		BOD$_5$	SS	Other Parameters			
		87 (215)	95			Fort Smith – Summer	Dawson et al. 1973
		53 (215)	72			Fort Smith – Winter	Dawson et al. 1973
	Combination short and long retention	73 (201)	91	99.99		Yellowknife – Summer	Dawson 1967
		60 (179)	88	97.7		Yellowknife – Winter	Dawson 1967
	Upgraded[5] Lagoons						
	Intermittent sand filter	67–94	19–66			Not yet applied to northern Systems. Effective sand size 0.17 mm Hydraulic loading 3700–7500 m^3/ha.-day	Middlebrooks & Marshall, 1975
		47–70	25–55			Effective sand size 0.72 mm Hydraulic loading 3700–5600 m^3/ha-day	

APPENDIX IV (CONT'D)

Treatment Type	Subclassifications	Treatment Efficiency [1,2]						Comments	References
		BOD$_5$	SS	Other Parameters					
				TOC	NO$_2^-$+NO$_3^-$	NH$_3$	Turbidity		
	Coarse, single media unstratified granular filters	68 (15)	79 (38)				82 (16 FTU)	No known northern applications to date	Cleasby, 1975
	Land Disposal	93–94 (59.3)	80–94 (64.6)	54–60 (53.6)	6 (0.02)	>95–98 (>10)		Effluent from second stage of four stage lagoon, loading rates of 8–18 cm per week.	Smith, 1975
Aerated Lagoons	Fine bubble	79 (175)	82 (188)					Winnipeg– avg. values over 21 month period, 2 cell operation	Pick et al., 1970
		81 (190)						Fort Greely – 2 cells	Christianson & Smith, 1974
		85 (238)						Northway – 2 cells	Christianson & Smith, 1974
		79 (173)						Eagle River – porous ceramic diffusers, 2 cells	Christianson & Smith, 1974

APPENDIX IV (CONT'D)

Treatment Type	Subclassifications	Treatment Efficiency[1,2]			Comments	References
		BOD$_5$	SS	Other Parameters		
	Coarse bubble	81 (175)	82 (188)		Winnipeg – avg values over 21 month period, 1 cell air gun	Pick et al., 1970
		79 (190)			Fort Greely – 2 cells	Christianson & Smith, 1974
		86 (281)			Northway – air gun, 2 cells	Christianson & Smith, 1974
	Surface aerator	78 (175)	79 (188)		Winnipeg – avg. values over 21 month period. Winter operation not practical due to ice build-up, 1 cell.	Pick et al. 1970
	Miscellaneous	79–91 (206–256)	58–91 (197–293)	TOC 64–74 (153–129)	Winnipeg – four aerated lagoons in series. Three aeration systems.	Girling et al. 1974
		92 (319)			Eielson AFB, fine bubble & coarse bubble aeration 6 cell operation	Christianson & Smith, 1974

APPENDIX IV (CONT'D)

Treatment Type	Subclassifications	Treatment Efficiency[1,2]							Comments	References
				Other Parameters						
		BOD$_5$	SS	Grease & Oil	Total Coliform	Fecal Coliform	COD			
Activated Sludge	Conventional	68-81 (76-81)	37-86 (46-51)						Pilot scale experiments at Kiruna, Sweden.	Balmer, 1970
	Extended aeration	78-80 (1750-2080)	69-75 (900-1339)	90-94 (414-419)	66-98 (4.8x10^6 – 3.5x10^9 per 100ml)	69-99.6 (3.9x10^5 –3.4x10^7 per 100ml)	60-61 (2505-3154) TOC 63 (827-887)		Drilling camps on artificial islands in Mackenzie Bay. T>20°C. Effluent chlorinated	Heuchert, 1974
		85 (200)							Alaska-recommended design removal at organic loading = 0.08 kg. BOD per kg MLSS – day and T<7°C	Coutts & Christianson, 1974
		90	85						Ft. Wainwright, Alaska. Single tank system with redwood stave tankage & tube settler	Buzzel, Reed & Wilbur, 1974

APPENDIX IV (CONT'D)

Treatment Type	Subclassifications	Treatment Efficiency [1,2]			Comments	References
		BOD$_5$	SS	Other Parameters		
	Oxygen, activated sludge (full scale)	90 (240)	91 (302)		Winnipeg, Man.	Suhr & Boyle 1975
		90 (153)	87 (105)		Newton Creek, N.Y.	Suhr & Boyle 1975
Rotating biological contactor		88-95 (183-133)	94-96 (199-96)	Grease & Oil 72-96 (36-386); COD 80-83 (342-243)	Yellowknife – Biodisc 1.14 m^3/day, Good treatment at 4 times design load also	Forgie & Christensen 1974
		75-96 (2000-500)	77-79 (314-2048)	66-72 (862-3315)	Arctic Red River work-camp-Biosurf – 57m^3/day design	Forgie & Christensen 1974
	Oxidation ditch	>90			College, Alaska T=2°C	Grube & Murphy 1969
		92			Glenwood, Min. T=1°C	Anonymous, 1965
		>87% 50% of time (92-342)			College, Alaska. No sludge wasting suspended solids in effluent from 8 to 2370 mg/l & BOD from 12 to 1380 mg/l.	Murphy & Rananathan, 1974

APPENDIX IV (CONT'D)

Treatment Type	Subclassifications	Treatment Efficiency[1,2]			Comments	References
		BOD5	SS	Other Parameters		
Activated Biofilter		90-99 (180-264)	70-94 (81-99)		Idaho Falls plant	Suhr & Boyle 1975
		~82 (170-200)	~93 (700-800)		Aberdeen, Idaho	Suhr & Boyle 1975
Physical-Chemical	Primary precipitation	BOD7 50-67 (40-120)		Phosphorus 83-86 (3-7); Nitrogen 20-50	Full scale plant at Kiruna, Sweden	Ulmgren, 1974
	Raw waste-water treatment	90-94 (115-153)	97 (137)	98 (3.3); COD 92 (222)	Pilot plant at Yellowknife – chemical precip/tube settlers/filltration/chlorination/carbon column. Plant downtime for repairs: ~40%	Forgie & Chistensen, 1974
Disinfection					Removal of bacteria & viruses depend upon: 1. Type of organism 2. Type of disinfectant 3. Temperature 4. pH	Chambers, 1974

APPENDIX IV (CONT'D)

Treatment Type	Subclassifications	Treatment Efficiency [1,2]				Comments	References
		BOD$_5$	SS	Other Parameters			
				Total Coliforms	Fecal Coliforms	5. Contact time 6. Residual 7. Other factors	
	Chlorination			>99.999 (2.7 x 10 per 100 ml)	>99.99 (4.8 x 10 per 100 ml)	Secondary effluent-60 min. batch disinfection at T<1°C Chlorine dosage = 5.4 mg/l as HOCl. Chlorine residual = 0.98 mg/l (orthotolidine method)	Gordon, Davenport, & Reid, 1974
	Lime Disinfection			>99.997 (~10 per 100 ml)	>99.997 (~3 x 10 per 100ml)	Secondary effluent, 90 min. contact time, pH = 12.0 T = 1°C	Morrison & Martin, 1974
	Ozonation			100%		10% wastewater (by volume) 10 min contact time pH 7.0, T=1°C, 3.5 mg/l O$_3$	Kinman, 1974
Incineration						Ft. Simpson - Dept of Public Works Workcamp - Black water incinerator.	Given & Chambers 1975

APPENDIX IV (CONT'D)

Treatment Type	Subclassifications	Treatment Efficiency [1,2]			Comments	References
		BOD_5	SS	Other Parameters		
					90 litres black water per hour at 700 to 830°C	
					No. 2 fuel oil rate=23 litres per hr.	
					Power consumption = 8 kwh per hour.	
Sludge digestion	Anaerobic digester				Normally heated to conventional Temperatures. Major diff. is heating cost.	
	Aerobic digester				City of Fairbanks is installing unit, 1975	
Land Disposal					Ft. Simpson & Norman Wells – No change in total vegetation productivity but improved growth of colonial species with up to 5cm of raw wastewater applied to cut-line	Fahlman & Edwards, 1974

APPENDIX IV (CONT'D)

Footnotes:

1. Values presented are generally average per cent removals over various time periods. For more details on their significance or applicability, consult individual references.

2. Values in parentheses are raw influent concentrations in mg/l unless otherwise specified.

3. Short retention ponds are considered to be those having retention times less than 10 days.

4. Long retention ponds are considered to be those having retention times greater than three months.

5. Values apply only to the individual system efficiencies and not to the overall efficiencies.

6. Final NO_2^- + NO_3^- concentrations in percolate averaged 5.28 to 8.48 mg/l.

ENVIRONMENTAL PROTECTION AND QUALITY ENHANCEMENT IN AN ARCTIC REGION

PART II - ENVIRONMENTAL EMERGENCY SITUATIONS

Herbert T. Doane

Environment Canada

Halifax, Nova Scotia, Canada

INTRODUCTION

Emergency situations creating a threat to the environment occur in the North just as they do elsewhere. There are reasons to believe that they occur more frequently than one might expect from experience elsewhere. This may be due to inexperience of operators in dealing with unique conditions of the North, difficulty of maintaining equipment in first rate conditions due to supply of spare parts, extreme temperature conditions causing failure of equipment which performs satisfactorily in less severe climate, and a number of other factors which you are no doubt familiar with and have been referred to already. Plans for dealing with emergency situations have traditionally lagged behind the development of plans for dealing with normal operations and this is understandable. There is, however, a growing concern with the accidents that are happening, particularly as they relate to petroleum spills in water and, I believe, petroleum spills in the sensitive Arctic environment. There are some obvious reasons for this growing concern. The generally accepted slow rate of biological activity at low temperatures has the effect of prolonging the effects of a spill on vegetation and animal communities and the apparent near absence of biological degradation of petroleum products in the North serves to greatly compound the problems. In conjunction with these effects, of course, the physical forces of erosion; melting, etc., will often operate unhindered or unaffected by the spill of petroleum or some other pollutant.

The still common view that the North country is an uninhabited
land of low environmental quality allows people to disregard the
environmental damage they may do, either deliberately or by accident.
Much of the activity in the North is carried out in a frontier
spirit where the environment and nature is seen as a great adversary
displaying no mercy toward man attempting to live there and wrest
resources from the area. Man's role then is often seen as one of
taking whatever is possible and disregarding anything which is not
immediately useful in order merely to survive. This attitude is
further enhanced by the temporary nature of residence by many of
the people operating in the Arctic environment. They often do
not feel they belong and therefore they do not feel much need to
preserve the nature of the environment as it is now or was before
man began to operate there.

GENERAL PHYSICAL PROBLEMS

As with all operations, environmental emergencies are very
much affected by the generally cold temperatures. If people and
machines and material are not prepared to deal with emergencies,
then the cold weather can seriously impair or even prevent any
effective action from being taken. For example, in southern areas
peat moss is commonly used to absorb oil; if this material is
stored at a temperature of less than $0^{\circ}C$, it will often become
a solid mass which cannot be used for absorbing oil. If workmen
who normally operate in a workshop are called upon to work on a
clean up which is exposed to the elements, they require special
protective clothing. The remoteness of most sites of operation
and consequently emergency situations often makes it difficult
or impossible to acquire materials, equipment, or manpower on
short notice.

EMERGENCY PROBLEMS

Environmental emergencies can take a variety of forms.
Forest fires are certainly environmental emergencies in the true
sense of the word. The need for dealing with forest fires has
long been recognized and the response mechanism is generally
quite functional. Land slides, wind storms and floods, may be
less well planned for natural emergencies, but we respond to them
in a positive way. There surely can be no reasonable doubt as to
the need for responding to man-caused environmental emergencies.
If during a well drilling operation, large quantities of hydrogen
sulfide are expelled or if a well catches fire, no reasonable
person would deny that the gas flow and fire should be stopped
so as to return conditions to nearly normal as soon as possible.

The almost universal use of Arctic diesel oil for fuel to
drive machinery and to provide heat has meant that this material

is very widely distributed throughout the North and, due to the relatively small size of packages, there is a high frequency of accidents and resulting spillage taking place. The use of pesticides and herbicides in the North in order to alter the environment and make it more acceptable to immigrant people poses both a possible continuing threat and certainly creates an accident hazard. While these materials are in storage or in transit or in the process of being applied, there is a possibility of their being spilled into the environment in an uncontrolled way.

Construction activities that are meant to provide some temporary facility such as a water supply or a road may very well create an environmental hazard after they have served their intended purpose. Road or dam embankments may eventually impound sufficient water to cause overtopping of the embankment and consequent quick washout and creation of a flood downstream.

This list of possible accidents is not exhaustive--it is intended to indicate that there is a variety of ways in which the environment may be accidentally damaged by people.

DAMAGE PREVENTION MEASURES

The nature of the threat, of course, determines the nature of precautions that should be taken. Since the most common threat is the spillage of petroleum products, it follows that preventive measures have been more highly developed to guard against this occurrence than others. If we exclude pipelines for now, storage facilities for petroleum, comprising large tanks, constitute the greatest risk of a large spill. Conventional measures to prevent spills include dykes or berms constructed around the tanks to prevent the loss of the tank contents to the general environment by confining it within the dyke. This has been frequently used in temperate climates and has one very great benefit which is generally recognized by fire prevention authorities and insurance companies. This is the prevention of the spread of burning petroleum and the consequent spread of fire. In adapting the dyking principle to the North, a number of problems have been identified. There is often considerable difficulty in finding material from which to make impermeable dykes, since sand and gravel are not very suitable and frozen material often has produced unsatisfactory dykes because of the voids and the differential settling that occurs. Impermeable liners of synthetic material are expensive and fragile and require skilled manpower to install properly and satisfactorily. Dykes of frozen material have been found to have some utility but refined petroleum products are only partially blocked by snow and ice. A useful review of petroleum spill containment was prepared for Environment Canada as Economical and Technical Review Report EPS 3-EE-74-1.

The design of piping for transfer of petroleum products requires more care than has been generally given. Piping systems should be well designed from the point of view of support and bracing and of course they should be protected from external damage from vehicles or natural hazards such as rock falls, falling trees, river flooding and forest fires. Subsidence due to thawing of permafrost of course should be prevented. As a general rule piping should be designed so that there is no interconnection of storage facilities (i.e., tanks), unless a transfer is taking place. At the completion of the transfer, it should be possible to close valves and effectively disconnect different stores of petroleum and this should be the established procedure. All of these criteria, of course, apply to temperate climates as well as the North.

Special conditions in the North require special design consideration. The selection of equipment and material needs to be done with consideration given to low temperatures, as this relates to operation and performance. The probability of rough handling in getting the material to site should be assumed. The scarcity of supplies, and difficulty of getting replacement parts on short notice, means that spares and reliability assume great importance. Recognition should also be given to the fact that many operators are not well trained or at least do not have the facility to call on an expert for advice in the event they run into trouble. The site conditions are likely to be rather different from standard conditions where designs are carried out and therefore site specific information, if adequately supplied to the design, is likely to result in less frequent emergency conditions. In the North small units are frequently used for a variety of reasons. Since this means more units, the probability of a spill is increased, however, the probable severity of the spill is decreased somewhat.

A major consideration is that the equipment should be simple to operate so that there is little possibility of faulty operations either due to carelessness or ignorance on the part of the operator. Standardization of units will reduce the training needs somewhat, and should be followed where possible.

This naturally brings us to another important factor in preventive measures and that is operator training. I am sure there will be wide acceptance of the fact that well trained operators are the best insurance against accidents and they are the best way to ensure good operation of any systems installed anywhere, including the North. There are various reasons for having poorly trained operators, some of which are barely tolerable but the existence and continued use of poorly trained operators should not be accepted. Training on site and in special courses continues to be a necessity.

A final consideration in preventive measures is the site selection for the facility. We find that most often the site is selected on the basis of resource availability, transportation, convenience, or a set of criteria which seldom include the associated environmental risk. Environment Canada is attempting to convince the federal government and others that an assessment of sites should be made in the light of environmental considerations before final site selection is made. This is not to say that environmental consideration should be the only factor, but we do consider that environmental protection ought to be one of the factors in selecting sites for any kind of industrial or institutional activity.

RESPONSE PLANS

When the site has been selected and equipment and facilities have been designed and installed, it is scill necessary to have a plan which will be followed in the event that emergency conditions arise. This is recognized, for instance, on board ship when they have lifeboat drill. It is recognized in many industries in relation to accidents which may threaten life, or property or processes. We consider that this should be equally recognized with regard to environmental protection.

The essential elements of an emergency response plan include:

1. Alerting: A means must be provided, and must be known by anyone in the vicinity who may detect an emergency, of sounding the alarm to the responsible authorities so that counter measures may be undertaken as quickly and effectively as possible. The alerting procedure should be well publicized so that people are familiar with it. It should be checked from time to time to ensure that it does function and it should be efficient and rapid. There is often good reason for combining the sytems or the communication channels used for sounding the alert of an environmental emergency with that for sounding other kinds of alarms. In fact, often the same communication facilities are used in normal operations. The essential points are that the system should work and that it should be workable by anyone who may possibly detect an emergency condition and that the system should be known by any possible reporter.

2. Responding: The organization to respond to spills needs to be carefully throught out, clearly set down and explained to all of the people who have a role in responding. An emergency situation is no time to begin sorting out personality clashes or vague areas of responsibility and division of authority. If all the people involved in a response know what their

responsibility is, to whom they are responsible, and what the responsibilities of others are, then they will be able to put their energies and talents to work solving the particular problem of the particular spill, they will already have settled the general procedures and the general relationships that are applicable to all emergencies.

3. Resources: Certain special resources are scarce but particularly necessary for responding to emergencies. These materials such as skimmers, oil booms, specialized pumps, and special containers should be procured and held in a state of readiness so that they may be put to use in short order. Frequent inspections to ensure that the materials are serviceable are needed. Other resources that are not particular to this kind of emergency response should be identified in the plan and their source or sources should be clearly noted. This would include sources of manpower, general construction equipment, and sources of food, drink, and accommodation, etc. Often a very important general requirement is for transportation equipment, for both pure transport and for inspection and surveillance needs.

The technology applicable to environmental emergencies that may be expected in the area should be identified in an appropriate appendix. Special aids such as electronic data searches, availability of special weather forecasts, and instructions on operation of specialized equipment should be included in appendices. Another appendix giving information on containment techniques that might be applicable in particular cases should also be included so that this may be studied by those people with responsibilities under the response plan. Included in another appendix might be a description of particular environmental sensitivities that should be considered when any counter measures are being conducted. Similarly there may be a need for an appendix to describe other social sensitivities that should be kept in mind when responding to an emergency.

RESEARCH NEEDS OF ENVIRONMENTAL EMERGENCIES

Since most of the environmental emergencies we have so far been called upon to deal with involve oil, we are aware that there is a need to determine the effects and the mechanisms of assimilating or dispersing oil in water, both fresh and seawater. This, of course, is a need which is common to environmental protection in general.

Another need is to research the effects of temperature on the behavior of oil in the environment and the temperature effect on the decomposition products of oil in the environment.

We also need to determine better ways to immobilize or contain
spills of hydrocarbons that are discharged onto land that is
permeable, onto land that is frozen, onto swamp or water surfaces
that are frozen, and onto open sea surfaces which are extremely
mobile and under ice.

The development of equipment to recover oil and other
similar materials from the environment still requires a good deal
of improvement. I would include in this category the development
of materials that can be used with equipment for recovery and for
containment of oil.

CONCLUSION

Consideration of the possibility of environmental emergencies
has not received very much attention from operators in the Arctic
environment. It is only recently that much consideration of the
Arctic environment under normal operating conditions has received
attention and many exploration and development concerns are com-
placent about environmental protection if they have begun to give
it some concern in relation to their day to day activities. This
situation requires a change of attitude and a selling job by
environmentalists concerned with Northern development activities.
The possibility of spectacular environmental damage looms larger
with continuing development of gas and oil and with the possibility
of larger quantities of hydrocarbons being moved by tanker, ice
breakers, or submarines. The fragility of the Arctic environment
has been only recently "discovered" by interested people and by
some laymen. It is possible that the true nature of Arctic
ecology is not well enough known for us to judge the adequacy
of protection measures that are presently considered appropriate.

In the circumstances, those of us who are in the very primitive
business of environmental emergency response consider that it is
only prudent to take whatever measures are possible to prevent
the discharge of any pollutant into the Arctic environment or,
after a spill, to remove as much of that pollutant as possible
with as little apparent disturbance of the environment as possible.

ENVIRONMENTAL ASPECTS OF CONTINGENCY PLANNING

AT PRUDHOE BAY, ALASKA

D. S. Braden

BP Alaska Inc.

Anchorage, Alaska, U.S.A.

LOCATION AND HISTORY

BP Alaska Inc.'s current operations are being conducted principally at Prudhoe Bay, Alaska, located on the Arctic coast of North America approximately 150 miles southeast of Point Barrow, Alaska, and approximately 200 miles west of the Canadian/United States border. Reconnaissance in this area was conducted by BP personnel as early as 1959. Subsequent to obtaining leases in the area, drilling started in 1963. BP Alaska Inc.'s discovery well - Put River No. 1 was spudded in on November 20, 1968 and an announcement was made on March 13, 1969 that oil had been struck. Atlantic Richfield was actually the first to locate commercial quantities of hydrocarbons in the area with Prudhoe Bay State No. 1 on January 16, 1968, predating BP Alaska's find by 13 months. Sag River State No. 1 confirmation well was announced June 25, 1968. Further drilling has defined the limits of the structures currently being developed which encompasses some 200 square miles. The spine of the producing area lies on a NW/SE trending axis and is approximately 20 miles long.

TOPOGRAPHY

The surface is tundra, most commonly a grassland, made up predominantly of grasses, sedges, lichens, mosses, and low-growing to prostrate woody shrubs growing on a thick, spongy mat of accumulated organic remains or peat. Summer thaw and poor drainage result in a wet, swampy, mushy substrate over most of the plain. Streams are shallow, have low gradients, are sluggish, and flow in contorted, braided, complex channels among the thousands of shallow lakes which reflect the polygonal pattern of ice wedges and produce

the polygonally patterned ground which is ubiquitous on the landscape. The polygons are normally a few tens of feet across and
are separated by troughs commonly several feet wide. Permafrost
presents critical problems in the management of the coastal zone as
it is readily eroded at shorelines by thermal and mechanical
processes. Any disturbance to the tundra surface resulting in
increased thaw produces characteristic thermokarst topography and
thaw lakes in ground containing large volumes of ice.

Prudhoe Bay is underlain with permafrost ranging in thaw
depth from 6" to 18" during maximum thaw back in September and
extending in depth to an average of some 1500 feet. The surface
topography is typical Arctic coastal plain. As described in
Geological Survey Professional Paper 482, the Arctic Coastal Plain
is a smooth plain rising from the Arctic Ocean to a maximum
altitude of 600 feet at its southern margin. The shoreline is
generally only one to ten feet above the ocean. Pingos, an Arctic
phenomenon, do occur and provide skyline relief to an otherwise
horizontal skyline. The Arctic Coastal Plain is poorly drained and
consequently becomes quite marshy during summer thaw. Bathymetry
work conducted by BP Alaska Inc. and Atlantic Richfield Co.
indicates an average depth of approximately 4-1/2 feet for those
lakes located in the proposed unit area. These lakes freeze solid
during the winter. There is one lake located near the Sagavanirktok
River most probably a former ox box cut off or meander loop that
has measured depths to 12 feet. The axial orientation of these
lakes lies in a NNW/SSE trend and they are believed to migrate.
Tentative evidence indicates they are enlarging and the energy is
probably wind. The long axis is normal to the prevailing winds
and downwind erosion accompanied by lateral erosion from wind
driven currents may cause this phenomenon.

CLIMATE

Climate data for this general area is sparse insofar as the
U.S. Government has only records from the Weather Bureau facility
at Point Barrow dating back to 1950 and the U.S. Military has
surface observations from Barter Island to the east, dating back
to 1951. Workers at the Naval Arctic Research Laboratory located
at Point Barrow have some climate data and this is being integrated
into other (U.S. Weather Bureau) data.

There is some reason to believe that the climate of the
Prudhoe Bay Area may be somewhat dissimilar to that at either Point
Barrow or Barter Island. The U.S. Tundra Biome Program conducting
work at Prudhoe Bay under the sponsorship of the proposed Unit
participants feels that subtle differences in local vegetative
species and abundance and certain dissimilarities in the ponding
distribution may be indications of this. Unpublished sources
indicate a mean annual snow fall in the area of somewhere between
20 and 30 inches with densities of 0.32 to 0.33 gm/cc. Mean

January minimum temperatures as shown by Watson range near (-22° F).
While total snowfall is not high, considerable wind drifting does
occur.

The maximum range of temperature may be about 130° F along the
coast. The highest temperature in summer is about the same number
of degrees above zero as the lowest temperature in winter is below
zero, that is roughly 65°. The summers are short and cool and the
temperature is likely to drop below freezing at times, especially
on clear nights. Temperatures rise above the freezing point only
occasionally from October through May. The ocean has the expected
effect in moderating temperature and both the high and low extremes
near the coast are surpassed at inland stations.

Recent data indicates that Prudhoe clearly has colder winter
months and warmer summers than both Barrow and Barter Island. The
net result is a colder annual mean at Prudhoe than for either
Barrow or Barter Island. The warmer summers are clearly indicated
by the consistently larger thaw indices. The two periods for which
winter freezing indices are available show larger values for Prudhoe
than the other two coastal stations. The significantly warmer
summer climate is most likely a reflection of the inland location
of the Prudhoe area station.

Annual regimes at Prudhoe parallel those for Barrow and
Barter Island. Both 1970 and 1971 were cooler than 1972 and 1973
across the entire coastal plain.

GOVERNMENT REQUIREMENTS

The Environmental Protection Agency of the United States
Government published regulations effective January 10, 1974
requiring owners and operators of facilities such as the proposed
Prudhoe Bay Unit to prepare Oil Spill Prevention Control documents
or as they are commonly called, Spill Prevention Control and
Countermeasure Plans. The authority for these regulations stems
from the Federal Water Pollution Control Act as amended (1972).
While it is not required to submit these plans for approval, they
must be available for inspection and have the approval and authority
of corporate management.

The Code of Federal Regulations Title 40 - Protection of the
Environment, Chapter 1 - Environmental Protection Agency, Sub-chapter
D - Water Programs, Part 112 - Oil Pollution Prevention, paragraph
112.3 - Requirements for preparation and implementation of Spill
Prevention Control and Countermeasure Plans, states in part:
"Owners or operators of onshore and offshore facilities in operation
on or before the effective date of this part that have discharged
or could be reasonably expected to discharge oil in harmful quantities,
as defined in 40 CFR Part 110, into or upon the navigable waters of
the United States or adjoining shorelines, shall prepare a Spill
Prevention Control and Countermeasure Plan (hereinafter SPCC Plan),
in accordance with 112.7."

40 CFR, Chapter 1, Sub-chapter D, Part 112, 112.7 Guidelines for
the preparation and implementation of a Spill Prevention Control
and Countermeasure Plan, states in part:
"The SPCC plan shall be a carefully thought-out plan, prepared in
accordance with good engineering practices, and which has the full
approval of management at a level with authority to commit the
necessary resources."
Perhaps somewhat obtusely 112.6 which precedes the guidelines for
preparation of an SPCC deals with Civil Penalties. Briefly it
states that owners or operators of facilities falling within the
purview of these regulations and who violate them shall be liable
for a civil penalty of not more than $5000 U.S. for each day that
such violation continues.
        Public Law 92-500 as passed by Congress provides, in part,
for liability with respect to discharge of oil hazardous substances
and states in part:
"Any person in charge of a vessel or of an onshore facility or an
offshore facility shall, as soon as he has knowledge of any
discharge of oil or a hazardous substance from such vessel or
facility in violation of paragraph (3) of this subsection,
immediately notify the appropriate agency of the United States
Government of such discharge. Any such person who fails to notify
immediately such agency of such discharge shall, upon conviction,
be fined not more than $10,000 (U.S.) or imprisoned for not more
than one year or both."

                        CONTINGENCY PLANNING

        For purposes of contingency planning, we might then consider
three separate seasons of the year rather than the traditional four.

Summer	Freeze-Up	Break-up
June 15	September	March 15
July	October	April
August	November	May
	December	June 15
	January	
	February	
	March 15	

During the summer, June through August, the tundra is quite wet and
marshy. Starting anytime during September freeze-up will occur;
however, as a rule of thumb it will take approximately ten
consecutive days at an ambient air temperature below -17.7°C to
freeze up the tundra surface well enough to allow off the road
travel by heavy equipment. Spring break-up may occur (anytime
after March) and while lasting for a relative brief time of
approximately two weeks, it is characterized by extremely rapid

snow melt and consequent high water runoff. Basically the SPCC plan addresses itself to three discrete elements: <u>Prevention</u>, <u>Containment</u> and <u>Countermeasures</u>.

Prevention

Prevention of spills is accomplished by a review of proposed engineering and construction plans to assure that adequate safe guards are installed and recurrent employee training sessions are conducted. At the Prudhoe Bay facility we have reviewed all installations designed to store and/or transfer fuels and have drawn up flow diagrams with operating instructions on how to accomplish fuel transfer. Valves on the diagrams have been labelled and then these valves on the facility have been tagged with metal tags. The flow diagrams with operating instructions have been printed and sealed in plastic and then posted at the facility. This is an ongoing requirement that is met by periodic updating of procedures as line and valves are changed or relocated. Normally this is accomplished by the Field Maintenance Department.

Recurrent employee training will be handled by the Personnel Department. As of this date it has not been finalized.

Company reporting forms and guidelines for their use have been drafted and made available to supervisory personnel. While spills may not be totally eliminated, procedures such as these have gone a long way towards minimizing them. With the high cost of petroleum and the even higher cost of products delivered to the Arctic for use, significant savings can be effected at the operational level by minimizing spillage.

At such time as the feeding and gathering lines are installed from the wells to the Gathering Centers and to Alyeska Pump Station Number 1, the control will be automated and operated by electronic console. This console will have flow diagrams and valving sequences appropriately shown for full understanding by the operators.

Containment

SPCC regulations call for a means of secondary containment in conjunction with any fuel storage facilities. This containment may take a variety of forms such as dyked structures, flow channels to sumps, or ability to contain a spill in a catch basin. The primary objective of the secondary containment is to prevent a spill from reaching navigable water as defined by the E.P.A.

The mode of secondary containment most commonly used at Prudhoe Bay for the more permanent storage facilities such as the 250,000 gallon diesel storage tanks is that of constructing gravel dykes and incorporating an impermeable membrane. This membrane is usually a cold adaptable, reinforced synthetic sheet which uses the gravel for structural support. Capacity of these containment systems is

designed and constructed to 110% of tankage volumes. Periodic
testing is called for; however, the fact that they contain water
during the thaw period is a reasonable indicator of integrity.

For the smaller or portable fuel storage requirements, pillow
or bladder tanks were originally used. Historical experience
indicates that beyond a three year life the reliability insofar as
integrity is concerned rapidly diminishes. The primary failure
mode is seam failure. These seam failures are not massive ones but
rather more discrete on the order of inches. BP Alaska Inc. is
retiring these bladders in favor of double hulled steel tanks.
The double or outside skin is constructed completely around the
inner steel container with an average 4" annulus and an open top
on both sides. This provides secondary containment in the unlikely
event of internal failure and also provides for containment in the
event of an overfill. These double hulled tanks are fabricated on
steel skids and nominally designed to fit into a C-130 or Hercules
type aircraft for mobility. Operational history of these tanks to
date has been extremely good with no reported problems in 16 months
of use.

A third method of secondary containment is used for exploratory
operations when conducted during the winter season. This consists
of using the natural elements of the Arctic and constructing
compacted snow berms or dykes and then spraying them with water to
form an ice punch bowl. Insofar as the fuel that might come into
contact with the ice in the event of a spill would be at ambient
temperatures, there should be no thermal erosion problems.

With regard to containing potential spills that could occur
in the event of a well blowout or major rupture of a large line,
BP Alaska Inc. has conducted two exercises. Horizontal and vertical
survey control was established throughout the proposed Unit Area.
This was done to second order survey standards. In conjunction,
high and low altitude, black and white aerial photography were
flown in stereo pairs. From this we currently have five (5) foot
contour mapping of the area conducted to National Map Standards.
By mid 1975 we will have two (2) foot contour maps of selected
areas around the drill pads and Gathering Centers. From these
contour maps we will be able to determine topographic catch basins
and areas most susceptible to damming to form various entrapments
for major spills. The survey and map information may then be fed
into computers to determine volumetric capacities of these basins.
With this information plotted against rate of flow on a blowout we
will be able to determine necessary timing for adequate containment.
The degree of relief within the proposed unit area was somewhat
surprising. Insofar as the Arctic Coastal Plain is quite flat
appearing it has come as a mild surprise that differences on relief
on the order of 50' are found. These large differences are caused
by pingos. Concurrently with this mapping exercise, we have had
conducted hydrologic studies of surface flow patterns to outline
ahead of time where spills might run should they occur.

Once the direction and path that a major spill might take is known, it is possible to pre-plot the optimum locations for rapidly constructing temporary dykes or dams. The gravel works that have been used for road, and pad construction can be considered for containment of crude oil such as found at Prudhoe Bay particularly with the high paraffin content rapidly gelling on contact with cold gravel.

As mentioned earlier, there are at least three seasons of the year that must be considered:

Summer Thaw Period. During the summer or thaw period with relatively high ambient temperatures we would be confronted with a situation consisting of a high or perched water table located on top of the uppermost level of permafrost. Standing water would be encountered; however, natural drainage courses are either dry or quite low in terms of water volumes carried. It is our feeling that with the high water table the spilled crude would essentially float on top of it providing separation from the soil and vegetative root mass in the main.

Winter Conditions. Given a freeze-up or winter condition the rate of spread would not be diminished due to rapidly increased viscosity with the lowering of temperature. Empirical data shows the following spread rates.

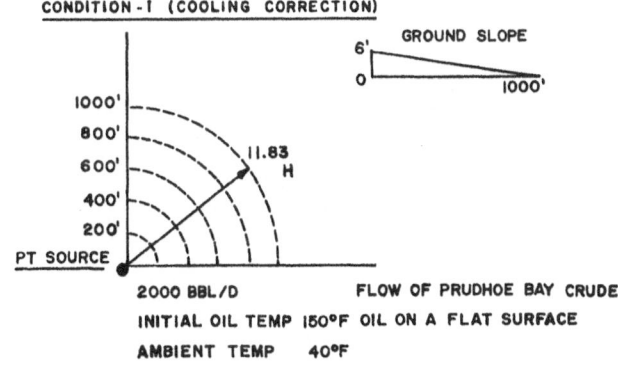

Figure 1 Spread rates for cooling conditions (cont'd on p. 184).

CONDITION-2 (COOLING CORRECTION)

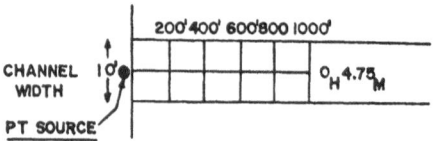

2000 BBL/D

INITIAL OIL TEMP 150°F   OIL IN A SHALLOW CHANNEL

AMBIENT TEMP   40°F

FLOW OF PRUDHOE BAY CRUDE

CONDITION-3 (NO COOLING CORRECTION)

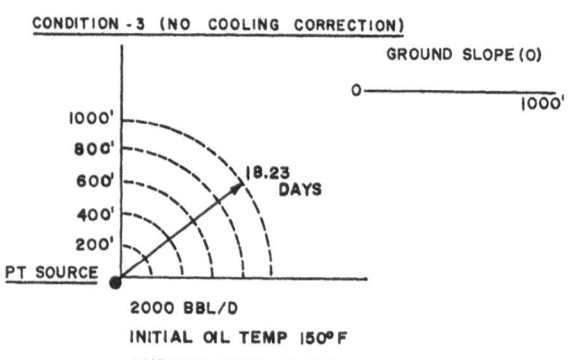

2000 BBL/D

INITIAL OIL TEMP 150°F

AMBIENT TEMP = INITIAL OIL TEMP (OIL DOES NOT COOL

DURING FLOW)

CONDITION-4 (NO COOLING CORRECTION)

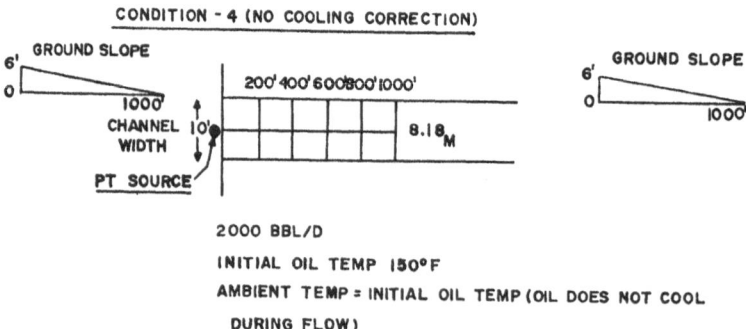

2000 BBL/D

INITIAL OIL TEMP 150°F

AMBIENT TEMP = INITIAL OIL TEMP (OIL DOES NOT COOL

DURING FLOW)

Figure 1 (cont'd) Spread rates for cooling conditions.

Concurrently during the winter with snow cover available, the snow will act as a wicking or absorbent agent. Arctic snow is quite low in moisture content, yet the manner in which fresh crude comes in contact with it seems to play a significant role in how well it will absorb. Field experiments conducted at Prudhoe Bay indicate in part that a layer of fresh crude poured on top of an existing crusted surface will penetrate to a depth of approximately 3/4". Disturbed or granular snow does not appear to absorb crude but rather to adsorb it. Knowing this, it becomes feasible to consider using snow as an agent from which to construct temporary dams or dykes. The best method here then becomes one of hauling or blowing snow into an area and spraying water on the outer surface to create an ice shell. Front end loaders and portable snow blowers could then be used cross country for rapid construction of snow and ice dams or dykes.

Spring Break-up. Spring break-up poses the third condition we must consider in containing large volumes of crude. During this time period of approximately two to three weeks, ground conditions are characterized by extremely fast surface snow melt with an attendant high volume, high velocity water run-off in all natural drainage areas. On the assumption that crude would be carried on top of the water surface, both standing and moving, our volumetric containment capacity is diminished with respect to holding crude in natural basins. While artificial dams or dykes of gravel, sand and sacks of drill mud could be constructed, we must be able to lower the rapidly rising water levels in the catch basins. This may be accomplished in several ways. Headboards can be constructed to contain the surface floating crude yet allow the run-off water to flow underneath. Using a straight containment approach we may choose to install large self-operating siphons. Large sea type containment booms may be deployed in the quieter waters of large lakes, or combinations of the above may be employed. Probably the most problems are associated with this season principally as a result of the rapidity with which surface waters will be flowing. Fortunately, this period is quite brief in duration.

Countermeasures

Countermeasure or Spill Cleanup techniques that have been used or developed in more temperate climates are applicable in the Arctic depending on the season. As with containment, the approach used to clean up a spill will depend in the main on the season and the location, that is whether they involve a terrestrial or aquatic spill.

Summer Aquatic. During the unfrozen season of summer conditions aquatic spills may be handled by the standard techniques of using booms for floatation and vacuum trucks or skimmers for primary recovery with sorbents used for secondary recovery or polishing. A variety of sorbents are available and generally the cheaper ones

are to be preferred. Due to the prevailing winds occurring at
Prudhoe Bay we have found that granular or loose materials are
difficult to deploy and handle. Consequently sorbents that come in
sheets or blankets are preferred by the field crews. Proper use of
containment booms can make a significant difference in spill
recovery time and amount of sorbent material needed for final
polishing. By slowly diminishing the size of the boom, oil on
water may be herded or crowded into a smaller area making it easier
to pick up by skimmer or vacuum. This also helps to decrease the
amount of final polishing or secondary recovery necessary. Oil on
water may also be herded by employing hoses.

Summer Terrestrial. Terrestrial spills occurring during the
summer or thaw period present a somewhat different problem. If
they occur in those areas where the ground water table is close to
or at the surface, damage to root stock appears to be minimized and
partial cleanup may be affected by sorbent materials. There appears
to be conflicting opinion on the use of burning; however, it has
been our experience that given the right conditions, i.e. a spill
on snow, this is a viable method of treatment. Another approach
that needs more work is that of water flooding an area to float as
much oil as possible off the ground and also to saturate the root
stock to prevent oiling. This has been tried at Prudhoe Bay on a
small scale and appears to have merit. Naturally, its application
may be limited to discrete areas where conditions such as topography
and water availability are met. On all terrestrial spills,
application of fertilizers appears to be highly desirable. On the
supposition that a percentage of the root stock survives oiling and
is in a depressed physiological condition, fertilizer application
would tend to strengthen the plant material helping to restore them
to full vigor. Experimental evidence indicates that as a result
of low soil temperatures even during the summer, nutrient transfer
between solid and plant root stock is only at about the 10% level.
We have applied fertilizer (10-20-20 and others) at the rate of
1000 to 1200 lbs/acre without any chemical burning occurring.
Application by hydro seeder appears more effective.

Spills occurring during break-up conditions may be treated
similar to those occurring during summer. The situation is
hindered on the one hand by high velocity water movement and also
by the increased spread that would occur as a result of containment
problems. However, root stock of plant material is in a natural
saturated condition most likely creating optimum survival. Again
with the diminishing of high water levels, fertilizer applications
would be indicated.

Winter. Winter or freeze-up condition spills present a quite
different situation. Experimental evidence indicates that the
quality of the snow cover and the manner in which the oil comes in
contact with the snow have a significant bearing on clean up
techniques.

Oil at temperatures up to those to be anticipated from a major
spill at Prudhoe Bay (85°C) when put into contact with hard crusted

snow by being sprayed on, will soak into the uppermost 3/4" and remain rather congealed. Clean up in this instance may be accomplished by removal using a front end loader and being careful to work from the outside edges inward, and not disturbing the contaminated snow until it is removed. Larger amounts of spilled crude will become quite viscous within a reasonable time and may be handled by mechanical removal to a safe disposal site. Disposal by burning once the spill has been removed to a safe area is probably the preferred method currently available. Depending on location, burning in place is desirable, with fertilizer applied in the spring.

In summary, Contingency Planning for the Prudhoe Bay Field has been predicated on the following factors:

I. Guidelines as established by the Government
   a. Prevention
   b. Containment
   c. Cleanup capability

II. Variations in techniques as dictated by seasons
   a. Break-up
   b. Summer
   c. Winter

III. Risk exposure to Environmentally Sensitive Areas

IV. Personnel Safety

V. Equipment Security

NATURAL GAS PIPELINES IN THE ARCTIC ENVIRONMENT

POLAR GAS PROJECT: PART I - "ENGINEERING RESEARCH"

L. M. Etchegary and Walter Hindle

Polar Gas

Toronto, Ontario, Canada

INTRODUCTION

The Polar Gas Project is one of the most challenging and
fascinating undertakings being contemplated in the world today.
It was formed in late 1972 with the mandate to plan the research
and engineering for the transportation of large volumes of frontier
natural gas from the Canadian Arctic Islands to markets in Canada
and the United States.

Participating in the Project are: TransCanada Pipe Lines
Limited, Panarctic Oils Ltd., Tenneco Oil & Minerals Ltd., Texas
Eastern Transmission Corporation and Pacific Lighting Gas Develop-
ment Company.

It can be seen from the reference map that the main sources of
gas are located well within the Arctic Circle in two general areas;
namely, The Sabine Peninsula on Melville Island and King Christian
and Thor Islands lying off the western side of Ellef Ringnes Island.

According to the latest published report of the Geological
Survey of Canada, ultimate recoverable reserves of natural gas are
estimated to total some 240 trillion cubic feet. To date, over
13 trillion cubic feet of these reserves have been established, an
amount fast approaching the minimum required threshold volumes of
20 to 30 trillion cubic feet.

The challenge facing Polar Gas is to devise a means of
economically moving trillions of cubic feet of Arctic Islands
natural gas reserves up to 2,000 miles southward and across up to
170 miles of Arctic ocean channels while minimizing impact on social
or biophysical environments.

Research programs are currently underway whose aim is to
obtain answers to the remaining technological and environmental
questions needed to enable feasibility determination and detailed

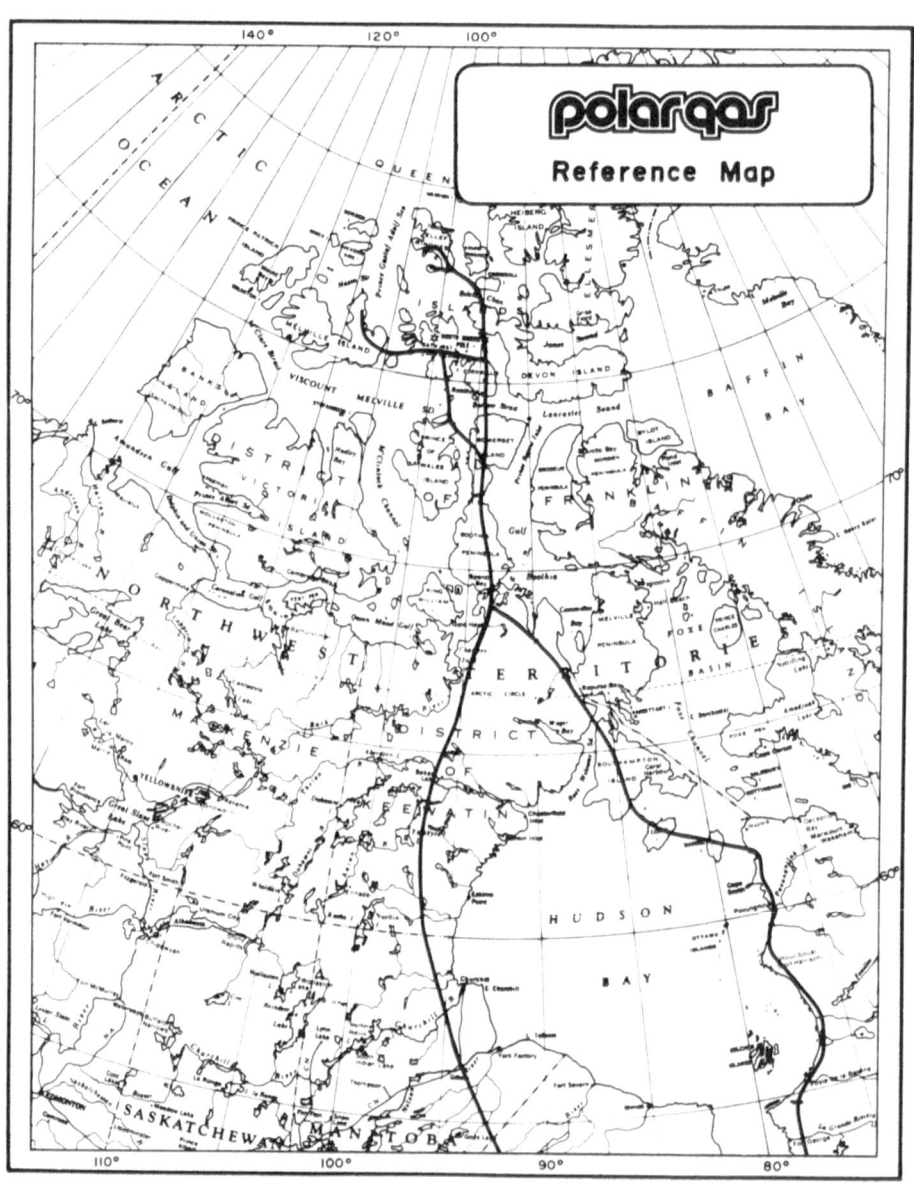

Figure 1

planning, design and routing of a pipeline. Polar Gas engineering
research, for example, is developing a number of methods which
would permit large diameter pipe to be laid during the winter,
across channels averaging 600 feet in depth.

Polar Gas currently plans to file regulatory applications to
construct the line in 1977. This date could be advanced should
threshold gas reserves be established as a result of 1975 drilling
programs in the Arctic Islands.

THE PIPELINE SYSTEM

In investigating a pipeline to deliver natural gas from the
Arctic Islands, Polar Gas is facing a unique technological
challenge - the construction of deepwater channel crossings in the
Arctic Islands.

Faced with the challenge of major pipeline crossings across
Arctic Island channels frozen for as much as 11 months of the year,
Polar Gas construction engineers developed a unique approach - one
which was based upon building the pipeline during a four-month
winter construction schedule. The intent is to utilize the strength
and stability of the ice as an extension of the land surface
whenever possible, preferably by modifying conventional pipelaying
techniques.

The proposed Polar Gas pipeline system would probably be
constructed in two phases - the first running south from Melville

Figure 2. Frozen Channel

Island and the second extending north at a later date to other
discoveries in the area of King Christian and Ellef Ringnes
Islands. Depending on routing, the system will involve some 2,200
to 3,200 miles of up to 48-inch diameter pipe to deliver in the
range of 2 to 4 billion cubic feet of gas per day. Marine
crossings will comprise a minimum of two lines of up to 36-inch
diameter pipe.

The major emphasis at this time is directed towards a 48-inch
diameter pipeline running from Melville and King Christian Islands
down the west side of Hudson Bay to markets in central Canada,
which would take maximum advantage of the available economies of
scale. With a capital requirement of 7 1/2 billion 1974 dollars,
this system would require reserves in the order of 30 trillion
cubic feet to support a minimum throughput of approximately 3
billion cubic feet per day.

An alternative route does exist via east Hudson Bay; however,
this would entail an additional 170 miles of water crossings, an
average of 630 feet deep, which, due to insufficiently strong and
land fast ice, would have to be constructed using lay barge
techniques. It is probable that even if a marine pipelaying system
were to be designed that could cope with the water depths, currents
and ice problems of the northern Hudson Bay area, the result could
be delayed timing and substantially greater cost for the Project.

A smaller 42-inch system from Melville Island, connecting with
TransCanada's system near Winnipeg, is also being considered.
The minimum throughput level for such a system would be in the

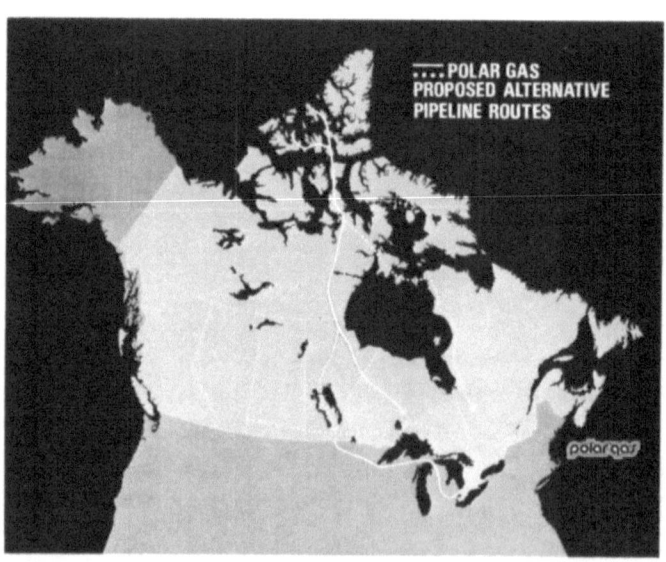

Figure   3.   Proposed Alternative Routes

order of 2 billion cubic feet per day, with capital requirements
of approximately 4 1/2 billion 1974 dollars and threshold reserve
volumes of about 20 trillion cubic feet.

Total compression requirements for a fully powered system
would approach the 3 million horsepower level. System
optimization studies of the compression facilities along the line
are in progress with the planning for compressor stations being
directed towards the use of turbines of high thermal efficiency
(about 33%) in the 20,000 to 35,000 horsepower range. In the
continuous permafrost zone in the north, it is generally planned
to cool the gas to 27 degrees F. by refrigeration systems and in
the south, outside permafrost zones, aerial cooling systems will
be used to maintain discharge temperatures of below 100 degrees F.

Low compression ratios will be used for stations adjacent to
channel crossings so that unit additions can be made in order to
increase the throughput capacity of the marine pipelines in the
event of loss or delay in the installation of one of the lines.

ROUTE DESCRIPTION

The Canadian Archipelago lies north of the Canadian mainland
to the polar ice shelf between Alaska and Greenland, comprising
approximately 500,000 square miles. Three of the islands found in
the Archipelago are among the 10 largest in the world.

As previously outlined, the main gas supply sources are located
on Melville Island, King Christian and Thor Islands. Lateral lines
from these areas will be constructed to meet at a junction point
on either Bathurst or Cornwallis Island and from this point, the
main line will proceed via the west or east Barrow Strait to
Somerset Island and then overland to southern markets.

Contrary to popular belief, the islands are not perpetually
ice-bound. In fact, less than 8% of the land area is permanently
ice covered. The central and western islands form a frozen desert
with low relief and from 2 1/2 to 5 inches of annual precipitation.
In the east, Ellesmere, Axel Heiberg and parts of Devon Island are
mountainous with higher elevations covered by permanent glaciers.

For one to three months in the summer, more than 90% of the
land is bare of snow. Temperatures range from 70 degrees above
zero on the odd day during the summer to more than 60 degrees
below zero in the winter. Mean temperatures at Resolute Bay on
Cornwallis Island are, for example, minus 28.5 degrees F. in
February and plus 40.3 degrees F. in July.

The ice increases in thickness to the end of May but breaks
up sufficiently in the east to permit ocean shipping with icebreaker support from July 15th to October 15th. Ice conditions
become progressively worse to the west. Polar Gas's Rea Point
base camp is generally accessible by sea between August 25th and
September 30th.

Figure   4.   Polar Gas camp at Rea Point

Figure   5.   Typical terrain on Boothia Peninsula

Surface characteristics vary considerably along the route, as might be expected with the pipeline passing through several physiographic regions. In the Arctic, the land surface consists mainly of sedimentary bedrock outcrops and unconsolidated materials. From Somerset Island to Chesterfield Inlet, the terrain is rough and hilly with a high frequency of bedrock outcrops. Permafrost and glaciation characteristics of this region are much in evidence.

South of Chesterfield Inlet, the line follows the margin between the wave-washed sediments which extend eastward to Hudson Bay and non-marine glacial deposits to the west. At the Seal River, located in north Manitoba, the zone of discontinuous permafrost is entered. This also marks the beginning of the tree line. In some areas of northern Manitoba, ponds and small lakes account for about 70% of the surface area. Alternating wet and rocky depressions persist further south into southern Ontario.

These terrain conditions are well known to Polar Gas through the experience gained by TransCanada PipeLines. The latter company has constructed large diameter pipelines during very cold weather and through rock and muskeg. In fact, TransCanada has actually been able to take advantage of such conditions to lay pipelines faster and more economically than in the summertime.

Figure   6. Aerial view of esker and muskeg country

INTER-ISLAND CROSSINGS: RESEARCH AND
INVESTIGATION PROGRAMS

In order to demonstrate the technical feasibility of
constructing the channel crossings, it has been necessary to
collect and analyze extensive data.   To date, three major
engineering research and investigation programs have been conducted
as follows:
a) An ice research program in the Spring of 1973;
b) A marine research program in the Summer of 1973;
c) A second-stage ice research program in the Spring of 1974.

A.   Ice Research Program - 1973

The principal objective of the 1973 programs was to begin
collecting ice and bathymetric data relevant to pipeline routing
and design, and to prove the feasibility of using the ice as a
support platform for winter marine pipelaying operations, including
verification that a conventional form of ditching machine could be
used to excavate a trench across sea ice.   Such a trench is a prerequisite for several possible crossing techniques and in total,
30 miles were successfully trenched during the trials.
The crossings chosen for this research were the Byam Channel
(Melville Island to Byam Martin Island) and Austin Channel (Byam
Martin Island to Bathurst Island).

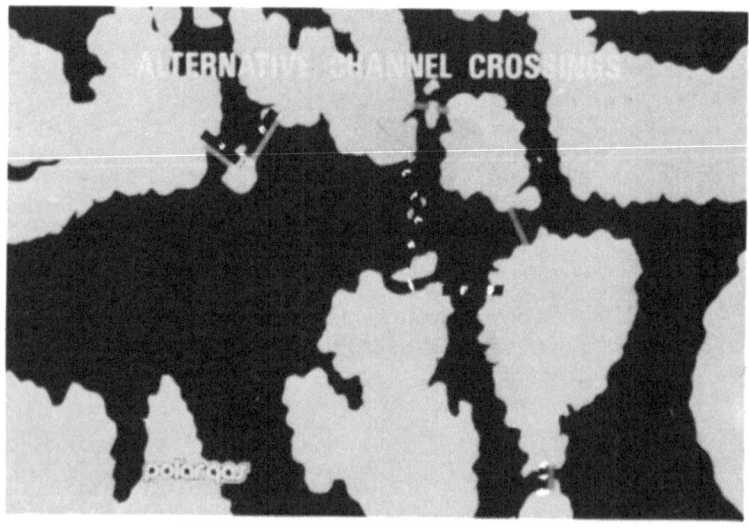

Figure 7. Map of channel crossings

Weather conditions during the program were typical of those which could be expected. In both April and May, the combination of temperature and wind produced wind-chill temperatures of minus 80 degrees F. These, however, are comparable to those experienced in pipelining across northern Ontario.

In March, 1973, a base camp for the program was established at Rea Point, Melville Island, near the base camp of Panarctic Oils.

During the month, men and equipment were flown to the site from Yellowknife, N.W.T., in preparation for trenching, which began in mid-April.

The equipment used for the trenching was basically standard land pipelaying trench equipment except that the diameter of the bucket wheel was increased to provide a capability for cutting ice up to a thickness of 15 feet. When excavating through 6 1/2 foot thick ice (the average thickness of sea ice), the maximum progress achieved by the machine in a twenty-four hour day was approximately 2 miles.

While these trenching operations were in progress, several other research and investigation activities were being carried out. They included, for example

Figure   8.   Polar Gas and Panarctic Oils base camps

a) Ice thickness measurements in several channels.   (Fig. 10)
b) Ice strength tests designed to provide basic
    parameters for further laboratory testing and
    analysis. These were full-scale load tests
    using large water tanks.   (Fig. 11)
c) Bottom sampling with piston gravity corers.   (Fig. 12)
d) Current measurements.   (Fig. 13)
e) Bathymetry, sub-bottom profiling and side-scan
    sonar observations through the trench.   (Fig. 14)
f) Scuba diving observations of the channel bottom
    to water depths of up to 200 feet.   (Fig. 15)
g) Environmental observations.   (Fig. 16)

In general, the results obtained from this overall program
were extremely encouraging and established the feasibility of cutting trenches through sea ice with only minor modifications to
conventional equipment. Moreover, the data obtained during the
program served as an excellent base for continuing research.

Figure 9 Ditching machine cutting a trench through the ice

Fig. 10   Augering through the ice to measure thickness

Fig. 11   Testing ice loading characteristics. Large tanks are
          filled with water to measure vertical deflection of ice.

Figure   12.   Collecting core samples

Figure 13. Installing current meters

Figure 14. Vehicle and equipment used to perform bottom
profiling and side scanning of ocean floor.

Figure   15.   Scuba diver

Figure   16.   Environmental observations on ocean floor

B.　MARINE RESEARCH PROGRAM - 1973

This program was designed with the main objective of providing marine data for the design of potential pipeline crossings in the Arctic Islands and northern Hudson Bay. The major part of this program was carried out in August, September and October from an ice strengthened, 2,400 gross ton vessel, the Percy M. Crosbie. The vessel had a total complement of 53 research staff and crew members.

Prior to commencing the program, the vessel was modified to meet the special requirements of the survey work to be undertaken. It was also equipped with 2 specially built 40 foot aluminum survey launches, helicopter and helicopter landing pad.

The launches were designed to handle the sophisticated electronic equipment required for bathymetry, sub-bottom and side-scan sonar data as well as the precise navigational systems required for research of this type.

Aided by this equipment, the following detailed activities were undertaken:

Figure　17.　The Percy M. Crosbie

Figure   18.   Survey Launch

a) Acoustics depth soundings to determine bathymetry and bottom
    roughness.
b) Seismic profiling to determine sub-bottom stratification.
c) Side-scan sonar surveys to detect ice scour, reefs, outcrops
    and other obstructions which could interfere with the pipeline.
d) Current metering, wave recording and tidal observations.
e) Collection and analysis of bottom samples.
    The Percy M. Crosbie travelled 15,000 miles during the program
and obtained data at selected locations in 19 different channels in
the Arctic Islands and Hudson Bay areas. The program was aided by
an excellent season from the standpoint of the great amounts of
open water available for research. However, this was offset by a
more than normal amount of fog which caused difficulty in the use
of air-borne transportation of personnel to support the program.
    Of particular interest was an independent air-transportable
system which was put together to collect similar basic data on
short notice in channels which only became ice-free for short
periods and which were inaccessible to the major vessel because of
the intervening ice conditions. The system comprised three 17 foot
aluminum launches coupled together, each being sufficiently light
for transportation from location to location by means of a small
helicopter.
    Ice reconnaissance patrols were also carried out from both
helicopters and fixed-wing aircraft. In addition to providing

short-term ice forecasts at designated crossings, valuable records
were obtained of the sea ice conditions prevailing in the channels
during the break-up season.

The data obtained during the Percy M. Crosbie's voyage
included approximately 800 miles of bathymetry, sub-bottom
profiling and side-scan sonar records. Most of the data collected
during this highly successful program has been analyzed and it
formed the basis for technical assessment studies of pipeline
installation methods which have been carried out on selected
channels. These marine studies also highlighted the gaps in
present knowledge and helped identify areas where research and
investigation should be concentrated in future.

Figure 19. Air-transportable survey vessel

C.   ICE RESEARCH - 1974

Though this was the general name of the program, it
encompassed far more than the title suggests and included work on
bathymetric and sub-bottom data collection from the ice.   This is
another area of innovation of our program at Polar Gas.   In the
past, as far as is known, such investigations on a production type
basis have, up to this time, only been carried out in open water
conditions.

Our general objectives in this program were briefly as follows:
a)   To conduct a research on ice strength in order to
     confirm and evaluate parameters predicted by laboratory
     tests and mathematical model computer studies.
b)   To research methods of preparing a right-of-way across
     the ice surface.
c)   To collect ancillary data on currents, bottom soils,
     ice thickness, ice movements and ice scour.
d)   To develop and test directly from the ice surface and
     obtain data in Byam Channel (Melville Island to Byam
     Martin Island) and West Barrow Strait (Bathurst Island
     to Prince of Wales Island).

For bathymetry and sub-bottom research programs, a variety of
sophisticated equipment was used which enabled us to acquire both
data simultaneously through the ice.   They are:
a)   A. 3.5 Khz high resolution system placed in a helicopter
     which was fitted with precise navigational equipment.
     Readings were taken at 300 meter intervals to provide
     quick evaluation of the bottom profile.   (Fig. 20)
b)   A similar high resolution system was placed in
     an enclosed housing on a large tracked vehicle
     which took readings at 15 meter spacings.   (Fig. 21)
c)   A Sparker 1500 Hertz system was placed in
     another enclosed housing on a large tracked
     vehicle which provided more detail for the
     interpretation of sub-bottom profile in sand
     and/or rock areas.   (Fig. 22)
d)   A precise navigational system placed in a small
     totally enclosed tracked vehicle which was
     fitted with flotation tanks as a safety measure.   (Fig. 23)

Information from all these systems was collected on magnetic
tape and fed into a computer housed at the Rea Point camp.   This
enabled the bathymetric data to be accurately plotted within hours
of the data arriving from the field.   It was then checked by a
pipeline design engineer to assess its suitability for pipelaying
operations, particularly as to expected stresses induced in the
pipe by the bottom roughness.   A total of 62 miles of detailed
bathymetric data (bathymetry profile, sub-bottom profile and sub-
bottom analyses) was collected in this manner during the program.
A complementary program of collecting data, consisting of both
through the ice and marine, provided maximum flexibility in
bathymetric research.

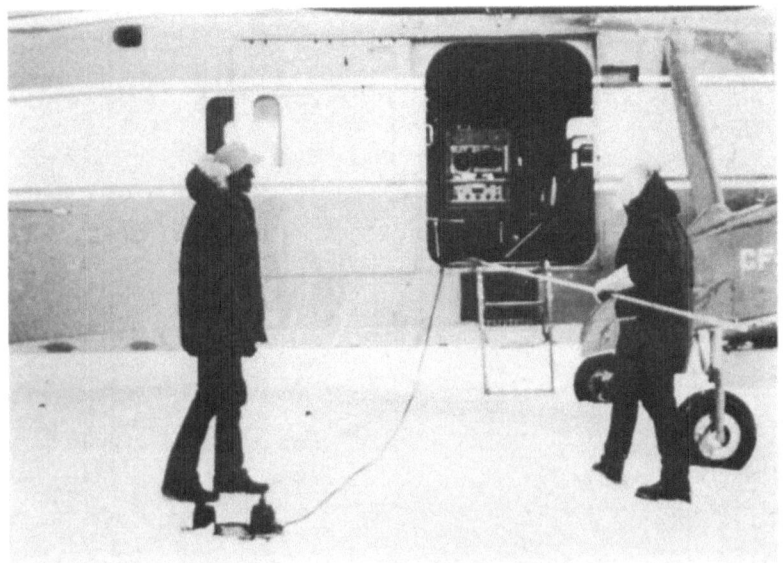

Figure 20.   S-62 helicopter fitted with resolution system
             and navigation equipment

Figure 21.   FN-110 Nodwell fitted with resolution system

Figure   22.   FN-110 Nodwell equipped with Sparker system.

Figure 23.   Navigation system in closed tracked vehicle
              with flotation tanks

The ice strength testing part of the program basically comprised a continuation of the initial work started in the previous year. Eighteen full-scale load tests with water tanks and 89 cantilever tests, under both long and short term loading conditions, were carried out.

For the right-of-way preparation part of the program, the work included:

a) Observations on the rate of build up of snow on cleared ice surfaces.

b) Thickening an ice sheet by flooding with sea water to obtain a stronger and smoother surface.   (Fig. 24)

c) Evaluation of the effects of snow fencing as a means of keeping the ice surface snow free.

d) Raising the surface of the ice sheet by adding compacted snow in order to assess the effectiveness of keeping the surface free from drifting snow by wind action.

e) Measurements of the tensile forces which can be sustained by anchors frozen or mechanically attached to the ice surface.   (Fig. 25)

Ancillary data collected during the overall program included such items as current velocities, bottom soils samples, ice thickness measurements, ice movement and ice scour observations. The techniques used to obtain these data were essentially the same as those employed during the previous year, with minor modifications and the addition of new pieces of equipment to improve efficiency.

Figure 24. Flooding the ice with sea water

Figure 25. Installing ice anchors for tests

Most of the data collected during the programs has now been
analyzed and is being utilized in the on-going technical
assessment studies of a number of pipeline installation
techniques for Arctic channels.

INTER-ISLAND CROSSINGS: CONSTRUCTION TECHNIQUES

Depending on the width, depth, ice conditions and other
characteristics of the channels to be crossed, a number of pipelaying methods could be required. In general, however, they fall
into two basic categories:
1. Those using the ice as a platform from which to lower the
    pipe (somewhat in the manner of a lay barge).

2. Those based on the conventional "bottom pull" technique -
    either from the ice surface or directly from one shore
    to the other.
    An additional option for one of the channels which could be
faced is to construct the pipeline crossing using a ship to pull
the pipe during the summer season. As currently conceived, the
"bottom pull" through the ice system would involve the building of
a series of thickened ice platforms or "ice islands" from which

Figure 26. Lay barge

Fig. 27. Model showing "bottom pull" from the ice technique

winches would pull the pipe assembled on shore. This method could
involve one or more "tie-ins" or welds of several mile long sections
of pipe which have been pulled into position. The technology to
carry out "tie-ins", working in chambers lowered to the bottom of
the channel or by lifting to the surface the two ends of pipe to
be joined, is available.

On shorter channel crossings, the "bottom-pull" shore to
shore technique would involve moving the pulling station from the
surface of the ice to the opposite shore.

OVERVIEW

The engineering research activities of the past three years
have been undertaken with the main aim of confirming the technical
feasibility of the pipeline. The results to date have confirmed,
to a high degree of confidence, that a natural gas pipeline from
the Arctic Islands is technically and economically feasible.

In future, research programs from a technical standpoint,
will be directed towards terrain and metallurgical studies, route
optimization analyses, the refinement of construction methods and
techniques and data collection for final pipeline design.
Needless to say, these programs present additional opportunities
for innovation and improving existing technology.

NATURAL GAS PIPELINES IN THE ARCTIC ENVIRONMENT

POLAR GAS PROJECT: PART II - "ENVIRONMENTAL RESEARCH"

R. J. Lamoureux and Walter Hindle

Polar Gas

Toronto, Ontario, Canada

INTRODUCTION

Since 1974 Polar Gas has been undertaking field studies of the High Arctic (North of Spence Bay) to assess the environmental implications of building a chilled gas pipeline, largely during the Winter, from fields on the Sabine Peninsula of Melville Island to the mainland of Canada (Figure 1). This research is a part of the phased program on environmental field studies which will eventually cover the entire Polar Gas Route from its northern to its southern extremities.

Field studies were carried out in 1974 on the Boothia Peninsula and on Somerset, Prince of Wales, Cornwallis, Little Cornwallis, Bathurst and Melville Islands under the broad categories of Landscape, Land Mammals, Birds, Fisheries and Marine Ecology. Studies of Land Mammals, Birds, Marine Mammals and general Marine Ecology are continuing in 1975 to further define the important biological parameters of the area.

This paper outlines some project-related environmental factors presently under consideration. Although continued refinement of our data base is necessary for detailed problem definition and solution, we presently have a sufficient grasp of the important physical and biological parameters to identify the key areas of interaction between the pipeline and the environment. The best solutions for potential environmental problems have yet to be worked out and in many instances we can only indicate our general direction of thought.

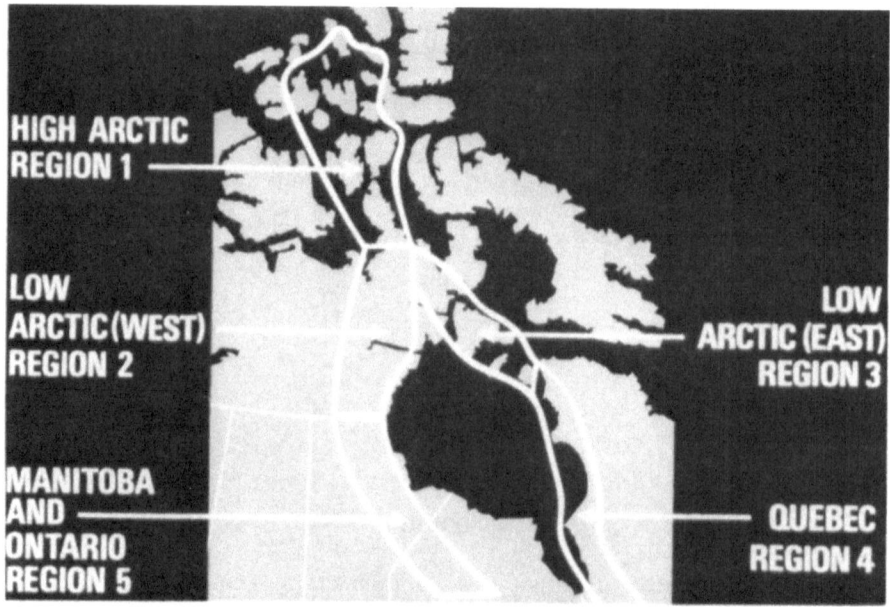

Figure 1.   Polar Gas route alternatives

ENVIRONMENTAL OVERVIEW

Climate

An extreme cold continental climate prevails over the study
area once the channels are frozen over. During winter, temperatures
fall to -30°F (-35°C) or below and the weather tends to be calm and
clear. In the summer period the weather becomes unstable when the
sea ice starts to melt, giving rise to cool maritime weather with
frequent fog and low cloud. Summer temperatures are typically in
the 5°C range, although they may be somewhat warmer on the Boothia
Peninsula.

Precipitation is typically low, the mean annual total
precipitation being only 5.35 inches (13.59 cm.) at Resolute on
Cornwallis Island. About half of the annual precipitation falls
as snow mainly in the months of May, September and October.
Normally the land is snow and ice free only in the months of July,
August and September.

Landscape

The study area can be divided into an igneous region which forms the central portion of Boothia Peninsula and the western portion of Somerset Island and a sedimentary region which forms the remainder of the study area. The landscape can be roughly sub-divided into six broadly defined land systems as follows:

(1) Coarse igneous rock uplands with or without frost-shattered rock fragments   (Figs. 2 and 3)

(2) Coarse sedimentary rock uplands with or without frost-shattered rock fragments   (Figs. 4 and 5)

(3) Continuous shallow tills and till veneers having gravelly and sandy surfaces with abundant angular rock fragments originating from underlying bedrock   (Fig. 6)

(4) Deep till and reworked till with rounded boulders of glacial origin   (Figs. 7, 8, 9)

(5) Coarse gravelly and sandy deposits (outwash valleys, eskers, kames, hummocky disintegration moraine)   (Figs. 10 and 11)

(6) Marine and lacustrine deposits   (Figs. 12, 13, 14)

The major portion of the route crosses land system types (1), (2) and (3). Land systems (4), (5) and (6) constitute a very minor portion of the total landscape.

Vegetation cover on land systems (1), (2), (3) and (5) ranges from sparse to extremely sparse. Plant cover, excluding the effect of crustose lichens, seldom exceeds 5%. Much of these areas would be classified as polar desert.

Marine and lacustrine deposits represent the opposite end of the vegetation spectrum. On these, vegetation cover may reach 100%. The communities are dominated by sedges and may also contain a significant component of mosses. Occurrence of thermokarst lakes is largely confined to these areas of fine-grained soils.

The vegetation of till and reworked till deposits which lie in areas intermediate between the sparsely covered uplands and the densely vegetated lowlands is the most difficult to characterize. Depending upon such factors as soil humidity, gradient drainage aspect, degrees of cryoturbation and activity of erosive forces, the composition and percentage cover of vegetation may vary considerably. Lichens, mosses, deciduous shrubs (willows) and evergreen shrubs (dryas, saxifrage) form components of varying importance. Plant cover generally exceeds 20% and may reach values greater than 50% in moist, protected areas.

Figure   2.   Coarse igneous rock uplands, Boothia Peninsula,
             ground view (Land System 1)

Figure   3.   Coarse igneous rock uplands, Boothia Peninsula,
             aerial view (Land System 1)

Figure 4. Coarse sedimentary rock upland with felsenmeer, Somerset Island, ground view (Land System 2)

Figure 5. Coarse sedimentary rock upland with felsenmeer, Somerset Island, aerial view (Land System 2)

Figure 6.   Sedimentary upland with thin till veneer,
             Somerset Island (Land System 3)

Figure 7.   Reworked till with typical elongated net pattern
             of vegetation, Somerset Island (Land System 4)

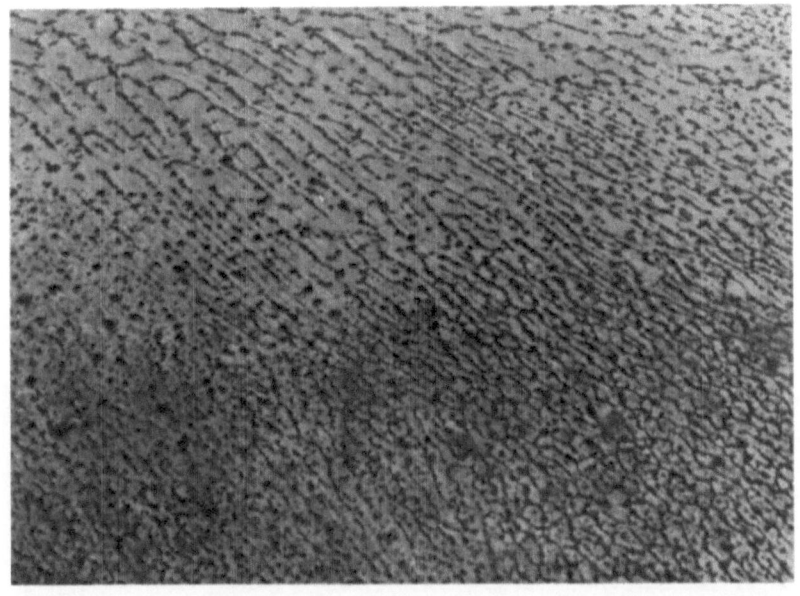

Figure 8. Aerial view of reworked till with typical elongated net pattern of vegetation, Somerset Island (Land System 4)

Figure 9. Heavily vegetated till pocket (Land System 4)

Figure   10.   Hummocky disintegration moraine south of
               Sanagak Lake, Boothia Peninsula (Land System 5)

Figure   11.   Aerial view of hummocky disintegration moraine
               south of Sanagak Lake, Boothia Peninsula
               (Land System 5)

Figure   12.   Well vegetated sedge meadow on raised delta,
              Boothia Peninsula (Land System 6)

Figure   13.   Closeup of sedge vegetation on raised delta,
              Boothia Peninsula (Land System 6)

Figure 14. Classic thermokarst lake area, Bracebridge,
Goodsir Inlet, Bathurst Island

Freshwater Bodies

The sedimentary portion of the study area contains few large
lakes and only limited areas of extensive thermokarst pond
development. The majority of the lakes are shallow enough to
freeze to the bottom. The igneous portion of the route contains
numerous small and large lakes, many of which are deep enough to
have unfrozen water in winter.

Both sedimentary and igneous regions have numerous streams.
Most stream runoff is derived from the annual snowmelt and there
is consequently a marked dropoff in streamflow by late summer.
Most, if not all, streams freeze to the bottom, but it is possible
that some streams, such as the Union River, which are lake
regulated may have some winter flow.

A number of lakes were tested in the spring of 1974 for winter
dissolved oxygen. It was found that most lakes had adequate
oxygen. However, a few instances of oxygen depletion were noted.

The bottom fauna of streams and lakes consisted almost
completely of species of larval midges (chironomids). A relatively
minor component of stoneflies and mayflies gives the bottom fauna
of Boothia Peninsula a somewhat greater diversity.

Saltwater Bodies

It will be necessary to cross Byam and Austin Channels to reach Bathurst Island. From Bathurst there is a choice of crossing by a west route via West Barrow Strait, Baring Channel and Peel Sound or by an east route via Crozier Strait, Pullen Channel and East Barrow Strait.

The channels of the east route break up earlier than those of the west route. In some years, such as 1974, Barrow Strait may open as far west as Griffith Island as early as March or April. In more typical years such as 1975 a stable ice front forms from the northeast corner of Somerset to the south shore of Devon Island and the deterioration of ice in East Barrow is a more gradual process taking place in June and July. Breakup of the ice on West Barrow, Austin and Byam Channel usually occurs late in the summer.

The presence of ice cover is a major factor influencing primary productivity. An important phytoplankton community utilizes the concentrated nutrients and limited light at the water-ice interface. Large zooplankters (amphipods) graze on the phytoplankton at the water-ice interface and are in turn grazed upon by ringed seals, and ducks and seabirds (through ice leads).

Terrestrial Mammals

The terrestrial mammal fauna is extremely simple, consisting of a total of only 10 species for the whole study area. The following species are present: caribou (peary and barren ground; Fig. 15), muskox (Fig. 16), arctic hare, arctic ground squirrel, collared lemming (Fig. 17), brown lemming, arctic wolf, arctic fox (Fig. 18) and ermine. Species diversity is lowest on Somerset Island with 7 species and highest on Boothia Peninsula with 9 species.

Suitable mammal habitat is restricted to those areas where moisture and soil conditions permit the development of significant vegetative cover (Figs. 19 and 20). Such conditions are usually found in areas of deep till, and marine lacustrine deposits. Within suitable vegetated areas, the thickness and physical properties of the snow cover greatly affect the availability of vegetation to grazing mammals. Consequently, windblown vegetated slopes and south-facing slopes are important winter and spring habitats.

Figure 15.   A small herd of Peary Caribou
              on Melville Island

Figure 16.   An adult cow Muskox on Bathurst Island

Figure 17. Lemming

Figure 18. Arctic fox in winter coat

Fig. 19.   Peary Caribou on well vegetated river valley slopes,
           west coast of Somerset Island

Fig. 20.   Muskox grazing in wet depression with
           sedge vegetation, Melville Island

Marine Mammals

The channels of the study area are frequented by ringed seals, bearded seals, harp seals, beluga whales, narwhal, polar bears and to a small extent, by walrus. Ringed seals and polar bears (inhabiting wide areas of fast ice) are permanent residents of the study area. The remaining marine mammals migrate into the area during the open water period. Their movements are greatly influenced by highly variable breakup events.

Perhaps the most spectacular phenomenon involving marine mammals is the annual Summer gathering of thousands of beluga whales and their calves in estuaries such as Cunningham Inlet (Fig. 21).

Fig. 21. Concentration of Beluga Whales in Cunningham Inlet

Birds

Many birds utilize terrestrial, freshwater and marine
environments during some part of their life cycle, although their
existence in the Arctic is limited to the Spring and Summer.

In the Spring the most important areas for birds are the
early leads which open up in the sea ice. In this period large
concentrations of snow geese, eider ducks and oldsquaw and a
variety of seabirds utilize the limited available habitat. During
the nesting season the limited areas of lacustrine and marine
deposits with their associated thermokarst development are critical
for nesting waterfowl and shorebirds. After nesting waterfowl
concentrate for moulting, at which time they are flightless (Fig.
22). Our information on moulting birds in the high arctic is
presently incomplete and the subject is being pursued intensively
during the 1975 program.

Seabirds require cliff nesting sites close to marine areas
where there are consistently recurring open leads, while peregrine
falcons generally utilize cliff-nesting sites along streams. The
numbers of important cliff nesting sites are limited and many of
these have already been identified.

The general statement can be made that very limited parts of
our study are capable of supporting abundant bird life. As a
result there is a very strong "oasis effect" and a judicious
combination of landscape analysis and field checks of promising
looking areas usually enables a fairly precise pinpointing of such
sites.

Fish

Arctic char (Fig. 23) is the most significant species in the
study area. In the islands it is the sole species present, but
on the Boothia Peninsula it is joined by lake trout and ninespine
stickleback.

It is known that arctic char must overwinter in freshwater as
they are not capable of enduring winter seawater temperatures less
than 32°F. Some char spend their entire lives in freshwater,
whereas others spend their summers in the sea and return to fresh
water in late summer. Our study area contains only a few streams
with sufficient late summer flow to sustain anadromous runs.
Significant exceptions are the Union River on Somerset Island and
the Lord Lindsay River on Boothia Peninsula. Most char found in
the study area are of the landlocked variety.

All char, whether anadromous or landlocked, require lakes of
sufficient depth to prevent freezing to the bottom. This allows
for the survival of fish and eggs. Such lakes are fairly frequent
on the igneous portion of the Boothia Peninsula but are rarer in
the sedimentary Arctic Islands portion of the study area.

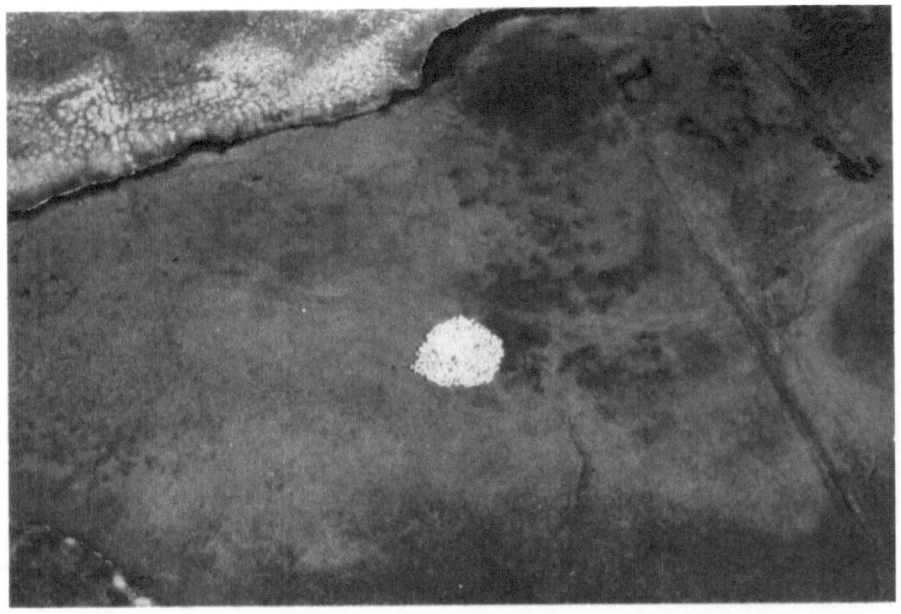

Figure 22. Concentration of 400 moulting Snow Geese
in a thermokarst pond near the Creswell River,
Somerset Island

Figure 23. Arctic Char

APPROACH TO ENVIRONMENTAL PROTECTION

We are using a rather simple but workable approach to
environmental protection consisting of four steps which can be
summarized as follows:
(1) Establish the nature and magnitude of potential disturbances;
(2) Establish the locations where potential problems may exist;
(3) If economically feasible formulate routing, scheduling or
     procedural measures to avoid potential problems;
(4) If potential problems cannot be totally avoided, carry out
     field research to establish the magnitude of the problem and
     safe limits of permissible disturbance.

At this stage in development of environmental science,
avoidance tends to be the basic approach to potential environmental
problems. As the experimental side of environmental science
develops there will likely be a shift to a more "engineered"
approach to potential environmental problems. However, at present
the cost of the experimental approach is often so prohibitive that
avoidance is a more economic strategy. Moreover an avoidance
strategy eliminates many subjective decisions concerning aesthetics
which would have to be made in seeking an "engineered" solution.

ENVIRONMENTAL CONCERNS

Landscape

Landscape disturbance factors which must be considered
include:
(1) Location and design of access roads, right-of-way and
     permanent facilities.
(2) Excavation and backfill operations.
(3) Effect of vehicular traffic.
(4) Alteration of subsurface drainage by the frost bulb of
     an operating gas line.
(5) Long term erosion problems.
(6) Disposal of solid waste.

All of these concerns overlap with basic geotechnical
investigations and engineering design. In many cases the optimum
engineering design results in the optimum environmental design.
However, this is not always true because the environmental input
introduces the added complexities of biological, cultural and
aesthetic factors.

In most cases, though, the types of foreseen changes in the
landscape would not likely have a significant influence on the
functioning of ecosystems and would largely be a matter of
aesthetics. Aesthetic evaluation does, however, require
considerable biological and cultural input.

Freshwater Systems

The basic objective is to maintain the freshwater environment in such a state that it will continue to support the existing fauna. The principal concerns that have been identified are:
(1) Siltation
(2) Pollution
(3) Water extraction
(4) Snow removal
(5) Blockage of winter flows

As indicated in the landscape section, some long term siltation effects are possible when pipeline construction causes terrain instability within the drainage area of a lake supporting over-wintering char. Routes are not sufficiently well established and drainage analysis is not sufficiently advanced to identify specific instances of this problem. However, our general analysis has indicated that the problem will be rare.

Siltation of streams will occur as a short term effect during actual stream crossing and it could occur as a long term problem if pipeline activity promotes terrain instability. Siltation is not considered as a major problem in the vicinity of our routes. Siltation could result in localized changes in relative abundance of the species forming larval midges (chironomid) fauna. This would not necessarily have any major effect on productivity. Regardless of the theoretical effects of increased siltation, the actual increase in stream siltation would likely be minimal because of the engineering precautions that would be taken to minimize cases of severe erosion.

We do not foresee any serious problems with pollution from poisonous substances. There are few probable uses for acutely toxic substances. Bulk storage of fuel for equipment and methanol for pipe testing (if used) will be protected by diking and drainage control. The greatest hazard exists during bulk transport where there is always some risk of spillage in an accident. This is an unavoidable risk that must be accepted and even this risk can be minimized by prearranged procedures for containment, recovery and disposal.

Water extraction from small lakes which are just barely deep enough to support char should be controlled to avoid damaging draw-downs and possible freezing to the bottom. Similarly, snow removal from such lakes should be minimized to prevent increasing the depth of ice cover. Potential problem areas will be easily identified with a minimum of field work at a more advanced stage of the engineering design.

In the likely small number of streams with winter flows, care must be taken to ensure that the frost bulb around the operating chilled pipe does not create a blockage of streamflow, resulting in icing and the freezing to the bottom of downstream reaches. Where this problem potentially exists, it will be necessary to do more detailed thermal studies to determine whether special measures such as deep burial or special insulation are required.

Marine Systems

The following concerns have been identified with regard to
the effect of a pipeline on the marine environment:
(1) Siltation resulting from nearshore trenching activities;
(2) Incidental fuel spillage associated with shipping.
Field work is being carried out to establish the nature of
bottom sediments and their relationship to the distribution of
benthic fauna. When this work is completed, it will be possible
to identify any potential problem areas. If this initial work
indicates the existence of such areas, further work will be
undertaken to establish the magnitude, duration and extent of the
siltation effects. On the basis of this information the environmental and engineering merits of alternative crossing sites will be
examined.
Fuel spillage from transfer at dockside is relatively easy to
control by the application of strict standards for fuel handling.
Hazards at sea are harder to control and can be considered as an
unavoidable hazard of development. This problem is not unique to
the project but is pertinent to all northern shipping. Effective
solutions to the problem include programs to increase safety of
navigation in Arctic waters and centrally coordinated emergency
procedures. It will not be possible for Polar Gas to tackle these
problems in isolation. Participation in a federally coordinated
effort would be a more reasonable alternative.

Atmosphere

Inversions during the winter months could result in ice fogs
and high concentrations of NO_x emissions in the vicinity of
compressor stations. It will be necessary, prior to final
location and design, to have a more detailed understanding of
local atmospheric conditions so that compressor station sites can
be located in sites favourable for disposal of emissions.

Terrestrial Mammals

The following concerns related to terrestrial mammals have
been identified:
(1) Disruption of normal behaviour patterns by noise, vehicular
     approach and human approach;
(2) Habitat destruction;
(3) Road kills;
(4) Increased hunting pressure.

Disruption of behaviour patterns of large ungulates is being carefully examined. Two mechanisms of disturbance occurring simply or in combination are postulated.

(1) Frequent disruption could interfere with feeding, procreation and rearing of young leading to population decline;

(2) Frequent disruption could lead to higher stress levels increasing energy demands of animals.

One has available the alternative strategies of either avoiding encounters or minimizing the impact of encounters.

The most promising approach to mammal protection is through procedural modifications to minimize impact of encounters. Such measures as activity restrictions at critical times of the year (such as during the calving period), rules to prevent the following of animals with aircraft and ground vehicles, controlled blasting and other such measures applied at the source of disturbance can go a long way to minimizing the impact of disturbance.

Animal density is an extremely important parameter in the assessment of overall impact of disturbance. In the High Arctic, mammal populations may be so small and dispersed that chances of encounters are low even during intensive pipeline construction activities. In many situations special protective measures may not be justified.

As outlined in the landscape section, interruption of ground-water flow through lowlands is the only conceivable situation that could significantly alter habitat values. This potential problem can be controlled if necessary by altering depth of burial and by the use of insulative materials.

Road kills do not appear to be a major problem in view of the low mammal densities and the lack of definite migration paths across the right-of-way.

It is likely that the pipeline will have some effect on hunting pressure of large ungulates. It is not yet certain whether total hunting pressure will increase or decrease. On the one hand employment of people from northern communities on work related to the pipeline will draw them away from hunting. Since guns will not be allowed in camp there will be no hunting opportunities on the job. However, increased exposure to potential new hunting areas may stimulate hunting of these areas when workers leave the project. An overall policy to prevent excessive hunting should be coordinated by authorities.

Marine Mammals

The following concerns have been identified with regard to the effects of a pipeline on marine mammals:

(1) Habitat Alteration;

(2) Direct damage to animals from shock waves from underwater blasting;

(3) Disturbance of behavioural patterns;
(4) Attraction of Polar Bears to campsites.

Habitat alteration has been discussed in the section on
Marine Systems. Changes in the bottom fauna are of major relevance
to marine mammals which directly or indirectly depend on this fauna
for their food intake.

The approach to minimization of disturbance will be essentially
the same as that applied to land mammals.

Prompt and positive disposal of garbage will be necessary to
discourage attraction of polar bears. Educational programs on
polar bear behaviour and training in special avoidance procedures
and in the use of scaring devices will be important. Cooperative
efforts with government agencies such as the Canadian Wildlife
Service may be necessary to develop effective measures to ensure
the safety of both men and bears.

Birds

The following concerns related to birds have been identified.
(1) Disruption of normal behaviour patterns;
(2) Habitat alteration;
(3) Increased hunting pressure.

Mechanisms of disruption of behaviour patterns of birds are
similar to those discussed for mammals. Disturbances during
nesting and moulting periods may be the most critical. In view of
the rather strong correlation between certain landscape features
and breakup phenomena, total avoidance of specified areas during
critical periods is a more workable strategy than for mammals.
Avoidance of important areas coupled with selectively-timed
procedural modifications should result in minimal disturbance.

Habitat alteration could take the form of drainage changes in
lowland habitats and siltation of marine feeding areas, subjects
which have been previously discussed.

The discussion of increased hunting pressure in the section
on Terrestrial Mammals applies equally to waterfowl. In addition
to control of hunting, strict policing measures will be necessary
to prevent the capturing of rare and endangered peregrine falcons
for export to falconers.

Fisheries

The primary consideration for protecting fisheries in both
the freshwater and marine situation is to avoid significant
environmental alteration as discussed in the sections on Freshwater
Systems and Marine Systems.

In addition precautions must be taken to minimize the following
direct disturbances:

(1) Blasting damage to fish;
(2) Creation of barriers to fish migration;
(3) Overfishing.
     It is a well known fact that fish are highly vulnerable to
air bladder damage from blasting shock waves. Blasting activities
will be timed in such a way to avoid fish concentrations.
Modifications in blasting technique may also be employed to
minimize the shock wave effect.
     Care will be taken in the design of temporary crossings of
streams to avoid the use of undersized culverts which cause high
stream velocities against which arctic char are incapable of
swimming.
     In view of the small char populations and their slow growth
rates, overfishing is viewed as a major environmental concern.
It is more difficult to control fishing than hunting because of the
practical difficulties involved in discouraging fishing rods in
construction camps. If even a small fraction of the pipeline
employees were to exercise their legal right to fish and catch their
legal limits, there would be a substantial local impact on the char
fisheries. It may be necessary to institute project specific
controls to prevent such an occurrence.

                              OVERVIEW

     Over the past two years Polar Gas has been conducting field
studies in the High Arctic to assess the environmental implications
of building a chilled natural gas pipeline from Melville Island to
southern Canada. Studies to date have been carried out on the
Boothia Peninsula and on Somerset, Prince of Wales, Cornwallis,
Little Cornwallis, Bathurst, and Melville Islands. Research on
Land Mammals, Birds, Marine Mammals and general Marine Ecology is
continuing in 1975 to further define the important biological
parameters of the area.
     The results of the environmental studies to date have provided
Polar Gas researchers with a sufficient grasp of the important
physical and biological parameters to identify the key areas of
interaction between the pipeline and the environment and to begin
assessing solutions to potential environmental problems.
     The research effort described in this paper is part of a
phased program of environmental field studies which will eventually
cover the Polar Gas routes.

AN ENVIRONMENTAL RESEARCH PROGRAM FOR DRILLING

IN THE CANADIAN BEAUFORT SEA

John Hnatiuk

Gulf Oil Canada Limited

Calgary, Alberta, Canada

This paper was presented at the 26th Annual Technical Meeting of the Petroleum Society of CIM in Banff, June 11-13, 1975. Permission has been granted to publish it in this Proceedings.

ABSTRACT

A multi-million dollar environmental research program consisting of thirty-three wildlife, biological, oceanographic, meteorological, sea ice and oil clean-up studies related to the southern Beaufort Sea is described. The studies are designed to provide ecological baselines, a better understanding of the physical environment, knowledge related to the consequences of a possible oil spill and means of oil clean-up in ice-infested waters. Seasonal offshore drilling in the Canadian Beaufort Sea from a floating vessel has been approved in principle and is scheduled to be undertaken during the open water period of 1976. The results of the studies discussed will be used to set operating constraints to safeguard the environment and minimize any adverse impact.

A unique feature of the program is its joint government-industry nature. The studies were specified by the Canadian Federal Government but the majority of the funding is by the oil industry. Government agencies co-ordinate the program with considerable management and scientific input from oil industry personnel. The two-year program will be completed by the end of 1975. The hostile environment is briefly described and problems arising during one of the worst ice seasons on record are discussed. The scope of the studies, research methods and progress to date in individual studies are outlined and, where possible, results are

provided. An important phase of the program is public interface
to inform northern natives, government, those with environmental
concerns and others about the program and the study results.

INTRODUCTION

The search for hydrocarbons in Canada's North has been
extended to the geologically attractive but environmentally
hostile southern Beaufort Sea. During the period, 1963 to 1969,
offshore permits to explore for oil and gas were issued for the
majority of the continental shelf which extends some 75 to 100
miles from shore, to water depths of approximately 600 feet,
shown on the location map, Figure 1. In mid-1972, one group of
ten companies and, later, two other companies applied to the
Federal Government for approval of floating vessels as exploratory
drilling systems for the Beaufort Sea. By this time, construction
of the first of many artificial islands in shallow water had
commenced, but these were in general regarded as being similar to
drilling operations on land. The first approval-in-principle of a
floating drilling system was granted in mid-1973, followed by other
approvals. These approvals were subject to completion of a number
of environmental studies which would serve as the basis for the
conditions and constraints to be included in a Drilling Authority.
This environmental program was stipulated in late 1973, and
consisted of twenty-nine studies, estimated to cost $5.3 million,
apart from government support. Following several months of
discussions and some misunderstanding, the basis for funding these
studies was resolved in February, 1974.

THE BEAUFORT SEA ENVIRONMENT

The southern Beaufort Sea is usually ice-covered nine months
of the year and it is during the remaining three or four months of
open water or thin ice that exploratory drilling from a floating
vessel is proposed. With the onset of winter in early October,
freeze-up typically progresses seaward with relatively smooth ice
extending for some 15 or 20 miles to an average water depth of 30
feet. Here, a very rough pressure-ridged band of ice grows
outward to the 60-foot water depth where the landfast ice mass
stabilizes early in the year. At this depth, a flaw or shore lead
of water, which can be miles wide and hundreds of miles long, will
recur under southerly winds. Beyond this lead, the ice is
fragmented into floes of varying thicknesses. The undisturbed
winter ice within the landfast zone reaches a thickness of 6 or 7
feet by May while the pressure ridges composed of broken blocks of
ice may reach up to 40 feet[1] above sea level and have keels
reported to exceed 100 feet.[2] Multi-year ice, 12 feet in thickness

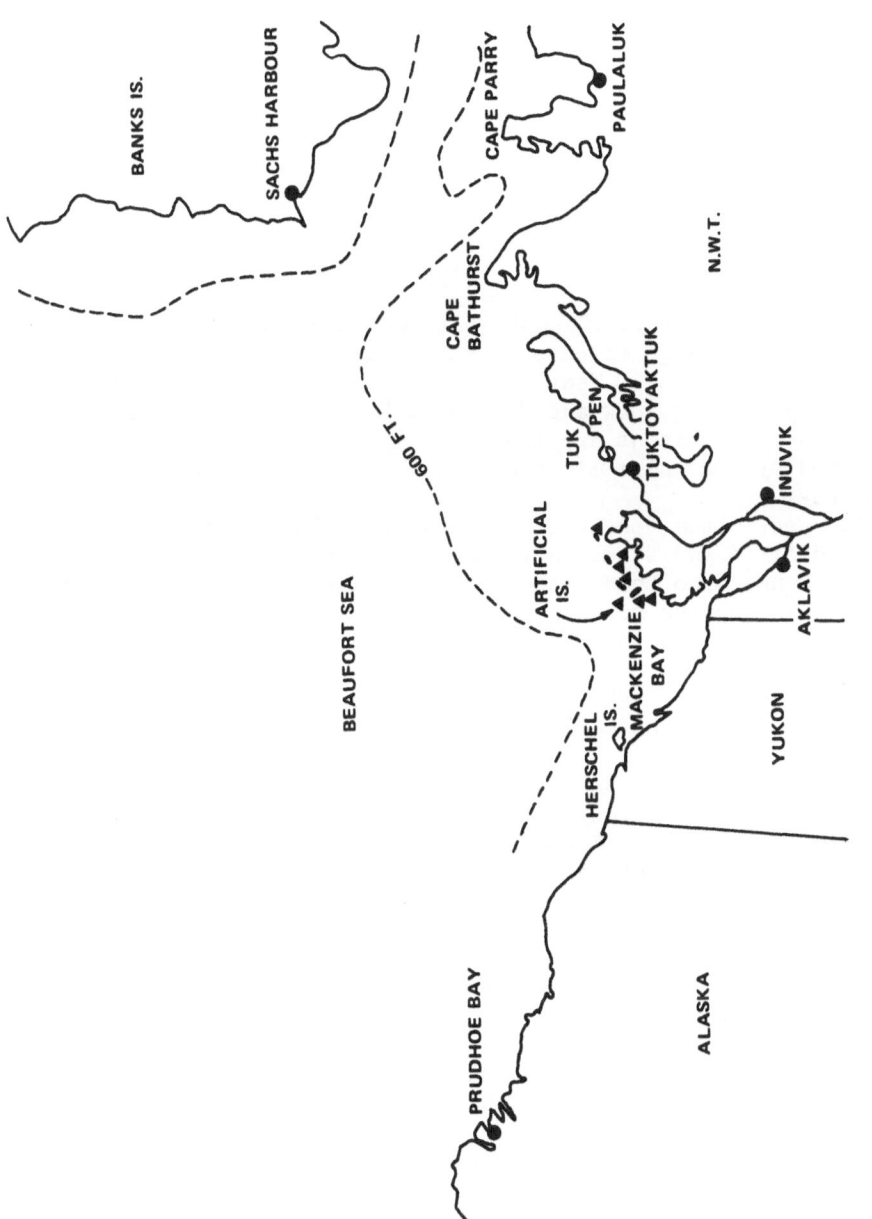

Figure 1. Location Map

and glacial ice islands may also be incorporated in the first year
sea ice. Beyond the ridged first year ice, a distance of 100 miles
or more, exists the permanent polar pack which comprises primarily
old ice and rotates in a clockwise motion.

Breakup commences in early July, influenced by the Mackenzie
River inflow, the Amundsen Gulf open water polynya, southerly winds
and increased temperatures. In severe years, such as 1974, the
area remains largely ice covered as northerly winds replenish the
ice in coastal areas. Figure 2 shows the 1953-74 variation in the
useful open water days at a proposed drilling location 45 miles
north of Tuktoyaktuk.

The average annual precipitation in the Beaufort Sea area is
6 inches comprising 4 inches of rain and 20 inches of snow.
Extreme temperatures range from -50°F to 80°F, with an annual mean
of 12°F. Winds average 10 knots but, on rare occasions, can reach
100 knots; when large fetches exist, these extreme winds can
generate waves up to 30 feet. Winter visibility is frequently
impaired by blowing snow and "whiteouts" at which time, depth
perception and horizons are lost. In mid-summer, visibility is
restricted by fog while limited daylight hours in winter reduce
efficiency.

The sea bottom consists of clays and silts in the westerly
areas and silts and sands in the easterly areas. Sea bottom
permafrost has also been detected during coring operations. The
sea floor is scoured or gouged by ice features, most commonly in
water depths from 50 to .150 feet. However, frequency of new scours
in specific areas is not known. Coastal currents are variable but,
in summer, average roughly 0.5 knots. Storm tides up to 8 feet have
been reported while normal tides are in the order of 1.5 feet.

The southern Beaufort Sea is inhabited by many species of fish,
ringed seals, some walrus, beluga and bowhead whales, seabirds,
polar bears and white fox. Since the native peoples in the
Mackenzie Delta area utilize this wildlife and marine life
extensively, their disturbance and reduction in numbers must be
minimized.

DRILLING SYSTEM

To date, offshore drilling in the Beaufort Sea has been
limited to eight artificial islands shown on Figure 1. These have
been constructed of gravel, silt, sandbag dikes and a barge in
water depths up to 11 feet. Moveable cone shaped or cylindrical
monopods which would sit on the sea bottom have been proposed for
intermediate water depths. Exploratory drilling from floating
vessels is proposed by Canadian Marine Drilling Ltd. for 1976.
Their system, which uses two ice-reinforced converted merchant
vessels, is limited to the open water season. An air cushion
vehicle and a vessel with a rotating ice cutting cylinder are
probably more applicable in the Sverdrup Basin of the Arctic

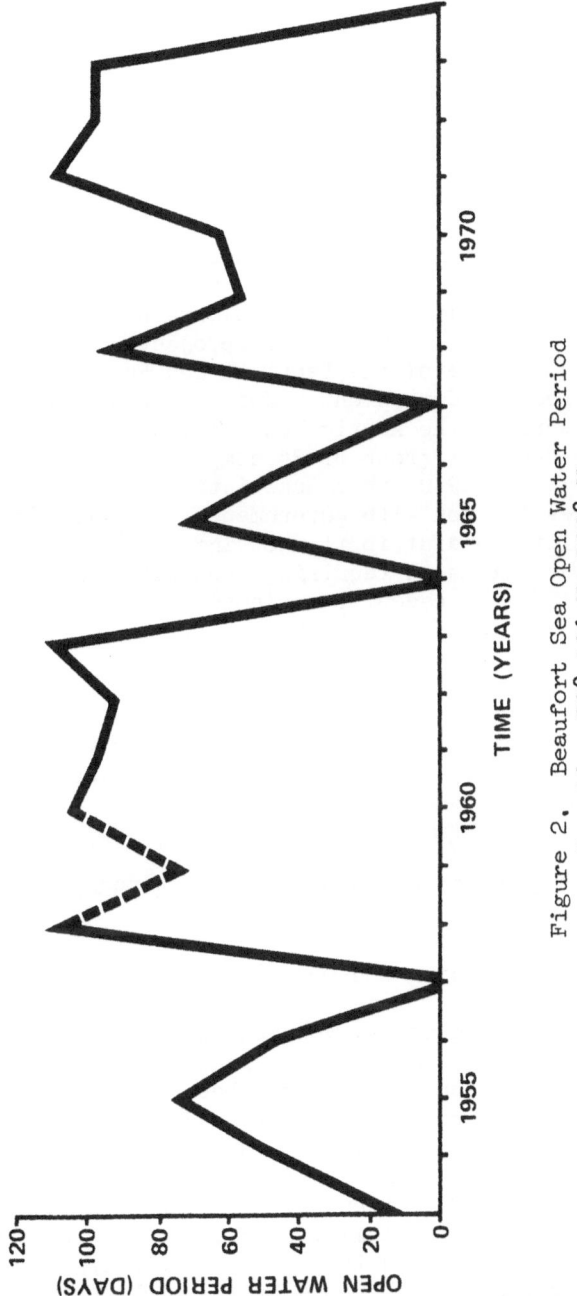

Figure 2. Beaufort Sea Open Water Period
Location 70° 10' N; 133° W

Islands. Ice movements in the Beaufort Sea are likely to be too
severe in winter to permit drilling from artificially thickened or
grounded sea ice.

Development drilling could be conducted directionally from
permanent artificial islands in shallow water and bottom-founded
cones in deeper water. Beyond some economic water depth, subsea
completions will be required.

INDUSTRY RESEARCH

In order to better understand and cope with the harsh
Beaufort Sea environment, joint industry research has been underway
since 1970. Approximately 10 million dollars have been expended
or committed for over 50 joint industry projects, the largest of
which is Industry's share of the Beaufort Sea Environmental Program
required by the Federal Government. Joint industry research has
been undertaken through the Arctic Petroleum Operators'
Association, a non-profit group of 34 companies holding permits in
the Arctic. Formed in 1970, this association promotes joint
research, improves liaison with governments, universities and other
groups, and facilitates distribution of information. When a project
is proposed, those companies requiring such data fund the study and
the results are, in most cases, restricted for up to five years to
the participants and the Federal Government.

Joint research projects conducted by APOA members include sea
ice strength, movement, distribution, thickness, physical
properties and modelling. Sea-bottom studies have included core
holes for soil composition and strength and sea bottom scouring by
ice features. Offshore drilling feasibility studies have also been
undertaken for seasonal and year-round systems.

JOINT INDUSTRY-GOVERNMENT RESEARCH PROGRAM

Funding and Organization

The unique joint industry-government Beaufort Sea Environmental
Program was necessary to supplement earlier government baseline
studies and industry research which had been geared toward offshore operations. This program was designed largely to accelerate
ongoing research so that a drilling authority could be granted for
offshore drilling in 1976 from floating systems.

Industry agreed to fund twenty-one of the required twenty-
nine studies by providing $4.1 million directly to the Federal
Government. These studies and their costs are shown in Table I.
An additional $400,000 was allocated by the eighteen funding oil
companies to cover the cost of industry co-ordination and a public
interface program. The Federal Government agreed to fund the

Table I. Beaufort Sea Environmental Program Industry Funded Studies

STUDY NUMBER	DESCRIPTION	COST – $
A3	SEA BIRD POPULATIONS	$ 389,400
A5	EFFECTS OF OIL ON MARINE MAMMALS	79,000
B4	NITROGEN FIXATION BY BACTERIA	28,500
B5	EFFECTS OF OIL ON MICRO-ORGANISMS AND INVERTEBRATES	63,400
B6	BIOLOGICAL PRODUCTIVITY	73,500
D1	MACKENZIE RIVER INPUT	105,000
D2	NEAR-BOTTOM CURRENTS	452,000
D3	OPEN WATER SURFACE CURRENTS	182,000
D4	PHYSICAL OCEANOGRAPHY	42,000
D5	STORM SURGES RELATED TO TOPOGRAPHY	49,000
E1	REAL-TIME ENVIRONMENTAL PREDICTION SYSTEM	630,000
E2	SYNOPTIC STUDY & WAVE CLIMATOLOGY	20,000
F2	BOTTOM SCOUR BY ICE	170,000
F3	SUSCEPTIBILITY OF SEA COAST TO OIL SPILLS	100,000
F4	SEDIMENT DISPERSAL	20,000
G1	ICE CLIMATOLOGY	50,000
G2	BEHAVIOR OF OIL IN ICE	550,000
G3	ICE THICKNESS AND DISTRIBUTION	45,700
H1, 2, 3	OIL SPILL DETECTION, CONTAINMENT, GATHERING AND DISPOSAL	265,000
Q1	SHIP CHARTER	480,000
Q2	POSITIONING SYSTEM	247,300
Z1	GOVERNMENT MANAGEMENT	58,200
		$4,100,000
	PUBLIC INTERFACE PROGRAM	200,000
	INDUSTRY MANAGEMENT	200,000
	TOTAL	$4,500,000

remaining eight studies totalling $1.2 million for outside contract
costs. Table II lists these eight studies plus four others added
later. In addition, government agreed to provide one of their
ships and in-house co-ordination.

An agreement between industry and the Department of the
Environment was entered into on May 15, 1974, covering the funding,
co-ordination and completion of the studies. The organization
chart for co-ordination of the program is shown on Figure 3. Both
government and industry formed steering committees and named
project managers. Each study has a primary government investigator
assigned to it who is responsible for co-ordination of the specific
study. In addition, an industry co-ordinator, particularly
knowledgeable in that study area, has been assigned to each study.
Contractors or consultants have been jointly selected to carry out
all or portions of individual studies.

The project managers provide the primary liaison between
government and industry. These managers, jointly with government
investigators and industry co-ordinators, discuss and modify the
terms of reference, monitor the progress of the studies, and re-
allocate funds. A major function of industry co-ordinators is to
bring existing information to the attention of government and
contract personnel and make suggestions based on experience.

The agreement required a review at the end of 1974 to determine
whether sufficient data was available for government to consider
issuing a drilling authority. This review indicated that
additional results, delayed by the 1974 adverse ice conditions,
would be required. All field studies are to be completed by
September, 1975, and all reports available by the end of 1975.

Over-expenditures on the program are the responsibility of the
Federal Government. Because of the adverse ice year in 1974,
addition of four new studies, extension of some studies and
additional management and reporting costs, there will be considerable over-expenditures. The latest estimate of the cost of the
completed program to the Federal Government is $7.5 million
including all in-house government personnel but excluding industry's
costs of $4.5 million. The total joint cost of the two-year
Beaufort Sea Environmental Program will thus be approximately $12
million.

It is of interest that a similar but longer term program
known as the Alaskan Offshore Program is now being considered.
Preliminary cost estimates are $54 million for a five- to ten-year
program to be co-ordinated by the U.S. Science Foundation. A
second $28 million program for U.S. Arctic waters commenced in the
spring of 1975, and will be co-ordinated by the U.S. National Ocean
and Atmospheric Administration.

Table II. Beaufort Sea Environmental Program Government Funding for Outside Contractors and Capital Costs

STUDY NUMBER	DESCRIPTION	COST — $
A1	DISTRIBUTION AND ABUNDANCE OF SEALS	$ 30,000
A2	DISTRIBUTION AND ABUNDANCE OF POLAR BEARS	80,000
A4	DISTRIBUTION AND MIGRATION OF WHALES	18,000
B1	POPULATIONS OF FISH IN COASTAL BEAUFORT SEA	120,000
B2	ANADROMOUS AND FRESHWATER FISH IN OUTER MACKENZIE DELTA	95,000
C1 & C3	BASELINES OF HYDROCARBONS AND PARTICULATES IN THE WATER AND ALONG COAST	311,500
F1	SEA BOTTOM PERMAFROST	388,000
G2a	HYDRODYNAMIC ASPECTS OF AN OILWELL BLOWOUT UNDER SEA ICE	194,500
G2b	LIGHT INTENSITY UNDER ICE CONTAINING OIL	13,000
M2	FISHERIES IMPACT STUDY	8,000
S1	SOCIO-ECONOMIC IMPORTANCE OF BEAUFORT SEA RESOURCE USE	7,600
		$1,265,600

NOTE: ADDITIONAL GOVERNMENT SHIP CHARTER COSTS AND IN-HOUSE SUPPORT COSTS TOTAL OVER SIX MILLION DOLLARS.

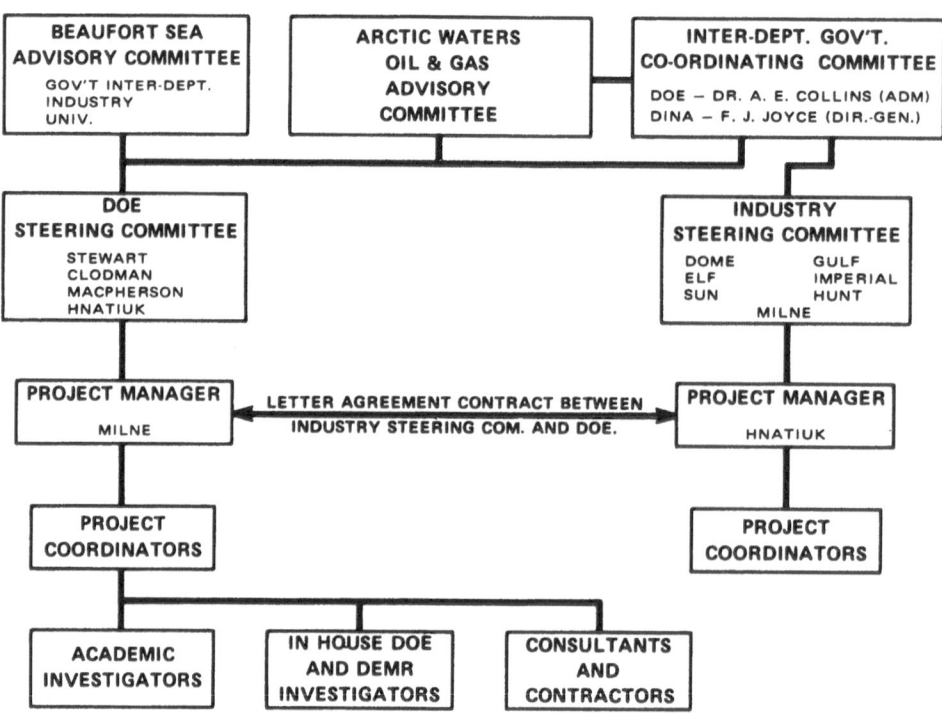

Figure 3.   Organizational Chart for Beaufort Sea
Environmental Program

Nature of the Program

As shown on Tables I and II, the program includes baseline studies of fish, birds and marine mammals and related biological studies of the marine ecosystems. These latter studies involve the effect of oil on marine mammals--in this case, the effect of oil on ringed seals, the effect of oil on nitrogen fixation in sediments and on invertebrates, and the biodegradation of oil. Baseline studies in which the waters are examined for pollutants are underway.

Physical studies include surface and bottom currents, river inflow, storm surges and waves. The sea bottom is being studied for the existence of permafrost and an effort is being made to learn more about frequency of scouring by ice features. The sea ice cover and thickness are also being studied.

A major study involving oil spills totalling 15,000 gallons deals with the behavior of oil under ice. Other studies deal with the detection, containment, gathering and disposal of oil in the Beaufort Sea.

Finally, a system to more accurately and rapidly predict weather, winds, waves and sea ice encroachment is being designed.

Objectives

The basic objectives of the Beaufort Sea Environmental Program are to better define the environmental conditions, establish base-lines, to improve forecasts and study means of coping with an oil spill. The program will identify sensitivities and document how drilling operations could damage the environment. At the same time, the program will identify how the environment could affect the drilling operation. More specifically, the environmental program and assessment of the results will provide a basis for the constraints to be placed on exploratory drilling operations in the Beaufort Sea.

Logistics

During the summer of 1974, the new 191-foot government vessel, "MV Pandora II", carrying the submersible "Pisces IV" was used to support marine geophysical studies and sea bottom investigations. The chartered "MV Theta" of similar size was also broguht from the West Coast to support oceanographic, fisheries and biological studies. These vessels spent one month in the Beaufort Sea in 1974 due to adverse ice conditions, but will both return during the summer of 1975. A small local craft, the "Pressure Ridge", was chartered for geophysical and sea bottom investigations in 1974. In 1975, it is planned to make part time use of the "MV Nahidik" from Hay River.

A Twin Otter was chartered full time and 206 and 205 helicopters used part time. These will again be used in 1975. A DECCA positioning chain is utilized.

The Federal Government's Polar Continental Shelf Project at Tuktoyaktuk served as a logistics base for some field operations, while camps were deployed elsewhere either by contractors or government agencies such as the Arctic Biological Station and the Geological Survey of Canada.

Although overall co-ordination is carried out from the government Project Office in Victoria, B.C., in-house government studies are underway from coast to coast with over 200 researchers involved.

Progress and Results[3]

Distribution and Habits of Wildlife and Fish (A1-4, B1 and B2)*
A total of 5,565 miles of transects representing 2,834 square miles were aerially surveyed to count seals. Vocalization was recorded and tissue samples obtained for analysis. Except for 6% bearded seals, the population of some 1 to 1.5 million seals in the southern Beaufort Sea are ringed seals.

The Polar Bear study was a continuation of a five-year program during which 300 bears were tagged, 35 of which have been recaptured or killed. Movements between Canadian and Alaskan areas have been recorded. The area north of Tuk Peninsula was found to be an important polar bear feeding area. It is of interest that white foxes move offshore to eat the remains of seals left by polar bears. 1974 was a difficult year for polar bears because of lack of water in the spring in which to hunt seals, presence of heavy ice all summer and early freeze-up.

The studies of sea birds commenced in April, 1974, but lack of open water, presence of sea ice, fogs and storms hindered these being conducted by aircraft and from vessels. Coastal waterfowl surveys were also carried out from Herschel Island to Cape Parry from June to September, including lakes and bays. The populations, migratory habits and moulting periods for various species were studied. Reproduction was very poor due to a late spring and a short cold season. In the case of snow geese, there were almost no young to migrate. A combination of radar, aircraft and ground observers will serve to fill data gaps in 1975.

Whales were counted using aerial reconnaissance. The White or Beluga whales, which migrate to the Mackenzie Delta area in June for calving and depart in September, are estimated to number 5,000, and are the largest known population in the Beaufort Sea-North Alaska area. Bowheads in the area are estimated at over 100 but are probably only the fringe of a larger population.

*Refers to Tables I and II

The studies of the movements, distribution, populations and food habits of coastal and anadromous fish were a continuation of earlier programs. During operations near Richards Island in the period, June to September, a total of 5,000 fish of 23 species were caught and 1,540 were tagged. Cisco were the most common and migration appeared to be limited. Using floating gillnets and trawls, other groups worked from the "MV Theta" and shore camps at Mackenzie Bay, Tuk Harbour and off Tuk Peninsula where Cisco are also the most common species. Generally fish appear to be concentrated near shore. There is ecological concern about an oil spill into lagoonal areas and bays which are important to fish fry.

Biological Studies (A5, B4-6). In order to study the effect of oil on mammals, six ringed seals were taken at Brown's Harbour on Cape Parry and immersed in oil-covered water for 24 hours, after which a number of parameters were investigated. Twenty seals were also taken to a southern university for additional studies which included the ingestion of fish containing capsules of oil. The study showed that the oil temporarily affected the eyes but most of the seals soon learned to stay under water to avoid the oil. Twenty-four hours after being oiled in the Arctic tanks, the seals appeared to be in good health and, by the fourth day, there was scarcely any evidence of the treatment. The seals transferred to the south were seriously affected by the stress of the move. In the spring of 1975, whitecoat seal pups in the Gulf of St. Lawrence were treated with oil but showed no permanent ill effects.

A second biological study is being conducted to determine the effect of oil on the nitrogen fixation capacity of bottom sediment samples from the Beaufort Sea and Eskimo Lakes. Norman Wells crude has not affected acetylene reduction under conditions studied to date.

In biodegradation studies, it was found that bacterial cultures obtained from the Beaufort Sea demonstrate a capacity to degrade the saturated aliphatic fraction of Norman Wells crude at 5°C and, in one case, at 0°C. In situ rates of biodegradation cannot be estimated at this time.

Studies of effect of oil on invertebrates using four different crude oils were undertaken. Eight biological productivity water samples were obtained off the ice using a helicopter and sixteen samples were obtained from the "MV Theta". Bottom samples for zooplankton studies were collected at the same sixteen sites. The submersible "Pisces" was also used for observations of productivity. However, biological productivity in 1974 was low due to the presence of ice as well as an increased thickness turbid water.

Pollutant Studies (C1 and C3). No tar balls, oil spills or natural oil seeps have been discovered although samples of particulates such as plastics were counted and examined.

In a baseline study of Beaufort Sea water samples taken from the "MV Theta", the water was found to be free of particulate pollutants and the level of hydrocarbons was very low.

Oceanography (D1 - D4). The flow of six channels of the Mackenzie River was measured at monthly intervals through the ice

and in open water. The combined flows peaked at over one million
cubic feet per second during runoff.

In studying bottom currents and tides, three current and two
tide gauges were were set through ice or in leads in October, 1973.
Ten current meters and nine tide gauges were deployed through ice
in May, 1974, using DECCA positioning. These were to be retrieved
from ships but, due to the presence of sea ice, only one tide gauge
and one tide gauge-current meter combination were recovered. Both
packages in 100 and 125 feet of water appeared to have been affected
by some physical force. They may have been struck by ice or the
flotation buoys which were 30 feet below ice level may have become
caught in a pressure ridge keel. It may be possible to recover
many of the instruments later in 1975, along with those being
deployed this spring.

It was intended that surface currents be measured by dropping
and tracking some 70 dispensable drift drogues containing radio
beacons. Due to severe ice conditions, the study was limited to
Kugmallit Bay and Atkinson Point although some beacons were also
placed on ice floes further offshore and tracked. Their positions
were monitored by frequent overflights. The results showed that
north winds tend to cause eastward currents in Mackenzie Bay and
along the Tuk Peninsula whereas easterly winds cause westerly
movement, increasing in velocity to the north. Current velocities
were generally less than 0.5 kts., but reached 1.3 kts. The
proximity of the ice in 1974 appeared to increase water flow to the
east into Amundson Gulf from the Mackenzie Delta.

The general physical oceanography study commenced with a
literature search of existing data. Later, in the field, the water
column was investigated for temperature, salinity, turbidity and
movement.

Real Time Environmental Prediction (E1, E2 and D5). A real
time prediction system has been designed in the form of a computer
model for rapidly updating forecasts of weather, winds, waves and
ice movement. A remote unmanned meteorological station, which
comprises part of this system, was tested on a floating ice island
in the summer of 1974. The prediction system will be tested during
the summer of 1975 using ground truth from a barge which will be
engaged in preparatory work at proposed drilling sites. Design
costs will be over one million dollars; installation costs, one-
half to three-quarters of a million dollars, and annual operating
costs similar to installation costs.

A hindcast study to the estimate probability of wind and wave
events has been undertaken and compared with that done for industry.
Structural icing, visibility and synoptic storm studies have been
undertaken.

A storm surge computer model for predicting the magnitude of
positive and negative surges has been prepared for incorporation
into the real time prediction system. Driftwood was examined on
beaches to obtain maximum surge levels.

Sea Bottom (F1 - F4). In their investigation of sea bottom permafrost, the Geological Survey of Canada drilled four holes off the ice in 1974 in Kugmallit Bay where seismic refraction had previously indicated some offshore permafrost. The discovery of permafrost in three holes agreed with these seismic predictions. In addition, 56 miles of shallow refraction shooting was undertaken from the small vessel, the "Pressure Ridge". Also, 20,000 miles of exploration reflection data from oil companies are being analyzed for indications of permafrost. These data supplement that from 11 core holes drilled by oil companies in 1970. The GSC interpretation suggests permafrost tops to be further below sea bottom than the 20 to 70 feet found in 1970 core holes.

In order to facilitate dating of sea bottom scouring, eleven areas were to have been surveyed with side scan sonar for construction of sea floor mosaics. Due to sea ice, only two were completed: one, 2 miles by 12 miles, in Mackenzie Bay and one, 4 miles by 6 miles, north of Pullen Island. The latter area overlapped work done earlier by APOA in 1971 and 1972, and provided a two- and three-year increment to see how many scours or gouges have been added over these periods. One 1974 mosaic has just been completed, and some additional scouring was noted. The areas inaccessible last summer will be covered in 1975. Additional precision fathometer and high resolution echosounding was done to determine scour depths and infilling of trenches. It has been suggested that the majority of scouring may occur in winter.[4]

A third annual coastal study was undertaken in the summer of 1974 to determine how coastal areas would be affected by an oil spill. Work extended along the Yukon coast in 1974. Effects of ice on the coast appear minimal but thermal and water erosion are severe in some areas.

One hundred soil samples obtained in 1970 were analyzed in 1974 in more detail to learn more about sediment dispersal in the Beaufort Sea.

Ice Cover and Thickness (G1 and G3). Historical ice cover data from aerial reconnaissance in the period, 1953-73, has been re-examined by computer methods on a 15 x 20-mile grid scale. Work is also underway on floe size distribution. Another interesting aspect of this study deals with the long range predictability of good and bad ice years.

A pulse radar method of determining ice thickness remotely is being developed. The distribution of ice types and open water from satellite imagery is being studied by several agencies.

Behavior of Oil in Ice (G2). This project is studying entrainment and migration of oil in growing ice, the spread of oil under ice by currents and biological implications of a spill. A remote little-utilized bay with a narrow mouth was chosen on Cape Parry as the site for eight successive discharges of oil under ice into retaining plastic skirts 100 feet in diameter. Norman Wells and Swan Hills crudes were both used. An open water spill was also simulated by removing the ice cover from a 50-foot diameter area.

The oil discharges are schematically represented in Figure 4 which
also includes dates and ice thicknesses. Solar radiation and
related parameters are being recorded. There had been no significant
upward migration of oil within the ice cover by early April. The
oil will be burned off just prior to ice breakup. Ice thickness
during discharges varied from one foot to nearly seven feet. Video
coverage of the oil migration and cores has been taken and
temperatures have been monitored.

Two discharges of 200 gallons each were made under ice in
April in offshore currents of just over 0.2 kts. off Cape Parry.
The oil was carried only 35 feet in one discharge and, in the
second, it accumulated to 4 inches in thickness in an upward
indentation. These sites will be monitored but it will be very
difficult to burn off the oil.

Biological and light penetration studies accompany the oil
discharge work. There have been no apparent adverse effects
beneath the oil discharge areas.

In Victoria, B.C., theoretical meteorological and oceanographic
studies have been undertaken which indicate that the effect of oil
from a single uncontrolled well would not have a major effect on
the volume of ice in the Arctic Ocean or on the global climate,
when compared to existing annual fluctuations.

Related to this study is definition of the scenario accompanying an uncontrolled well. This work has included discharge of
compressed air on the sea bottom on the West Coast to simulate
natural gas. Other smaller scale laboratory work will be undertaken
with oil in test tanks to study the underwater plume.

Oil Cleanup Studies (H1 - H3). Engineering feasibility studies
related to containment, gathering, transport and disposal of oil
from a spill are underway for various ice conditions. Another
related study is being conducted to determine methods for remote
detection of oil under ice. There are no firm plans for field
tests as yet within this program.

FUTURE FIELD STUDIES

Field studies affected by the adverse ice conditions in 1974
will be completed in the summer of 1975, and should more closely
represent average conditions. The "MV Pandora II" with the
"Pisces IV" submersible, the "MV Theta", fixed wing and rotary
aircraft and DECCA positioning will again be used. The "MV Nahidik"
will be added on a part time basis.

The first studies will involve physical oceanographic
measurements through ice in March. These will be followed by water
sampling for biological studies in late spring from helicopters and
later from vessels. Eight tide gauges and current meters will be
placed through ice in May and these, plus existing packages, will
be later retrieved from vessels. Six tide gauges will be set near
shore. Drift buoys will be deployed and tracked by the Twin Otter.

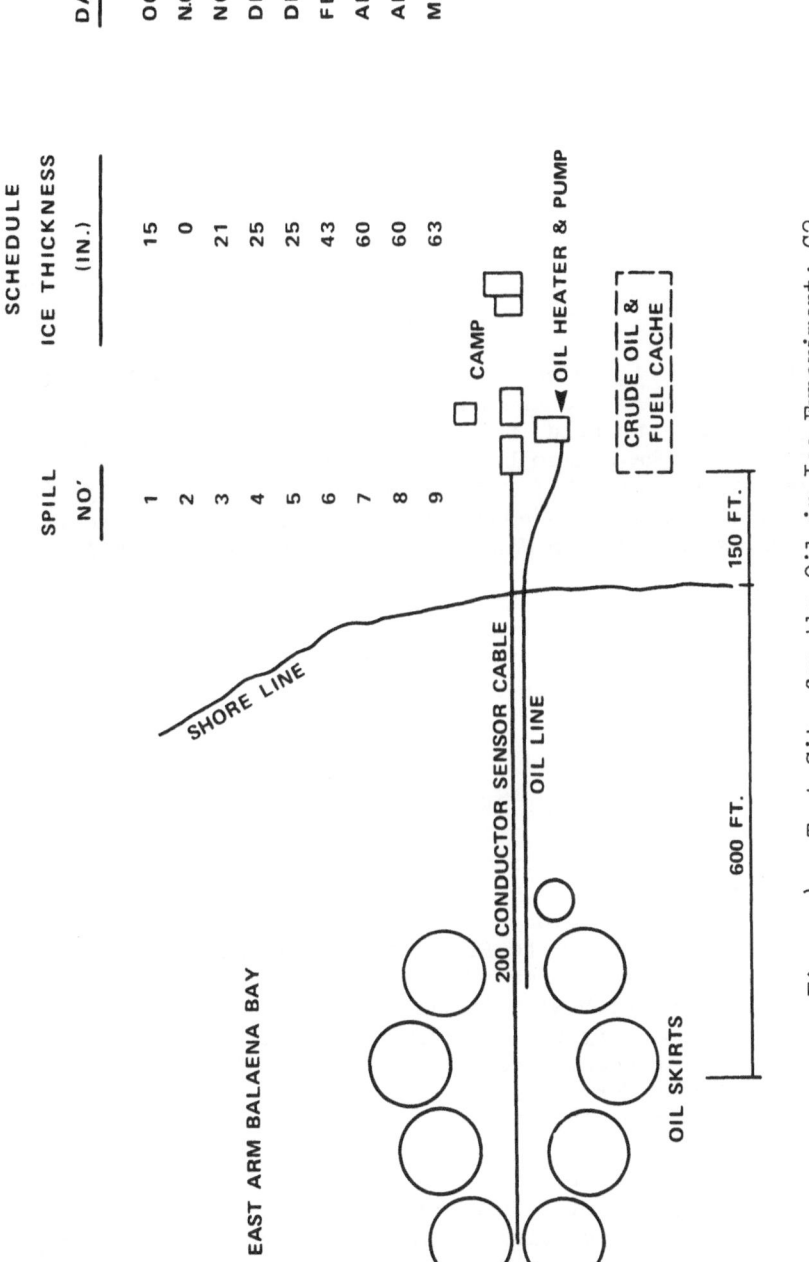

SPILL NO'	SCHEDULE ICE THICKNESS (IN.)	DATE
1	15	OCT.24
2	0	NOV.1
3	21	NOV.14
4	25	DEC.7
5	25	DEC.9
6	43	FEB.15
7	60	APR.12
8	60	APR.15
9	63	MAY 15

Figure 4. Test Site for the Oil in Ice Experiment: G2

Polar bear tagging and surveys will continue from May to November. A census of sea birds and migratory birds will be undertaken using radar. Studies to fill data gaps relating to fisheries will be undertaken through ice and throughout the summer. The Arctic Biological Station will continue their field activities.

Surveys related to sea bottom scouring and permafrost will also be undertaken in additional areas.

Field testing of the wave and ice forecasting equipment and system will continue this summer.

Other ongoing field work such as the oil in ice study will be completed during the summer.

REPORTING

Interim reports on each study have been made available to government, industry and the public. A major conference to exchange results was held in January, 1975, in Calgary. Monthly progress reports have also been made available.

Near the end of 1975, final reports on each study will be available. These will be condensed into six overview reports by subject areas as shown in Figure 5, and will serve as an interpretive assessment of pertinent scientific and technical data as it relates to offshore drilling. One simplified summary report covering the entire program will also be issued early in 1976.

As results become available, they could be used in considering a Drilling Authority rather than waiting for completion of all reports.

PUBLIC INTERFACE PROGRAM

A consultant was retained by industry in mid-1974 to assist in communicating information on the research program to northern peoples, environmental groups and other publics. A movie film entitled, "Understanding the Beaufort Sea", was prepared showing activities during the summer of 1974. Four issues of a "newspaper" have been produced with wide distribution. Visits by the consultant and industry and government representatives have been made to the Arctic coastal communities of Inuvik, Aklavik, Tuktoyaktuk, Sachs Harbour and Paulatuk in October, 1974 and January, 1975 to describe the research program and provide results to date.

Visits will again be made to these communities plus Holman Island in May, 1975. Slide presentations and movies have been shown to public groups and schools. Film strips have also been prepared which will be left at schools.

Reception by northern natives has been very gratifying. They have complained that, through the years, they see many researchers

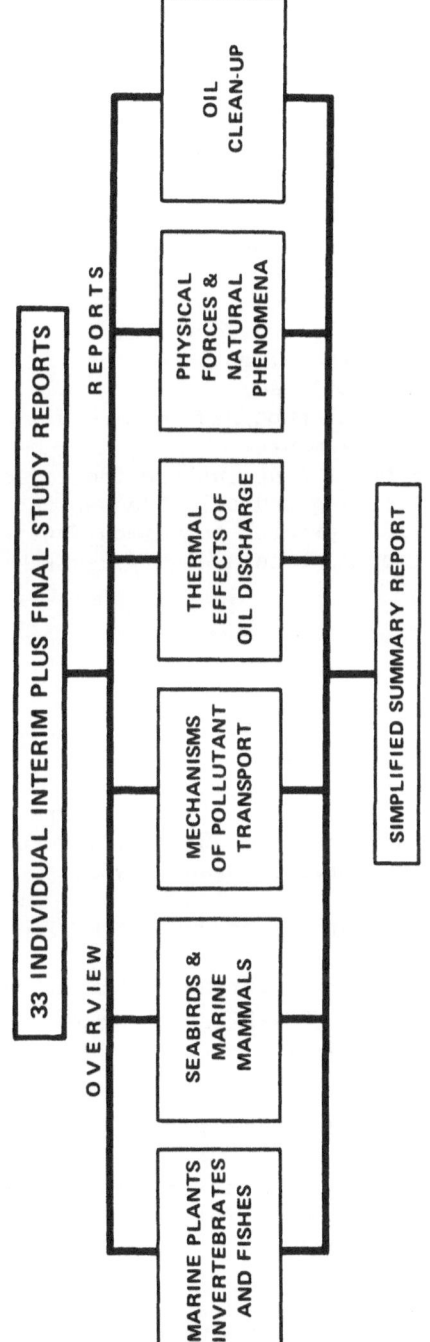

Figure 5. Beaufort Sea Environmental Program Reporting Mechanism

but are not made aware of the nature of their activities nor are they provided with results. In view of this response, a public interface program is now considered an essential aspect of any major activity in the Arctic.

CONCLUSIONS

1. A jointly-funded environmental research program has been undertaken in the Beaufort Sea with good co-operation between industry and the Federal Government.
2. Thus far, no alarming results have come to light which would preclude drilling in the Beaufort Sea.
3. The concern about the climatic effects of a major discharge of oil has been largely alleviated.
4. Adult seals and seal pups have not been significantly affected by oil during experiments.
5. Currently, the major concerns are the effect of oil on fisheries, sea bottom scouring and oil clean-up techniques.
6. A public interface program to communicate with northern natives and other publics has been a highly desirable addition to this research program.

REFERENCES

1 Kovacs, A., "On Pressured Sea Ice", International Sea Ice Conference, Reykjavik, May, 1971.

2 Weeks, W.F., Kovacs, A., and Hibler, W.D., "Pressure Ridge Characteristics in the Arctic Coastal Environment", POAC Conference, Trondheim, Norway, August, 1971.

3 Interim Reports from Beaufort Sea Project Office, December, 1974.

4 Shearer, J., and Blasco, S., "Further Observations of the Scouring Phenomena in the Beaufort Sea", Geol. Surv. Can., Paper 75-1, Part A.

ACKNOWLEDGEMENTS

The author wishes to express his appreciation to the funding members of the Arctic Petroleum Operators' Association for permission to present this paper.

ARCTIC OIL CLEANUP SYSTEM – SUBSYSTEM REQUIREMENTS*

Lt. J.H. Getman

United States Coast Guard

Washington, D.C., U.S.A.

NATURE OF THE PROBLEM

The U.S. Coast Guard has the responsibility for preventing and controlling oil spills in and along the coastal waters of the United States from authority granted by the Water Pollution Control Act. The coastal waters of arctic Alaska are in the direct area of Coast Guard responsibility.

Cleanup and disposal of oil spills are the responsibility of the person or persons causing the discharge. If the responsible party is taking appropriate action, the role of the Coast Guard is relatively passive, consisting primarily of monitoring progress and providing advice. In the event that the responsible party cannot be determined or does not take appropriate action, the Coast Guard institutes the necessary cleanup and disposal actions.

The requirement for direct Coast Guard response or active assistance to the responsible party is more probable in arctic Alaska than in similar spill situations in the "lower 48" states. The geographic expanse, remoteness and underdeveloped nature of Alaska minimize the probability that the required equipment, personnel and facilities will be available in the immediate area of the spill. Logistic support will be a prime factor in all potential spill areas. The Coast Guard has the only large and established air-sea emergency response capability within Alaska. Direct Coast Guard participation in the event of marine spills is highly probable due to the requirements for very rapid response and a general lack

---

*The opinions contained herein are those of the writer and are not to be construed as official or reflecting the views of the Commandant or the Coast Guard at large.

of marine transportation systems within the state.

The attendant problems of containment, recovery, storage and disposal associated with oil spillage will vary widely, depending upon

(1) the volume and type of petroleum products released,
(2) the geographic and environmental setting of the spill, and
(3) the availability of equipment and personnel to conduct the spill cleanup.

The potential for crude oil production from the Alaskan mainland and adjoining continental shelf is truly immense, in both volume and areal extent. Developed fields in Cook Inlet and the probable reserves in Prudhoe Bay field (estimated at 10 billion barrels by industry) will provide approximately 10% of United States crude oil demands in the early 1980's. The preliminary assessment of the reserves in other areas such as the Naval Petroleum Reserve #4 (NPR 4) on the North Slope and the presence of promising geologic formations on virtually all of the accessible continental margins surrounding the Alaskan mainland bespeak a potential for eventual production several times the 2-3 million barrels per day projected for the Trans-Alaska Pipeline System (TAPS).[1] Declining domestic production, uncertainties in the future reliability of foreign imports, the "Project Independence" goal for United States energy selfsufficiency by the mid-1980's, and recent acceleration of Outer Continental Shelf (OCS) schedules for sales in new areas are indicators that Alaskan arctic and sub-arctic petroleum resources will be developed within a shorter time frame than previously assumed.

A recent Outer Continental Shelf (OCS) tentative lease sale schedule includes sales of tracts in the following areas of Alaska before the end of 1978:[2]

SUB-ARCTIC

Lower Cook Inlet
Gulf of Alaska (eastern)
Bering Sea (St. George)
Gulf of Alaska (Kodiak)
Outer Bristol Basin (Bristol Bay)
Gulf of Alaska (Aleutian Shelf)

ARCTIC

Beaufort Sea
Bering Sea (Norton Basin)
Chukchi Sea (Hope Basin)

The areas tabulated above encompass the coastline of the Alaskan mainland, thus extending the marine waters potentially threatened by oil spillage to virtually the entire coastal margin of the mainland.

The major oil spill threat in Alaska is expected to become and remain that related to crude oil development, production and transportation. However, expanding construction activity and industrial development will continually increase the threat of releasing refined petroleum products. The bulk of Alaskan crude oil will continue to be exported for refining outside the state in the foreseeable future, as it has in the past.

The frequency and volume of oil spillage are commonly predicted by correlation to ongoing exploration, production, transfer and transportation activities; which can in turn be projected into the future on the basis of past development. Oil spill statistics from past petroleum activities in the United States and other countries are commonly used to predict oil spill frequency and volume resulting from future Alaskan development. Such statistics are virtually the only source available. However, it must be remembered that these statistics were primarily derived from operations in more temperate climates and, as such, may have limited relevance to arctic operations. Prediction of oil spills in Alaska is fraught with uncertainty because of the extremely limited past experience with crude oil production (Cook Inlet). Also, the scant number of established transportation systems for either crude or refined products , and worldwide lack of experience in arctic petroleum development and production.

Spills in arctic regions may result from casualties to conventional, icebreaking, or submarine tankers or barges; from drilling accidents (blowouts); from pipeline breaks; or as a result of oil transfer operations from offshore terminals. Spills from these sources will not result in a uniform slick as in spills on ice free water. Spilled oil in arctic regions can be located on, under, or sandwiched in between solid ice, in leads between ice floes, mixed in with broken ice, or along ice-locked shorelines.

The potentially hostile environment and lack of development in arctic regions necessitates a re-evaluation of priorities established for oil spill response in temperate climates. The widespread availability of equipment and facilities and complex technology developed for coping with spills in the "lower 48" states does not exist or may have little application in the Arctic. Logistics is a severe problem. It must be assumed that all equipment, personnel and support functions necessary to conduct cleanup and disposal operations at any specific site will have to be brought to the site from locations up to one thousand miles away. Improvisation, which is so common in temperate area spills, is nearly impossible throughout the arctic because of the remoteness and general lack of nearby equipment, materials and facilities.

The most serious present limitation to the evaluation of systems and procedures for recovery, storage, and disposal of spilled oil is a general lack of knowledge regarding the engineering properties and behavior of petroleum products in the arctic environment.

Limited testing has been conducted with Prudhoe Bay crude and diesel
oil and some information is available from oil spills in other
arctic regions of the world (primarily Canada and the Scandinavian
countries). However, the bulk of arctic oil fields have yet to be
developed, so knowledge regarding the behavior of many arctic
crudes must await the development. Crude oils with pour points
above approximately -10°C may require completely different systems
for recovery, storage, and disposal than those which remain fluid
in the arctic environment.[3]

CONDITIONS UNIQUE TO THE ARCTIC

Constraints on the recovery, handling, temporary storage and
ultimate disposal of arctic oil spills will exist that render many
of the conventional equipment and procedures developed for more
temperate regions ineffective or greatly reduce their efficiency.
The conditions which produce the constraints must be accommodated
and can sometimes be used advantageously if recognized and worked
with rather than avoided. The constraints imposed by the arctic
environment suggests that equipment and procedures for arctic spill
cleanup must be developed specifically for arctic conditions, rather
than by adaptation of conventional systems.
Some of the more significant factors considered unique or exaggerated for arctic marine spills and the resultant constraints on
oil spill response operations are tabulated below:

Arctic Environment

Extreme Cold. Temperatures down to -50°C reduce efficiency of
personnel up to 90%; drastic alteration of the properties and
behavior of petroleum products; causes failure or non-performance
of equipment and materials.
Winter Darkness. Limited extent or nonexistence of daylight
severely hampers field operations and reduces personnel efficiency.
Wind. Winds in excess of 100 knots are not uncommon along
coastlines which can result in structural damage and greatly
increase wind chill factors.
Oceanographic. Sea ice is the dominant feature effecting
arctic spill response; storm surges may raise the water level up
to 5 meters; nearly constant water temperatures of -1.5°C.
Ecological. Abundance of fish and wildlife accentuates the
potential damage from spills, relatively sensitive shoreline
vegetation (tundra) easily damaged by field operations; large
animals such as bears can destroy storage systems (notably pillow
tanks).

Permafrost. Seasonal instability of shoreline sediments containing moisture complicates travel overland and virtually precludes
sound footings for structures.

Geographic and Physiographic

Vast Expanse. Long, largely uninhabited and inaccessible
coastline seriously complicates logistics.

Lack of Ports and Harbors. The combination of extremely shallow
coastal waters and virtually complete lack of suitable ports and
harbors along the Bering Sea and arctic coastlines complicates
shoreside access and logistics.

Low Shorelines. Practically the entire Bering Sea and arctic
coastline is low-lying and therefore susceptible to storm surges
and wind-driven pack-ice.

Industrial Development

Remoteness. Population and industry concentrated in areas of
south-central mainland and Southeast Alaska; few facilities available in Arctic or westward; population centers not necessarily in
areas of high oil spill potential.

Lack of Transportation Systems. Virtually no roads in coastal
areas, overland travel difficult in summer, air transport almost
mandatory which restricts size and weight of equipment.

Lack of Widespread Coast Guard Facilities. Coast Guard concentrated in Southeast Alaska and south-central portion of mainland
which necessitates response over long distances.

Lack of Communications Systems. Established communications
systems inadequate or nonexistent.

Petroleum Industry Development

Limited Previous Experience. Development restricted to Cook
Inlet, limited arctic experience.

Potential for Rapid Expansion. Enormous probable reserves will
lead to rapid development of new fields and modes of transportation
which complicates planning for prevention and control of future
spills.

Varied Locations of New Fields. Promising geologic formations
virtually encompass the entire mainland opening nearly one million
square miles to development which severely affects logistics and
planning; properties of crude oils unknown.

Physical Fate of Arctic Oil Spills

On Ice. The primary factor effecting the final size of a quantity of oil spilled on ice is the surface roughness. Neglecting absorption into the ice and surface tension and viscosity effects, the oil will spread out from the spill source filling one "pocket" after another until the volume spilled is contained. The spill size on ice will be much smaller than a spill of the same volume on temperate water. A spill of 150,000 tons of oil on temperate water will spread to an area of approximately 300 square miles. In comparison, the same volume of oil spilled on ice is expected to spread over a 1/4 to one square mile area depending on the surface roughness.[4] Aging, the evaporation of volatile components, will occur on arctic spills exposed to air, although at a slower rate than occurs in temperate areas. Blowing or falling snow will·mix with and eventually cover a spill causing a further reduction in aging rate. The resulting oil/snow mulch can contain up to 80% snow by volume,[5] and represents a quadrupling of the amount of material which must be handled during the spill cleanup.

On Water. Oil spilled on temperate waters will spread out until it reaches monomolecular thicknesses and is controlled by three separate phases of oil spreading - inertia, viscous, and surface tension. Field experiments have shown that the surface tension phase of oil spreading is absent on near freezing water. Spreading on arctic waters therefore ceases when a certain limiting thickness is reached on the order of 1 cm.[6] Oil spilled in the arctic may become contained in leads. If these leads close, the oil will be driven beneath or on top of the ice depending on the density of the oil and ice. If the pressure which caused the lead to close continues, forming a pressure ridge, the oil will very likely become entrapped in the ice rubble of the pressure ridge.

Under Ice. Oil released under ice will spread in a manner similar to spills on ice, filling interconnecting pockets until the volume released is contained. Since the under ice roughness is greater than the surface roughness, the ultimate spill size will be less than that of a similar sized spill on the ice surface.[4] During the fall and winter, when ice is actively growing, ice will continue to grow beneath the oil layer forming an entrapped oil lens in the ice.[7]

Arctic Cleanup System

The function or purpose of an arctic oil cleanup system will be to remove oil from arctic spills in a manner which results in the minimum possible damage to the environment. The physical fate of oil spills in the arctic is much more complex than the fate of spills on temperate waters. Arctic spills may be located on or under the ice, trapped in pressure ridges, or frozen as oil lenses

within the ice. While the fate of arctic oil spills differs
greatly from temperate spills, the cleanup subsystem requirements
are identical.

Many of these subsystem requirements are taken for granted in
temperate areas. In the arctic nothing can be taken for granted.
An oil spill cleanup system, both arctic and temperate, must provide
for the following functions:
- Containment
- Recovery
- Storage
- Disposal
- Transfer of recovered materials
- Detection and Surveillance
- Logistical Support

Arctic Cleanup System - Subsystem Requirements

The following is an attempt at identifying and defining the
subsystem requirements for an arctic cleanup system. References 3
and 8 were relied upon heavily for the subsystem breakdown and
corresponding requirements. The total cleanup system includes all
the subsystems listed. However, the degree to which each subsystem
is utilized during a spill response will depend on the specific
situation.

Containment.* Limit initial and further spreading of released
oil. Concentrate oil to facilitate recovery. Serve as immediate
storage.

Recovery. Remove spilled oil and oil stained debris (ice and
other debris) on, under, or amongst ice, entrapped in ice, or on,
near freezing water.

Storage. Provide for immediate (1 - 3 days) and temporary
secure storage (3 days to 1 year) of recovered oil and debris.

Disposal. Dispose of recovered oil in an environmentally safe
or acceptable manner.

Transfer. Provide for transfer of liquid, semisolid, and
solid product and debris from recovery to storage to disposal.

Detection and Surveillance. Detect spills when they occur and
monitor movement of the spill and progress of the cleanup effort.

Logistics. Provide for movement, maintenance, and disposition
of equipment, supplies, facilities, personnel, and services required
to support an arctic oil spill cleanup effort. (All equipment suitable for arctic use and packaged for ease of transportation and
assembly).

---

*The first function in any pollution response action is to
remove the source of the spilled oil to prevent further contamination.
For the purposes of this discussion, the objective of a cleanup
system will be to detect, recover, and dispose of spilled oil.

Support Bases

<u>Logistics Center</u>. Store and maintain spill response equipment and all equipment and supplies required to sustain field efforts for extended periods. Access to all primary transportation modes is mandatory.

<u>Staging Area</u>. Terminus near spill site to provide for transfer and support of personnel and equipment from logistics center to field camp.

<u>Field Camp</u>. Located at spill site for immediate support of personnel and equipment.

Transportation

Provide for movement of equipment and personnel from logistics center to staging area (primary) and from staging area to spill site (secondary).

Life Support

Coupled with support bases, provides for food, protective clothing, and shelter for spill response personnel.

Communications

Long-range communications to link the staging area and spill site to the logistics and command center are required along with short-range communications for coordination of the cleanup effort.

<u>Ancillary</u>. Provides for additional support activities such as weather and ice forecasting and establishment of temporary navigational aids for support aircraft and vessels.

Emergency

<u>Evacuation</u>. Provides for immediate evacuation of personnel from hazardous ice conditions and provides access to medical facilities.

CONCLUSION

An arctic oil spill cleanup system has not, as yet, been developed. The form such a system will take is therefore not known. It could be built around an icebreaker or arctic surface effect vehicle incorporating the major subsystems, or may consist of many

modules designed for ease of transport by all probable means of
transportation. In either case, the total system will be composed
of the same subsystems--containment, recovery, storage, disposal,
material transfer, detection, surveillance, and logistic support.

REFERENCES

1. Swift, W.H. et al., "Geographical Analysis of Oil Spill
Potential Associated with Alaskan Oil Production and Transportation
Systems," U.S. Coast Guard Report CG-D-79-74, Washington, D.C.,
1974. AD 784-099

2. U.S. Department of the Interior, Bureau of Land Management,
"Proposed OCS Planning Schedule," November 1974.

3. Peterson, P.L., "Temporary and Ultimate Disposal of Oil
Recovered from Spills in Alaska," Draft Final Report of Contract
DOT-CG-23, 223-A, Prepared for U.S. Coast Guard by Battelle Pacific
Northwest Laboratories, May 1975.

4. Hoult, D.P. et al., "Oil in the Arctic," U.S. Coast Guard Report
CG-D-96-75, Washington, D.C., 1975. ADA 010 269

5. McMinn, T.J., "Crude Oil Behavior on Arctic Winter Ice,"
U.S. Coast Guard Report 734108, Washington, D.C., 1972. AD 754 261

6. Glaeser, J.L., and Vance, G.P., "A Study of the Behavior of Oil
Spills in the Arctic," U.S. Coast Guard Report 714108/A/001,
Washington, D.C., 1971. AD 717 142

7. Brown, R.F., Interim Report of Oil in Ice Studies Conducted by
Norcor Engineering and Research, Ltd. for Beaufort Sea Project,
Presented at Oil in Ice Workshop, Yellowknife, N.W.T., April 1975.

8. Peterson, P.L. et al., "Logistic Requirements and Capabilities
for Response to Oil Pollution in Alaska," U.S. Coast Guard Report
CG-D-97-75, Washington, D.C., 1975.

MULTISPECIES MODELS OF AN EXPLOITED SEA

Jan E. Beyer   Hans Lassen

Technical University   Danish Institute for Fishery
of Denmark   and Marine Research
Lyngby, Denmark   Lyngby, Denmark

1   INTRODUCTION

   Man's impact on the sea has increased to the point where it is
obvious that the living resources are by no means inexhaustible.
This shows clearly in the present year's commercial fisheries of
the North Atlantic where catches have been constant or even
decreased in spite of more effort put into operation. Even on a
world scale landings have tended to stabilize on the present level
of about 60 million tons. This figure should be compared with the
FAO estimate of 100 million tons as the maximum sustainable yield
on a world scale if each stock could be properly managed. It is
expected that only 80 million tons is achievable due to mixed
fisheries problems, etc. The crux of the matter is that careful
management of the fish stocks is required if these 80 million tons
as a stable long term yield is the objective.
   The use of the sea as a wastedump is in conflict with the use
as a production unit, as can be seen from coastal waters of the
industrialized countries. Pollution introduces major changes in
the ecosystem but little is known of the long term consequences of
such encroachment.
   It is obvious that the very slow biological turnover of arctic
systems makes these very vulnerable to man's interference. It is
therefore essential to these systems that the biological consequences
of any interference by man are investigated before operation is
taken. While the short term effects to some extent are elucidated,
little is known about the long term result.
   The forecast of short term changes of the marine ecosystem
requires quantified knowledge of the present state of the system.
Prediction of the long term changes requires in addition a
theoretical understanding of the biological and economical mechanisms

of the system.

Fishery biologists have assessed and forecasted changes in
fish stocks using the Beverton & Holt assessment model [1], while
economists and administrators have used this work for further
analysis. However, a thoroughly integrated biological and
economical analysis requires a description of the interaction
between stocks as the exploitation has become so serious that
stocks virtually are obliterated. The consequences of such
interference can not be analysed using the B & H model or any other
single-species model.

It is necessary to provide a proper basis for analysing the
interaction between species, i.e. a multispecies ecomodel. The
model builder (System scientist) meets several problems when
attempting to fulfill this goal.

System ecology or ecomodelling is a new branch of theoretical
biology found in a trial and error phase which means that the
skeleton for building models is weak. Much biological information
is available in qualitative terms but is required in quantitative
terms. Furthermore, the technical problems in handling the models
are very pronounced.

The language of the model is chosen according to the desired
type of answer. Three main classes of models may be distinct as

    <u>a</u> Time-continuous deterministic models
    <u>b</u> Time-discrete deterministic models
    <u>c</u> Time-discrete stochastic models

If one is only interested in an average situation, models of
type a and b will probably suffice while extra information, and
labour, about variance of the average situation is gained using a
model of type c.

The present paper describes a multispecies extension to the
Beverton & Holt assessment model. The model has been formulated
both as type a, time-continuous deterministic, described in part 1,
and as the time-discrete type b described in part 2.

The model of part 2 is the deterministic equivalence of a
stochastic model (type c) which presently is under investigation.

The deterministic model has been used to analyse the fishery
[4] and the eutrophication situation of the North Sea [3]. It has
also been formulated to investigate the problem of persistent
chemical compounds [5].

2   OUTLINE OF THE BASIC MODEL

The model is a multispecies extension to the Beverton & Holt
assessment model and is described in detail in [2]. The model has
been further investigated since 1973, the results are given in [3],
[4] and [5].

In contrast to the B & H model, this model highlights the food
turnover of the system. This requires a decision of where to
"begin" the multispecies extension. Andersen, Lassen and Ursin [2]

start at the algae production as function of sun radiation and
inorganic phosphorus in the sea. The model of primary production
is that of Steele [6] and [7] as adopted by Lassen and Nielsen [8].
The algae is used as food to higher tropical levels causing growth
to other species. The growth model is that of von Berthalanffy,
see [1], with the refinements adopted by Ursin [9]. The predation
is developed in [2] by requiring that the system is closed, i.e.
      Food intake = animals dying from predation.
The food selection is based on the weight ratio between prey and
predator (Ursin [10]).
      The inclusion of pollutants in the system is the latest
extension of the model (Lassen [5]). The mechanisms are absorption
of the pollutant from the food and the water and eliminated by
metabolits and chemical decomposition. The knowledge of the physiological mechanisms involved is limited and the description is often
chosen by the principle of the simplest possible description.
      The state of the system is given by
A. State vector of the system
      1. Content of nutrient salts of the sea
      2. Content of pollutants of the sea
      3. For each species and age group
            a Weight of an animal
            b Number of animals
            c Accumulated fishery yield
            d Burden of pollutants
The system is subject to some forces which are controlled from
outside, i.e. these forces appear as explicit functions of time.
These are

B. Input-Natural and man-controlled functions
      1. Sun radiation and influx of nutrient salts
      2. Discharge of nutrient salt
      3. Discharge of pollutants
      4. Fishery effort, i.e. fishery mortality
The evolution of system is determined by the imposed forces and
the mechanisms within the system. These are

C. Mechanisms
      1. Primary production
      2. Growth and feeding of animals
      3. Predation
      4. Uptake and elimination of pollutants
      5. Death caused by pollutants, parasites, illness, etc.
      6. Spawning
The system is extensive and complicated to work with. However,
it is only in rare circumstances that the entire model is required.
Submodels which may be extracted from the above system will often
suffice. This is a lucky situation as the technical problems of
handling the entire model are enormous.

PART 1
THE TIME-CONTINUOUS DETERMINISTIC MODEL

3   Model Description

The present section is intended to give a fairly detailed description of the model. The exposition follows mainly [2]. The formulation is then used to set up an explicit "sea" in section 4 where the results of a run on a computer are shown.

The animal has at time t the weight w and there are N specimens in the same group, i.e. of the same species and of the same age. The burden of a pollutant on the animal is P. The accumulated fishery yield of the group is Y. We shall use the subscript j, j_1, j_2.. to indicate the group.

3A   Primary Production

The primary production of algae is controlled by nutrient salts and sun radiation. The problem of eutrophication in the sea has been investigated by Ursin and Andersen [3] using the model of Steele [6] and [7] as adopted by Lassen and Nielsen [8] combined with the present model. In the runs of the model of the present paper, the primary production is handled as an input function outside control of man.

3B   Growth and Feeding of Animals

The growth model must provide a fair description of growth all the way from egg to adult and lead to a realistic size at old age considering the repeated loss of mass for reproduction purposes. Also it must represent growth as a function of food consumption. Beverton and Holt used the von Berthalanffy growth equation in the form of

$$\frac{dw}{dt} = Hw^{\frac{2}{3}} - kw \tag{3.1}$$

Although recognizing its analytical value in the description of metabolic processes they used it mainly to provide empirical expressions for size-at-age. In the present context a realistic description of metabolism is necessary which requires some refinements. The description below largely follows Ursin [9].

The crux of the matter is the concept of feeding level, $0 \leq f \leq 1$. By definition, an animal having access to unlimited food has f=1. When not fed at all it has f=0 whereas an animal eating a fraction f^o of what it would eat if food were in unlimited supply has $f=f^{o}$. The determination of f as a function of food availability is deferred to the next section on predation, see (3.10). The rate of food consumption is determined by:

1. A species-specific rate constant h
2. The feeding level f
3. The area of the intestine assumed proportional to w^{m} where w is the body weight. If specimens of different size but of the same species are similar bodies, then $m=\frac{2}{3}$.

For each species we then have with R, the quantity of food consumed, and t, the time

$$\frac{dR}{dt} = hfw^{m} \tag{3.2}$$

not bothering for the moment about the functional relationships of f. A fraction β of (3.2) is digested. The rate of shedding of excrements therefore is

$$\frac{d\Phi}{dt} = (1-\beta)\frac{dR}{dt} = (1-\beta)hfw^{m} \tag{3.3}$$

A fraction α of the digested food is lost in energy transfer (costs of eating, passage through semipermeable membranes, synthesis of large molecules, etc.) The "anabolic term" of the von Berthalanffy equation therefore becomes

$$Hw^{m} = \beta\frac{dR}{dt} - \alpha\beta\frac{dR}{dt} = (1-\alpha)\beta\frac{dR}{dt} \tag{3.4}$$

The "catabolic term" $-kw^{n}$ is what remains when f=0 and therefore represents the metabolic rate of a fasting animal. We now have using (3.2)

$$\frac{dw}{dt} = (1-\alpha)\beta hfw^{m}-kw^{n} \tag{3.5}$$

There is evidence from observations of weight loss in fasting animals that n ~ 5, rather than one. Using the physiologically estimated value $\frac{5}{6}$ of n makes k independent of body size. It is large in lively species like the mackerel and small in sluggish ones like plaice. In the present paper α and β are both considered "universal constants" although there is some evidence that they are both functions of f. The rate of excretion, mainly as urine, is the sum of the negative terms in the growth equation (3.4)

$$\frac{du}{dt} = \alpha\beta fhw^{m} + kw^{n} \tag{3.6}$$

3C   Predation

Ursin [10] investigated the prey size preference by cod and dab and found that on encounter with a prospective prey the predator evaluates the size of the prey relative to its own size and that the evaluation is symmetrical on a percentage scale so that, e.g., "half as big" and "twice as big" differ by the same amount from "just fine". In cod the preferred predator/prey weight ratio was estimated as approximately 100:1, and in dab about 1000:1. Formally, the prey selection can be described by a function of form as a log-normal density adjusted to a maximum value of one. With w_{j_1} the predator body weight, w_{j_2} the prey body weight and η the value of $\log_e \frac{w_{j_1}}{w_{j_2}}$, corresponding to the ratio referred to above as "just fine", we have

$$g(w_{j_1}, w_{j_2}) = \exp\{-\gamma(\log_e \frac{w_{j_1}}{w_{j_2}} - \eta)^2\} \tag{3.7}$$

which can be used as a weight on estimating the available amount of food for a specified age group and species of predator.

It may be desirable also to admit empirical information on the value of any one species as food for another, given the right size proportions. This is achieved by multiplying a constant into g to give an index of suitability of the j_2th group as food for the j_1th group

$$G_{j_1}^{j_2} = \xi_{j_1}^{j_2} \, g(w_{j_1}, w_{j_2}) \tag{3.8}$$

An obvious use for ξ is the separation of a dermersal phase from a planktonic phase.

Now define the quantity of food, ϕ_{j_1}, available to the j_1th group as the biomass of all groups j_1 each weighted by its index of suitability, $G_{j_1}^{j_2}$

$$\phi_{j_1} = \sum_{j_2} G_{j_1}^{j_2} w_{j_2} N_{j_2} \tag{3.9}$$

The principle of Beverton and Holt adopted above of allotting the same body size to all members of an age group of a given species proved unsatisfactory in trial runs with few species and age groups because some groups became temporarily starved when no food of suitable size happened to be present. This could be overcome by decreasing arbitrarily the γ of (3.7) or by replacing the power 2 by some higher (but even) power. When building the model it was attempted, however, that each parameter should have a simple interpretation and be separately estimable. The problem therefore had to be dealt with differently. This led to the realization that the natural size distribution should be incorporated for the purpose of computing the G values and for that only. As shown recently by Mr. K.P. Andersen (personal communication) the result is a natural broadening of the G values without affecting radically the behaviour of the model because the calculation of biomass is not affected. The time discrete model shown in part 2 of this paper, has the weight distribution as an essential feature.

The feeding level f should be related to the available food ϕ and in [2] the relation is suggested as

$$f_j = \frac{\phi_j}{\phi_j + Q_j} \tag{3.10}$$

where Q is a species dependent constant.

A point of paramount importance in the present model is the close functional relationship between the rates of food consumption and of predation mortality: the process of group j_1 eating out of group j_2. The predation, or grazing, mortality coefficient of group j_2 is the fraction consumed in one time interval of the biomass of j_2 by the biomass of all groups preying upon j_2

$$\hat{M}^{j_2} = \sum_{j_1} \frac{G_{j_1}^{j_2} h_{j_1} w_{j_1}^m N_{j_1}}{\phi_{j_1}} f_{j_1} \tag{3.11}$$

3D   Uptake and Elimination of Pollutants

The animals in the system are able to uptake and eliminate a persistent (non-degradable) chemical compound, which may be thought of as f.ex. DDT or PCB. The uptake takes place either directly from the water or from the food. The elimination occurs either through metabolits or as the original compound f.ex. in faeces. The compound follows the biomass in general, not any specific part of the tissue, e.g. the pancreas or the fat.

The pollutant is dissolved in the seawater with the concentration c. The contents change due to

1. Absorption in animals
2. Chemical decomposition
3. Water exchange with neighboring waters
4. Excretion from animals
5. Release from detritus
6. Man made production

The uptake in the animals is directly from the water through the surface of the animal and thus proportional to some surface of the animal most likely the gillnet, i.e. to w^n (Ursin [9]). When working with the simple von Berthalanffy equation $n = 1$, the uptake is proportional to the concentration of the pollutant c. Summing over all animals in the sea

$$\frac{dc}{dt} = - \frac{c}{V} \sum_j \rho_j^{(1)} w_j N_j \qquad \text{(V volume of the sea)} \qquad (3.12)$$

The pollutant absorbed is found in the individuals as burden of pollutant to the animals

$$\frac{dP_j}{dt} = \rho_j^{(1)} w_j c \qquad (3.13)$$

and when the predator j_1 eats some prey j_2 the compound is transferred to the predator

$$\frac{dP_{j_1}}{dt} = h_{j_1} f_{j_1} w_{j_1}^m \sum_{j_2} \frac{G_{j_1 j_2}^{j_2} N_{j_2} P_{j_2}}{\phi_{j_2}} \qquad (3.14)$$

The contents of pollutant in dead animals and eggs are transferred to a "detritus" compartment from where it is later released to the sea. The burden to detritus is P_D

$$\frac{dP_D}{dt} = \sum_j \overset{V}{M}_j N_j P_j - \rho^{(5)} P_D \qquad (3.15)$$

$$\frac{dc}{dt} = \rho^{(5)} P_D / V$$

$$P_D = P_D + (E_j - R_j) P_j$$

where E_j is the number of eggs spawned and R_j the number of recruits.

The organism is able to eliminate the pollutant either as metabolits or through faeces. In the first situation some amount of the pollutant leaves the system, while in the later it is immediately recirculated to the water. The metabolism is assumed to be

$$\frac{dP_j}{dt} = - \rho^{(4)} \frac{P_j}{w_j} \qquad (3.16)$$

Decomposition of the compound dissolved in the water is often an oxydation. It is assumed to be a 1. order reaction

$$\frac{dc}{dt} = - \rho^{(2)} c \qquad (3.17)$$

The exchanges of water with neighboring seas make some of the pollutant to disappear from the system. The process is similar to a chemical decomposition. If the incoming water is clean:

$$\frac{dc}{dt} = -\rho^{(3)} c \qquad (3.18)$$

The system is fed by man, this is an input function, an explicit function of time $U(t)$.

The physical processes of exchange of pollutants with the sediment and problems of splitting the sea into boxes of homogeneous hydrodynamic properties are ignored. The extension is in principle not very complicated if a proper hydrodynamic model of the sea is available but the model then overflows any computer.

3E Death Caused by Pollutants and "Other" Causes

The mortality caused by any pollutant is included in the "other causes" mortality $\overset{v}{M}$. When analysing contamination problems the mortality caused by the pollutant must be explicitly specified. The mortality $\overset{v}{M}$ is constituted by

$$\overset{v}{M} = \overset{v}{M}_o + \overset{v}{M}_1 \frac{P}{w} \qquad (3.19)$$

the mortality due to pollution is assumed proportional to the concentration of pollutant in the animal. The M_o mortality is caused by illness, parasites, etc. The fishery leads to a mortality F which is a man-controlled function. The total mortality equation becomes

$$\frac{dN_j}{dt} = - (F_j + \overset{v}{M}_j + \hat{M}_j) N_j \qquad (3.20)$$

3F   Spawning

Spawning of a species occurs with discrete time intervals beginning at the age t_m of first maturity, m being the number of immature age groups. Spawning causes the weight w of the spawning age groups to be reduced by the fraction π

$$w_j \text{ (after)} = (1-\pi)w_j \text{ (before)} \quad , j > m \qquad (3.21)$$

Because the model as it stands does not distinguish males from females all animals are considered females, π being adjusted accordingly. With $\underline{a}$ the number of age groups and ω the egg size for the species considered, the number of eggs spawned becomes

$$E = \frac{\pi}{\omega} \sum_{j=m}^{a} N_j w_j \text{(before)} \qquad (3.22)$$

Eventually, the eggs will have to constitute the youngest age group. Putting the number of recruits to E proved disastrous because nothing like a stationary solution could be found. A simple, consistent and in the applications effective means of controlling the population is to introduce at the moment of spawning a Beverton & Holt density dependent recruitment model. The reasoning is as follows.
We have

$$\frac{dN}{dt} = -(\mu_1 + \mu_2 N)N$$

$\mu_1 = 0$ gives $dN = \mu_2 N^2 dt$ which with N=E for t=0 gives

$$R = \frac{1}{\mu_2 t + 1/E}$$

where $\mu_2 t$ = constant when independence of time is required. Adopting this argument even in the limit t = 0 and putting $\mu_2 t = 1/E_{max}$ leads to

$$R = E \frac{E_{max}}{E_{max} + E} \qquad (3.23)$$

This means of recruitment control is extremely rigid and may be considered an emergency solution to an intricate problem. It may be fairly realistic in a few cases like herring spawning large quantities of eggs on a small area of sea bottom: the possibility that only the topmost eggs would hatch is easily perceived. When

(3.23) is the only stock-recruitment controlling mechanism it is necessary, however, to put E_{max} so small that a large part of the larval mortality is transformed into hatching mortality, particularly for big, long-lived species.

All age groups can now be moved up one number, R entering group 1. The contents of age group a-1 is added to the contents of age group a (the last one).

The eggs have a heritage of the pollutant from the parent stock. The eggs are born with the burden

$$P = \sum_j \pi_j P_j N_j / E \qquad\qquad (3.24)$$

and the eggs which do not hatch are together with their pollutant burden added to the "detritus" compartment.

4 Examples of Runs of Models

The object of this section is an illustration of the calculation procedure and an exposition of some feature of the present model. We shall use "example a" of Andersen, Lassen and Ursin [2], pp. 33-35, as this is the simplest system with all the principal elements of the total model.

This example is extended by inclusion of a single pollutant. The "sea" considered holds the following species

a	Primary producer "algae"	1 age group
b	"Euphausiacean"	4 age groups
c	"Herring"	4 age groups
d	"Cod"	5 age groups
	total	14 age groups

The last age group of each species comprises mature animals.

The spawning takes place twice a year 1. April and 1. October in the following pattern

a	Primary producer "algae"	1/4 and 1/10	
b	"Euphausiacean"	1/4 and 1/10	
c	"Herring"		1/10
d	"Cod"	1/4	

Sufficient age groups are established to allow agetransformation every half year, 1/4 and 1/10. This leaves some empty age groups

c	"Herring"	1 and 3 empty	1/4 - 1/10
		2 empty	1/10 - 1/4

and

d	"Cod"	1 and 3 empty	1/10 - 1/4
		2 and 4 empty	1/4 - 1/10

Table 1. Numerical Values of Parameters Used in All Model Runs

General Constants

$(1-\alpha)\beta$	$= 0.4$	Conversion factor: food intake to anabolism
m	$= 2/3$	Exponents in the growth equation.
n	$= 1$	
γ	$= 0.108574$	Parameters in the G, the food selectivity
η	$= 4.60517$	
ξ	$= 1$	
Q	$= 50000$	Food searching parameter
V	$= 5_{10}{}^{12}$	Sea volume
M_o	$= 0.1$	Death by "other" causes
Ψ spring	$= 1_{10}{}^{9}$	Primary production
Ψ autumn	$= 1_{10}{}^{9}$	
U	$= 5_{10}{}^{9}$	Yearly production of pollutant
$\rho^{(2)} + \rho^{(3)}$	$= 0.7$	Water exchange and chemical decomposition
$\rho^{(5)}$	$= 3$	Recycling from detritus to the water.

Species dependent Parameters

Species	Growth h	k	"Egg" weight ω	Spawning E_{max}
a	0	0	.001	–
b	10	2.5	.01	$2_{10}{}^{6}$
c	18	1.75	.02	$1_{10}{}^{4}$
d	35	0.4	.05	2

Table 1. (continued)

Fishing Mortalities applied in the time-continuous model.

Species	Age group 2	3	4	6
a	-	-	-	-
b	0	0	-	-
c	1	1	.2	-
d	0	0	0	.5

Pollutant Mortalities, Uptake and Elimination

Species	Run A+B+C $M^{(1)}$	Run A $\rho^{(1)}$	$\rho^{(4)}$	Run B $\rho^{(1)}$	$\rho^{(4)}$	Run C $\rho^{(1)}$	$\rho^{(4)}$	Run 0 $M^{(1)}$
a	0	200	175	2000	175	2000	175	0
b	70	200	175	2000	175	0	0	0
c	.7	500	325	5000	175	0	0	0
d	.7	500	325	5000	175	0	0	0

Initial Stock at 1. april after spawning used to start calculation in the time-continuous models.

Species	var.	Agegroup 1	2	3	4	5	6
a	w	.001	-	-	-	-	-
	N	$1_{10}{}^9$	-	-	-	-	-
b	w	.01	.23	.814	-	-	-
	N	$8.6_{10}{}^5$	$1.42_{10}{}^5$	$1.17_{10}{}^5$	-	-	-
c	w	.02	1.26	1.26	14.2	-	-
	N	0	3660.	0	2140.	-	-
d	w	.05	.05	54.7	54.7	295.	917.
	N	2.	0	1.14	0	1.	1.55

The fishery yield Y and the pollutant burden P are initialized
to zero. The concentration of pollutant dissolved in the sea
c is initialized to 0.00143.

Table 2. Results from run of time-continuous models (section 4), numerical values of the parameters and initial stocks are given in Table 1. Results after 20 half year; 1. April before spawning

Age-groups / Species	1 biomass	c**	2 biomass	c	3 biomass	c	4 biomass	c	6*) biomass	c	total biomass
a Run 0	6.28×10^4	–	–	–	–	–	–	–	–	–	6.28×10^4
a Run A	8.36×10^4	.0016	–	–	–	–	–	–	–	–	8.36×10^4
a Run B	1.04×10^6	.0163	–	–	–	–	–	–	–	–	1.04×10^6
a Run C	9.16×10^6	.0162	–	–	–	–	–	–	–	–	9.16×10^6
b Run 0	3.75×10^4	–	3.91×10^4	–	7.32×10^4	–	–	–	–	–	1.50×10^5
b Run A	3.98×10^4	.0017	4.03×10^4	.0017	6.87×10^4	.0017	–	–	–	–	1.49×10^5
b Run B	5.13×10^4	.0173	3.99×10^4	.0171	3.17×10^4	.0170	–	–	–	–	1.23×10^5
b Run C	$1. \times 10^{-6}$	.0620	$5. \times 10^{-7}$	.0804	$8. \times 10^{-8}$	.102	–	–	–	–	1.6×10^{-6}
c Run 0	4.63×10^3	–	–	–	6.65×10^3	–	2.59×10^4	–	–	–	3.72×10^4
c Run A	4.88×10^3	.0022	–	–	6.91×10^3	.0022	2.62×10^4	.0022	–	–	3.79×10^4
c Run B	8.08×10^3	.0224	–	–	1.13×10^4	.0222	3.37×10^4	.0222	–	–	5.31×10^4
c Run C	1.16×10^4	.0543	–	–	1.99×10^4	.0820	5.04×10^4	.188	–	–	8.18×10^4
d Run 0	–	–	66.1	–	–	–	311.	–	1.83×10^3	–	2.20×10^3
d Run A	–	–	65.8	.0022	–	–	307.	.0022	1.80×10^3	.0022	2.17×10^3
d Run B	–	–	71.8	.0222	–	–	298.	.0221	1.66×10^3	.0221	2.03×10^3
d Run C	–	–	92.2	.0741	–	–	244.	.226	598.	.441	934.

*) Agegroup 5 is empty for all species.

** Concentration of pollutant in the animals $c = \dfrac{p}{w}$

Table 2. (continued)

Fishery Yield accumulated over 1. october to 1. april.

Species		Agegroups 3	4	6	total
c	Run 0	$3.58_{10}{}^3$	$2.62_{10}{}^3$	-	$6.20_{10}{}^3$
	Run A	$3.73_{10}{}^3$	$2.65_{10}{}^3$	-	$6.38_{10}{}^3$
	Run B	$6.13_{10}{}^3$	$3.46_{10}{}^3$	-	$9.59_{10}{}^3$
	Run C	$1.03_{10}{}^4$	$5.20_{10}{}^3$	-	$1.55_{10}{}^4$
d	Run 0	-	0	465.	465.
	Run A	-	0	459.	459.
	Run B	-	0	425.	425.
	Run C	-	0	163.	163.

Weights and number of animals in the stationary state in Run 0 at 1. april before spawning.

Agegroup

Species	param.	1	2	3	4	5	6
a	w	.001	-	-	-	-	-
	N	$1.28_{10}{}^7$	-	-	-	-	-
b	w	0.221	0.620	1.08	-	-	-
	N	$1.70_{10}{}^5$	$6.30_{10}{}^4$	$6.81_{10}{}^4$	-	-	-
c	w	1.23	-	8.43	18.7	-	-
	N	$3.77_{10}{}^3$	-	788.	$1.39_{10}{}^3$	-	-
d	w	-	57.1	-	304.	-	1163.
	N	-	1.16	-	1.02	-	1.57

Account of the pollutant is kept by introducing two extra
groups
  e Pollutant contents in detritus
  f Pollutant dissolved in the sea.
The single group (species-age) is described by
  1. Weight; dry biomass
  2. Number of animals
  3. Accumulated fishery yield: dry biomass
  4. Burden of pollutant
A total of 4 independent variables per group.

The burden of pollutant to the sea and to detritus adds two
more differential equations resulting in a total of $2 + 4 \times 14 = 58$
coupled ordinary first order differential equations to be solved.

The discontinuity points of the model are the spawning moments.
The spawning for species b, c and d is of ordinary Beverton & Holt
type. The "algae" produces a fixed number of new specimens every
half year (here 10^9) which are added to the primary producer group.

The numerical values of parameters are given in Table 1. They
are extracted from [2]. The parameters describing mortality due
to the pollutant compound is chosen such that a concentration of
1 ppm in the two fish like species causes a yearly death of 50%,
$\overset{v}{M}_1 = 0.7$ while the value for "euphausiacean" is fixed to 100 times
this value, $\overset{v}{M}_1 = 70$. No mortality due to pollution is applied to
the "algae".

The rate of uptake and elimination was guessed from data given
by Jensen and Renberg [12] for shrimp and stickleback for penta-
chlorphenol (PCP, Lindane). These values are used in Run B. In
Run A the uptake is divided by 10, while in Run C the uptake and
elimination is ignored except for the primary producer where the
values of Run B are used. The uptake and elimination rate of the
primary producer were arbitrary put equal to that of the
"euphausiacean".

The initial state of "sea" was chosen as the equilibrium state
found in "example a" and the concentration of pollutant in the sea
as a good guess of the equilibrium value. All other initial
states were put to zero. The actual initial state is shown
together with the parameters values in Table 1.

A programme in BASIC using a standard Runge-Kutta-Merson
routine [11] was set up for the RC 7000 computer of the Danish
laboratory. The calculations were carried through with an accuracy
of four significant digits. The computer time required is
extensive, about 12 hours for 10 years.

The first Run 0 no pollutants exist in the system. The mean
weights, number of animals and fishery yields are given in Table 2
in the stationary state. Three other runs Run A, Run B and Run C
are shown. They are all concerned with the influence of a single
pollutant in the system. The results are shown in Table 2.

Run A is the situation with relatively slow uptake from the water compared to the elimination from the organisms. Run B has a fast uptake compared to the elimination while in Run C only the primary producer is able to absorb and eliminate the pollutant. The runs reach a stationary state after integration over five to ten years. The results are given in Table 2 as biomass and concentration of pollutant in the animal of each of the four species.

The result of the calculation example, Table 2, illustrates the competition between "euphausiacean" and "herring". The ratio between the two species is changed due to the extra mortality. imposed on "euphausiacean". It is noted that the surplus of "algae" grows with the increase of mortality in the other species. The catastrophy to "euphausiacean" in Run C does not effect the system very seriously. This is due to the unrealistic broad food selection curve applied in this example. The "herring" takes advantage of the extra mortality on "euphausiacean" and the fishery yield shifts from "cod" to "herring". In fact the total biomass fished increases markedly.

PART II
THE TIME-DISCRETE DETERMINISTIC MODEL

5   Some Model Considerations

In the present part interest is focused on modelling the long-term variations of the fish-stocks in "a generalized sea". As in the examples of the preceding section, the body of water considered is assumed to contain a plant species and three animal species.

In the multispecies models which have been described so far animals belonging to the same species/age group grow in exactly the same manner. This means that animals of group j have had identical life-histories. The idea has been to model the development in time of the average body weight in the different age groups. But, to what extent is it possible to give an adequate description of nature by means of "average animals"?

The basic model, i.e. the time-continuous model that does not take nondegradable pollutants into account, operates with high egg mortalities. If this mortality mechanism is not introduced the model explodes, i.e it results in large populations consisting of extremely small animals. It appears from Table 1 that egg mortality is placed upon species d in particular. In each spawning season a maximum of two larvae will survive, i.e. join the first year class of species d. This, of course, is not what we believe to occur in nature. Over a long period of time, it is true that each female adult animal alive on average will be replaced by precisely one female from the following generation because the size of such animal populations in nature as we are interested in

observing, neither to increase nor to decrease permanently. However, this does not mean that a one-for-one replacement occurs
between two successive generations.

The populations in the sea show relatively great fluctuations
in size. The North Sea herring for example is frequently estimated
to change from 0 to 100% from one year to the next. However, much
greater variations in population size occur every now and then.

In the 25 years period from 1945 to 1970 an extremely large
North Sea herring year-class has occurred twice. This is the
famous herring bulge of 1956 and 1960. Unpublished work from the
ICES North Sea Herring Working Group gives the following estimates
of the number of herring in billion:

1954	4.46
55	4.56
56	2.41
57	13.19
58	3.01
59	4.13
1960	1.14
61	10.55
62	4.70
63	4.00
64	5.46

This is the number of herring recruits (age approx. 2 years)
estimated from fishery data. It should be stressed that the
number of recruits always is negligible compared to the number of
eggs spawned by the parent stock of mature herring. The population
booms are therefore not caused by birthrate bulges in the winters
of 1955/56 and 1959/60. Obviously, an unusual survival of herring
larvae must have occurred in these two years. But, why?

To date, a North Sea herring bulge has not occurred in the
seventies. On the contrary, the number of recruits has been very
small in three consecutive years. Is this dangerous situation due
to a heavy or misplaced fishery effort?

We need adequate models first of all to aid in our understanding
of what determines the long-term variations in the numbers of
animals in the sea.

It appears from part 1 that multispecies modelling involves
four levels of biological organizations corresponding to organ
systems, organismic systems, population systems and, the ecosystem.
Our main interest is located to the ecosystem level where the
interactions between different populations take place. Although
there exists great uncertainty about the different mechanisms
involved in ecosystem development and control, we do not wish to
incorporate a stochastic description of each mechanism for the
sake of stochasticity. The basic model-problem relates to the
finding of relationships that need a stochastic description. That
is, to describe the uncertainty of those mechanisms that greatly
affect the long-term behaviour of the population sizes.

We need a multispecies model of great flexibility which can
serve as a basis for investigations of the importance of applying
the stochastic device. The model which is described in the
following section represents the first trial. It is a time-
discrete, extended edition of the time-continuous basic model of
part 1 (i.e. "example A" of [2], see section 4). The model is
formulated as a stochastic model in [13]. In the present paper
we consider the deterministic equivalent. The model incorporates
a weight grouping of animals. This means that the ontogeny of one
larva may differ from the ontogeny of another larva from the same
batch of eggs. Another new model mechanism is larvae dying of
starvation. If only a small amount of food is eaten by a group of
larvae then this food will not be distributed equally among the
larvae alive. A larva dies if it can not maintain its body weight.

In section 7, the behaviour of the model will be compared with
the behaviour of the time-continuous model. We shall furthermore
see what happens when the egg-mortality mechanism (see eq. (3.23))
is removed.

6. Model Description

The basis of the model is a classification of animals
according to species, age and weight. Time is treated as a
discrete variable corresponding to a step length of τ, equalling
one week in the present four-species version of the model. The
resulting tableau of groups is depicted in Fig. 6-1. There are
17 weight classes (index i) and 14 species/age classes (index j).
This gives a total of 238 weight/species/age groups which shall
be labelled (i,j): $i=1,\ldots,17$; $j=1,\ldots,14$. However, some groups
are defined to be empty. It appears from the figure that the
model consists of 135 groups. The state of the total system at
time t is thus given by specifying the number of animals (or
plants) in all groups $\{N_{i,j}(t); (i,j) = (1,1),\ldots,(17,14)\}$ in that
particular week to which t refers.

The species are similar to those described in the time-
continuous model, section 4. Species a, the phytoplankton input
to the system, is represented by age group j=1. It suffers death
due to predation from the animal groups. Species b (zoo-plankton)
covers four classes, j=2,3,4 and 5. This species spawns regularly
with intervals of half a year, starting when one year old. It
stays half a year in the first and the second age class. Mature
animals are now represented by two j-classes. The first class
comprises animals that have not yet spawned and the second, animals
that have spawned, in a given spawning season. Adult animals
therefore do not have to spawn punctually on 1st April and 1st
October in the model. A more or less arbitrary time-distribution
of spawning in the spring and in the autumn season can be applied.
However, such a spawning distribution should not extend over too
many weeks because the age shift necessarily must occur immediately

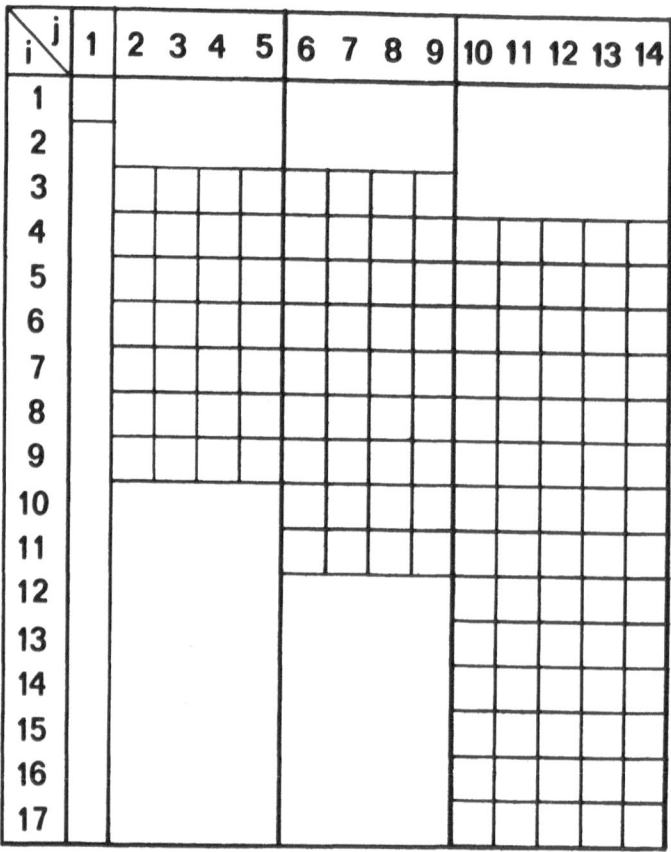

Figure 6-1   The group tableau of the time-discrete
             deterministic model. Index i denotes weight
             classes while index j mainly specifies species
             and age.

before the first week in which spawning is permitted, making the
account of age less precise.

Species c ("herring") and species d ("cod") spawn once a year.
They are therefore represented by whole year classes until they
get matured. The last two j-classes for each species are reserved
for differentiation between mature animals that are ripening and
mature animals that are spent. The j-index should therefore be
read as follows:

Species a: j=1   phytoplankton

Species b: j=2   0 - 1/2 year, immature
          3   1/2 - 1   - ,   -
          4   1 - ∞   - , mature: non-spent
          5   1 - ∞   - ,   - : spent

Species c: j=6   0 - 1 year,   immature
          7   1 - 2   - ,   -
          8   2 - ∞   mature: non-spent
          9   2 - ∞   - ,   - : spent

Species d: j=10   0 - 1 year, immature
         11   1 - 2   - ,   -
         12   2 - 3   - ,   -
         13   3 - ∞   - , mature: non-spent
         14   3 - ∞   - ,   - : spent

The weight classification is logarithmic with a weight factor
of $10^{1/2}$ between neighbouring classes. Minimum weight and maximum
weight for each species are fixed by the growth constants applied
(see Table 1). The weight in class 1 is $w_1 := 1$ mg. Class 17
corresponds to the weight 10^8 times w_1, i.e. 100 kg. In general

$$w_i = 10^{1/2(i-7)} \text{ gram,} \quad i=1,2,\dots,17 \qquad (6.1)$$

The selection of the factor $10^{1/2}$ is a proposed compromise. A
rough weight grouping spoils the possibility of obtaining an
adequate picture of the weight distribution. A fine grouping
results in complicated predation patterns making computer time
restrictive. Furthermore, maximum weekly growth of animals in
class i, calculated from the growth eq. (3.5) with f=1 and $w=w_i$,
should not exceed $(w_{i+1} - w_i)$.

In the following subsections interest is focused on the interactive mechanisms which cause the group population size to change
as time goes by. Each animal group serves as prey to all animal
groups but is also a predator itself. Throughout, subscript 2
denotes a prey group while subscript 1 characterizes a predator.

6A   Death

We consider the number of animals in an arbitrary group (i_2,j_2) during a timestep, starting at time 0 with $N_{i_2,j_2}(0)$ animals alive. The predator groups $(i_1 j_1)$ that graze on group (i_2,j_2) are determined by the index of suitability $G_{i_1}^{i_2}$ which, according to eq. (3.7) and (3.8), with $\xi=1$, exclusively is a function of $\log(w_{i_1}/w_{i_2})$, i.e. from (6.1)

$$\log\frac{w_{i_1}}{w_{i_2}} = 1/2(i_1-i_2)$$

The constants of the G-index are chosen as in section 4. The preferred predator/prey weight ratio is therefore put to 100:1, meaning that all animals in weight class $i_2 + 4$ find animals in group (i_2,j_2) "just fine". A maximum of nine predator weight classes can eat of group (i_2,j_2). These are

$$i_1 = i_2, i_2+1, i_2+2, \ldots, i_2+8$$

Note, that the predator/prey selection is a matter of size exlclusively. All prey-animals in weight class i_2 are thus exposed to the same predation pattern.

The <u>predation mortality</u> coefficient of group (i_2,j_2) is calculated in exactly the same manner as in the time-continuous model, section 3c. The difference is only that the group now is specified by two indices. According to eq (3.11) and (3.10) we obtain

$$\hat{M}^{i_2 j_2} = \sum_{i_1,j_1} \hat{M}^{i_2 j_2}_{i_1 j_1} \tag{6.2}$$

$$= \sum_{i_1,j_1} G^{i_2}_{i_1} h_{j_1} w^m_{i_1} N_{i_1 j_1}/(\phi_{i_1 j_1}+Q)$$

where N and ϕ refer to the start of the time-step (t=0). We <u>assume</u> the predation mortality coefficient to be constant during the entire step of length τ.

The <u>"other causes"</u> mortality coefficient is $M_v^{i_2 j_2}$ while the <u>fishery</u> leads to a mortality coefficient $F_{man}^{i_2 j_2}$. The total mortality coefficient of group (i_2,j_2) is therefore

$$Z = \hat{M} + \overset{\vee}{M} + F \quad , \quad 0 \le t \le \tau \tag{6.3}$$

where the group index has been omitted. Since Z is constant in the time-interval $[0,\tau]$ the number of animals in group (i_2,j_2) is governed by the differential equation:

$$\frac{dN(t)}{dt} = -ZN(t) \quad , \quad 0 \le t \le \tau \tag{6.4}$$

giving

$$N(\tau) = N(0)\exp(-z\tau) \tag{6.5}$$

as the number of animals in the group, alive at the end of the time-step. The total number of deaths is

$$N(0) \cdot [1 - \exp(-z\tau)].$$

It follows from (6.3) and (6.2) that the predator group (i_1,j_1), at time τ, has eaten the fraction

$$\hat{D}_{i_1 j_1}^{i_2 j_2} = \hat{M}_{i_1 j_1}^{i_2 j_2} \frac{1-\exp(-\tau Z^{i_2 j_2})}{Z^{i_2 j_2}} \tag{6.6}$$

of the $N_{i_2 j_2}(0)$ animals present in group (i_2,j_2) at time 0.

6B   Growth

Group (i_1,j_1) has consumed the food

$$T_{i_1 j_1} = \sum_{i_2 j_2} w_{i_2} N_{i_2 j_2}(0) \, \hat{D}_{i_1 j_1}^{i_2 j_2} \tag{6.7}$$

in the time-interval $[0,\tau]$. In the beginning the food is distributed evenly among the animals _alive_ in group $(i,j) = (i_1,j_1)$ at time τ, i.e.

$$R_{ij} = T_{ij}/N_{ij}(\tau) \tag{6.8}$$

is the food consumed per animal. Animal growth is computed from the growth eq. (3.5) as

$$\Delta w_{ij} = (1-\alpha)\beta R_{ij} - \tau k_j w_i^n \tag{6.9}$$

If $\Delta w_{ij} = 0$ the $N_{ij}(\tau)$ animals stay in weight class i. But, if $\Delta w_{ij} \neq 0$ a redistribution of the food (T_{ij}) must take place due to the weight grouping. The animals are classified in two groups. One group comprises animals which just maintain their body weight, i.e. this animal group stays in weight class i. The second group comprises animals which move to the neighbouring weight class.

If $\Delta w_{ij} > 0$

$$\frac{\Delta w_{ij}}{w_{i+1}-w_i} N_{ij}(\tau)$$

animals grow into weight class i+1.

If $\Delta w_{ij} < 0$

$$\frac{\Delta w_{ij}}{w_{i-1}-w_i} N_{ij}(\tau) \qquad\qquad (6.10)$$

animals fast and are moved down to weight class i-1. Numbers of animals are obviously considered as continuous variables. The two cases, corresponding to maximum and minimum weight of species/age class j must be dealt with separately.

If $\Delta w_{i_{max}j}$ is positive we <u>assume</u> that the metabolism has been increased by $\Delta w_{i_{max}j}$ in the time-interval $[0,\tau]$, meaning that all animals alive stay in the maximum weight class.

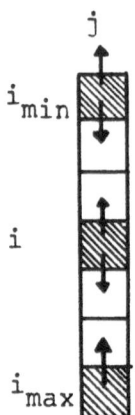

Figure 6-2. The possible movements of animals which can occur at the end of a time-step. It appears from Fig. 6-1 that the j-class belongs to species b.

Note: The top arrow represents deaths of starvation.

6C   Starving to Death

We <u>assume</u> that the small animals in the minimum weight class suffer death if they can not maintain their bodyweight. In the case $i=i_{min}$ eq. (6.10) is replaced by the following:

If $\Delta w_{i_{min}j} < 0$

$$\frac{-\Delta w_{ij}}{\tau k_j w_i^n} N_{ij}(\tau) \qquad , \; i=i_{min} \qquad\qquad (6.11)$$

animals of group $(i_{min,j})$ will die of starvation.

To see this more clearly, let

$$N_o = N_s + N_d \equiv N_{ij}(\tau) \qquad , \qquad i=i_{min}$$

N_d is the number starving to death and N_s the survivors. For short, introduce $v=(1-\alpha)\beta$ and K for the "catabolism" term, i.e. the denominator of (6.11). Multiplying (6.9) by N_o we obtain

$$N_o \Delta w = (vT - N_s K) - N_d K \qquad\qquad (6.12)$$

The total food consumed, T, is used to maintain the survivors body weight, i.e. the bracket equals zero, so

$$N_s = \frac{vT}{K}$$

and consequently

$$N_d = \frac{-N\Delta w}{K}$$

In accordance with (6.11).

This starvation mechanism has been introduced in the model because larvae alive grow very fast as has been reported by Rodney Jones [14]. It is therefore the first group of each species which have our interest, i.e. the groups (2,3), (6,3) and (10,4) – see Fig. 6-1. The food intake requirement for a larva to stay alive can easily be strengthened. To require in (6.12) that

$$(vT - N_s K) = pw N_s$$

means that a larva must $(p-1)$ double its weight in the period of time considered (τ) if it shall stay alive. This leads to

$$N_s = \frac{v T}{K + pw}$$

and   (6.14)

$$N_d = \frac{-\Delta w + pw}{K + pw} \cdot N_o$$

instead of the equations (6.13) for $p=0$. An appropriate p-value for the first week of the larval development must at least be one. But, as the larva grows it tolerates fasting better. The parameter p should therefore be a decreasing function of the body weight. However, such features have not yet been incorporated in the model. The present model operates with the $p=0$ requirement (6.13).

The amount of food lacking in order to keep all the N_o animals alive is

$$\Delta T = \frac{(-\Delta w + pw)N_o}{v}$$

according to (6.14) it causes the biomass

$$N_d w = \Delta T \cdot \frac{vw}{K + pw}$$

to die, i.e. to leave the system.

6D   Spawning

Immature animals are moved one age class up immediately before the start of a spawning season. Maturity is considered to be a function of <u>age as well as of body weight</u>. Animals of "mature age" ($\geq t_m$) with "mature weight" ($\geq w_{im}$) spawn a fraction π of their body weight. Eq. (3.22) gives the number of eggs spawned. The situation is illustrated for species b in Fig. 6-3.

7.   Example of Runs of Model

The object of this section is to illustrate some features of the time-discrete model. We shall use "example A" of Andersen, Lassen and Ursin [2], pp. 33-35 as a standard of reference. The numerical values of model parameters are therefore essentially those of section 4 given in Table 1.

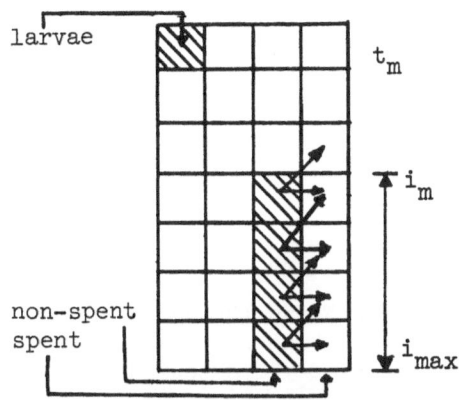

larvae

t_m

i_m

non-spent
spent

i_{max}

Figure 6-3.
Species b under spawning. The
four mature groups are shaded.
Animals which have spawned are
immediately transferred to the
"spent class" with reduced body
weight. The first group (shaded)
is supplied with larvae corresponding to the biomass spawned
unless, of course, egg mortality
is introduced. When the spawning
season terminates, all "non-spent"
mature groups will be empty.
Animals in the "spent" mature
groups are transferred to the "non-spent" groups simultaneously with
the age shift before the next spawning season. The start of a
spawning season is therefore characterized by empty groups in the
spent domain and in the first age class.

We confine our attention to a few important problems
illustrated by the following questions.

Question 1: Is the stationary* solution of "ex A"
comparable with the "average behaviour of the model"?
This question arises because the model operates with growth of
single animals and not with growth of "average animals" as in
example A, cf. section 5.

In ex A it was necessary to keep the stock of "cod" small
enough for them to be fed. This is achieved by a heavy restraint
on E_{max} for species d, cf. the discussion in section 3F. Perhaps
such artificial restraints can be avoided in the present model.

Question 2: Does a reasonable stationary solution
exist when there is no egg mortality at all?
The next question relates to mature animals. In section 6D,
maturity was considered to depend on age as well as body weight.

---

* Apart from seasonal variations, of course.

Question 3: To what extent is the behaviour of the
model affected by the weight-requirement for maturity?
Species a, phytoplankton, constitutes the food supply for larvae in
particular.
Question 4: Is relative model behaviour influenced
by the amount of phytoplankton available?

7A Fixed Variables

The model has been programmed in FORTRAN and the calculations
were carried out on an IBM 370/165 computer. The step length of τ
was fixed to one week. Computer time for a 50 years period
amounts to 1 min. (cpu).

The mortality M by other causes is zero for the algal group
(1,1) and .1 per year for all others. We attempt to choose fishing
mortalities in accordance with Table 1, i.e. heavy "young herring"
fishery ($F=1.$ year^{-1}), modest "adult herring" fishery ($F=.2$ year^{-1})
and medium fishery for adult "cod" ($F=.5$ year^{-1}). The groups
selected for fishery are shown in Fig. 7-1.

The predator patterns are fixed by the G-function, see
section 6A. Parameter values are adopted from Table 1.

i \ i	1	2	3	4	5	6	7	8	9	10	11	12	13	14
1														
2														
3														
4														
5														
6														
7														
8														
9						1.	1.	.2	.2					
10						1.	1.	.2	.2					
11						1.	1.	.2	.2					
12														
13													.5	5
14													.5	5
15													.5	.5
16													5	.5
17													5	5

Figure 7-1.
Fishing mortalities per year.
Fishery is placed upon 10-100
gram "herring" and age-mature
"cod" with minimum body weight
1 kg.

Parameters for growth are given in Table 1 (h and k in year^{-1}). The mechanism of starving in eq. (6.11) is applied. Spawning occurs 1. April (species b and d) and 1 October (species b and c) - see section 4. Species a is poured into the sea in equal quantities on 1 April and 1 October. What is left after half a year disappears.

The initial state of the "sea" is principally the same as in part 1, i.e. as the equilibrium state of ex A - see Table 1. However, a translation must take place due to the weight grouping. Table 3 shows the numbers in each group on the starting day which is a 1st April after spawning. The time-horizon is set to 50 years.

7B Trial Runs

Five parameters are chosen as independent variables for the experimentation conducted with the model. The first being $N_{1,1}$, the number of phytoplankton units present 1 April and 1 October. The second variable is an indicator of egg-mortality defined as

$$\hat{E} = \begin{cases} 0 & \text{"no egg-mortality"} \\ 1 & \text{"egg-mortality"} \end{cases}$$

Thus, if $\hat{E}=0$ all eggs spawned are hatched and join the first-group as larvae; R=E. If, alternatively, $\hat{E}=1$, the number of recruits is governed by eq. (3.23). The last three independent variables, i_m (species) denote for each species the minimum weight class of mature animals. The lowest possible value of i_m is $i_{min}+1$ because some of the animals from class i_m are moved down to class i_m-1 when they are spent, cf. Fig. 6-3. The five independent variables are listed in Table 7-1.

The dependent variables at time t correspond to the group sizes, i.e. $N_{i,j}(t)$. The aim of the experimentation is to provide information about the variation of the dependent variables as a function of different values of the independent variables. The design of experiments is shown in Table 7-2.

Presentation of Results

The results are shown as graphs on Figures 7-2 to 7-5 and placed at the end of this section. Each run is represented by 18 graphs: 10 for the average weight in each animal age-class, the last two classes of each species being combined; 6 for total population sizes and number of age-mature animals of each species; and 2 for animal biomass and total biomass. Figure 7-2 gives a total picture of the first three runs while Figures 7-3 to 7-5 show what is believed to be the stationary solutions. The graphs are made with automatic scaling meaning that <u>different units are applied</u>.

Table 3. a) Initial state used in all model runs of part 2. (1. April after spawning)
b) Equilibrium state for run 1 of part 2. (1. April after spawning in year 50)

Table 7-1

Independent Variables

Variable	Symbol	Type	Domain of variation
Phytoplankton present 1 Apr. and 1 Oct.	$N_{1,1}$	Real	$[0,\infty]$
Indicator of egg-mortality	$\hat{E}$	Integer	0 or 1
Minimum weight of maturity for animal species	$i_m(b)$ $i_m(c)$ $i_m(d)$	Integer	$[4, 9]$ $[4, 11]$ $[5, 17]$

Table 7-2

Design of Experiments

RUN	Independent Variables				
	$N_{1,1}$	$\hat{E}$	$i_m(b)$	$i_m(c)$	$i_m(d)$
1	10^9	1	4	4	5
2	10^9	0	6	8	13
3	"	"	"	"	14
4	10^8	"	"	"	13
5	"	"	"	"	14

Comments on Run 1. This run represents a situation very close to "ex A". It appears that the system has been started near its equilibrium state, meaning that the answer to Question 1 is yes. The most marked difference between the start situation and the equilibrium state occurs for species d. The number of "cod" increases but their body weight decreases. However, we could not expect to obtain exactly the same equilibrium state as in ex A because the same food selection curve was applied. The food available for large animals will therefore be greater in the time-discrete model (weight grouping, i.e. many animal groups) than in ex A (few "average" groups). But it would have been difficult to predict in which way the discrepancies would occur.

The more detailed picture in Fig. 7-3 shows that average growth can be followed from one age shift to the next. Table 3 gives the population sizes at 1 April after spawning in year 50. A brief investigation shows that the logarithmic weight distributions are close to Normal distributions.

Model runs have been carried out with grossly different initial data but with the same parameter values. The stationary solution given in Table 3 was always reached after runs of 20-30 years.

Comments on Run 2. Compared with run 1, two changes have been introduced. However, the situation in run 2 in principle remains unaltered if the weight requirements for maturity are omitted. With no egg-mortality species b and c become extinct approximately at the 10. year. But, the "cod" manage in the long run through cannibalism.

The first generation of cod comprises 2 recruits, cf. Table 3. However, in the next year all the eggs spawned are hatched which results in a generation consisting of approximately 7000 recruits. When this generation, three years later becomes age-mature there are only 110 animals alive; 20 animals are matured. They give rise to a large 5. generation consisting of nearly 100,000 recruits - see the first peak on the graph.

It appears that the population of cod becomes stationary after 20 years. The equilibrium state is reasonable because it consists of an animal population with appropriate body weights in all age classes.

Comments on Run 3. It is now more difficult for the cod to manage because its body weight shall be greater than approximately 3 kg before it becomes matured. The picture changes drastically when compared with run 2. The "herring" is eaten. Species b and d show Lotka-Volterra oscillations. The period of cycle being approximately 25 years. Note that the population of cod begins to increase in the 50. year. The answer to Question 3 is obvious. Note that the cod is exposed to fishery when its body weight is 1 kg, i.e. one weight class below the requirement for maturity.

Comments on Run 4 and Run 5. With reduced food input we obtain 10 years oscillations of herring & cod while species b becomes extinct (run 4). When the weight-requirement to cod is raised one class, the herring is the only species that survives

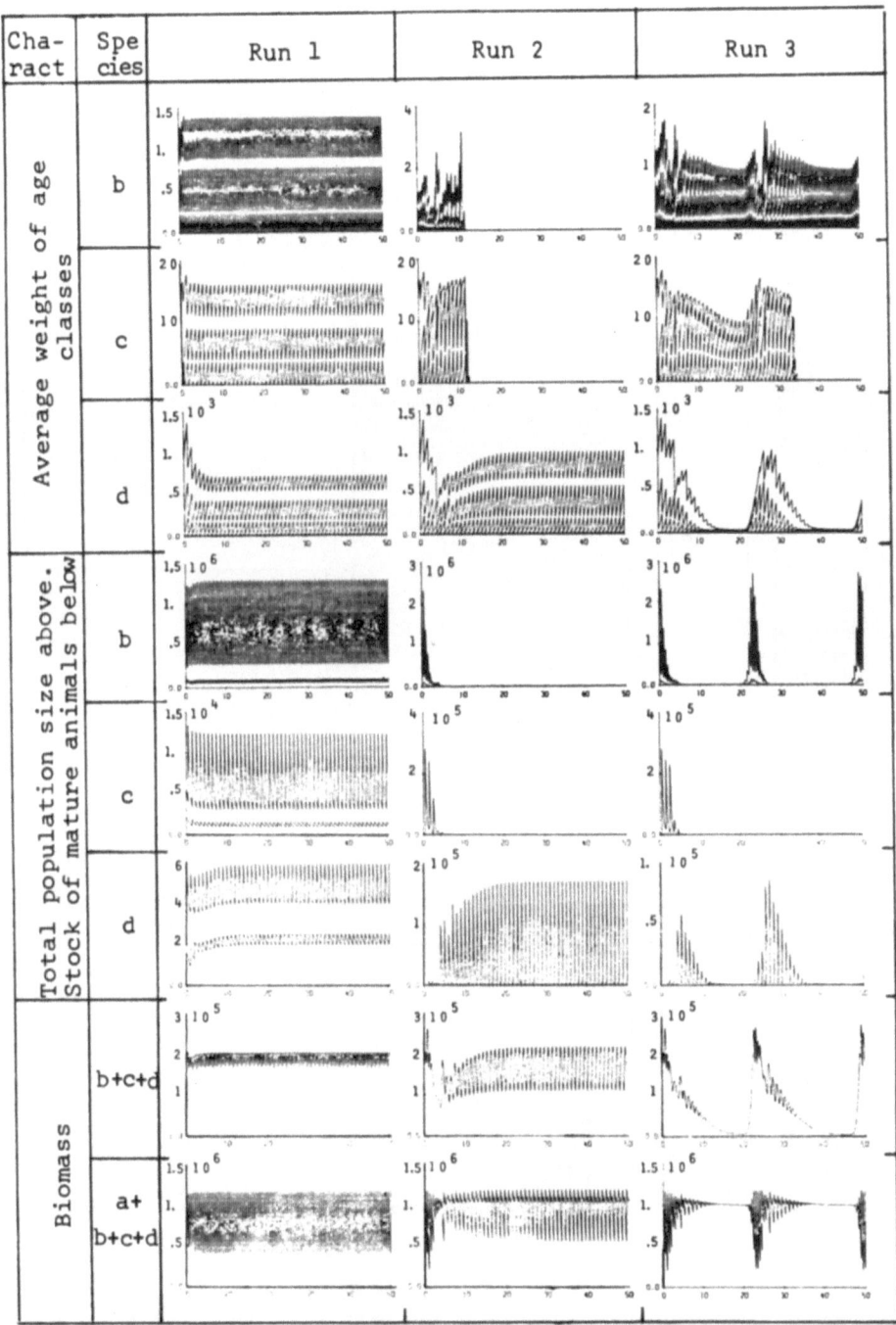

Figure 7-2. Three runs over fifty years - see the run-specification in Table 7-2. Note that different units are used in particular for the number of "cod," i.e. species d.

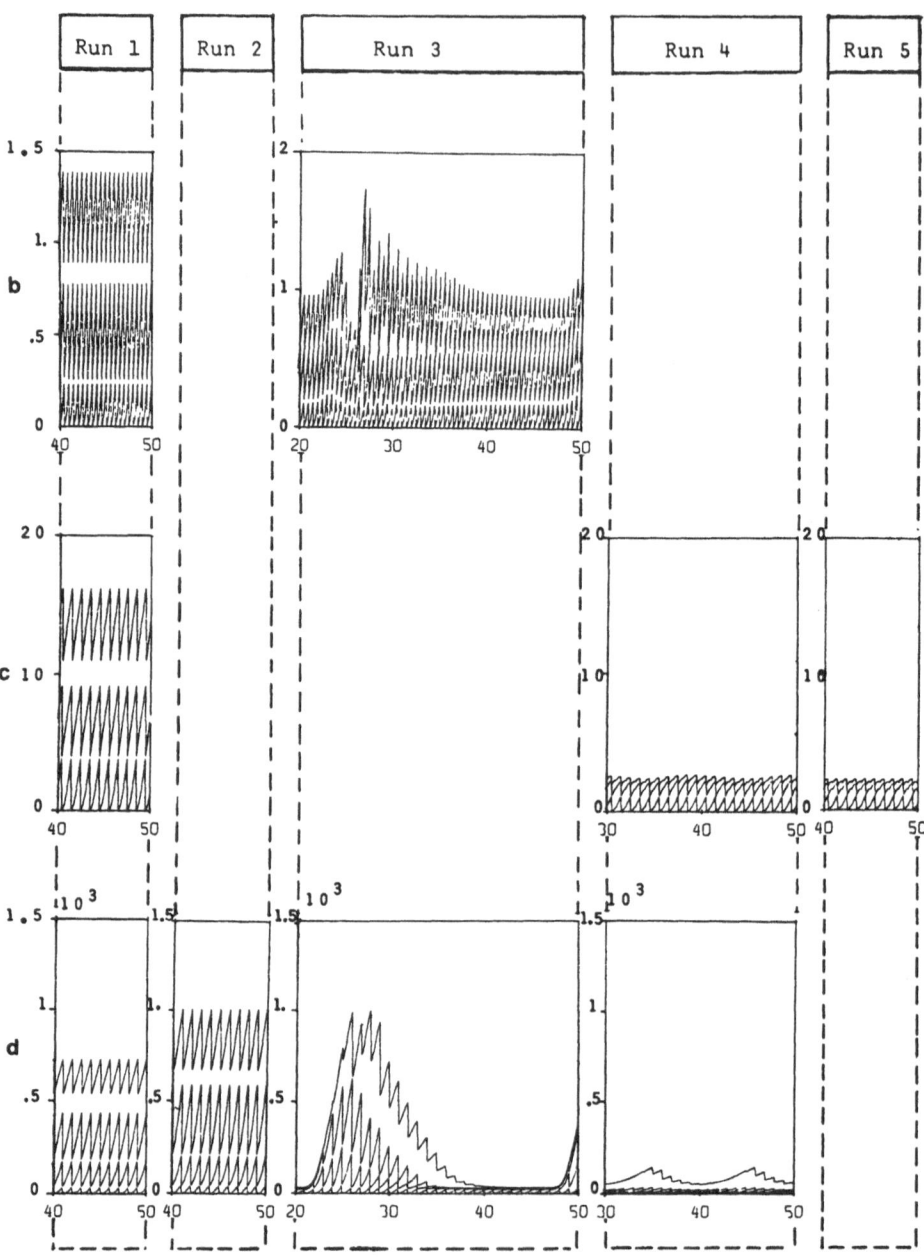

Figure 7-3. Average weight of each animal species/age class in steady state. Note the different units for species b.

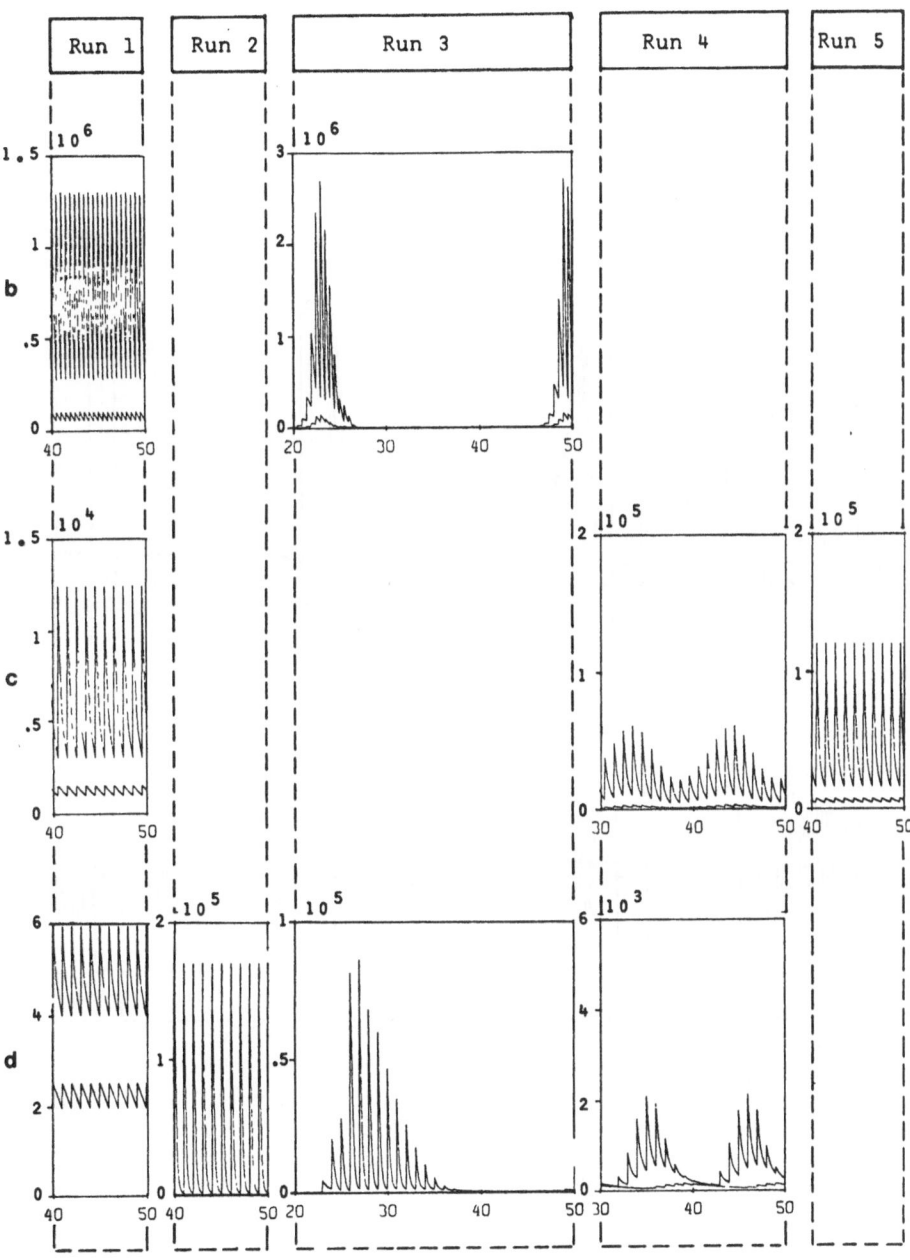

Figure 7-4. Population size of each animal species and the stocks of age-mature animals in steady state. Note the different units.

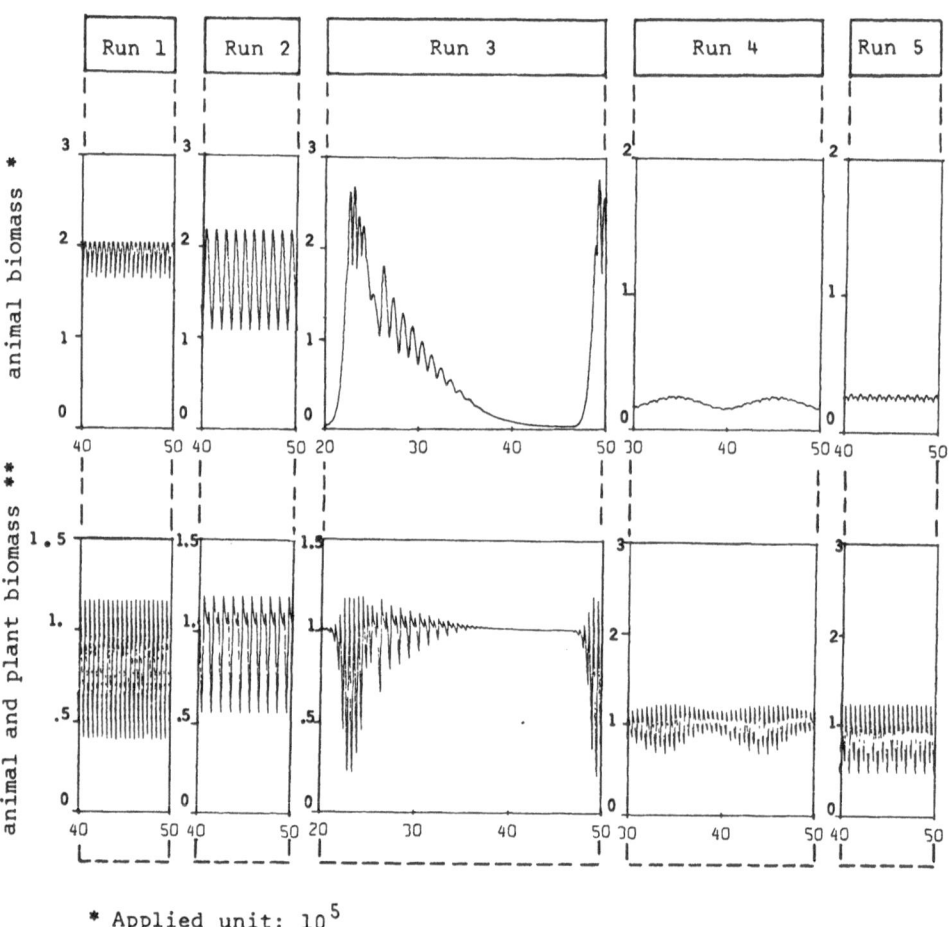

* Applied unit: 10^5
**   -   - : 10^6

Figure 7-5. Total biomass in steady state.

(run 5). These runs show completely different model behaviour compared with the runs No 2 and 3.

The mechanism of starving to death has been totally unimportant in the first three runs. In runs 4 and 5 the mechanism has been active for species b, but only in the first year where 30,000 larvae starve to death. However, this number corresponds to only 5% of total larval mortality.

In the present five runs all spawning occurs at 1 April and 1 October. In nature, however, spring and autumn plankton outbursts are not always synchronized with the spawning seasons. It may be disastrous to the recruits if the plankton outburst is delayed for several weeks. The mortality "starving to death" is presently being investigated with phytoplankton poured into the sea according to more realistic rules than those applied in the present paper.

Symbol Glossary

t	Time
τ	Length of step in the time-discrete model
w	Body weight
N	Number of animals in a group
N_d	Number of animals in a group starving to death during a step of time τ
Y	Accumulated fishery yield
P	Burden of pollutant to a single animal
P_d	Burden to pollutant to detritus
C	Concentration of pollutant dissolved in the sea-water
$\check{M}$	Total instantaneous mortality coefficient due to "other" causes: $\check{M} = \check{M}_o + \check{M}_1 P/v$
$\check{M}_o$	Instantaneous mortality coefficient due to "other" causes
$\check{M}_1$	$M_1 P/w$ represents instantaneous mortality coefficient due to pollution
$\hat{M}$	Instantaneous mortality coefficient due to predation

$\hat{D}$ Mortality coefficient in a time-step due to predation

Z   Total instantaneous mortality coefficient

F   Instantaneous fishery mortality

U   Man made production of pollutant per year

$\alpha,\beta,h,$ Coefficients in the growth equation
k,n,m

v $(1-\alpha)\beta$

f   Feeding level

T   Food consumed of a group in a time-step τ

R   Food consumed of an animal; Number of recruits

Φ Shedding of excrements

u   Excretion

ϕ Biomass available as food to the group

G,g   Food selection functions

γ,η,ξ Parameters in the G-function: $G = \xi g(\gamma,\eta)$

Q   Food searching function: $f = \phi/(\phi+Q)$

V   Volume of the sea

$\rho^{(i)}$ Parameters in uptake and elimination of pollutant

E   Number of eggs spawned

E_{max} Maximum number of eggs hatched

ω Weight of a single egg

μ_1,μ_2 Parameters in the recruitment model

i   Weight class index used in the time-discrete model

i_m Minimum weight for mature animals

j   Species/age class index

Acknowledgement

Most of Jan Beyer's work was carried out with support from Danish Natural Sciences Research Council. He wants to thank N.H. Hansen at the Institute of Mathematical Statistics and Operations Research for useful discussions. The authors want to thank K.P. Andersen and E. Ursin at the Danish Institute for Fishery and Marine Research for much criticism and inspiration.

References

1   Beverton, R.H.J. and Holt, S. (1957): On the Dynamics of Exploited Fish Populations. Her Majesty's Stationary Office.

2   Andersen, K.P., Lassen, H. and Ursin, E. (1973): A Multispecies Extension to the Beverton & Holt Assessment Model with an Account of Primary Production. C.M. 1973/H:20. Revised 1.9.1973.

3   Ursin, E. and Andersen, K.P. (1975): A Model of the Biological Effects of Eutrophication in the North Sea. Presented to the ICES Symposium "The changes in the North Sea Fish Stocks and their Causes", No. 44 (1975).

4   Andersen, K.P. and Ursin, E. (1975): A Multispecies Analysis of the of the Effects of Variations of Effort upon Stock Composition of Eleven North Sea Fish Species. Paper No. 45 (1975) presented as (3).

5   Lassen, H. (1975): A method of predicting the biological effects of synthetic pollutants in the North Sea. Presented as (3).

6   Steele, J.H (1958): Plant Production in the Northern North Sea. Mar. Res. Scot. 7 36 pp.

7   Steele, J.H. (1962): Environmental Control of Photosynthesis in the Sea. Limnol. Oceanogr. 7 137-150.

8   Lassen, H. and Nielsen, P.B. (1972): Simple Mathematical Model for the Primary Production as a function of the Phosphate Concentration and incoming Solar Energy applied to the North Sea, C.M. 1972/L:6

9   Ursin, E. (1967): A Mathematical model of some aspects of fish growth, respiration and Mortality. J. Fish. Res. Bd. Canada, 24 (11) p. 2355-2453.

10   Ursin, E. (1973): On the Prey Size Preferences of Cod and Dab. Meddr Danm. Fisk.- og Havunders. N.S., Vol. 7, pp. 85-98.

11.   Sparre, P. and Lassen, H. (1972): Ordinary First Order
       Differential Equations. Runge – Kutta – Merson Method.
       User's Manual 4/9 1972. Danish Institute for Fishery and
       Marine Research.

12.   Jensen, S. and Renberg, L. (1974): Uptake and Elimination
       of some Organochlorine Hydrocarbon Compounds in Marine
       Organisms. Presented to Oslo Convention Working Group on
       the Degradability of substances in annex I, §§ 1,2. Paper
       No. 4. Lowestoft Spring 1974.

13.   Beyer, J.E. (1975): Ecosystems – An Operations Research
       Approach. The Institute of Mathematical Statistics and
       Operations Research, Denmark. To appear.

14.   Jones, R. (1970): Density dependent regulation of the
       numbers of cod and haddock. The ICES symposium on "Stock and
       Recruitment" No. 27 (1970).

A VISION OF A FUTURE HOME RULE CULTURAL POLICY FOR GREENLAND COMPARED WITH PRESENT CONDITIONS

Father Finn Lynge

Head of Greenland Radio

Godthaab, Greenland, Denmark

FUTURE POLICY	PRESENT CONDITIONS
A bi-lingual nation, ruled according to the principles of representative democracy, with close ties to the social and cultural traditions of the Scandinavian nations.	Present status.

Language policy

The West Greenland dialect of the Inuit tongue is the nation's official first language.	Generally speaking, this will not be contradicted, but in actual fact Greenlandic is far behind in asserting its position as the nation's first language.
Danish is the nation's second language, and until the Greenlandic language has evolved to meet the demands of modern civilization, it will be the main language of administration as well as of higher education.	Same.
The evolution of Greenlandic to a stage corresponding to the other Nordic languages is to be encouraged by all responsible authorities.	Same.

The orthography will be chosen so
as to meet the demands of computerized data-processing as well as
to provide the best possible
vehicle for pan-inuit
communication.

The new standard orthography
meets these demands.

Policies of media

A. Printed media

One weekly nationwide publication
edited in Godthab.

Present status.

Three weekly district publications, one edited in a southern
town, one in Godthab for the
central part of the west coast
and one in a northern town.

Same.

Local papers up to the private
initiatives that may manifest
themselves in the various towns
and settlements.

Same.

The editorial policy of the one
nationwide and the three district
papers will be laid down by a
Committee for Information and
Culture directly responsible to
the Home Rule governing body.

There is no unified editorial
policy for the existing nationwide and district papers, and
there is no Committee for
Information and Culture.

B.   Electronic media

The evolution of Radio and TV
will be based on the principle
of technical decentralization and
the broadest possible contact
to the people, whereas the
responsibility for the contents
of the transmissions will be
centralized in the non-political
Committee for Information and
Culture.

An evolution of this kind is
possibly under way.

Regional radio studios in north and south - besides the central broadcasting station in Godthab - with built-in technical possibilities of simultaneous broadcasting to the whole nation from each and any of the regional studios for the purpose of creating mutual understanding and closeness between the different and geographically isolated parts of the population.

An evolution of this kind seems definitely under way.

Privately owned cooperative cable TV nets in the various towns and settlements - for the purpose of involving people's own sense of responsibility - linked together by a state-financed UHF-chain which serves as a vehicle for simultaneous TV broadcasting along the coast of programmes edited in the central TV studios in the capital - the UHF-chain opening possibility for nationwide broadcasting also of TV programmes produced locally in the various towns.

This may possibly be realized in the years to come, labeled "solution no. 4" in the report from the Greenland TV Commission now being published.

Responsibility for editing of Radio and TV programmes placed in the capital with the Home Rule Committee for Information and Culture.

Editorial responsibility for the Radio centralized in Godthab under a Radio Committee responsible to the Ministry of Greenland. At the moment, no kind of centralized editorialship for TV, all cable TV nets being owned by private coops.

C. Media language policy

Greenlandic the first language, Danish the second language. Both languages will be used when communicating about matters of internal national importance to both language groups. Greenlandic alone will be used as a vehicle of general adult education and in the coverage of affairs external

There exists no coordinated media language policy. In the one nationwide weekly paper, the language distribution is 50-50, each and every item being printed in both languages. In at least one local paper, practically everything is printed in Greenlandic only.

to Greenland, the Danish language
group having already sufficient
access to Danish and foreign
language sources of information
in these matters.

In the Radio, presently 70%
of the broadcasting is in
Greenlandic and 30% in Danish.
In TV, sporadic efforts have
been made to produce and/or
version some programmes in
Greenlandic, otherwise everything is being tapped from
the Danish TV tel quel. No
general rules of the type
mentioned have ever been
formally set down as the basis
for the media language
distribution.

Book publishing policy

In close cooperation with the
printed press and the electronic
media, the Greenlandic Publishing
Company will provide the broadest
possible spectrum of reading
material in comics and coloured
picture magazines, in recognition
of the fact that unless the
people is reached where it is
found, no foundation can be laid
for any future improvement of
the general reading habits of
the Greenlandic middle class.

The Greenlandic Publishing Co.
provides a rather narrow
spectrum of reading material
and definitely shys away from
comics and coloured picture
magazines. Publications are
traditionally limited to
fiction and poetry, recently
however with a growing interest
for children's books also.
The publishing policy is also
understandably limited by lack
of funds and manpower.

Art support policy

The Home Rule governing body as
well as the municipal governments
will in their budgets set money
apart for the purpose of buying a
number of Greenlandic art works
every year and will see to it
that they are kept within the
boundaries of the country and
placed so as to be seen by the
greatest possible number of people.

Nothing of the sort is being
done systematically, but
attempts are being made in
this direction.

Direct financial support to
worthy artists will be
administered from several sources
and by a special board appointed
by the Committee for Information
and Culture. All the members of

The Ministry for Greenland and
the Greenland Council set
apart a certain small amount
of money for cultural purposes
in their yearly budgets for
which individual artists may

this board will be periodically changed.

The artists will have their say in every town planning committee and in every major building project and will be kept responsible for the aesthetic quality of the human made environment.

apply. Direct financial support is also given occasionally by the artists or artistic activities.

All town planning and major building projects are being run by the Danish "Greenland Technical Organization" with complete planning worked out in Copenhagen and with no kind of consultation of local artists. Consultation of the local municipal governments does take place, but seems all too often to leave no real room for local influence on the plans.

Educational policy

In the children's school, the language of instruction will be Greenlandic. At the high school level, part of the subjects will be taught in Greenlandic, part in Danish, the demarcation line between the two languages being set by the internal capacity of the Greenlandic tongue to cope with the subject matter in question. At the college and university levels, all instructions will be given in Danish or in foreign languages, except for subject matters touching directly upon the Inuit culture and heritage.

As far as the children's school and the high school levels go, Greenlandic politicians have settled for the policy outlined, the general problem being only the lack of a sufficient number of native teachers. As far as the college and university levels go, things haven't evolved that far as yet.

The academic as well as the professional training will aim at a level of knowledge and skills suited to the requirements of Greenlandic society, not dictated by traditional university and unions' demands in Denmark.

This seems to be the trend that's favored especially by the younger generation of Greenlandic poloticians. The issue, however, has never been formally debated, especially as far as the academic formation goes.

The graduation level at which   Same.
terminates the academic and
professional training of this
country will however contain a
built-in basis suitable for
voluntary further training of a
kind that terminates at the
traditional Danish graduation
level for the subject matter in
question.

At all levels, from first grade   The basic and urgent need for
to university courses, the basic   a formation aiming above all
aim of the formation of the   at flexibility and adaptability
Greenlandic mind will be,   is only dawning upon us all.
regardless of the IQ-level, to
cultivate its flexibility and
adaptability to the ever-
changing current of evolving
standards and demands of the
outside world, preparing future
generations to a world, the only
foreseeable feature of which is
an ever accelerating aculturation.

Church policy

The Home Rule will favor an   The Evangelical-Lutheran Church
independent diocesan status for   of Greenland has no bishop of
the traditional Evangelical-   its own, only a dean, who is
Lutheran Church.   subject to the bishop of
                                     Copenhagen.

The Evangelical-Lutheran Church   Present status.
of Greenland, having been
mothered by the Danish Protestant
state church, is however only
half-way Protestant itself, in as
much as the Christian believers
of this country - which was
christianized only after the 16th
century - have never gone through
the protest against the universal
church. Among the believers of
Greenland, this fact makes for a
special openness and lack of
prejudice toward other Christian
denominations, which sets them
apart among the Lutheran churches
of the Nordic countries as

ecumenical minded in a special
way.

This ecumenical charisma will be
cultivated as a growing awareness
of the Christian values embodied
also among believers other than
Scandinavian Protestants, and the
diocese of Greenland will on its
own accord aim at sacramental
fellowship with the universal
church.

At the present, the believers
at large are not much aware of
spiritual values elsewhere in
the world.

The church will foster a sense
of social involvement and
responsibility among the believers
and will represent a source of
personal renewal and universal
spiritual outlook for the
generations to come.

The church doesn't seem to
sufficiently foster a sense of
social involvement, although
initiatives in this direction
are under way.

Social policy

Regardless of the stage of
general disintegration of a given
individual, family or local
community, nobody will be allowed
to lack the basic necessities for
life: food, clothing and lodging.

Present status - although
severe housing problems are
making themselves felt.

Social welfare will be constructed
after the principle: help to self-
help, and the welfare system will
be constructed so as to prevent
people from taking advantage of
the public subsidies out of sheer
laziness or lack of cooperation
to the above principle.

Present status - although in
practice some abuse is taking
place.

Various levels of social worker
training programs will be laid
out in cooperation with the public
school system and will be
conducted under the auspices of
the teachers' college training
center, thus securing a mutual
contact between the social
welfare training and humanistic
schooling, fostering a wider

Present status

outlook and a greater flexibility
of thought in both groups.

Tax policies will be constructed so as to favor regular marriages and protect family life.	Tax policies are constructed so as to discourage regular marriages and endanger family life.
The central welfare administration office under the Home Rule governing body will ensure a fair and even policy in all municipalities.	The present welfare administration office is doing its best in this direction.
The same office will be given full authority to demand and effectuate a total ban on alcohol consumption during working hours in all public institutions and workshops.	Authorities have an extremely wishy-washy attitude the way things are at present - words, words, words and no determination to face the real problems where they are found.

General policy of cultural belonging

The Davis Strait constitutes the frontier between the Scandinavian family of nations and the North American civilization.	Present status.
On its own premises, Greenland will cultivate its ties to Denmark and the other Scandinavian nations so as to ensure a national fellowship eastward across the North Atlantic basin.	Whichever direction the internal evolution in Greenlandic society takes, this will probably be the momentum of the need for contact to the outside world.
Programmes of exchanges to the west will be conducted so as to cultivate a state of neighborhood friendship with Canada and the U.S. on the special historical premises of the family of the Inuit.	This has been taking place for a number of years already.

DYNAMIC MODELS OF THE SOCIO-ECONOMIC ROLE

OF TELECOMMUNICATIONS TECHNOLOGY

J. C. Beal and L. E. Peppard

Queen's University

Kingston, Ontario, Canada

INTRODUCTION

During the past three years a multidisciplinary research team at Queen's University has been engaged in a study of the usefulness of dynamic models in the evaluation of the impact of telecommunications on developing regions. These dynamic models are similar in concept to those introduced by Forrester [1]. When based solidly on multidisciplinary investigations, they enable a wide diversity of information about a region or community to be brought together into a dynamic whole that has the potential to illustrate clearly the way in which the many facets of its socio-economic behaviour are interacting over a period of time. The model's behaviour is, of course, no more than the end result of all the hypotheses built into its structure by its builders, with the major advantage that such structure is explicitly described and hence open to critical inspection and possible modification by the builders and the users. A key feature of the models is the incorporation of a large number of "feedback" loops to describe these socio-economic interactions over a period of time.

Since the focus of this project is on telecommunications technology, the models developed include explicit descriptions of such items as telephone installations, telephone usage, and broadcasting. At an early stage, however, it was recognized that an adequate description of their roles in a region or community would inevitably involve an overall view that included, as well, such aspects as the economy, demography, transportation, social services, and, particularly in the case of certain communities, socio-cultural components.

The region chosen for investigation in this study at an Ontario university was northwest Ontario, including the districts

of Kenora, Rainy River, and Thunder Bay, as shown in Figure 1.
While not strictly an arctic community, it does exhibit some
features of the sub-arctic, having a substantial native population
in rural areas and an economy based largely on the exploitation of
natural resources (mining, forestry and tourism). Like many such
developing regions in Canada, it has its own metropolis, Thunder
Bay, and a large sparsely populated hinterland that includes many
small communities, some of which have entirely native populations.
In 1966, 144,000 of the total regional population of 223,000 lived
in Thunder Bay.

This region is currently experiencing an introduction of major
new developments in telecommunications technology: "ANIK", the
domestic communications satellite, will provide modern telephone
facilities in several of the small outlying communities; new local
broadcasting outlets have recently been introduced into some areas;
and a novel inter-community H.F. radiotelephone network has been
put in operation in 25 small native communities.

It was decided that it would be useful to attempt to build a
model of this region as a whole, as well as models of two small
communities within the region, whose characteristics would be quite
different from those of the region treated as an entity. This
paper describes these models and relates them to the contributions
of the several disciplines involved (biology, business, electrical
engineering, political studies, psychology, and sociology).

It is perhaps important to stress at this point that the whole
subject of socio-economic modelling should be considered to be
still at an early and experimental stage in its development. As
with any experimental science, concepts are introduced, models are
constructed, tests of their usefulness are performed, and new and
modified approaches are formed; with the whole process taking on
an iterative nature on its way towards a rising level of
confidence in its usefulness. The essence of these models is not
that they yield exact predictions of the future; they should rather
be considered to be forecasters of possible alternative futures,
subject to all the assumptions and hypotheses included in them by
their builders.

SYSTEM DYNAMICS MODELLING TECHNIQUES

Before discussing the regional and community models in detail,
it will be useful to describe the system dynamics approach to
modelling a system. Figure 2 shows a diagrammatic representation
of the socio-economic system of northwestern Ontario. This
"system" is, of course, completely interactive with the rest of
Canada and the World. Hence, in defining the system "boundary",
certain outside influences or exogenous variables must be taken
into account.

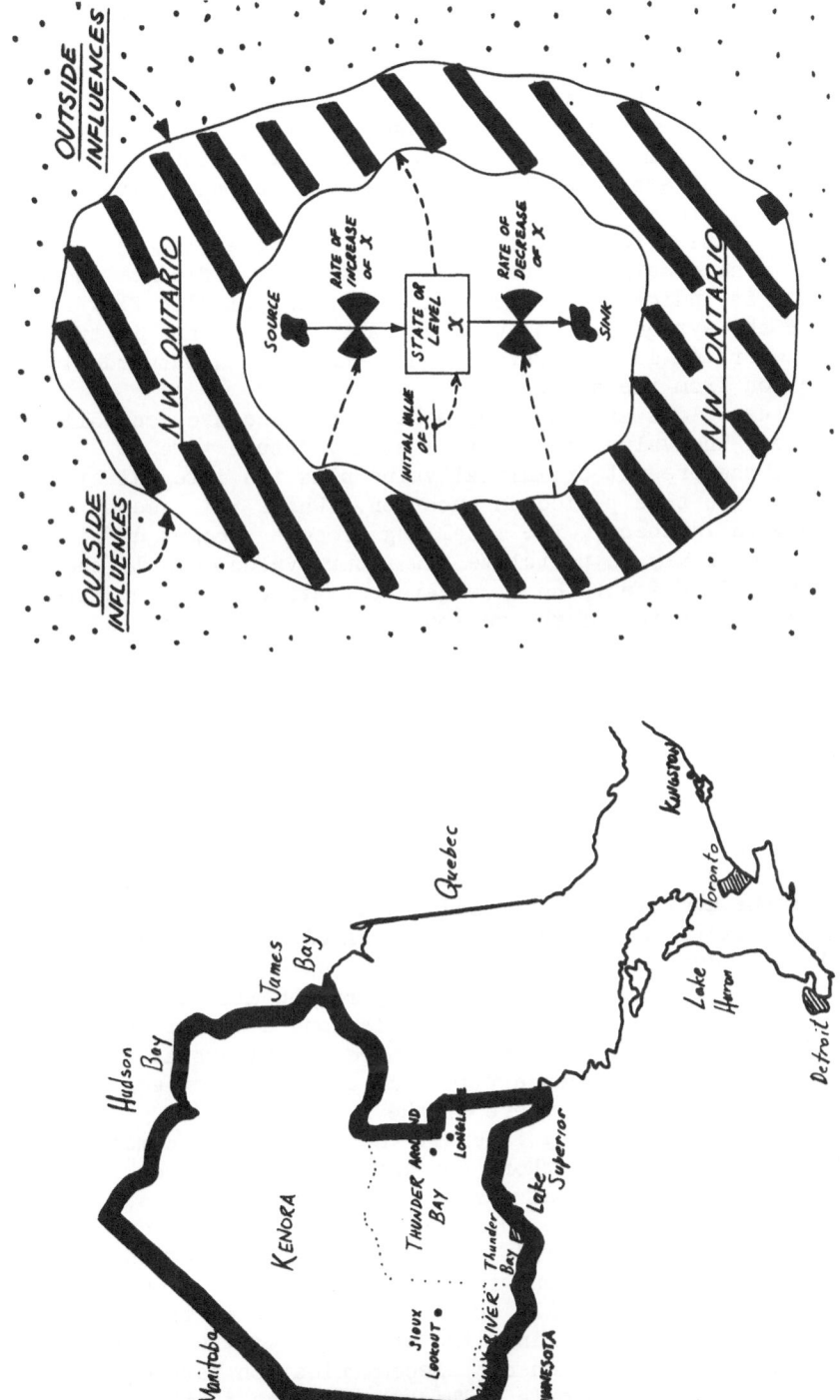

Figure 2. Basic Concepts of Dynamic Models.

Figure 1. Map of Northwest Ontario.

In modelling the system, a number of states or "levels", the knowledge of which describes the state of the system at any time, are defined. There are any number of possible choices for a set of states depending on what features of the system are deemed to be of interest. One such state is shown in Figure 2. Since the state is dynamic (time-varying), it has associated with it a rate of increase and a rate of decrease as indicated by the valve symbols. The difference between these rates gives the instantaneous rate of change of the state. In modelling the system, it is necessary to determine what factors influence these rates. This is indicated by dashed lines of information flow from the rest of the system to the two flow rates. The state in turn affects other rates in the system; hence there is an outward flow of information from the state.

The actual flows are shown by solid lines and are controlled by the rates. The value of the state at any future time is dependent on the present or initial value plus the integral of the net flow over the time period in question. While this seems straightforward in theory, the modelling process becomes difficult since it is necessary to postulate how future values of all the system states will affect the flow rates. This will become more clear in the discussion of the regional and community models which follow.

THE REGIONAL MODEL OF NORTHWEST ONTARIO

Introduction

Figure 3 is a much simplified block diagram of the regional model and shows some of the main interactions. An intensive study was made of the available statistical data on the region and formed the basis of quantification of the items named. In addition, substantial fieldwork was conducted in the region and in associated government offices, with very useful cooperation also being obtained from Bell Canada, who supply telephone services in the region. Inevitably, however, the data were not always as full as desired and it was then necessary to rely upon the knowledge and experience of the various members of the multidisciplinary team.

While there is a growing interest in socio-economic modelling, generally (e.g. [2]), the authors are not aware of any working models designed with a special focus on telecommunications. Many studies have, of course, been published on the socio-economic role of telecommunications (e.g. [3], [4]) and, while some of these have provided useful background for this work, it has been very difficult to find any generally agreed conclusions on this role that could be built directly into the structure of the model. For example, consider the question of telecommunication between people versus transportation of people. The substitution of the latter by the former is often the view put forward. Yet, if anything,

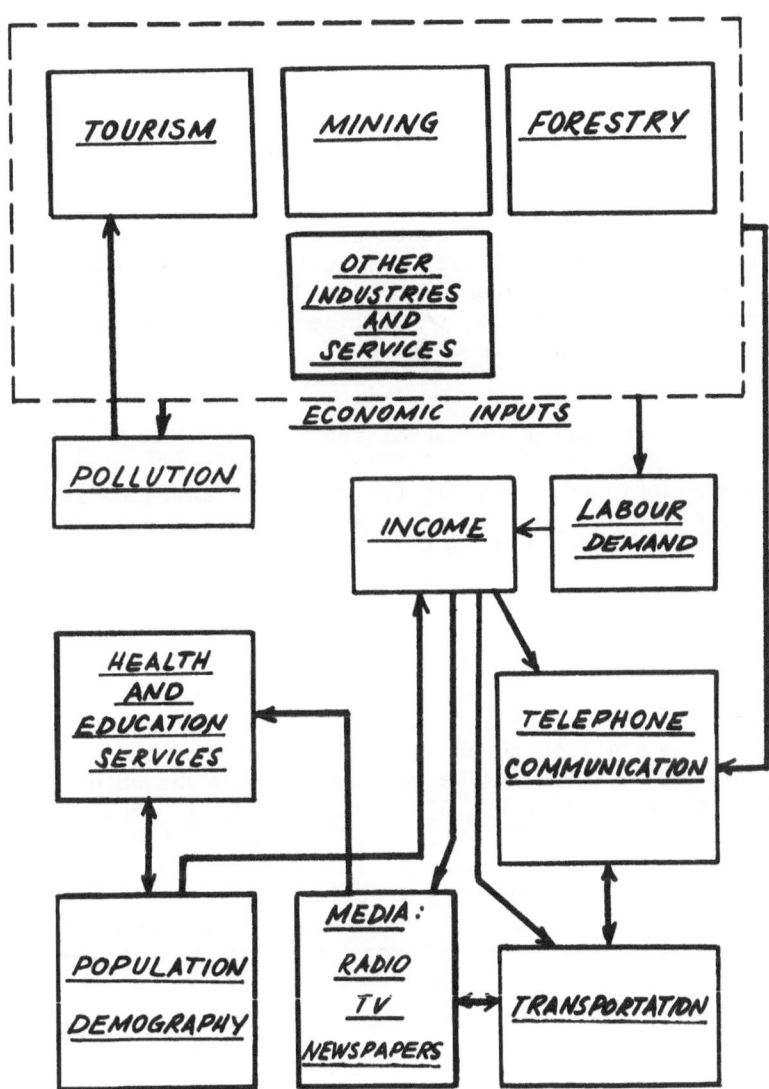

Figure 3.   Block Diagram of Regional Model of Northwest Ontario.

the evidence available suggests a process of reinforcement rather
than substitution; improved telecommunications facilities give rise
to more travel rather than less.

    Structure and data are then the major factors in the synthesis
of a dynamic model. Wherever possible this model is based on
reliable data and the structure is derived from the best available
background work, subject to the accumulated wisdom and experience
of the members of the multidisciplinary team. The entire model
includes 18 "levels" and 60 "variables", with associated "rates"
and "multiplier functions", as well as 16 exogenous inputs that
enable the user to postulate possible future conditions acting on
the region from outside, such as the world price for the products
of the region, the possible presence of Government subsidies to
economic activity, and the levels of health and education services
available in the rest of Ontario.

    In the following sections several sectors of this regional
model are examined in greater detail.

The Demographic Sector of the Regional Model

    The demographic behaviour of the region is modelled in outline
form in Figure 4. The population is split into the customary
divisions and each forms a "level" in the dynamic model.
Associated with each group is a death rate, an aging rate (transfer
to the next higher age group), a migration rate which can be into
or out of the region, and, in the case of the youngest group, a
birth rate. Only the most significant relationships with the rest
of the model are indicated. Migration, for instance, is considered
to be a function of the average wage rate offered in the region,
which thus becomes a link to the economic sector of the model, as
does the total population in the working age groups, which
constitute the labour supply. Similarly, the total population is
a factor contributing to the total usage of telephones and
transportation, as well as, detrimentally, to the tourist industry
via pollution and to the forestry industry via fire hazards.
Health services obviously affect the birth and death rates, while
the former is considered to be affected also by the average level
of education.

Mineral Resources

    In the demographic component of the regional model described
above, the dynamic behaviour (i.e. the rate dynamics) was largely
determined by factors originating in other parts of the system.
The levels and rates associated with the model component describing
mineral resource discovery and extraction are shown in Figure 5.
All mineral resources are aggregated into two levels: potential
(undiscovered) resources and proven resources. The dynamics are

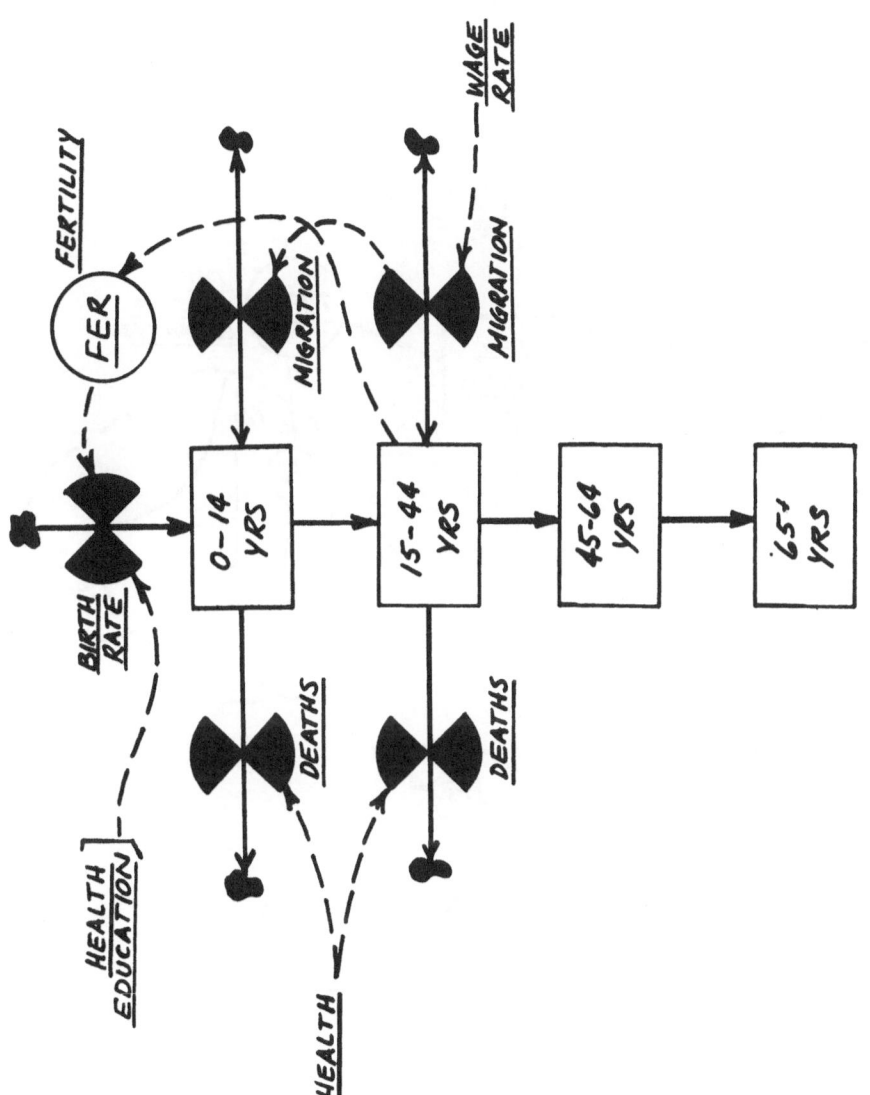

Figure 4.   Basic Demographic Model for the Region.

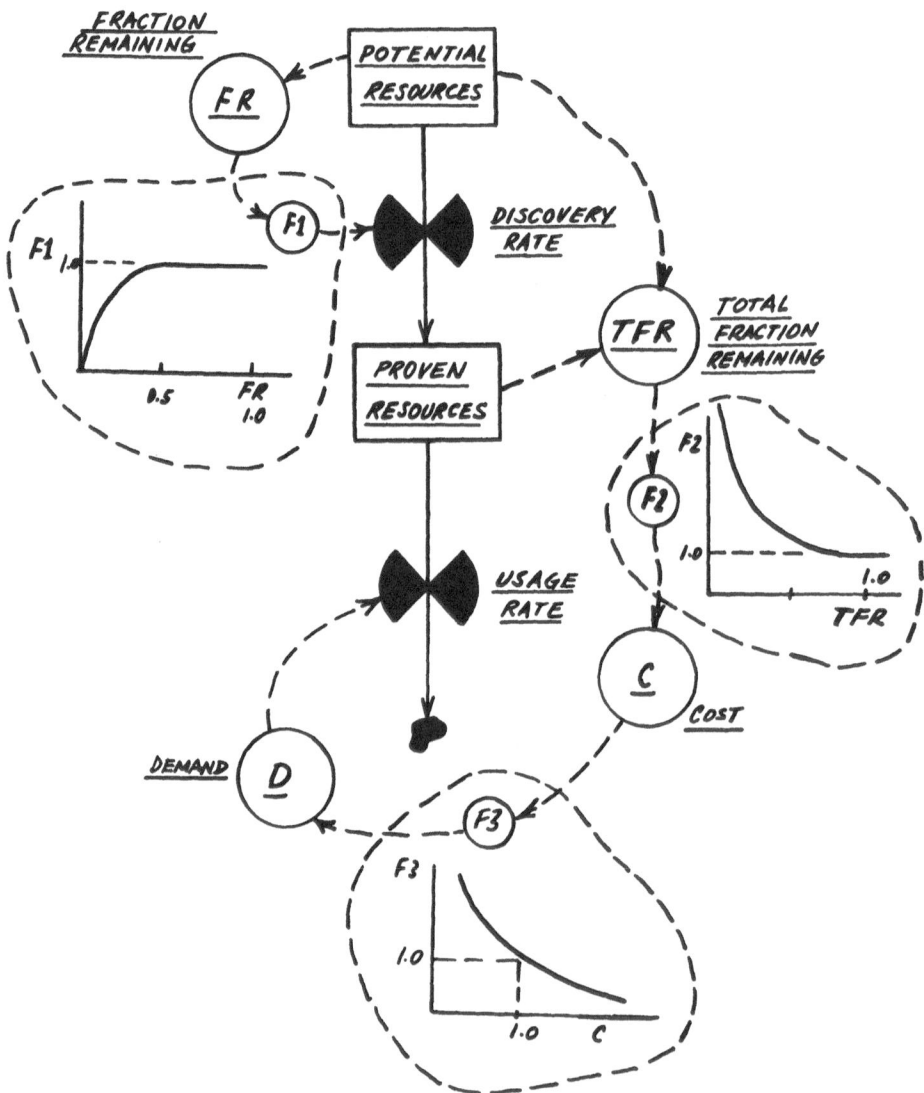

Figure 5.   Basic Mineral Resources Model for the Region.

determined by the resource discovery rate and the extraction or usage rate. For purpose of illustration, a number of feedback loops are shown in the figure.

First consider the loop from potential resources to the discovery rate which describes how, as resources grow scarce, their rate of discovery declines. The variable FR (fraction remaining) is calculated from knowledge of the initial level of potential resources (based on a best estimate). Function F1 multiplies the initial discovery rate (1971 value) and is shown in the figure. As FR declines beyond 0.5, the discovery rate declines to zero. Information used to estimate the shape of function F1 is difficult to obtain since few resources have yet been totally exhausted. Also, the discovery rate of most resources is dependent on many other factors such as current price, government incentives, etc., and the effect of resource depletion alone is difficult to extract. Function F1 is one of many functions in the model chosen to reflect a reasonable and educated estimate.

Next, consider the feedback loop from total fraction of resources remaining (TFR) to production cost (C), to demand for resources (D), and finally to the usage (production) rate. Function F2 describes how the production cost per unit resource varies with the fraction of total resources remaining in the region. The form of F2 was based on data obtained on a gold mine operating for 30 years in the region. Extraction cost per ounce was compared with the fraction remaining of the total production over the 30 years. Function F3 describes how demand for regional resources varies with the production cost (closely related to price). Again, data from the gold mine was used as a guide to the shape of this function.

Figure 6 illustrates one of the links between the two components discussed above. The mineral resource usage (production) rate is one of several inputs to the variable describing labour demand (LD). Labour supply (LS) is a function of the population. These two variables determine the employment ratio (ER) which influences the wage rate (WR). The wage rate (average) in turn influences the migration rate which completes the feedback loop. In operation, this part of the model (given that all other factors are constant) will produce a migration rate which responds to the labour demand relative to the available labour in the region.

The Model of Telephone Usage in the Region

In a model focussed on telecommunications it is appropriate for telephones and their usage to be treated in some detail, as shown in Figure 7. To provide more carefully defined links with the economy and with other factors, telephones are divided between those installed for residential use and those for business use (appropriate statistics were available). Each of these levels has growth and decay rates (combined in one symbol) influenced by

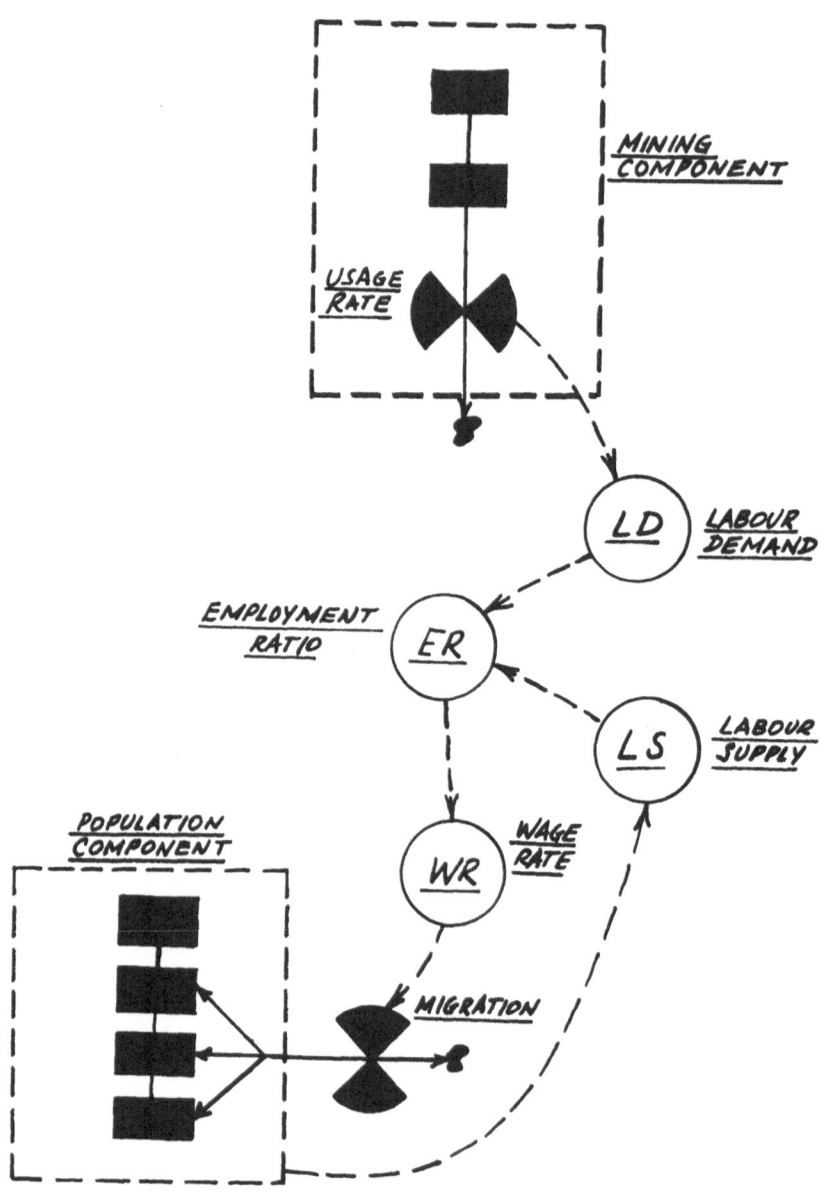

Figure 6. A Link Between Mineral Resources and Demography.

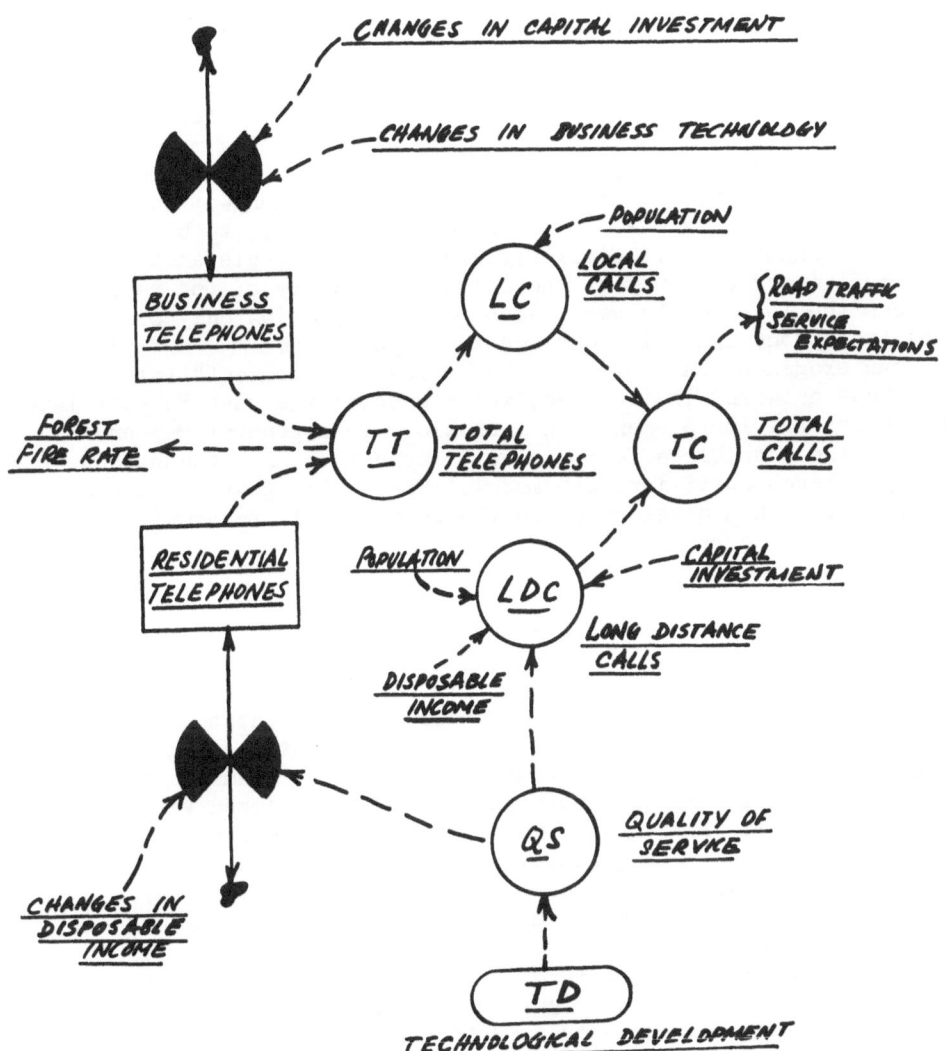

Figure 7. Basic Telephone Model for the Region.

different factors. Not shown in the figure are links between
business telephone installations and the tourist industry, and
between residential telephone installations and the regional level
of health services. An increase in the total number of telephones
is considered to lower the forest fire rate by the consequent
general improvement in rapid communications.

Telephone usage is divided into two categories: local calls
and long-distance calls originating within the region (for which
statistics were available). The total telephone traffic was taken
to influence road passenger traffic, in a reinforcing mode, as well
as the expected, or desired, levels of health and education
services. The division of usage into local and long-distance
enabled the latter alone, and thus more accurately, to be linked
to air transportation. Similarly, the total long-distance calls
were influenced directly by the economic activity of the region,
whereas the local calls were taken to be a function only of the
total telephone installations and the population.

An exogenous variable, "Technological development" (of the
telephone network) is included, which enables the user to postulate
future general improvements in technology in a general way and
observe the possible effects, always subject to the assumptions
and hypotheses built into the model. The dialling system was an
example of such a development in the past; satellite ground
stations are a current instance.

Use of the Regional Model

The entire model is programmed in FORTRAN on a dedicated
medium-sized computer (PDP-15). Considerable attention has been
paid to the question of how such a model can best be used,
particularly by those unaccustomed to direct involvement with
computers. An interactive terminal has been developed with the
output displayed directly in the form of graphs on a video screen.
After three short opening statements have been typed on the usual
typewriter keyboard, a special operating console takes over
completely. This enables the user to select for display any one
of a wide selection of graphs of the model's behaviour.

First, however, the list of 16 exogenous variables appears
on the screen, as shown in Figure 8 but with the right-hand column
initially blank, and the user selects appropriate annual growth or
decay rates, in steps of ±1% up to ±6% (in constant 1971 dollars,
where appropriate). Figure 8 shows that all but two of the 16
have been set to have neither growth nor decay ("CONS"), while the
world prices for mining products (1) and for forestry products (50)
are considered each to be growing at an annual rate of 2%. When
all 16 have been specified, the previously selected graph is then
traced directly on the video screen almost instantaneously. The
behaviour of certain aspects of the model over 30 years from the
base year (1971) is thus indicated.

```
  1  -RESOURCES WORLD PRICE    2 EX
  2  -MINE TECHNOLOGY AVAIL    CONS
  3  -MINE OPERTN SUBSIDY      CONS
  4  -MINE EXPLORTN SUBSIDY    CONS
 50  -WORLD PRICE FOREST RE    2 EX
 51  -FORESTRY SUBSIDY         CONS
 52  -FOREST TECNLGY AVAILA    CONS
200  -TEL INDSTRY PROG INDX    CONS
201  -BUS TECHNLGY PROG IDX    CONS
300  -SUBSIDY-IND +SERVICC     CONS
350  -DISPOSABLE INCOME        CONS
351  -TOURISM CAP INVESTMNT    CONS
400  -EDUCATION LEVEL SOUTH    CONS
401  -EDUCATION QUAL SOUTH     CONS
450  -H SERVICE LEVEL SOUTH    CONS
451  -H SERVICE QUAL SOUTH     CONS
```

Figure 8.   Exogenous Variables and Growth Settings.

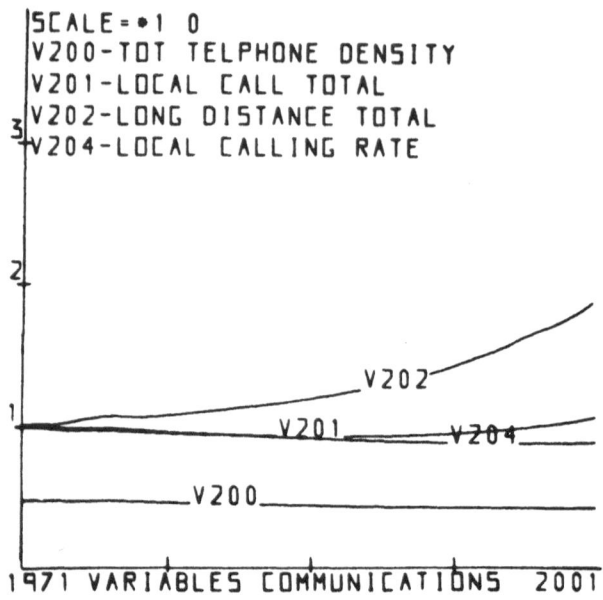

Figure 9.   Telephone Behaviour of Regional Model for Constant Exo-
            genous Variables.

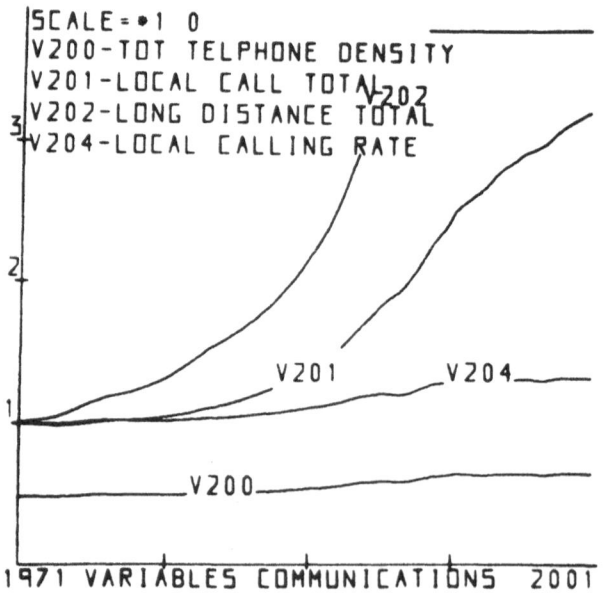

Figure 10. Telephone Behaviour of Regional Model for 2% Annual
            Growth Rates in World Prices for Products.

For simplicity, only two of the large number of possible graphs are shown in this paper (Figures 9 and 10). Figure 10 shows the behaviour of some of the telephone variables for the particular set of exogenous variables given in Figure 8. Figure 9 shows the behaviour of these same telephone variables when all 16 exogenous variables are held constant. It is evident that dramatic increases occur in telephone usage, both local and long-distance, over the 30-year period, most probably as a result of the general economic prosperity of the region when the world price for its resources is steadily increasing (in constant dollars). The density, or number of telephones per capita, however, does not change much, nor does the rate of local calls made per capita. Other curves in fact show similar dramatic increases in the total population of working age, due to migration into the now prosperous region, and it is this factor that is primarily increasing the total calls made (not per capita).

All the curves are normalized to their 1971 values, with V200 on half-scale for ease of viewing. Further refinements include the possibility of stopping a run at some time during its course, altering the exogenous variables, perhaps to simulate a change in government policy or other external factors, and then continuing the run to the end of the 30-year period.

Many tests have been made on this model to test its "reasonableness of behaviour" and also its sensitivity to changes in some of the structural parameters. The general question of validation of such models will be discussed briefly in a later section.

COMMUNITY MODELS

Introduction

Before discussing the two community models in detail, it will be useful to consider some differences in the approach taken in modelling a community as opposed to a region. These can be summarized as follows:

(a)   The economy of a community is most directly linked
      to that of the region as a whole; the regional economy
      is linked to that of the nation, continent and the world.

(b)   The public services and resource industries present in a
      community may differ from the regional norm and must be
      individually studied.

(c)   Social measures such as cohesiveness, attachment, kinship,
      cultural attitudes, etc., are difficult to define for an
      entire region as diverse as northwestern Ontario; for a
      community they can be more meaningful.

(d) Telecommunications and transportation services differ widely
     from community to community and must be treated individually
     for each community.

     The above points stress the need for community field studies
to gather information on community economy, services, and social
structures. These studies serve to augment the data obtained from
census and other government records.

Ojibway Community

     Figure 11 shows a block diagram of the community model for the
village of Aroland which was the subject of field studies during
the summer of 1973. The community is an independent native
settlement of about 260 persons located on leased land. Wage income
arises from woodcutting and loading operations near the community.
The main feature of the model is the inclusion of variables
describing cultural attitudes toward integration with and rejection
of the dominant (white) culture. These are closely linked to
communications and transportation and to the cultural variables
describing language retention, ownership, traditional religion,
and acculturative stress.
     Figure 12 shows a system dynamics flow diagram relating one
cultural variable (ownership) to the attitudes toward integration
and rejection. As can be seen, these variables form complete
feedback loops such that, while the degree of ownership influences
the attitudinal variables, they in turn affect the rate of change
in ownership.
     The Aroland model has not been simulated on a computer since
it was felt that insufficient data and information were available
at this time to proceed further than the detailed flow diagram.
This will, however, serve as a prototype for future community models
when the process of acculturation is of prime interest.

Service Community

     The second community studied is that of Sioux Lookout, which
is a predominantly white frontier service community of about 2,500
persons. A significant body of data concerning social structure
related to communications and the economy of the community was
obtained from field studies carried out during the summer of 1974.
     Figure 13 shows a block diagram of the Sioux Lookout community
model. Rather than the process of acculturation, the main feature
of this model is a description of how various measures of community
attractiveness affect migration. Economic attractiveness is a
function of employment generated by various economic sectors.
Social attractiveness is a function of social mobility and other
measures such as community cohesiveness, degree of citizen

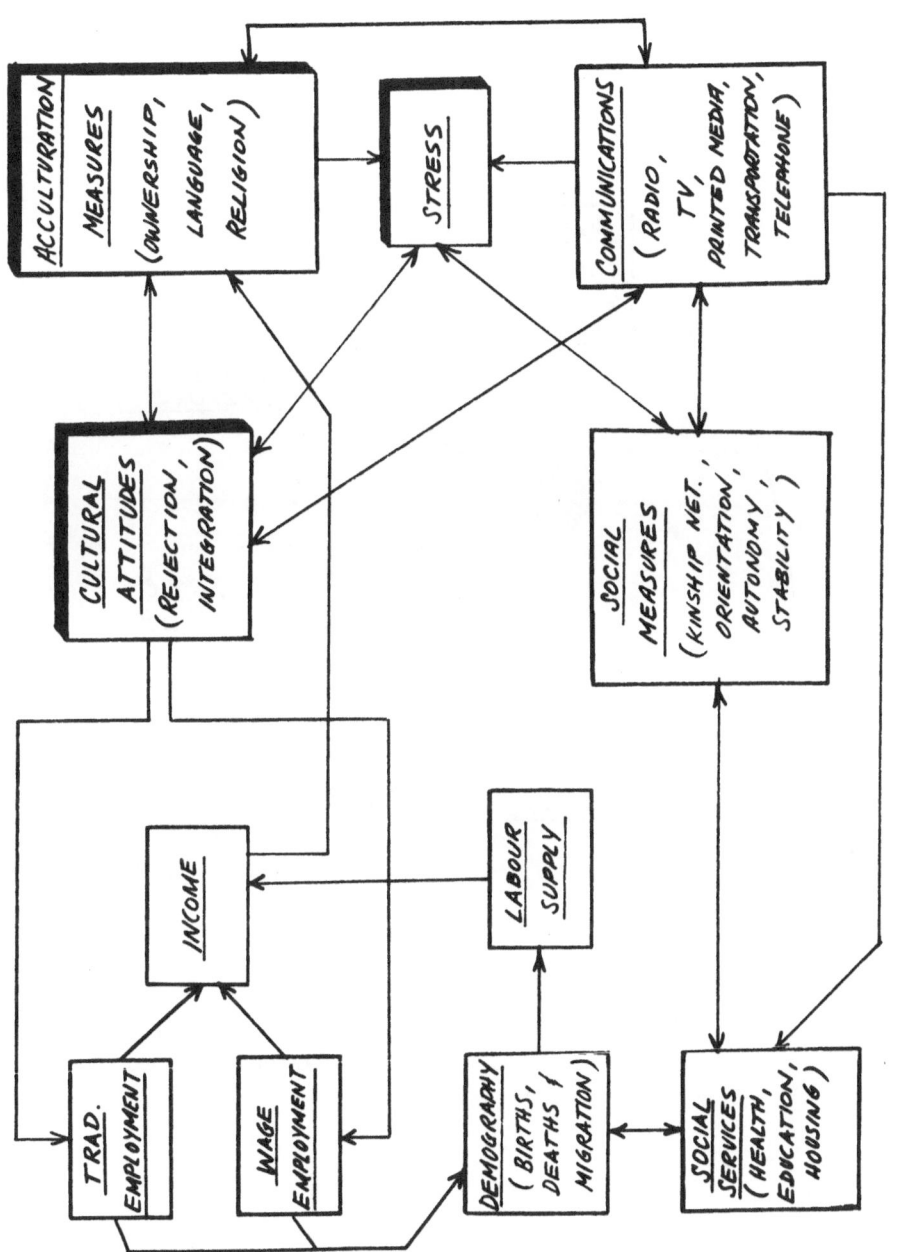

Figure 11. Aroland Community Model Block Diagram.

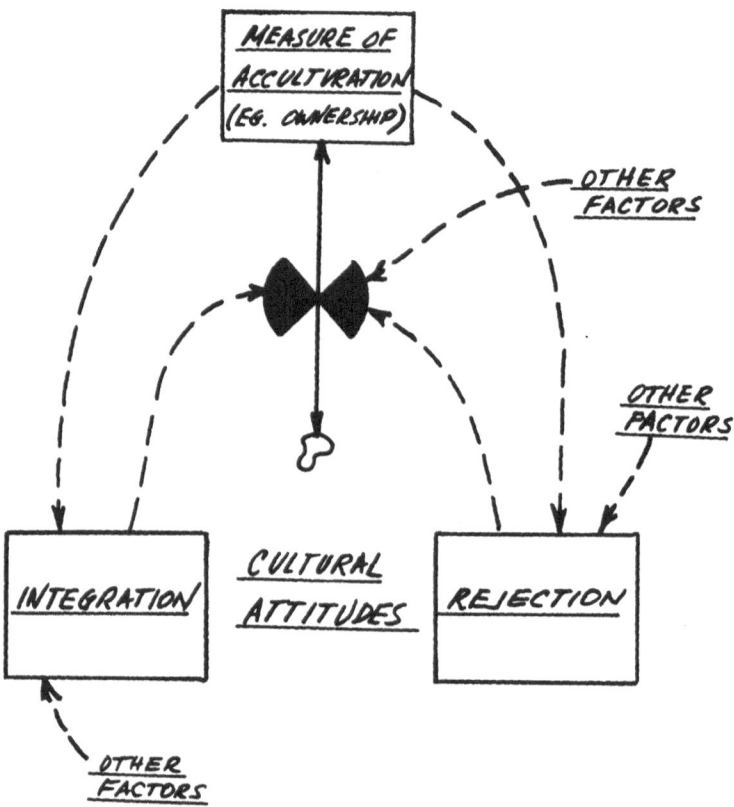

Figure 12.   Some Cultural Aspects of a Community Model.

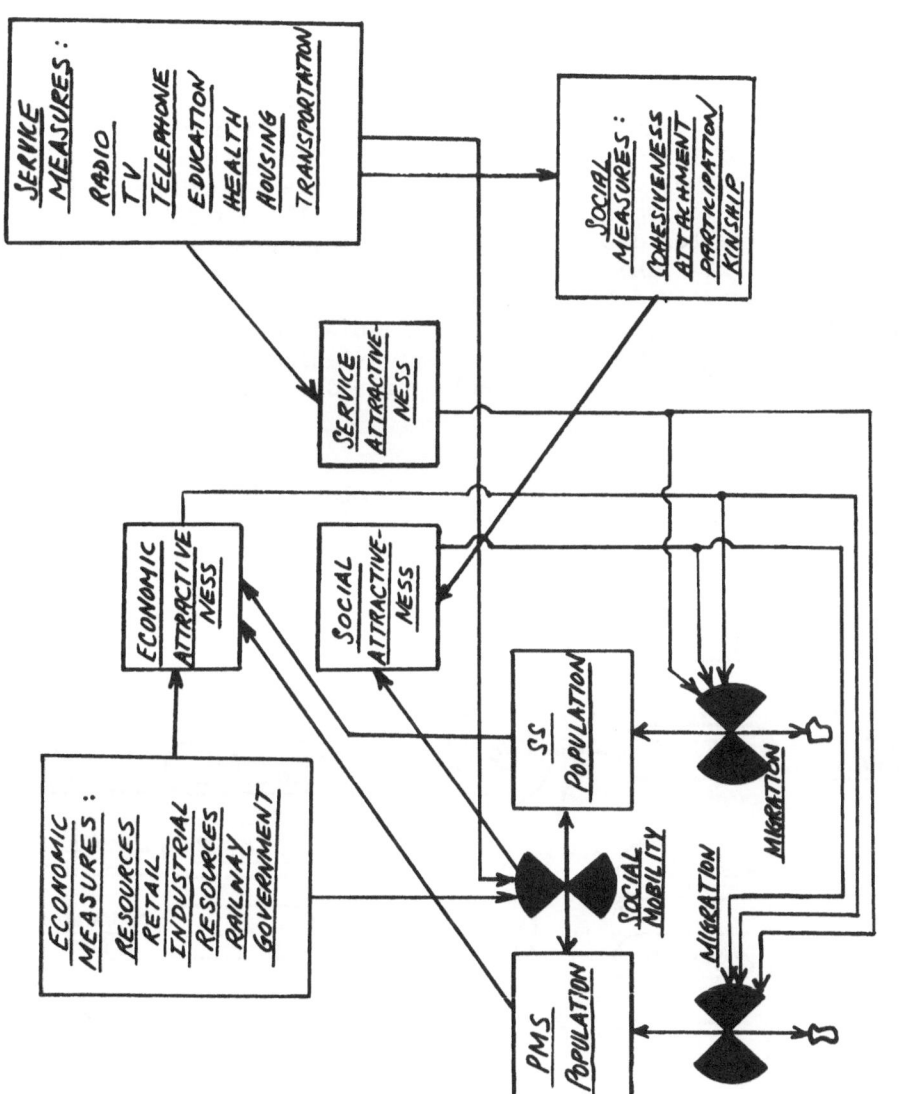

Figure 13. Sioux Lookout Community Model Block Diagram.

participation in community activities, etc. Service attractiveness
is a function of the level of various services in the community,
including telecommunications.

The demographic sector of this model is somewhat more complex
than that of the regional model described above. Each of six
population age groups is represented by a level representative of
families involved in professional, managerial or skilled
occupations (PMS), and by a level representing families involved in
semi-skilled or unskilled occupations (SS). A social mobility rate
describes the processes of upward and downward mobility.

The state of telecommunications services in the community is
described by a number of levels, rates and variables related to
radio and television services and usage, as well as telephone
usage. The variables describing radio and television services are
shown in Figure 14. These variables are treated as exogenous in
that they are largely determined from outside the community itself
via government policy, etc. The two variables describing how these
services are used are average TV contact (viewing) hours per person
per day (TVCH) and a similar variable for radio hours (RCH).

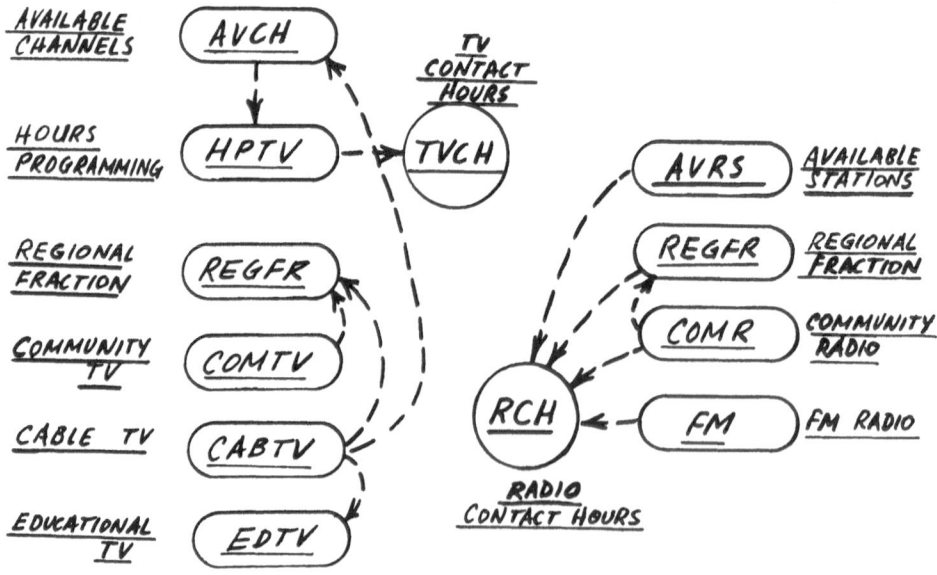

Figure 14. Radio and Television Services in a Community.

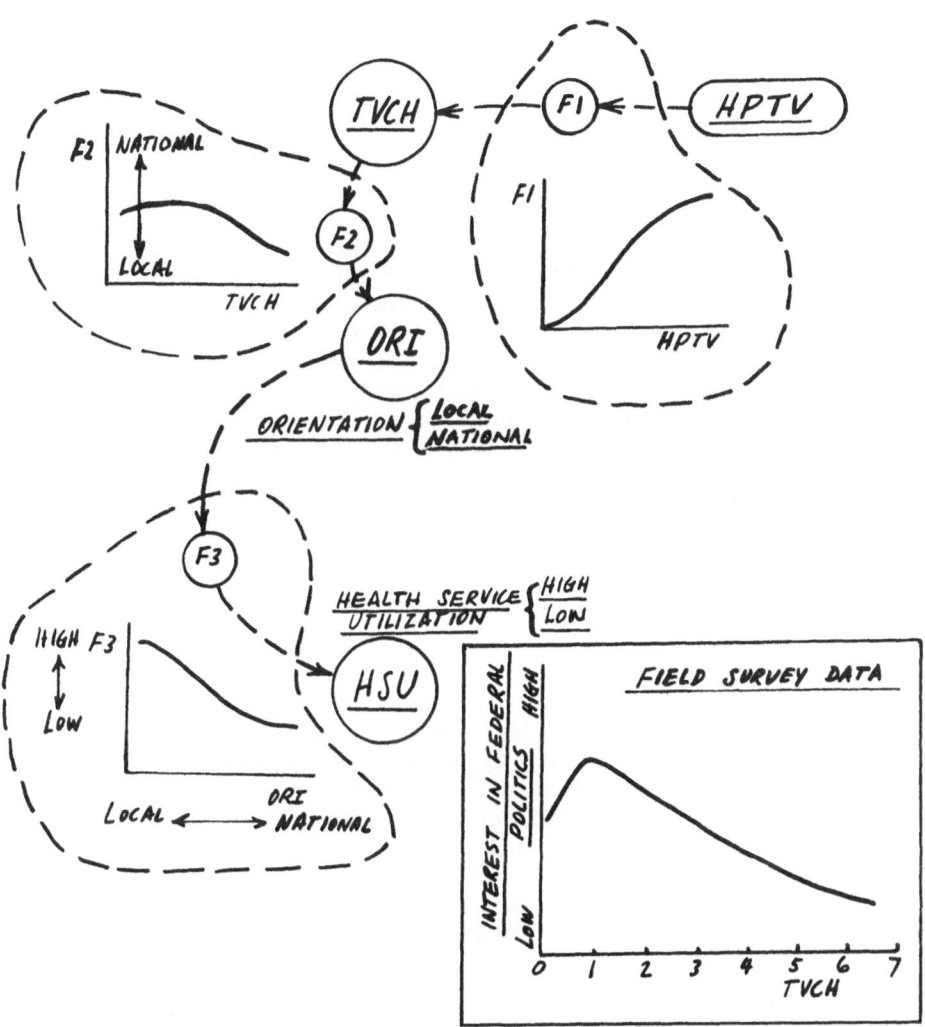

Figure 15. Possible Model of Impact of Television Contact Hours.

Figure 15 illustrates how TV contact hours is linked to other variables in the model and how field survey data can be used to guide the choice of function shape. At the top right hand corner of the figure, TV contact hours (TVCH) is seen to be a function of the hours of TV programming (HPTV) available as described by function F1. A qualitative variable describing the orientation of the community (ORI), tending to be either local or national, is believed to be affected to some extent (along with other factors) by TV contact hours. Field survey data relating TV contact hours to interest in federal politics is shown in the figure and is used as a guide to the choice of the form of function F2. It appears that those who watch a great deal of television tend to have a more narrow, local orientation despite their greater exposure (via TV) to national and world affairs. One link from orientation is to the degree to which local health services are utilized in preference to regional (Winnipeg) facilities. Health service utilization (HSU) is related to orientation (ORI) by function F3.

In the above discussion, only single factor causal effects were described. In fact, there are many factors influencing a variable such as HSU that are included in the model. At the present time, the model is partially simulated and is undergoing tests on a digital computer. A great deal of further data analysis must be carried out before the large number of model functions can be described with sufficient confidence to make complete simulation possible.

THE VALIDATION OF DYNAMIC MODELS

It is sometimes said that only when dynamic models have been fully "validated" can they be used with confidence as predictors of the behaviour of the subject modelled. Such "validation" is then defined as the comparison with known historical data - "If you started the model with the conditions that existed 30 years ago, would it predict the current reality?" This argument is very dubious in as much as the actual future never repeats the actual past. While a model could in principle be refined to follow precisely the behaviour of the past 30 years, this would still be no more than a guiding indication, rather than a guarantee, of its ability to predict the future behaviour during the next 30 years [5]. Furthermore, in practice it seems to be almost imposs-ible to get reliable and consistent historical data over any substantial length of time for anything as complex as a region or community, because of continual changes in policy regarding the keeping of statistics, if at all, by various bodies.

What then can be done? The models can be thoroughly tested for the "reasonableness of their behaviour". For instance, no large and complex social system in practice would suddenly exhibit a massive migration rate, or a colossal fall in the number of telephone installations, over a very short time period, with

the possible exception of utterly unpredictable catastrophes on an
apocalyptic scale. Therefore, any such grossly unexpected behaviour
of a model should be carefully investigated for possible flaws in
the structure and hence in the underlying hypotheses. An associated
test is that of the sensitivity of the model to small changes in
its structural parameters. If such a small change produces a large
change in the model's behaviour, then one should be very sure of
the data on which those parameters and their structure are based;
if one is not sure, the structure should be modified to reflect
this.

What, finally,is the ultimate test of their usefulness but
their use? By this is meant their experimental use over a
substantial period by persons having a full understanding of their
structure and the ability to make modifications in the light of
experience. Then, gradually, an increased confidence in their
usefulness can be developed in a realistic planning environment.

CONCLUSIONS

Dynamic models have been constructed of a developing region
of Canada and of small communities within that region. Wherever
possible these models have been based on statistical data and field
work, conducted and interpreted by appropriate specialists within
the multidisciplinary team. The regional model is complete and
has been investigated in some depth while those of individual
communities have been taken to various degrees of completion. All
these models allow for exogenous inputs under the control of the
user, which can be used to simulate the effects of planning and
policy decisions as well as other factors external to the models.

The work described here should be considered experimental and
to be at an early stage in the history of socio-economic modelling.
In the history of the physical sciences, the mathematical models
that we find so useful and reliable today usually bear but little
resemblance to the first tentative steps towards mathematical
models of physical reality that were made by the earliest workers.
Despite the complexity and ability of the modern computer, it is
perhaps wise to look on current attempts to build dynamic models
of socio-economic systems as only the first, but nevertheless
necessary and useful, steps along a difficult road of continual
experimentation and modification. Only then can there arise a
gradually increasing confidence in their usefulness as forecasting
aids to those who have to make plans for the future development
of various aspects of our society.

ACKNOWLEDGEMENTS

This work is supported by the federal Department of
Communications in Canada.

This paper summarizes the work of the entire multidisciplinary team and the authors wish to acknowledge with deep appreciation the contributions of their co-investigators, including: J.W. Berry, C.E.S. Franks, R. Harmsen, K.A. Herman and J.G.M. McKirdy; and of the many students and assistants who worked on the project, including: R.J. Adomavicius, J.R. Kane, J.A. Love, T.A. Mawhinney, L.N. Mombourquette, M. Rose, R. Rumberg, H.K. Tang, Y. Vautrin, V.G. Weinreb.

REFERENCES

[1]   J.W. Forrester, "World Dynamics", Cambridge, Mass., Wright-Allen Press, 1971.

[2]   Special Issue on Social Systems Engineering, Proc. I.E.E.E., Vol. 63, No. 3, pp. 340-534, March 1975.

[3]   C. Cherry, "World Communication: Threat or Promise", New York, Wiley-Interscience, 1971.

[4]   G. Gerbner, L.P. Gross, and W.H. Melody (eds.), "Communications Technology and Social Policy", New York, John Wiley, 1973.

[5]   K. Chen, "An Evaluation of Forrester-Type Growth Models", I.E.E.E. Trans Systems, Man, and Cybernetics, Vol. 3, No. 6, pp. 631-632, November 1973.

THE MAN IN THE ARCTIC PROGRAM (MAP): ECONOMIC, DEMOGRAPHIC,

AND SOCIOCULTURAL EFFECTS OF DEVELOPMENT IN ALASKA

Judith Kleinfeld

University of Alaska

Fairbanks, Alaska, U.S.A.

Alaska is in the midst of a major period of economic growth, stimulated largely by oil and gas development.[1] The 800-mile trans-Alaska pipeline, which is now under construction, will have a peak employment level of 16,000 workers, many of whom have migrated to Alaska from other states. Commercial production of Prudhoe Bay oil is expected to begin in 1977. While direct employment in the oil industry will decline substantially at this point, royalty payments and taxes to the state government will provide a substantial new economic stimulus, creating new jobs and bringing new migrants to Alaska. Alaska presently has about 350,000 residents. Even if no major new oil developments occur, by 1990 Alaska's population is projected to exceed 600,000.

What is more likely, however, is that major new oil developments will occur. Prudhoe Bay and the oil pipeline may be only the beginning of the oil boom in Alaska. After the oil pipeline is completed, a gas pipeline will be built either following a Canadian route to the midwest or an Alaskan route to a tidewater port in south-central Alaska. Congress has authorized exploration of the Naval oil reserve (Pet 4). The federal government is planning an Outer Continental Shelf lease sale in the Beaufort Sea, Bering Sea, and Gulf of Alaska. The state government may hold additional lease sales in Prudhoe Bay and Cook Inlet to bridge the gap in the state budget until North Slope oil production gets under way.

---

[1]This section summarizes information presented in G.S. Harrison, Alaska Growth Policy Issues. Summarized from the Growth Policy Symposium held at University of Alaska, Fairbanks, December 3-5, 1974.

Native corporations formed under the Alaska Native Claims
Settlement Act of 1971 have selected land with oil potential as
part of their settlement, and six corporations have already
entered into exploration agreements with oil companies. Such
future oil and gas development would have tremendous impact on
economic development, population growth, and sociocultural change
in Alaska.

Alaska must make a number of major policy decisions over the
next few years which will affect the rate of economic development,
the quality of the natural environment, and the social well-being
of Native and non-Native residents. Many of the developmental
forces in Alaska obviously are out of the control of the state
government, but the state has significant leverage on the issue of
the rate of economic development through its petroleum leasing
policies and the rate at which it spends state oil revenues. The
state can also influence well-being in Alaska by the ways in which
oil revenues are used. The state, for example, could seek to
mitigate the costs of oil and gas development in Alaska, attempting
to preserve natural wilderness or the traditional lifestyles of
Native and non-Native residents. Or it can elect to bear these
costs of development and allocate its revenues toward achieving
other goals such as reduced taxes or a higher level of education
or health care.

The Man in the Arctic Program (MAP), a major research effort
in progress at the Institute of Social, Economic and Government
Research at the University of Alaska, is examining economic,
demographic, and sociocultural effects of development in Alaska
with the objective of providing information for such policy choices.
MAP, funded primarily by the National Science Foundation, is
organized into two phases of three years each. The first phase,
which will come to a close at the end of 1975, is concerned with
broad economic and demographic effects of oil and gas development.
I will very briefly describe the approach used in this research
and a few of the preliminary findings. Research on social and
cultural effects of development in the north will be done primarily
in Phase II, beginning in 1976. During Phase II, the economic and
demographic research program will focus on the distribution of
economic welfare across different population groups and in
different regions of the state. Since the sociocultural research
program is now being developed, I cannot, of course, present findings.
However, I will discuss the types of social problems and policy
issues in Alaska that stimulated the sociocultural research effort
and the Arctic social and cultural systems model we are using to
explore the social effects of development in the north.

MAP ECONOMICS AND DEMOGRAPHIC RESEARCH[2]

MAP in its Phase I program has developed computerized models of Alaska's economy and population structure and has carried out studies of certain key development policy issues. The models are used to estimate the impact of different policy options on such variables as employment, gross state product, and personal income. While MAP is examining a wide range of specific policy options, many of the important policy decisions fall within two broad categories:

1. The rate of petroleum leasing in Alaska -- choices, for example, between very limited leasing to preserve Alaska's environmental quality and rapid leasing to promote accelerated economic development.

2. Alternative uses of the state revenues generated through North Slope oil production -- uses of these funds, for example, to reduce taxes on personal income or corporate profits or to provide various public services, social programs, and welfare payments.

The policy analyses in Phase I are being carried out on a fairly aggregate level while the more detailed, distributional aspects will be studied in Phase II. Some of the preliminary results of MAP research provide estimates of the impact on Alaska's economy of alternative state policies concerning the rate of petroleum development and the rate of state expenditures. The effects of three broad alternative growth policy sets are compared:

1. A "limited growth" policy, defined by no additional petroleum leasing and placing most of petroleum revenues in an investment trust fund.

2. A "moderate growth" policy, defined by moderate additional petroleum leasing and the use of half of the petroleum revenues for current spending.

3. A "rapid growth" policy with extensive lease sales opening several new fields and the majority of petroleum revenues used for current spending.

These analyses indicate that rapid growth policies can markedly increase gross state product, employment, and population, but available policy options cannot substantially curtail Alaska's growth. While the size of Alaska's economy increases substantially under rapid growth policies, at the same time population grows as

---

[2]This section summarizes papers prepared by Dr. David Kresge, Director of the MAP Phase I program, especially, D. Kresge, <u>Alaska's Growth to 1990: Policies and Projections</u>. Fairbanks: Institute of Social, Economic and Government Research, University of Alaska, November 21, 1974.

new workers migrate to the state. The result is virtually no
change in the unemployment rate or in real per capita income. In
sum, what this analysis forcefully points out is that under a wide
range of government spending and petroleum leasing policies,
unemployment will not be significantly reduced nor real per capita
income increased.

MAP SOCIOCULTURAL RESEARCH

Although certain northern population groups may derive
substantial economic benefits from development, rapid development
in itself not only has little effect on achieving the social goals
of reduced unemployment and increased income. Rapid development,
at least in certain phases, also may cause serious social and
cultural dislocations in both Native and non-Native communities.
In other phases and to some population groups, of course, rapid
development may bring social benefits. The fundamental purpose of
MAP sociocultural research is to develop a model predicting the
effects of rapid economic growth on social problems in different
types of communities and on the quality of life experienced by
different northern population groups.

Social Problems

Concern with the social and cultural consequences of growth
has intensified in Alaska, in part as a result of the serious
social problems occurring through construction of the trans-Alaska
pipeline. While little systematic empirical research has been
done, community agency reports[3] and interviews with informed
observers suggest a wide range of social problems. A housing
crisis, for example, has occurred in communities along the pipeline
corridor. In major pipeline service centers, such as Fairbanks
and Valdez, rental housing is virtually unavailable and is
extremely high in price. Indeed, the newspapers have carried
advertisements offering rewards for locating a vacant apartment.
Many long-term tenants are subject to eviction or to large rent
increases, which force out many of those on fixed incomes. The
scarcity of housing has precipitated a number of other serious
problems. Substandard housing is brought into use which in turn
may be leading to increased accidents, health hazards, fire dangers,
and stress on families.

---

[3]The information on social problems in Fairbanks discussed in
this section is drawn largely from the series of pipeline impact
reports done by the Fairbanks North Star Borough, Fairbanks,
Alaska.

The housing crisis is probably a short-term dislocation which will end with pipeline construction. Another social problem resulting from rapid growth, however, which may have long-term social consequences is family breakdown and defective socialization of children. During the pipeline construction period in Fairbanks, for example, the divorce rate has rapidly increased. The juvenile arrest rate, especially among 11-12 year olds, has substantially risen. Rates of serious child neglect have risen. Runaways have also increased with a new pattern of groups of children floating from house to house while parents are away at remote construction camps. Social workers and probation officers suggest that these problems are caused in part from a changed employment structure. Parents may take high-paying pipeline jobs requiring them to leave town for a period of several weeks or may work in town on the six-day, ten-hour work schedule of pipeline employment. At the same time, there is a shortage of domestic labor at low-wage rates, so children may be left unattended.

While pipeline construction may have brought social costs to such groups as juveniles, wage levels in union pipeline jobs are extremely high. A dishwasher working a 40-hour week in Fairbanks, for example, makes $638 a month while a diswasher on a pipeline job, working 60-70 hours a week, makes $3061 a month.[4] Many people are using this income to extricate themselves from financial difficulties, to buy houses and land, and to live the lifestyle they prefer. Another social benefit of pipeline construction is the increase in jobs available to teenagers and other marginal groups. Indeed, what may be occurring is a general shift in the social class structure with the professional and government worker classes being edged out, at least in the short run, by the working and business classes. Some professionals explicitly voice the anxiety that they are descending into the lower middle class.

Native villages also appear to be undergoing severe social dislocations as a result of pipeline construction. To fulfill minority hire provisions, pipeline companies are making some attempt to employ Native workers. In many villages a large proportion of men are leaving for pipeline jobs. Again, I must stress that no systematic empirical research has yet been done, and these comments are based primarily on interviews with members of Native organizations which serve these villages. Native organizations, however, are concerned about such problems as food and fuel shortages in the villages. Air transport may be deflected to pipeline service, and men may not be available to perform subsistence tasks. A substantial increase in drinking and violence has been reported when men return home with large pipeline paychecks. Native organizations are especially concerned that pipeline

---

[4]Pipeline Impact Information Center Report, Number 10, Fairbanks North Star Borough, Fairbanks, Alaska, May 27, 1974.

construction may undermine the movement for Native self-
determination and the implementation of the Native Land Claims
Settlement Act because qualified Natives prefer the high-paying
pipeline jobs. For example, when the Tanana Chiefs Conference
offered training for the position of business manager of the
village corporation, in only 14 out of 44 villages was a person
available.

The types of social problems that may be occurring in Alaska
"boomtowns" as a result of pipeline construction may be repeated
with other anticipated oil and gas development in the north, such
as development of Alaska's Outer Continental Shelf, development of
the Naval oil reserve (Pet 4), and construction of other gas and
oil pipelines across Alaska or Canada. A rather obscure but quite
useful literature on boomtowns was stimulated by the impact of
large military and industrial installations in nearby communities
in the World War II period. Much of this research consists of
studies of impact situations which have parallels in current and
anticipated oil and gas developments in Alaska -- small rural towns
overwhelmed by the sudden emergence of an installation involving
large sums of investment, a large construction force but
comparatively smaller long-term labor force, and a large influx of
population.[5] While much of this research is limited to single case
studies, some work has been done in identifying characteristic
impact patterns, relating them to social problems, and suggesting
policies for dealing effectively with impact in its various stages.[6]
The litany running through this literature is the predictability of
these social effects combined with the repeated lack of appropriate
response of policy-making institutions. As Breese[7] concludes in
an analysis of five such impact situations:
"Thus the familiar pattern is repeated over and over again whenever
the boom conditions which accompany large plant industrialization
and rapid urbanization are experienced...
On the basis of this and past experiences, one thing is un-
mistakably clear. The social and economic costs of misdirected

---

[5]See, for example, R. Havighurst and H. Morgan, The Social
History of A War Boom Community. New York: Longmans Green, 1951;
L. Carr and J. Sterner, Willow Run: A Study of Industrialization
and Cultural Inadequacy. New York: Harper Brothers, 1952.

[6]G. Breese, The Impact of Large Installations on Nearby Areas:
Accelerated Urban Growth. Beverly Hills: Sage Publications, 1965;
F. Hoehler, Efforts at Community Organization In Boom Towns of
Defense, Journal of Educational Sociology, 15(8), 1942, pp. 445-505.
Many other studies in this area may be found in government agency
reports and other sources not readily available.

[7]Breese, op. cit., p. 557.

development are prohibitively high -- high in dollars and cents,
but higher still in the countless irritations induced by the
congestion, noise, confusion, and the added hazards of fire and
traffic which such development engenders. In the face of countless
and continuing examples of communities being subjected to unduly
severe strains as a result of defense connected 'boom' conditions,
one might ask why these situations are perpetuated..."
Similar social consequences of growth have been found in boomtowns
arising from energy production in such states as North Dakota and
Wyoming.[8] As Kohns[9] points out in reviewing Wyoming's experience:
"The history of power production synonomous with 'boom development'
in Wyoming is a dismal record of human ecosystem wastage ...
Almost every community in Wyoming has experienced boom, e.g.,
Newcastle, Cheyenne, Laramie, Hanna, Salt Creek, Casper, Gilette,
and now Rock Springs and Douglas. There has been little change in
the social consequences over the past one hundred years ... Divorce,
tensions on children, emotional damage, and alcoholism were the
result. The pattern of depression, delinquency, and divorce was
so well documented that the consequences were predictable."

The sociocultural impact of rapid economic growth may differ
in the north due to such factors as the cultural backgrounds of
Eskimo and Indian populations and the extreme small size and
isolation of many northern communities. However, this literature
taken as a whole can be quite helpful in developing a model of the
social effects of rapid economic growth because it suggests the
emergence of a similar pattern of problems in quite different
communities, time periods, and cultural contexts. This literature
also suggests the strategic role of federal, state, and local
institutions. In some boomtowns, public agencies implemented
policies which took advantage of a temporary boom to make long-term
community improvements while, in others, no such policy intervention
occurred and the boom led to long-term community problems.

Quality of Life and Northern Lifestyles

MAP sociocultural research is also examining the effects of
development on the "quality of life" and lifestyles of different

---

[8]L. Nellis, What Does Energy Development Mean for Wyoming?
Human Organization, 33(3); E. Kohns, Social Consequences of Boom
Growth in Wyoming. Paper presented at the Rocky Mountain American
Association of the Advancement of Science Meeting, April 24-30,
1974, Laramie, Wyoming; R. Engler, The Politics of Oil: Private
Power and Democratic Directions. Chicago: University of Chicago
Press, 1961.

[9]Kohns, op. cit., p. 3.

northern population groups. Concern that rapid development would
erode or end the lifestyles characteristic of the north has
intensified in Alaska, in part as a result of pipeline construction.
The image of change held by many northerners prior to the pipeline
was "traditional Alaska grown rich."[10] Northern lifestyles were
not expected to change; everybody would just be better off. Again,
pipeline construction has led to increased population, increased
traffic, higher levels of air pollution, and mushrooming trailer
parks. The types of satisfactions that brought many migrants to
the north -- the absence of crowding, the natural beauty of the
wilderness, the personalized social and political relationships,
and the opportunity to fulfill significant social roles that is
characteristic of a small semi-isolated society -- are, for many,
no longer as available. Again, however, rapid economic development
may be having important redistributional effects. While pipeline
construction may be leading to decreased availability of northern
lifestyles for some groups, it has increased this opportunity for
others. Indeed, in pipeline construction camps a course on how to
build a log cabin shared first rank with a course on how to avoid
income taxes.[11] Workers see the high-paying pipeline jobs as a
way to make enough money to live a northern lifestyle -- building
a log cabin and retiring in the bush for a while.

      MAP sociocultural studies will explore quality of life issues
by extending research defining and measuring quality of life which
has been done at the University of Michigan's Institute of Social
Research. The survey instruments developed to assess quality of
life for nationally representative population groups will be
adapted to include the satisfactions characteristic of the north.
The relative importance and levels of satisfaction in different
life domains, such as natural environment and family life, will be
compared for northern and nationally representative samples.
Satisfactions will also be compared for northern population groups
experiencing different forms and rates of development.

                    Social and Cultural Systems Model

      To explore the social effects of rapid economic development
in the north, we have developed the following heuristic model (see
Figure 1). Such a model is useful in early research stages because

----

      [10]Mike Carey and David Halprin, unpublished interviews
conducted in Fairbanks, summer 1974.

      [11]Alyeska Report, Alyeska Pipeline Service Company 1(1),
January, 1975.

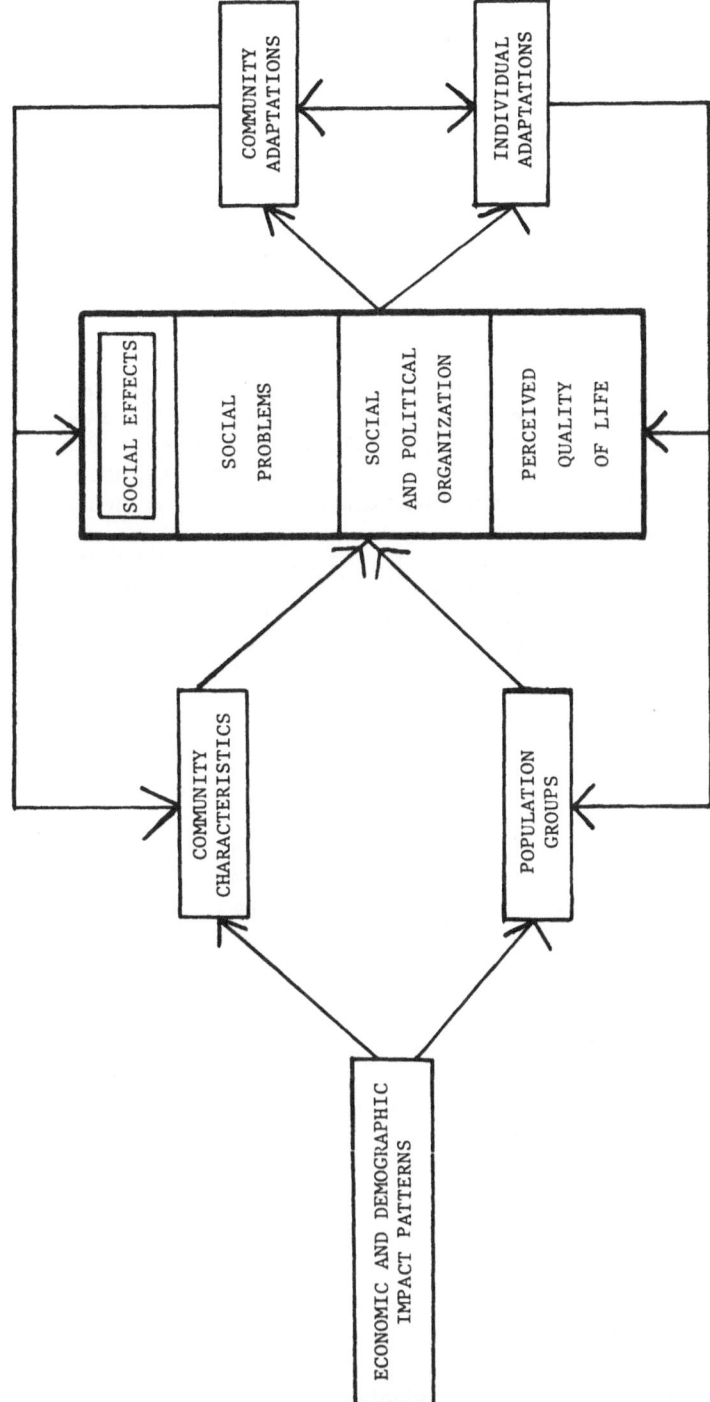

Figure 1.   Heuristic Model: Sociocultural Effects of Oil and Gas Development.

it identifies classes of variables considered important and suggests
that certain types of phenomena are related. As MAP sociocultural
research proceeds, this initial model should be replaced by one
suggesting predictive relationships.

The heuristic model developed to explore social effects of
development in the north identifies five classes of significant
variables:
1. Social Problems
2. Social and Political Organization
3. Perceived Quality of Life of Different Population Groups
4. Individual Lifestyle Adaptations
5. Institutional Adaptations.

These types of social effects are examined primarily at the
community level. First, individual perceptions of well being are
often organized around conditions in the community where one lives.
Second, policy actions designed to ameliorate negative effects of
development are often directed at communities, for example, impact
aid funds. In addition, social statistics tend to be available by
communities.

The sociocultural effects occurring in a community affected
by rapid development are conceptualized as a joint function of the
particular pattern of economic and demographic impact and the
community's social, political, and economic structure. The model
suggests that different types of communities will adopt different
patterns of institutional response to changes occurring through
development and that these institutional adaptations will have
feedback effects on levels of social problems, community social
and political organization, and the quality of life of different
population groups. Similarly, the model suggests that different
types of individuals will adopt different patterns of individual
response to changes occurring through development and that these
individual adaptations will have feedback effects on community
social problems, social and political organization, and quality of
life. These community and individual adaptations, of course, will
also be influenced by state, federal, and industrial policy
decisions, which structure the options available to the community.
The collective result of this economic and social change over time
may be to dramatically alter the economic, political and population
group structure of the community. When stabilization occurs, the
old society may be replaced by a new society, where quality of
life may be based on a very different set of values and where very
different types of lifestyles may dominate.

Research resources available under MAP will not, of course,
permit the development of predictive theory regarding all the sets
of relationships suggested in the heuristic model. The first
priority of MAP sociocultural base studies will be to conceptualize
and examine linkages between patterns of economic and demographic
impact and community types (viewed as causal and intervening
variables) and social effects (viewed as dependent variables) in
the areas of social problems, quality of life, and individual
behavioral adaptations.

1. Patterns of Economic and Demographic Impact: Research on
boomtowns suggests that the economic impact pattern precipitating
serious social problems can be conceptualized in terms of such
factors as a rapid, multifold increase in employment and population,
change in the dominant economic sector, and substantially increased
price and wage levels. In Alaska, another community impact pattern
prevalent in Native villages, appears to be a sharp decrease in
adult population caused by migration to nearby employment
opportunities. Community impact patterns will be conceptualized
in so far as possible in terms of the economic and demographic data
projected by MAP economic and demographic models in order to link
projections of social change under different policy sets with
projections of economic and demographic change.

2. Community Types: Different types of communities may respond
quite differently to a similar pattern of economic and demographic
impact depending on such factors as: 1) degree of isolation,
2) cultural background of population, 3) economic structure,
4) community size, and 5) social and political institutions. For
example, in communities where the dominant industry is fishing,
large numbers of residents may attempt to maintain this traditional
occupation and lifestyle, although it becomes increasingly less
viable with oil development.[12] In a community where large numbers
are employed in seasonal construction, in contrast, resident labor
may shift over fairly easily to new employment opportunities, which
will result in fewer social dislocations. A task of MAP sociocultural research will be to develop a typology of Alaska
communities based on the key factors leading to differential social
responses to economic and demographic impact patterns. Such a
typology will be developed through a pattern analysis of community
characteristics, following procedures used by Barth (1963) in
analyzing the effects of community structure on the types of
relationships which develop between air force bases and their host
communities.[13]

3. Population Groups: MAP sociocultural research will examine the
effects of development on the quality of life of different
occupational, age, sex, and cultural groups. In addition, the
possibility of defining northern population groups in terms of
dominant value orientations and lifestyles will be explored since

---

[12]G. Rogers, Offshore Oil and Gas Developments in Alaska:
Impacts and Conflicts, Polar Record, 17(108), 1974, pp. 255-275.

[13]E. Barth, Air Force Base/Host Community Relations: A Study
in Community Typology, Social Forces, 41, 1963, pp. 260-264.

such orientations may be key mediators in changes in satisfaction levels resulting from development.

4. Social Problems: The range and level of social problems occurring in different types of communities under different economic and demographic impact patterns will be identified, such as housing shortages, overload on municipal services, family disintegration, child neglect, alcoholism, depression, etc. Research on boomtowns suggests, for example, that communities anticipating a boom almost invariably expect significant increases in crime and devote major attention to building up the police force. Yet, in some situations, crime dramatically increases; in others, crime increases in only a few categories; and, in others, there is an "amazing dearth of crime."[14] A major task of MAP sociocultural research will be to develop testable hypotheses suggesting the specific conditions and processes through which such problems arise.

5. Perceived Quality of Life: The conceptualization and measurement of perceived quality of life for Native and non-Native population groups will be a major research task. Pilot interviews and analyses of lifestyles of people holding highly divergent orientations toward the satisfactions and dissatisfactions of living in the north will be carried out. On the basis of such pilot work, a northern quality of life survey instrument will be constructed and hypotheses concerning the effects of rapid growth on satisfactions levels and lifestyles will be developed.

6. Individual Lifestyle Adaptations: A major research task will be to identify the forms of individual adaptation to the economic and social consequences of development and to develop hypotheses suggesting the conditions under which particular adaptive choices occur. The major lifestyle adaptations explored will be:
    1. migration decisions -- for example, migrating from the north, migrating to regions of new employment opportunity;
    2. employment decisions -- for example, changing occupation, entering the labor market;
    3. allocation of income -- for example, the disposition of increased income resulting from high wage levels in impact-related employment, the adjustments in purchasing made to deal with impact-related inflation on a fixed income;
    4. pursued activities -- for example, changes in participation in community organizations, changes in the uses of leisure;

---

[14]Carr and Stearner, op. cit.

5. allocation of time -- for example, changes in time devoted to
   subsistence versus wage-earning pursuits, changes in seasonality
   of employment and leisure.

CONCLUSION

Economic development in the north poses complex issues and
difficult choices. MAP economic research calls into question the
widespread belief that economic growth in and of itself will
achieve such social goals as increased income or decreased
unemployment. MAP sociocultural research may identify serious
social problems resulting from rapid development, at least in the
short term. At the same time, the substantial state oil revenues
could be used to mitigate these social costs and perhaps to achieve
central social goals in such areas as health and education. My
own preference, as a northerner who values the satisfactions and
lifestyles resulting from low population and natural wilderness,
is for a moderate rate of development. But here I am reminded of
a point made in research literature on World War II boomtowns --
that local populations use traditional lifestyles and values as a
screen for their real concern, that development will mean the
redistribution of economic power and social status to new groups.
The Man in the Arctic Program Phase II effort will focus on this
issue of distributional effects. Through this type of research,
hard policy choices can be informed by the best available
understanding of the range of consequences.

THE DEVELOPMENT OF PETROLEUM RESOURCES IN THE CANADIAN ARCTIC:

PERSPECTIVE ON THE EVOLUTION OF ENVIRONMENTAL AND SOCIAL POLICIES

Douglas H. Pimlott

University of Toronto

Toronto, Canada

It is difficult to deal with cultural, ecological and aesthetic factors so they will be meaningful components of cost-benefit ratios or of models which relate to social and environmental systems. It is, however, important that they be included in meaningful ways because considerations related to them should be important elements of recommendations which are made on the mitigation of the social and environmental impact of resource development policies and programs.

The principal objective of this paper is to show the difficulties that are being encountered in achieving a balance in Northern Canada between the development of non-renewable resources on one side and the protection of the environment and the continuation of the traditional uses of the land by the native people on the other. This goal was enunciated as part of a policy statement entitled Northern Canada in the '70's (Chreffian, 1973). In attempting to accomplish my objective I will trace the history of the development of petroleum resources in the Canadain Arctic during the past twenty years. I will relate aspects of the development to some environmental and social issues and I will discuss some of the inter-departmental problems which have arisen.

THE APPROACH TO DEVELOPMENT

During the past twenty years the Government of Canada has made a very intensive effort to encourage the multinational resource corporation of the world to undertake exploration programs for oil, gas and minerals in the Northwest Territories and the Yukon Territory.

The effort to find and exploit these resources was based primarily on the growth-progress syndrome not, at least until very recently, on a sense that they were required to meet Canadian needs.

Although the pace of northern development began to pick up immediately following World War II, in retrospect many Canadians think of 1958 as an important bench mark. It was an election year and the Conservative party was making a strong bid to convince Canadians that it was the party of the future. The campaign focused strongly on what a Conservative Government would do to promote resource development. The Vision of the North was a prominent campaign slogan of the party. It is often referred to now as "Diefenbaker's Vision of the North" after the venerable politician who lead his party to victory in the election and who is still a member of the House of Commons.

Subsequent events show clearly that the stage was being set for a program of intensive development. Land and geological surveying was greatly intensified and in 1960 greatly improved geological maps were published. These maps showed the location of formations likely to bear petroleum and minerals. Prior to 1964, the average number of claims staked north of '60 was less than 6000 a year. During the next five years there were five major staking rushes, reaching a peak of 52,000 claims staked in 1968 and representing a tremendous level of activity in exploration and development (Passmore, 1970).

In 1964, a railway was completed to Hay River on Great Slave Lake and to Pine Point where a large lead-zinc mine was being developed. Late in the decade intensive studies were conducted to determine the feasibility of establishing a deepwater port at Herschel Island on the Beaufort Sea.

In 1966, a Northern Exploration Assistance Program was established. It provided up to 40% of the exploration costs of Canadian citizens, or of companies incorporated in Canada. Added to this, were many speeches extolling the virtues of northern development and referring in glowing terms to the rapid evolution of the "petroleum play".

In the case of petroleum the only field in production was a small oil field at Norman Wells which had been discovered in the 1920's. However, the acreage held by oil companies under exploratory permits had increased rapidly after the publications of the geological maps in 1960. Much of the interest was of a speculative nature and there was only little exploration of permit areas in the early part of the decade.

The announcement of a major discovery of oil and gas at Prudhoe Bay came just after the Canadian Government had joined a consortium to form Panarctic Oils Ltd., in which it held 45% of the equity. The new company immediately began an active exploration program on its holdings of approximately 44 million acres in the Arctic Islands. It has constituted the most successful operation in northern Canada. To date, it has discovered approximately 15 trillion cubic feet of gas and undisclosed quantities of oil. But, at the same time Panarctic achieved an unimpressive record in the conduct of its operations. It had made two major gas discoveries out of the first 8 wells drilled. Both had blown out - one on Melville Island and one on King Christian Island. Fortunately for the environment, neither discovery was crude oil (Woodford, 1972).

After drilling wells on the Arctic Islands, the company commenced exploration of its offshore holdings in 1974 and 1975 with the drilling of wells to delineate the extent of one of the major gas fields which it had discovered (Pimlott, 1974).

Many of the large multinational corporations also became interested in the potential of the Arctic and by 1971, 460 million acres were under permit (Yates, 1972). Earlier, the oil industry had been invited to draft the oil leasing regulations so they would provide maximum incentives for exploration. The leasing arrangements were much more favourable than those which prevailed in Alaska and in terms of petroleum speculation they had certainly accomplished their purposes (Thompson, A.R. and M. Crommelin, 1973). The areas held under exploration permits represented virtually all the potential oil bearing formations north of the 60th parallel. It also included large offshore holdings. These covered the Beaufort Sea to the edge of the Polar Pack, and virtually all of the waters enclosed by the Arctic Islands including the Northwest Passage. It even included McLure Straits at the Northwest end of the Passage which is normally free of ice only one year in ten.

In the post-Prudhoe Bay climate, expenditures on exploration increased from $24 million in 1967 to $170 million in 1971. They remained at a high level through 1974. They are said to have dropped during 1975 because of uncertainty caused by the Government's failure to clarify royalty and other taxation policies.

There have been several discoveries of gas and oil in both the Mackenzie Delta region of the Western Arctic as well as in the Arctic Islands. In the case of gas, the combined total of all the finds still falls short of that discovered in the Prudhoe Bay Reservoir. Although no announcements have been made on the size of the oil discoveries it is believed that they still fall far below the volume required for commercial production.

The present situation is that much of the focus on exploration is changing from the land to the sea. It appears that the results of exploration programs have indicated that elephant-sized discoveries are unlikely to be made on land. However, the industry remains hopeful that offshore regions contain reservoirs of the magnitude of Prudhoe Bay.

Offshore operations will be of quite a diverse nature. Up to the present time they have included drilling from artificially constructed islands in shallow water areas at the mouth of the Mackenzie Delta; the use of land rigs on thickened ice platforms in the Arctic Islands and the use of a semi-submersible rig in Hudson Bay. In 1976, two drill ships are scheduled to begin

operations far off the land in the Beaufort Sea. Approval in
principle has also been given for the operation of a drill ship in
Lancaster Sound in 1975; however, the company involved has decided
not to proceed with the operation this year. The well there was
to have been drilled in 2500 feet of water.

Soon after the Prudhoe Bay discovery the Canadian Government
became intensely interested in the possibility of both oil and
gas pipelines being constructed from Alaska through the Mackenzie
Valley to markets in the United States and southern Canada. A
network of interlocking committees was formed and in 1970 as a
result of their efforts the Government promulgated a generalized
set of guidelines for pipeline construction.

Progress toward a rapid and smooth integration of the energy
resources of the Canadian Arctic into a Continental Pool was
profoundly disturbed by the voyages of the Manhattan into and
through the Northwest Passage in 1969 and 1970. One result was
that the importance of the Arctic came into perspective for many
Canadians for the first time. However, the passage created a
chaotic situation in the echelons of the Federal Government.
Although they were never publicly admitted, there were strong fears
that Canadian sovereignty over the Northwest Passage was being
challenged by the United States. The Government reacted by
enacting the Arctic Waters Pollution Prevention Act in 1970. Its
enactment was primarily a matter of international politics. It
was a clear case where an expression of concern for the environment
was used as a means of strengthening Canada's jurisdiction over
the ice-bound waters enclosed by the Arctic Islands. The United
States objected strongly to some of the provisions of the Act
which it claimed were contrary to international law. However, the
Bill was passed unanimously by the House of Commons in June 1970
and came into force in 1972 when Regulations were promulgated under
the Act (Pharand, 1973).

The 'qualified' success of the Manhattan in traversing the
Northwest Passage, the legal barriers which were being posed to
the Trans Alaskan Pipeline by U.S. environmental groups and the
fears in British Columbia that major oil spills would occur off
the B.C. coast from tankers operating from Valdez to Washington
encouraged pipeline proponents in Canada. The two departments
(Indian Affairs and Northern Development and Energy, Mines and
Resources) which were promoting the construction of both oil and
gas pipelines moved rapidly to take advantage of the promising
situation. Early in 1971, Canadians became aware of the fact that
the Federal Government was offering a Canadian route as an
alternative to the Trans Alaskan pipeline. The offer was being
made in a series of speeches by Ministers of the Federal Cabinet.
However, we also believed that it had been made directly at an
informal meeting with high ranking members of the U.S. administration in Washington by the Minister of EMR. Significantly, these
activities followed shortly after the TAP's hearings in Washington
and Alaska at which the Department of the Interior's preliminary

environmental impact assessment was reviewed and strongly criticized.

However, environmental concerns over the development of oil and gas had also surfaced in Canada over the construction of an oil pipeline. Under pressure in the House of Commons and in the press, the Government felt it necessary to qualify statements which had been made about Canada's willingness to accept an oil pipeline. Top executives of multinational oil companies jetted in and out of Ottawa; by the late spring of 1971 it had become evident that the industry would continue its fight to gain approval of the Trans Alaskan Pipeline. The Federal Government recognized that while there would probably be a gas pipeline through the Mackenzie Valley, there was no immediate hope for an oil pipeline. The Federal Government then went on and attempted to ensure that nothing would stand in the way of the construction of the gas pipeline.

In the meantime two consortia, Arctic Gas Ltd. and the Northwest Project Study Group, had been developed as competing organizations to prepare feasibility studies, engineering plans and social and environmental impact assessments. The Government brought strong pressure for the amalgamation of the two consortia and this was achieved in 1973. The new organization, Canadian Arctic Gas Study Ltd. continued EPB but de-emphasized its role. The job of preparing the environmental data for the public hearings was given to the more traditional environmental section which had come into CAGSL from the Northwest Project Study Group.

These hearings are now going on before Justice Thomas Berger who was appointed in January 1974 to study and recommend on the application for the right-of-way of a gas pipeline through the Northwest Territories and the Yukon Territory.

ENVIRONMENTAL: ISSUES AND PROBLEMS

During the 1960's few Canadian conservationists, outside of the Federal Government, realized the extent of the developments which were shaping up in the Arctic. Virtually no questions were being raised about the potential impact of exploration on Arctic ecosystems in Parliament, the press or even in the publications of conservation organizations.

In retrospect it seems evident that Canada simply did not have either the individuals or the organizations which could, or would take the initiative to alert the country to the dangers posed by the massive thrust by government and industry in the Arctic.

The development of powerful national conservation organizations has proven to be difficult in Canada. During the 1960's, the Canadian Audubon Society (now the Canadian Nature Federation) and the Canadian Wildlife Federation were the principal national conservation groups. Both operated at submarginal levels throughout the decade; bankruptcy was an ever present threat. As a result, neither had much capability to monitor the policies, programs or activities in the North. As will be seen, the CWD did an important job in 1970 of bringing the problems into focus in spite of the limitations imposed on it by the shortage of money and human resources.

In the case of governmental agencies, the Canadian Wildlife Service was the principal one which was interested in the protection of Arctic ecosystems. Members of CWS were actively engaged in studies of Arctic foxes, caribou, muskoxen, waterfowl, seabirds, and other species in the postwar period; however, the main thrust of these studies was over by the early 1960's. Local government was evolving in the Yukon and N.W.T. Game management was one area which was assumed by the Game Branches when territorial governments were formed. Thick walls soom developed between the two organizations and CWS encountered strong resistance to its request for funds for Arctic research.

During the critical pre-development period, CWS had little influence on the formulation of environmental policies for the Arctic. I became aware of this in 1969 as leader of a study for the Science Council of Canada (Pimlott, et al., 1971); we attempted to make an analysis of the situation for our report. However, CWS was extremely sensitive about its relationships with the Territorial Game Branches and refused to give our research team access to the files which would have provided us the knowledge we needed to review the situation and report on it to Government and to the public.

The responsibility for the aquatic ecosystems of the North came directly under the Department of Fisheries (which developed into the Department of the Environment in 1972). In this case there were no jurisdictional conflicts involved with either Territorial Government. However, the department was very ineffective in representing Arctic environmental interests. The Fisheries Research Board established an Arctic Biological Station in 1955. It was formed to be responsible for "Canadian research on marine and anadromous fish and biological oceanography in northern waters, as well as marine mammals on all coasts". In terms of the two territories alone, its area of responsibility represented 30% of the area of Canada; however, up until 1972, "the full-time scientific staff of the station has never exceeded nine!" (Sprague, 1973). The interest of the Department, and of FRB, was primarily in the commercial fisheries of the two coasts; as a result it did even less than the CWS to promote the development of an ecological conscience among the agencies who were so actively promoting the development of oil, gas and minerals in the Arctic.

The result of the lack of environmental purview began to be evident to Canadians in 1969, the same year that Panarctic Oils drilled its first well on Melville Island and had it blow out of control. It turned out that virtually nothing had been done to enact environmental legislation to control any phase of exploration activities. As in the case of the Naval Petroleum Reserve in Alaska, the heavy equipment involved in seismic operations had been allowed to work on the tundra during the summer. It had resulted in severe erosion of the tundra on the Tuktoyaktuk Peninsula and in at least one other area of the N.W.T.

Dr. W.A. (Bill) Fuller's interest in Arctic conservation began in 1963 when he had "... observed cat tracks on the North slope of Alaska west of point Barrow". This caused him to become concerned about the potential effect of exploration activities on the Arctic (Fuller, 1970). He worked over a five-year period within the International Union for the Conservation of Nature and succeeded in convening a Conference on Productivity and Conservation in Northern Circumpolar Lands in October 1969 at the University of Alberta (Fuller and Kevan, 1970).

It was a significant event. The activities leading up to it stirred interest at many levels, including the Department of Indian Affairs and Northern Development. Nine days before the Tundra Conference Mr. Chrétien, the Minister of the Department, gave his first "We will protect the Arctic" speech. His department made a modest grant to support the conference. He sent a telex message which was read on his behalf at the opening of the conference. It stated that his department would establish land use regulations, support a broad program of research and propose water conservation legislation to Parliament (Chrétien, 1970). A member of his department presented a paper which filled in some of the detail (Naysmith, 1970).

The Conference passed a resolution welcoming his statement but while "... endorsing the program, deep concern is expressed that it should be implemented on a scale and with a speed and determination that will adequately cope with the thrust of development now penetrating the north." (Fuller and Kevan, 1970, p. 326).

The Conference was conducted as a very circumspect, unemotional scientific activity; however, it served as a focus for Canadians to develop perspective on what could happen to the Arctic if exploration crews were allowed to run rampant in the North.

The new perspective was sharpened by the course of events in the United States and in Canada in 1970. The preliminary injunction against the construction of TAP which was granted under the mandate of the National Environment Policy Act in April 1970; the sinking of the tanker Arrow off the coast of Nova Scotia brought out how little was known about the effect of oil and how impossible it would be to clean up a spill in the Arctic (McTaggart-Cowan, 1970); Panarctic's second natural gas well blew out in October and burned with a deafening roar for 91 days; in February the Canadian Wildlife Federation sent out a 'Crisis in the North'letter. It stated a succinct case on the nature of the hazards and argued a strong and rational case for a partial moratorium on oil exploration, lasting until 1974.

The proposal never got to first base but it, and many other events of 1970, intensified the discussion among the public. This speeded up the enactment of environmental legislation and made it evident to government that people were finally aware of what was happening in the Arctic.

Before the end of the year the Arctic Waters Pollution Prevention Act and the Northern Inland Waters Act were passed by Parliament. The Land Use Regulations were also promulgated under the Territorial Lands Act which had been revised by Parliament so that land use activities could be controlled under it.

The happenings and the expressions of concern for Arctic environments during the early 1970's were too numerous to review in an account of this nature. They are documented in the proceedings of a federal-provincial wildlife conference, a conference of the Royal Society of Canada, a special report by Pollution Probe at the University of Toronto and many articles and editorials in newspapers and in television documentaries. To a large extent, the focus was on the construction of pipelines but other types of activity were occasionally involved. The period was epitomized by very sharp contrasts. Background studies for the assessment of environmental impact of the construction of the gas pipeline were quite intensive. They were conducted on behalf of industry by an environmental section of the Northwest Project Study Group and by the Environment Protection Board.

The Department of the Environment had many scientists involved in research along the route and the Arctic Land Use Research (ALUR) program provided grants for a variety of studies.

Offshore drilling in the Beaufort Sea and elsewhere in the Arctic provided an example of the contrast. In this case, DIAND had been encouraging exploration for years. However, in 1973 when the Cabinet was requested to give approval-in-principle for off-shore drilling operations, no specific research had been undertaken and the proposal to Cabinet did not make any mention of the need for an environmental impact assessment. In addition, the plans for the operation were shrouded in secrecy. To write a report on them for a native organization, the Committee for the Original Peoples Entitlement (COPE), I had to search out restricted, confidential and secret reports and other documents within govern-ment agencies and industry (Pimlott, 1974).

A decision was finally made to conduct a research program in the Beaufort Sea. It was an entirely inadequate program conducted in a single season in one of the worst ice years in a decade. In other Arctic areas, offshore drilling has been conducted without even the pretence of research programs. In one case, approval was based on an environmental impact assessment which made a mockery of scientific processes (Canadian Arctic Resources Committee, 1974a).

In another case, where the construction of artificial islands was involved, an environmental impact assessment was made of the effect of construction of an island on beluga whales; however, the terms of reference for the study specifically excluded the consultant from considering any other impacts of the exploration process- The most obvious one was the effect of oil spills (Pimlott, 1974). This topic would have been particularly relevant because the island was constructed at the interface of the Mackenzie Delta and the Beaufort Sea, an area which was very critical to many species of birds as well as to whales.

One of the most controversial happenings was the announcement by Prime Minister Trudeau in April 1972 that an immediate start would be made on the construction of a Mackenzie Valley Highway. It would terminate at Tuktoyaktuk on the Beaufort Sea and open up the entire valley to traffic and people from the South (Trudeau, 1972).

When the announcement was made the preliminary engineering surveys had not been done. In addition, no research had been conducted, and none was ever undertaken, to determine either the potential social or environmental impact of the project. This in spite of the fact that by then the Government had required the pipeline consortia to embark on such a program in preparation for public hearings on the pipeline.

On the side of industry one of the most positive initiatives was taken by Arctic Gas Ltd. It contracted to have its environmental studies and its environmental impact assessment prepared by a specially constituted body, the Environmental Protection Board. It seemed like an important innovation because EPB included some of the best-known ecologists in Canada; most important was the fact that the results of all studies were to be made public at the discretion of the Board. However, EPB barely survived the amalgamation of the consortia. In addition, it has become clear that the petroleum industry considers that the EPB was a dangerous experiment. Unfortunately, there seems to be no likelihood that the endeavour will be considered as a precedent for the industry to follow in the future. Rather, it appears that the model will be the Arctic Petroleum Operators Association. This organization has hosted annual arctic environmental conferences and has sponsored studies on various topics of interest to exploration and production systems. These studies are, however, held confidential among the sponsors and are unavailable to environmental organizations, and even to many government agencies.

Perhaps the most significant development on the public interest side was the formation of the Canadian Arctic Resources Committee (CARC) in 1971. It sponsored a workshop in early 1972 on social, environmental and legal aspects of northern development (Pimlott, et al., 1973). I stated why we had decided to form CARC in these terms:

"By the end of March (1971) we were convinced that Canada badly needed an organization which would provide a pair of eyes to look in on the North in a more perceptive way than any existing citizens' organization was capable of doing; which could act in an

Honest Broker capacity to attempt to ensure that the things that
needed to be done in advance of development of whatever type,
got done; which could help to bring to the surface the question of
what was to be done about the claims of the native people; and
which could help to overcome the barrier to factual information
existing between the Canadian people and the Government on matters
that pertained to development, the native people and the environment." (Pimlott, 1973).

We solicited financial support of both the government and the
oil companies for the Workshop. It was our objective to emulate
the low-key approach of the Tundra Conference, while at the same
time dealing frankly with the problems. But, even though we were
very thorough in our approach to fund raising, no oil company nor
any department of government would give us financial assistance
for the Workshop. This is the way I spoke of CARC and the
situation it faced in my talk at the opening of the Workshop:

"In the first place, it was an attempt to form an organization
which would further public participation in quite different ways.
Those of us who conceived of the organization had often taken
adversary positions on government policies on resource and
environmental matters. But we recognized that significant things
appeared to be happening to attitudes and approaches towards both
the environment and the native people. We reasoned, if attitudes
in government and industry had changed, perhaps it was possible
to achieve results through processes of reason rather than of
confrontation.

In the second place, there are not many organizations where
people with a wide diversity of backgrounds work together to further
the cause of the native people or the environment. We hoped to
add a little to the development of that type of organization.

But the experiment goes beyond that. It seemed to us that
there are at least two fundamental problems which must be faced if
Canadian society is to come to terms with socio-cultural and
environmental problems. One is the need for much greater day-to-
day participation by people in decision-making processes. The
other is honouring the right of people to know. (I apologize for
the cliche - I know of no phrase that states the case better.)
Government in Canada is not finding it easy to come to terms with
these problems. Part of the experiment has been to determine if
CARC could do anything to speed up the process of letting the
people in on decisions in the North. As far as I am concerned that
is what the Canadian Arctic Resources Committee is mostly about."
(Pimlott, 1973).

CARC has been very active on Arctic problems and issues since
1972. It has sponsored a series of publications including a
regular letter, Northern Perspectives, a technical report on Land
Use Regulations, the proceedings of two additional conferences.
It is a far cry from the doldrums of the 1960's.

In terms of the evolution of environmental policies and
programs for the Arctic the greatest disappointment has been the
role of the Department of the Environment. A second disappointment
has been the inability of the Department of Indian Affairs and
Northern Development to overcome the conflicts of interests which
are represented by having the dual role of protecting the environment and promoting the development of non-renewable resources.
This is the way I reflected on the problem at a CARC conference a
year ago:

"The formation of Environment Canada delighted many conservationists. At last, we said, there is going to be a department of
Government which has a broad mandate to protect the Canadian
environment. Its formation seemed to bode particularly well for
the Arctic since there were no competing governments in the North
to develop confusion and jurisdictional squabbles. But since 1970,
the Government has increasingly removed the responsibility of
protecting the Arctic environment from the jurisdiction of Environment Canada and placed it under the Department of Indian Affairs
and Northern Development which is 'seeking the exploitation of non-
renewable resources'. Environment Canada was not named as an
administrative or a cooperating agency for the enforcement of either
the Arctic Waters Pollution Prevention Act or the Northern Inland
Waters Act. In addition, during the past winter I have realized
that its prerogative to enforce the Fisheries Act in the NWT is
being progressively reduced and taken over by DIAND. The
Environmental Protection Service of DOE has been, and is, operating
at less than a marginal level both in terms of staff and budget.
It even has unwritten orders to clear all of its actions to enforce
Section 33 of the Fisheries Act with DIAND. EPS is so ineffective
that it might as well be withdrawn from the Territories."

To further widen the credibility gap, the responsibility for
environmental protection has not been given to the Conservation
Branch of DIAND but to the branch which until recently was called
The Northern Economic Development Branch. Of course, if the job
was being done in an adequate way it would not matter what agency
or department did it. But there is much evidence that DIAND has
not resolved the conflict of interests between development and
protection. The examples are legion.

The Land Use Regulations have been authoritatively criticized
on many occasions by Andrew Thompson and others. More recently
Peter Usher and Graham Beakhurst have brought the weakness in
implementation of the regulations into sharp focus. But in terms
of perspective on environmental policy one of the most significant
things was the fact that mineral exploration was virtually excluded
from the regulations - by carefully worded definitions.

DIAND's and DOE's approaches to preparation for offshore
drilling in the Beaufort Sea is another example of the Government's
lack of capability to bring environmental protection into perspective with the development of non-renewable resources. That
case history has been documented in some detail in a report which

I wrote for COPE (the Committee for the Original Peoples
Entitlement) and for the last issue of Northern Perspectives so I
will not detail it here. A few points will serve to bring it into
focus:

(1)   DIAND has been issuing permits for petroleum exploration in
      the Beaufort Sea since the early 1960's. It made no attempt
      whatsoever to foster or encourage a program of research to
      assess the potential impact of exploration and development
      on the environment prior to submitting a memorandum to
      Cabinet to obtain approval-in-principle for offshore drilling.

(2)   The memorandum to Cabinet failed to give Cabinet members any
      appreciation of the environmental hazards and risks which
      would be involved in petroleum operations in the Sea. DOE
      had two representatives on the Committee which reviewed the
      memorandum so must share some of the responsibility for its
      inadequacies.

(3)   DIAND allows the petroleum industry to 'shelter' potentially
      important reports on environmental topics, such as meteoro-
      logical and ice conditions, under proprietary interests
      arrangements so that they are not even reasonably available
      to members of government. It even allowed these proprietary
      interests to dictate that a major conference on offshore
      drilling in the Arctic be held in camera and the distribution
      of the report restricted to participating oil companies and
      government agencies.

(4)   The Government tried very hard to prevent information on the
      project becoming publicly known. In fact, all the important
      documents associated with the plans were rated as either
      Restricted or Confidential. In retrospect, the most obvious
      effort at subterfuge was the steps taken to disguise the
      construction of artificial islands as normal extensions of
      land-use operations rather than to identify them as the first
      stage of offshore drilling operations.

"There are many more examples of DIAND's inability to incorporate environmental considerations in decisions on resource
development. Even a short list would include the limited research
undertaken to determine the impact of seismic and other exploration
work on the tundra and animals in the Delta and the Arctic Islands;
approval of Panarctic's first offshore drilling operation after
only a crude and limited study had been made of environmental
considerations; the recent approval-in-principle by Cabinet of major
expenditures to support the development of mining operations in the
Strathcona Sound area of Baffin Island, again before DOE had conducted any environmental research; the decision to promote the
development of the hydro potential of Great Bear Lake before any
program of environmental assessment had been undertaken; the

failure of the Government to undertake even preliminary studies of much of the animal-resource base of the people of Victoria Island and Resolute Bay, even though oil companies were granted exploration permits for large areas on the island some time ago."

The role of DIAND in protecting Arctic environments is of course seen in a much more positive way by members of the department (e.g. Chretien, 1973; Naysmith, 1971, 1973a, 1973b).

In retrospect, experience has shown that the concerns reflected in the resolution passed at the Tundra Conference in 1969 and by the Canadian Wildlife Federation in 1970 were justified. However, concern shown by Canadians over the possible impact of Arctic development has as yet only had a modest influence on government policies and programs for the Arctic.

THE RIGHTS OF THE NATIVE PEOPLE

The aboriginal rights of the Indian and Inuit (Eskimo) people of the N.W.T. and the Yukon were ignored during the thrust for the development of Arctic Petroleum resources in the 1960's.

The most striking documentation of that statement is provided in the report of a study of the people of Banks Island. Usher (1971) traced how the government had issued petroleum exploration permits for the entire island (28,000 sq. miles) and the adjacent waters without informing the Bankslanders of what was happening. In June 1970 representatives of two oil companies visited Sachs Harbour to tell the people that they held exploration permits and would conduct seismic exploration during the winter of 1970-71. There appeared to be no recognition of the fact that the people, who lived almost entirely from trapping, would feel that the wildlife and, as a result, their interests were threatened by the operations.

The case history gives a thorough account of the callous way the Bankslanders were treated. Exploration did come to the island against the objections of the people.

However, as a result of the confrontation, they have been able to maintain some control over the time of year when exploratory operations will be conducted (Sachs Harbour Hamlet Council, 1974).

There have been some improvements in the approach of government since 1970 but the changes have not been profound ones. In the early 1970's, when government and industry were intricately involved in planning for offshore drilling operations, native people were not consulted and were not told about plans until they had reached the fait accompli stage. For example, DIAND held a Northern Canada Offshore Drilling Meeting in December 1972 as one stage in the preparation of a submission to Cabinet to get approval-in-principle for the project. Officials from oil, gas and consulting companies were included but no native organizations were invited to send representatives. The Committee for Original People's Entitlement (1974) reflected on the situation in one of its news releases.

The rationale for the situation which existed through the 1960's was perhaps given by Prime Minister Trudeau, who in referring to Indian land claims stated the government "would not recognize aboriginal rights" (Trudeau, 1969).

The negative attitude of the government, knowledge of the moratorium on development in Alaska until native claims were settled, the rapid expansion of exploration activities in the North and the discovery of oil at Atchinson Point on the Tuktoyaktuk Peninsula in 1970 caused a great deal of concern among native people. One result was the formation of five organizations to represent their interest between 1969 and 1972. These included the Committee for the Original People's Entitlement (COPE), Indian Brotherhood of the N.W.T., Inuit Tapirisat of Canada (national Eskimo organization), Yukon Indian Brotherhood, the Metis Association of the N.W.T., and the Federation of Natives North of Sixty.

The pressures on the government from these organizations, and from public opinion in southern Canada gradually forced the government to adopt a more flexible policy. All the organizations eventually received financial support from the government to support their programs; by mid-1971 the Minister of DIAND had begun to make statements such as "... I am ready at any time to sit down to discuss the settlement of their treaties" (Chretien, 1972). On August 8, 1973, the Minister issued a communique on the Claims of Indian and Inuit People which stated that the government had changed its stance and was willing to negotiate a land settlement.

It remains to be seen how profound the change in policy really is. In the first place, the government has never considered a moratorium on development, such as was imposed in Alaska, while claims are being negotiated; secondly, the communique appeared to exclude the possibility that the native people would gain major control of land and resources. The key statement seemed to be "An agreed form of compensation or benefit will be provided to native peoples in return for their interests" (DIAND, 1973). Thirdly, a proposal for a settlement of Indian claims in the Yukon, made in a white paper earlier this year, appears to fall far short of the hopes and expectations of the native people of that territory (the legal basis of land claims and the historical record of settlements has been reviewed by Cumming (1973).

But, there are also things to be positive about. The government provided quite liberal funding for land use and occupancy studies to establish the basis for negotiation of both Indian and Inuit claims. In the case of the N.W.T. the Inuit studies are virtually complete; the Indian studies for the N.W.T. will probably be completed within a year.

Another positive action was the appointment of Justice Thomas Berger of the British Columbia Supreme Court to conduct the hearings to determine the social and environmental impact of a gas pipeline through the Mackenzie Valley. When in law practice Justice Berger had represented the interests of an Indian Band on a land claim in British Columbia; as a jurist he has a reputation of being an able arbitrator.

In retrospect an action taken by the Indian Brotherhood of the N.W.T. was probably of paramount importance in stimulating the government to put out the communique stating that it was willing to negotiate on land claims. The action was the filing of a caveat in the Territorial Court early in 1973. It warned developers to beware of going ahead with major programs, like pipeline construction, because the native people had a legal interest in the land and resources. In September 1973, Mr. Justice William Morrow found that the Indian people had sufficiently established their case to give them the right to file the caveat (Wah-shee, 1974).

PERSPECTIVE ON MAN AND RESOURCES IN THE ARCTIC

Each year since 1962, when I ended the last of four periods of my life spent in government service, I have become increasingly involved in the public-interest side of conservation and environmental affairs. For some years my attention focused on the preservation of natural areas and endangered species. My interest in these is still very strong; however, my energies are devoted more and more to broader areas of environmental concern, such as the one I am talking about today. The evolution was prompted by the realization that preservation of bits and pieces of species, parks and wilderness will have little meaning unless they are part of a beautiful, diversely patterned fabric which is represented by a whole healthy earth.

The increasing level of my activities has resulted from a growing conviction that democratic societies will not be able to come to terms with problems of environmental degradation unless 'people' become much more involved in the day-to-day decision-making processes of government.

Intricate symbiotic processes involving government and industry have developed over the years as the growth/progress syndrome evolved. At the same time the growth of massive bureaucracies has resulted in government and politicians becoming more and more remote from the people they represent.

Events documented in many environmental case histories demonstrate that the system is not working well. There are many examples which indicate that when large investments are involved it simply does not seem to be possible for industry, and particularly those working in remote areas, to give environmental protection adequate consideration in the scheme of things. It is also often the case that the priorities and conflicts within governments result in either inadequate laws and regulations or inadequate enforcement of those which exist.

In Canada, the pipeline situation is a positive one because
public hearing processes are involved. In the case of offshore
drilling, the potential environmental hazards and risks are
probably greater than in the case of the pipeline but there is no
requirement for hearings so the processes involved have been very
shoddy ones. A great deal of dust is being swept under the rug.

In the United States, public process is much more firmly
established so it would seem probable that a better balance could
be achieved. However, I am aware that limited financial resources
always pose serious problems to the groups involved. Organizations
with million-dollar annual budgets are pitted against corporate
giants which spend millions of dollars on advertising and public
relations programs every day.

My conclusion from all of this is that in Canada we must
continue to fight for a much more open process than we now have.
I think that in both countries we must try very hard to convince
governments that they should support environmental organizations
so they can prepare for and participate effectively in public
hearings. Professor Andy Thompson, present Chairman of CARC, has
been a consistent proponent of such an approach for Canada
(Thompson, 1973).

The Mackenzie Valley pipeline hearings have established some
precedents which I hope will have meaning for the future in Canada.
In this case the native organizations have received funding through
the front door (directly from DIAND) while the environmental groups
have been funded through the back door (by the Berger Commission).

However, the government demonstrated recently that it does
not like the process at all, at least as far as the environmental
organizations are concerned. The Northern Assessment Group (NAG,
a consortium of environmental and conservation organizations) had
received money from the Berger Commission to review the application
which the Canadian Arctic Gas Ltd. had made for the pipeline
right-of-way. But when NAG, as directed, applied to the government
for additional funds to cross-examine and present its evidence at
the hearings, it was twice refused by Treasury Board, which is the
Court of Last Appeal when federal funds are involved. The problem
was finally resolved by some intricate juggling which permitted
the Commission to again fund NAG's operations.

The implication seems very clear - the federal government does
not want to establish the direct precedent of providing the money
so that public hearings can become more meaningful processes. It
is obvious that we still have a lot of convincing to do in Canada.

A sense of perspective on the use of Polar regions should
surely be based on a sense of humility which our species has seldom
achieved. But we must somehow find both the humility and the
perspective if we are to change the frontier approaches which we
seem compelled to use in commandeering resources for our use.

LITERATURE CITED

Canadian Arctic Resources Committee. 1974. Offshore drilling in the Beaufort Sea and Arctic offshore drilling. In Separate issues of Northern Perspectives, both entirely devoted to the offshore topic.

Chretien, J. 1970. Statement from the Minister of Indian Affairs and Northern Development. In Productivity and Conservation in Northern Circumpolar Lands. p. 7-8.

Chretien, J. 1972. Mackenzie Corridor: Vision becomes reality. Keynote address to the 18th Annual Convention of the Pipeline Contractors Association of Canada. Montreal, May 11. Mimeo, 15 pp.

Chretien, J. 1973. Northern development for northerners. In Arctic Alternatives, p. 26-35 (see Pimlott, et al., 1973).

Committee for Original People's Entitlement, 1974. Drilling for oil and gas in the.Beaufort Sea. Press Release. Feb. 8. Inuvik, N.W.T.

Cumming, P. 1973. Our land, our people: Native rights North of '60. In Arctic Alternatives, pp. 86-110.

Department of Indian Affairs and Northern Development (1973). Claims of Indian and Inuit people. Communique 1-7339. August 8. 8 pp.

Fuller, W.A. 1970b. Opening address. In Productivity and Conservation in Northern Circumpolar Lands. p. 7-8 (see Fuller and Kevan).

Fuller, W.A. and P.G. Kevan. 1970. Productivity and conservation in northern Circumpolar lands. IUCN Publications new series. No. 16.

Naysmith, J.K. 1970. Conservation in Canada's North. In Productivity and Conservation in Northern Circumpolar Lands.

Naysmith, J.K. Canada North - Man and the Land. Department of Indian Affairs and Northern Development, Ottawa. 44 pp.

Naysmith, J.K. 1973a. Management of Polar lands. Proceedings of 12th Technical Meeting, International Union for the Conservation of Nature and Natural Resources, pp. 295-317.

Naysmith, J.K. 1973b. Toward a northern balance. Department of Indian Affairs and Northern Development. 30 pp.

Passmore, R.C. 1971. Environmental hazards of northern development.
Transactions Federal-Provincial Wildlife Conference. Canadian
Wildlife Service, Ottawa.

Pharand, D. 1973. The law of the sea of the Arctic with special
reference to Canada. University of Ottawa Press. Ottawa. 367 pp.

Pimlott, D.H. 1973. People and the North: Motivations, objectives
and approach of the Canadian Arctic Resources Committee. In
Arctic Alternatives, pp. 3-24.

Pimlott, D.H. 1974a. Offshore drilling in the Beaufort Sea: A
report to the Committee for the Original People's Entitlement
(COPE) Mimeo. 16 pp.

Pimlott, D.H. 1974b. Delta gas: Time and environmental considerations In Gas from the Mackenzie Delta, now or later? Canadian
Arctic Resources Committee. pp. 93-107.

Pimlott, D.H. 1974c. The hazardous search for oil and gas in
Arctic waters. Nature Canada Vol. 3, No. 3, pp. 20-28.

Pimlott, D.H., C.J. Kerswill, and J.R. Bider. 1971. Scientific
activities in fisheries and wildlife resources special study No.
15. Science Council of Canada, Ottawa. 191 pp.

Pimlott, D.H., K.M. Vincent, and C.E. McKnight. 1973. Arctic
Alternatives: A national workshop on people, resources and the
environment North of '60: Canadian Arctic Resources Committee.
Ottawa, Canada. 391 pp.

Settlement Council, Sachs Harbour, N.W.T. Press Release, March 11,
1974. Exploration for oil and gas in the summer on Banks Island.

Sprague, J.B. 1973. Aquatic resources in the Canadian North:
Knowledge, dangers and research needs. In Arctic Alternatives,
pp. 168-189.

Task Force - Operation Oil, 1970. Report to the Minister of
Transport on the Clean-up of the Arrow oil spill in Chedabucto Bay.
Vol 1. Information Canada, Ottawa.

Thompson, A.R., and M. Crommelin, 1973. Canada's petroleum
leasing policy: A cornucopia for whom? In Proceedings of
Conference by same name. Canadian Arctic Resources Committee.

Thompson, Andrew R. 1973. Public Hearing for Northern Pipelines in
Canada. Arctic Institute of North America.

Trudeau, P.E. 1969. Transcript of remarks at a Liberal Association dinner in Vancouver, August 8. From Violated Vision by J. Woodford.

Trudeau, P.E. 1972. Press release of notes for remarks by the Prime Minister to a public meeting. Edmonton, Alberta. April 28.

Usher, P. 1971. The Bankslanders: Economy and ecology of a frontier trapping community. Vol. 3. The Community. Department of Indian Affairs and Northern Development. Ottawa, 88 pp.

Wah-shee, J. 1974. A land settlement: What does it mean? In Gas from the Mackenzie Delta: Now or later? Canadian Arctic Resources Committee, Ottawa. pp. 83-92.

Woodford, J. 1972. The violated vision: The rape of Canadian North. McClelland & Stewart, Toronto. 136 pp.

Yates, A.B. 1972. Energy and Canada's north: The search for oil and gas. Nature Canada. Vol. 1, No. 3. pp. 9-14.

ARCTIC RESOURCE TRANSPORTATION:

PRESENT SYSTEM AND FUTURE DEVELOPMENT

J. L. Courtney

Ministry of Transport

Ottawa, Ontario, Canada

This Workshop is to examine a finite number of Arctic systems, drawing upon various techniques to examine present status and to identify problem areas requiring further research. In the transportation field, a number of quite recent government policy initiatives and scientific studies have provided new insights. If we define systems analysis as the rational analysis of complex problems, we find that a number of Arctic resource transportation systems have received systematic scrutiny to determine ENGINEERING FEASIBILITY AND FINANCIAL VIABILITY.

Mr. Nitzki of A.G. Weser outlined in his keynote address a number of the modal characteristics and the design parameters required for effective development. I would like to examine more closely the structural characteristics of present systems and the research being carried out for various potential resource systems. In particular, while this Workshop no doubt is most concerned with problem definition, study design and methods, I would like to inject an element of what is needed from you, the system analysts to assist the decisionmaker in the specification and evaluation of alternatives.

1. BACKGROUND

Systems serve as means of denoting diverse communalities integrated to deal with certain specified problems and sets of goals. In the case of the Canadian Arctic, all the systems being examined at this Workshop should be set amidst the goals defined for its development. Problem definition should be slanted to dealing with those issues that hinder the realization of each of these National Northern Objectives:

1. To provide for a higher standard of living, quality of
    life and equality of opportunity for northern residents
    by methods which are compatible with their own
    preferences and aspirations.

2. To maintain and enhance the northern environment with due
    consideration to economic and social development.

3. To encourage viable economic development within regions
    of the Northern Territories so as to realize their
    potential contribution to the national economy and the
    material wellbeing of Canadians.

4. To realize the potential contribution of the Northern
    Territories to the social and cultural development of
    Canada.

5. To further the evolution of self government in the
    Northern Territories.

6. To maintain Canadian sovereignty and security in the North.

7. To develop fully the leisure and recreational opportunities
    in the Northern Territories.[1]

     This policy statement also gives guidelines and priorities for
the attainment of these objectives. In essence it describes a form
of balanced development where the needs of people are given pride
of place and the ecological balance is maintained.
     The Ministry of Transport in Canada recently completed a
major transportation policy review (TPR), which concluded that
future government strategies for the development and regulation of
transport systems must vary with their "maturity". The Yukon and
Northwest Territories that make up the Canadian North are truly an
immature system, lacking in infrastructure and real inter-modal
competition. Bulk commodity markets for the Arctic fall into
quadrant "C" below, characterized by low competition (Y axis) and
immature systems:
A series of "northern extensions" to the present Canadian east-west
transport system will have a significant impact on the extent and
distribution of industry development. This should be borne in mind
when route selection is being made for the systems discussed in
this paper.

_____

1. Government of Canada, Policy for Northern Development, Ottawa,
    1972.

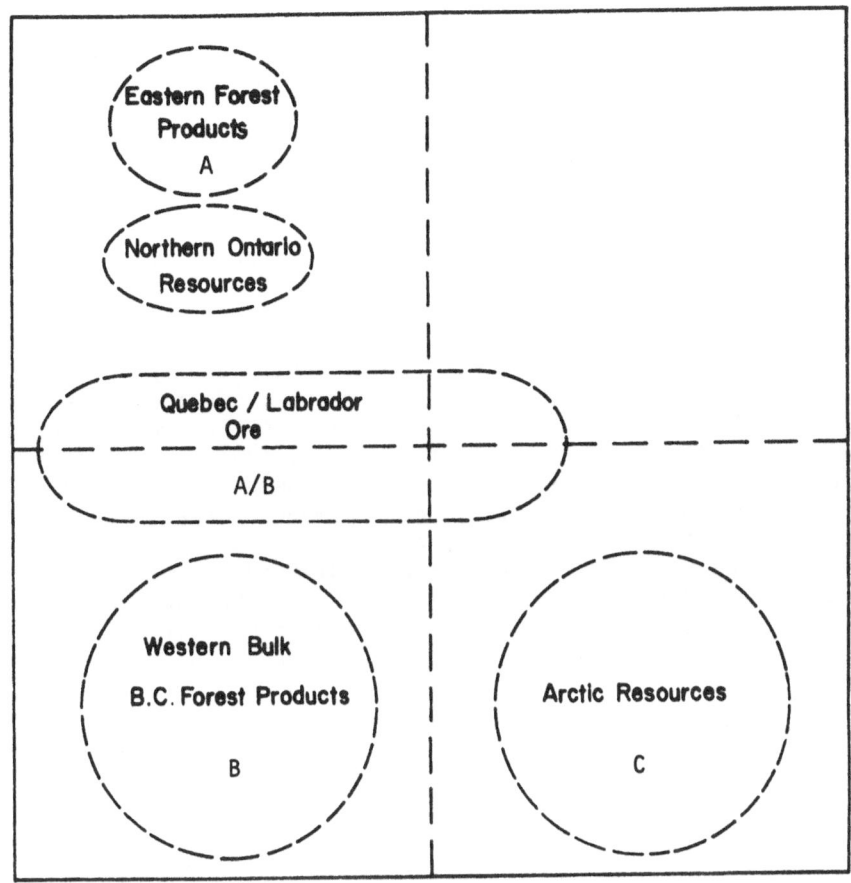

Bulk Commodities Markets

The role of the transport system is a varied one. Support
for resource exploration is provided, as is the means to actually
transport resources to market once the development stage is
attained. Both these themes will be explored for each mode.

People priorities do take precedence and affect the pattern
of resource exploitation. Communications are vital. This includes
not only telecommunications systems, but also the use of the air
mode to enable people to "communicate" with their peers, in the
south (for resource development workers) or in other communities
(for the native people).

2. GEOGRAPHIC SETTING

That part of the Canadian North with which I am concerned
lies north of the 60th parallel of latitude and comprises the
Yukon and Northwest Territories. A small portion of Nouveau-Quebec
also lies in about the 60th parallel but this part of the north,
of course, comes under provincial jurisdiction. The Yukon and
N.W.T. comprise 1½ million square miles and has a population of
less than 60,000. This region contains less than 1/4 of 1% of the
Canadian population. The distance east to west is 2,660 miles
whilst the distance north to south is 1,575 miles. When speaking
of such a vast region, one might call it a country, it is dangerous
to be too specific about the climatic and other geographic
conditions. The north generally, however, is characterized by
(a) long winters and short summers, (b) very little precipitation -
the region is often called the Arctic desert, (c) permafrost, both
continuous and discontinuous. This is that part of the earth
which is perennially frozen; it is caused more by the length of
the winter than the cold temperatures.

Climatically the Territories can be divided into two
divisions: the Arctic above the tree line and the sub-Arctic
below the tree line. The tree line extends in a generally south-
easterly direction from the mouth of the Mackenzie River near
Inuvik to the west coast of Hudson Bay just north of Churchill.
It is a relatively sharp demarcation line.

The impact of this geographic setting upon modal systems is
variable. The ice-infested waters of the Arctic restrict vessel
passage to a few months of the year. Permafrost inhibits road,
rail and pipeline construction. This leaves air transport as the
most flexible mode, restricted only by the change of seasons.

3. ARCTIC RESOURCES

As a guide to the places discussed in this paper, Figure 1
shows the frontier hydrocarbon discovery areas that are the subject
of so much recent attention. The question of alternate transport
systems hinges upon the most feasible, economic means to move
large volumes of oil and gas south from the Arctic Islands and the
Mackenzie Delta to markets in Canada and the United States. Each
area offers different logistic and system design challenges.

Other minerals are also found in this region. In particular,
Strathcona Sound (site N) and Little Cornwallis Island (site M)
mentioned by Mr. Nitzki, are shown in Figure 2. The locations
identified have potential for export by sea. There are also a
number of other sites in the southern part of the Yukon Territory
that are now developed - such as the Anvil mine. Around Great
Slave Lake, a number of mines are in operation, Pine Point is
connected by railway to the south and produces more than 400,000
tons of concentrates annually. On the north side of the Lake,

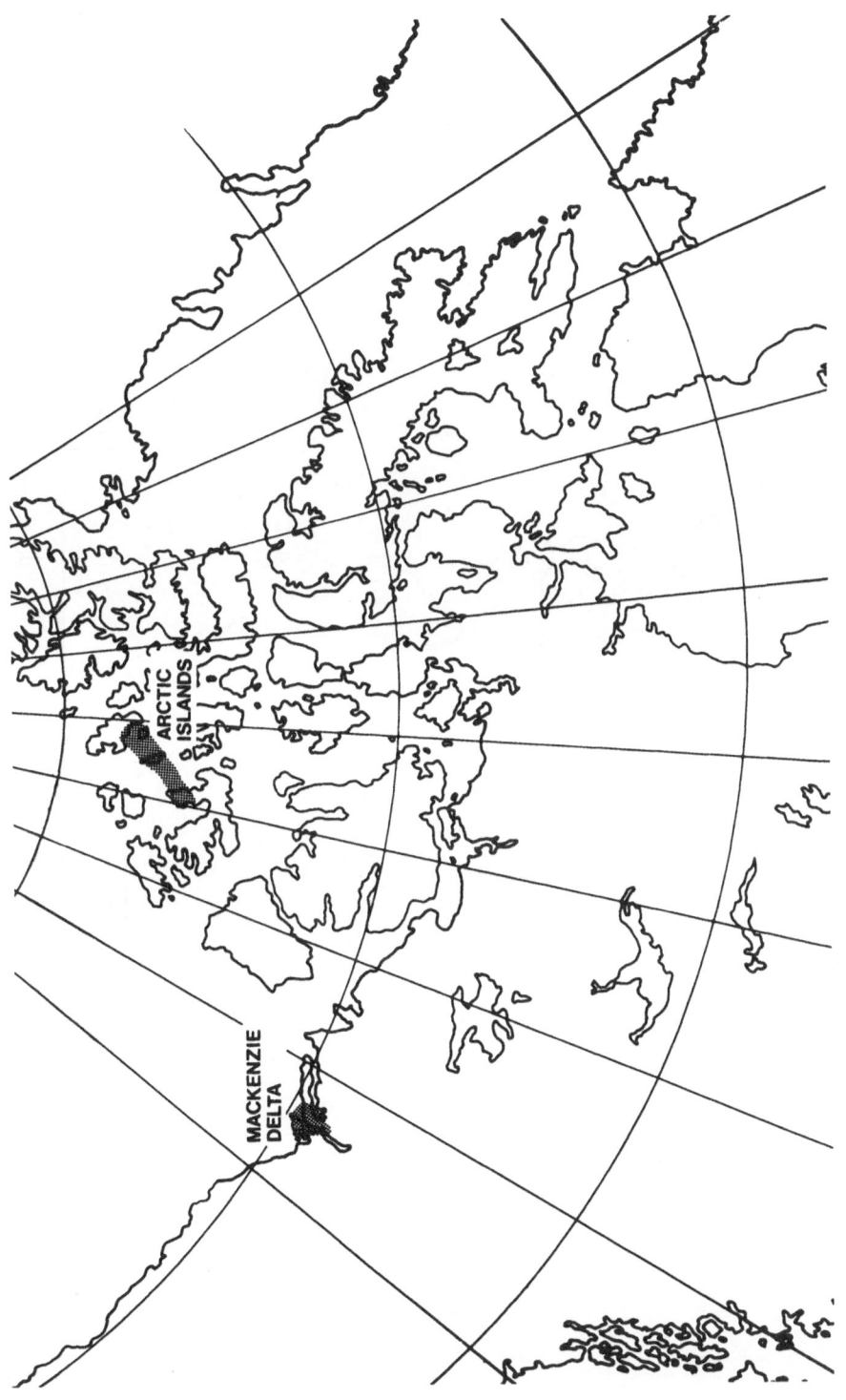

Figure 1. Frontier Hydrocarbon Discovery Areas

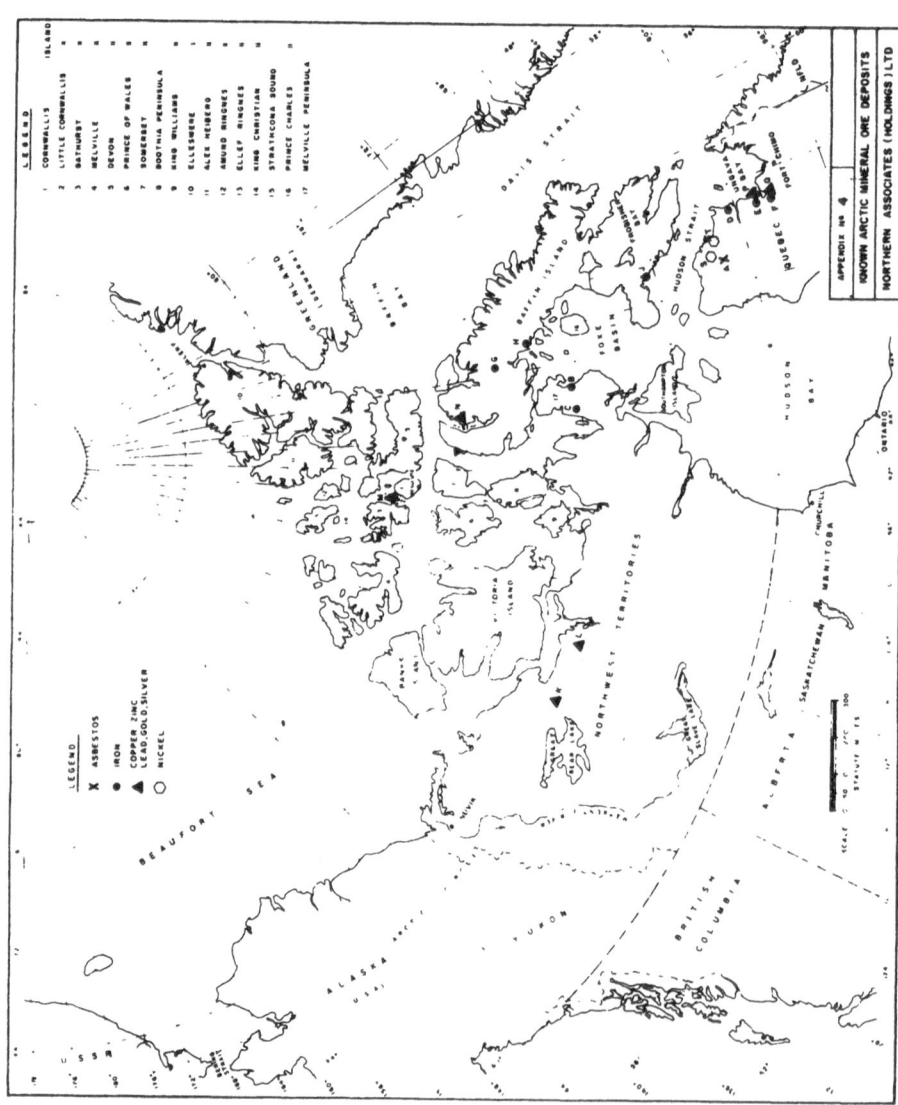

Figure 2.   Known Arctic Mineral Ore Deposits [Source: Northern Associates (Holding) Ltd.]

Yellowknife has been producing gold since 1936.

4. MODAL SYSTEMS

This section will examine the present systems with an indication of short term plans for their continued improvement. Research conducted to date on alternate modes for resource transportation are described with particular regard to engineering feasibility and economic viability.

While access to port facilities is a major consideration for such minerals as lead, copper and zinc, hydrocarbon development is perhaps more flexible. Figure 3 outlines the system parameters for various modes that have been suggested for gas movement in the North. It is taken from the Canadian Arctic Gas submission.

4.1 Air

The air mode is of prime use in logistical support for resource exploration and development. An extensive network of aerodromes and air navigation facilities exist in northern Canada. Figure 4 indicates all communities and with the exception of some located on highways all are served by some type or class of air service. In many of the smaller communities the air transportation facilities may not support regular, reliable air services. The standards for this program are noted in Figure 5.

We are now in the second year of a multi-year facilities program. While the policy itself represents the Government's desire to accelerate the provision of an adequate system of air transportation and navigation aids in the Territories, in other respects it can be said to represent a more enlightened approach in that, as a sociological objective, it recognizes that Canadians everywhere are entitled to safe, reliable and regular transportation services, in this case by air.

Without aircraft such as Lockheed Hercules which has been used extensively for resource exploration, we would not have achieved our present knowledge of our northern resources. Now proposals are coming forward to utilize large aircraft such as the Boeing 747F to market these resources. One example is the use of such aircraft to transport those components of natural gas which cannot be transported by a gas pipeline. This proposal by Dome Petroleum Ltd. is to use 747F aircraft operating up to 60 flights per day from the Mackenzie Delta to Hay River where these gas liquids would be transferred to a surface mode such as rail. The alternative is to reinject them into the reservoir until such time as an effective transportation system is available.

Another similar proposal has been put forward by Dr. Vern Atrill. He proposes this same aircraft be utilized for transporting liquified natural gas from the Sverdrup Basin fields to a location

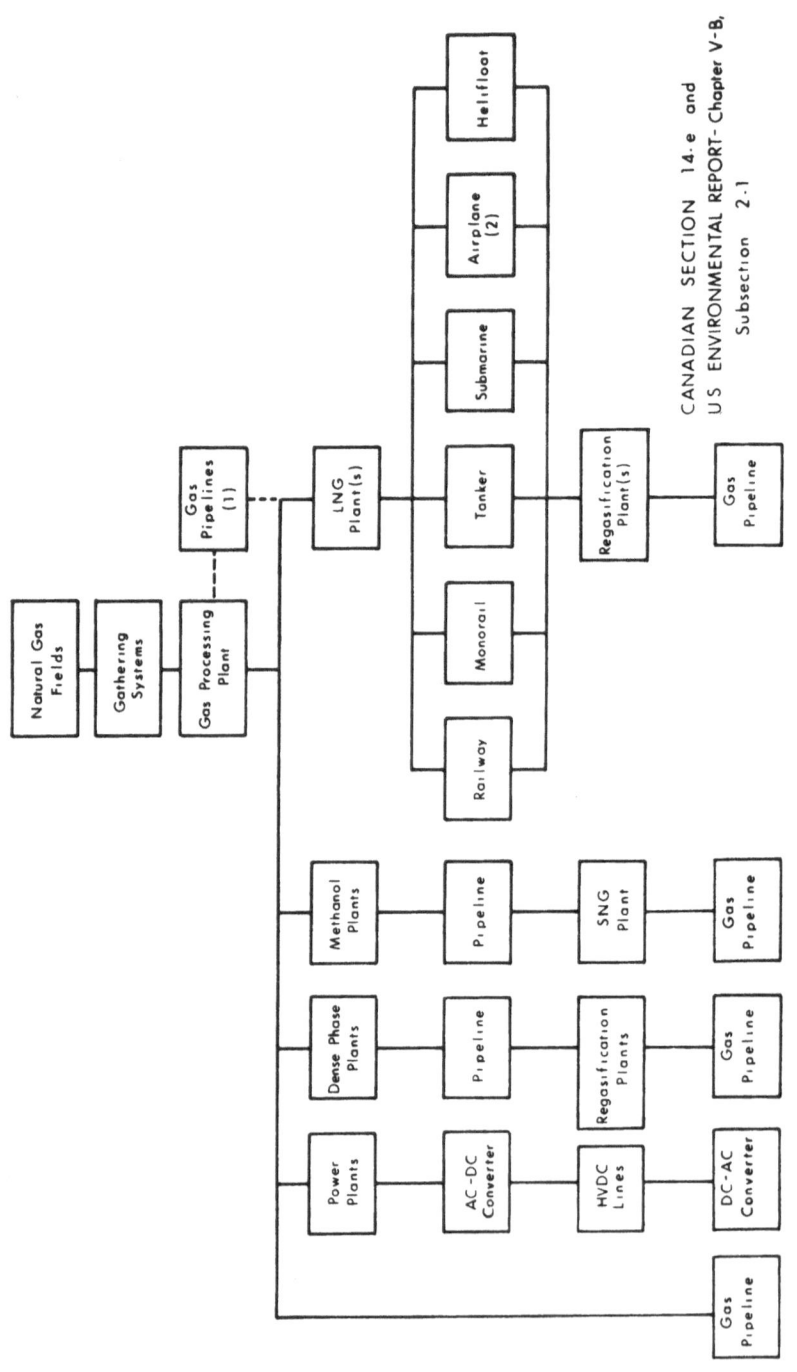

(1) For the tanker or submarine system, pipelines are required to move the natural gas to a central LNG Plant

(2) Short range and long range alternatives were estimated

Figure 3. Principal Process Steps in Delivering Energy to Market by Various Transportation Systems [Source: Canadian Arctic Gas Limited]

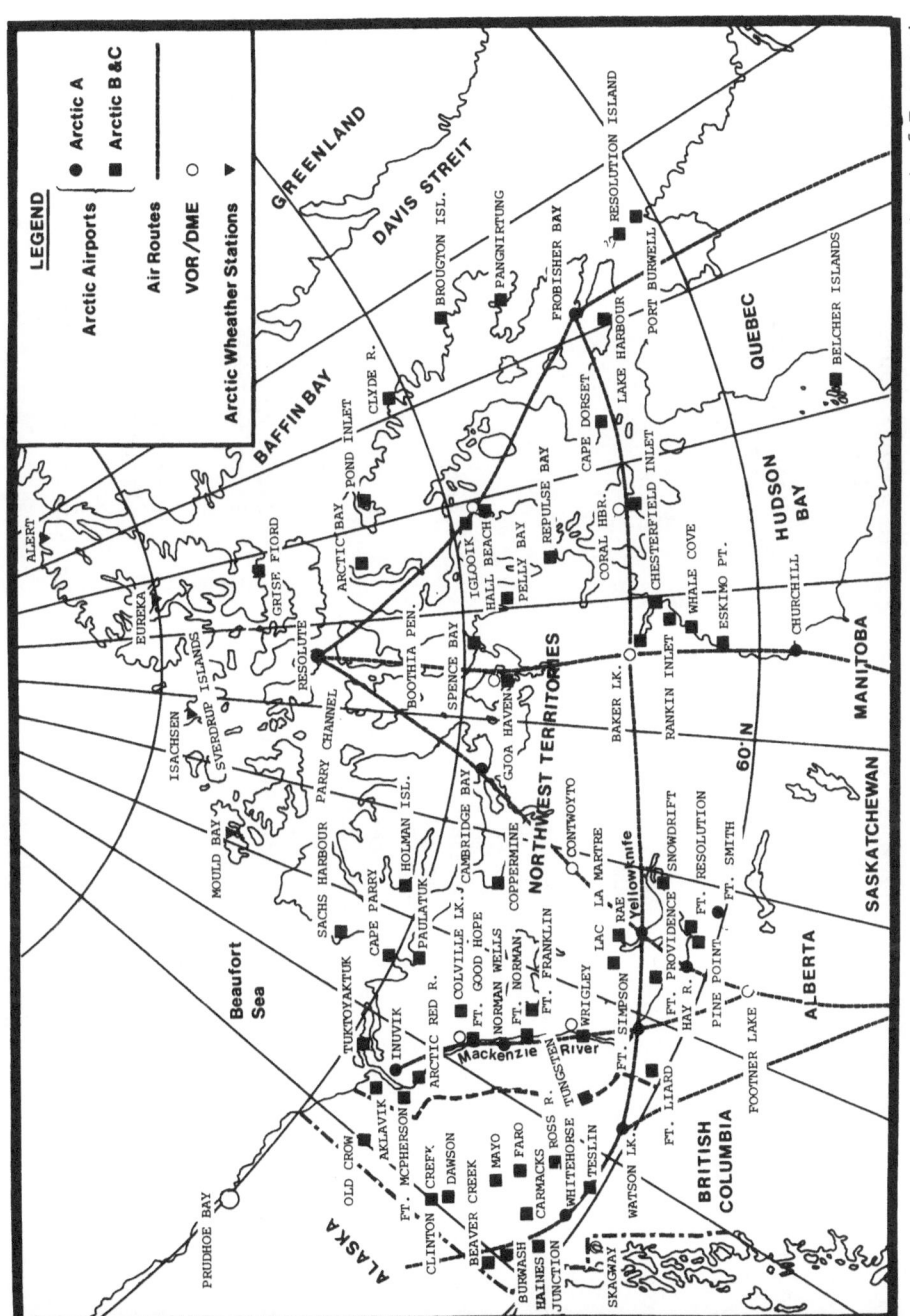

Figure 4

Department of Transport

	Arctic A	Arctic B	Arctic C
Runway	6000' x 150' Paved	5000' x 150' Gravel	3000' x 100' Gravel
Lighting	High Intensity Approach & Runway	Medium Intensity Approach & Runway	Low Intensity Approach & Runway
Approach Aids	ILS NDB	NDB	NDB
Navigation Aids	VOR/DME	NDB	NDB
Passenger, Aircraft & Airport Facilities	Terminal Apron & Access Rd. Fuel Maintenance Bldg. Equipment	Same	Same
Communications	Aeradio Station	Air Ground Point to Point	Air Ground Point to Point
Meteorological	Full Time Program	Part Time Program	Part Time Program

Figure 5. Minimum Standards for Airport Facilities

on the east side of the Arctic Islands or the south side of
Lancaster Sound/Barrow Strait from where it could be transported
by marine.

It would appear that the development of a super-large 12-engine
aircraft for moving hydrocarbons is now seen as beyond present
technology and not competitive economically with more conventional
modes.

4.2 Marine

Present use of this mode is generally restricted to resupply
of coastal and river settlements. Generally, for safe delivery of
cargo to communities, except those on the coast of Hudson Bay,
support from Canadian Coast Guard icebreakers is required. At the
present time, Government cargo carried northward in the annual
Eastern Arctic resupply is of the order of 100,000 tons. Non-
government commercial cargo tonnages are now in excess of those
shipped by and for the Government. Total tonnage moved into the
Eastern Arctic is now approximately 300,000 to some 60-70
settlements and exploration sites, and involving some 80 ships and
7 icebreakers. In the Western Arctic the annual resupply is
carried out by the Northern Transportation Company Limited, a
Crown Corporation. Total tonnages delivered in this part of the
Arctic are now of the order of 300,000 to 400,000 tons, to
Mackenzie River points and to settlements along the Arctic coast
as far east as Spence Bay. Oil exploration has accounted for
large tonnage increases since 1968. Again the commercial vessels
must be provided with Coast Guard icebreaker support along the
Arctic Coast.

In the wake of Prudhoe Bay discoveries, considerable
attention was given to transporting oil to markets by large
tankers through the Northwest Passage. A major event in this
investigation was the transitting of the Northwest Passage in 1969
by the U.S. tanker "Manhattan", accompanied by a Canadian icebreaker.
Probably as a direct result of this voyage it became clear to the
Canadian Government that ice-strengthened super-tankers might well
be plying the ice-infested waters of the Canadian Arctic in the not
too far distant future. With very little known - still true today -
about the possible irreparable damage that a major oil spill could
have on the delicately balanced ecosystems in ice-covered Arctic
waters, Canada acted promptly to invoke safeguards to minimize the
pollution hazard that will always exist where oil tankers are
concerned. The Arctic Waters Pollution Prevention Act was
proclaimed and the attendant Regulations governing water-borne
traffic in the Arctic were produced. The Act and the Regulations
divided the Canadian Arctic into a total of 16 zones as shown in
Figure 6, and provided a classification of vessels to conform with
specified operating seasons in each of the zones. The regulations
contain direct and appropriate requirements for vessel

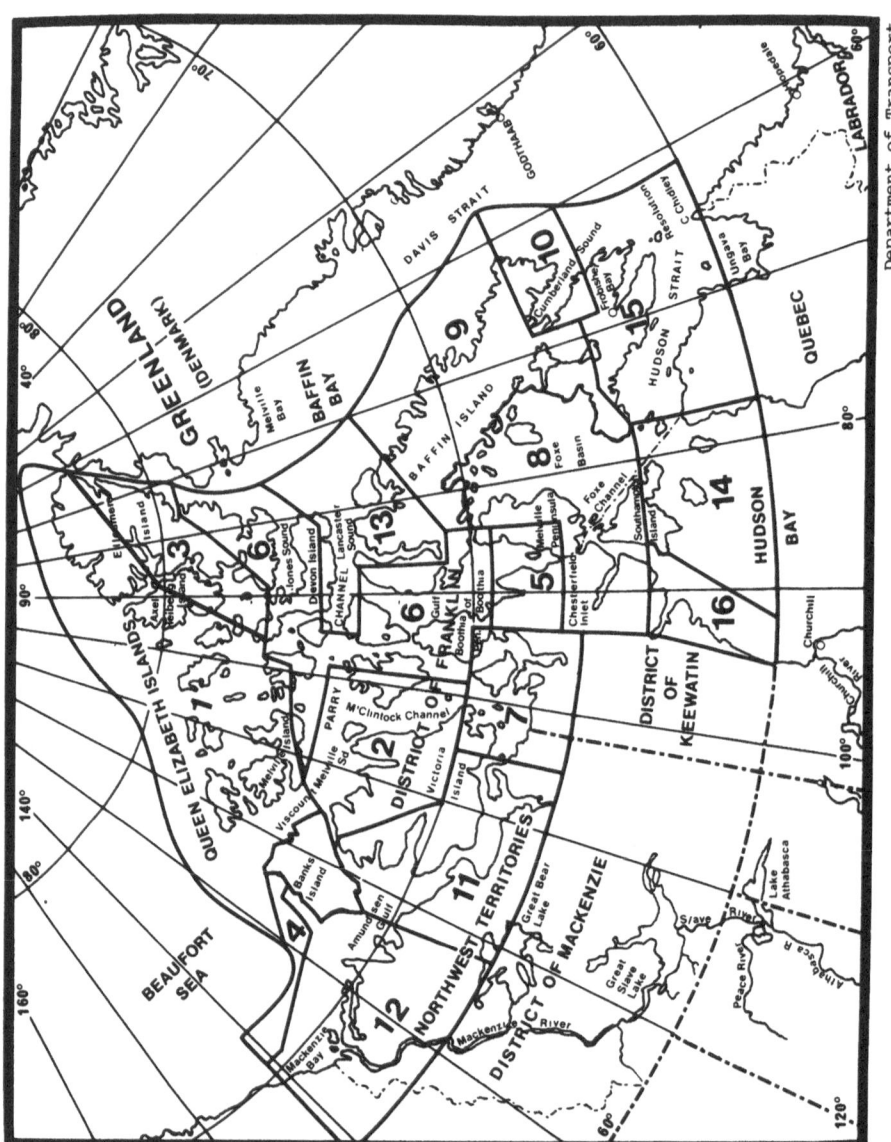

Department of Transport

Figure 6 Canadian Arctic Divided into 16 Zones for Classification of
Vessels to Conform with Specified Operating Seasons

construction, specifying Arctic vessels in classes from 1 to 10
with the number of class corresponding to the ice thickness through
which the vessel can make steady progress.

Present marine facilities are limited to temporary loading
areas for resupply, to the major harbour facilities at Churchill
on Hudson Bay, and to the major NTCL installations at Hay River,
Inuvik and Tuktoyaktuk. All Mackenzie River facilities would be
expanded to handle the more than 1 million tons required for pipeline and gas plant construction should either of the two
applications before the National Energy Board be successful.

Year round Arctic marine shipping will have to contend with
an ice mantle for which, as an impediment to shipping, large
knowledge gaps exist. In preparation for the likely resources
development in the Arctic and the likely role of the marine mode,
two recent actions by the Canadian Government are of significance.
In 1973 the Canadian Government announced publicly the intent that
Canada develop within 5 years an internationally recognized
excellence in operating on and below ice-covered waters; and even
before this announcement, the Minister of Transport had
commissioned a comprehensive study to a leading Canadian consulting
firm, for which the terms of reference were:

"To examine and make recommendations as to the suitability,
applicability and relevant economics of marine surface, sub-
surface and submarine vessels for the transportation of bulk oil,
gas and minerals from the Canadian Arctic to the main world
markets for these commodities."

The report entitled, Arctic Resources By Sea, by Northern
Associates provides a valuable inventory of the present state-of-
the-art. One important economic conclusion is that, based upon
the 150,000 ton dwt. vessel size, cargoes can be moved from the
Arctic to the North American east coast at estimated shipping
rates not exceeding 1/2¢ per ton mile. Ice zones limit the type
of shipping (Figure 7) and proposed design has been adjusted for
the various ice environments (Figure 8).

One recent study on the transport of oil from the Delta bears
noting. Imperial Oil Ltd. has studied the feasibility of
transporting crude oil from the Mackenzie Delta to east coast
ports, possibly Point Tupper via the Northwest Passage.

Vessels in the 30,000 dwt. class will be carrying concentrates
from Strathcona Sound to Europe in 1977. For subsequent years,
the Ministry of Transport is now committed to the construction of
an Ice Class II cargo vessel, as a joint industry-Government
venture, and this vessel may, as one of its first commercial
applications, participate in the transport of Strathcona Sound
concentrates. The $3.6 million dock at Strathcona Sount is the
first to this degree of sophistication in the High Arctic.

A major area of concern is ice research, which others are
dealing with during the Workshop. Considerable research is required.
Questions of labour practices, new training, "Canadian content",
user charges for icebreaker services, and construction cost-sharing

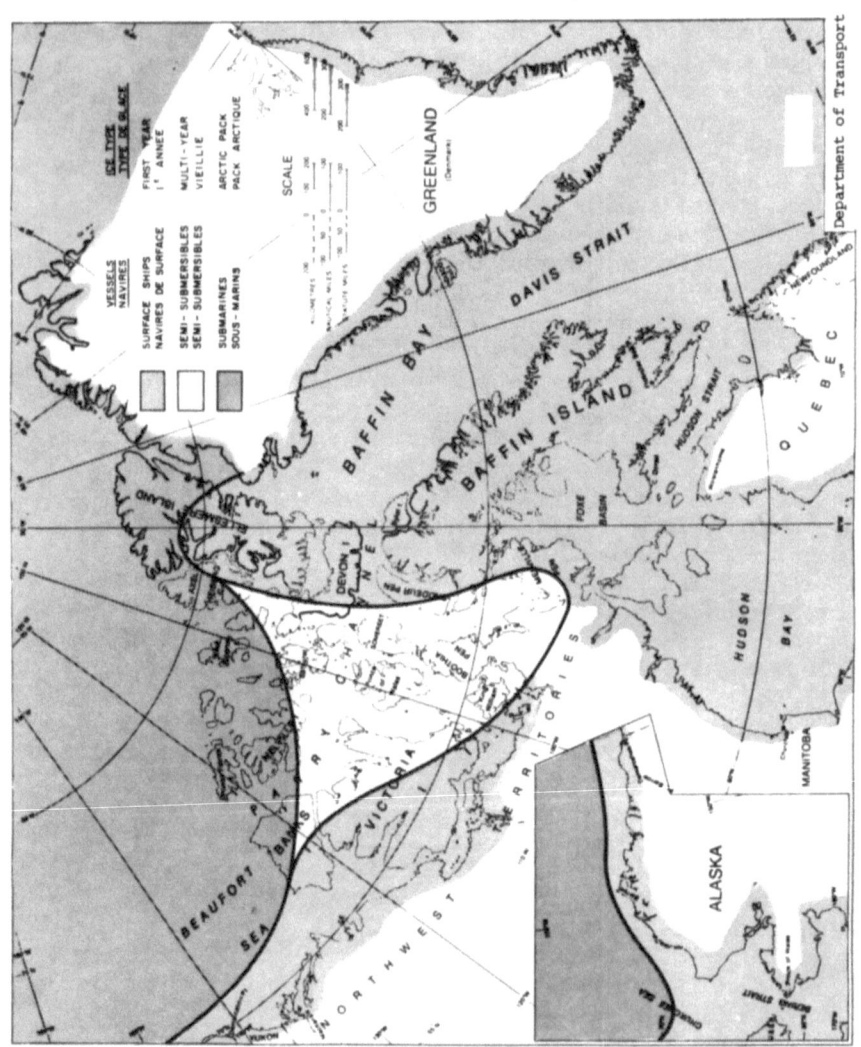

Figure 7   Map Showing Ice Zones Limiting the Type of Shipping

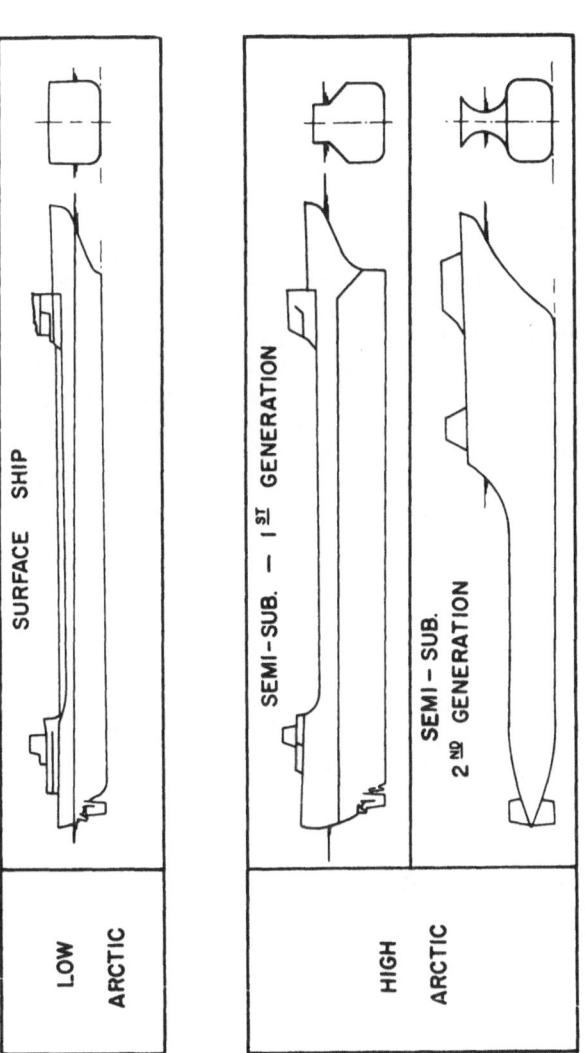
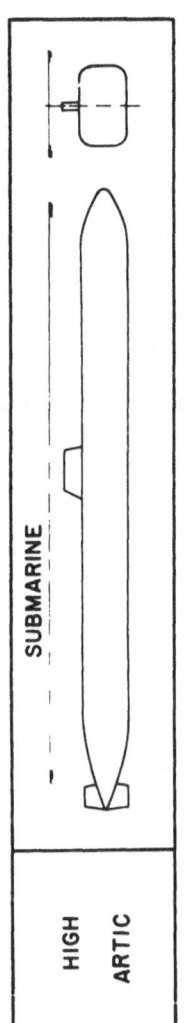

Figure 8   Proposed Designs of Transport Vessels for Various Ice Environments

for single-purpose docks all have to be confronted by government and industry.

Systems research is needed, and the number of papers both in this session and others attest to the growing importance attached to the marine mode in Arctic resource exploration and development.

4.3 Rail

Rail transport has linked the mineral resources of the Yukon Territory to tide-water at Skagway since the beginning of this century. More than half a million short tons of zinc, lead, asbestos and copper concentrates are handled annually. This is projected to exceed one million in the early 1980's, and this volume could double by the year 2000. In the northern Yukon, there is an extensive iron ore deposit at Snake River which could produce more than 10 million short tons annually after the year 2000. A separate railway would probably be built to this last deposit. Extensions to the existing railway could be built over the next twenty years north-east or north-west of Whitehorse to move a variety of minerals. Cost estimates are now being developed and would appear to range from $40 million to $200 million depending upon the trackage involved. Presently, trucking is used for the mines north of Whitehorse, but it is possible that potential deposits could warrant the construction of one or more of these railway extensions.

Mention has already been made of the railway to Pine Point. Churchill is also linked to the south by rail. Major new rail projects outside the Yukon relate to the movement of hydrocarbons to the southern extent of permafrost to connect with conventional pipelines. CN-CP have recently completed a study for the Ministry of Transport concerning the enginerring feasibility, cost and time required to construct a railway for the transportation of frontier oil and gas from the Arctic. This massive systems study, running to more than 1310 pages, deals with route selection, construction, maintenance and financial return.

The railway would serve both the Mackenzie Delta and Prudhoe Bay areas. The route would lie along the Arctic coast from Prudhoe Bay to Arctic Red River where it would join the branch originating near Swimming Point in the Delta. The preferred route would be south, along the west side of the Mackenzie River, crossing north of Norman Wells or at Wrigley, and again at Fort Simpson, terminating at a hypothetical point at Trout River, where transshipment to pipeline would take place. The Trout River point is chosen as a point south of permafrost and not as an optimum transfer point. A feeder rail line would be built from Trout River to Enterprise for connection to the continental rail system for movement of supplies other than oil and gas (Figure 9).

The line would comprise three main sections: Prudhoe Bay to Arctic Red River, 507 miles; Swimming Point to Arctic Red River,

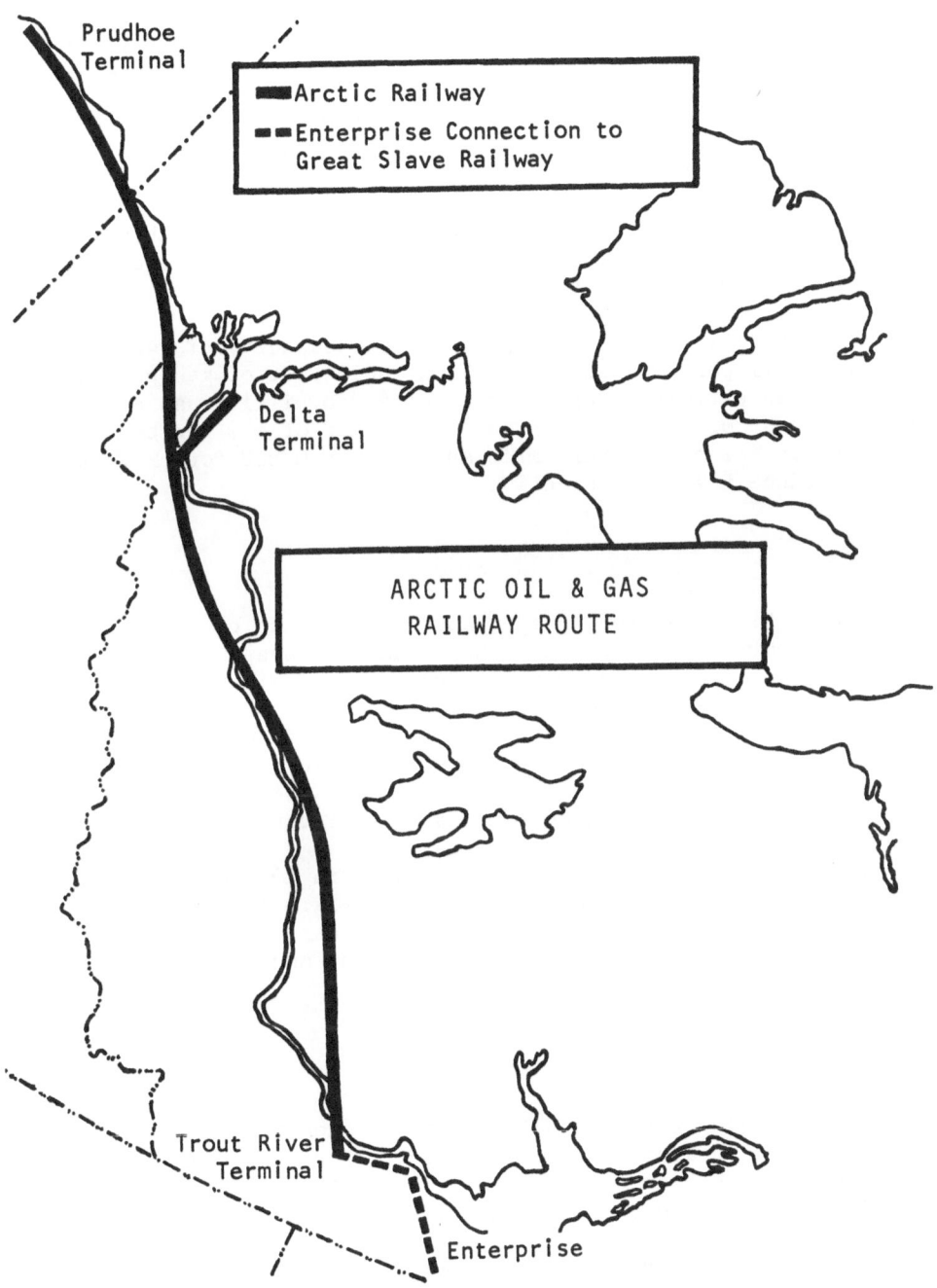

Figure 9

60 miles; Arctic Red River to Trout River, 507 miles.

Full capacity operation would require up to 70 trains per day in each direction, each comprising 225 cars and 7 locomotives for oil, and 110 cars and 4 locomotives for gas, necessitating a double track line. Natural gas would be shipped as LNG. A liquefaction and a regasification plant would be required at the north and south terminals respectively.

The CN-CP Study concludes that a railway system devoted to the transport of either oil or gas alone achieves reasonable transport cost levels only if large quantities are involved. A combined oil and gas operation would offer the lowest transportation costs (Figure 10).

The full oil and gas system, a maximum capacity of two million barrels of oil per day and six billion cubic feet of gas per day, would have an estimated capital cost of $12.8 billion, an annual operating cost of $430 million, an oil tariff of 64¢ per barrel and a gas tariff of 44¢ per MMBTU and 93¢ per MMBTU for ex Delta and ex Prudhoe Bay gas respectively.

4.4 Road

As Mr. Nitzki stated, road transport tends to be limited to local transportation. However, the Yukon road system, which is relatively well-developed, permits transport of mine concentrates from as far away as Dawson to the rail-head at Whitehorse (Figure 11). In this case, road serves as the forerunner to rail for resource transport. In the Mackenzie Valley, road access would be useful for future pipeline construction. Other areas of the Arctic, particularly the High Islands, are of course quite unsuited for either road or rail transport.

Construction plans call for completion in 1977 of the Dempster Highway, linking Whitehorse to Inuvik. This would permit surface access from the Pacific to Arctic Oceans in Canada. The future of the Mackenzie Highway is at present uncertain, due to rapidly escalating costs.

4.5 Pipelines

The oilfield at Norman Wells in the Mackenzie Valley has been producing small quantities of oil products for more than forty years, which have been transported by barge to a number of communities down-river. A pipeline was completed in 1972 to tap a natural gas field at Pointed Mountain in the south-west corner of the Mackenzie District. Two applications are now before the National Energy Board that could see up to 4.5 BSCF of natural gas transported by pipeline along the Mackenzie Valley to southern markets.

Mackenzie Delta	Gas	Oil & Gas
Oil (Million BBL's/Day)	-	2
Gas (Billion SCFD)	3	3
System Capital Cost ($Millions)	5,294	6,869
Annual Operating Cost ($000)	161,252	281,786
Operating Employment	3,883	5,721
Oil Tariff (per BBL)	-	78¢
Gas Tariff (per MMBTU)	73¢	53¢

Mackenzie Delta	Gas	Oil & Gas
Oil (Million BBL's/Day)	-	2
Gas (Billion SCFD)	3	3
Prudhoe Bay		
Oil (Million BBL's/Day)	-	-
Gas (Billion SCFD)	3	3
System Capital Cost ($ Millions)	11,801	12,880
Annual Operating Cost ($000)	305,928	430,127
Operating Employment	6,971	8,965
Oil Tariff (per BBL)		
Mackenzie Delta	-	64¢
Prudhoe Bay	-	-
Gas Tariff (per MM BTU)		
Mackenzie Delta	59¢	44¢
Prudhoe Bay	97¢	93¢

Figure 10. Arctic Oil and Gas Railway

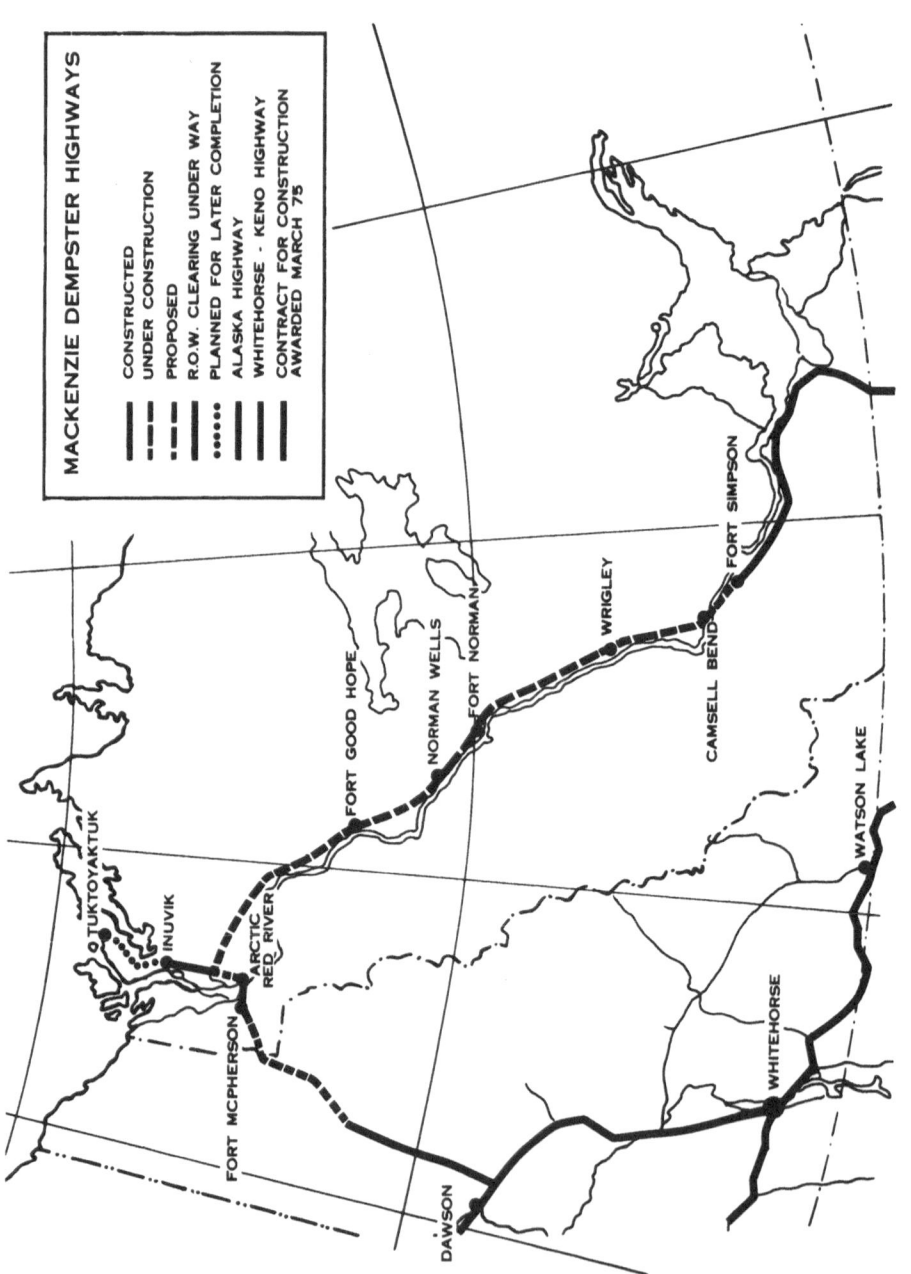

Figure 11

The Canadian Arctic Gas Pipeline Ltd. (CAGPL) is a consortium of seven companies of majority Canadian ownership, five companies of minority Canadian ownership and nine non-Canadian companies. The consortium proposes to construct and operate a pipeline for the transport of natural gas from Prudhoe Bay, Alaska and the Mackenzie Delta, to southern markets in both Canada and the United States (Figure 12).

One half of the capacity of the line would be devoted to Prudhoe Bay gas, destined for the U.S. market. The remaining half would carry Delta gas for domestic use and possibly export.

The pipeline would originate at Richard's Island in the Mackenzie Delta at the Taglu gas plant. It would be joined by the Parson's Lake pipeline at a point about 20 miles north of Inuvik. At Travaillant Lake, immediately south of the Delta, it would be joined by the 492 mile branch from Prudhoe Bay. It would proceed south along the east side of the Mackenzie River, recrossing the River near fort Simpson. It would then leave the River and proceed southward through Alberta and Caroline, a point north of Calgary, where it would divide into two branches, the eastern branch going via Empress to Monchy, a point on the Saskatchewan-Montana border, and the western crossing the B.C.-Idaho border near Kingsgate.

The Canadian portion of the line would be a total of 2,404 miles in length, of which 1,727 miles would be 48 inch diameter and 66 miles, those from Caroline to the two border crossings, would be 42 inch diameter.

The capacity of the line when in full operation would be approximately 4 billion cubic feet per day. The capital cost, originally estimated to be in the order of $7.5 billion, is now expected to be increased to $11 billion or more owing to the effects of inflation.

Construction of the line would require three years, with two years of pre-construction preparations of stockpile sites. Construction north of 60 would be performed in the winter, and logistics supplies would be transported in the summer navigation season.

An alternative to the CAGPL proposal is offered by Foothills Pipe Lines Ltd. who submitted an application in March of this year. The Foothills proposal is for an exclusively Canadian system to transport Mackenzie Delta gas only to Canadian markets. North of the 60th Parallel, the line would follow the same route as the CAGPL proposal but would omit the Prudhoe Bay. Laterals would be provided to serve the northern communities of Inuvik, Fort Good Hope, Norman Wells, Fort Norman, Fort Wrigley, Fort Simpson, Fort Providence, Hay River, Pine Point, Rae-Edzo and Yellowknife. The line would be 40 inch diameter, having a full capacity of 2.4 billion cubic feet per day. The capital cost is estimated at $4 billion. Construction of the line would require two winter seasons and could be ready for operation in 1979, if approval to start is given.

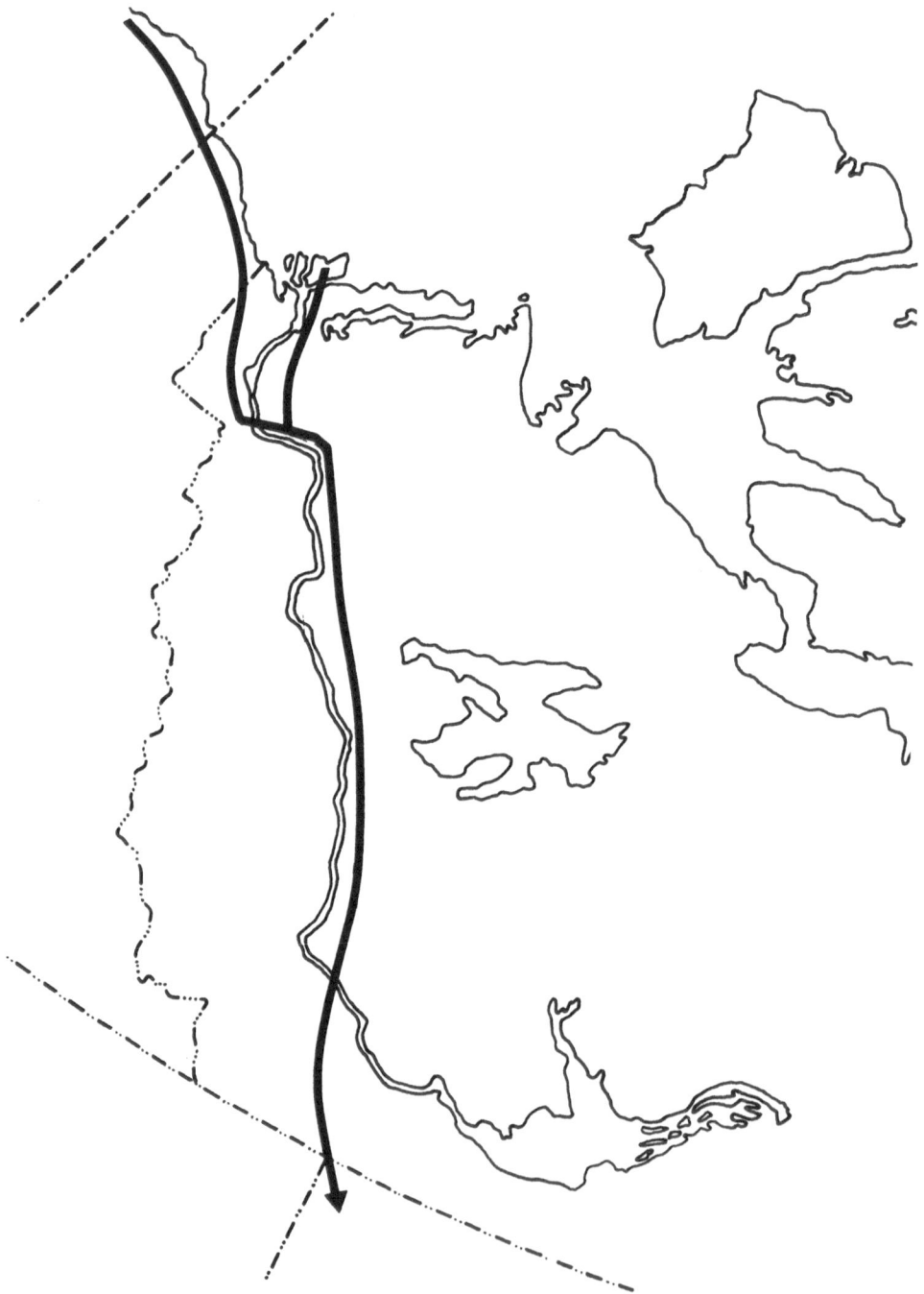

Figure 12. Mackenzie Valley Natural Gas Pipeline Route

5. INTER-MODAL CONSIDERATIONS

Inter-modal systems analysis is required, in some cases to sort out the alternative technologies, such as those outlined in Figure 3.

The most immediate comparison that comes to mind is pipeline versus railway in the Mackenzie Valley. While both have been studied individually, there are difficulties in making direct economic comparison between the two. The following standardizing assumptions are required:

(a) The pipeline costs are based on gas movement to the 49th parallel, whereas rail costs are only to Trout River. It is necessary to estimate the incremental values to the 49th parallel.

(b) Both systems throughput will be gas only.

(c) North slope gas will be available.

(d) Return on investment is the same for both cases.

(e) Equal depreciated life of 25 years.

The costs and tariffs for the Foothills and CAGPL pipelines are compared to the rail equivalents below in Figure 13. It is seen that the capital costs of the rail mode are approximately twice the pipeline equivalent, annual costs of rail exceed those for the pipeline by 2 to 3 times and the estimated tariffs reflect these differences.

		Volume BCFD	Capital Cost Billions	Annual Cost Millions	Tariff
Pipeline	Foothills	2.25	3.7	72.4	.65
	CAGPL	4.5	5.74	93	.695
Rail		3.0	5.3	161.3	.98
		4.5	11.64	299	1.359

Figure 13 Mackenzie Delta and Prudhoe Bay Gas Pipeline vs. Rail Comparison Summary

High Arctic oil and gas development is still a number of years away and it is uncertain whether pipeline or marine, or both will be used. The Polar Gas Project was formed in late 1972 to investigate the feasibility of a natural gas pipeline or other modes of gas transmission from the High Arctic Islands. Participants in the project are: Trans Canada Pipelines Ltd., Panarctic Oils Ltd., Tenneco Oil and Minerals Ltd., Texas Eastern Transmission Corporation and Pacific Lighting and Gas Co. Until recently, Canadian Pacific Investments Ltd. was also a member.

Panarctic Oils Ltd., a member of the project, was established to conduct major oil and gas exploration in the Arctic Islands. It is 55 per cent industry and 45 per cent government owned. To date, five major gas discoveries have been made. They are: Drake Point and Hecla on Melville Island, Kristoffer Bay on Ellef Ringnes Island, Thor Island and King Christian Island.

While the project is examining all modes of transportation, major emphasis is currently directed to the pipelines, for which three alternatives are proposed:

1. 48 inch pipeline via the west side of Hudson Bay and thence through Ontario to eastern markets.

2. 42 inch pipeline via the west side of Hudson Bay connecting to the Trans Canada Pipeline system at Winnipeg.

3. 48 inch pipeline via the east side of Hudson Bay through Quebec to eastern markets.

These alternatives are now being examined with regard to technical feasibility, environmental impact and overall costs.

The line requires a threshold of 20-30 trillion cubic feet of which 12 trillion has been proved to date. Depending upon the alternative chosen, the line would be 2,200-3,200 miles in length, 42 or 48 inches in diameter (36 inches at marine crossings), having a capacity of 2-4 billion cubic feet per day and cost from $4.5 to $7.5 billion (1974). The group plans to make an application for a pipeline in 1977 and expects first gas delivery in the early 1980's.

A study to investigate and evaluate the relative economics of transporting hydrocarbons by various modal combinations from the Arctic Islands to southern markets, specifically Montreal, has recently been completed. Costs have been developed on a 21-year time base, from 1980 to 2000, and include both capital and operating costs. Tariffs were developed using a standard flow, commencing at 0.1 trillion cubic feet per year in 1980 and increasing to 1.5 trillion cubic feet by 1986 and thereafter (equivalent to 4.5 billion cubic feet per day).

The alternatives have been reduced to four cases utilizing combinations of pipeline, rail, air and marine modes as follows:

Case I: Pipeline - Ellef Ringnes to Montreal

```
        Case II:    Air      - Ellef Ringnes to Devon Island
                    Marine   - Devon Island to Canso
                    Pipeline - Canso to Montreal
        Case III:   Pipeline - Ellef Ringnes to Devon Island
                    Marine   - Devon Island to Canso
                    Pipeline - Canso to Montreal
        Case IV:    Pipeline - Ellef Ringnes to Somerset Island
                    Rail     - Somerset Island to Montreal
```
(Figure 14).

The pipeline route examined was that route via the western
side of Hudson Bay, through Ontario to Montreal. The railway
considered in the study would follow the same route as that of the
pipeline. For the marine mode, LNG tankers of ice class 6 would be
required. Marine mode costs include terminal, liquefaction and
regasification. The air mode costs are based on use of the Boeing
747F aircraft but do not include infrastructure costs such as
navigation aids, communications, air traffic control, etc.

Comparisons of the four cases are shown below as Figure 15.
It is to be noted that the tariffs for the pipeline-marine
alternative, Case III, are very competitive with those of the
pipeline, Case I, and other factors, such as energy efficiency and
time scheduling may be more important factors in the final choice.

While the major interest has been the transport of hydrocarbons
from the Mackenzie Delta and the High Arctic Islands to markets in
southern Canada and the United States, research has also been
conducted into the feasibility of various forms of conventional and
new technology to move a wide range of commodities. In fact, it
would seem that every form of "exotic" technology has been explored
to varying degrees by "experts" both in Canada and abroad. Such
"exotics" include aerial tram-ways, super submarine tankers,
dirigibles and twelve-engine super-heavy aircraft. These rely on
scale economics for effectiveness, but generally involve untested
technology with capital costs estimated in the tens of billions of
dollars. Resource transport research by and large has concentrated
on costing the relative advantages of conventional technology for
application in relatively short-term.

This review of present and potential systems for the transport
of oil, gas and other mineral resources from the Arctic leads us to
a number of general observations:
(1) very large reserves (deposits) are needed before development;
(2) extensive research is still required before conventional
    technology can be adopted to Arctic environment;
(3) high capital and operating costs limit the range of alternatives.

                6. RESOURCE DEVELOPMENT SCENARIOS

World market forces generate a continuing requirement for
petroleum and metals exploitation. Various trading alignments have
encouraged Canadian resource development - such as coal and metals

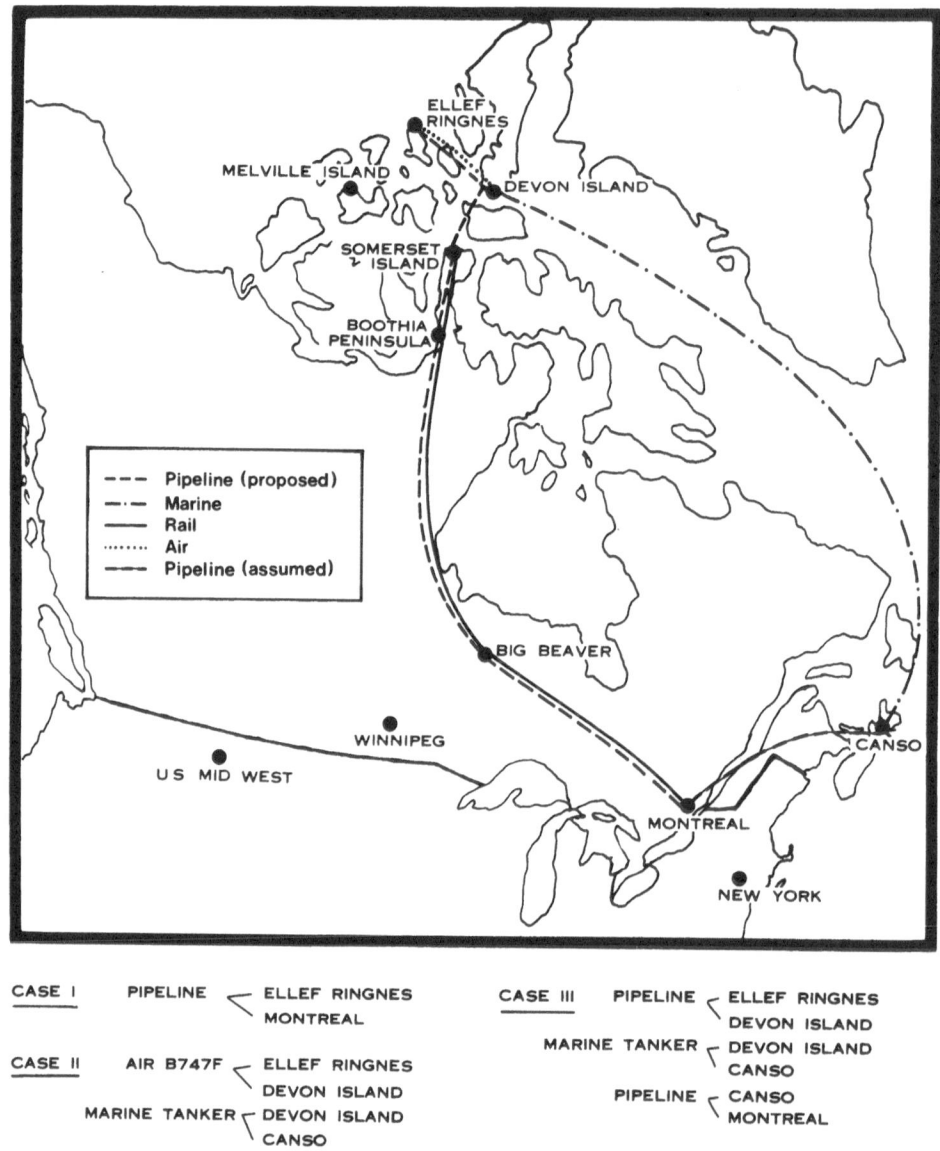

Figure 14   Proposed pipeline route via westernside of Hudson
               Bay, through Ontario to Montreal

Systems	Pipeline	Air-Marine	Pipeline-Marine	Pipeline-Rail
Tariff ($ per mcf)	1.35	2.00	1.40	2.90
Capital Cost ($ millions)	7,150	8,140	7,390	15,140
Annual Operating Cost ($ millions)	160	1,630	640	440
Energy Efficiency (present)	90	67	76	86
Fleet Size		- 66 Aircraft 26 Ships	26 Ships	84 Trains

Figure 15   Arctic Islands Gas Transportation Alternatives

to Japan and petroleum to the United States. The course of the
Mackenzie Valley Gas Pipeline Hearings in Washington and Ottawa
could see a new definition of pipeline systems, with "northern
extensions". These could be followed in ten to twenty years time
with either tanker routes or another major pipeline from the High
Arctic. While the development corridor in the Mackenzie Valley is
quite well known, that for the High Arctic is not. Railway and
road systems will continue to move northward, especially in the
Yukon.
    While pipeline transport in the Mackenzie Valley should be
able to surmount technical difficulties, the solutions for the High
Arctic are not quite as apparent. Inter-modal analysis will have
to be conducted for the following should sufficient reserves be
found:
(i) pipeline versus marine for gas, with inter-island pipeline
     transport the critical factor;
(ii) marine versus air (?) for oil, in that industry concedes a
     pipeline to be out of the question.
Social and environmental impact, issues central to the proceedings
of the Berger Commission now holding community hearings in the
Mackenzie Valley cannot be dismissed and will form an important

part of decisions for development of future resource transport
systems.

The enactment of a bill to create Petro-Canada, a government
body to explore and develop petroleum resources could well influence
the types of systems developed. This in turn must be related to
national energy policy, which may take one of two directions with
regard to the North:
  (i) promote the export of petroleum to balance international
       trade, particularly with the United States and Europe,
       using the "big project" approach;
 (ii) develop Canadian self-sufficiency based upon the gradual
       or incremental exploitation of reserves, using flexible
       means of transport.
The divergency of these two approaches is evident in the two
applications before the Canadian National Energy Board.
Deliberations on the Mackenzie Valley gas pipeline proposals will
have an important bearing upon how various transport scenarios
unfold over the next twenty years.

7. RESEARCH IMPLICATIONS

A gauge of the requirement for research comes from government
evaluation of the relatively well known Mackenzie Valley gas pipeline route. The United States Department of the Interior reported
on July 29, 1975, that for the American section of the CAGPL line
there are "eighteen faults" in the submitted plan. These include
fracture toughness, pipeline safety factors, and particularly, the
problem of burial for a chilled gas pipeline. These engineering
problems associated with environmental impact are aside from socio-
economic problems such as labour force instability, cultural shock
and drug abuse.

In this paper I have reviewed the current state of the art for
resource transport systems research, but have not touched as much
as I would like on the people aspects of these systems. A coherent,
inter-disciplinary approach is needed to solve these Arctic transport
problems, involving both the physicist and the psychologist.
Training of crews is just as important as vehicle design. I would
not like to see research confined wholly to the economist or the
engineer working in isolation.

The Arctic Transportation Agency was founded three years ago
to co-ordinate systems analyses similar to those I have so briefly
outlined and to co-operate with industry in its endeavours in the
same field. This NATO Workshop will, I am sure, make a very
meaningful contribution to our efforts to ensure that, for the
Canadian Arctic, the right transport system is available at the
right time - when it's needed.

ICE RESISTANCE WITHOUT ICE:

SIMULATION OF SHIP RESISTANCE IN ICE FIELDS

Fen-Dow Chu

State University of New York, Maritime College

Bronx, New York, U.S.A.

INTRODUCTION

This paper deals with a series of six ice-breaking ship models
with commercial hull forms, which have been systematically tested
with the intention of gathering more information about resistance
in ice. In contrast to conventional ice resistance tests,
performed in real ice towing tanks and followed by complicated
data regression, the tests discussed in this paper were performed
in a regular towing tank using an artificial ice substitute.

The proposal for the systematic tests was initiated in the
summer of 1970 by F.D. Chu and was submitted to Professor Dr. Ir.
J.D. van Manen, director of the Netherlands Ship Model Basin in
Wageningen. The proposal being approved, a model test plan,
outlining techniques to be used and specifying further details,
was made in the summer of 1971. In the following winter six wooden
models were made and in the summers of 1972 and 1973 a series of
systematic tests were done in the deep water towing tank of the
NSMB, using paraffine to simulate ice.

The present paper is a brief of the theory underlying these
experiments and of the methods used in carrying them out. A more
detailed description of theory, methods, and results of the
experiments can be found in the author's doctoral dissertation,
entitled Ship Resistance in Homogeneous Ice Fields: Theory,
Systematic Tests, and Estimation of Resistance and Effective Power
at Constant Speed (1).

THE SEPARATION OF THE TOTAL RESISTANCE AND THE
CLASSIFICATION OF THE COMPONENTS

The resistance encountered by a ship in an ice field has long been observed and accurately described. Observation and correlation of ice resistance data by various experts in the field have led to the following conclusion: The total resistance of a ship in a continuous ice-breaking process is the sum of the components of (1) breaking the ice, (2) moving through the ice-clogged channel which is formed by the broken pieces, and (3) the resistance of the open water. In short:

$$R_{total} = R_{breaking} + R_{ice\text{-}clogged\ channel} + R_{open\ water}$$

simply:

$$R_t = R_b + R_i + R_o \tag{1}$$

whereby:

R_b is the resistance required to crack the ice sheet into characteristic breaking patterns. More precisely, it is the resistance of breaking the ice without any motion and gross displacement involved. This resistance is directly related to the structure failure of the ice sheet: it is the summation of the energy required for the individual break, which is a function of the thickness and strength of the ice and of the extent of the area that is broken, as well as of the form of the bow and its frictional coefficient against the ice.

R_i is the resistance of the hull and the broken pieces. Once the ice sheet is broken, energy is required to push the ship through the ice-clogged channel. R_i is the resistance caused by pushing the broken pieces sideways or downwards, and also by the friction between the hull and the ice pieces, and the ice pieces between themselves. The energy of R_i can be divided into kinetic, potential, and frictional energy. The kinetic energy consists of the speed the ice pieces gain from the ship; it can be in either linear or rotative form. The potential energy is the energy necessary to submerge the ice; it is expressed in ρgh form. The frictional energy is the sum of the energies caused by friction between hull and ice and between ice and ice. Except for the frictional energy, R_i is not only a function of the volume of the ice, but also of the speed of the icebreaker.

R_o is the open water resistance of the ship, with which we are familiar. It is the sum of the skin resistance and the wave-making resistance. As the ship in a continuous ice-breaking mode is operated at a very low speed-length ratio (ca. 0.1), the wave-making resistance can safely be ignored.

The separation of the different resistances also corresponds to the fact that each type of resistance consists of individual

phenomena and leads to its own scaling law, even though these
resistances act simultaneously in the full-scale operation.

THE LAW OF COMPARISON FOR THE RUPTURE OF THE ICE SHEET BETWEEN PROTOTYPE AND MODEL TESTS

The variables involved in the rupture of the ice sheet are
the following:

E = modulus of elasticity
ρ = density
t = thickness
σ = strength of the ice
P = downward force

By applying Buckingham's theory, the above variables yield:

$$\frac{\sigma}{E} = \frac{\sigma_m}{E_m} \qquad\qquad (2)$$

or: $\quad \dfrac{\sigma}{\sigma_m} = \dfrac{E}{E_m}$ $\qquad\qquad (2)a$

and: $\quad \dfrac{P}{Et^2} = \dfrac{P_m}{E_m t_m^2}$ $\qquad\qquad (3)$

whereby: subscript m = model

Equation (3) can be rearranged as:

$$\frac{P}{P_m} = \frac{Et^2}{E_m t_m^2} \qquad\qquad (3)a$$

For two systems of geometric similarity with a scaling factor of
$L/L_m = \lambda$,

$$P \propto L^3$$

Thus:

$$\frac{P}{P_m} = \frac{L^3}{L_m^3} = \frac{Et^2}{E_m t_m^2} = \lambda^3 \qquad\qquad (3)b$$

Equation (3)b is the most commonly adopted form of simulation law
in model breaking tests. Because downward force P has a linear
relation with breaking thrust T_B, equation (3)b can also be written
as:

$$\frac{T_B}{T_{Bm}} = \frac{Et^2}{E_m t_m^2} = \lambda^3 \qquad\qquad (3)c$$

In the case of a single rupture of the ice sheet, neither the thickness nor the modulus of elasticity have to maintain the proper geometric similarity λ, as long as the functional relation in equation (3)c is fulfilled. In a perfect model test, however, both equations (2) and (3)c have to hold true simultaneously. Moreover, in a continuous ice-breaking mode (multiple ruptrues), the total breaking thrust is directly proportionate to the number of breakings. This leads to only one conclusion, namely:

$$\frac{t}{t_m} = \lambda \qquad\qquad (4)$$

$$\frac{E}{E_m} = \lambda \qquad\qquad (5)$$

$$\frac{\sigma}{\sigma_m} = \lambda \qquad\qquad (6)$$

$$\frac{\sigma}{E} = \frac{\sigma_m}{E_m} \qquad\qquad (2)$$

The geometric similarity is easy to achieve. In order to maintain the relationship λ, the model ice strength and the model ice elasticity have to be scaled down according to λ. Usually the strength of fresh water model ice is greater than that of natural ice. By making high-salinity ice and using special seeding techniques, the strength of the model ice can be lowered down. However, if we really would manage to fulfill equations (4), (5), (6), and (2) simultaneously, the model ice would be so weak that the tests would be no longer performed in the elastic stage but in the plastic stage. In that case, the entire breaking theory, that we developed out of the theory of elasticity, would be no longer applicable. In conclusion we can say that the law of comparison, even though it is unshakable, cannot in general be applied to ice-breaking model tests. We consequently question whether it is useful to try to improve model ice-breaking techniques, as long as there is no guarantee that the results will be correct. Perhaps we should use the same logic that is used for the treatment of skin friction in ship model resistance, where Reynold's number presents practical difficulties, and simply calculate it.

THE LAW OF COMPARISON OF THE
ICE-CLOGGED CHANNEL RESISTANCE

The origin of Froude's number is the correlation between the resistance and the waves made by the moving ship. In the ice-clogged channel, the ice pieces are moving up and down while the icebreaker is moving forward. The ice pieces will form a wave train with the wave speed of V relative to the ship and the wave height of the average submersion of the ice pieces (see fig. 1). This creates the ideal condition for the application of Froude's law of comparison.

The frictional resistance, whether of ship and ice, or ice and ice, is directly associated with the frictional coefficient (not viscosity). It will follow any law of comparison, as long as the geometric proportion is maintained.

The resistance from kinetic energy loss, either linear or angular, is not different from any other energy form: as long as the geometric similarity is maintained, it follows Froude's number.

SHIP MODELS - NSMB SERIES 80

In earlier systematic model tests for arctic shipping, the models were chosen according to the projected prototype. Unlike those model series, Wageningen series 80 was designed with special regard to the variation of bow and spread angle, two unique features that have an absolute priority in the design of an ice-breaker.

Upon consideration, it was decided that the models should have a block coefficient of 0.80 and a smaller beam-draft ratio. A set of basic lines from series 60, $C_b = 80$, was chosen and the forebody design was varied according to Table 1.

The maintaining of α and β at the same value throughout the fore-body was considered as the only means to evaluate the true effect of these two variables in the tests. It appeared later that, when α was small, some of the combination were not practical.

The models had a long parallel mid-body. The lines of the aft-body were drawn for a twin-screw installation. There were six bow sections and one aft-body section. These were constructed separately and were interchangeable (see fig. 2).

MODEL BROKEN ICE PIECES (FILLING OF THE ICE-CLOGGED CHANNEL)

We have examined the laws of comparison for various component resistances. It is clear that, unless the ideal model ice can be made (cf. equations (2) and (3), the possibility of combining breaking and ice-clogged channel resistance in a single test must, on theoretical grounds, be ruled out.

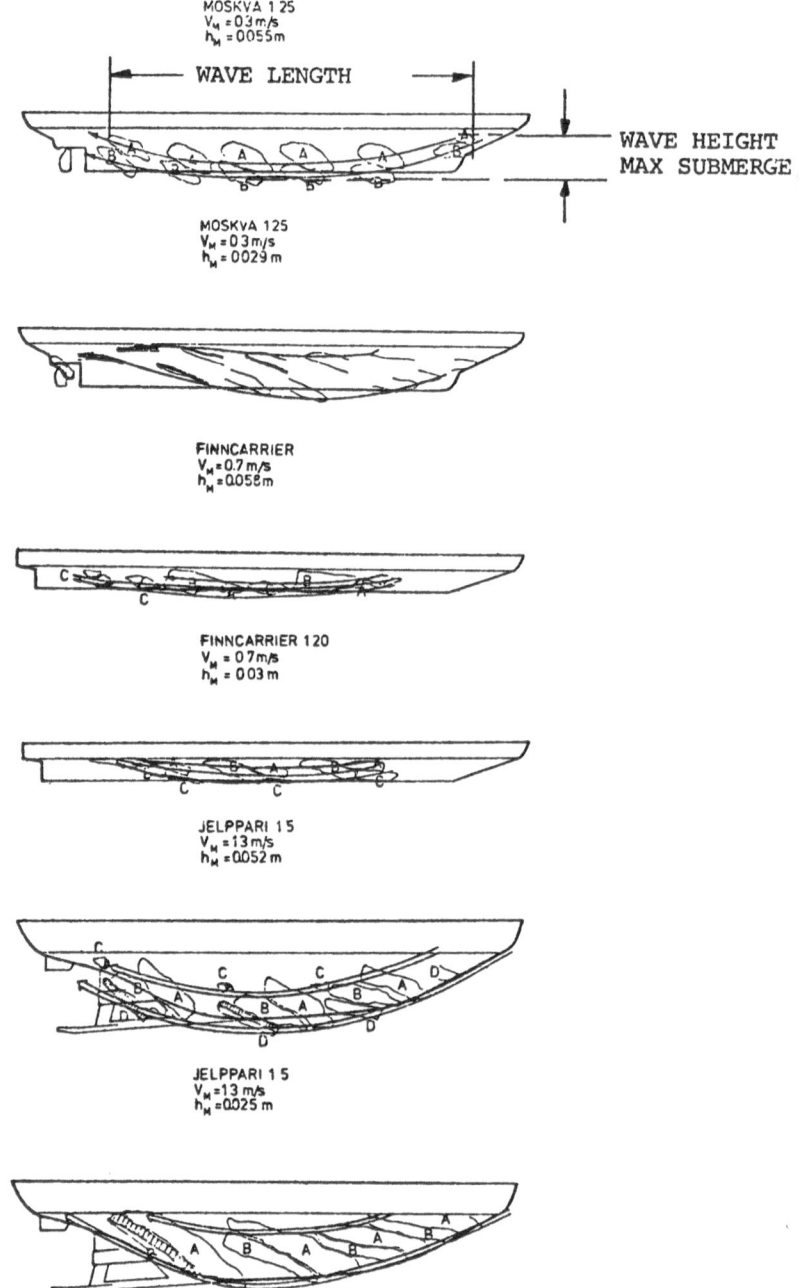

Fig. 1. A perfect example of the ice-sector train wave (a reinter-
pretation of Enkvist's ice streamline patterns) (see ref, 2).

Table 1. Wageningen series particulars; model scale 1 : 67.936; length = 4.500 m; beam = 0.5625 m; draft = 0.2813 m

MODEL No.	BOW ANGLE	SPREAD ANGLE	C_b	DISPL. kg	WET SURFACE m^2
4231 f. + a.	30°	10°	0.838	596.63	4.4080
4287 f.	30°	20°	0.830	590.87	4.3745
4288 f.	30°	30°	0.819	580.82	4.2330
4289 f.	30°	50°	0.789	560.06	4.1780
4290 f.	20°	10°	0.801	570.41	4.2971
4291 f.	20°	30°	0.769	547.16	4.1267

f = forebody; a = aft-body

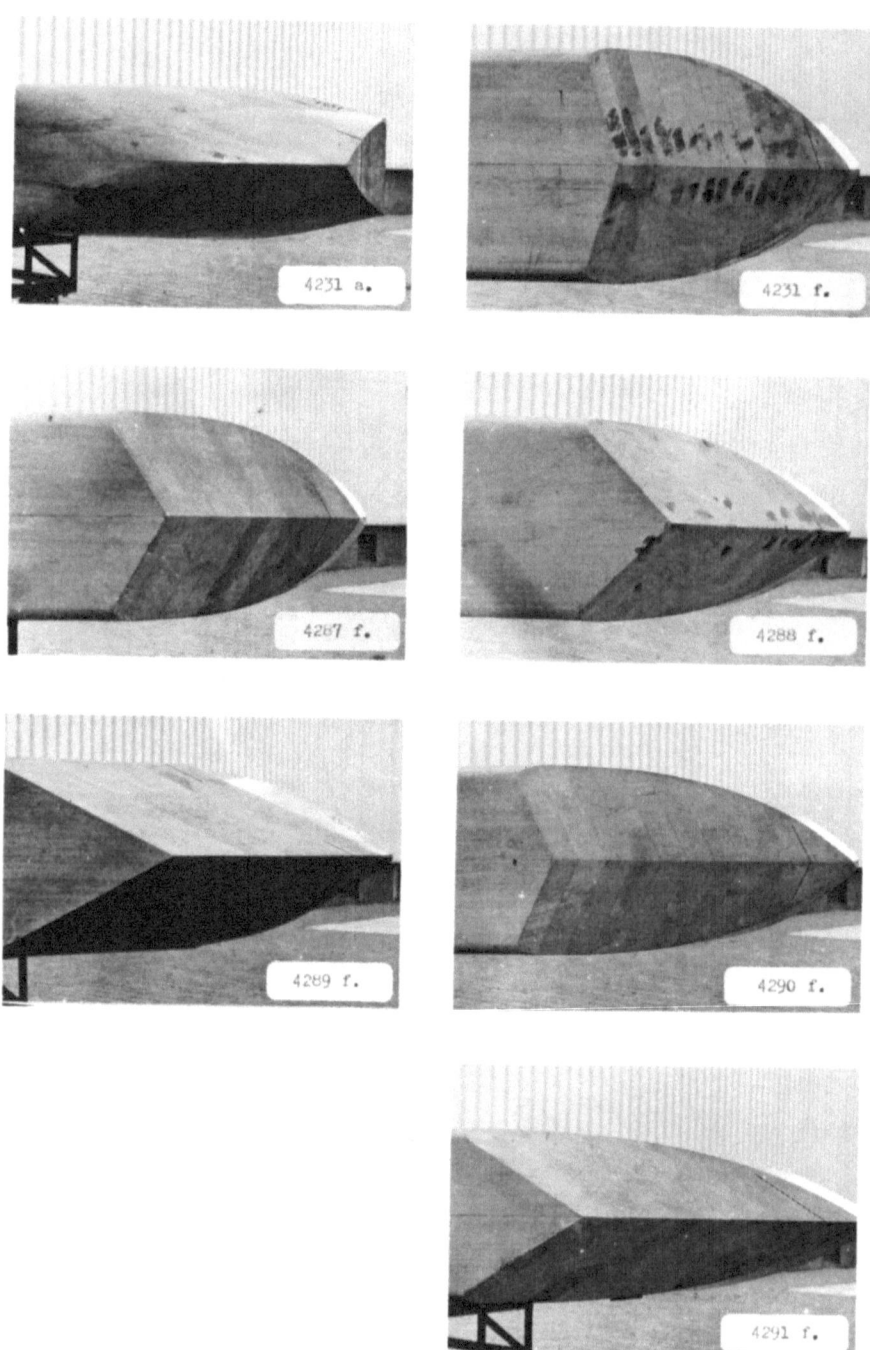

Fig. 2. Icebreaker ship models

Test voyage data of full-scale icebreakers show that, in practice, the breaking resistance only makes up a small percentage of the total resistance. In model tests the proportion of each component of the total resistance turns out to be inaccurate (see fig. 3).

The breaking resistance is relatively so small that it is difficult to simulate in model tests, but it can be calculated accurately. The remaining resistance components constitute 90% of the total resistance. They all follow the same law of comparison and they all can be simulated in a single ice-clogged channel test.

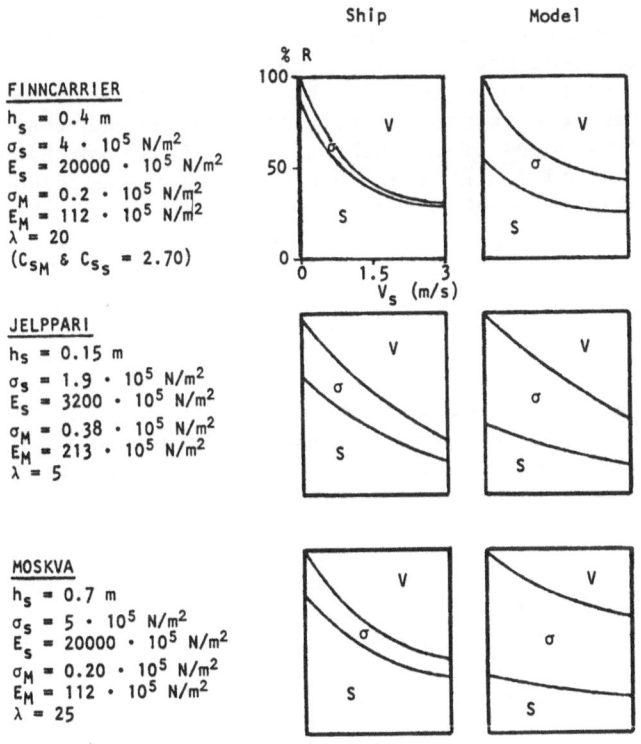

S = Submersion Resistance Percentage
σ = Breaking Resistance Percentage
V_s = Velocity Resistance Percentage

Fig. 3 Cases illustrating the relative importance of the main resistance components. (after Enkvist) (2)

For the entire breaking process the variables are the following:

density of the water	= ρ
density of the ice	= ρ_i
elasticity of the ice	= E
strength of the ice	= σ
thickness of the ice	= t
static friction coefficient between ice and contact surface of hull	= μ_s
kinetic friction coefficient between ice and contact surface of hull	= μ_k

In an ice-clogged channel where no breaking takes place, the variables E and σ can be left out of account. The thickness of the ice, t, is governed by the geometric scaling factor. The only variables that remain to be examined are the density of the ice, ρ_i, and the frictional coefficients μ_s and μ_k.

The use of real ice in model tests involves so much labor, expense, and time (the author can recall a ten-seconds test that was preceded by 24 hours of ice making), that it is worth while trying to find a good substitute. Since in the ice-clogged channel resistance chemical elements play no role, any material that matches the physical properties of natural ice (which depend on water salinity, weather condition when freezing occurred, and the age of the ice) can be used as a substitute in ice-clogged channel resistance model tests. In the Wageningen tests paraffin was used. The density of paraffin wax (ρ = 0.86 to 0.91) is so close to that of sea ice (ρ_i = 0.88 to 0.92) that no adjustment is necessary.

The frictional coefficient between the ice and the icebreaker hull surface depends on whether the friction is static or kinetic, whether the contact surface is wet or dry, and whether or not the surface of the ice is covered with snow. It varies from 0.05 to 0.6. By adjusting the roughness of the model hull surface, the same frictional coefficient can be achieved between the paraffin wax and the model surface. Paraffin wax, moreover, has the advantage that it is inexpensive, easy to work with, and reusable.

Besides their density and frictional coefficient, other characteristics of the broken ice pieces in the ice-clogged channel are their shape and size. Theoretically, the basic shape of a broken piece after a characteristic break is approximately triangular. In reality, the shape of the broken pieces is determined by a combination of basic breaks and the fractions caused by secondary breaks. In the model tests, the shape of the pieces has to be generalized and idealized, but at the same time they should physically resemble the real ice pieces. In order to correlate the model ice piece with its prototype, the randomly broken ice piece, the author has examined many photographs of randomly broken ice fields as well as fields broken open by icebreakers. Appendix 1 shows part of the photographs used (collected from various sources).

When we carefully observe the ice segments (see especially the outlined parts of the detailed pictures, we find that, by generalizing, we can reduce the shapes of the broken pieces to the four types shown in fig. 4.

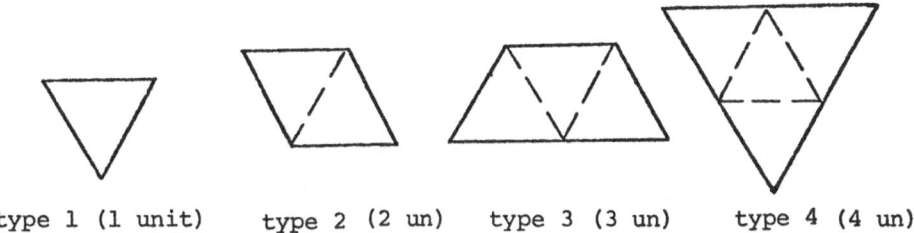

type 1 (1 unit)   type 2 (2 un)   type 3 (3 un)   type 4 (4 un)

Fig. 4   Basic shapes of the broken ice floe

These four types of shapes, combinations of the basic unit equilateral triangle, can form any configuration within the ice-clogged channel (see fig. 5 (a) to (e)).

MODEL TESTS AND THEIR REPRESENTATION

For the tests the high speed towing tank in the NSMB was used. The channel was made according to fig. 6 (a) and (b).

The installation of the whole channel was fastened by frames to the tank wall. Each frame had a set of screws which could adjust the channel width and the alignment. The length of the entire channel was 40 m. It could be installed for the test in one day.

The preparation of the tests consisted of the installation of the channel and the instrumentation of the model. Both preparatory actions could begin simultaneously and took about two days, depending on how long it took to adjust the channel width and to align the channel. The paraffin wax pieces were prefabricated and stockpiled. They were laid out in the channel in the manner of a brick pavement. The actual test run was of very short duration. The team was standing on the carriage, so that one could very well observe what happened during the test. It seemed that all the pieces were pushed downwards and then sideways. After the run the pieces of wax were fished out of the channel and the procedure could start again. About four to six test runs could be performed within one day (see fig. 7).

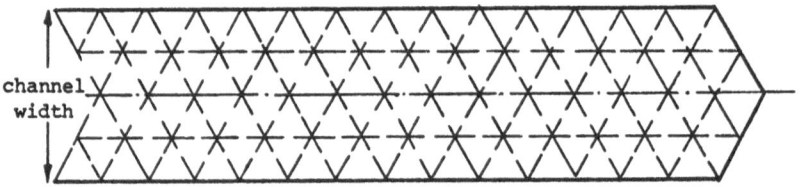

(a)   Basic sketch of an ice-clogged channel.

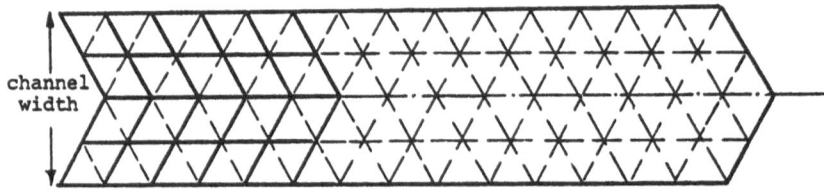

(b)   Type 2 ice-clogged channel formation

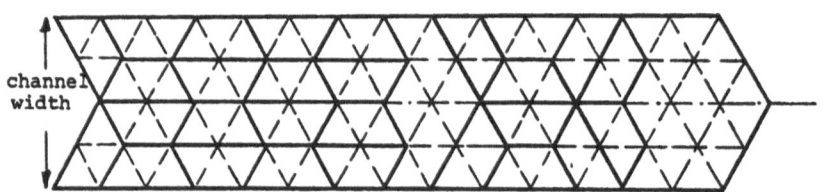

(c)   Type 3 ice-clogged channel formation

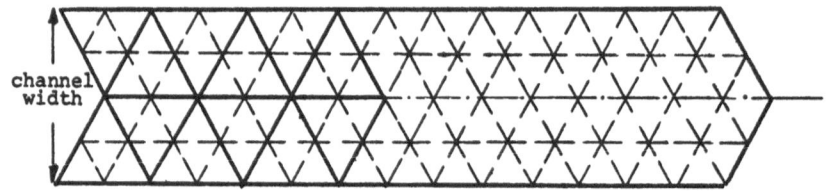

(d)   Type 4 ice-clogged channel formation

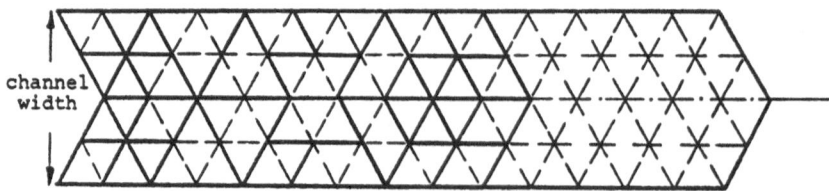

(e)   Ice-clogged channel formation of all four types

Fig. 5.   Configurations within the ice-clogged channel

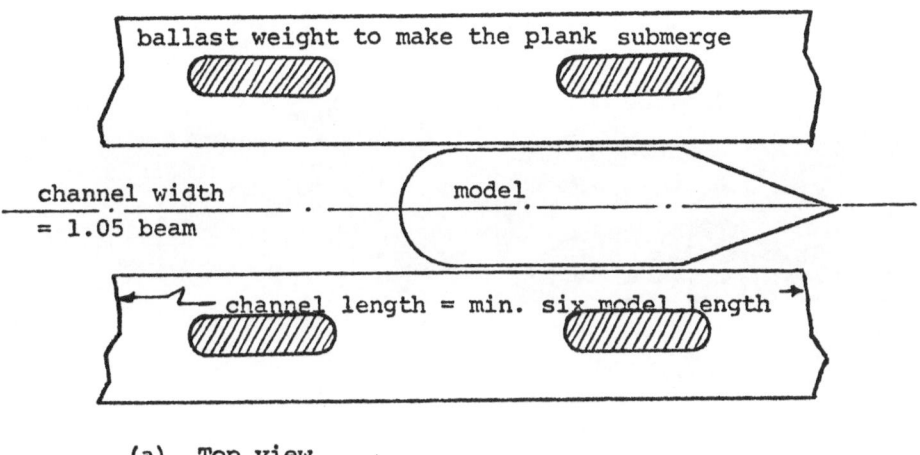

(a)   Top view

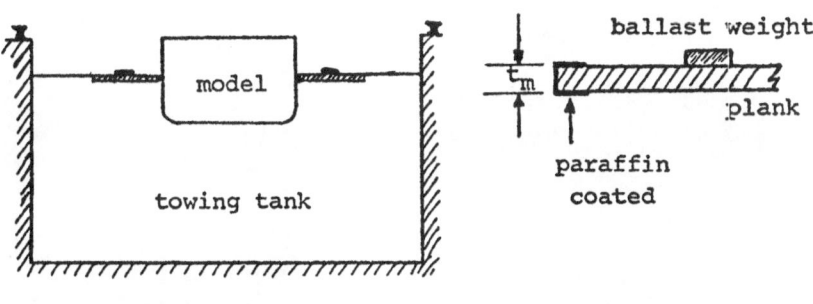

(b)   Sectional view

Fig. 6   Schematic diagram of a channel

Packing of the channel   Paraffin wax sectors in stock

Open Channel   Clogged Channel

Fig. 7.   Ice-clogged channel

THE MODEL ICE-CLOGGED CHANNEL TEST AND NSMB SERIES 80 CONTOURS

The test was divided into two parts:

Part 1. The first part of the test served to determine whether or not the resistance was influenced by the size and the shape of the ice pieces. If it could be proved that at a given (low) speed the resistance in an ice-clogged channel is independent of the shape and the size of the ice pieces and only affected by the unit volume of the ice displaced by the icebreaker, one would be free to choose ice pieces of any one size and shape for the second part of the test. If the result of test part 1 were negative, however, the ice pieces for test part 2 would have to be customarized for each individual run, which would be costly as well as time-consuming. As test part 1 was designed, the model went through a 2 cm thick ice-clogged channel at variable speed. Further specifications were as follows:

Bow angle   : $\alpha = 30°$
Spread angle   : $\beta = 30°$
Speed range   : ν from 0.1 to 0.6 m/sec
Model ice thickness : $t_m = 2$ cm
Ice field (total 3 types)
     1. uniform one unit size channel
     2. uniform two units size channel
     3. uniform three units size channel

The total amount of runs was 12. The test data are plotted in fig. 8 and form the experimental evidence for the conclusion.

Part 2. It was proved in test part 1 that the size and the shape of the ice pieces had little or no effect on the ice-clogged channel resistance, so that throughout test part 2 pieces of uniform size and shape (preferably type 1 or 2) could be used. The only factor that varied, of course, was the thickness of the pieces. Test part 2 covered the operational and speed ranges that are desirable at this moment. Six interchangeable bows were used at four different speed ranges through an ice-clogged channel of three different ice thicknesses and various types of ice pieces. The total amount of runs was more than 150. The following matrix served as guideline for the test:

ice thickness (in cm)	size and shape of ice-clogged channel	bow section	speed v_p, knots ($\lambda=68$)		range v_m, m/sec	
3	type 1 or 2	all six variations	1.6	9.6	0.1	0.6
2					0.1	0.6
1					0.1	0.6
0 (open water test)					0.1	0.8

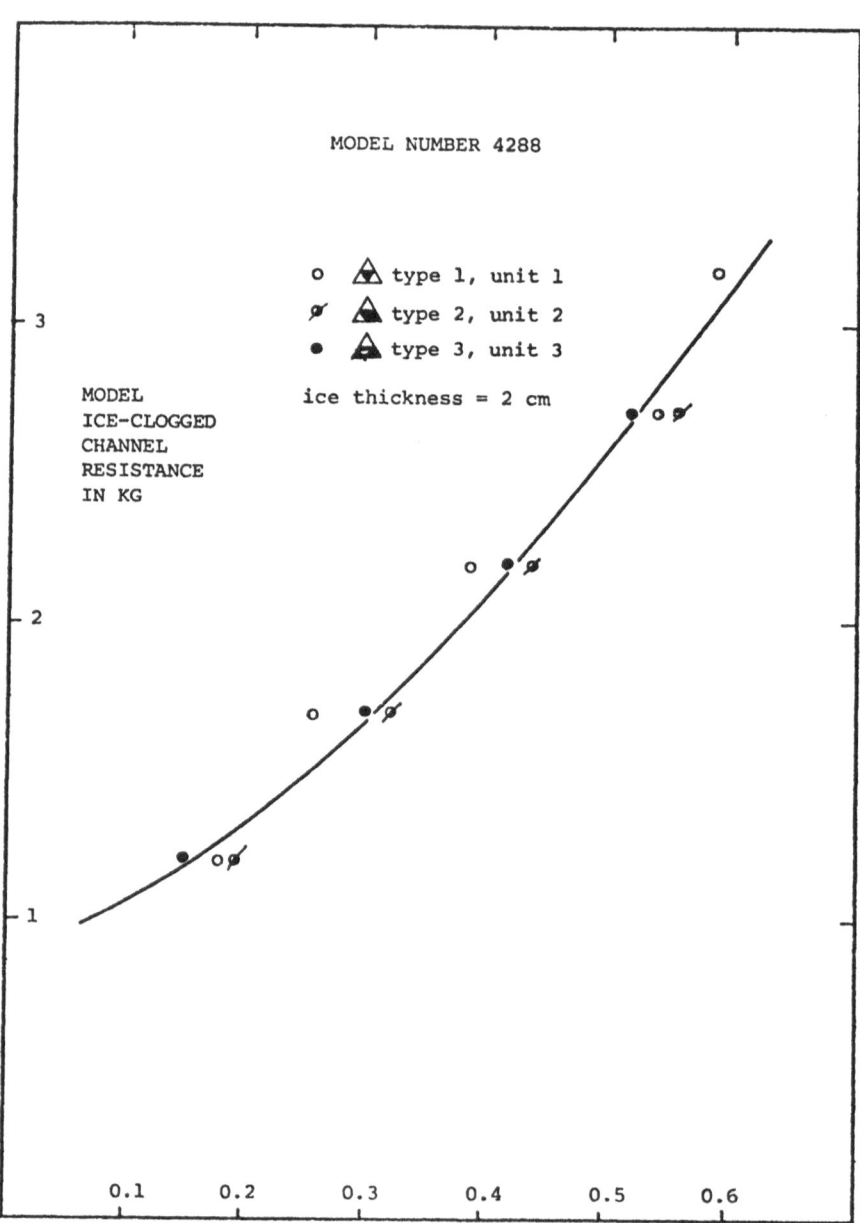

Fig. 8. Model ice-clogged channel resistance vs. model ice size and shape

For some of the tests zero speed was tried, but the results were too inconsistent to be recorded. The original data were in speed (m/sec) versus thrust (kg). The speed was changed according to Froude's number, $V/\sqrt{gL}$, for a non-dimensional representation. The thrust, R_{im}, was divided by the model displacement, Δ_m. R_{im}/Δ_m was defined as ice-clogged channel resistance coefficient in kg/ton. In appendix 2 the data are plotted. These charts are used for the thrust estimation.

RESISTANCE AND EFFECTIVE POWER ESTIMATION FROM NSMB SERIES 80

When a ship traveling at V speed encounters an ice field of t thickness, the total resistance can be written as:

$$R_t = R_b + R_i + R_o \qquad (1)$$

R_o can be subdivided into:

$$R_o = R_w + R_f \qquad (7)$$

whereby:

$\quad R_w$ = wave-making resistance

$\quad R_f$ = skin frictional resistance

Substituting equation (7) into equation (1), we get, by regrouping:

$$R_t = R_b + (R_i + R_r) + R_f \qquad (1)a$$

or:

$$R_t = R_b + \frac{(R_i + R_w)}{\Delta}(\Delta) + \frac{R_f}{S}(S) \qquad (1)b$$

whereby:

$\quad$ b = beam

$\quad \Delta$ = displacement

$\quad$ S = wetted surface

The first term of equation (1)b can be written as (see ref. 1):

$$R_b = \frac{1.05\ b}{(1/2)\pi r^2}(W_B + W_{FB}) \qquad (8)$$

whereby:

$\quad W_B$ = breaking energy

$\quad W_{FB}$ = frictional energy

The numerical values of W_B, W_{FB}, and r, can be read off in Appendix 3.

The second term of equation (1)b is:

$$\frac{(R_i + R_w)}{\Delta} (\Delta)$$

We know that:

$$\frac{R_i + R_w}{\Delta} = C_i = \text{ice-clogged channel resistance coeff.}$$

The value of C_i is represented by the contours in Appendix 2. By the application of Froude's number, the reading, for a given thickness $t = \lambda t_m$ can be found easily.

The third term:

$$\frac{R_f}{S}(S)$$

can be subdivided into R_f/S and S. R_f/S is a rather familiar term in naval architecture: it is the frictional resistance per wetted surface. In order to simplify the calculation of R_f/S, a nomograph has been provided (see Appendix 4).

The practical procedure of calculating the individual terms of equation (1) and their summation, is more time-consuming as well as error-inviting than the simplicity of the equation would make us suppose. Three standard sheets have been introduced (Appendix 5), which present the interpolation of the data as well as step-by-step chart calculation. Fig. 9 is the numerical example of a full-scale prediction resulting from the table calculation in Appendix 5.

FULL-SCALE COMPARISON

In order to correctly evaluate the above-described model tests, one should not judge them on a theoretical basis but by means of a full-scale comparison.

There are few existing icebreakers that match the models used in the systematic tests, which have a long and narrow hull, a deep draft, and a full block. The only existing icebreaker than can be correlated with the models used is the T.S. MANHATTAN.

The test voyage report of the T.S. MANHATTAN probably contains the most valuable full-scale ice-breaking data as yet available. On board of the ship were sixty scientists and about two hundred tests were performed. The result of this was a very detailed as

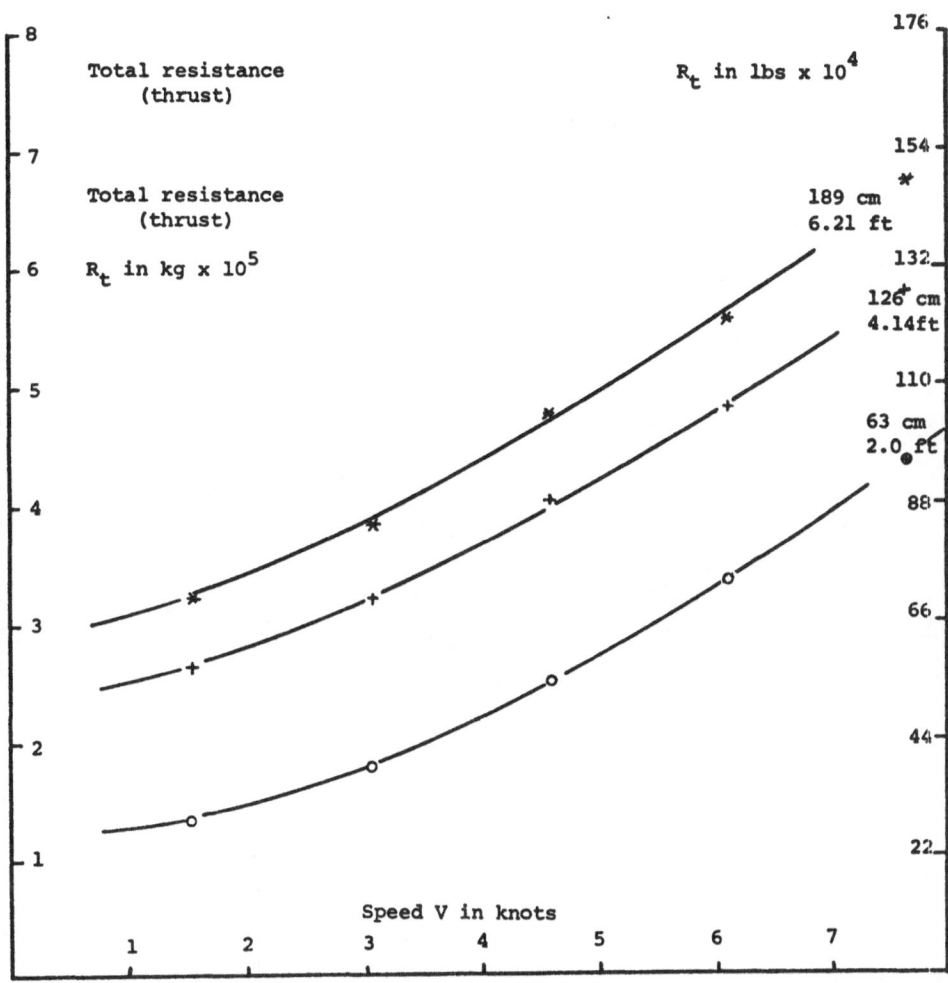

Fig. 9. Full-scale prediction of speed versus thrust in a continuous ice-breaking operation

well as expert report entitled: <u>Test Voyage Report 1970 Manhattan Full-Scale Ice-Breaking Tests</u>. Unfortunately, the report has not been made public and is only accessible to selected individuals. Some information about the specifications of the MANHATTAN and its tests voyage can, however, be gathered from other sources (which are described in detail in ref. 1). This information is plotted in fig. 10, which is a direct estimation from the chart calculation of Appendix 5. In the calculation the basic ship dimensions are chosen as close as possible to the T.S. MANHATTAN.

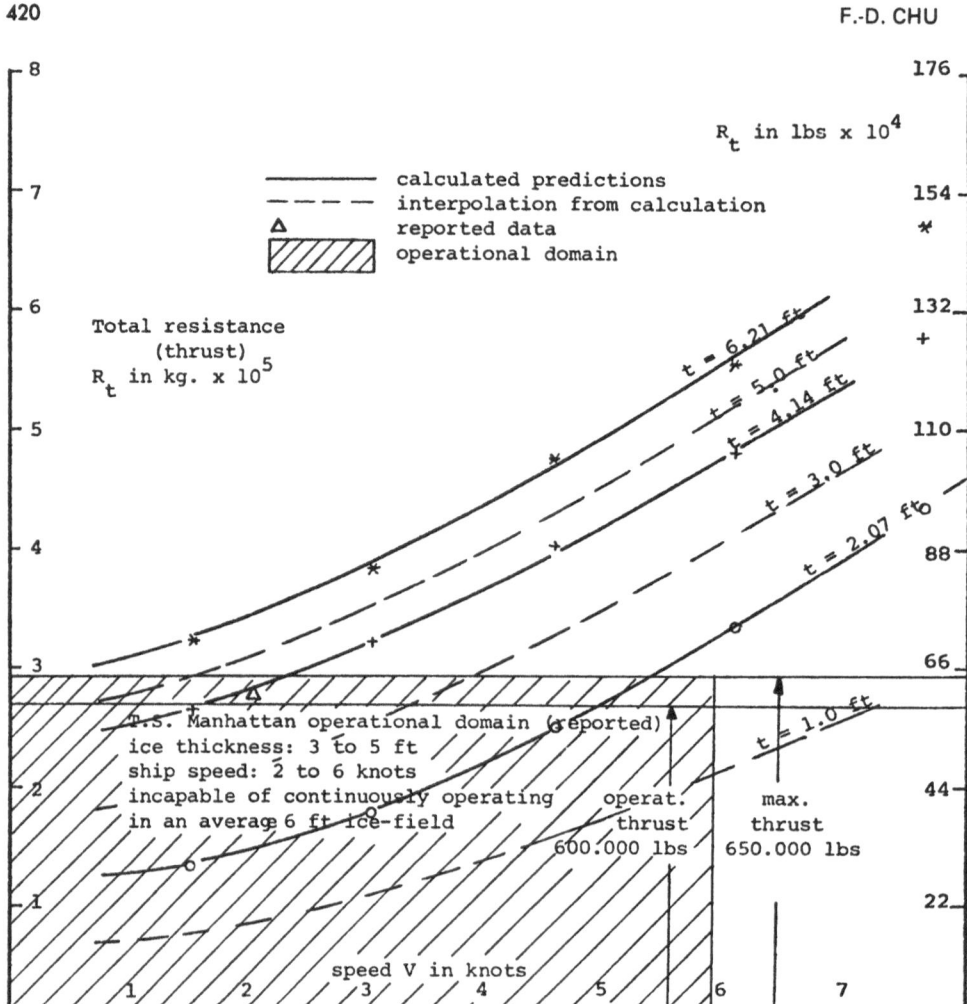

Fig. 10. Calculated full-scale prediction for the T.S. Manhattan
compared with her voyage performance in the Northwest Passage,
September, 1970

CONCLUSION

Out of the foregoing we can distill the following major points:

(1) The application of Froude's principle to the model test of
continuous movement in the ice field is focussed on the manner of
displacement of the ice pieces. For the ship's hull, Froude's
number is too small to be effective.

(2) The separation of the resistance of a ship operating
continuously in an ice field is unavoidable. Theory has confirmed
that the total resistance consists of different components with
conflicting physical characteristics, so that, with the present
technological means, it is impossible to simulate the resistance
in one single experiment.

(3) It has been confirmed that the ice-clogged channel resistance
is independent of the size and the shape of the broken pieces, as
long as the total volume to be displaced remains the same.

(4) It is advisable to perform ice-clogged channel resistance
tests in regular ship towing tanks, not only because it is more
economical but also because the results are satisfactory.

(5) The diversive character of the natural ice and the narrow
range of full-scale data allows individual researchers to propose
different data correlations. We should keep in mind, however,
that the perfection of the correlation techniques is not the goal
of ice-breaking resistance research. It is the predictability over
a wide range of ship proportions that we are after.

REFERENCES

1. Fen-Dow Chu, Ship Resistance in Homogeneous Ice Fields: Theory,
    Systematic Tests, and Estimation of Resistance and Effective
    Power at Constant Speed, 1974. Doctoral Thesis, Helsinki
    University of Technology.

2. Ernst Enkvist, On the Ice Resistance Encountered by Ships
    Operating in the Continuous Mode of Icebreaking, The Swedish
    Academy of Engineering Sciences in Finland, Helsinki, 1972.

Appendix 1

Photographs of Randomly Broken Ice Fields

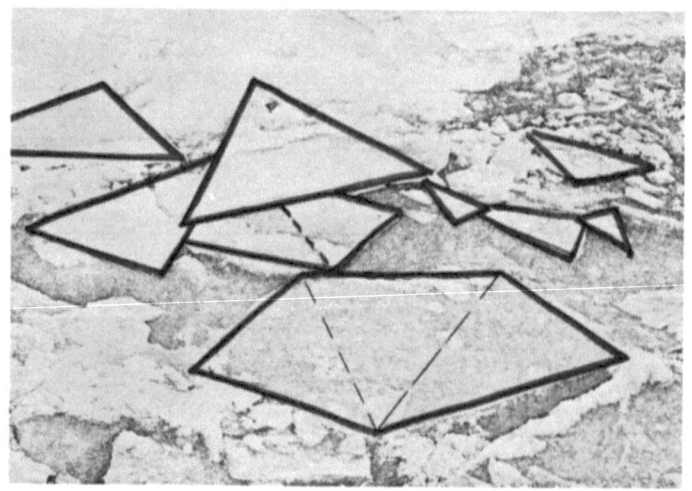

(a) Natural ice and the broken pieces

(b) Ice breaker USCG WIND 282 in an ice field

(c) Enlargement of right corner of (b)

(d) St. Lawrence Seaway in winter time

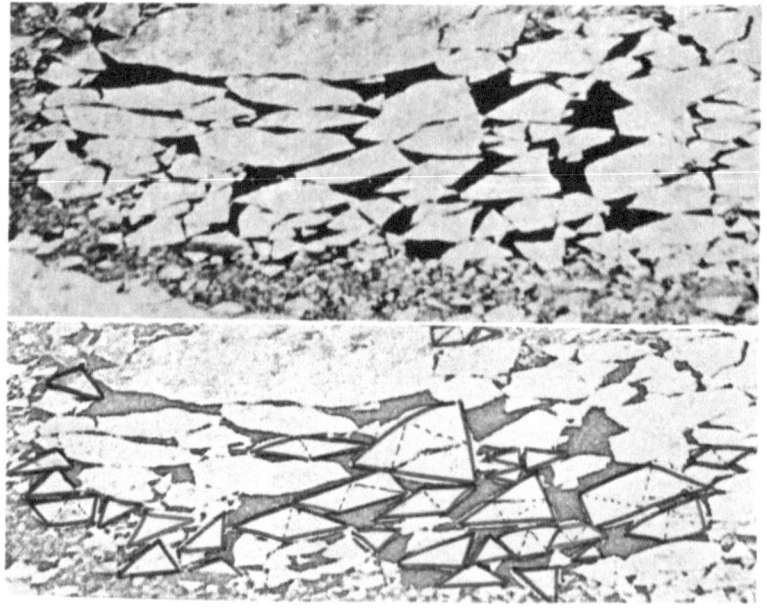

(e) Enlargment of lower part of (d)

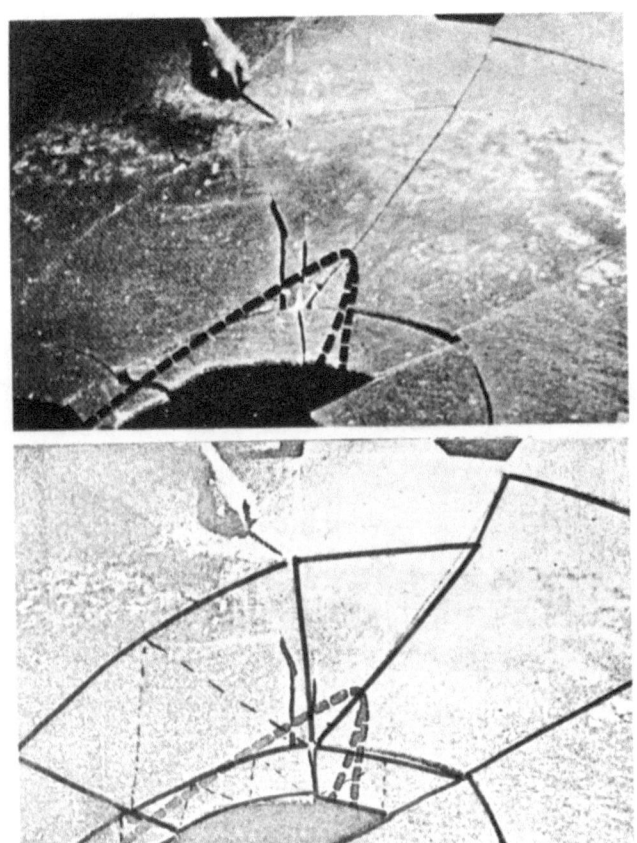

(f) Model break of paraffin wax ice sheet

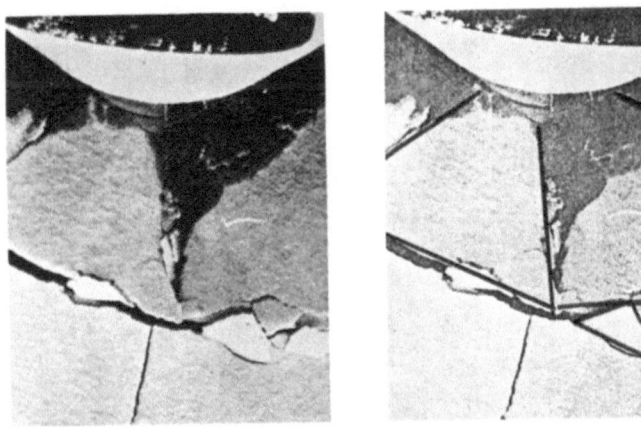

(g) Crack formations around the bow of CCGS Wolfe

(h) Canadian ice breaker in an ice field

(i) Enlargement of (h)

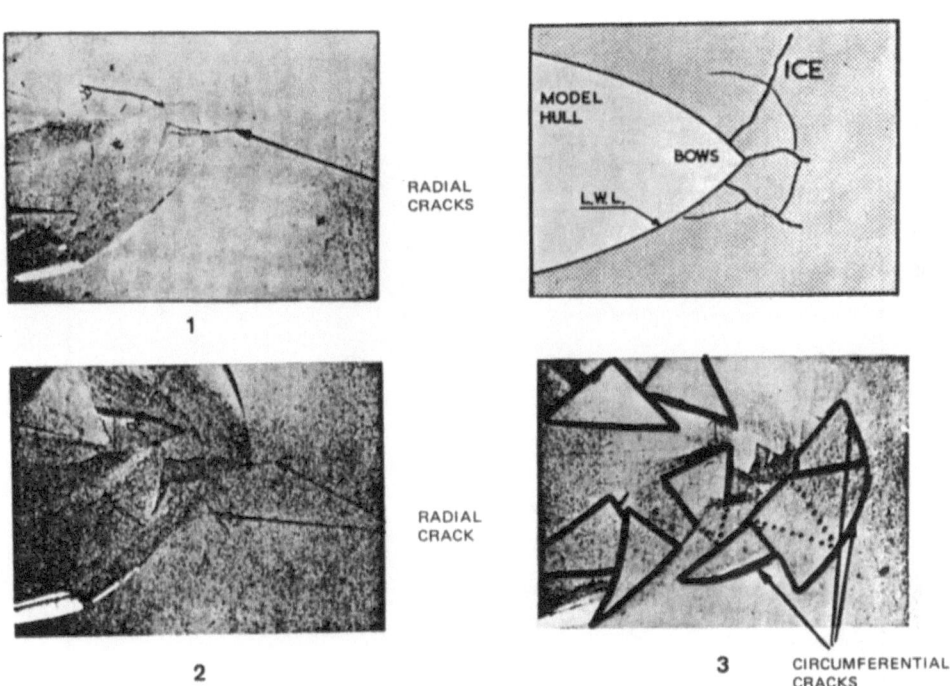

(j) Sequence of frames from underwater cine film of model, illustrating crack formation

Appendix 2

Contours for Ice-Clogged Channel Resistance

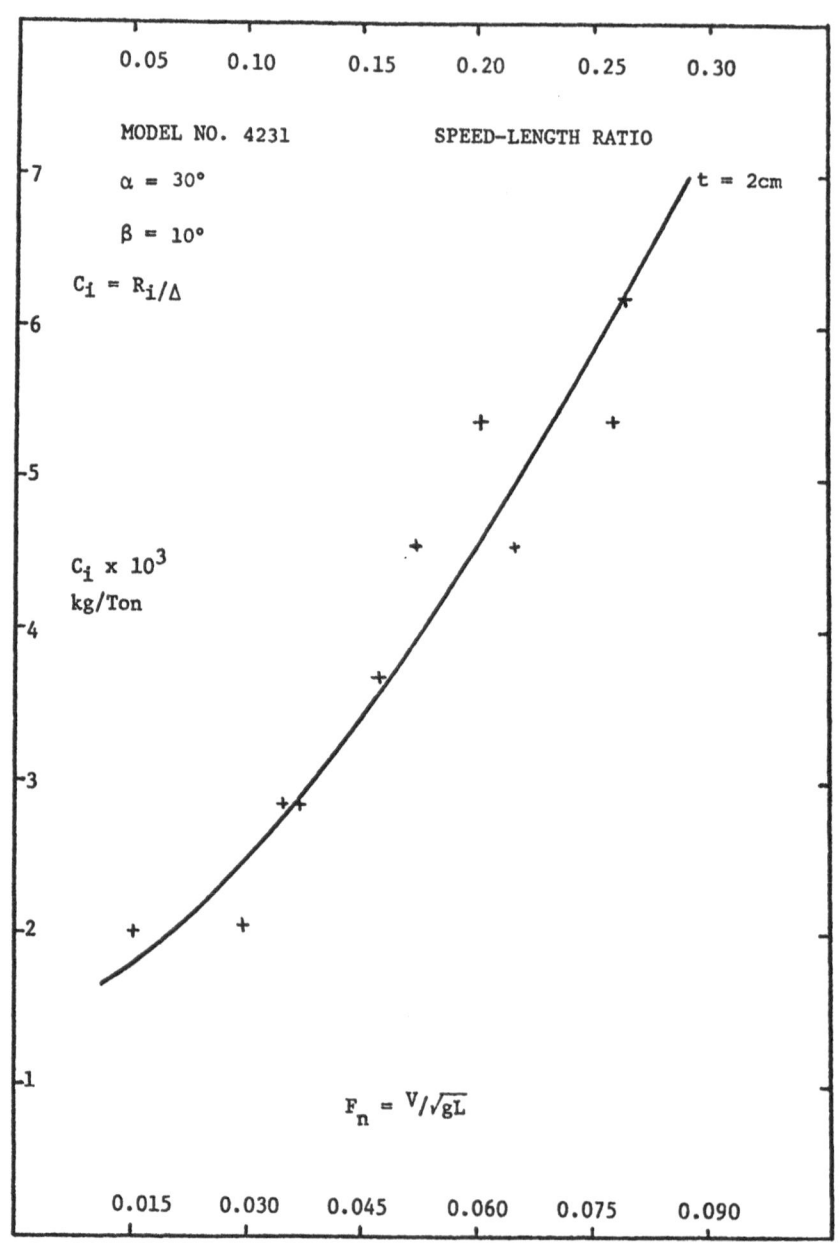

(a) Contour C_i for ice-clogged channel resistance

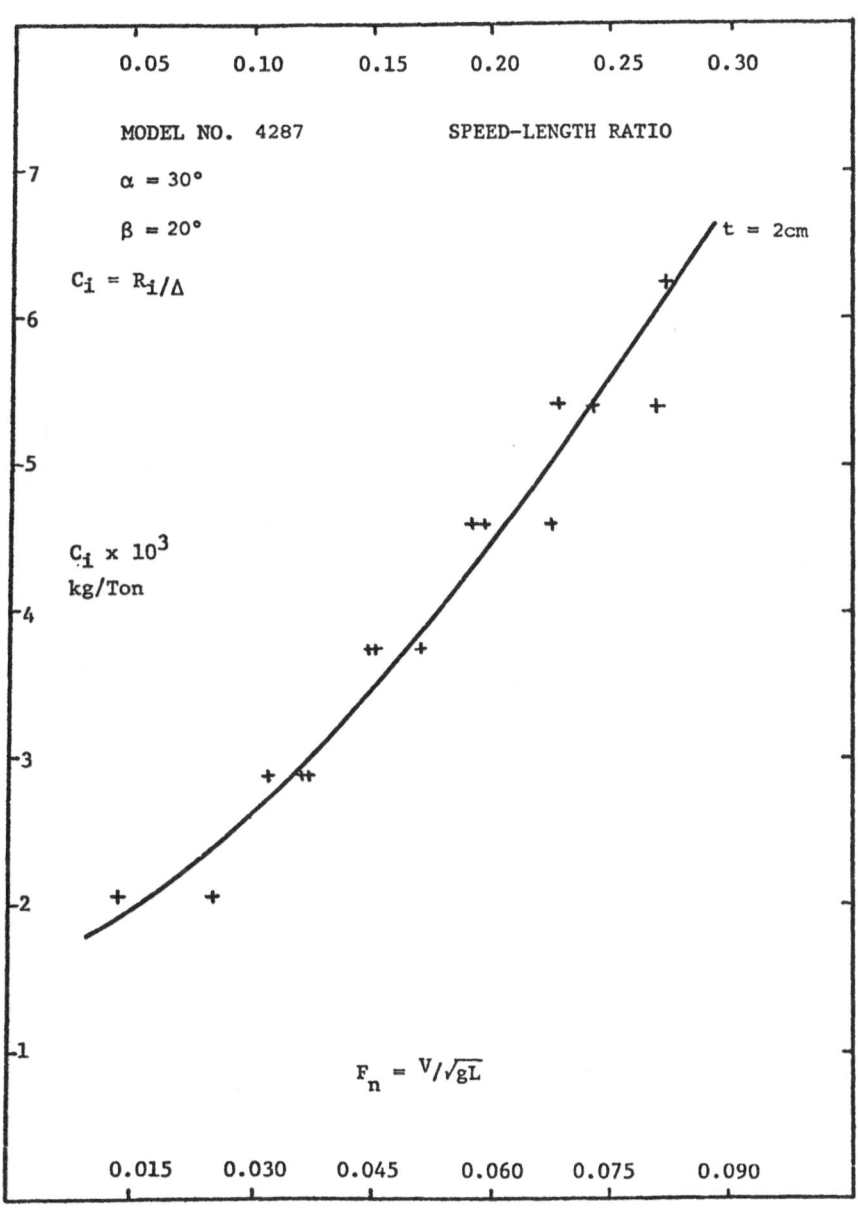

(b) Contour C_i for ice-clogged channel resistance

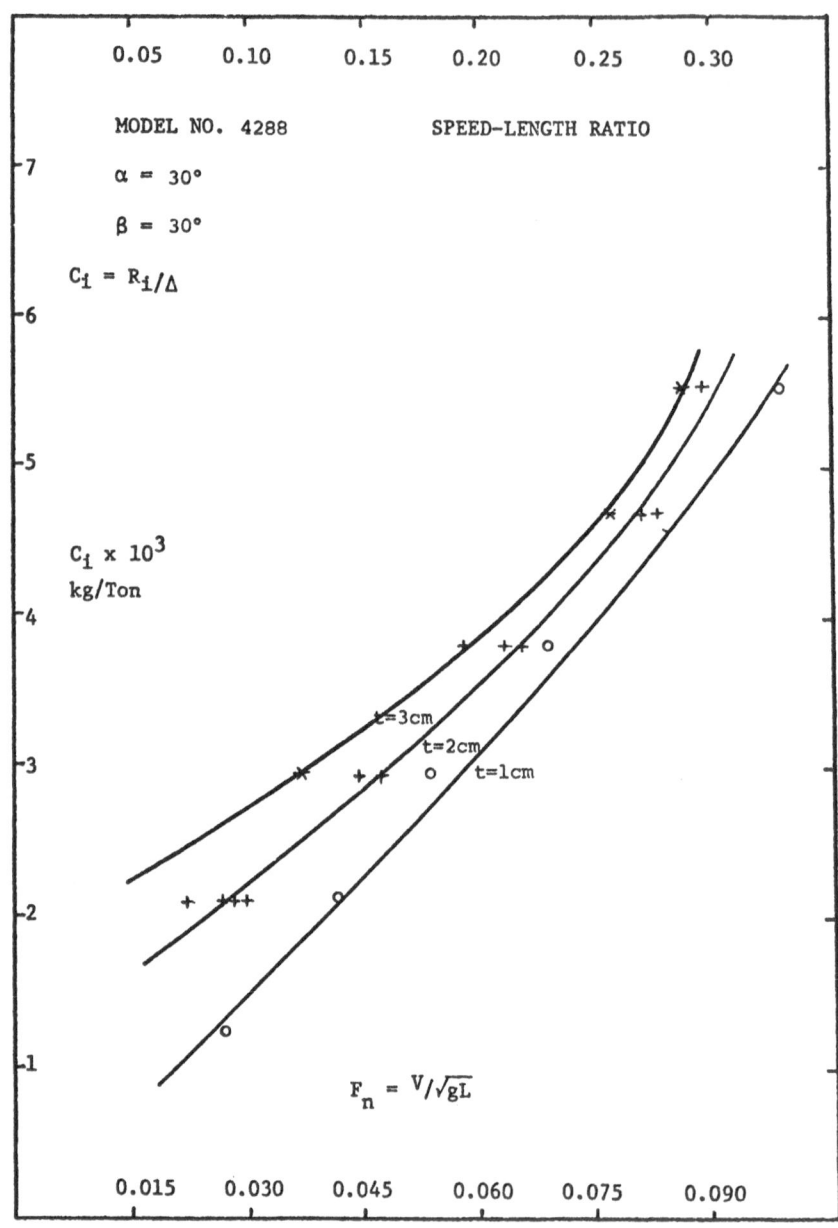

(c) Contour C_i for ice-clogged channel resistance

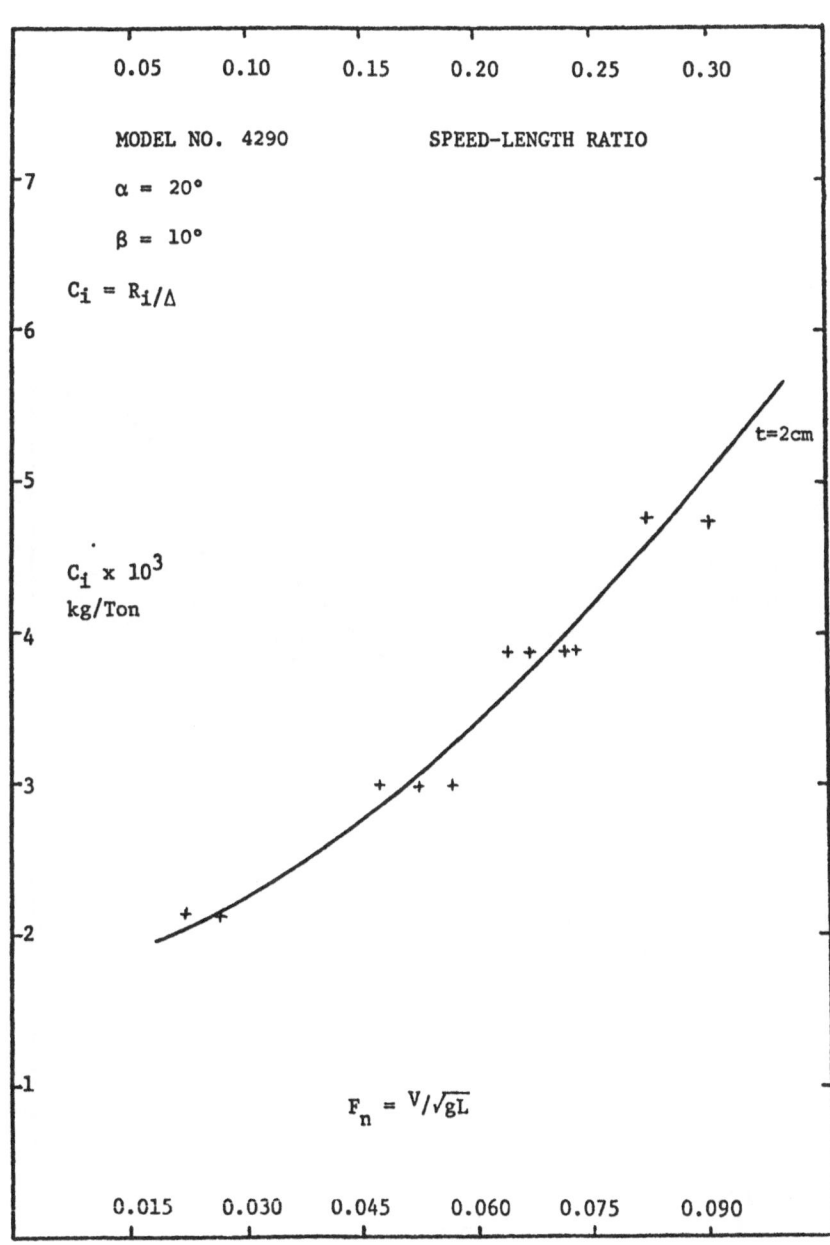

(d) Contour C_i for ice-clogged channel resistance

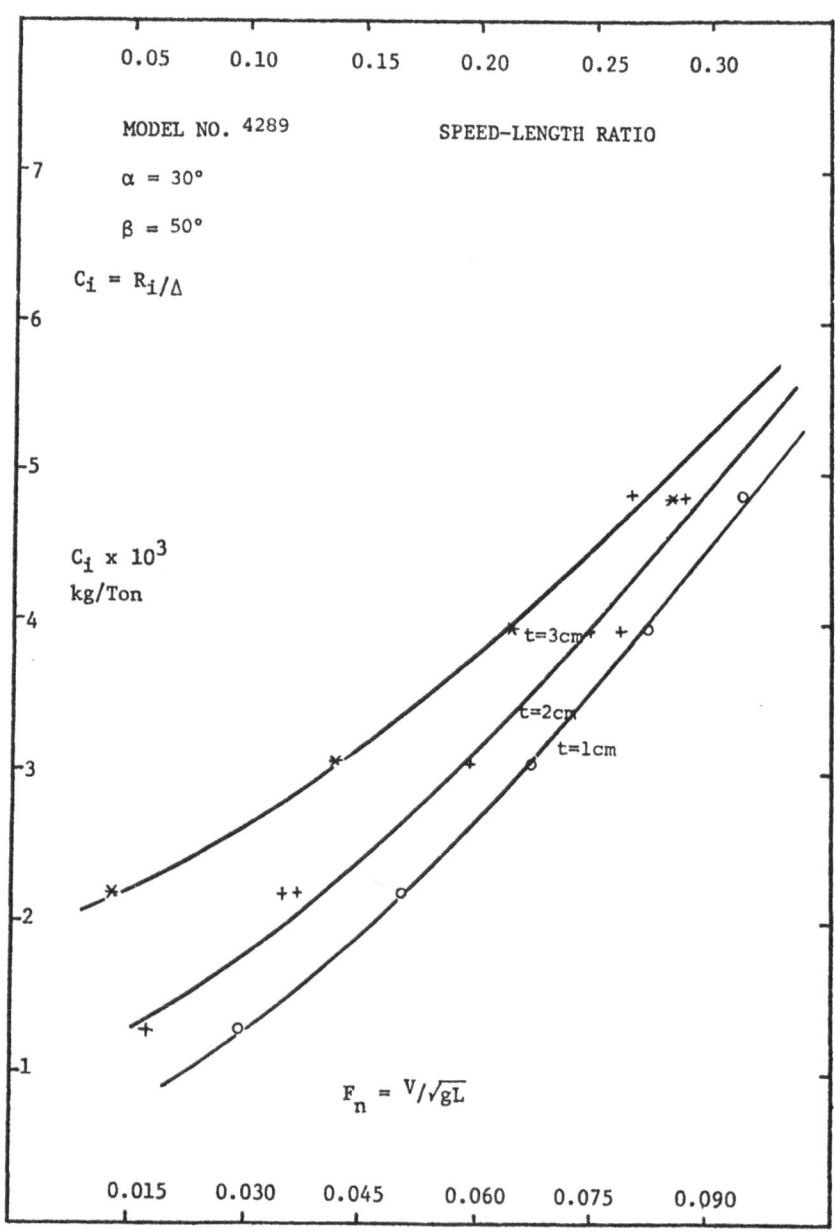

(e) Contour C_i for ice-clogged channel resistance

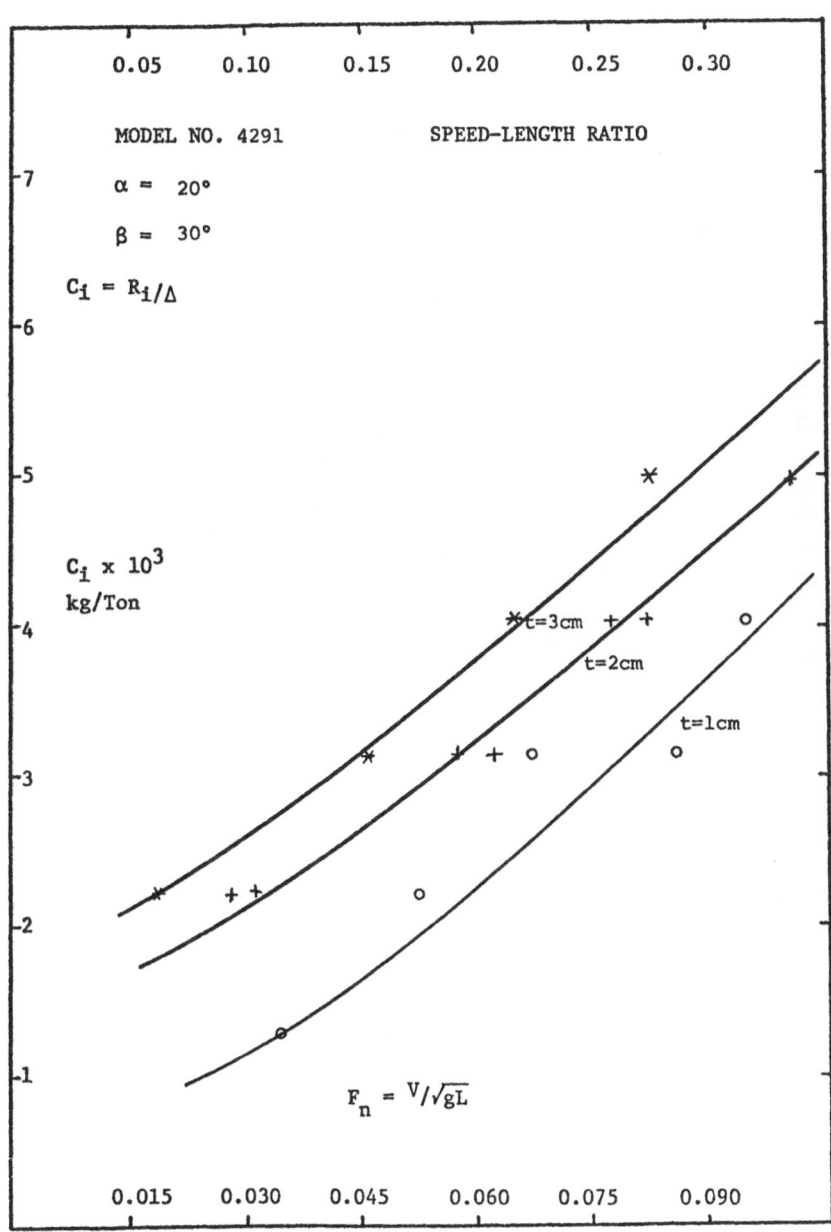

(f) Contour C_i for ice-clogged channel resistance

Appendix 3

Charts for Calculation of Breaking Resistance

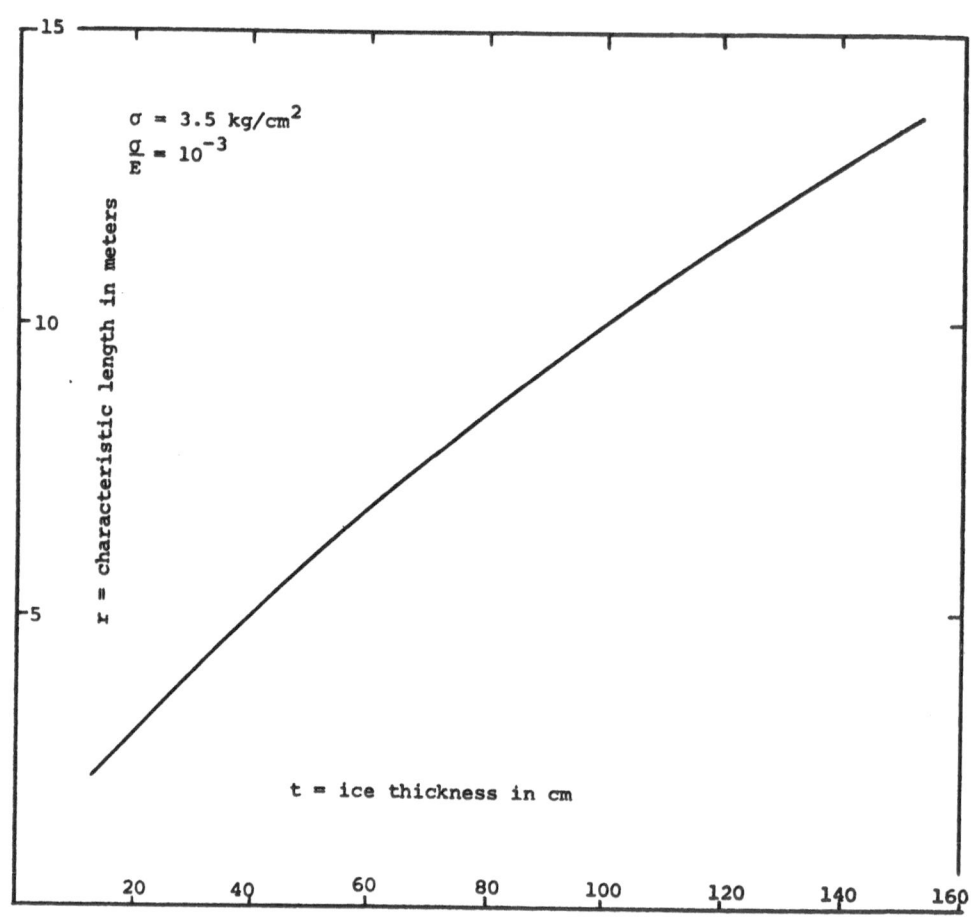

Characteristic length vs. ice thickness

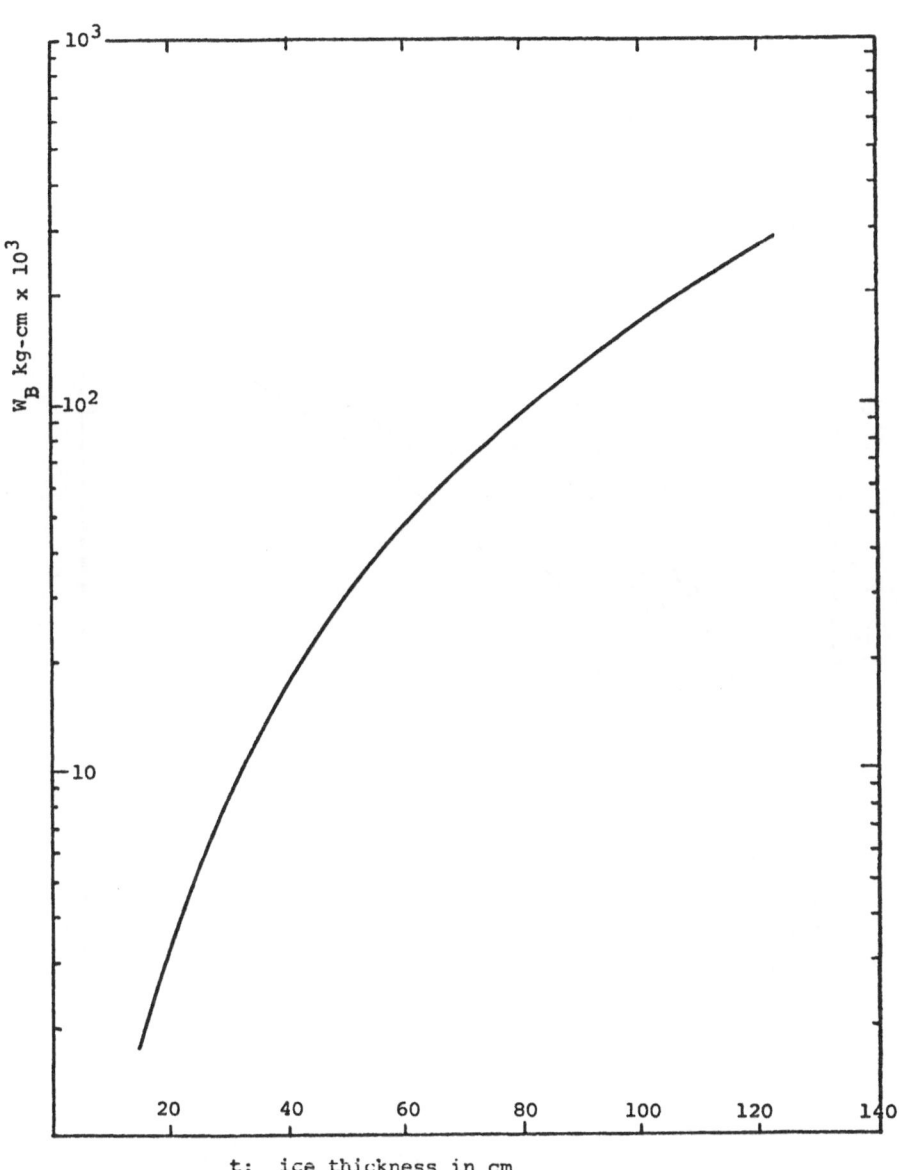

t: ice thickness in cm

Downward force breaking energy vs. ice thickness

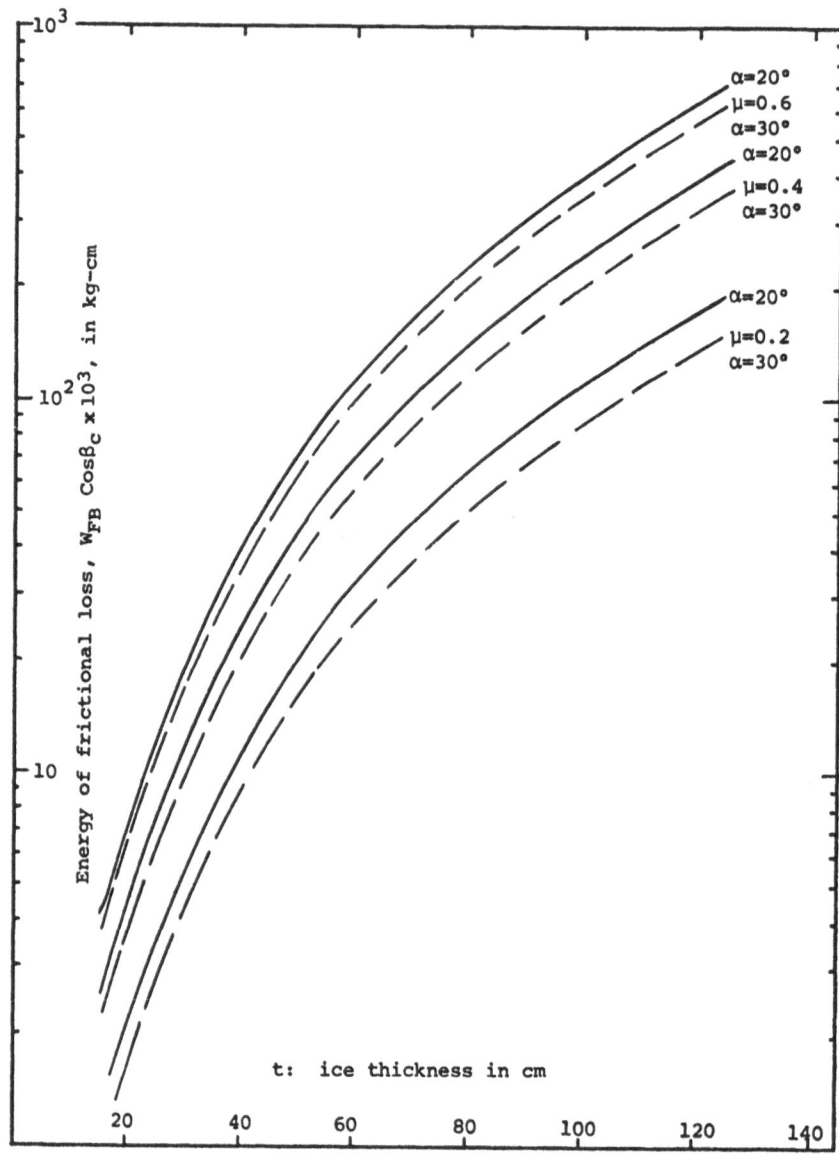

Energy of a frictional loss at a downward
bow breaking vs. ice thickness

Appendix 4

Nomograph for R_f/S Computation (After Series 60)

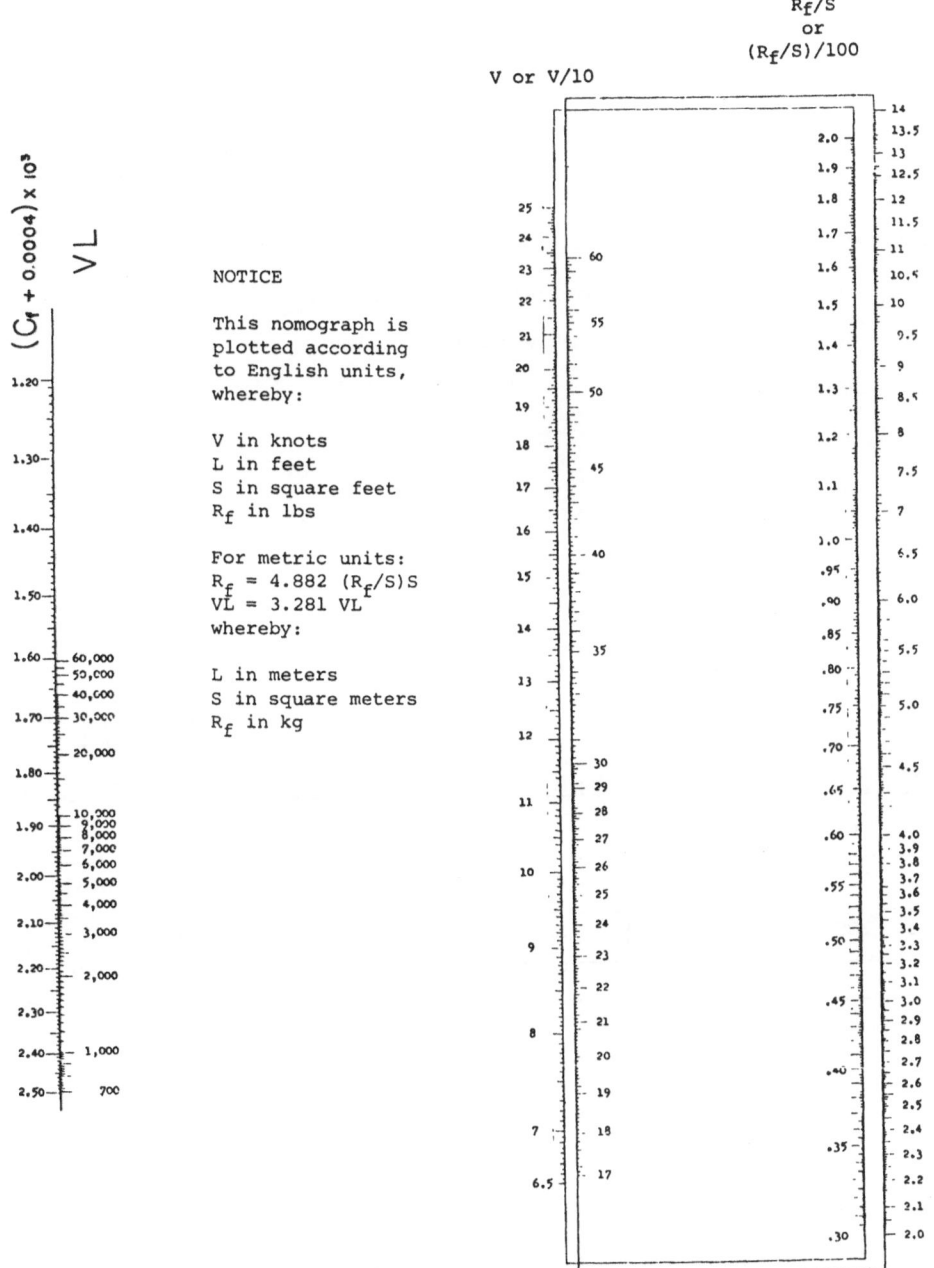

NOTICE

This nomograph is
plotted according
to English units,
whereby:

V in knots
L in feet
S in square feet
R_f in lbs

For metric units:
$R_f = 4.882 (R_f/S)S$
$VL = 3.281 VL$
whereby:

L in meters
S in square meters
R_f in kg

Appendix 5

Numerical Example for a Total Resistance
Estimation in the Ice Field

Breaking resistance (R_b) estimation

L_{WL} = _____   modulus of elasticity E   = 10500 kg/cm^2

L_{BP} = ___283m___   average ice strength   = 3.5 kg/cm^2

beam = ___45.13m___   average frictional coeff.   = 0.4

α = ___18°___   channel-width/beam factor k_{cb} = 1.05

β = ___25°___   channel width = k_{cb} x beam = 47.4m

β_c = ___17.65°___

$\cos \beta_c$ = ___0.95___

λ ___63___

C		.L	M	N	O	P	Q	R
ice thickness cm.		W_B	W_{FB} $\cos \beta_c$	W_{FB} M/$\cos \beta_c$	r	L + N	k_{cb} x b / $\frac{1}{2} r^2$	R_b
t_m	t	fig. I-12 kg-m	fig. I-14 kg-m	kg-m	fig. I-15			Q x P KG
1	63	500	720	758	7.01	1257	0.61	767
2	126	3000	4400	4631	11.79	7631	0.22	1678
3	189	7794	11223	11813	15.98	19607	0.12	2352

Ice-clogged channel resistance estimation
from NSMB series 80 R_i/Δ contours.

L_{WL} = _____

L_{BP} = _____

α = _____

β = _____

λ = _____

interpolation
of readings
$x_\beta = (\beta - 30°)/20°$

$a = (D - F)/2$

$b = (F + D - 2E)/2$

$G = E + ax_\beta + bx_\beta^2$

interpolation
of readings
$x_\alpha = (\alpha - 20°)/10°$

$H = G_1 + (G_1 - G_2)/x$

A	B	C	$D_1 E_1 F_1 G_1$				$D_2 E_2 F_2 G_2$				H*
$V/\sqrt{gL}$	V	ice thickness cm	R_i/Δ for α = 20° spread angle°				R_i/Δ for α = 30° spread angle°				R_i/Δ α = 18° β = 25°
		t_m | t	10	30	50	β =	10	30	50	β =	
0.015		1 | 63									0.91
		2 | 126									1.80
		3 | 189									2.20
0.030		1 | 63									1.20
		2 | 126									2.20
		3 | 189									2.60
0.045		1 | 63									1.66
		2 | 126									2.70
		3 | 189									3.20
0.060		1 | 63									2.20
		2 | 126									3.20
		3 | 189									3.70
0.075		1 | 63									2.80
		2 | 126									3.80
		3 | 189									4.45
0.090		1 | 63									3.60
		2 | 126									4.45
		3 | 189									5.05

*   The bow shape of T.S. MANHATTAN very much resembles Model No. 4291

Continuous ice-breaking resistance and effective power
estimation from NSMB series 80.

Ship T.S. MANHATTAN   Model No. 4291

ENGLISH OR METRIC UNITS   ENGLISH UNITS   METRIC UNITS

L_{WL} _____ $C_{B(L_{BP})}$ _0.80_ R_i in lbs=(R_i/Δ) R_i in kg=0.4464
                                                                      $(R_i/\Delta)\Delta$(m.tons)
L_{BP} _283 m_ R_f in lbs=$(R_f/S)S$ R_f in kg=4.882
                                                                      $(R_f/S)S$ (sq m)
B   _45.13m_ L_{BP}/B _6.27_   V in knots=$(V/\sqrt{gL})\sqrt{gL}$ V in kn.=1.811
                                                                      $(V/\sqrt{gL})\sqrt{gL}$ (m)
H   _14.02m_   B/H   _3.22_   power = $\dfrac{V \cdot R_t}{325.6}$ power in metric un.
                                                                      $\dfrac{V \cdot R_t \ (kg)}{145.7}$
Δ _144426 ton_ α _18°_

S   _185559 ft^2_ β _25°_ $V \cdot L = V \cdot L_{WL}$ $V \cdot L = 3.281 \cdot V \cdot L_{WL}$ (m)

L_{WL} _____

λ _53_

A	B	C			H	I	J	K	R	S	T
		ice thickness cm			R_i/Δ	$R_i=$	R_f/S	$R_f=$	R_b	R_t	power
$V/\sqrt{gL}$	v				Tab. IV-1	H×Δ	fig. IV-1	J×S	Tab. IV-2	I+K+R	S×B
		tm	t								
0.015	1.53	1	63		0.91	131	1.53	1.29	0.792	133	
		2	126		1.80	260	1.53	1.29	1.700	263	
		3	189		2.20	318	1.53	1.29	2.377	321	
0.030	3.06	1	63		1.20	173	5.3	4.5	0.792	178	
		2	126		2.20	317	5.3	4.5	1.700	323	
		3	189		2.60	375	5.3	4.5	2.377	381	
0.045	4.59	1	63		1.66	239	12.5	10.5	0.792	291	
		2	126		2.70	390	12.5	10.5	1.700	402	
		3	189		3.20	462	12.5	10.5	2.377	474	
0.060	6.12	1	63		2.20	317	21	17.8	0.792	336	
		2	126		3.20	462	21	17.8	1.700	481	
		3	189		3.70	534	21	17.8	2.377	554	
0.075	7.64	1	63		2.80	404	32	27	0.792	432	
		2	126		3.80	549	32	27	1.700	577	
		3	189		4.45	642	32	27	2.377	672	
0.090	9.18	1	63		3.60	520	45	38	0.792	558	
		2	126		4.45	635	45	38	1.700	675	
		3	189		5.05	729	45	38	2.377	769	

ARCTIC MARINE STRUCTURES:

SOME ASPECTS OF SYSTEMS APPROACH

W. H. German

German and Milne

Montreal, Quebec, Canada

I purposely chose the title of my paper in order to give me the widest latitude possible when collecting my thoughts for this presentation.

The theme of this conference is the Arctic, and the coupling word is Systems - a word that has a specific meaning in modern terms and in this context. But when it comes to ships or marine structures, the meaning behind the word has been with us since the beginning.

I think it is probably true to say that nothing made by man does not eventually find its way into a ship, and it is therefore equally true to suggest that the ship as a system is probably as complicated and as expensive as anything else one can imagine with the possible exception of the Soyez, the Apollo, or perhaps even the human body.

At any rate, the interrelationship between man, material, action and existence in ships has been with us for a very long time. Therefore, to discuss systems as something new is not that easy.

In my title I also chose the expression "Arctic marine structures" rather than "ships" because whilst our first real entry into the Arctic was in ships, the fact we must face today is that there are many other types of offshore engineering that will be applied to the Arctic, and all of which will have similar problems; that is, the protection of man against the elements (hostile under the best of circumstances), and a machine that will perform the physical function expected of it.

Ships in the Arctic have been around for a very long time and we are all familiar with the impressive pictures of naval heroes, explorers, people who penetrated to unbelievable degrees of distance and human endurance, and people who lost their lives.

441

After the explorers (and the whalers) came the first
commercial venturers to the Arctic, and here I specifically refer
to the Canadian sector. They were generally the Newfoundlanders
and the Maritimers who started to man the supply missions and so
forth. The first major activity in what was then considered the
rim of human activity was the Hudson's Bay Supply Ship called the
NASCOPIE. When that gentle lady eventually became too old in the
late 1940's, she was replaced by the RUPERT'S LAND who carried on
serving the Hudson's Bay Company for many years and by the C.D. HOWE,
a Government vessel which then took over specific Government
functions instead of merely chartering space in a commercial vessel.

I do not propose in this paper to get into any significant
detailed mathematics or physical facts, largely because the subject
is much too broad and also because the audience, to the best of my
knowledge, covers a very wide spectrum of disciplines, and nothing
could be more boring to such people than to have one member of a
sector pounding out remote facts, figures or theories.

Furthermore, my thoughts today lead not to the development of
special points of routine engineering but into areas which are
undoubtedly well known, but seldom approached. Ships themselves
in the Arctic carried on a slow process of improvement and development from the early 1900's up to the time that DEW Line was
commenced. The Military Sea Transport Service, in supplying the
defence requirements for DEW Line, produced phenomenal efforts in
carrying great quantities of material to the north into areas that
had not been serviced before, and under conditions and during
seasons that were at that time inconceivable.

The record is impressive and my personal belief is that it
has not received publicity appropriate to their efforts.

However, what they did do was leave a core of experienced men
about, who were available for the next step forward which was the
Prudhoe Bay "discovery" and which resulted in the meridian of
transportation in the Arctic. You all know that event as the
MANHATTAN.

MANHATTAN, with the impetus that she gave to transport and
research, is the meridian of transport in the Arctic and today we
are all moving as quickly ahead in this area as possible. Competent
consultants, learned scientists, capable testing institutions are
all heavily occupied, and the process is grinding out improved
hull forms, improved structures, improved materials and improved
propulsion systems, all to the extent that improvements now are
no longer quantum leaps forward but are short steps; but steps
that must not be discredited because every step, no matter how
small, is a move in the right direction.

Regarding machinery, we have been examining everything from
the traditional steam, diesel, electric, through fixed propellers,
C.P. propellers, on up to the use of gas turbines and, of course,
nuclear.

Since there are many combinations of machinery with potential
application in Arctic icebreakers and structures, it would probably

be helpful to provide a range of powers and the preferred arrangements to produce those powers.

Propulsion Systems for Ships Operating in Ice

0-5,000 SHP Total Single Screw	Geared Diesel Direct Drive Diesel DC Diesel Electric
5,000-10,000 SHP Total Twin Screw	Geared Diesel DC Diesel Electric AC Diesel Electric AC HD GT Electric
10,000-20,000 SHP Total Twin Screw	Geared Diesel DC Diesel Electric AC Diesel Electric AC HD GT Electric
20,000-40,000 SHP Total Triple Screw	Geared Diesel DC Diesel Electric AC Diesel Electric AC GT Electric Geared GT Steam Electric
40,000-75,000 SHP Total Triple Screw	Geared Diesel-GT (CODOG) AC-DC GT-Diesel Electric AC-DC GT-Diesel Electric Geared Steam Turbine Steam Turbine Electric Nuclear
75,000-120,000 SHP Total Triple Screw	Geared GT AC-AC GT Electric Nuclear
Above 120,000 SHP Total Triple Screw	Geared GT Nuclear Geared Steam Turbine

A discussion of systems related to the foregoing prime movers would be too large a task for this paper, but it is fair at this time to point out that the election of machinery is qualified to an extensive amount by the characteristics of that machinery relative to the requirements of the hull into which it is fitted.

For example, all ice-capable ships, whether commercial or Government, must have a machinery power that bears a delicate relationship to the mass of the hull into which it is fitted.

But the ratio of power to mass must be higher in a working ice-breaker due to the need for higher acceleration forces.

Similarly, icebreaking ships, commercial or Government, must have the ability to apply maximum power when going astern. This was clearly demonstrated in MANHATTAN, a ship fitted with conventional turbines with sufficient turbine blading only to provide 10% power astern, ample in temperate waters, but not enough in the ice. Since the vessel must be capable of going astern, obviously the Government icebreaker requires a higher capability than a MANHATTAN-type ship.

In order to achieve the greatest effect through the use of ramming, the time from full ahead to full astern at the propeller must be to an absolute minimum, and therefore careful attention must be given to the inertia of the driving train, i.e. propeller shafting, reduction gears, etc.

Traditionally, icebreaker propulsion systems have also required an unusual characteristic not found in other vessels, which is the ability to provide maximum torque at zero RPM (hence the popularity of electric drives).

In the early days of icebreaking, when propulsion was inevitably steam, usually from Scotch marine boilers through triple expansion engines, the arrangement was ideal since if a propeller was stalled due to ice jamming, torque to the ultimate in boiler pressure permissible could be applied.

Another approach to achieving maximum effect at minimum speed of advance is via the use of the controllable pitch propeller, a device that has been unreasonably long in securing recognition in the type of service we are discussing.

My personal view is that the U.S. Coastguard is to be highly recommended for fitting C.P. propellers in the Polar Star. In view of the horsepowers and the service involved, this is a significant step forward.

C.P. propellers, whilst not as capable in comparison with fixed wheels/electric drive in some areas, offer one significant advantage - when the ship has forward momentum with no power being transmitted through the propellers, fixed wheels at zero RPM can produce a dragging or hoe effect with a consequential increased opportunity for propeller damage.

Alternatively, under the same circumstances, the C.P. wheel at zero pitch can be left rotating with less possibility of such damage.

A step forward in the debate over the use of propellers will take place when the giant ships are produced for the Arctic, since the propellers in these ships will in all probability be sufficiently deep in the water to avoid the danger of serious impact with the ice or the jamming of ice between the propeller tip and the hull; and furthermore, being of such large size, i.e. strength, all the time realising that the strength of the ice is constant, the possibility of damage becomes increasingly less.

Conventional land-based power distribution systems have evolved over many years under conditions where generating plant size and weight are eclipsed by the demand for low loss power distribution over vast areas. Here voltages must be low enough to prevent corona discharges and severe insulation leakage, and frequencies must be low to prevent AC power induction losses in the long lines.

Power systems for Arctic ships do not generally operate under the conditions of a city or rural supply. Shippower distribution, for example, hardly requires a system that maintains low AC losses over distances of many miles and insulation for the shorter runs can be economically upgraded to permit higher distribution voltages.

When traditional restraints are removed, electrical distribution systems and machinery for Arctic vessels and other installations should be designed for maximum efficiency while minimising bulk and weight. The weight of an electrical transmission cable is a function of the current and permissible voltage drop. As the voltage is increased, the current falls accordingly but the voltage drop for the same power loss increases. A high voltage cable is thinner and lighter than its low voltage counterpart carrying the same power. The design limits are the practical limitation on insulation and the maximum voltages that electrical machinery or transformers and switchgear can handle.

In order to reduce the size and weight of the electrical machinery, the supply frequency has to be increased to permit lower pole and field inductances and consequently less iron. This also increases operating speeds allowing lower torques and lighter shafting for comparable power. Changing from a 60 cycle 400 volt to a 400 cycle 1,000 volt motor for example, results in a typical weight and volume reduction of 4 to 1 and cuts the weight of the supply cable in half.

Systems warranting investigation are the newer types of cryogenic machines and the homopolar systems which require high currents at low voltages to produce extraordinarily compact high speed units with high efficiencies.

Regarding hull form, tank tests have been done with great success and much brilliance in Finland, North America (U.S.A. and Canada), Russia and Germany, with some support from other leading maritime areas.

Significant recent advances have related to the bow form and the traditional bow profile, that is the 30° angle of attack, has now been displaced by the so-called concave bow, or popularly known as the White bow, which was fitted on the MANHATTAN. The bow is a major advance over the traditional.

Tank tests done for ships' design since that point have continued to improve the vessel's performance, perhaps not to a spectacular degree but certainly to a degree sufficient to warrant the costs of such research.

The world has a large number of very inventive people, and since the MANHATTAN affair with the publicity that accompanied

that experiment, hundreds of people have developed ideas on how
icebreaking performance can be improved. Some of the ideas are
incredibly imaginative and hopelessly impractical, but the common
denominator of all seems to be a genuine interest in attempting to
solve the problem.

Of all the ideas that have been submitted for consideration,
only a very few will survive, and only one or two will in my view
make **any real headway**. I think the White bow is the most
effective and will continue to dominate hull design for these
ships for many years to come.

The upward breaking bow concept known as the Alexbow, had a
brief and fiery piece of publicity, but it is my view that in
certain specific applications, the upward breaking bow will have
its advantages.

Similar comments apply to devices built into the ship that
give the ship a reciprocating or orbital action, not for
commercial vessels or working icebreakers but quite possibly for
specific application under particular circumstances. Rotating
saws, high pressure jets, concussions - they have all been
proposed; some of them will survive.

Much of the mystery of icebreaking hulls is no longer a
mystery and there is no need to discuss procedures which are now
recognised, such as bilge keels, sea-suction inlet, cooling water
loops, etc., but I believe that one or two possibilities are
worthy of mention. The first one is much more than a possibility.
It is a fact, and that is the bubbler system developed by the
Finns.

It is indisputable that this system has substantial advantages
under given circumstances and the concept will undoubtedly become
a standard fitting on most Arctic vessels.

Another concept that warrants close examination is the
opportunity for heated sides. Major resistance to the ship in ice
is provided from two sources: one is pressure, which no one has
properly solved as yet; the other is abrasive or adhesive
resistance at the ship's side in the area of the ice belt.

The bubbler, mentioned above, is one solution. Another
solution is ship-side flooding and heated sides in that area, and
there are presently arrangements worked out to achieve the results.

We can afford to re-examine hull design with particular
reference to the stern of the ship, and give serious thought to
modifying outlines to encourage buttock flow rather than waterline
flow, particularly for deep draft ships. If we have buttock flow
coming into the propellers under deep draft circumstances, it is
probable that an absolute minimum of ice will find its way to the
propellers as opposed to the amount that will arrive in that area
if the flow to the propellers is waterline flow. If buttock flow
is combined with the potential for annular rings or Kort nozzles,
the ultimate result is likely to be very successful indeed.

Whilst I am in this portion of the ship, I would like to add
some further comments on propellers. Any damage sustained by a

ship in temperate waters will, of course, be bothersome and costly
to the owners, but if it cannot be fixed by the ship's complement
the simple expedient of calling for a salvage tug will bring the
ship home to drydock and safe haven. Unfortunately, this is not
the case in the Arctic, and the vessel, with what would normally
be considered a very simple ailment, could discover that ailment
to be a terminal disease; in other words, the ship could be a
constructive total loss in the ice as a result of a breakdown or
failure that otherwise might be relatively unimportant.

For example, should an ice-capable vessel, particularly if it
is single screw, suffer a propeller fracture or failure under
heavy ice conditions, and if a search and rescue icebreaker of
adequate strength and capability were not available, that ship
could be in real difficulties.

Some years ago, when the Royal Canadian Navy had an Arctic
cruiser, the HMCF LABRADOR, this possibility was anticipated and
exercises were conducted which developed an ability in the ship to
change her fixed propellers at sea, in the ice, using scuba divers
and ship-carried equipment.

With the implications of the foregoing, it makes good sense
for the architect, when designing such vessels, to satisfy himself
that propeller damage can be repaired in the Arctic by ship's staff.

If the propellers are controllable pitch, it should be
possible to arrange the ship's hull in way of the propellers in
such a fashion as to permit divers to go to the propellers from an
underwater aperture in the hull above the propeller and with
suitable tools and equipment remove the defective blade and replace
it with a new one.

Fixed propellers, or solid propellers, of course, present a
different problem, largely due to their weight and at present also
due to the way they are connected to the tail shaft. However,
designs are now available which suggest that the conventional
fixed-blade propeller can now be produced as a built-up, thus
providing circumstances under which propeller blades can be
replaced. This concept will in the future be receiving a great
deal more attention than to date.

Steel in ships' hulls for cold weather service has gradually
improved over the years, and at the moment has arrived at a
situation where Lloyd's type grade E steel is considered to be the
best available, all circumstances being considered. A normal
specification for an Arctic vessel requires that all steel in the
ship, unless otherwise specified, should be at least Lloyd's A
quality (ASTMA-131-58).

Plating for the shell should be grade D or E quality, and the
specification for grade E plate is as follows:

When stainless steels are used for structural work they are
of the AISI austenitic type in the 18.8.3MO group, 316L or equal.
There shall be no cast iron in the ship whatsoever, unless it is
nodular iron or better.

Hull castings such as the stern frame shall be nickel vanadium steel of the following description:

Mechanical Property Requirement

Property

Tensile strength (psi)	76,000 min.
Elongation in 2 in. (%)	25 min.
Reduction of Area (%)	35 min.
Bend Test (Specimen 1 in. x 3/4 in. in cross section bent around 1 in. (degrees) diameter bar on 3/4 in. side)	160 min.
Charpy V-notch impact strength at -40°F (ft.-lb)	15 min.

Chemical Requirements

Element	Percent
C	0.20 max.
Ni	1.5 min.
V	0.07-0.12
Total Al	0.10 max.
S	0.02 max.

Other Properties

Sea Water corrosion:

Static i.p.y.	–	0.005
27 f.p.s.i.p.y.	–	0.035

Heat Treatment

The heat treatment shall be double normalize with a temper following the final normalizing. A suggested procedure is as follows:

(i)	Normalize 1675°F	– 5 hours
(ii)	Normalize 1575°F	.– 8 hours
	temper to obtain mechanical properties as specified under "Mechanical Property Requirement". Minimum time at temperature – 8 hours.	

The casting shall be in the normalized and tempered condition prior to machining.

We have often considered the use of high tensile steels for this application. However, apart from the much higher material cost, the scantlings cannot be reduced sufficiently to make the substitution cost effective. For structural members, deflection is often the governing consideration. With the use of the H.T. steel, it therefore becomes necessary to maintain the section inertia high enough to reduce deflections to a tolerable limit. This means that the high tensile steel members would be under-stressed and provide larger factors of safety in the structure than required, and at more cost.

The actual reduction of the hull weight should high tensile steel be used, would be relatively unimportant and, in fact, reduces the weight at a low part of the body which would possibly reduce the stability of the vessel slightly.

There therefore seems to be no strong advantage to using high tensile steel, and there are disadvantages:

(a) higher cost
(b) more difficult repair procedures
(c) risk of substituting a mild steel plate, where a high tensile plate had been used
(d) the quality control in repair work would be more difficult to maintain.

As a matter of interest, it is understood that modern Finnish icebreakers built for the Soviets use essentially the same grades of steel. The subject was considered extensively by them recently prior to the building of the latest series of icebreakers and the outcome was the use of comparable grades of steel to those we use ourselves.

Professional icebreakers are developing in themselves. Recent publicity has been mostly on the commercial ice-capable ship; indeed, vessels on the surface, under the surface and surface piercing in sizes up to 250,000 tons and horsepowers on a 1:1 basis have been prominently mentioned. Not so much has been heard of professional icebreakers, the most notable of which has, of course, been the Russian nuclear ship LENIN. The American U.S. Coastguard Polar Stars with powers of 60,000 are the current interest, and we now have on our design boards the new ship which, for want of a better name, is called POLAR 7. The principal particulars of the POLAR 7 are as follows:

Length overall (excluding stern notch)	182.00 m
Length on design W.L.	173.00 m
Length between perpendiculars	167.50 m
Breadth moulded	32.20 m
Breadth at design W.L.	30.50 m
Depth moulded to upper deck	18.25 m
Draft, moulded, design, S.W.	12.20 m
Displacement, full, S.W. at 12.20 m	33,000 Tonnes
Deadweight, total	12,260 Tonnes
Power, normal continuous at propellers	67,115 Kw total
Speed, service at 11,185 Kw at propellers, total	31 Km/hr

Complement:

Deck officers and crew including Flight personnel	55
Engineer, officers and crew	32
Catering officers and crew	15
Met. observers	2
3rd and 2nd year cadets	56
RCMP officers	8
Detainees	4
Spares	11
Total persons on board	183

Function

The vessel will be designed for icebreaker service in the Arctic with primary and secondary missions as follows:

Primary Missions:

(a) to prevent marine disaster, and
(b) to maintain the flow of marine transportation, by providing icebreaker assistance capable of:
    (i) preventing beset ships from being carried by ice into shallow water or other dangerous situations;
    (ii) releasing ships that are beset;
    (iii) assisting ships that have sustained damage to a port of refuge or to open water.

Secondary Missions:

(a) to provide a policing function, ensuring compliance with Canadian Regulations.
(b) Search and Rescue.
(c) Pollution control.
(d) to provide a vehicle for hydrographic and marine science investigations on a limited scale in otherwise inaccessible areas.
(e) to contribute to the provision of environmental intelligence on the route.
(f) to provide emergency medical assistance to ships on the route and remote settlements in the area.
(g) to provide emergency logistic support to remote areas.

Environmental Conditions:

The vessel and all its machinery equipment shall be designed for the following environmental conditions:
(a) Outside air temperature -51°C in winter
(b) Wind velocity 185 km/hr
(c) Water temperature -2°C in winter; 30°C in summer
(d) Ice pressures - as given in Arctic Waters Pollution Regulations for Arctic Class 7 vessel
(e) Roll of vessel 60° out to out, cycle frequency 10 seconds. Permanent list 15° port or starboard.

(f) Pitch of vessel ± 12°, cycle frequency 5 seconds
    5° fore and aft trim.
(g) Shock loading 3g Horizontal and Vertical
(h) Icing conditions 150 mm thickness of ice on all exposed
    surfaces.

Environmental conditions to be applied shall be specially
considered for certain equipment such as cranes and elevators.

It will come as no surprise to anyone that all of these new
ships involve major improvements in all systems. But to the best
of my knowledge, a factor which is equally important has received
little attention, and that factor is the ergonomics or the human
engineering of such voyages.

Up to now, voyages into the Arctic have been on a one-trip
basis. The people crewing those ships, in leaving on the voyage
know full well that they'll be back for the rest of the year in a
few short weeks. The ships or structures we are talking about
today will be expected to stay out for extended periods, and how
do you keep personnel aboard such ships under the very difficult
living conditions of the Arctic? I say difficult because the
conditions are mostly in the mind. Lond periods of darkness;
small colour contrast; a virtually unending vista; and when you
add to that the noise to be found in an icebreaking ship, close
confinement of living, there is no question that an important
consideration in the design of Arctic systems must include human
engineering.

There are several basic factors to take into account - space,
colour, temperature, and humidity, noise, sound, smell, motion and
vibration, exercise and mind-challenging activities.

The term "mind-challenging activities" will be quite obvious.
When personnel are off-duty, they must be occupied with sleeping,
eating, exercise and mental occupation. Unfortunately, under the
present dismal conditions in the Arctic, the more intelligent, or
perhaps the word should be imaginative, members of the crew are
the ones most susceptible to this need for mental activity, lacking
which they tend to brood over possible problems at home or elsewhere, and become a problem.

Our friends the Russians have very directly tackled this
problem in that much of the leisure time of the ship's complement
is taken up by lectures, information exercises, propaganda and
self-improvement routines. I think it would be fair to suggest
that such sessions are compulsory, and are sufficiently taxing of
the person's mind to satisfy his need for mental involvement and
at the end of the session, fatigue.

In our society where the problem has been handled somewhat
differently, it would be necessary to provide these ships with the
equivalent of a modern lecture hall, and a fully stocked library
of books and audio-visual equipment, giving complete courses,
training, information, etc. on a wide range of subjects that have

been very carefully selected by the education specialists.

Such mental involvement could be divided into three parts: direct teaching of difficult subjects to those who wish to improve their status or achieve higher degrees; matters of general information and general knowledge; and matters that verge on or are in fact pure entertainment.

Modern methods of teaching, using highly sophisticated audio-visual techniques, offer a completely successful tool in this area: indeed, sometimes one would suspect even more successful than the personal attention of some teachers available to us today!

The same comment applies to physical fitness and condition. The ship should carry either a medical officer with specialist qualifications, or an athletic director with a medical background in order that the crew will be kept in top physical condition.

Once again in our society it is difficult to insist, at least outside of the military, on activities of this nature. But I feel that some disciplinary agreement must be established which will make proper physical and mental conditioning an assured activity.

The old expression of an "idle sailor is an unhappy sailor" is still a true one.

Ergonomics or environmental engineering to which I have referred is also referred to as environmental psychology, a discipline that has been concerned with two major topics. The emotional impact of physical stimuli and the effect of physical stimuli on a variety of behaviours such as work performance, or social interaction. It is my belief that a great deal of attention should be paid to this particular discipline and its recommendations when accommodation in the ships or offshore structures is being laid out.

Messrs. A. Mehrabain and James A. Russell, in their book An Approach to Environmental Psychology clearly indicate that physical stimuli of all sorts have significant and predictable effects on humans. As a result, human behaviour under controlled conditions can be manipulated or "engineered".

1. Colour and pleasure. There have been numerous studies of emotional reactions to colour which clearly indicate impact on the human psyche.

Brightness: there is in general a positive relationship between brightness and pleasure - white and black are both pleasant, greys are unpleasant. For "cool" colours, pleasure increased with brightness; for "warm" colours, increase in pleasure was a positively accelerated function.

Saturation: Similar reactions exist for saturation, but for most values of hue and brightness, saturation was a direct correlate of pleasure.

Hue. Pleasure associated with hue varies with the value of saturation and brightness. It may be interesting to the reader to realise that the degree of pleasure for a specific colour can be easily ascertained from extensive charts already in existence.

2. Sound simulation. Numerous studies have constantly indicated that arousal level increases with noise level, but for pleasure purposes a different yardstick must be considered.

For example, noise is unpleasant, and music is pleasant. The problem is, of course, to select music for a large group that will appeal to all individuals in that group, and have a predictable and desired effect.

3. Food: is a point that should be kept in mind, and the Americans have long recognised this problem. I refer not only to the military but more so to the oil exploration industry, where wild-catters, roughnecks and people who live and work on the frontier are always plentifully supplied with food. Indeed, it has been my experience that quantity has been the alter rather than quality.

Today, and this is particularly true on a ship, quantity is not the only requirement. A wide range of well-cooked, well-presented food is fundamental. And at the same time, access to alcoholic beverages to accompany the meal should never be restricted, albeit under control.

4. Smell. In the matter of ventilating, heating and air conditioning accommodation and work spaces, if complete economy of energy is to be maintained resulting in a system that would be a closed system, then some effort must be made to introduce into the air being circulated, carefully selected chemical agents with identifiable smell characteristics designed to have the desired psychological effect upon the crew.

5. Studies indicate similar predictability on pleasure and arousal by other factors such as tactile stimulation and fantasizing.

All of the foregoing is, I am sure, generally acknowledged, and is indeed practised in many areas, sometimes on a hit-and-miss basis, other times on a more positive programme approach.

Our friends in the icebreaking service of the Russian Government are a case in point. Their ships are extremely well appointed with large cabins and very large public rooms, tastefully decorated with wooden furniture, drapes and the lot.

These people obviously recognise the point that I have been trying to make, and have done something about it (See Figures 1 a, b,c,d,e and f).

I have included them merely for comparison purposes. These are pictures of accommodation in the new U.S. SPRUANCE Class destroyers; and whilst the U.S.N., in recognition of the problem presently under discussion, have made considerable efforts to improve living conditions by going in for colour combinations and printed vinyl coverings, it is pretty evident that there is more to the solution of this problem than merely slapping a few colours up on adjacent flat surfaces.

It is fairly easy to criticise the U.S.N., but I have considerable sympathy for them since any move in the direction that I am advocating will at the same time be a move backwards

Figure 1a

Figure 1b

Figure 1   Recreational Facilities in U.S. SPRUANCE

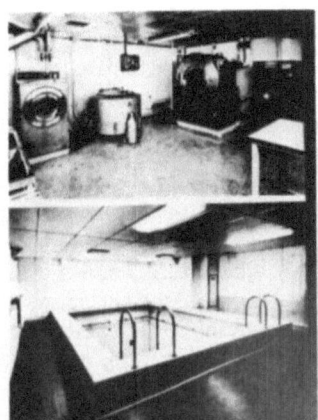

Figure 1c   Figure 1d

Figure 1e

Figure 1f

from the requirements of the damage control manuals, since we have
unfortunately not yet been able to find fire-proof materials that
meet our standards of comfort, and I would like, therefore, to
enlarge somewhat on this point with particular reference to
insulating materials as well as fittings and furnishings.

As mentioned elsewhere in this paper, improved insulation is
necessary, but the ability to achieve the ideal is restricted
because of our laws respecting the use of combustible materials.
Either the acceptable material is too heavy for this service, or
it has other disadvantages such as non-structural rigidity or the
inability to foam.

Whilst I hold no argument with the authorities in the need
for fire-proofing passenger ships, I begin to wonder if such rigid
adherence to these regulations is necessary in commercial vessels.
Presupposing the crew are just as human as passengers, I will give
you that point. But when the design of a ship and its ultimate
function in a severely restricted area such as the Arctic is likely
to be affected by such regulations, then I feel that it is time to
re-examine the regulations and, if necessary, issue supplementary
regulations applicable only to the Arctic in order to make possible
other desirable characteristics.

Why, for example, when we must conserve energy in an Arctic
vessel are we required to have fifteen air changes per hour in a
crew member's toilet attached to his cabin? Assuming he uses that
particular cubicle for one to a maximum of four times in any given
24-hour period, and assuming that he is the only user, why is it
necessary to have 15 air changes an hour or, in the 24-hour period,
360 air changes, and so on?

Cocooning

The present design of icebreakers is traditionally that of a
hull with superstructure, but the character of the superstructure
has been changing radically in recent times from the usual
collection of graceful curves, open decks, catwalks, rails and
bulwarks, to something that is becoming very much more functional.
The ERMAK is a good case in point (Figure 2) leading to an even
better example, the ATLE (Figure 3 a and b). But even the ATLE
does not go far enough, and I submit for your examination Figure 4
which is a conceptualised presentation of what I am about to say.

Traditional superstructure in these vessels suffers from many
disadvantages. It encourages the collection of superstructure
icing in those areas where icing may be a problem. It presents all
kinds of equipment, fittings, etc. to the elements with the
possibility of their becoming inoperative due to freezing.

Areas such as this are extremely difficult to insulate in
effective terms, and they are costly to maintain. Because of this
insulation problem and other inherent characteristics, they are
costly to heat and to ventilate, and they are also difficult to

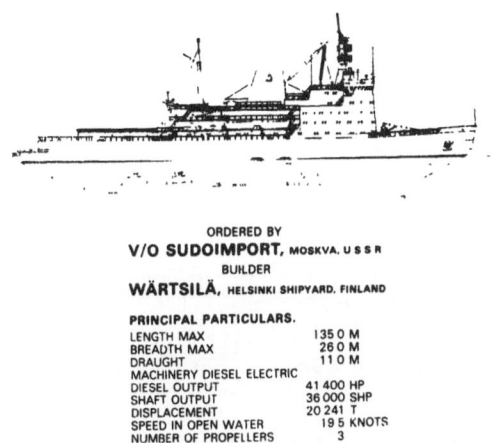

ORDERED BY
V/O SUDOIMPORT, MOSKVA. U S S R
BUILDER
WÄRTSILÄ, HELSINKI SHIPYARD. FINLAND

PRINCIPAL PARTICULARS.

LENGTH MAX	135 0 M
BREADTH MAX	26 0 M
DRAUGHT	11 0 M
MACHINERY DIESEL ELECTRIC	
DIESEL OUTPUT	41 400 HP
SHAFT OUTPUT	36 000 SHP
DISPLACEMENT	20 241 T
SPEED IN OPEN WATER	19 5 KNOTS
NUMBER OF PROPELLERS	3

Figure 2   Polar Icebreaker ERMAK

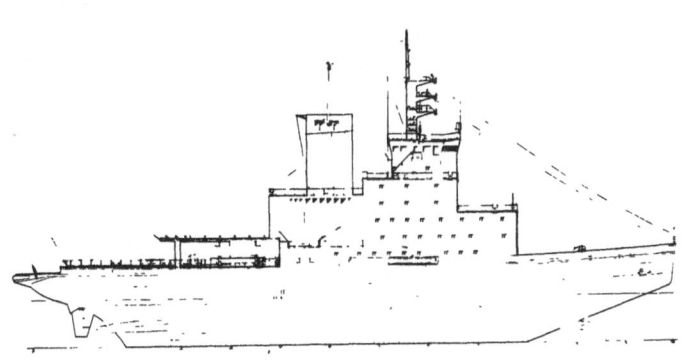

Figure 3a.   Polar Icebreaker ATLE

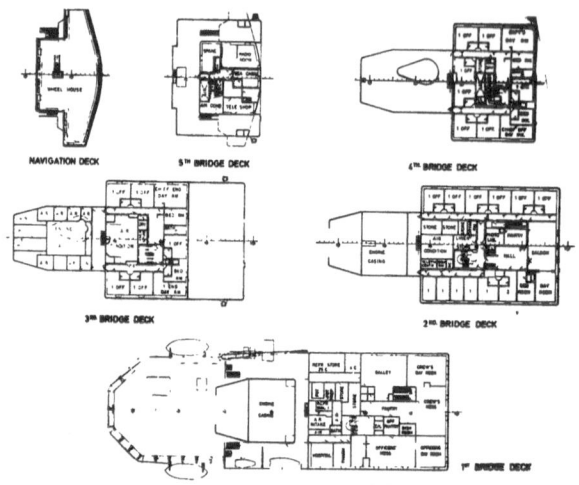

Figure 3b.   Polar Icebreaker ATLE

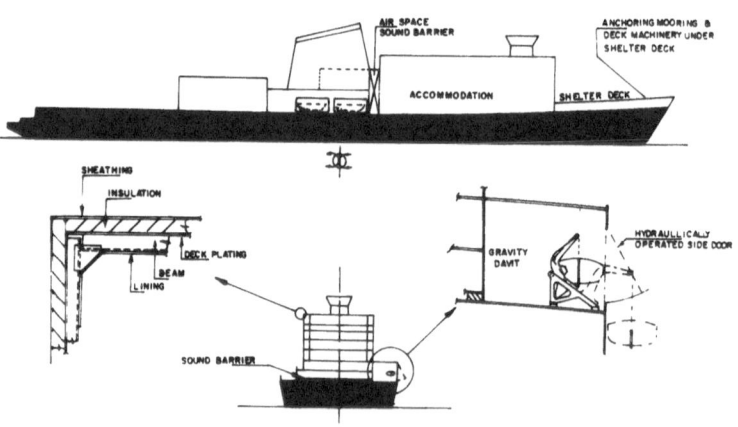

Figure 4   Proposed Design of vessel with deck
machinery under cover

isolate from hull and machinery noises.

However, if the vessel were built in what I choose to call a completely cocooned condition, where all deck machinery would be under cover such as the suggestion in Fig. 4 at the forward end, and the lifeboats in the midship area, there is no reason why properly designed equipment cannot be contained under cover at all times, and controlled either remotely or from a vantage point such as the wheelhouse. I suggest this matter of control because even if the equipment is designed to withstand the rigours of Arctic conditions on an unprotected deck, the deckhand who may be required to stand by that machinery for some time could be extremely uncomfortable under adverse conditions, unless, of course, he is adequately supplied with heated clothing.

If the superstructure is built in cocooned elements, it is reasonable to assume that all the machinery uptakes and related services above the strength deck can be isolated physically and acoustically. The accommodation and working section of the cocoon can be isolated from the ship's hull for acoustic and vibration reasons, thereby removing the crew from one of the most serious noise pollutions in the Arctic, the sound of breaking ice which, over long periods of time, can be extremely fatiguing.

By cocooning I mean that the superstructure, having been built in straight lines and planes, would then be covered externally in a layer of insulation and that insulation would then be protected by external sheathing. With no obstructions, no deviations, the insulation and vapour barrier should be completely lacking in holidays.

At this point I would like to suggest the conventional insulants are not really appropriate to this work. Any insulation made from wool, mineral or otherwise or from foam or material derived from hydrocarbon sources is not the ideal, either because they are structurally inadequate or they are subject to fire or the production of gases under high temperatures.

What must be developed is a new material that will neither burn nor produce such gases, that is of a foamed type preferably capable of being foamed in place rather than laid on slab by slab, and should have sufficient structural flexibility to discourage the possibility of breakdown due to vibration or low frequency concussion.

Inside the superstructure, under the cocoon, one would expect to see the general layout completely changed from the traditional; that is, corridors inside and cabins on the outboard side, to the reverse. All cabins and human occupancy rooms would be inside and the external spaces, that is, those adjacent to the outside structure, would be passageways, store rooms and the like. This latter group of spaces does not require temperatures above, shall we say, 60°, whereas the accommodation and operating areas would require temperatures up to 72°F. By having this two-stage temperature situation, much would be saved in heat loss.

It is not really necessary for cabins to have outside port-
holes or windows, particularly if the ship is well supplied with
public rooms, lecture halls, gymnasia and the like. The occupant
of each cabin will spend most of his leisure time not in the cabin
but in the public rooms to which I have referred elsewhere. Cabins,
therefore, do not require windows. The public rooms may or may
not be fitted with windows depending upon the circumstances.
Certainly the gymnasium doesn't require them. The theatre or
projection room does not, and neither do the lecture rooms, which
significantly cuts down the need for windows, therefore heat loss,
and other related inefficiencies.

Air Conditioning

Air conditioning, which is defined as the process of treating
air in order to control simultaneously its temperature, its humidity,
its cleanliness, its transmission of noise and provide distribution
at the same time, is the second most important system in connection
with the maintenance of crew comfort, and therefore warrants some
examination.

Air conditioning systems are required to maintain compartments
under conditions described as "comfortable". As all humans are
physically different, the term comfort is purely an individual
feeling. At the present time most air conditioning systems are
designed to satisfy the majority of persons within the space.

Air conditioning is expensive to operate, particularly in the
Arctic due to the requirements of the Regulations which are 15
cubic feet per minute of fresh air per man. The greater the
sophistication that the system offers, the more costly it is. And
finally, since ventilation is normally considered to be a
supplementary supporting system, it normally has to take second
place when the vessel is being built and where prime spaces are
being fought for by other competing systems.

In Arctic vessels, because ventilation is so important to
crew comfort, it is suggested that all of these situations be
reversed, that the ventilation and heating system be given first
priority and the utmost in sophistication to ensure that all
requirements are met.

To obtain the nearest to the individual control system,
several methods have been used, perhaps the most successful being:
(a) a reheat system where the air from a central system is
distributed to each space at a temperature which is cool enough to
meet the lower end of the comfort scale, and the reheater is
located at the terminal outlet with sufficient capacity to heat
the air to meet the higher point of the comfort scale;
(b) an individual fan coil system where possibly only the amount
of air required for dilution is supplied, and the room air is
circulated through a fan coil unit where it is either heated or
cooled according to the conditions to be met.

One important requirement of any air conditioning system is the need to design it in order to eliminate completely the transmission of sounds that are normally conducted through such systems.

Traditionally, odour-producing areas such as toilet spaces, galleys and the like have been fitted with air extraction systems as opposed to the other areas where the air is supplied. Since extraction is yet again expensive to operate in cold latitudes, consideration could well be given to introducing completely self-contained 100% recirculation systems, all suitably fitted with grease extractors, using scrubbers and activated charcoal filters; humidity reducers or increasers, activated by electrical sensing units and temperature differential coils; and by other filtration equipment designed to extract CO_2, CO, SO_2 and various dusts and particles using a combination of electrostatic filters and filters containing activated charcoal and other hydrocarbon filaments.

Contrary to popular opinion, a complete recirculation is neither difficult nor is it expensive, and it is certainly healthy. The attached graph (Figure 5) listing ventilation requirements indicates that the amount of air required to prevent CO_2 concentrations from rising above 0.6% is less than 4 cubic feet per minute, and the equivalent figure for necessary oxygen is about 1.5 cubic feet per minute, both as indicated on curves A and B.

Curves C and D indicate the amount of air required under normal conditions; that is, lacking any type of scrubbing or filtering network.

The difference in air volume required will be clearly apparent and will suggest the reason why closed circuit systems will save considerably in fuel costs. The real disadvantage of a closed system is, of course, the unreality that is achieved because the air is so pure it carries no odours at all, and this is a condition to which the human frame is not accustomed to meeting. However, as suggested elsewhere in this paper, normality can be re-established by introducing into the air chemicals formulated to produce odours appropriate to the occasion, time of day, circumstances and, to a degree, the imposition of the Master's will on the mood of the crew.

The introduction of such additives will not be new to most seafaring personnel in this room since any lengthy period at sea usually finds the crew drinking a degree of distilled water, considered by most to be highly unpalatable simply because it has no flavour.

Following on my remarks about the air conditioning system also being designed to eliminate noise transmission, the attached tables of indoor design goals (Tables 1 and 2) for air conditioning system sound control, together with Russia's latest figures, should be of interest due to the remarkable similarity in requirements between the two countries.

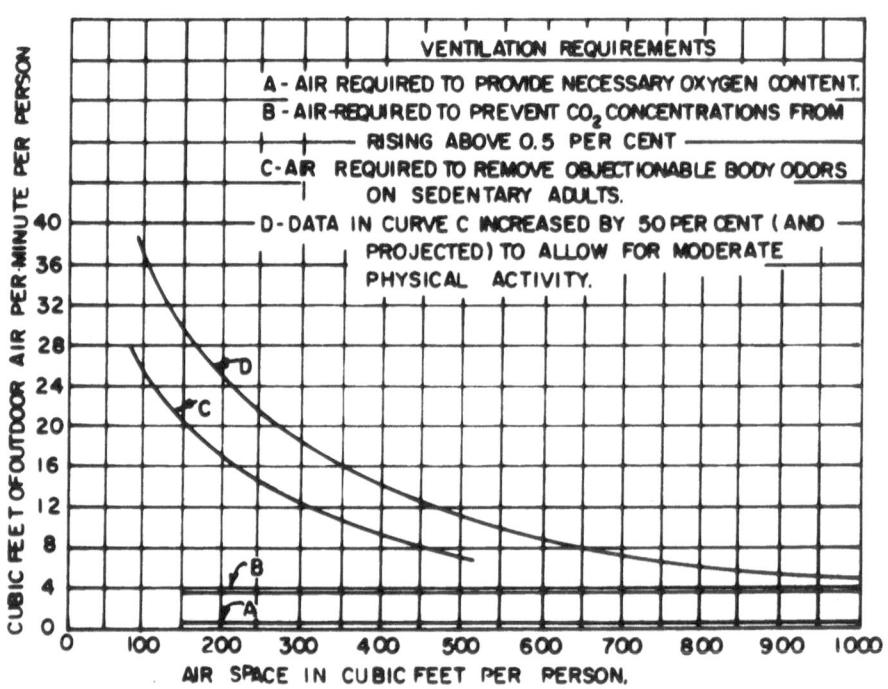

Figure 5   Ventilation Requirements

TABLE 1

New Russian Shipboard Noise Limits

Maximum permissible noise levels in Soviet ships set out in
1962 are shortly to be replaced by a new set of provisional norms
drawn up on the basis of international recommendations and
agreement with Iron Curtain countries. The accompanying table
includes a column which shows a proposed further stage in the
lowering of noise levels on board ship.

Area onboard	Prov regulations	New norm proposal
	dBA	dBA
Engine-rooms:		
with permanent watch	85	80
with periodic service	95	90
unmanned	105	95
Sound-insulated control rooms	70	65
Service areas	70	65
Galley	65	55
Wheelhouse & navigation spaces:		
sea-going ships	55	55
river ships	60	60
Public rooms:		
sea-going ships	55	55
river ships	60	60
Radio cabins:		
river ships	55	50
other river ships	60	-
Accommodation:		
Ships which are at sea for a period longer than 24 hours and fishing vessels etc which sail more than 20 nautical miles from land	55	50
other ships	50	45
Sick-bays	45	40

Svensk Sjofarts Tidning 13/1975

Shipbuilding & Marine Engineering International,
June, 1975

TABLE 2

Ranges of Indoor Design Goals for
Air-Conditioning System Sound Control

Type of Area	Range of A-Sound Levels Decibels	Range of NC Criteria Curves
RESIDENCES		
Private homes (rural and suburban)	25-35	20-30
Private homes (urban)	30-40	25-35
Apartment houses, 2- and 3-family units	35-45	30-40
HOTELS		
Individual rooms or suites	35-45	30-40
Ball rooms, Banquet rooms	35-45	30-40
Halls and corridors, Lobbies	40-50	35-45
Garages	45-55	40-50
Kitchens and laundries	45-55	40-50
HOSPITALS AND CLINICS		
Private rooms	30-40	25-35
Operating rooms, Wards	35-45	30-40
Laboratories, Halls and corridors, Lobbies and waiting rooms	40-50	35-45
Washrooms and toilets	45-55	40-50
OFFICES		
Board room	25-35	20-30
Conference rooms	30-40	25-35
Executive office	35-45	30-40
Supervisor office, Reception room	35-50	30-45
General open offices, Drafting rooms	40-55	35-50
Halls and corridors	40-55	35-55
Tabulation and computation	45-65	40-60
AUDITORIUMS AND MUSIC HALLS		
Concert and opera halls, Studios for sound reproduction	25-35	20-25
Legitimate theaters, Multi-purpose halls	30-40	25-30
Movie theaters, TV audience studios	35-45	30-35
Semi-outdoor amphitheaters, Lecture halls, planetarium		
Lobbies	40-50	35-45

TABLE 2 (Continued)

Type of Area	Range of A-Sound Levels Decibels	Range of NC Criteria Curves
CHURCHES AND SCHOOLS		
Sanctuaries	25-35	20-30
Libraries	35-45	30-40
Schools and classrooms	35-45	30-40
Laboratories	40-50	35-45
Recreation halls	40-55	35-50
Corridors and halls	40-55	35-50
Kitchens	45-55	40-50
PUBLIC BUILDINGS		
Public libraries, Museums, Court rooms	35-45	30-40
Post offices, General banking areas, Lobbies	40-50	35-45
Washrooms and toilets	45-55	40-50
RESTAURANTS, CAFETERIAS, LOUNGES		
Restaurants	40-50	35-45
Cocktail lounges	40-55	35-50
Night clubs	40-50	35-45
Cafeterias	45-55	40-50
STORES RETAIL		
Clothing stores, Department stores (upper floors)	40-50	35-45
Department stores (main floor) Small retail stores	45-55	40-50
Supermarkets	45-55	40-50
SPORTS ACTIVITIES INDOOR		
Coliseums	35-45	30-40
Bowling alleys, gymnasiums	40-50	35-45
Swimming pools	45-60	40-55
TRANSPORTATION (RAIL, BUS, PLANE)		
Ticket sales offices	35-45	30-40
Lounges and Waiting rooms	40-55	35-50
EQUIPMENT ROOMS		
8 hr/day exposure		
3 hr/day exposure	90	
(or per OSHA requirement)	97	

Note: NC curves are shown in Fig. 4. For a discussion of the relations between NC curves and A-sound level, see Chapter 6, 1972 Handbook of Fundamentals.

Lifeboats

Any life-saving gear that a ship carries, and obviously there must be a satisfactory quantity, must be very carefully considered, since conventional life-saving equipment would be hopelessly inadequate.

Whatever is put over the ship's side for the purpose of saving human life must be capable of three major functions. It must be unsinkable and so designed that in the event that it is pinched between two opposing forces, ice in this case, that it will rise rather than sink.

Secondly, it must be painted a brilliant international orange for easy visual discovery. It must be coated with a suitable material to make it an excellent radar reflector, and it should carry an automatic radio homing device.

Thirdly, and most important, self-propulsion of any type is not necessary, but what must be provided is complete protection for the occupants, including a source of heat generation in order that the passengers will not die of the cold.

After these requirements are met, then other needs can be incorporated.

Cargo and Hull Access Devices

All above hull external access devices must be specially considered.

Hatch access, shipping ports, lifeboat protection doors, etc. must be specially considered in order to:
(a) reduce to a minimum damage of equipment from ice build-up;
(b) ensure "watertight" sealing at extreme low temperatures;
(c) reduce to a minimum man-handling of equipment including wires, cleats, blocks, etc.
(d) provide maximum protection from elements to deck machinery, life boats, etc.

Icing

Equipment must be designed to withstand a coating of 1/2 inches of ice. All moving parts should be shrouded to prevent ice build-up on wheels, rollers, cleats and the mechanical driving mechanisms.

All tracks or rails should be heated in order to prevent ice build-up and ensure easy operation. Heating can be provided by means of electrical tracers or heated glycol circulated through hollow structures or flexible tubes. If flexible tubes are used they should be stainless steel or the equivalent, not fabric or rubber.

Sealing Arrangements

Most materials used for watertight seals are suitable only for temperate water conditions. Special materials must be selected for low temperature operation.

Care must be taken in heating sealing bars. Temperatures of over 95°F are not to be presented to rubber or neoprene materials as this would tend to create a permanent set in the material and reduce the compression necessary to provide adequate watertightness. When linear seals are fitted, a built-in hydraulic method of forceably raising the cover off the seal in the event that it is frozen thereto should be fitted. A half inch rise should be sufficient.

Automatic Operation

In cold temperatures, automatic operation or remote operation of opening, closing, cleating and dogging should be considered. Care must be taken that the automatic operation is simple and presents a minimum of maintenance to be carried out to external parts.

Wire ropes and cranes can be used but since their handling during low temperatures invites awkward and dangerous situations, they are to be avoided wherever possible.

All deck machinery, lifeboats and any other gear normally exposed to the elements should be protected.

Traditional fabric covers (canvas) normally acceptable in temperate water conditions should be avoided. The most effective form of protection is to place all deck gear, lifeboats, etc. either below decks or behind bulkheads, all protected by automatic doors and access.

Automatic and remote operation must, of course, be backed up by local hand operation in the event of failure.

Wheelhouse Arrangements

More efficient Bridges are necessary and much work is being done on this subject in several countries.

The wheelhouse design in icebreakers requires a good deal more than an efficient standard design for a normal merchantman. For example, Mr. Per Larsen of the Ship Research Institute of Norway has suggested, amongst other things, that "there should be an unobstructed view from the manoeuvring position from dead ahead to 115° to each side". This recommendation is simply inadequate for icebreakers. Ships of this type require all-around view. The nature of their activity is almost akin to that of an offshore supply vessel, and as a result all-around view is as necessary to the icebreaker, regardless of size, as it is to the offshore supply vessel.

Figure No. 6 is a page extracted from a study made on this subject.

In addition to the all-around requirement, the Master or Officer of the Watch of an icebreaker must be able to look directly over the ship's side at the ice thus making it mandatory that he be able to move from wing to wing as well as operate from the centre line position at his discretion.

A traditional requirement, no longer as forcefully demanded as heretofore, was that at least the Bridge wings should be open, it being claimed by the older ice mariners that under certain navigating conditions they could "smell" the ice or they could feel the cold of a close but possibly invisible ice mass on their faces, all presupposing that visibility is obstructed by blowing snow, white-outs, fog or possibly just plain darkness.

This latter requirement I rather think is no longer applicable due to younger Masters and modern electronics. But it serves to illustrate the particular special conditions reserved to wheelhouse design in icebreakers.

In conflict with the foregoing is the other belief that since an icebreaker, when operating in heavy ice, performs very largely in accordance with the skill of the ship handler, it is important that the person handling that ship be as unified with the ship as is possible. In other words, he should be handling the controls himself (see Figure 7 and Table 3).

The pilot of a 747 is a professional charged with the task of handling a very large piece of structure with enormous power built

Table 3

Key to Equipment in Figure 7

1. 10-in. Diameter Radar Display
2. FFU & NFU Helm Control
3. Power Control Quadrants, Shaft RPM and Power Readouts
4. Communications and Alarms
5. TV Monitor
6. TV Pan and Zoom Control
7. Navigation Instruments (Fathometer Readout, Radar Ranging Unit, Gyro Repeat)
8. Crew Seat on Flush-Mounted Rails
9. Fold-Down Chart Table
10. 10-cm. Radar Transceiver
11. 3-cm. Radar Transceiver

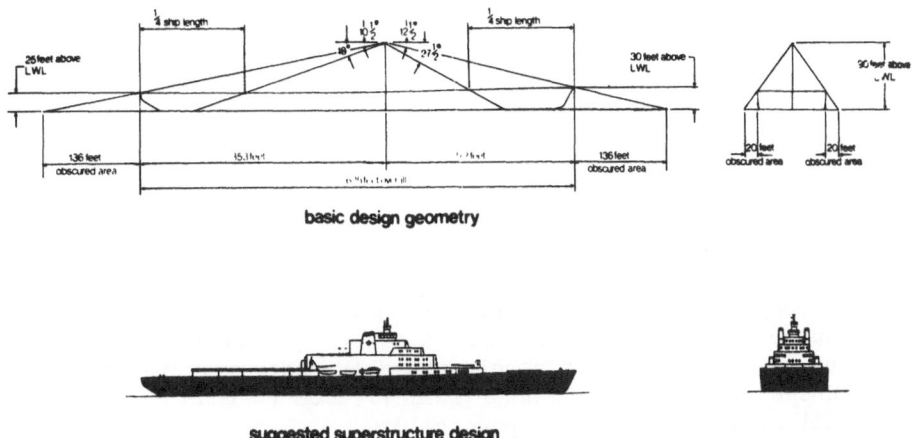

basic design geometry

suggested superstructure design

Figure 6 Overall design on small bridge and stepped superstructure

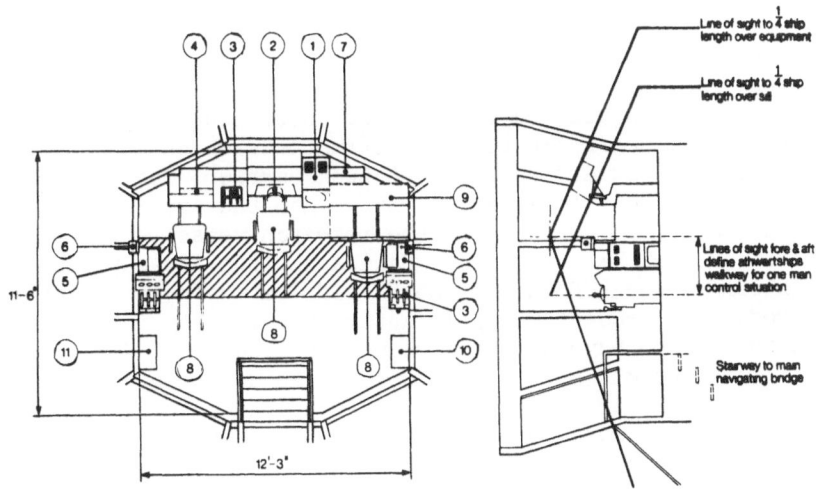

Figure 7 Proposed layout of four-place conning position

into it and with heavy responsibilities, all depending upon the man's skill. I am not aware of any pilot who will turn the landing of his aircraft or the take-off of his aircraft over to the co-pilot, excepting under unusual circumstances. In other words, he chooses to have the craft under his direct fingertip control.

When an icebreaker is being operated under particularly difficult or heavy ice conditions, it would make good sense to assume that a control position of a 747 type would be more effective in melding the Master/ship relationship than the conventional way of the Master strutting about the deck calling out commands to a quartermaster, who then imparts the signal into the ship's helm, or to a supporting officer who rings up instructions on the engine room telegraph. Whether the Master himself or whether a new class of officer, shall we say an ice navigator/ship handler is developed, is not germane to this paper. But the suggestion is seriously made that such a new class of officer be developed who can handle the ship under very personal circumstances, allowing the Master full freedom to roam about and inspect all areas under his responsibility.

Immediate, sensitive, instinctive handling of the ship is important. I can recommend some very excellent work done in this respect by the Defence and Civil Institute of Environmental Medicine, a branch of the Department of National Defence in Canada.

Conclusion

I mentioned in the early part of this paper that my references would be directed not only at ships but at offshore marine structures, and whilst it is correct that the general direction of my remarks has been pointed towards the Government or professional icebreaker, it also refers to other structures which include not only the ice-capable dry bulk or wet bulk carrier, but also other craft that are likely to find their way into the Arctic.

First, and most obvious, is the research vessel which at the moment is somewhat hard to find. Canada has two ice-capable research craft, the HUDSON and the BAFFIN, but even they must be very careful and prudent in their activities if they intend to stretch the season beyond nominal summer period.

Military vessels are a class apart. No doubt there will be requirements for displacement-type surface craft, but it is the author's view that outside of submersibles such surface craft would be restricted to the roles of support and auxillary vessels. The prime fighting units are most likely to be a marriage of the light frigate using highly sophisticated rocketry and a large hovercraft-type hull (see Figure 8).

The major commercial activity in the Arctic at this time is, of course, oil and gas exploration, and support fleets are going to be necessary in this area. Pipeline layers for inter-island hopping have been designed and the attached general arrangement

Figure 8   German and Milne's conception of Hovercraft-
              destroyer for service in the Arctic

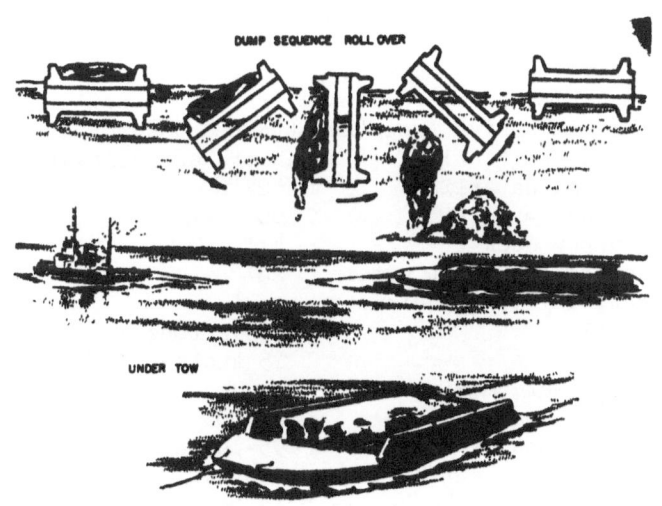

Figure 9   Dumping sequence for man-made ocean bottom
              plateaus

indicates the latest thinking in this respect. The principal
particulars of this craft are:

Length overall	167 metres
Breadth at the upper deck	26.2 metres
Depth at the upper deck	12.5 metres
Deadweight total	8,600 tons
Speed service	31 kilometres per hour
Propulsion power installed (maximum continuous rating)	11,186 Kw
With a complement of	100 persons

A vessel of considerable power, and even greater cost.

As exploration is forced to go into deeper and deeper water,
so will the drilling devices require increased sophistication.
At the moment, industry is using artificial islands, but their
use is limited to shallow water. The next step out is likely to
be the large monopod or conical-shaped drilling platform, and
this device can be expected to survive depths up to, say, 150 feet
of water. Beyond that point they may still continue to be used by
being strategically located on plateaus of fill placed on the ocean
bottom at that particular location.

Since the man-made bottom plateaus are costly to produce and
must be done both accurately and quickly, a special type of barge
for carrying and dumping the fill has been devised which provides
the advantages of exact positioning (with the aid of accurate
electronic positioning devices), rapid turnaround and minimum
damage to structure from heavy fills such as large rocks (Figure 9).

Offshore drilling beyond this point will be conducted by
either drilling ships or by residence drilling. Drilling ships
will not only be expensive (see the attached drawing of one design
recently produced, particulars of which are as follows:

Length, design waterline	522'6"
Breadth at upper deck	86'
Depth at upper deck	41'
Draft, design mean salt water	28'
Deadweight, drilling equipment and materials	3,912 tons
Service speed	15 knots
Propulsion power, maximum continuous rating	24,000
With a complement of	95 persons),

but Canada's pollution control regulations will increase the cost
since the requirement is that every hole being drilled must have
a back-up drilling vessel, fully capable in all respects, and each
of the two drilling vessels must have two fully capable offshore
supply vessels supporting them.

The cost of such a fleet reaches incredible proportions, and
at this point one wonders whether the cost of the drilling is not
likely to be prohibitive.

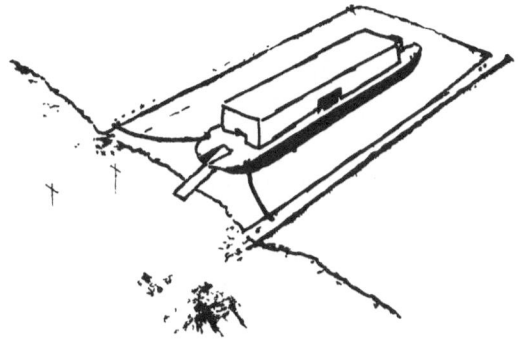

Figure 10   Winter Layup for Arctic

Figure 11   Sheltered Anchorage

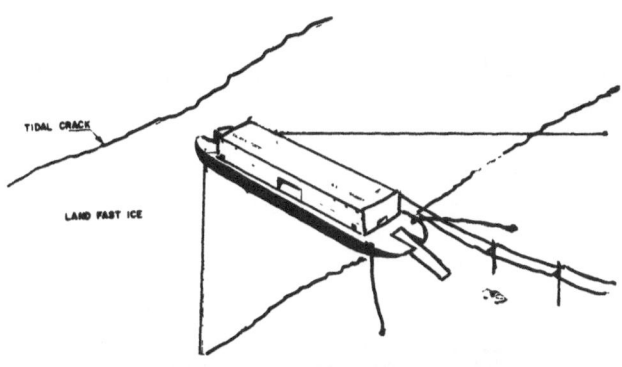

Figure 12   Cargo Discharge on Open Beach

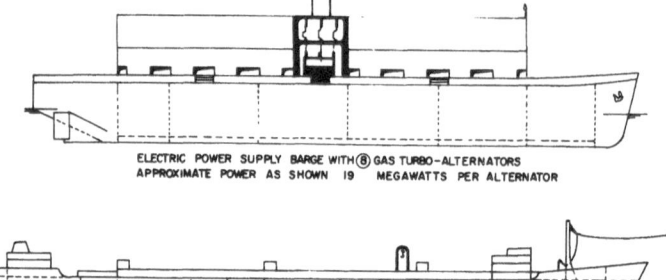

ELECTRIC POWER SUPPLY BARGE WITH ⑧ GAS TURBO-ALTERNATORS
APPROXIMATE POWER AS SHOWN 19 MEGAWATTS PER ALTERNATOR

TANKER CONVERSION TO ELECTRIC POWER SUPPLY BARGE WITH ⑥ DIESEL ALTERNATORS
APPROXIMATE POWER AS SHOWN 11 MEGAWATTS PER ALTERNATOR

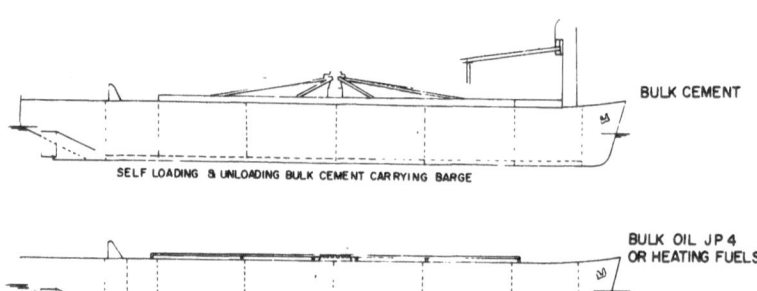

BULK CEMENT

SELF LOADING & UNLOADING BULK CEMENT CARRYING BARGE

BULK OIL JP 4
OR HEATING FUELS

BULK OIL CARRYING BARGE WITH SELF CONTAINED PUMP ROOM

Figure 14   Bulk cargo

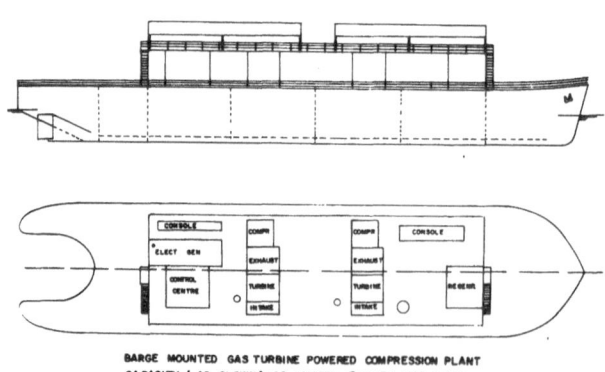

BARGE MOUNTED GAS TURBINE POWERED COMPRESSION PLANT
CAPACITY (AS SHOWN) 145 MMSCFD @ 80 TO 1600 P.S.I.G
INSTALLED POWER 40,000 B.H.P

Figure 15   Natural Gas Station

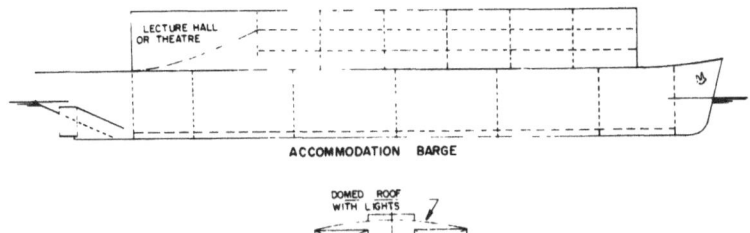

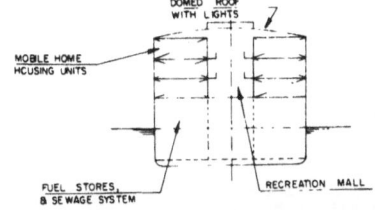

CONSTRUCTION — BARGE HULL BY SHIPBUILDER OR ANY STEEL FABRICATOR
HOUSING UNITS BY HOUSE CONSTRUCTOR, BUS BODY OR TRAILER
MANUFACTURER, OR AIRCRAFT CONSTRUCTOR

Figure 16 Accomodation

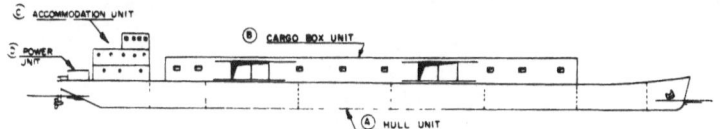

HARBORMASTER PROPELLED BARGE FOR EMERGENCY CONSTRUCTION FOR DISTRESSED AREAS

4 - UNIT CONCEPT SHOWN

CONSTRUCTION BREAKDOWN AS FOLLOWS,
Ⓐ HULL UNIT BY SHIPBUILDER OR ANY STEEL FABRICATOR
Ⓑ CARGO BOX UNIT BUILT SEPARATELY BY SHIPBUILDER OR ANY STEEL
FABRICATOR (BOLTED OR WELDED TO UNIT - A)
Ⓒ ACCOMMODATION UNIT BUILT SEPARATELY BY SHIPBUILDER OR
ANY STEEL FABRICATOR (BOLTED OR WELDED TO UNIT —A)
Ⓓ POWER UNIT, HARBOURMASTER, SCHOTTEL OR SIMILAR,
SELF CONTAINED AND BOLTED TO UNIT - A

Figure 17 Emmergency service

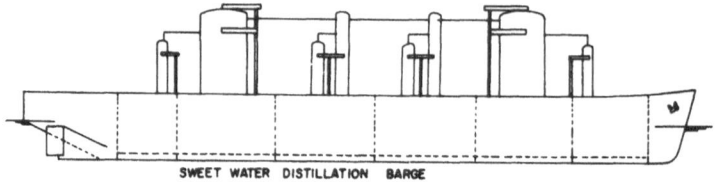

SWEET WATER DISTILLATION BARGE

STORAGE BARGE WITH REPAIR SHOPS, DEPOT, HOSPITAL, ETC

Figure 18   Fresh water supply

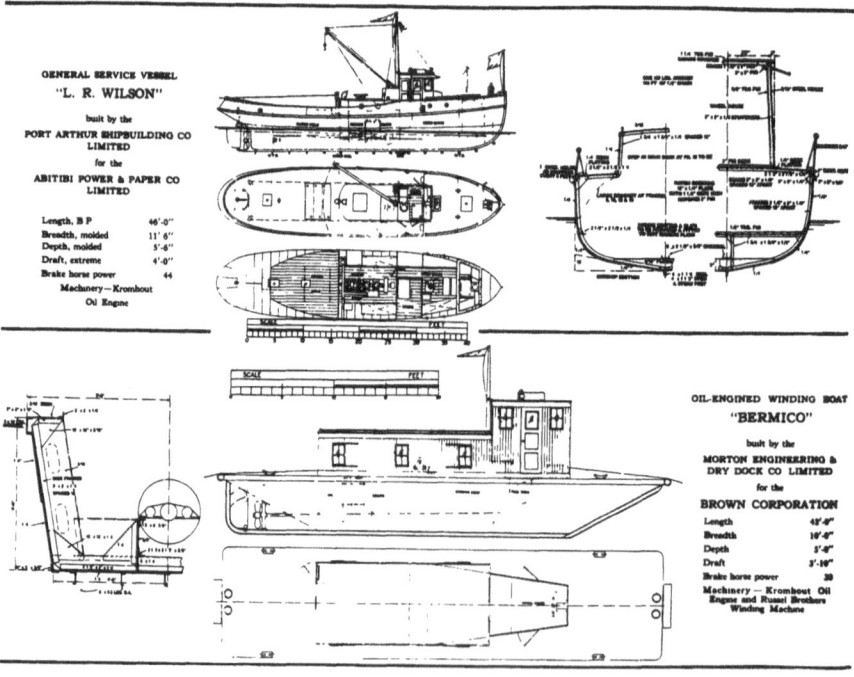

Figure 19   General service vessel - L.R. WILSON

The real problem to be faced in the Arctic in exploration drilling is in the deeper waters, particularly those waters where there is an arrogant bush by Arctic ice. Here no man-made object can withstand the opposing forces, and the economics will not permit the drilling of holes when a major portion of the time is spent avoiding the ice. Therefore, it becomes necessary to consider underwater or residence activities.

Major concrete structures recently being produced for the North Sea provide great encouragement for the design of such underwater structures for oil and gas production, though they possibly may be considered as a trifle expensive for oil and gas exploration. In this latter case, it is probable that a craft operating on the FLIP principle, but modified to resemble a submarine of about 250 feet in length, might be the answer. This craft would proceed to the drilling site horizontally, either on the surface or under the surface. At the drilling site it would nose down until it was standing vertically. The equivalent of the drilling mast would be a large centre tube within the submarine and through the nose doors drilling would take place.

There are already in operation some very sophisticated and competent underwater control completion and treatment systems for well-head operations which suggest that this very much larger version of an offshore activity is not unreasonable.

Any construction that must be done in the Arctic will be expensive and tedious. Therefore it behooves those organisations with construction problems ahead of them to examine the distance of their construction site or operating centre from salt water, and if they are close enough to salt water, then strong consideration should be given to mounting all their equipment requirement or needs on barges.

This concept has become well recognised in recent months and to date, to the author's knowledge, the following barge-mounted equipment has been designed:

A complete village.

A complete natural gas liquifaction plant.

A complete minehead treatment plant for separating and primary beneficiation of lead-zinc-silver ore.

And there are many more concepts presently being considered (Figures 10 to 19).

Requirements can be built in their entirety in the civilised and more competitive part of the country. They can be completely tested before delivery to ensure that they are in working condition. On delivery, it is simply a matter of lodging them into a pre-engineered stowage location and with an absolute minimum of delay, they are onstream.

Process handling or treatment plants are largely matters of engineering design, but barge-mounted villages are something yet again and much of what I have said in the foregoing regarding human engineering can be given the greatest prominence in this particular area.

SYSTEM ANALYSIS OF LOGISTICS PLANNING FOR NATURAL GAS PIPELINE CONSTRUCTION IN THE ARCTIC REGION[1]

E. Buchholz, T.E. Kingsbury and G.E. Bushell

Canadian National Railways

Montreal, Quebec, Canada

INTRODUCTION

This paper describes an application of the technique of Linear Programming to a large transportation planning problem. The work is part of a project concerned with planning the movement of materials and supplies required to build a large diameter pipeline from the North Slope of Alaska to northern Alberta.

The development of the LP model was carried out by the authors as members of the Operational Research Branch of Canadian National Railways. In what follows, emphasis will be placed on the identification of all elements of the problem and on the application of the linear programming model which was developed.

The first section gives the background to the study and a description of the problem. This is followed by brief sections on the formulation and development of the model and the processing steps involved in solving it. The final section illustrates the input requirements and the usage of the model.

---

[1]This paper was also presented at the Canadian Operational Research Society Annual Conference, Toronto, 1972 under the title of "Transportation Planning Model for the Construction of a Northern Gas Pipeline".

PROBLEM IDENTIFICATION

Since the discovery of major oil and gas reserves at Prudhoe
Bay on the North Slope of Alaska, several groups have announced
pipeline projects for the transportation of natural gas to
southern markets. Canadian National is involved with such a
consortium, known as the Gas Arctic Systems Study Group. The
other members are Alberta Gas Trunk Line Company Limited, The
Columbia Gas System Incorporated, Northern Natural Gas Company,
Texas Eastern Transmission Corporation, and Pacific Lighting Gas
Development Company.

The Gas Arctic Project Group is concerned with planning the
construction of approximately fifteen hundred and fifty miles of
forty-eight inch diameter, high pressure gas pipeline from
Prudhoe Bay to northern Alberta. The proposed route (Figure 1)
runs south-east from Prudhoe Bay through the Brooks Range of
Alaska, swings east through the Yukon north of Old Crow to the
Mackenzie River valley, and follows the Mackenzie upstream to
near Fort Simpson where it heads south to connect with an existing
gas system in northern Alberta. The estimated capital cost for
the initial system is two and one half to three billion dollars.

For a construction project of such scope, the transportation
of materials and equipment necessary to construct and maintain
the line requires advanced and specialized planning. The immense
volume of supplies to be transported, the lack of established
transportation routes, and the harsh environmental conditions of
the Arctic combine to complicate the logistics problem.

Canadian National became involved in the advanced
transportation planning for the project in mid 1970 when a small
group was set up in Edmonton. Once the complexity of the problem
became apparent, this group recognized the need for more
sophisticated planning tools in addition to the manual methods
they were using. At this point, early in 1971, the Operational
Research Branch was approached to assist in the study.

The construction of the pipeline will extend over
approximately two and one half years with most of the work being
carried out during the winter months. The exact timing of the
construction has not yet been determined and it depends partly on
the building of the trans-Alaska oil pipeline if it is approved.
For the proposed route, equipment and material requirements have
been estimated at each of several stockpile locations close to
the pipeline right-of-way (Figure 1). The major item is steel
pipe but considerable volumes of fuel, cement, contractor's
equipment, and coating and wrapping materials must be moved as
well.

Due to the remoteness of many stockpiles from major existing
transportation routes, it is apparent that the majority of
supplies will move multi-modally from supply locations to
requirement stockpiles. The team in Edmonton has spent a great
deal of time examining the separate modes and routes available,

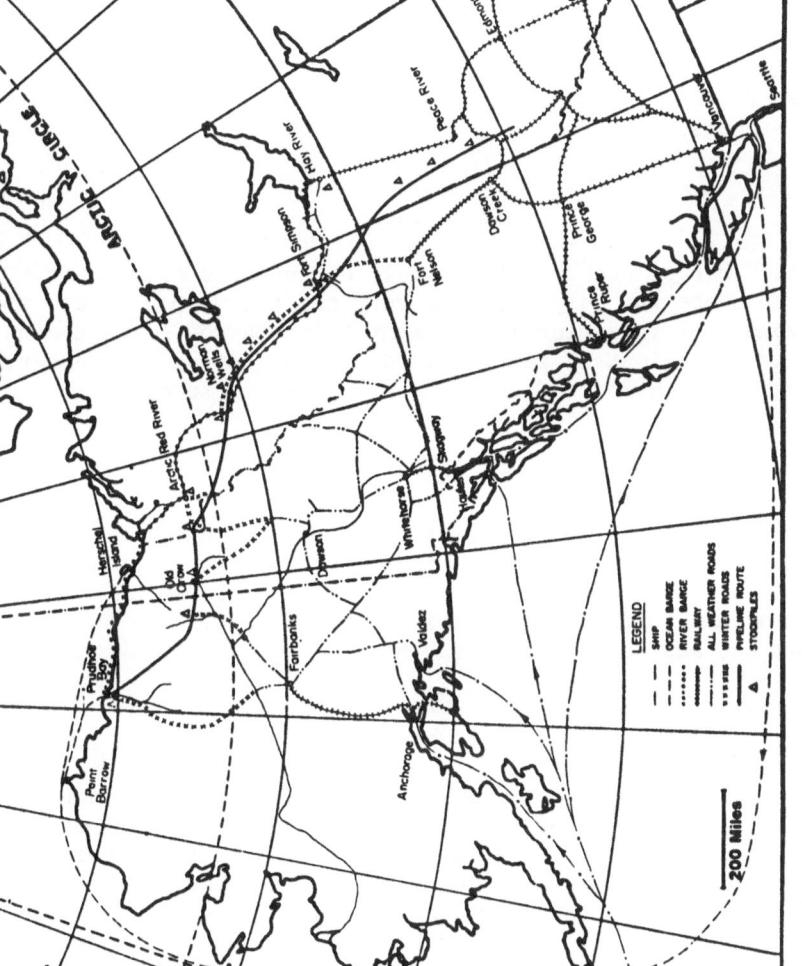

Figure 1

and acquiring an understanding of the capabilities and restrictions
of each mode. The following summarizes the information they have
obtained and illustrates the principal transportation modes and
routes being considered (see Figure 1).

Ocean Ship

The movement of supplies to west coast ports from off-shore
supply origins is virtually a year round operation. Supplies from
Japan can be shipped to the ports of Seattle, Vancouver, Haines,
Skagway, Valdez, and Anchorage where they are in turn transshipped
to other modes.

Ocean Barge

Ocean barges can be used to deliver supplies from Seattle to
the North Slope but only during a very short time period due to
the movement of the polar ice pack at Point Barrow. The season
length is approximately forty days during August and September
restricting the number of trips per barge train to one.

Rail

Several railroads have existing track and equipment which
would allow them to participate in the movement of supplies, and
all are year-round operations. These are the Alaska RR, the
White Pass and Yukon RR, the Pacific Great Eastern RR, Canadian
National RR, and Canadian Pacific RR.

Mackenzie River Barge

The major barge fleet, run by Northern Transporation Limited,
operates out of Hay River to points downstream on the river and as
far as the North Slope. The shipping periods to the Arctic
Coastline, Lower Mackenzie, and Upper Mackenzie are approximately
seventy, one hundred, and one hundred and twenty-five days in
duration respectively. Barge trains are made up to six barges
plus a tug with each barge designed to carry a maximum load of
fifteen hundred tons.

All-Weather Trucking

Limited all-weather roads in Alaska, the Yukon, and the
Northwest Territories extend north from the terminals of rail and

water modes. Still others, such as the Dempster Highway which
will eventually extend from near Dawson to Inuvik, are being
developed and may be completed in time for the movement of some
supplies.

Winter-Only Trucking

Winter roads or trails are available for limited use during
short periods and under severe conditions. On the average, the
availability period is three or four months in the December to
March interval.

For more information on all the transportation alternatives,
I refer you to a paper by Mr. J.R. O'Rourke of Canadian National
Railways ("Logistics Planning For a Northern Pipeline", The
Journal of Canadian Petroleum Technology, July-September, 1971).

Given this background, the problem of interest is to
determine the least-cost combination of routes and modes to be
used in each time period subject to supply, demand, and transportation restrictions. Initially, the main concern was with the
supply of steel pipe since it represents the largest proportion of
the total materials to be moved. From the outset, however, the
importance of considering other supplies was recognized and the
model was developed so that it would be relatively easy to alter
the number of commodities and time periods being represented.

FORMULATION

The problem can be visualized as a multi-modal, multi-
commodity, and multi-time period constrained directed graph. For
each commodity in the problem, the transportation network is
described by nodes and arcs for each distinct time period, with
additional arcs joining one time period to the next. Constraints
may apply to a single commodity in one or more time periods or to
a group of commodities in the same time period. An example of
the latter type of restriction would be an equipment availability
constraint limiting the total volume of all commodities carried
over a route.

In association with each commodity, nodes in the graph
represent locations in various time periods and arcs represent
activities between places in space and time. Such activities are
transportation by a certain route and mode, direct transshipment
between two modes, transshipment to or from a stockpile, or
stockpiling inventory from one time period to another. For each
commodity, origin or supply nodes are identified as are destination
or requirement nodes.

The objective is to determine what quantities of each
commodity should be moved over what routes to minimize the total

cost of materials, transportation, transshipment, and stockpiling without violating any of the constraint conditions. These apply to supply and requirement volumes, stockpile capacities, and transportation equipment availability. It was felt that all constraints and the objective functions could be adequately represented by linear equations and the technique of linear programming used to solve.the problem. The multiple arc constraints, such as those required to model the Mackenzie River barge equipment, were recognized to be the most difficult to model but they could be handled in a LP formulation.

In the linear programming model, the directed graph is represented mathematically as a system of simultaneous linear equations. Each variable in the system represents the flow over a specific arc and the relationships among the variables are described as linear equations. The objective function includes all variables and their associated per unit costs. The constraint equations are defined in the following manner.

Supply Nodes - (Figure 2). For certain commodities and supply locations, an upper bound on the supply reflects a production capacity for one time period.

Requirement Nodes - (Figure 2). At destination stockpiles during construction time periods, the flow on arcs into the node must equal the demand plus any flow on arcs out of the node.

Intermediate Nodes - (Figure 2). At all transshipment nodes, conservation-of-flow equations require the flow on arcs into the node to equal the flow on arcs out of the node.

Stockpile Capacities - (Figure 3). At all stockpiles, the summation of all commodities being carried from one time period to the next must be less than or equal to the storage capacity of the facility.

Inventory Requirements - (Figure 3). In order to meet the construction schedule, the plans require a certain percentage of

1.   SUPPLY NODES

$\sum$ flows out' $\leq$ production capacity

2.   REQUIREMENT NODES

$\sum$ flows in $-\sum$ flows out $=$ demand

3.   INTERMEDIATE NODES

$\sum$ flows in $=\sum$ flows out

Figure 2. Node constraints for all comodities and all time periods

1. STOCKPILE CAPACITY

   $\sum_i$ tonnage stockpiled between two time periods $\leqq$ stockpile capacity

   all commodities

2. INVENTORY REQUIREMENT

   tonnage stockpiled through to a construction time period $\geqq$ percentage of total

   requirement

3. EQUIPMENT CAPACITY (MACKENZIE BARGES)

   a. $\sum$ (8 week flows) $\leqq$ 8 week capacity

   b. $\sum$ (8 week flows) + $\sum$ (10 week flows) $\leqq$ 10 week capacity

   c. $\sum$ (8 week flows) + $\sum$ (10 week flows) + $\sum$ (14 week flows) $\leqq$ 14 week capacity

ALL FLOWS NORMALIZED BY COEFFICIENTS FOR CYCLE AND VOLUME DIFFERENCES.

Figure 3. Arc constraints

the material requirements to be at the stockpiles at the beginning of the construction intervals. This constraint is modelled by imposing lower bounds on the arcs representing stockpiling from the previous time period to the construction time period. The availability of the required tonnage is thereby ensured.

Transportation Equipment Capacity. These constraints express the capacity of pools of transportation equipment. They are more difficult to model since the equipment can be used over multiple routes with different cycle times and often different season lengths. As well, it is necessary to consider the different volume characteristics of the commodities competing for the equipment.

To illustrate such restrictions, consider the constraints on the pool of Mackenzie River barges (Figure 3). In each summer time period, the different shipping seasons (approximately eight, ten and fourteen weeks) to points along the river are represented using three equations. In each equation, the variables representing barge movements have associated coefficients to reflect the differences in cycle times to each destination and the different volume characteristics of the commodities. Thus the equations reflect the amount of barge capacity actually used up by each move.

One equation sums the normalized flows to all locations in the eight week shipping season and limits this total to the tonnage that could be moved if the entire fleet of barges were used in this service. The second equation includes flows to both the eight and ten week periods and requires the sum to be less than the increased capacity. The final equation includes all variable flows out of Hay River, and relates the total to the tonnage that could be moved in a fourteen week season.

COMPUTER PROCESSING

Having formulated the problem as a linear program, a standard LP package was chosen to solve the problem. IBM's Mathematical Programming System (MPS) was selected due to prior familiarity with the system and compatibility with in-house computers. Since the initiation of the project, IBM has issued an improved Mathematical Programming System Extended (MPSX) and this package is currently being used.

MPSX requires a standard input format to represent the problem. The LP is defined in matrix form by naming the rows and columns, and specifying all non-zero coefficients, objective function costs, right hand side values, and bounds on any variable. Several processing steps are involved in setting up a problem in this form, solving for an optimal solution, and producing useful output. These are illustrated in Figure 4.

Without going into detail, it is sufficient to say that one master file is maintained which contains the basic model structure and all cost and capacity information. For any particular

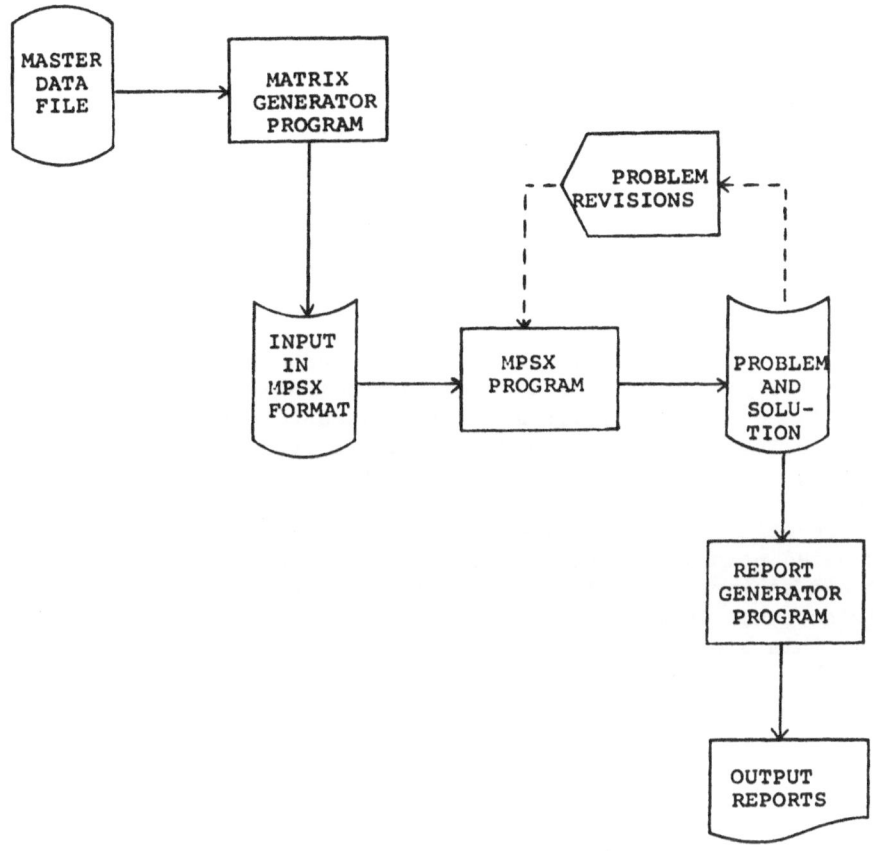

Figure 4.   Computer Processing Steps

formulation, this file is processed by a FORTRAN computer program
which generates a matrix in MPSX format for the required number of
commodities and time periods. This is passed as input to the
MPSX program which solves the problem using standard procedures.
Once an optimal solution has been found, a Report Generator
program manipulates the results to provide output reports which
are meaningful to the users.

For each major formulation, the problem and its solution are
saved on disk. The stored problem can then be easily revised
without going through the matrix generator step, and solved for a
new solution using the previous solution as a starting point.
This allows for numerous sensitivity runs to be made with only a
limited increase in computer usage.

The major programming development work was carried out using
a remote terminal and the Conversational Remote Job Entry system

on the IBM 360/Model 75 computer at McGill University. Presently
the model is run on an IBM 370/Model 165 computer at Canadian
National headquarters in Montreal. As an example of the time
required to solve the LP, a two commodity formulation of 900 rows
and 3000 columns was solved in a little over one minute of C.P.U.
time on the Model 165. A seven commodity formulation of 3000 rows
and 10,000 columns required approximately six minutes of computer
time.

APPLICATION OF THE MODEL

The model is a useful planning tool since it allows users to
rapidly examine the optimal transportation scheme for various sets
of conditions. The computer model is suited to easily determine
effects on the transportation system caused by changes in network
structure, costs, capacities, etc. Due to the complexity of the
problem, the impact of such changes might otherwise be overlooked.
This type of detailed logistics planning is important to ensure
coordinated efforts of the various modal participants in planning
plant and equipment requirements, and to ensure efficient and
timely movement of materials from source to destinations.

In this section we will explain the typical data required to
set up the model, and the type of results made available to the
user. To do this, it would seem appropriate to consider one
particular formulation that has been studied by the planners,
mentioning the type of sensitivity analysis carried out.

The sample formulation involves the movement of two
commodities during five defined time periods (Figure 5). The two
commodities are low temperature 48" diameter pipe, required for
the most northern part of the pipeline, and normal temperature
48" diameter pipe, required for the remainder of the line. The
time periods represent the winter 1973/74, summer 1974, winter
1974/75, summer 1975, and winter 1975/76 with the final two winter
periods being the main construction intervals. To fully describe
the formulation, the following information is required as input to
the computer model.

1) Material Supply (Figure 6)
   The potential origins, both off-shore and domestic, are
   identified for both types of steel pipe. Japan is the only
   source of low temperature pipe while several Canadian sources
   are able to supply the normal temperature pipe. For each
   Canadian source, a limit on the production in any time period
   is given.

2) Material Requirements (Figure 7)
   In conjunction with the pipeline construction plan, fifteen
   stockpiles close to the pipeline right-of-way are identified
   as destinations for the two types of pipe. The number of
   miles of pipe required, expressed as a total tonnage, is
   specified for each construction period and each stockpile.

```
COMMODITIES:
                LOW TEMPERATURE PIPE
                NORMAL TEMPERATURE PIPE


TIME PERIODS:
                WINTER 1973/74
                SUMMER 1974
                WINTER 1974/75
                SUMMER 1975
                WINTER 1975/76
```

Figure 5. Sample Formulation

```
LOW TEMPERATURE PIPE:
                JAPAN


NORMAL TEMPERATURE PIPE:
```

SOURCE	MAXIMUM TONS/TIME PERIOD
EDMONTON	38,250
CAMROSE	60,000
CALGARY	65,000
REGINA	19,250

Figure 6. Material Supply

LOW TEMPERATURE PIPE:

LOCATION	WINTER 74/75	WINTER 75/76
PRUDHOE BAY	69,000	67,000
.	.	.
.	.	.
.	.	.
FORT NORMAN	39,000	–

NORMAL TEMPERATURE PIPE:

LOCATION	WINTER 74/75	WINTER 75/76
SALINE RIVER	58,500	–
.	.	.
.	.	.
HIGH LEVEL	–	55,900

Figure 7. Material Requirements

Only one type of pipe is required at each stockpile, and of the total of approximately one million tons, thirty per cent is normal temperature pipe and the remainder is low temperature pipe.

3)   Network Arc Costs (Figure 8)
Each arc in the network has an associated per unit cost to express the material, transportation, transshipment, or inventory carrying cost. The cost information is gathered by the team in Edmonton, through contact with suppliers, transportation companies, and others. While it is not possible to say at present what exact rates individual suppliers and carriers will charge, estimates can be made on the basis of informed opinions and historical costs for similar types of projects.

Inventory carrying costs at intermediate and final stockpile locations are estimated on the basis of a per annum interest charge. To calculate the dollar per ton cost associated with a stockpiling arc, an approximate delivered cost at the stockpile is found and then the interest charge is applied over the appropriate length of stockpiling time.

4)   Equipment Capacity Restrictions (Figure 9)
For each mode of transportation, equipment capacity restrictions are represented in the model by limiting flow over single arcs or groups of arcs. Based on the most reliable information, estimates have been made of the number of units of equipment that could be provided for the movement of pipe in each time period. These estimates are translated into maximum tonnage bounds, taking into account the capacity of the equipment, the routes covered, the cycle times over the routes, and the length of the time periods.

For each transportation facility, one or more pools of equipment are available for use over various routes. For example, three truck "pools" have been defined for representation in the model. These are the Northwest Territories, Yukon, and Alaska pools, and due to the geography and national boundaries involved, it is thought that each one is more or less independent of the other's resources. Each movement of material within the boundary of a pool uses up a portion of the total capacity.

Similar pools of equipment are available for transportation by ship, ocean barge, Mackenzie River barge, and railroad. As well, the model structure includes restrictions so that the model will not, for example, move material on the Mackenzie River barges in a winter period or use a winter-only road in a summer period.

5)   Stockpile and Inventory Restrictions (Figure 10)
Certain locations have limitations on the amount of material that can be stockpiled due to the physical plant available. The model restricts the total tonnage of the two types of

SAMPLE COST ESTIMATES

TRANSPORTATION:

       EDMONTON TO HAY RIVER (RAIL) $30.00/TON
       HAY RIVER TO WRIGLEY (RIVER BARGE) $26.00/TON
               .
               .
               .
             . ETC.

TRANSSHIPMENT:

       RAIL TO STOCKPILE (ENTERPRISE) $ 2.00/TON
               .
               .
               .
             . ETC.

MATERIAL:

       PIPE AT CALGARY $337.00/TON
               .
               .
               .
             . ETC.

INVENTORY CARRYING:

       WINTER TO SUMMER (ENTERPRISE) $13.00/TON
               .
               .
               .
             . ETC.

Figure 8

SAMPLE EQUIPMENT CAPACITY CONSTRAINTS

MAXIMUM AVAILABLE UNITS:

      MACKENZIE RIVER BARGE   ...   6 BARGE TRAINS

      VANCOUVER TO ENTERPRISE (CN) ... 140 RAIL CARS

      ALASKA RAILROAD   ... 25 RAIL CARS

      YUKON TRUCK POOL   ... 50 TRUCKS
        .
        .
        .
       . ETC.

Figure 9

OTHER RESTRICTIONS

STOCKPILE CAPACITY:

      ENTERPRISE   ...   200,000 TONS

      HAY RIVER   ...   25,000 TONS
       .
       .
       .
      . ETC.

INVENTORY REQUIREMENTS:

      FIFTY PER CENT OF PIPE IS REQUIRED

      TO BE IN STOCKPILE BEFORE THE START

      OF CONSTRUCTION.

Figure 10

pipe stockpiled to the specified upper limit. On the other
hand, lower bounds are imposed on some stockpiling arcs to
ensure that a given amount of material is available at the
beginning of construction. By imposing a lower limit on the
amount of pipe stockpiled at a destination location, the model
ensures that a certain percentage of pipe will be available
before construction actually begins.

The linear programming model finds the least-cost method of
satisfying all pipe requirements subject to the plant, inventory,
and equipment restrictions. As direct output from the MPSX
package, the results are the non-zero flows over each arc in the
network for the optimal solution. Both graphical output and
summary reports are then produced for the users.

Figure 11 illustrates the sample formulation results for a
single time period (the winter of 1973/74) and one commodity
(low temperature pipe) in graphical form. The following
observations can be made on the movement of low temperature pipe
to various destinations.

Coleen River. The solution has 36,900 tons of low temperature
pipe moving by ship to Anchorage, by rail to Fairbanks, and then
by truck to the Coleen River stockpile. This move must take place
one year prior to the first construction season in order to meet
the condition that fifty per cent of the total requirement be in
the stockpile at the start of construction. Since winter road is
the only means of reaching this destination, half the requirement
of 73,900 tons must be moved in this first time period.

Old Crow. Since this stockpile can only be reached by
winter road also, 25,300 tons of Japanese pipe moves in via
Whitehorse in this time period.

Arctic Red River. The Arctic Red stockpile is in a different
situation since it can be reached in the summer by Mackenzie River
barges. Looking ahead to the following summer, the solution shows
that only 13,000 tons are barged to Arctic Red before the capacity
of the barge fleet is used up. This forces the move of 37,900
tons by winter road in the winter of 1973/74.

It should be pointed out that of the pipe moving from Japan
to Whitehorse, 57,500 tons move over the White Pass and Yukon
railroad and the remaining tonnage moves by truck via Haines.
This happens because the total capacity of the cheaper route (the
railroad) is used up, thus the remaining volume is forced in by
the alternative route at a higher cost.

Enterprise. The remaining 81,500 tons of steel from Japan
moves via Vancouver to Enterprise where it is stockpiled until the
next summer. It is then moved in to Hay River by truck to supply
the Mackenzie River barges in the summer of 1974. The solution
shows that the pool of rail equipment for this move is msed up to
capacity in the following time period, and this results in the
large volume being moved in this earlier time period.

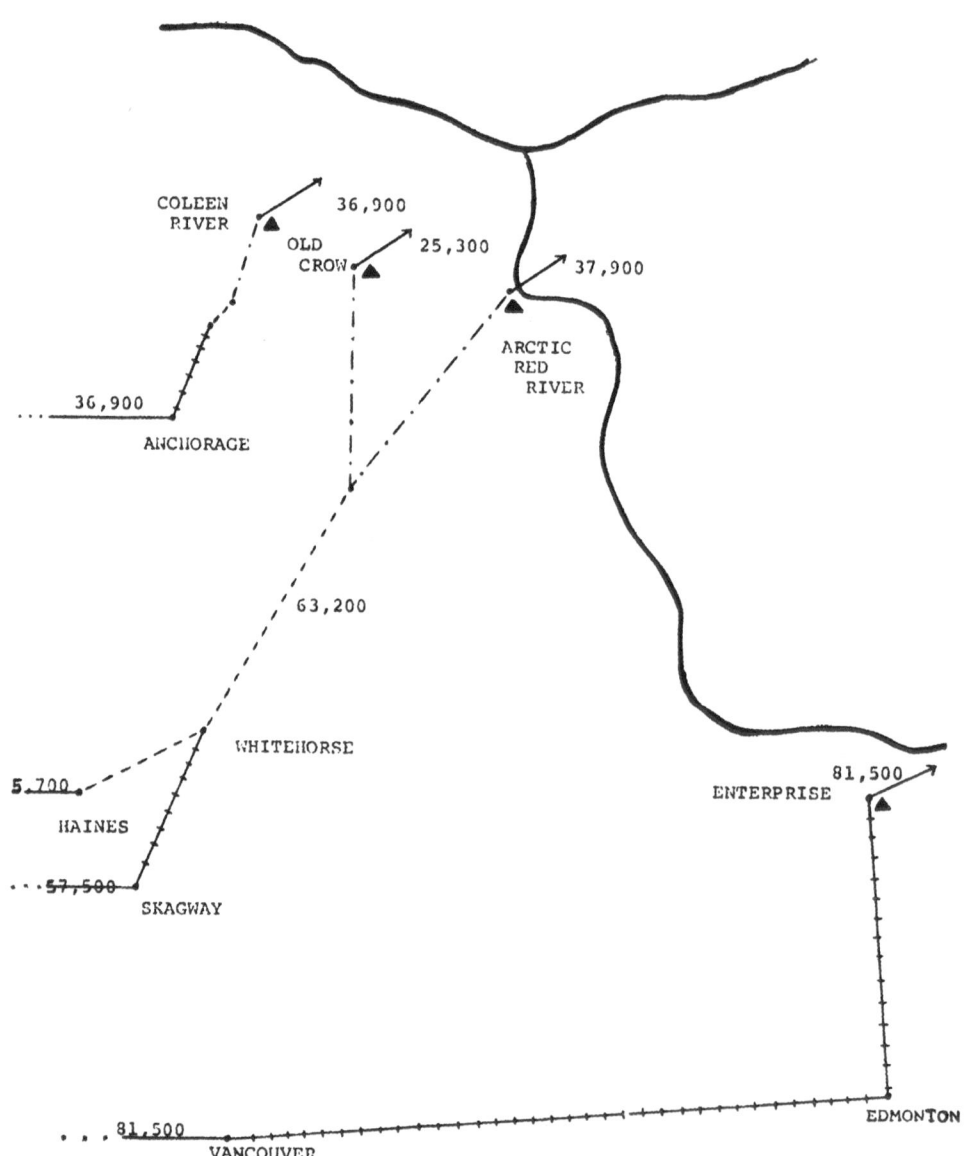

Figure 11

The above illustrates only a small portion of the solution
but indicates the complexities and interactions of the various
transportation modes in the problem. Along with such diagrammatical
output, the following types of reports are generated giving the
details of the solution.

1. Transportation Volumes and Equipment Requirements.
    For every mode of transportation and each commodity, a
    detailed report of the tonnage moved over the network links
    in each time period is given. The total number of loads and
    the transportation equipment requirements are calculated and
    listed as well. Figure 12 is an example of such a report.

2. Pipe Supply.
    For each type of pipe, the supply locations, volumes, and
    costs are summarized.

3. Inventory Stockpiled.
    The amount of each commodity stockpiled from one time period
    to the next at all intermediate and final stockpiles is
    given.

4. Summary of Costs.
    The total cost of each transportation mode is calculated and
    listed by commodity and time period. Also included are the
    material costs and inventory carrying costs as shown in
    Figure 13.

5. Summary of Equipment Units Required.
    This report summarizes the amount of equipment utilized in
    each time period for all "pools" of transportation equipment.
    From such a report (for example, Figure 14) one can readily
    examine how even the utilization of equipment is over the
    five time periods and whether the total capacity was used up.
    For example, notice that the 36 river barges were used to
    capacity in both summers.

        As previously mentioned, one of the major benefits to this
type of model is the ability to test the sensitivity of the
solution to changes in the input. This is useful to study the
effect of major decisions which would have an impact on the
transportation of pipeline materials and supplies. For example,
one of the questions posed concerns the effect on the over-all
transportation scheme if the number of barge trains available for
moving pipe on the Mackenzie River is changed. The model is
easily revised to reflect different barge capacities and solved
again for a new solution.
        In order to illustrate the effect of such a change, we can
examine the movement of pipe to meet the first construction period

SUMMARY OF TONNAGES (IN THOUSANDS OF TONS)
AND
EQUIVALENT TRANSPORTATION EQUIPMENT REQUIREMENTS

COMMODITY - JPN. PIPE

MODE - RAILROAD   UNITS - CARS

TIME PERIOD

		W 73/74		S 74		W 74/75		S 75		W 75/76		TOTAL TONS	TOTAL NO. OF LOADS
		TONS	UNITS	TONS	UNITS	TONS	UNITS	TONS	UNITS	TONS	UNITS		
VANCOUVER	TO ENTERPRISE	81.5	139.6	81.5	139.6	81.5	139.6	81.5	139.6			326.0	8467.53
VANCOUVER	TO FORT NELSON					5.8	6.4					5.8	149.89
SKAGWAY	TO WHITEHORSE	57.5	27.9	9.9	4.8	57.5	27.9			50.6	24.6	175.5	6750.85
ANCHORAGE	TO FAIRBANKS	36.9	16.9			54.1	24.7			54.1	24.7	145.1	3768.83
RAIL	TOTAL	175.9	184.4	91.4	144.4	198.9	198.7	81.5	139.6	104.7	49.3	652.4	19137.10

Figure 12

SUMMARY OF COSTS (IN THOUSANDS OF DOLLARS)
COMMODITY - JPN. PIPE

TIME PERIOD

	W 73/74	S 74	W 74/75	S 75	W 75/76	TOTAL
OCEAN SHIP	6720.8	4742.9	8704.9	3015.5	3955.3	27139.5
OCEAN BARGE		4271.9				4271.9
RAILROAD	5767.8	4252.8	6366.4	4114.9	1955.2	22457.2
RIVER BARGE		6447.9		6610.7		13058.6
TRUCK	8205.4	588.9	13779.2	456.4	9507.6	32537.4
TRANSSHIPMENT	2012.6	3693.2	2483.8	1731.2	1500.3	11421.1
TRANSPORTATION	22706.7	23997.5	31334.3	15928.8	16918.4	110885.7
INVENTORY		2749.3	6013.6	2571.0	4946.4	16280.3
SUBTOTAL						127166.0
COMMODITY						262013.5
TOTAL						389179.5

Figure 13

SUMMARY OF EQUIPMENT UNITS REQUIRED

		TIME PERIOD					
		W 73/74	S 74	W 74/75	S 75	W 75/76	MAXIMUM
OCEAN	SHIP	3.2	2.3	4.0	1.5	1.8	999.0
OCEAN	BARGE		3.7				999.0
RAIL	CN.JPN	139.6	139.6	139.6	139.6		140.0
RAIL	CN.CDN	18.3	30.0	30.0	30.0	30.0	30.0
RAIL	CN.EDM	79.6	108.3	108.3	106.7	105.3	120.0
RAIL	PGE			6.4			35.0
RAIL	WPY	27.9	4.8	27.9		24.6	28.0
RAIL	AR	16.9		24.7		24.7	25.0
RIVER	BARGE		36.0		36.0		36.0
TRUCK	N.W.T.	10.9	4.0	20.0	7.0	18.3	20.0
TRUCK	YUKON	50.0	3.2	50.0		35.0	50.0
TRUCK	ALASKA	42.0		100.0		87.8	100.0

Figure 14

requirement at Prudhoe Bay for the case with six barge trains
available and that with four barge trains available. Figure 15
shows the solution for the six barge train alternative and we note
that of the 69,000 tons required for the winter of 1974/75, the
following moves are made:

- 34,300 tons shipped by Mackenzie River barge in the
  summer of 1974.

- 28,000 tons moved by ocean barge around Point Barrow
  in the summer of 1974.

- 6,700 tons transported by truck over a winter road in
  the winter of 1974/75.

This solution is explained by examining the delivered costs
by different routes in the model. The delivered costs to Prudhoe
Bay show that the Mackenzie River barge move is the least
expensive, the winter road move intermediate, and the ocean barge
move the most expensive. These costs include the necessary
stockpiling costs from summer to winter when the sea or river
routes are involved. Thus, if there was unlimited river barge
capacity available, all the pipe would move by this mode. However,
other stockpiles are also competing for the six barge train
capacity and in the optimal solution only 34,300 tons are moved
to Prudhoe Bay by this mode. Likewise, trucking capacity of the
Alaska pool is limited, and only 6,700 tons are moved in this way.
The remaining requirement moves by the most expensive route or
the ocean barge.

When the Mackenzie River barge capacity is reduced to four
barge trains, the movement of pipe to Prudhoe Bay in this new
solution is shown in Figure 16. In this case, to minimize the
total logistics cost, none of the Mackenzie barge capacity is
made available for moving pipe to Prudhoe Bay, and the tonnage
formerly moved is shifted to the ocean barge route. Again, this
example has only shown the effect on a minor portion of the entire
network but serves to illustrate how this type of sensitivity
analysis is carried out. It should also be stressed at this time
that the costs used in this example are only for illustrative
purposes.

The model has recently been used to study the impact on the
pipeline logistics of the building of new all-weather roads in the
far north. Only recently, the Canadian Government announced plans
to go ahead with the construction of the Mackenzie Highway from
Fort Simpson to Tuktoyaktuk. The savings that could accrue to the
project if selected road links were completed before pipeline
construction have been estimated with the aid of the model.

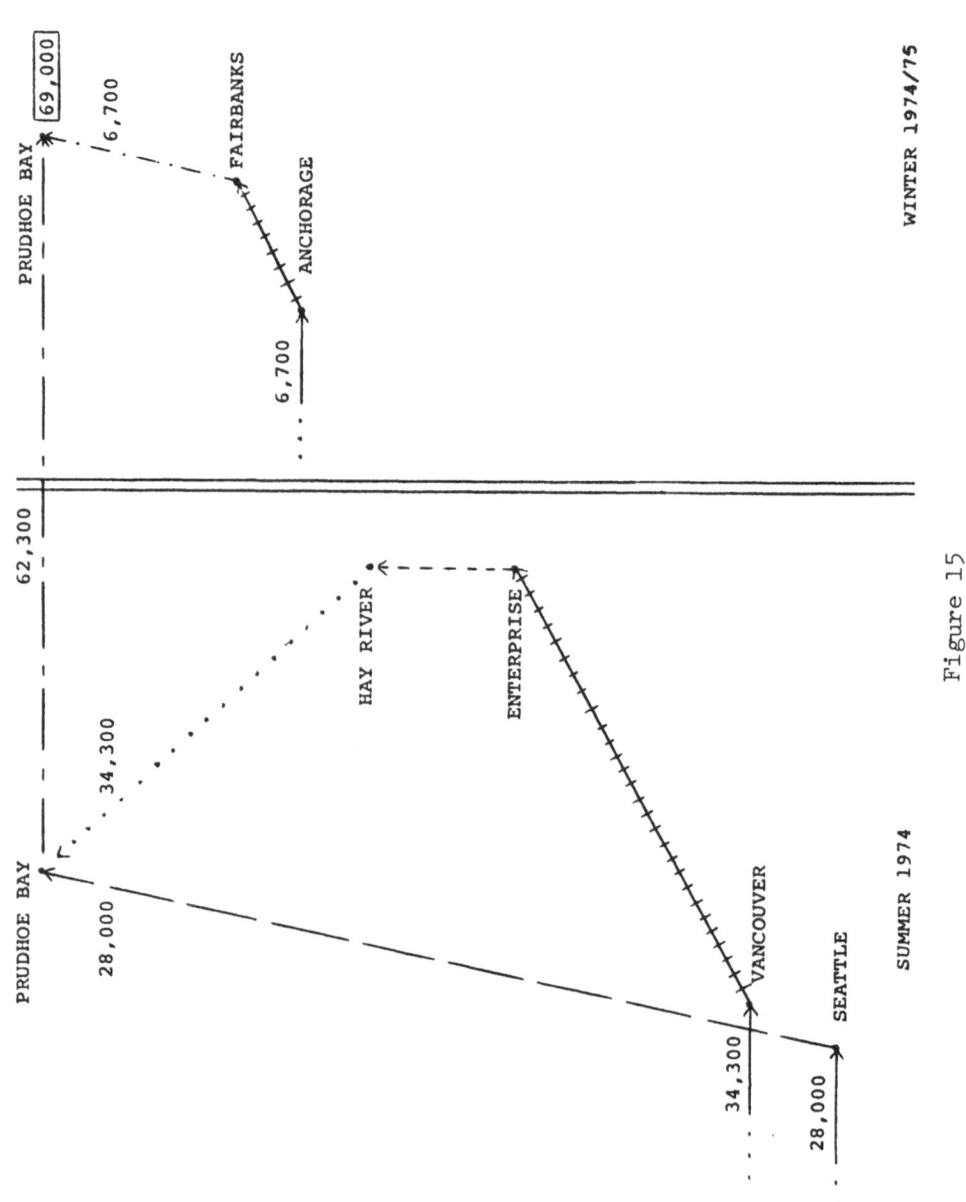

Figure 15

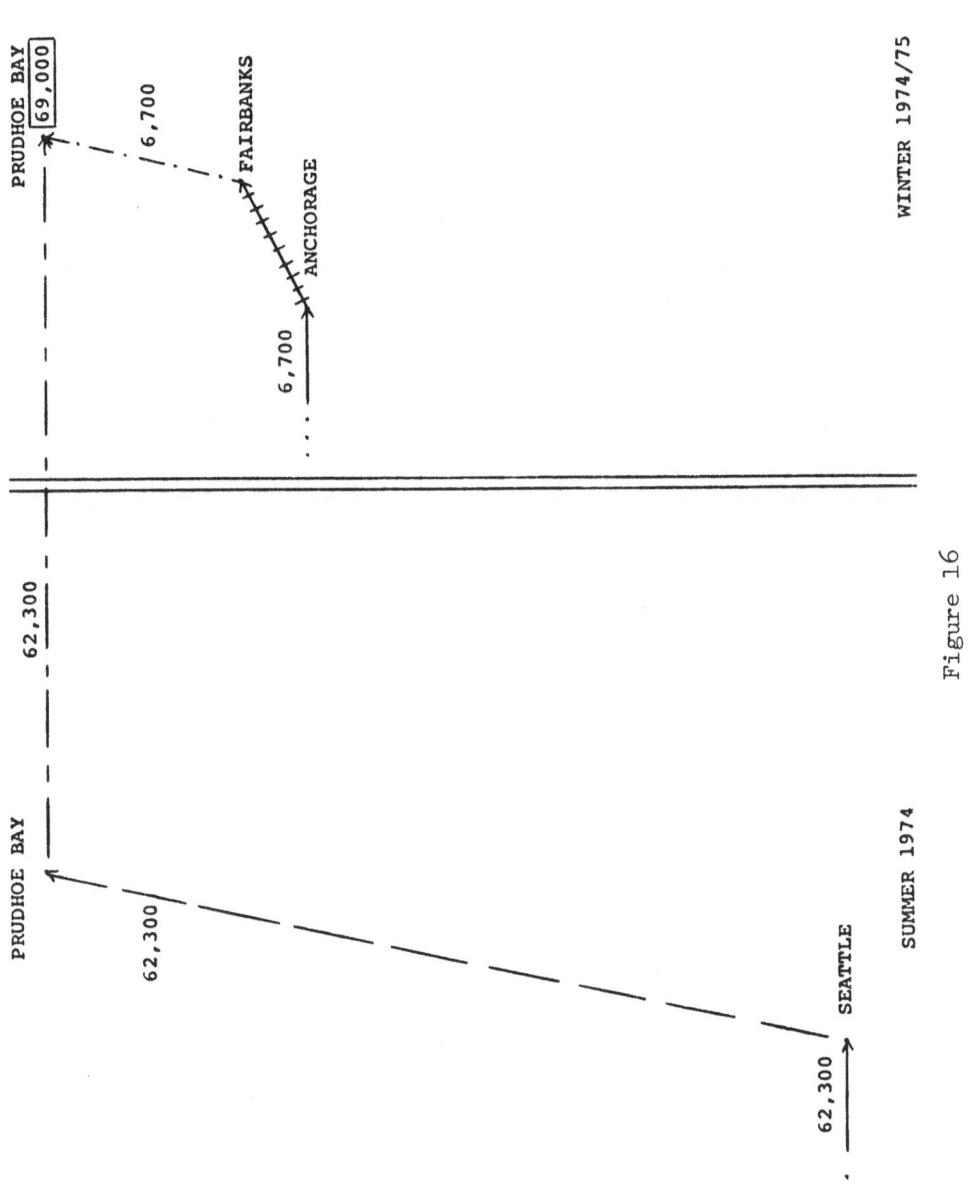

Figure 16

CONCLUSIONS

A successful application of a well known OR technique has been described. The LP model continues to be used in planning the transportation of materials associated with the pipeline. The extent and detail of the model is still being expanded as the planners become more aware of the usefulness of this tool.

Another benefit has been the education of the users in the type of analytical support available to them from the Operational Research Branch. As a result of our involvement in this study, we have consulted on other aspects of the pipeline project and have become involved in related studies.

DISCUSSION CONTRIBUTION REGARDING

IMPROVEMENT POTENTIAL OF ICE-NAVIGATION

F. J. Legerer

Memorial University of Newfoundland

St. John's, Newfoundland, Canada

PREFACE

Already, 300 years ago Gottfried Wilhelm Leibnitz complained that it was not any more possible to read all books.

In recent years, the author of this discussion contribution has heard two public complaints by eminent scholars that it is not any more possible to follow the publications in their respective, very specialized, fields.

It is, therefore, not surprising to find a communications gap between two disciplines which would mutually benefit if an exchange of thought and concepts took place.

This to initiate is the purpose of this discussion contribution as it is within the aim of this conference.

SYNOPSIS

The discusser submits that current analysis of ice-breaking neglects the progress in applied mechanics of approximately the last forty years. This is essentially caused by a communications gap between designers and scientists in mechanics. By closing this gap, substantial benefits can be derived: energy costs of ice transportation may be reduced by as much as 50 percent of present levels.

INTRODUCTORY AND DISCUSSION STATEMENTS

This discussion contribution deals with some aspects of the present state of the art of ice-navigation. Any improvement is

equivalent to a cost reduction of ice transportation with
corresponding ramifications into other branches of Arctic systems.
    The basis of any engineering design is an engineering analysis;
in the case of ice-going vessels, the analysis is rooted in applied
mechanics as a scientific discipline. While the above statements
are undisputedly in agreement with everybody else, this discusser
presents herewith to this respected audience the following
submission for their consideration:

1st Discussion Statement: "The Present analysis of ice breaking
    (even in most recent publications) is exclusively based on
    elementary mechanics and it neglects relevant advances in
    applied mechanics of almost the last forty years."

2nd Discussion Statement: "While advanced analytical methods need
    not be applied for their own sake, substantial cost reductions
    of ice breaking can be expected by implementing into actual
    design those conclusions which can be derived by means of
    recent scientific progress."

SUBSTANTIATION OF DISCUSSION STATEMENTS

    First Statement: For the present state of the art, it is
referred to references 1-6 listed at the end; the references listed
therein have been used as further guidelines. As there were
repeated references to Russian researchers, this discusser checked
in PMM (Prikladnaja Mathematika u Mekhanika) for authors known in
icebreaking literature and for icebreaking-related subjects. As a
result of this literature search it was concluded that even in
Russia where substantial effort has been put into icebreaker
research there is simply a communications gap between experts in
applied mechanics and engineering designers of ice navigating
vessels.
    Second Statement: The second discussion statement can only be
supported by an incomplete list of heuristically expected benefits
due to advanced analytical concepts. It is beyond the scope of a
discussion contribution to verify every detail; however, even a
cursory list of applicable methods and corresponding potential
benefits is impressive.

A. ICE RESISTANCE VERSUS POWER CAPACITY AND ENERGY CONSUMPTION

    The utility industry is well aware of cost reductions due to
'peak shaving' if the load demand fluctuates widely. Similarly,
in ice navigation the load demand varies; therefore, we regard it
as one objective of analysis to pinpoint methodically these areas
which can contribute to 'peak shaving of load demand'. Several,

not all, ice conditions are considered in the following along with
more recent methods of analysis in view of a reduction of the peak
load demand.

(a) Resistance in brittle, elastic, fast ice.

Resistance in brittle, elastic, fast ice has drawn most
attention of researchers. Despite many efforts and differences
between the results of various investigators all formulations of
the ice resistance belong to one particular class of mathematical
models -.as this discusser has already pointed out elsewhere (7).
We show first that this particular class of mathematical models is
unsuitable at the limit of the ice negotiating capacity of a ship.
According to current analysis, the ice resistance is defined as a
sum of contributory terms

$$R_i = R_f + R_v + R_w \qquad\qquad\qquad (1)$$

(The total resistance R_i is thought as the sum of the
resistance due to fracture R_f of the resistance and dipping,
turning over and pushing aside ice floes R_v, and of the open water
resistance R_w). Any resistance is in that case defined as energy
expended per unit distance of progress. We apply this definition
to the limit when the vessel is just stuck. The fracture resistance
although possibly the decisive contributor remains finite but
indefinite per definitionem $(R_f \to 0)$ while the viscosity resistance
becomes infinite $(R_v \to \infty)$. This 0 formulation certainly cannot
reflect properly an inability to break ice, thus creating a pond
in the wake of the propeller with an almost closed flow circle
(see figures 1a and 1b).

This leads to the following conclusion: 'The resistance
expressed as a sum of contributory terms implies a mutual
independence of these terms which does not hold at the limit of
the vessel's capability'.

Therefore, it seems preferable to split the resistance,
similarly to the method efficiencies are split, into a product of
contributing terms; we will elaborate on this later.

A further examination of the customary method to express the
ice resistance exhibits that the contributory terms to sum (1)
have been established by some reasoning of elementary mechanics
combined with statistical regression analysis of actual tests.
However, the confidence interval of any statistical regression
analysis is widening at the ends of the range over which it is
performed (cf (8) or even (9)). The widening of the confidence
interval may, therefore, be regarded as the second reason against
using resistance formulation of the type of equ. (1) as a basis
of design for peak shaving.

Finally, the relative magnitude of the fracture resistance
versus the resistance of turning over and pushing aside of ice
floes need to be examined regardless of whether the total resistance
is formulated as a product or a sum.

Fig. 1a. Under continuous progress, ice transport is governed by
          the vonKarman vortex configuration: small piece experi-
          ence translational and angular acceleration, while for
          large floes translational movements are almost negligible
          rotation becomes predominant.

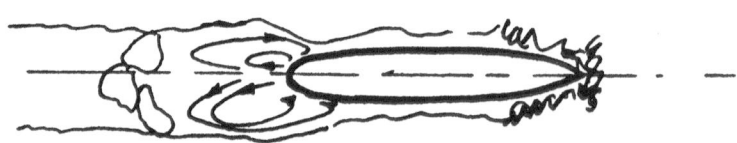

Figure 1b. Under stalled conditions, the broken channel behind the
            ship tends to become clogged due to the von Karman
            interaction of water and ice; in the vicinity of the
            propeller, a pond is created of almost circular flow
            pattern: the propeller becomes a 'giant kitchen
            blender' for ice.

The customary formulation of the fracture term is most aptly summarized by Enkvist (10) according to Kashteljan (11) which we discuss in the following: Accordingly, failure of the ice is assumed to occur in bending; consequently, fracture of the ice sheet occurs at some multiple of the characteristic length; with increasing ice thickness, the broken ice floes increase in size and, therefore, the number of breaks per unit distance decreases. As a result of this theory, the fracture resistance is proportional only to the first power of the ice thickness. Bending failure, as considered by this analysis, is typical for a simplified engineering analysis with quasistatic assumptions.

In contrast, the analysis of a transient load (impact) on a thick plate or beam results in a combination of bending and shear loads (the influence of shear increases with increasing thickness (cf (12) and (13)). Accordingly, the ice floes do not increase in size with increasing ice thickness; as the shear strength of the ice is lower than the flexural strength, the failure criterion of the load combination is rather difficult to establish; as our first approximation the second power of the thickness seems to be justifiable. The term dipping, turning over and accelerating of ice floes has been denoted here as the 'viscosity term' in reference to A. Einstein's doctoral thesis (14). This work dealt with the viscosity of heterogeneous fluids. It is suggested to use the same method to derive the 'viscosity' of the heterogeneous mixture of ice and water and to consider a ship moving in a liquid of an increased viscosity (see also reference (7)), because this will permit us to unify the approach of determining the ice resistance for fast ice as well as mush ice and frazil ice. With this approach, it will be easier to determine under which practical conditions Froude's number becomes less important while Reynolds' similarity conditions have to be considered.

Summing up, we suggest therefore to establish the ice resistance in the form

$$R_i = (1 + a_f)\, R_v + R_w \tag{2}$$

with a_f being an amplification factor for fracture and based on impact mechanics; equ. (2) is more general than equ. (1) because it holds as well for pack ice, mush ice and frazil ice. While at the limit of the fast ice capability the resistance increment with incremental ice thickness is constant in equation (1), it becomes according to equ. (2):

$$dR_i = \frac{\partial a_f}{\partial h}\, R_v + (1 + a_f)\, \frac{\partial R_v}{h}\ dh \tag{3}$$

Along with previous remarks on the influence of transient loads on thick plates equ. (3) permits the conclusion that for the objective

of 'peak shaving' fracture is more important than generally assumed
due to an incorrect application of equ. (1).

Despite the critique on equ. (1), it may be remarked that it
serves its purpose well for the range for which it was established
and verified.

(b) Mush ice and frazil ice.

The formulation of the principal ice resistance as a viscosity
based term permits to include mush ice and frazil ice resistance as
above mentioned. If peak loads occur in mush or frazil ice, peak
shaving must be based on laminar boundary layer theory.

(c) Resistance in pressurized ice.

Pressurized ice or ice pressure due to onshore winds as an
obstacle to navigation is well known. In the Baie d'Exploits
region on the Northeast coast of Newfoundland, onshore winds cause
a total stop of all navigation (including icebreakers) for about
15 percent of the ice season.

The practical effects of ice pressure have been extensively
investigated by D. Bradford (15). The physical phenomenon can be
mathematically modelled as a visco-elastic material. The visco-
elastic material properties are well reported in the literature on
the physics of ice (cf (16)). The mechanics of visco-elastic
bodies (i.e., forces and deformations) is a relatively new branch
of applied mechanics.

The discusser submits herewith:
"The conventional wedge shaped prow of ships which has evolved in
the development of navigation in water is almost the worst
conceivable tool for negotiating a visco-elastic plate." A strict
proof of the above submission would be beyond the scope of this
discussion contribution. However, we illustrate in the following
way the basic elements heuristically; for simplicity, the example
contemplates fast ice, although open sea ice could be considered
accordingly.

Typical for visco-elastic mechanics are slow, time delayed
movements and residual internal stresses which recede only after
the external forces have vanished for a considerable time. Even
after the onshore wind has receded, a hole punched in the ice plate
will close due to internal forces (in contrast, a hole in a brittle,
elastic plate - no ice pressure - will just stay open and refreeze).

In Figure 2 the upper half illustrates a ship progressing
through brittle, elastic ice while the lower half illustrates
progress through pressurized ice. Two consecutively broken cusps
of ice are considered: In the first case, the travel distance of
breaking two consecutive cusps is only determined by the ship's
geometry; in pressurized ice, the ice closes again while the ship
moves forward; hence, the distance of two consecutive cusps
depends also on the velocity of the ship and of the ice closing.
Therefore, more cusps have to be broken. In addition to the
increased fracture resistance, the substantial side friction has to

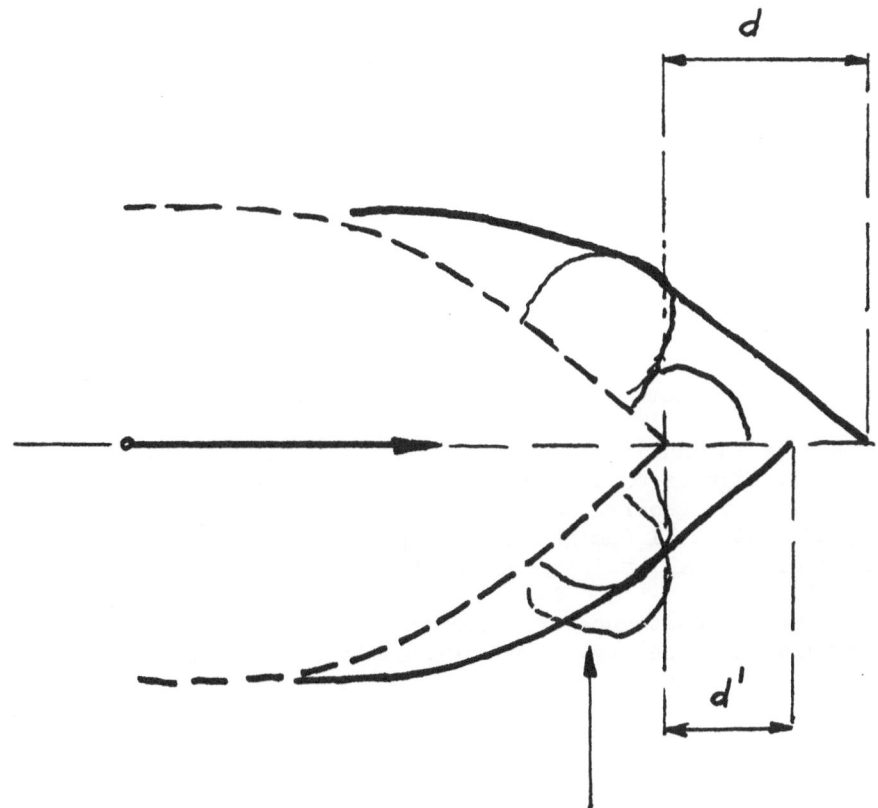

Figure 2.   Scheme to illustrate the effect of the wedge shaped bow
            in visco-elastic (pressurized) ice: the upper half il-
            lustrates the travelling distance d between the fracture
            of two consecutive cusps in brittle elastic ice, while
            the lower half shows the effect of simultaneous closure
            of the gap in pressurized ice, which leads to a shorter
            fracture distance d'.

be considered on top of it. Compared to the wedge shaped prow a
more or less blunt prow may be more efficient in pressurized ice.
This discusser suggests therefore a development programme for prows
and other means which permit an improved efficiency in pressurized
ice without carrying an unacceptable penalty under other environmental conditions of ice-going vessels.

(d) Peak shaving by improving the manoeuvrability.
     Studies have been performed on the potential benefits of
remote sensing permitting ice navigating vessels to follow leads
or, generally, routes of least resistance (17).

However, information on minimum ice resistance is less valuable if a ship is unable to utilize such information due to a lack of manoeuvrability (depending upon the size of an ice navigating cargo vessel, the radius of the turning circle in ice is about 20-50 times that in open water). It is obvious that ice forces should form the basis of steering equipment because these forces are much higher than fluid dynamic forces on which rudder and bow thruster are based. A reduction of the radius of the turning circle in ice to about the open water radius may under typical conditions result in approximately 25 percent reduction of operating costs.

B.   HULL RELATED IMPROVEMENTS

Costs of the hull account for about half of the capital costs of a vessel, figures on maintenance costs are difficult to assess. However, there is general agreement that they are high. We contemplate only hull damage due to impact which accounts for about one-third of the hull maintenance costs for ships operating in the Arctic (18).

The impact resistance of armoured military vehicles has been successfully increased by introducing laminate sandwich structures. The corresponding theory of dampening transient load spectra has been developed extensively during the last twenty years. Certainly, a successful method against the impact of bombshells is likely to be adaptable for resisting the impact of ice-floes - despite a necessary adjustment for a different type of load spectra.

To our knowledge, there was only one publication reporting research into a particular composite material regarding its applicability for strengthening the hull of an icebreaking vessel (19). It was not in a shipbuilding related journal and its wider consequences were seemingly not even recognized by the author.

We suggest, therefore, a development programme for both, the design and the manufacturing methods of suitable sandwich hull platings, which are bound to require less maintenance costs and which are also a lesser pollution hazard. The capital costs of such hulls need not be more expensive than present comparable ice-strengthening conventional hulls.

PRACTICAL CONCLUSIONS

Our contention, substantial cost savings, can be expected if design analysis is based on more recent progress of applied mechanics was substantiated in the following ways:

Requirements of power capacity and energy consumption can be reduced by a method known in the utility industry as 'Peak Shaving'. This means, translated to ice-navigation, extreme load demands should be met by means of suitable auxilliary devices rather than by larger power machinery and ships.

However, the success of such devices depends on a correct assessment of the operating conditions of a vessel and an improved analysis of the actual operation. We gave three different reasons why present analysis of the resistance in fast ice is not suitable to reflect a vessel's operation at the limit of its capability. Due to impact theory, a more recent discipline of mechanics, it can be concluded that fracture is more important than presently recognized. Advances in laminar boundary layer theory and multi-phase fluid flow can provide the basis for equipment design to reduce mush ice resistance. Also, a new discipline within applied mechanics, the mechanics of visco-elastic bodies can provide the means for reducing the losses due to pressurized ice conditions.

Stochastic methods of structural stability theory can bring about the means for designing ice-force based steering equipment which could drastically improve the manoeuvrability of vessels in ice.

Design criteria for hulls which should be less costly than conventional ice-strengthened hulls can be obtained by means of the theory of transient loads on sandwich structures.

The total benefit to be derived by the above mentioned means can be estimated from statistical information on ice conditions. Adding up possible incremental improvements for typical Eastern Arctic conditions results in a reduction of the energy requirements (fuel costs) to about one-half of their present level. There exist some design concepts (for instance (20) and some others) which have been developed in view of the thoughts presented in this discussion contribution.

However, the full potential benefit in terms of engineering hardward design and actual economic results will only accrue if on a broad scale the gap is reduced which exists at present between design analysis and applied mechanics as a scientific discipline.

REFERENCES

1. Chu, F.D., Ship Resistance in Homogeneous Ice Fields, Doctoral Thesis, The Helsinki University of Technology (1974).

2. Enkvist, E., Report No. 24, Swedish Academy of Engineering Sciences in Finland, Helsinki (1972).

3. Lewis, J.W., and Edwards, R.Y., Trans. SNAME, Volume 78 (1970).

4. Milano, V.R., Doctoral Thesis, Stevens Institute of Technology (1972).

5. Vance, G.P., Doctoral Thesis, University of Rhode Island (1974); also, SNAME Symposium Ice Tech 75, Montreal, April (1975).

6. White, R.M., Doctoral Thesis, M.I.T. (1965).

7. Legerer, F.J., Discussion Contribution to G.P. Vance, SNAME,
    Ice Tech 75, Montreal (1975).

8. Linder, A., Statistiche Methoden, 4th Edition, Birkhaeuser,
    Basel (1964).

9. See Reference 4, p. 40.

10. See Reference 2, pp. 109-120.

11. Kashteljan, V., et al. Ice Resistance to Motion of A Ship,
    Sudostroenie, Leningrad (1968).

12. Abramson, H.J. Plass, Ripperger, E.A., Advances of Applied
    Mechanics, Vol. 5 (1958).

13. Goldsmith, W., Impact, the Physical Behaviour of Colliding
    Solids, Butterworth, London (1960).

14. Einstein, A., Ann d. Physik, Volume 19, pp. 289-306 (1906) and
    Einstein A., Ann d. Physik, Volume 34, pp. 591-592 (1911).

15. Bradford, D., Proceedings, Sea Ice Conference, Reykjavik (1972),
    pp. 154-158.

16. Weeks, W.F., and A. Assur, Fracture of Lake and Sea Ice, CRREL,
    Report 264 (1969).

17. McQuillan, Benefits of Remote Sensing in Sea Ice, IARCS
    Subcommittee, December (1973).

18. Crosbie, C., Oral Communication, April (1975).

19. Vetter, M.F., ASME Petroleum Division, Petroleum Engineering
    Conference, Los Angeles, California, September (1973).

20. Canadian Pat. No. 964 527
    U.S. Pat. No. 3 878 804, etc.

A PROPOSED FORMULATION AND ANALYSIS PROCEDURE FOR A NATIONAL POLICY

ON CANADIAN ICE-CAPABLE OCEAN TRANSPORT MANAGEMENT

T. W. Kierans and P. J. Amaria

Memorial University of Newfoundland

St. John's, Newfoundland, Canada

This submission on policy formulation for Canadian Ice-Capable Ocean Transport Management is intended by the authors solely to demonstrate a proposed procedure. It is not a complete analysis. It is estimated that a complete policy formulation analysis, evaluation and selection procedure would require thirty to thirty-six months with a staff selected and trained to rigorously follow the iterative and filtering process indicated. The cost of this process is estimated at about $6,000,000. It should be noted that the economic value of Canadian ice-capable ocean transport for the Eastern Arctic alone by 1990 is estimated at more than $600,000,000 per year.

INTRODUCTION

Purpose and Terms of Reference

The purpose of this submission is to suggest and to demonstrate a multi-discipline procedure for:

1. formulating optional Canadian ice-capable ocean transport management policies and associated plan alternatives which could be used to implement the stated national objective;

2. indicating how this procedure may assist those responsible in evaluating and selecting that policy and plan development which conforms most acceptably to all national objectives, design criteria and natural and man-caused constraints;

3. indicating the management organizations and engineering systems required for implementation of the selected policy option; and

4. recommending steps and estimating costs to immediately initiate the proposed full policy formulation and analysis procedure.

This presentation is based on a paper prepared by Professors T.W. Kierans and P.J. Amaria of the Faculty of Engineering and Applied Science, Memorial University of Newfoundland at the request of Dr. A.A. Bruneau, the University's Vice-President of Professional Schools and Community Services. The general approach to policy formulation for Canadian ice-capable ocean transport management has been based on policy engineering concepts developed by the authors, at Memorial University, over the past two years. It is also an exercise demonstration of the multi-discipline methods and procedures that have been proposed for the formulation, analysis and evaluation of policy alternatives for public, corporate, institutional and cooperative groups. This process seeks to avoid intuitive judgments on the one hand and single purpose task force objectivity on the other. It seeks to provide comprehensive information in an acceptable format to those who have the authority and responsibility to select a final policy from the options considered. The basic concept in policy engineering is the multi-discipline, definition of objectives and the iterative screening, filtering and evaluation of all possible policy options in order to produce rational choices. A basic goal is the avoidance of the creation of new peripheral problems. It seeks comprehensive consideration of all parameters and a step by step testing procedure. It emphasizes concern for the inherent objectives of all the groups involved. It proposes policy statements having regard for time, space (location), matter (resources), energy (consequences) and unknowns.

In addition to describing the proposed policy formulation and
analysis procedure, reference is also made here to the management
organization required to implement a policy option, the engineering
systems which the organization should use in such implementation
and a brief description of the interaction between policy
formulation and its required management organization and engineering
systems. Using the general procedure described a very much
condensed sample procedure is developed for the subject policy
problem. The basic principles of policy engineering are described
briefly in an appendix to this paper. An even more condensed
description of the simple procedures used in developing policy and
analysing and evaluating policy options is included in the paper
along with an estimate of the requirements for a full analysis and
evaluation of the subject policy options.

GENERAL STEPS IN POLICY ENGINEERING PROCEDURE

Step 1. State specific concern and the primary objective
requiring policy formulation. State these in terms of: time -
space (location) - matter (resources involved) - energy (anticipated
consequences of policy) - unknowns.

Step 2. Define groups involved in policy formulation:

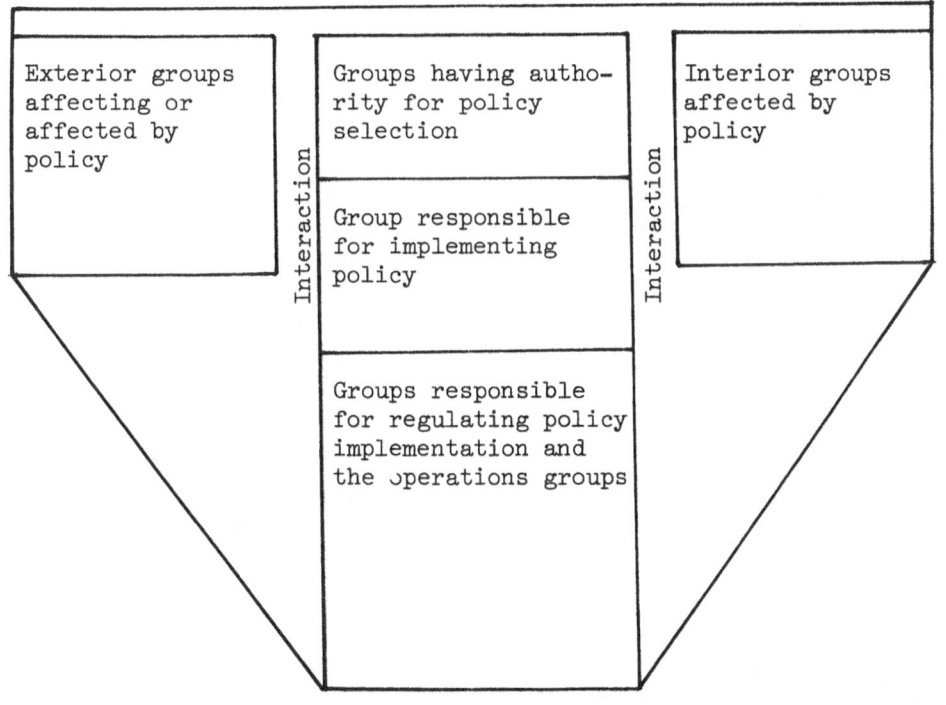

Each of the above groups should be defined as to: function -
jurisdictional authority and areas of influence - capability in
regard to economic, technical, manpower, resources - composition,
compatibility and interaction with other groups - its decision
making procedures - motivation in regard to cultural, economic and
life style interests - character and reliability regarding policy
implementation - numbers of personnel involved - special interests
and conflicting interests between involved groups - terms of
office - potential for unforeseen changes in the above factors -
other factors.

All the above should be stated in terms of time, space, matter,
energy and unknowns and in a format which permits policy option
screening and filtering as to conformity with group objectives,
design criteria and natural and man-caused constraints.

Step 3. Establish the design criteria for the proposed study,
e.g. cost, functions, readiness, quality, reliability, flexibility,
maintainability, etc. State in terms of: time - space (location) -
matter (resources) - energy (consequences) - unknowns.

Step 4. State all possible policy options, in terms capable
of rough screening through Step 3. Describe each option in terms
of: time - space (location) - matter (resource dependability and
requirement) - energy (consequences) - unknowns.

Step 5. Prepare preliminary and alternative implementation
plans for each policy option including:
- management organization for each option
- engineering system needs for each option, e.g. economic forecasts,
  scheduling, technology, R & D needs, infrastructure, support
  services, monitoring and information requirements, etc.

The interaction between policy, plans and the required
organization and engineering systems should be described.
Activities in plans should be described in terms of: time - space
(location) - matter (resources) - energy (consequences) - unknowns.

Step 6. Prepare lists of all policy selection and implementation constraints in terms of time, space, matter, energy,
unknowns.
- natural constraints, e.g. climatic, hydrographic, visibility,
  etc. in terms of time, space, matter, energy, unknowns.
- man-made or man-caused constraints, e.g. law, existing
  facilities, required facilities, labour, finance, technology
  management, jurisdictional, markets, R & D needs, etc.

Step 7. Each policy option and each activity in the
associated detailed plans are screened and filtered through (a) the
inherent objectives of the group involved in policy making (Step 2),
(b) the policy design criteria (Step 3), and (c) the policy
selection constraints, both natural and man-caused (Step 6).

Step 8. Select policy alternatives which successfully survive
Step 7 and prepare detailed implementation plans for these
policies with special reference to unknown elements of each plan.
Watch for possible modification to policy options, plans or
constraints which may enhance the value of certain policy options

and plans.

Step 9. Simulate actual policy implementation for all policies selected in Step 8 - modify as required and recycle modified plans through Steps 1 to 8.

Step 10. Develop limited, on-site tests of selected policy or policies to conform to design criteria and constraints under actual performance, before full implementation:
- limitation on tests may be
  (a) as to time or test period for total group involved
  (b) as to a portion of total group involved
  (c) as to a selected area for test
  (d) other limited tests, e.g. voluntary and imposed.

Monitor all relevant data and provide information to implementing and regulatory authorities.

Step 11. Select final policy and fully implement it with suitable flexibility to permit necessary adjustments for which the early implementation period indicates the need. Regulatory bodies develop and maintain suitable monitoring and surveillance systems as required. Conduct scheduled operations.

STEPS IN POLICY FORMULATION, ANALYSIS, EVALUATION, SELECTION AND IMPLEMENTATION FOR CANADIAN ICE-CAPABLE OCEAN TRANSPORT MANAGEMENT POLICY

Step 1: Statement of National Concern and Primary Objective

The National Concern. Canadian interest in the North, stimulated by an urgent world concern for hydro-carbon fuels and mineral ores, and their occurrence in the Arctic Islands and off the northeast coast, has now grown to the extent that the managed development of the resources of these areas is a national objective. Full achievement of this objective is constrained by severe natural conditions, concern for the total environment, the welfare of indigenous and resident people, the extremely heavy economic and social commitments that will be required and the timing, technology needs and go-no go consequences to all Canadians. A basic part of the total regional policy requirements is that related to all in-going and out-going transportation by air, land and sea. Of these three modes of transport, this paper is concerned only with a proposed policy formulation process for the management of ice-capable marine transport in Canada's ice-affected Arctic Islands and northeast coastal jurisdictions. This is an essential prerequisite for effective and long term achievement of the national objective. As such, the possible policy options require comprehensive consideration by all the groups involved before a final policy is selected and implemented. The alternative to such consideration is the creation of many new peripheral problems and concerns. The area of these seas and associated land areas is equivalent to about two million square miles or more than half of

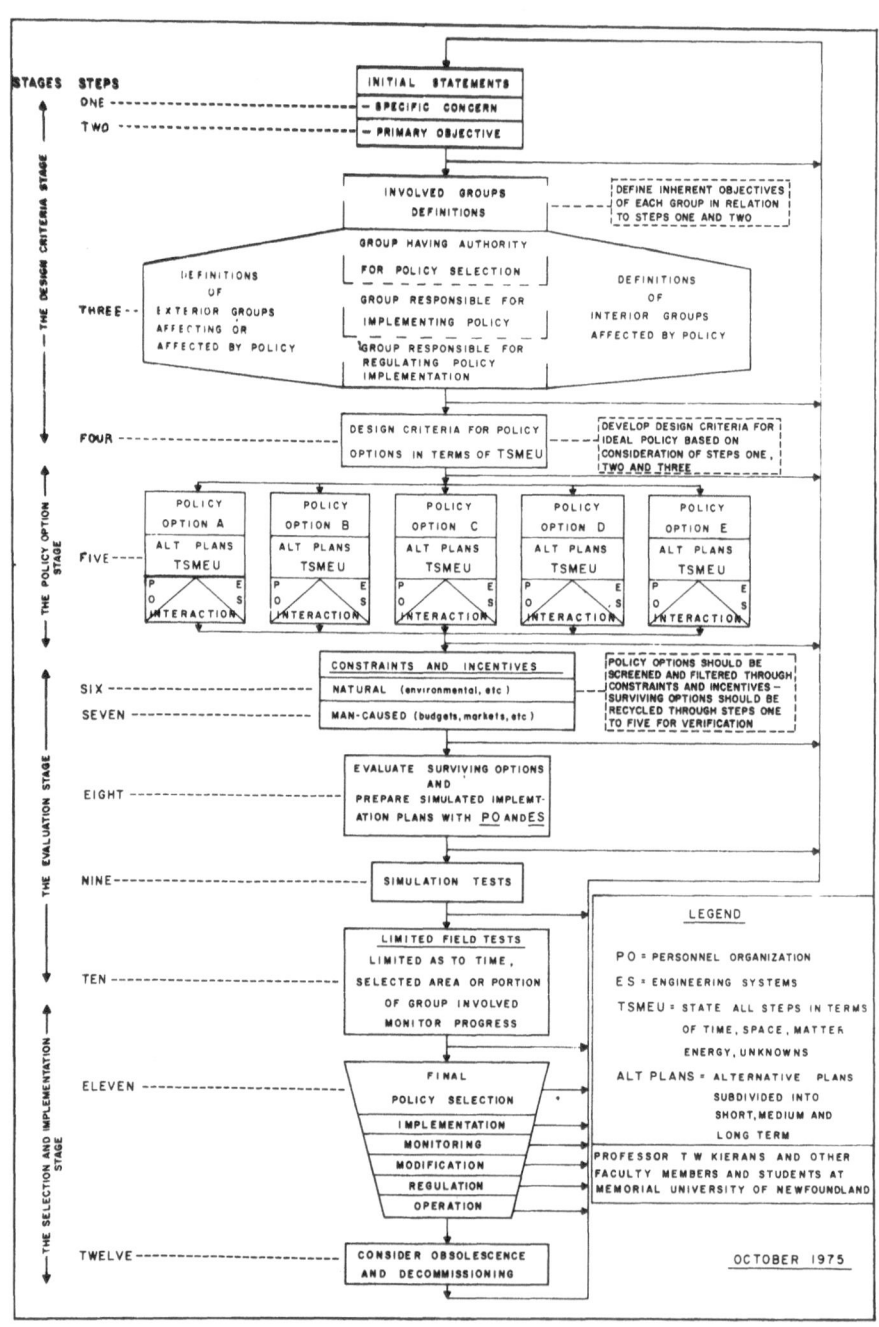

Outline of General Policy Formulation Procedure

Canada's total land area.

The National Objective. Based on the above general considerations as well as official and semi-official statements regarding Canadian governmental goals, the national objective of interest to Canadians in this submission may be stated as follows:

Canadians should establish, as soon as possible, a national policy for an ice-capable ocean transport management system for its ice-affected Arctic and northeast coastal areas. This policy should be designed to effectively maintain our Canadian sovereignty, and good relations with our neighbours and between our internal governments. It should serve our cultural, social, political, environmental, and economic objectives and provide efficient and lowest possible cost marine transportation to regional marine transport users. It should be designed to apply to the exploration, development, production and depletion phases of the region's currently estimated natural resources. It should ensure the provision of on-going values for the area following such anticipated depletion.

Time. In this division we should consider the period for which policy or parts of the policy would be applicable. This parameter describes existing and current plans, short-term plans, mid-term plans and long-term plans, all of which should be compatible.

The ocean transport policy should include provision for the research and development period, transport vessel design and construction, engineering systems, mining and oil and gas exploration periods, environmental and marine investigations including ice-research, production and bulk-cargo usage period and activities during the mineral and oil and gas exploration, development, production and depletion period, port design and construction, manpower training and all other management organization and support activities.

With regard to time involvement, it has been stated as a policy objective that Canadian resources should be directed towards establishing and assuring Canada's marine transport pre-eminence and sovereignty in ice-affected oceans as soon as possible. Therefore, in the scale of time as related to policy formulation, we have assumed that an implementation schedule should be established which would effectively begin at once and proceed with full vigour to the achievement of its objective.

Space or Location. In this parameter we should describe the geographical or other limits to which policy or parts of the policy would apply. This relates to both macro-structural limits, that is, the outside limits of the organization's policy influence, as well as the micro-structural limits, including that of all sub-groups within the organization.

Geographical Coverage of Policy. To some Canadians the north
begins a few hundred miles north of the 49° parallel, whereas to
others the north begins at the 60° parallel. For this policy
proposal we have drawn largely upon the definitions of L.E. Hamelin
of Laval University. The conceptual area of concern here is shown
in Figure 1 superimposed on Hamelin's north. His definition is
based on a "Scale of Nordicity" and includes such factors as
latitude, climate, type of ice, precipitation, vegetation,
accessibility, population and degree of economic activity.
Hamelin's area of the north is roughly triangular in shape and
includes Ellesmere Island in the high Arctic, the Strait of Belle
Isle on the east and Prince Rupert on the west. The southernmost
point of its boundary is approximately one hundred miles south of
James Bay and includes that portion of the east coast of Canada
and its adjoining seas which extends southwards from the eastern
portion of the high Arctic Islands to the northeast coast of
Newfoundland, where the Arctic pack ice drifts. For all or a
substantial part of the year, much of the area is covered with
locally formed sea-ice, drifting Arctic pack and icebergs. These
conditions severely restrict the operation of the area's marine
transport and resource development. In our proposal we have
excluded fresh water river areas such as the Mackenzie River and
the St. Lawrence River and the Great Lakes. However, we have
included those areas of the Continental Shelf offshore from the
above land masses which are currently claimed as Canadian waters.
The area referred to here is about two million square miles or
more than half of all of Canada's land areas.

Matter. With regard to the material aspects of the policy,
we have assumed that the effort should be one in which the maximum
possible application of all Canadian resources and technology will
be employed. This parameter defines all the resources that would
be developed. It also includes resources that will be required,
such as the human, material, equipment, capital and technical
know-how, as well as the locations, numbers, types of persons and
organizations involved, the rate of expenditure of funds, and
other types of materials for various support activities.

Resources. In the 730,000 square miles of Canada's Eastern
Arctic alone, the short marine transport season, long periods of
darkness, extreme weather and terrain conditions, great distances
and difficult water and ice crossings between the islands, present
major transportation challenges. Nevertheless, this area contains
about one-third of this country's potential oil and gas reserves
as well as high grade iron ore and other minerals. If we include
the northeast coast hydro-carbon exploration area the estimate of
potential economic resources is more than doubled. In the Arctic
Islands current exploration and governmental as well as other
economic activity is in the order of two hundred million dollars
per year. As a basis of rate of growth comparison, we have used a
suitably modified western Canadian oil field growth pattern. Using
this as a guide we have estimated that just before production is to

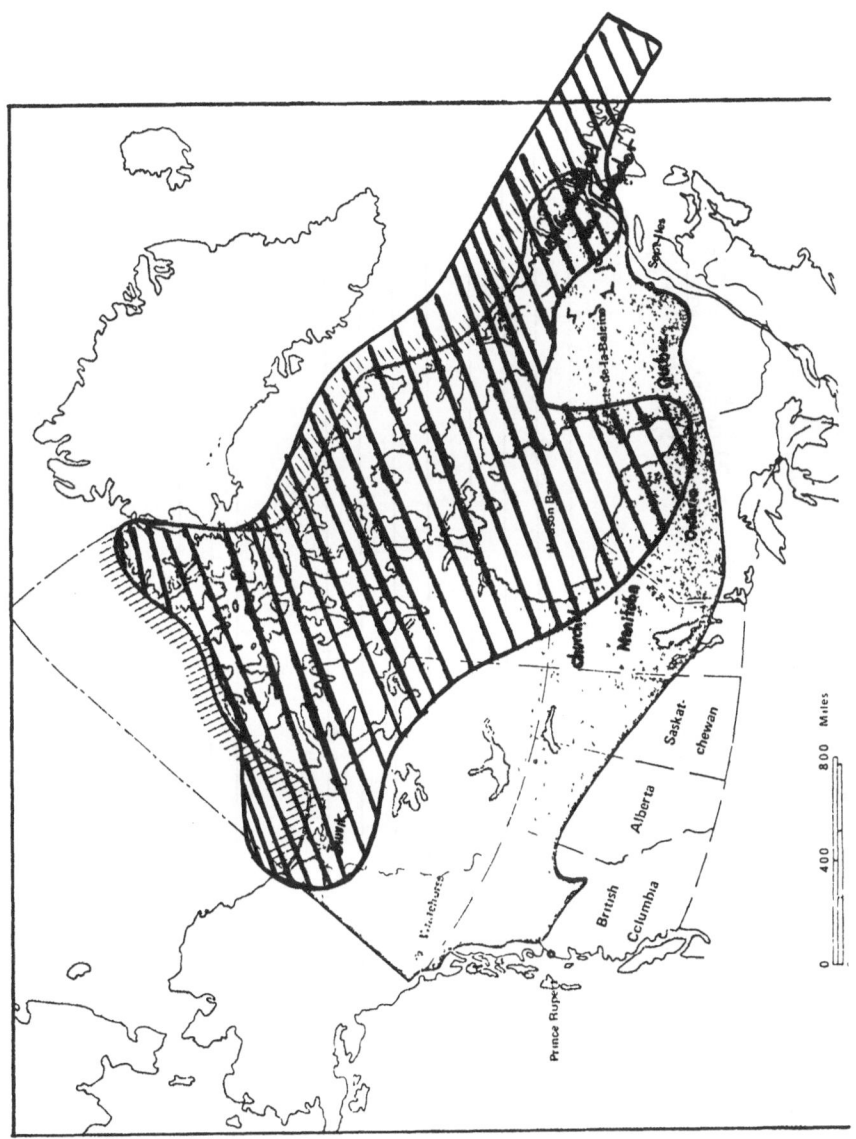

Figure 1. Canada's Ice-Affected Ocean Areas

Table 1. Eastern Arctic

Summary of Estimated Activity and Transportation Costs for Selected Years
($ millions per year 1972 prices)

	TOTAL VALUE OF ECONOMIC ACTIVITIES (In millions of $/yr.)				VALUE OF TRANSPORT PORTION OF ECONOMIC ACTIVITIES																
	1972 Expends. include Exploration only.	1980 Expend. include Development and Exploration	1990 Expend. here include Development and Exploration	1990 Produc. Revenue.	190,000 tons 1972 (In only) Sea, Air, Land. In millions of $/yr.	Air	Land	Total	700,000 tons 1980 (In only) Sea, Air, Land. In millions of $/yr.	Air	Land	Total	800,000 tons 1990 (In only) Sea, Air, Land. In millions of $/yr.	Air	Land	Total	67,000,000 tons 1990 (Out only) Sea, Air, Land, Pipeline. In millions of $/yr.	Air	Land	Pipeline	Total
OIL & GAS																					
1972 Expl.	85				10	15	5	30													
1980 Dev. & Expl.		300							30	25	22	77									
1990 Dev. & Expl.			330										20	22	15	57					
1990 Produc. Revenue				1250													469	–		141	610
MINING																					
1972 Expl.	1																				
1980 Dev. & Expl.		100							5	4	4	13									
1990 Dev. & Expl.			100										5	3	5	13					
1990 Produc. Revenue				70													30	1	4	–	35
GOVERNMENT & NATIVE POPULATION																					
1972	52				15	3	2	20													
1980		100							20	6	4	30									
1990			200										35	9	6	50					
Total	136	500	630	1320	25	18	7	50	55	35	30	120	60	34	26	120	499	1	4	141	645
GRAND TOTALS Millions $/Yr.	136	500	1950		50				120				765								

Source: Data compiled by T.W. Kierans

Table 2

Comparison of GSC, CPA, and CSPG Estimates
(based on 1973 GSC regions)

	Oil Potential				Gas Potential			
	CPA '69	GSC'72 (Est.I)	GSC '73 (Est.II)	CSPG(A) '73	CPA '69	GSC'72 (Est.I)	GSC'73 Est.II	CSPG(A) '73
	Billions of Barrels				Trillions of Cubic Feet			
1. Arctic Islands and Coastal Plain (North)	43.5	49.3	20.3	28.6	260.7	327.4	242.0	208.8[1]
2. Beaufort Mackenzie		14.7	6.2	8.0		117.2	93.5	64.0
3. Western Canada	47.4	28.6	22.2	25.0	283.8	207.4	120.3	155.5
4. Offshore East Coast	24.8	38.5	47.5	22.1	149.9	229.6	307.1	132.6
5. Hudson Platform	2.9	1.5	1.5	[1]	17.4	8.7	7.3	[1]
6. Eastern Canada Onshore	2.3	1.8	1.5	1.5	13.0	15.8	12.7	16.6
Totals	120.9	134.4	99.2	82.5	724.8	906.2	782.9	577.5

[1] Potential for Hudson Platform included in Arctic Islands and Coastal Plain (North).

NOTE: Estimates all have slightly different basis in that areas considered, limit of depths of sediments or depth of water on continental slopes are not the same in every case. The derivation of each of the above sets of estimates is discussed in Appendix A.

Source - Canadian Energy Policy 1973.
Canadian Department of Energy, Mines and Resources.

Table 3. Summary of Known Mineral and Oil and Gas Locations in the
          Eastern Arctic Excluding Labrador Coastal Area

Commodity	Area	Estimated Reserves (Short Tons)	Estimated Annual Production (Short Tons)	Life of Reserve	Probable Best Transportation Mode
Iron Ore	Mary River	600 million tons	8 million tons	+30 Yrs.	Surface Ship.
Copper	Coppermine	4 million tons	20 thousand tons Concentrate	+10 Yrs.	Surface Ship or Barge
Lead/Zinc	(Little Cornwallis (Island	+20 million tons	300 thousand tons Concentrate	15 Yrs.	Surface Ship.
	(Strathcona (Sound	12 million tons	60 thousand tons Concentrate	15 Yrs.	Surface Ship.
Sulphur	Axel Heiberg Island	10 million tons	500 thousand tons	20 Yrs.	Surface Ship.
Copper/Nickel	Tehek Lake	25 million tons	500 thousand tons Concentrate	13 Yrs.	Ground Conveyor and Surface Ship
Oil/Gas	Arctic Islands & Coastal Plain	43 billion bbls.oil 250 trillion cu. ft. gas	50 million tons 2 million tons liquefied gas		Pipeline and Surface Ship.
Oil/Gas	Melville Island	500 million bbls. 3 trillion cu. ft.	3 million tons 2 million tons liquefied gas		Pipeline and Surface Ship
Oil/Gas	Axel Heiberg	500 million bbls. 3 trillion cu. ft.	3 million tons 2.4 million tons liquefied gas		Pipeline and Surface and Ship

Source: Author's Estimates Using Available Sources of Information

begin in 1980 the value of economic activity will reach at least
five hundred million dollars per year. By 1990, when production
is in full swing in the Arctic Islands, this will increase to a
total of two billion dollars per year considering both ingoing and
outgoing ocean transport. Using experience only as a guide in
estimating the value of the transport portion as a percentage of
the total economic activity for the years shown the economic value
of in and out marine transport has been estimated to be about $560
million per year by 1990.

    <u>Energy (Consequences)</u>. Here we should consider the dynamic
and static effects or the consequence of each of the proposed
policy alternatives, as they affect other organizations, both
outside and inside the policy making group. This includes the
economic, technological, political, environmental and social
interaction between people of the Northern Territories, the people
from the south, the developers, the users, the researchers, the
explorers, the financiers, governments and the outside world
organizations, law of the sea and the environmental regulatory
bodies. With regard to the consequences, or in this context the
"energy" of a chosen policy, it is apparent that our most important
concern should be to consider the consequences to Canadian
sovereignty of having no effective policy or a policy error. This
potential consequence is unfortunately already evident as we see
marine development steps taken by the other nations which will have
a direct bearing on Canadian sovereignty in these areas.

    <u>Unknowns</u>. It is vital to recognize the limitations of
knowledge. Even though advancing science and technology often
permit intelligent and quantitative forecast of future events,
there is always the unexpected, and the unknown event that might
take place and this should be emphasized in the synthesis of policy
alternatives. It is possible that other sources and/or other
locations of hydro-carbons may be found and therefore could have
adverse effect on the policy. Flexibility should therefore be an
important consideration.

    With regard to unknowns, it is evident that the major unknowns
are those associated with Canadian motivation to develop and manage
our own resources. In addition we have economic factors in terms
of exploration, production and production scheduling, pricing and
markets; as well as contingency and escalation factors in budgeting.
These, in turn, should be considered in the light of both production
costs and current and future sales value and reserves allocations
for foreign and domestic markets. We have included an estimate of
eastern Canadian Arctic resource development and other economic
activities, as well as the requirement for other transport in the
north. Anticipated growth in the Arctic is based on an evaluation
of 1972 Arctic exploration activities and results in which these
are related to growth rates from the 1946 Alberta oil and gas
discoveries to the present Alberta production rates and revenues.
This evaluation includes consideration of:

(a) the current world hydro-carbon demand, costs, and market
    prices, and
(b) the relative difficulties associated with the
    exploration development, production and transport in the
    Arctic environment as compared to the 1946 Alberta
    environment.
In this first step the estimates developed should be refined to
include high, low and medium calculations.

Step 2: Define Groups Involved in the Policy Formulation Procedure

     Groups Involved. In general terms the group or groups of
people who will be involved in Canada's ice-affected ocean
transport policy alternatives are:
1. Public Groups
    Federal and provincial governments, and many of their numerous
    agencies and regulatory bodies. External governments and
    international bodies such as the United Nations, etc. would
    be included in this sub-section. The native people and
    residents of the Northwest Territories and the provincial
    areas having ice-affected ocean transport requirements.

2. Corporate Groups
    Resource Corporations - Canadian, multi-national and others,
    Ocean transport vessel and equipment manufacturers,
    Bulk and general cargo carriers,
    Engineering firms,
    Private consultants,
    Finance companies,
    Marine Insurance companies,
    Construction companies.

3. Institutional Groups
    Research and development institutions, universities,
    conservation organizations, etc.

4. Cooperative Groups
    Production and consumer cooperatives, labour unions, etc.

              Step 3: Basic Design Criteria of an Ideal Canadian
                 Ice-Capable Ocean Transport Management System

     An integrated Canadian ice-capable ocean transport management
system should be considered as the base for the development of a
total transportation management system for Canada's ice-affected
ocean regions. It is not intended that the system would cause
economic injury to existing sea, air or land carrier companies
currently engaged in the various stages and modes of transport.

However, its basic goal should be, in due course, to integrate all transport modes into a coherent, efficient whole, designed to provide <u>maximum efficiency and reliability in the delivery and pick-up of goods on behalf of all users of transport from point of origin of material or equipment to point of use or point of delivery. It should seek to encourage competition among competent long-term carriers.</u> The system should be representative of social and economic interests in that order of priority. A separation of transport coordination management and carrier ownership and management seems to be indicated in order to avoid conflicts of interest.

<u>Proposed Area of Operations for the Ice-Capable Ocean Transport System.</u> The area of operations of the proposed transport system would, in general, be all locations in Canada as shown on the geographical coverage map, Figure 1.

<u>Specific Requirements.</u>

1. Lowest possible transportation costs for delivery of inbound and outbound goods via all necessary transportation modes from any of several specified locations in Canada and elsewhere to the ice-affected regions and at specified and dependable rates and schedules.

2. Full amortization of the most suitable and efficient types of ocean transport equipment within a contract between manager and carrier.

3. Provision for high quality transportation service to native communities and small settlements in order to ensure social as well as economic benefits to the natives and residents of the regions under development.

4. Provision of suitable inter-modal transfer bases in the north and coastal areas as well as system insertion bases in Canada close to the points of origin of supply of goods to be delivered to the Arctic. This would include sea-air bases, warehouses, lightering craft, wharves, heavy-lift facilities and containerization and freight consolidation facilities.

5. Standards of transportation suitable to the protection of the northern environment.

6. Standards of packaging suitable for Arctic marine transport.

7. Information service to users regarding the movement and location of their goods within the ocean transport system.

8. Insurance, forwarding and customs clearance facilities as required.

9. Flexibility in freight consolidation so that reasonable changes

in priority of delivery of a user's goods may be effected.

10. Warehousing at trans-shipment locations of certain staple goods such as fuel, cement, etc.

11. Communication facilities for transportation management purposes.

12. Adequate hotel accommodation, at trans-shipment locations.

13. Up-to-date forecasts and studies of future transportation requirements and plans in order to see that growth requirements are promptly and efficiently served.

14. Coordinated and efficient development of airports, docks, townsites, and campsite facilities.

15. The provision of coordinated manpower education and training facilities for personnel.

16. Priority as to the employment of indigenous workers and residents.

17. Protection of public interest by means of suitable regulatory bodies.

Step 4:   Some Policy Options for a Canadian Ice-Capable
          Ocean Transport Management System:
          Possible Alternatives

Policy Alternative A (continue current policy).
Continue the present system with Department of Transport taking responsibility to improve ocean transportation services through contracts with carriers and load consolidation services to native and other communities. Other users such as resource corporations and large communities will be allowed to continue to make separate contracts with individual carriers. Ice-capable services for the latter groups will be arranged with the Department of Transport on the existing availability basis. Ports and other infra-structure services will be provided by a combination of local and/or corporate and federal government cooperation.
Policy Alternative B (carrier managed systems).
Establish a number of separate ocean transport systems with exclusive franchises for specific areas of operation, each controlled by a marine carrier or joint venture of carriers. Each group would manage its own port facility, transfer location, ice-capable and infra-structure services in cooperation with other competitors, resource corporations or cooperatively with the federal government.

Policy Alternative C (resource corporation managed systems).

Establish multiple ocean transportation systems, competitively directed or contracted for by major resource corporations, possibly including some associated municipal users. Each transport system would manage its own port facilities, transfer locations, ice-capable and infra-structure services in cooperation with other competitors, or cooperatively with the federal government.

Policy Alternative D (government managed system).

Establish an ocean transport corporation controlled and managed by a government agency or agencies or departments. This agency would own or contract for transport vessels, port facilities, transfer location, ice-capable and infra-structure services and all other support services required for the ocean transport operation.

Policy Alternative E (cooperative user managed system).

Establish a voluntary, user controlled, user directed Canadian ice-capable management authority. This authority would contract with carriers for transport by ice-capable ocean transport vessels. It would own or manage port facilities, transfer locations, infra-structure services and all other support services. It would co-ordinate all contractual activities required for ice-capable ocean transport operation for its members in Canadian ice covered oceans. Under suitable government regulation it would be granted exclusive transport management authority in specified areas with the right to negotiate long-term, open-tendered contracts with carriers with the view to full amortization of the most suitable and efficient marine transport equipment. It would be regulated to avoid conflict of interest by having financial investment in carrier operations.

Step 5: Preliminary Implementation Plans

Policy - Plan Activities: refer to Step 5 Outline (Figure 2)

1-4: Data input from groups involved and their objectives.
Gather data on groups involved and their objectives:
   i) Native people and residents,
   ii) People of Canada,
   iii) Governments - Regional and Federal,
   iv) North West Territories Government,
   v) Resource Corporations,
   vi) Bulk and Cargo Carriers (i.e. shippers),
   vii) Construction Companies,
   viii) Research Institutions - Universitite, Trades and
         Technical Colleges, Arctic Institute
   ix) Environmental protection agencies - government & private
   x) Natural Resource Development and Management agencies
   xi) Transportation Development agencies
   xii) External Affairs
   xiii) Arctic Transportation agencies
   xiv) Ministry of Transportation - air, sea, land
   xv) Ministry of State, Science and Technology
   xvi) Ministry of Defence
   xvii) Department of Industry, Trade and Commerce

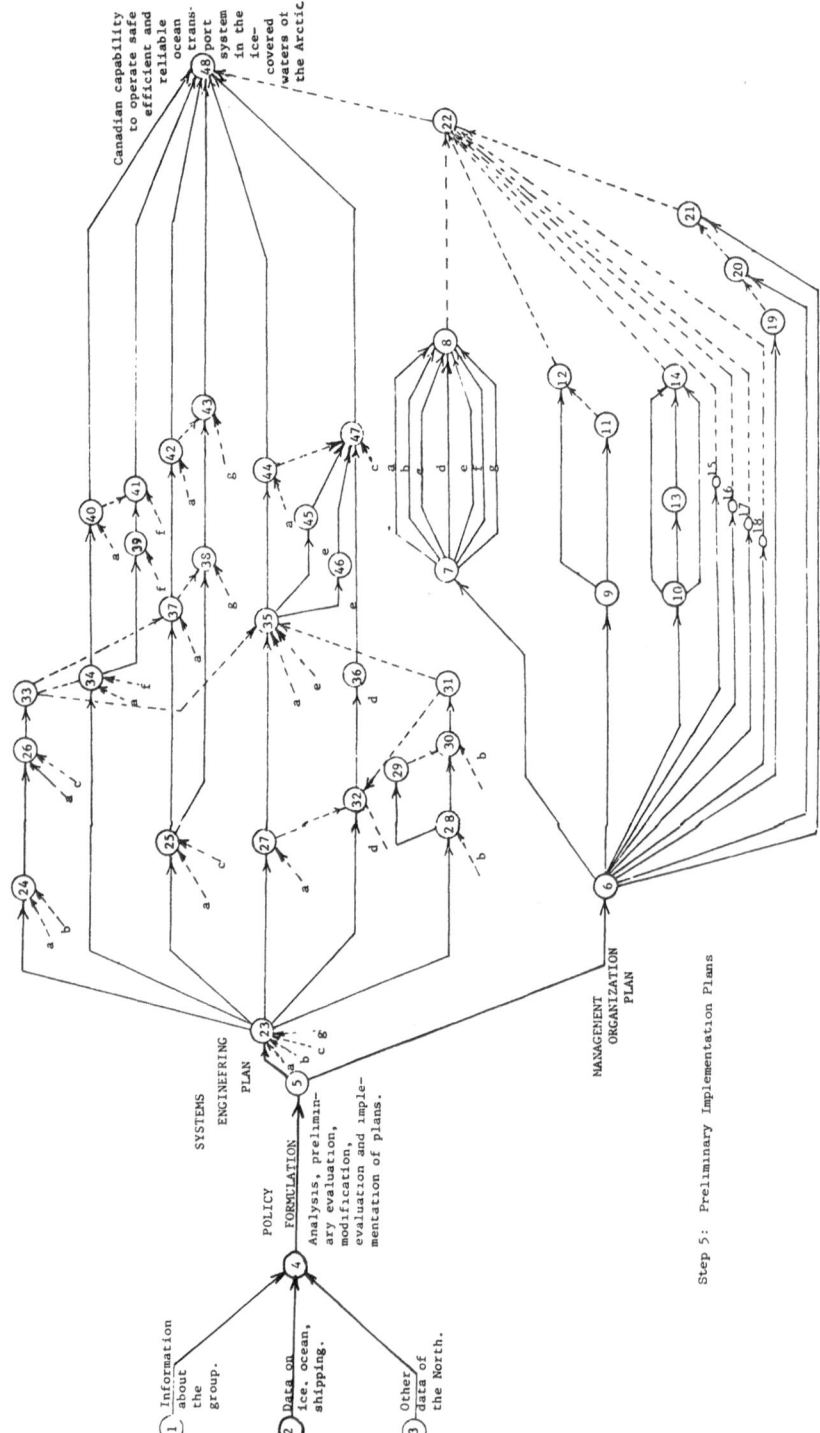

Figure 2.   A Proposed Canadian Ice-Breaking Ocean Transport Policy – Management Organization and Systems Engineering Plans.

xviii) Other governmental regulatory agencies directly
or indirectly involved in the Northern transportation
   xix) Private Finance Companies
    xx) Marine Insurance Companies
   xxi) Other interested groups directly or indirectly
involved - private, public, foreign.

2-4: Gather all existing data input and research on ICE,
OCEAN, SHIP, CLIMATIC, GEO-TECHNIC, TOPOGRAPHIC,
OCEANOGRAPHIC, HYDROGRAPHIC and other information
broken down to (a) High Arctic, (b) Sub-Arctic,
(c) Offshore , and (d) Inshore.

3-4: Other data input and research information of the north.

4-5: CANADIAN ICE-CAPABLE OCEAN TRANSPORTATION POLICY -
formulation, analysis, preliminary evaluation,
preliminary policy recommendations, test implementation,
evaluation, modification, final policy collection,
final plan, and continued updating, review and
modifications.

5-6: SELECTED MANAGEMENT ENGINEERING PLAN APPROVAL.

   6-7: Economic and Financial Management Plans.

      7-8(a): Estimate capital for research and data input.
      7-8(b): Estimate capital for manpower training
         and education.
      7-8(c): Estimate capital for field studies.
      7-8(d): Estimate capital for ice breaker construction.
      7-8(e): Estimate capital for bulk and cargo carriers.
      7-8(f): Estimate capital for exploration and
         extraction gears.
      7-8(g): Estimate capital for other systems support
         and management support.

   6-9: Resource Development and Management Plans.

      9-12: Non-renewable resource management.
      9-11: Renewable resource management.

   6-10: Environmental Protection Management Plans

      10-14(a): Marine protection plans.
      10-13: Pollution control plans.
      13-14: Oil spill clean-up plans.
      10-14(b): Animal life protection plans.

6-15: Information services.

6-16: Indigenous and resident population services and benefits.

6-17: Navigation management.

6-18: Legal and Jurisdictional management plan.

6-19: Air-Sea Rescue management.

6-20: Naval Support management.

6-21: Administrative support management

5-23: SELECTED SYSTEMS ENGINEERING PLAN APPROVAL

23-24: Data input on manpower with scientific know-how.

    24-26: Set up new or expand existing education and
            training.
    26-33: New graduates.

23-34: Data input and research for exploration for extraction
gear technology.

    34-40: Further data input and research for extraction
            gear technology.
    40-48: Continuation of further data input and research.
    34-39: Design of extraction gear.
    39-41: Construction of extraction gear.
    41-48: Trial test of extraction gear.

23-25: Data input and research for port design.

    25-37: Further data input and research for port design.
    37-42: Continuation of further data input and research
            for port design.
    42-48: Continuation of further data input and research
            for port design.
    25-38: Preliminary analysis of port locations and routes.
    38-43: Design of ports, docking and materials handling
            facilities.
    43-48: Construction of ports, docking and materials
            handling facilities.

23-27: Data input and research for vessel design.

    27-35: Further data input and research for bulk and
            cargo carriers.
    35-44: Continuation of further data input and research.
    44-48: Continuation of further data input and research.
    35-45: Design of bulk carriers for the north.

45-47: Construction of bulk carriers.

35-46: Design of cargo carriers for the north.

46-47: Construction of cargo carriers.

23-32: Field study and research in the north with existing icebreakers to test breaking capabilities.

32-36: Design of new icebreakers based on new data.

36-47: Construction of new icebreakers.

47-48: Field study and test of new icebreakers and carriers in the northern ice-covered waters.

23-28: Data input on manpower with technical know-how for vessel design and construction.

28-29: Set up education and training for trade personnel.

28-30: Set up education and training schools for marine navigation and naval architecture.

30-31: New naval architects, marine engineers, technologists and technicians.

Step 6:   Natural Constraints

Major natural constraints are shown in the chart (Figure 3),
others should be listed.
Man-made Constraints
   - Jurisdictional and BNA Act
   - Resource markets - resource prices - values - costs of
     production - value added processes
   - existing terminal facilities
   - existing equipment
   - see Arctic Resources by Sea by Northern Associates (Holdings)
     Ltd.
   - etc.
Supply Bases. Fundamental to Canadian ice-capable ocean
transport management policy is the vital linkage between marine
supply bases in the south and points of delivery to marine air
bases in the north. These southern Canadian ports which currently
serve the transport supply bases for the far north such as Montreal,
St. John's and Churchill (Manitoba) and in the north marine air
bases such as Resolute, Frobisher and even Eureka should be
considered. Further, in discussing ocean transport to the north,
it is necessary to understand the close relationship between
seaport facilities and on-going air transport needs. Thus, it is
almost a sine qua non of Arctic port facilities that they should
be closely associated with good air bases. The investigation of
new northern terminal sites such as Radstock Bay on Seven Island
is an important consideration in developing implementation type
policies.
Transport. Currently Arctic transport delivery is highly
irregular and, notwithstanding the excellent efforts of many
carriers, difficult natural conditions can only be satisfactorily
dealt with by means of a substantial increase in new transport and
terminal facilities. These are difficult to finance and insure,
particularly in the face of short-term carrier competition. Firm
long-term contracts with carriers on a "chosen instrument" basis
designed to insure amortization of the most suitable equipment for
specific routes or transport modes provide the logical answer to
this problem and should be welcomed by competent carriers. In
addition an integrated transportation management system with
proper sea-air bases for necessary inter-modal transport inter-
changes is required. With load consolidation and warehousing·
facilities at suitable transport system insertion points and inter-
modal transfer bases, substantial improvements in rates and
schedules to benefit Arctic transport users should be expected.
In order that this may be done effectively, and without concern on
either user or carrier, a voluntary user management system should
be considered as a possible alternative. This is described in
more detail in this proposal.

Step 6: Natural Constraints

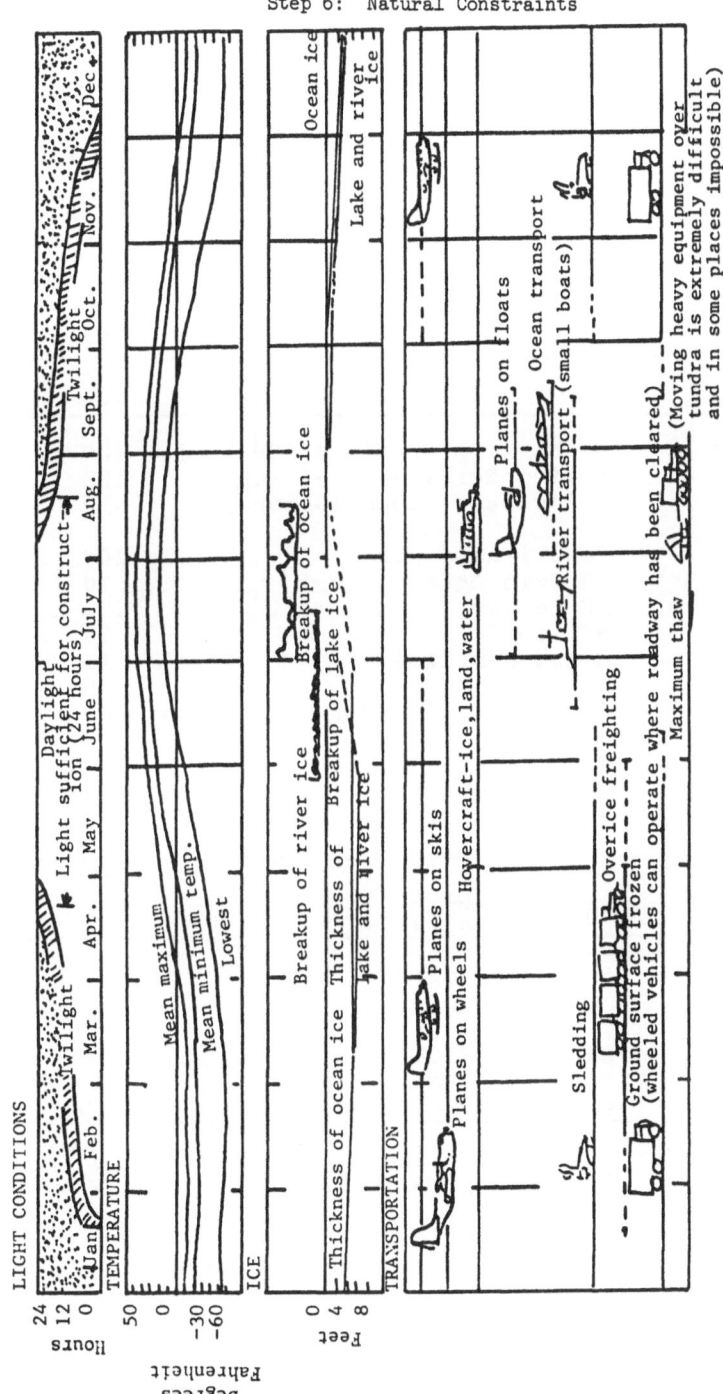

Source: Pacific Hovercraft Ltd., Vancouver.

Financial Post Nov. 7, 1970.

Figure 3. Arctic Transportation Problem. – U.S. and Canadian Experts Investigate New Cargo Moving Methods

Table 4

Comparitive Ton-Mile Rates for Different Transportation Modes

Mode	Vehicle or Craft	(1972 estimates) Cost/Ton Mile (in cents)
Air Cushion Vehicles	SNR 6	100
	Hovermarine HM$_2$ (sidewall type)	50
Tractor Train		100
Aircraft	G. Helicopter (load 500 lbs)	400
	Helicopter (load 3-4 tons)	125
	Helicopter (load 10 tons)	50
	DeHavilland Otter	100
	DH Twin Otter	70
	DH.Beaver	130
	Bristol	30
	C 130 Hercules	15
	Boeing 747 f	5
	Boeing tilt wing (20 ton load)	30
Truck	Std. highway vehicle on winter road	20
	Std. highway vehicle on highway	8
Rail	Std. equipment (northern railways)	7
Marine	Ships on northern supply operations	5
	Barge in Mackenzie	4
	Super tanker in Arctic	?
	Cargo submarine in Arctic	?
Pipeline	Standard in south	1
	Northern pipeline	?

Note: the ton-mile rates for aircraft and ACV's would
      double if the craft returns with no load.

Source: North Magazine June 1969

<u>Current Status</u>. At present, Canadian ice-covered ocean
transport to re-supply eastern Arctic communities is arranged by
DOT chartered vessels with the assistance of government icebreakers.
While some load consolidation takes place at the southern ocean
terminals, it is not fully integrated. Many industrial users such
as oil and mining companies arrange their own transportation. The
current status of ocean transportation in the eastern Arctic has
been described by Captain Michel Dussault, Director of Marine
Pilotage for DOT at a meeting before the Society of Naval
Architects and Marine Engineers of eastern Canada on April 18,
1972 as follows:

"Looking at commercial marine transportation in the Arctic,
past and present, very definite facts stand out. Excepting a few
modern and ice strengthened tankers and cargo vessels of the
coasting type, the fleet involved is a non-descript mixture of all
tonnages, structure and power. Several vessels are old and poorly
equipped for the trade in which they are engaged. Insurance rates
are rising fast and are becoming a serious problem to some owners.
Many vessels and some of the services strike me as somewhat more
of an opportunist operation than a genuine and special trade with
proper and experienced personnel, good ships and equipment at a
fair price to the shipper and a reasonable return to the operator.
However, serious marine operators are showing interest and, most
encouraging, some experimentation has taken place in the last few
years such as containerization and sky crane operation. But, on
the whole, too little and in the future much more will have to be
done."

In 1972 marine costs to users from Montreal to the beach at
Resolute were about $175.00 per ton. The bulk of fuel supplies
goes to Resolute. General cargo is unloaded on beaches by
lightering barge. On-going transportation to another point of
delivery or to point of use is arranged by the users of transportation services on a charter basis. More and much improved sea-air
transfer locations are needed.

The key to ensuring that northern economic growth and the
solution of that area's difficult transport problems on behalf of
the natives, residents, developers, and Canadians at large, seems
to lie in the establishment of an effective user-oriented ice-
capable ocean transport management system, capable of ensuring
Canadian sovereignty and leadership within its jurisdiction.

Step 7: Screening and Evaluation of Policy Options

In this proposal we are not attempting to make firm conclusions
on final policy selection. This brief discussion is inserted in
order to assist an understanding of the complexity of the problem
and to indicate the need for much more detailed and systematic
consideration of all policy options.

If we accept the premise that existing principles for selecting
an ice-capable ocean transportation management system are inadequate
to deal effectively with the far north's transport needs now and
in the future, then, without going into a complete analysis at
this time, the merits and demerits of the five proposed policy
options are briefly discussed.

Alternative A: This alternative would merely attempt to
revise the existing inadequate system without coming to grips with
the fundamental requirement that the system should be designed to
serve both social and economic users as the system expands to 1990
needs, and that it clearly establish Canadian sovereignty.

Alternative B: This alternative has the major defect that it
would be essentially designed to provide profit to carriers not to
users. Because of the competitive allocation of geographical
areas, the long term financing of the most effective equipment for
a specific route or task could be difficult and the selection of
equipment could be directed towards greatest short term carrier
profit rather than total user service.

Alternative C: This alternative would have the major defect
that it would be designed to efficiently serve high economic
priorities of resource corporations but not necessarily the social
priorities, native populations or long term national interests.

Alternative D: This alternative has the major defect that it
would be designed to serve the political interests of all of
Canada and not necessarily the natives, residents and developers
of the eastern Arctic. It would also suffer from interdepartmental
communication problems frequently evident in government controlled
public service activities.

Alternative E: While no system is perfect, this alternative
appears to have sufficient built-in safeguards to prevent abuse by
any section or group of transportation users. It also covers a
sufficiently broad area so that the financing by both the system
itself and the carriers involved should have the least possible
difficulty. It would require approval and support by the Federal
Government on a trial basis for a relatively small area. It could
then expand to cover all the eastern and northern Arctic as
experience indicates. It could also be introduced on a voluntary
self interest basis.

Step 8: Preliminary Policy Alternative Selection

Within the terms of reference of this submission which are
essentially to indicate a procedure for formulating, analysing and
evaluating alternative Canadian ice-capable ocean transport
management policies, we recommend, at this preliminary stage, that
Policy Alternative E would appear to most effectively serve all
groups of users.

Alternative E (modified): Establish a voluntary, user-
controlled, user-directed ice-capable ocean transport management

authority. This authority would contract for all ice-capable
ocean transport vessel activity and may own or contract for the
required port facilities, transfer locations, infra-structure
services and all other support services. It would manage the
coordination of all activities required for ice-capable ocean
transport operation for its members in Canadian ice-affected oceans.
It would establish suitable standards for vessel design and
activity, goods packaging and the loading, unloading and intermodal
transfer to and from ice-capable ocean vessels. It would be
regulated to avoid conflict of interest with ownership of carrier
interests and other interests such as environmental authorities.
The authority would recommend to appropriate governmental
regulatory bodies, northern and southern ice-free or relatively
ice-free bases, intermodal transfer bases, navigation aids, and
ice-affected ocean shipping routes.

As stated elsewhere, this proposal is intended by the authors
solely to demonstrate a policy formulation and analysis procedure.
It is not a complete analysis or evaluation. A detailed study of
such a policy analysis is estimated to require approximately three
years with a staff selected and trained to rigorously follow the
iterative and filtering process indicated. On the basis that
there may be six policy alternatives to investigate in detail, the
estimated cost could be approximately $6,000,000.

A detailed simulation of this modified policy option
implementation should be undertaken with new implementation plans
prepared as in Step 5.

Step 9: Simulate Implementation of Selected Policy

Preliminary Schedule and Costs for Formulation, Simulated
Testing and Implementation Phases for Policy Formulation and
Analysis Procedure for Canadian Ice-Capable Ocean Transport
Management. (See Figure 4)

Step 10: Conduct Limited Tests

These policy implementation tests should be conducted on a
voluntary basis as follows:
(a) for a specified test period of time,
(b) for a selected geographic area,
(c) for a selected portion or type of marine transportation
(d) for some other arrangement.

These limited tests should be carefully monitored to provide
information for modifications to the policy as it is more widely
extended.

TIME = 0.00   Jan 1976   Step 9: Simulate Implementation of Selected Policy   1990

Gather Data on the
Policy Making Groups,
the internal Groups
affected by the Policy
and the external
groups affecting the
Policy

Canadian Ice-
Breaking Ocean
Transport Policy
Formulation, and   Dec. 1978
Preparation of
Management Organ-
ization and
Systems Plans

All Existing Data
on Ice, Ocean,
Ship

All other Exist-
ing Relevant Data

Evaluation   Obtain   Test Imple-
and Pre-   Voluntary   mentation,
liminary   support of   Evaluation
Selection   Users   and Modifi-
                                    cation

                                                    Final
                                                    Policy
                                                    Selection

                                                    Provide Capital
                                                    for carrying out
                                                    certain activities
                                                    for action.

Field Study - Existing ice-breaker
trial run to test capability in
Norhtern ice-covered waters.

On Going Work & Research

Further Data for Vessel Design - ice
breaker, bulk & cargo carriers.

On Going Work & Research

Further Data for Port Location,
Design, Docking & Handling facilities.

On Going Work & Research

Further Data for Exploration and
Extraction Gear Gechnology

On Going Work & Research

Plan Manpower
Education and
Training Programmes

Expand existing and or
set up new institutions
for Naval Architecture,
Marine Engineering, Ice
Navigation, Research
Scientist Technologist
and Trades Personnel.

Figure 4.   Data Gathering Steps in the Formulation and Analysis of Policy Alternatives Procedure
for Canadian Ice-Breaking Ocean Transport Management.

Step 11:   Implementation - Regulation Operations

Select final policy and fully implement it with suitable
flexibility to permit necessary adjustments for which the early
implementation period indicates the need. Regulatory bodies
develop and maintain suitable monitoring and surveillance systems
as required.

Proposed Steps to Initiate and Carry Out Full Policy Study

If Canada is to establish her.pre-eminence and sovereignty in
the north, where ice-capable ocean transport facilities are
currently inadequate, then Canada should act now, for already there
is evidence that development steps have been taken by other
countries which may endanger Canadian sovereignty and leadership
in the area. The development of this Canadian ice-capable
transport management policy formulation system with its analysis
and preliminary evaluation procedure has been outline only.
Suggested phases for initiating this policy study are briefly
stated as follows:
Phase I - Schedule of Proposed Activities and Costs to
          Initiate Support for Full Policy Study
          (July 1975-January 1976)

Phase II - Schedule of Proposed Activities and Costs
           for a Full Policy Study (January 1976-
           December 1978)
The steps and costs for each of the above phases are shown
briefly in the following two schedules.

Table 5

Canadian Ice-Capable Ocean Transport Management Policy

Stage I – Schedule of Proposed Activities and Costs to Initiate Support for Full Policy Study
(July, 1975–January, 1976)

Step	1975 Dates		Costs $
1	July 1975	Submit revised draft to selected governmental and industrial groups with suitable discussion arrangements to permit revisions as required	$ 3,000
2	Sept. 1975	Conduct Canadian ice-capable ocean transport management policy meeting in St. John's. From this develop plans for full joint study by government and industry.	$15,000
3	Dec. 1975	Revise proposal and seek joint support and approval from federal and provincial governments and from industry involved for a full study	$ 5,000
			$23,000

Stage II – Schedule of Proposed Activities and Costs for a Full Policy Study (Jan. 1976–Dec. 1978)

Steps	Dates		Costs $
1	Jan. 1976	Final definition of objectives - scope - terms of reference with federal and provincial governments.	
2		Definition of groups involved - decision making - internal - external with their objectives relative to Step 1.	
3		Conduct resource inventory in area of concern	
4		Conduct ice-capable ocean transport environment assessment	
5		Gather existing information - forecasts, schedules, volumes, technology, man-caused constraints - natural constraints. Existing ports - aids - vessels - technology etc.	
6		Define function of management of CICOT	
7		Define design criteria of management system with requirement schedule for readiness - capabilities. Budgetary conformation.	
8		Outline policy alternatives in preliminary formulation in terms of TSMEU.	
9		Outline alternative plans within each policy alternative.	
10		Outline plans in terms of management organization and engineering systems according to TSMEU.	
11		Screen and filter plans (a) within boundaries, (b) by moving boundaries, (c) by moving plans.	
12		Evaluate successful plans according to design criteria.	
13		Simulate successful plans under distance, 'service' and extreme load conditions.	
14		Formulate and present successful policies and plans in detail for selection by those with decision making authority.	
15		Conduct simulation and limited test of selected policy or policies.	
		TOTAL	$6,000,000

Summary: Stage II

Estimated Time: 36 months to study in full detail concurrently six policy options with a staff of 40 - 50 trained technical experts

Estimated Cost: $6,000,000
Staff of 50 technical experts

APPENDIX
BRIEF DESCRIPTION OF POLICY ENGINEERING
AS IT IS BEING DEVELOPED AT
MEMORIAL UNIVERSITY OF NEWFOUNDLAND

Introduction

One of the major characteristics of the modern world is the
steadily accelerating integration and interweaving of its variant
cultures, social systems and philosophies within the political,
economic and communal organizations which comprise human society.
This blending process is common to virtually all governmental,
institutional, corporate and cooperative groups. Constraints on
the process may be natural such as distances, climate or human
behaviour patterns (e.g. self-preservation, the search for
knowledge). There are also manmade constraints, such as legal,
jurisdictional or economic.

Bearing in mind our dimly understood origins, history and
ultimate destiny, the intermingling process seems to be inevitable
and, in the long term, beneficial. In the short term, as social
organizations attempt to develop, and implement goals and policies
related to their own welfare, we can see disturbing confusion and
incompatibility. In addition, there are conflicts of interest
which in some cases cause serious social unrest and even violence.
Witness our current confusion with regard to world economic value
systems, natural resource sharing systems, quality of life
problems, management and labour relations, offshore jurisdictional
conflicts, and population density and aging difficulties. This is
not to mention resulting armed conflict and human deprivation and
oppression in many parts of the world.

The modern, fast pace of this process seems to have resulted,
to a large extent, from recent engineering developments particularly
in the fields of communication, industry, transportation and
education.

Much has been accomplished in the science of achieving common
social objectives since policy was the prerogative of the tribal
chief, or the "divine right" of a king or the "right" of a
physically dominant group. Spiritual and cultural development
resulting from special environments or historical occurrences has
often been the basis for policy in a particular society. The more
modern, "51% of the voting strength" influence has ensured a
democratic approach to policy making within certain jurisdictions.
However, this has not guaranteed that the best policy for any
social group will be formulated or implemented, or even clearly
stated.

The availability of single brilliant minds capable of naturally
or intuitively resolving policy conflicts seems to have been out-
paced by the number of social policies which now urgently require
clear, compatible, and progressive formulation. What appears to
be lacking is some universally acceptable process for stating

objectives inherent to a group of people and for developing,
implementing and evaluating alternative policies applicable to the
objectives and compatible with the objectives of other groups.
As well, there is need for a set of guidelines for the education
and training of people competent to apply such a process. For
example, good progress has been made in such areas as the development of universally accepted cost accounting and data storage and
retrieval systems. Therefore, even though policy engineering may
be far more complex, it seems reasonable to look towards the
ultimate, multi-discipline development of a universally acceptable
system for social policy formulation, evaluation and selection by
those having authority.

The most important aspect of policy engineering, as it is
currently developing at Memorial University of Newfoundland, Canada,
is that it represents the multi-discipline application of the
natural law, as defined by the scientific method, to the synthesis,
analysis, evaluation, implementation and modification of policies
for any group of people sharing common primary aspirations and
objectives. Here objectives are assumed to be the inherent natural
goals of the group. These objectives simply require clear
statement, not evaluation. They must precede policy; they have no
alternatives since they are inherent to the particular group for
which policy at a given time is being developed.

Policy engineering should not be confused with some of its
related engineering technologies or its components such as
"management engineering", which deals with the organization of
people designed to implement policies, nor with "systems
engineering" which refers to the technical tools and systems which
management uses in implementing a policy. The basic elements of
natural law, namely time, space, matter, energy, and entropy (or
the measurement of uncertainties or unknowns) are the basic
elements of policy engineering.

The goal of policy engineering is to present in a universally
accepted format, the policy details and alternative policies and
plans which relate to the objectives of any social unit. The policy
engineer should be able to do this clearly and with adequate
distinction between known and unknown elements. In some ways, it
could be comparable to the almost universally accepted format in
which auditors present the financial statements, or economic facts,
of social groups and organizations.

Policy engineering should represent the practical application
of the collective knowledge of all professional, social, scientific
and technical disciplines. As such it must be multi-discipline in
character.

Much more basic work is required to advance policy engineering
to the status of a universally accepted technology. Nevertheless
even at this early stage of development it may be stated with
reasonable certainty that both the challenge and the promise will
ensure that increasing numbers of engineers will apply their best
skills, imagination and effort toward the continuing and future

development of policy engineering.

The Policy Engineering Thesis

Animals, humans, the earth, the universe and the entire cosmos, both microscopically and macroscopically, appear to function in accordance with a totality of order which we call the natural law of the cosmos. For the purpose of policy engineering the natural law of the cosmos is defined as the totality of scientifically demonstrable order. This order includes that which relates to time, space, matter, energy and the unknown, including uncertainties. It includes incompletely understood observations of order which have recognizable and measurable effects on the motivation or behaviour of humanity singly and in groups. This includes an understanding of the relationship of personal faith in man's existence. Above all, man's greatest concern in unravelling the secrets of the natural law is the search for knowledge about himself and his relationship to his fellow humans.

Man came into existence and is sustained in existence, within the cosmos, in accordance with and in conformity with the natural law, including the above mentioned unknown forces. Were he to proceed contrary to the natural law, for example with respect to gravity, he would most likely perish or at least be unsuccessful in his efforts.

Because of his existence within, as opposed to being outside, the cosmos, it does not seem possible for him to observe its totality in his present human form. However, he has demonstrated his capability to (a) expand his objective knowledge of the natural law, and (b) choose to apply this knowledge, or part of it, in both individual and collective enterprises. To the extent that man has successfully learned and effectively applied this knowledge, he has expanded his presence in the cosmos and appears to have improved his ability to further expand his knowledge. Where he has failed to understand and effectively apply natural laws, the opposite effect appears to have resulted.

It is therefore reasonable to conclude that the interests of men and groups of men are best served by (a) choosing objectives and policies which reflect his best knowledge of the natural law; (b) studying such objectives and policies with maximum possible competence to ensure that they conform to this natural law; and (c) seeing that policies are presented for group evaluation, selection and implementation in a format in which all natural law may be ultimately defined - that is, in terms of time, space, matter, energy and the unknown.

Thus, the foundation upon which policy engineering is based is the natural law as defined by the scientific method. To be effective and universally respected, policy engineering must be especially sensitive to "unknowns", particularly as the "unknown" may relate to the cultural, behavioral or spiritual unknowns or faith, including negative processes, which motivate all of us in one way or another. Where faith or a part of faith can be demonstrated to be a part of the natural law, it should be included amongst the "known", where it cannot it should be recognized and respected as an "unknown".

The basic objective of the policy engineer should be to ensure that a policy is itself clearly defined so that there can be as little doubt as possible as to its effects in terms of its time, or period of application, its space or ambit of influence, its material aspects, its energy or consequences, and its unknowns. The energy portion of the policy is concerned with its effects as they are recognized to act inwardly upon the group making policy, and outwardly upon other groups affected by the policy's implementation. The goal of policy engineering should be to permit, insofar as possible, total objective and pragmatic evaluation and comparison of one policy with other policies with the purpose that groups which will work and live together will find compatible and effective policies. Policy engineering should recognize the totality of the natural law as the base line, or value measurement for the evaluation of a policy.

Basic Rules of Policy Engineering

From the above, certain basic rules should follow, for example:

(a) The natural law of the cosmos has developed and continues to sustain and transform human life. It follows that the best interests of humanity and groups of persons are best served by

1. Choosing objectives and policies which reflect the most advanced scientific knowledge of the natural law.
2. Seeing that objectives and policies are studied with maximum competence by those specially trained to ensure that the stated policies conform to the natural law and that public safeguards will be established to ensure that such multidisciplined competence is in fact applied.
3. Seeing that the group objectives and the policies under review are presented for group evaluation, selection and implementation, in a format similar to that in which the natural law is ultimately defined, namely in terms of time, space, matter, energy and the unknowns.

(b) Group policies should be designed to protect the group as it functions in harmony with the policies of other legitimate sectors of society.

(c) When a group no longer serves the common objective of its component members it should transform itself, by termination of itself or by separation or by integration in relation to other groups.

These rules follow the basic rules of the cosmos itself
which has developed, and transforms and sustains life.

Clearly, additional rules and standards will require to be
stated as the structure of the technology or discipline of policy
engineering develops. Rules should conform to the natural law.

Definition of Policy Engineering

Policy Engineering is defined here as a comprehensive, multi-
discipline, professional technology which may be used to assist
those in public, corporate, institutional or cooperative authority
within a democracy to formulate, analyze, evaluate, select, modify
and implement rational policies. It has fundamental regard for the
basic components of the natural law, namely: time, space (location),
matter (resources), energy (consequences), and unknowns. Policy
engineering should not be confused with management organization or
engineering systems which themselves form essential parts of
policy engineering. The process used seeks to avoid reliance on
personal intuitive judgement on the one hand or single-purpose
task force objectivity on the other, both of which may lead to the
achievement of given objectives but with the consequent development
of many peripheral new problems. Policy engineering emphasizes
concern for the inherent objectives of all the groups who are
involved or who may become involved. It moves iteratively and
sequentially through statements of concern and group objectives.
It then sets down design criteria for the policy followed by
suitable policy option outlines, each of which include appropriate
management organizations and engineering systems. Natural and
man-caused constraints on the implementation of each of the options
are then set down and the screening and filtering of options begins.
Surviving policy positions are then analyzed and evaluated for
final selection. The process then continues through simulation,
limited testing, modification, implementation and operation. It
is clear, of course, that for any given primary objective a complex
network of secondary policy searches will be required.

Essential Factors in Policy Engineering

In the development of policy engineering technology we should
consider that all phases of policy development should include
reference to the five basic factors in natural law, namely: TIME,
SPACE (LOCATION), MATTER (RESOURCES), ENERGY (CONSEQUENCES) and
the UNKNOWN. These should be stated in a format which permits
them to be used as screens or filters for policy option evaluation.
A very brief description of these five basic factors in policy
engineering follows:

(a) <u>Time</u>. In this factor we should consider the period for which the policy will be applicable. For example, in the oil industry policies must include provision for the exploration period, development period, production period and the depletion period. In other words the time involvement of any policy or goal should be defined clearly and in detail within the policy.

(b) <u>Space</u>. In this factor we should set out the geographical or other limits to which the policy shall apply. This will relate to both the macrostructural limits, that is the <u>outside</u> limits of the organization's policy influence, as well as the microstructural limits. In this division we should specify, define and describe all of the other organizations which lie within the area of influence of the particular organization which is making policy. In this context space should be thought of as, at least, three dimensional.

(c) <u>Matter</u>. In this division we should carefully define the material resources with which the policy to be considered is concerned in any way. Since engineering seeks to quantify, we will, insofar as it is possible, attempt to define matter in terms of numbers. This obviously is not always possible. For example, if we are seeking to develop a public policy with regard to the training of personnel for special skills it may be difficult to quantify in some aspects of the problem. On this subject selection and motivation are the main or guiding principles. However, much more can be stated. The location, numbers and types of suitable persons, the ratio of expenditure of funds for such purposes as compared to others should be defined. The requirements and cost of educational facilities and other related statistics should be listed in a format suitable for evaluation and selection.

(d) <u>Energy</u>. In this division we should be concerned with the study of the dynamic and·static effects, or the energy of each of the proposed policy alternatives as they affect other organizations, both <u>outside</u> and <u>inside</u> the group developing the policy. For example, what would be the effect of a policy change by a government in regard to offshore oil leasing? Here we should sequentially examine each of the outside governments affected, the oil industries concerned and then examine each of the interior groups within the population of the government making new policy. These highly detailed studies will have particular reference to the effects, <u>outward</u> and <u>inward</u>, of the government's policy changes. In addition, the feedback effects which will result in further changes and effects should be stated in a format suitable for use as a screen or filter. It should always be borne in mind that the effects of a policy unimplemented may be static: if implemented, policy effects may be expected to be dynamic to the extent that they will change the character of the group making the policy.

(e) <u>The Unknown</u>. In the affairs of man it is vital to recognize the limitations of knowledge. Advancing science and technology often permit intelligent and quantitative forecasts of future events. The study of modern forecasting methods is important.

It is not expected that man will ever be free of the unknown.
Nevertheless with actuarial experience in the insurance business
we intelligently and statistically make usable policy forecasts.
For example, unmanned flights to distant planets should precede
manned flights until man can proceed with adequate knowledge to be
reasonably assured of the safety of the astronaut.

Policy Analysis and Preliminary Evaluation

In recent years there has been an increasing development and
use of sophisticated quantitative techniques for the solving of
complex problems. These methods have been developed under labels
of systems analysis, operations research, statistical decision
theory, etc. However, it is important that the students of policy
engineering understand the capabilities and limitations of such
techniques. Quantitative systems for decision making are of great
value in the solving of many problems; however, they offer little
prospect of serving as final arbiter of the conflicts of interest
involved in large policy problems. Here final selection by
authority may be in error. However, the chances of error are less
if the policy engineering is carried out according to an acceptable
detailed screening process. Such procedures may well be too costly
for certain policy matters. For others they may be much less
costly than simple trial and error processes.

Definitions

The usefulness and endurance of policy engineering will
depend on the clarity of the definition of its basic language or
words. In the importance of their relationship to each other,
these structural terms can be compared to the structural members
of a building. In turn, the strength of the foundation upon which
this structure will rest will depend on the understanding by those
taking part in this work of the various organizational groups for
which policy formulation alternatives are prepared for consideration.
With this in mind, the following definitions are proposed. They
are obviously neither complete nor final.

Group. In policy engineering this term will be used to
indicate an organization of people for whom a policy is being
formulated. It is intended at this time that it will be confined
to organizations operating within the democratic type of structure.
The groups in mind are, specifically, the following:
(a) Public groups, i.e. national, regional (provincial) or
     municipal
(b) Corporate groups, i.e. investor-owned enterprises
(c) Institutional groups, i.e. groups designed for specific type
     of public service such as hospitals or universities

(d) Cooperative groups, i.e. unions, various consumer
     cooperatives, etc.

The above arrangement recognizes the possible need for many sub-
groups within a group. Therefore good group definition should
precede statements of objectives or policies.

Policy. A policy is a considered statement of a plan of
action designed to achieve a given objective by a group of people.
It includes plans of action which will initiate or operate new
projects or maintain existing activities or respond to developments
outside the group, or within the group. Policies may be selected
from a number of alternative policies. They are distinct from
objectives which have no alternatives. They may be components in
a complex set of policies all designed to achieve a higher level
or a specific objective.

Objectives. Objectives are clear, concise statements of the
inherent natural goals forming, in whole or in part, the basic
intentions of the group. They are frequently the reasons for the
group's existence and continuance. Objectives must precede policy.
There are at least two types of objectives:

(a) Primary objectives are statements of inherent purpose
     regardless of time, place, matter or consequence. They
     admit to no alternatives for the group for which policy
     is under consideration.

(b) Secondary objectives are modifications of primary
     objectives. They are designed, because of constraints
     of time, place, matter or consequences , to provide
     limited goals which will ultimately permit the
     achievement of the primary objective. They also are
     distinct from policy statements which they should
     precede.

Just as policy should result from objectives, policy may be
expected to generate new objectives, each requiring new policy.
They all form a complex of policies and objectives designed to
achieve a high level primary or secondary objective.

Policy Engineer. A policy engineer is one who is competent
by virtue of his fundamental education and training to apply the
scientific method and the natural laws of the cosmos to the
synthesis, analysis and solution of problems related to achieving
the objectives of organizations of people. He should be able to
recognize deficiencies in his own training so that he may work
jointly with other professional disciplines having such training.
He should be able to formulate alternative policy proposals and
then analyze and compare these alternatives with the view to
recommending or selecting the most appropriate towards the
achievement of the specific group objective. He should be able
to recognize the appropriateness of a given policy within a complex
set of policies and objectives.

REFERENCES

1. Kierans, T.W. Notes on Policy Engineering, July 1973.

2. Amaria, P.J. and Kierans, T.W. Policy Engineering. Paper
    presented at Technical Workshop session at CIE-APEN meeting,
    October 1973.

3. Kierans, T.W. Eastern Arctic Transport Management. Paper
    presented at The Engineering Institute of Canada, October 1973.

4. Steltner, H.A.R. EOS Routing Study - 1972. Summary Report of
    the work conducted in the Arctic.

5. Hersey, W. Arctic Transportation Study. Report prepared for
    Department of Indian Affairs and Northern Development.

6. Edgington, A.N., et al. Exploration and Production in the
    Canadian Arctic. Archipelago. J.C. Sproule and Associates
    Limited, Calgary, Alberta.

7. Dussault, Capt. L.M. Cargo Ships of Hudson Bay and the Canadian
    Arctic. Presented to Eastern Canadian Section of the Society
    of Naval Architects and Marine Engineers, April 1972.

8. Proceedings of the Arctic Transportation Conference, Yellowknife,
    December 1970.

9. Science and the North. A seminar on guidelines for scientific
    activities in Northern Canada, October 1972.

10. Canada Science and the Oceans. Science Council of Canada.
    Report No. 10. 1970.

11. A Study on Marine Science and Technology. Science Council of
    Canada. Special Study Report No. 16. 1971.

12. The Churchill Falls Power Development, its Transport and
    Transportation Information System, August 1972.

13. Northern Associates Ltd. Arctic Resources by Sea, November
    1973.

14. Arctic Institute of North America. Arctic Marine Commerce
    Study, August 1977.

15. Science Council of Canada. Background Study: A Case Study of
    East Coast Offshore Petroleum Exploration.

PROPOSAL FOR AN ARCTIC SERVICES JOINT OPERATIONS CENTRE*

L. W. Morley

Canada Centre for Remote Sensing

Ottawa, Ontario, Canada

D. J. Clough

University of Waterloo

Waterloo, Ontario, Canada

1.   INTRODUCTION

The Canadian arctic environment is hostile to man and to the works and machines of man. But the environment is also delicate, easy to damage and slow to heal.

In large measure, the future of the arctic will be shaped by the exploitation of its rich natural resources, particularly oil and gas.

The future of Canadian arctic is subject to the pressures of economic development, but within the control of the federal government. The government must not only formulate laws and regulations, but also create infrastructures for orderly development to meet the nation's needs.

Laws and regulations often arise out of cases of political crisis, and may crystallize quickly. But infrastructures are evolutionary and take time to build. It is necessary to plan and build an arctic infrastructure in the last half of the 1970's to meet the requirements of arctic economic development and to cope with the potential crises of the 1980's.

This paper deals with a proposal to develop an Arctic Services Joint Operations Centre (ASJOC) as a linchpin for the arctic infrastructure. ASJOC would provide civilian and military support services related to ship surveillance, environmental monitoring, communications, navigation, Coast Guard icebreaker support, air-

---

*This paper is solely the opinion of a public servant and a professor, based on publicly available information. It does not, in any way, represent government opinions or positions.

sea rescue, RCMP operations, research activities, and other
activities.

ASJOC is envisaged to be an arctic "cybernetic centre",
providing a kind of central nervous system for the arctic
infrastructure. The term "cybernetic" refers to processes of
communication and control in machines and biological organisms,
including human beings, their organizations and their technologies.
The term is in keeping with the technological and organizational
thrust of the proposal.

ASJOC is also envisaged to be an evolutionary centre that
would adapt to the requirements of various government agencies for
the implementation of laws and regulations, and to the requirements
of civilian agencies for support of social and economic developments.

The paper is arranged in the following order, to provide an
initial basis for discussion:

2. <u>Arctic oil and gas potential</u> - a description of the basic
economic pressures on the arctic.

3. <u>Arctic development and law of the sea</u> - a description of
Canada's recent position on the law of the sea, and the status of
issues such as the use of international straits (e.g., the Northwest Passage).

4. <u>Proposed objectives of ASJOC</u> - a list of proposed initial
objectives, limited for purposes of initial discussion.

5. <u>Some operational programme requirements</u> - a description
of some operational requirements to implement the proposed
objectives.

6. <u>Conclusions about organization and implementation of ASJOC.</u>

2. ARCTIC OIL AND GAS POTENTIAL

Various studies have indicated that the arctic will ultimately
become one of the world's largest oil and gas producing areas. At
the Ninth World Petroleum Congress in 1975, potential recoverable
reserves of more than 100 billion barrels of oil and 100 trillion
cubic feet of natural gas were estimated for the Beaufort Basin,
Alaskan North Slope and Sverdrup Basin. These quantities are
enormous compared with the North Sea estimates of 15 billion
barrels of oil and 18 trillion cubic feet of natural gas, for
example, reported at the same Congress.

The economic pressure for development of U.S. and Canadian
arctic oil and gas resources, and transportation via the Northwest
Passage, will continue to increase if world supply and demand
trends continue and if the OPEC nations maintain their policy of
high prices.

Conventional oil and gas reserves in the producing provinces of Alberta and Saskatchewan are declining, at a time when world oil prices are artifically maintained above $10 per barrel by the OPEC nations. Intensive oil and gas exploration is underway in three frontier regions of Canada: (1) the onshore and offshore arctic area of the Mackenzie Delta and Beaufort Sea, (2) the east coast offshore area of the Newfoundland and Labrador continental shelves and slopes, and (3) the north-and-eastern continental shelf and island areas of the Arctic Archipelago.

It has been estimated that arctic oil costs will be in the order of $8 or $9 per barrel at the wellhead, and that oil will be available from the western arctic by 1983 and from the east coast and north-eastern arctic by 1985. But a great deal of exploration and development work will have to be carried out, and a collosal infrastructure built before that time.

Wells have to be drilled, production platforms installed, collector pipelines and storage vessels constructed, and terminal facilities built. Settlements, roads, airfields, utilities and logistical support services have to be created. There will be a growing need for support services such as aircraft and satellite all-weather ice reconnaissance, communications, navigation aids, icebreaker support, traffic control, and settlement administration.

How will all the arctic oil and gas be transported to markets across the vast expanse of North America? Various combinations of pipelines and surface tankers seem technologically and economically feasible, though stringent environmental safeguards are required and political conditions are uncertain.

Whatever the long term cost-benefit estimates may indicate, surface shipping may prove to be the most attractive transportation mode for some arctic offshore and island production areas. Heavy weight may be placed on short-term financial and political considerations such as the following: (a) necessary threshold reserves and initial capital costs are less for a variable-scale tanker system than for a fixed-scale pipeline system, (b) tanker shipments can start sooner than pipeline shipments, generating cash flows to finance further development under difficult financing conditions, and (c) early tanker shipments will contribute to continental security of supply during a period of adjustment to new world price and supply conditions, particularly with respect to Canadian and U.S. eastern demand centres.

Numerous studies have been carried out by various agencies to determine the technological and economic parameters of surface and underwater ship and pipeline designs and routes. Plausible scenarios indicate that one million barrels per day of Canadian oil will probably be shipped by surface vessel out of the Baffin Bay-Davis Strait area by 2000 A.D.; 600,000 bpd will probably be transported out of the arctic islands by surface vessel or pipeline; about 1.2 million bpd will probably be carried by pipeline out of the Beaufort Sea, though it may prove to be technically and economically feasible to send it by surface tanker via the Northwest

Passage route to eastern markets. Recent studies for MARAD
(Maritime Administration, U.S.A.) have postulated a "potential
requirement" for the U.S. to ship 4 million bpd by surface tanker
from Alaska through the Northwest Passage by 2000 A.D. From
Canada's viewpoint, the pollution hazards of such a venture loom
very large at present, and it seems likely that the use of the
N.W. Passage would be forbidden. However, future economic,
technological and political conditions may make the N.W. Passage
route acceptable as a joint venture of Canada and the U.S.

3. ARCTIC DEVELOPMENT AND LAW OF THE SEA

Much of the future oil and gas development is expected to be
offshore in the arctic and off the Labrador coast. Since water
transportation will be vital to arctic economic development,
attention must necessarily focus on the sea.

The future development of Canada's arctic resources will depend
on both the international law of the sea and a fine balancing of
national laws and regulations to provide (a) incentives for economic
development and (b) constraints against environmental and social
damage.

The law of the sea is that part of international law which
determines how much control a coastal state has over how wide a sea
area off its coasts, and, as a corollary, what the rights of other
states are in the same areas.

At the present time, Canada claims and exercizes the following
rights:
(a) full sovereignty over a 12-mile territorial sea, subject only
    to the rights of "innocent passage" extended to other nations;
(b) exclusive fishing rights in the Gulf of St. Lawrence, the Bay
    of Fundy, Queen Charlotte Sound, Hecate Strait and Dixon
    Entrance, as well as within the 12-mile limits;
(c) pollution control in the same bodies of water as in (b).
(d) pollution control north of latitude 60° north, throughout the
    Arctic Archipelago and out to 100 miles from the Canadian
    coastline;
(e) sovereignty rights over the resources of the continental shelf
    out to the edge of the continental margin;
(f) full sovereignty over Hudson Bay and Strait (historic Canadian
    waters).

The future development of Canada's arctic sea resources must
take account not only of its present maritime rights, but, more
appropriately, the much more demanding ones that will arise as a
result of ongoing negotiations at the Third Law of the Sea
Conference (LOS-3). It can already be assumed that the LOS-3
treaty will be structured upon the "economic zone", a radically
novel concept in international law which involves aspects of both
the territorial sea, falling within the sovereignty of the coastal

state, and the <u>high seas</u> together with their concomitant traditional freedoms.

On the basis of progress so far achieved at LOS-3, the coastal state will in all likelihood be vested with sovereign rights over the living and mineral resources within the 200-mile zone as well as with lesser powers with respect to the preservation of the marine environment and for the control of scientific research. Agreement upon the economic zone is not expected to affect the legal rights of the coastal state over that part of its continental margin lying beyond the 200-mile seaward limit of the zone. As a corollary, the coastal state will have the duty not to interfere with legitimate activities, including navigation, carried out by foreign states in the zone.

<u>Territorial Sea</u>. In 1970, by amending the Territorial Sea and Fishing Zones Act, Canada moved the outer limit of its territorial sea from three to twelve miles. Although the international community had failed to agree in 1958 and 1960 on a precise breadth for the territorial sea, the Canadian action was clearly in line with the prevailing practice of other coastal nations. As provided for in the 1958 Territorial Sea Convention, the coastal state exercizes sovereignty over the territorial sea, the air space above and the subsoil below. However, there is one important limitation placed on this exercise of sovereignty: foreign vessels have a right of 'innocent passage' through the territorial sea, that is, a right of passage which is not prejudicial to the peace, good order and security of the coastal state. If it deems such action essential to its security, then the coastal state may, after giving prior publicity, temporarily suspend innocent passage on a non-discriminatory basis.

At LOS-3, all items under consideration are intimately interrelated and no one issue can be settled without taking into account the others. It is, however, practically a foregone conclusion that the Conference will agree on the breadth of the territorial sea as being twelve miles. As to "innocent" passage, the Conference may decide to set out objective criteria for determining whether passage is innocent or not, rather than leaving that judgement to be made by the coastal state, as is now the case.

<u>Straits used for international navigation</u>. The 1958 Territorial Sea Convention also touches upon straits used for international navigation and defines these as either linking two parts of the high seas or one part of the high seas and the territorial sea of a foreign state. As within the territorial sea, foreign vessels have a right of innocent passage through such straits, except that in the straits case this right may under no circumstances be suspended by the coastal state.

<u>Northwest Passage and Arctic Archipelago Waters</u>. The question may be asked whether the straits articles of the 1958 Convention apply to the Northwest Passage. Canada has consistently maintained that the Passage cannot be said to have been used for international navigation and as such is not one of those straits referred to in

the Convention. Additionally, Canada has repeatedly stated that
the waters of the Arctic Archipelago are internal waters wherein
passage of foreign vessels is subject to its control for the
purposes of protecting the marine environment. Some of the major
powers have not acquiesced to such claims; on the other hand, at
least one of them has supported Canada's stand with respect to the
Arctic. Passage in other bodies of Canadian waters or the channels
thereof are regulated from an environmental point of view under the
Canada Shipping Act.

It is at this stage virtually impossible to predict how LOS-3
will be able to work out an accommodation of the straits issue, as
it still sharply divides the states bordering these narrow waters
and the Maritime Powers. The issue is such a pivotal one that it
could virtually "make or break" the Conference. The yet unresolved
debate may be summarized as follows: the strait states are pressing
for the present regime of innocent passage to be retained for
straits used for international navigation; the maritime powers, on
the other hand, are promoting acceptance of an unimpeded and un-
suspendable transit passage regime. In fact, the latter states
attach such importance to their thesis that they have indicated
time and again that, in the event it were not endorsed by
Conference, they would find it impossible to become parties to the
future treaty. Another related question concerns the precise
definition of the straits to be covered in any new regime devised
by the Conference.

Contiguous Zone. The 1958 Convention on the Territorial Sea
contains as well a section on the contiguous zone. It provides
that such a zone, beyond the territorial 3-mile limit, should
extend at a maximum distance of 12 miles from the coast of
appropriate baselines and allows coastal state jurisdiction in
relation to customs, immigration, health and fiscal matters only.
Canada's contiguous zone jurisdiction was subsumed under its
sovereignty when the territorial sea was set at twelve miles in
1970.

The issue of contiguous zone jurisdiction is not one of the
most controversial at the Conference and consequently not much
attention has been paid to it, to date. One proposal, apparently
acceptable to a majority of states, would have a 12-mile contiguous
zone lying beyond the coastal state's territorial sea. The outer
limit of the contiguous zone could thus be delineated at a maximum
distance of 24 miles from the baselines for measuring the
territorial sea.

Continental Shelf. The 1958 Continental Shelf Convention - to
which Canada is a Party - describes in a clear and precise manner
the rights of the coastal state with respect to its Continental
Shelf. These rights may be summed up as follows: (a) the coastal
state exercizes over its continental shelf sovereign rights for the
purposes of exploring it and exploiting its resources, i.e., not
only the mineral resources but as well the sedentary species of
living resources which at their harvestable stage either are

immobile or found in the subsoil or are in constant physical contact
with the seabed; (b) these rights are exclusive in the sense that
the resources may only be exploited through the authorization of
the coastal state; and (c) research in respect of the continental
shelf may be conducted only if the coastal state has given its
prior consent. Altogether these rights provide the coastal state
with considerable power and control over any activity conducted
with respect to its continental shelf. The Convention, on the
other hand, lays down that the status of the waters over the shelf
is that of high seas.

The difficulty with the Convention lies in the fact that it
gives an ambiguous definition of the shelf thereby creating some
confusion as to the extent to which states rights apply. Article
1 states that the Continental Shelf begins with the outer limit of
the territorial sea (on the breadth of which it proved impossible
to reach agreement) and extends to the 200-meter isobath, or beyond
that depth, to where the waters will admit of exploitation. A
liberal interpretation of the "exploitability" criterion could have
opened - theoretically - coastal state claims to the sea floor out
to the middle of the ocean had the world community not recognized
in 1970 the seabed beyond national jurisdiction as the "common
heritage of mankind", subject to no appropriation by State or
individual. Fortunately the Convention's definition was clarified
by the 1969 decisions of the International Court of Justice - in
the North Sea Continental Shelf cases - describing the shelf as the
natural submerged prolongation of the land-territory of the coastal
state. On the basis of the natural prolongation thesis, this meant
that the coastal state had exclusive sovereign rights to the whole
of its continental margin, comprising the physical shelf, slope and
rise, but none beyond. The practice of a number of broad-shelf
states has also endorsed the Court's definition.

Basing its position on the terms of the Convention, the ICJ's
decision and state practice Canada has over the years laid claim to
exclusive sovereign rights over its continental margin. The
Canadian margin is asymmetrical on the Pacific coast and it barely
reaches a distance of 70 miles; off the Atlantic, it is broad and
even reaches out over 600 miles, in some instances.

At LOS-3, the 200-mile economic zone concept, applicable to
renewable as well as to non-renewable resources and endorsed by a
majority of states, has come into open conflict with the "natural
prolongation" approach for the continental margin, as propounded by
some 35 to 40 states with extensive margins. An accommodation may
be devised that will confirm the rights of states, such as Canada,
to their continental margin beyond the 200-mile seaward limit of
the economic zone.

Fishing. Codified international law, as it appears in the
1958 Conventions on the Living Resources and on the High Seas, does
not endow the coastal state with realistic rights for the harvesting
of living resources, let alone their management and conservation.
The Conventions in effect limit the coastal state's exclusive

fishing zone to the confines of its territorial sea. Since the
1960's, fisheries operations have considerably increased and become
sophisticated to the extent that over-fishing is presently endanger-
ing certain species. The negative impact of such operations is
particularly felt on Canada's Atlantic coast where communities
heavily depend on fishing for a living.

In order to provide proper management and protection to its
most significant stocks, Canada, in 1970, sealed off the Bay of
Fundy, the Gulf of St. Lawrence, Queen Charlotte Sound, Hecate
Strait and Dixon Entrance as exclusive fishing zones. To that end,
phasing-out agreements were concluded with six of the European
nations which had traditionally conducted fishing activities in
those areas.

Considering the degree of agreement reached on the fisheries
issue at the 1975 Session of LOS-3, it may well be that Canada will
wish to establish its exclusive fishing zone within the next few
years. By such a move, Canada would not necessarily be putting an
end to all foreign fisheries within the zone but would simply be
limiting them to that part of the total allowable catch which
Canadian fishermen could not harvest themselves, i.e., the surplus.
In addition, Canada will probably be authorized to subject those
foreigners in the zone to its management and conservation regulations.

At LOS-3 Canada is endeavouring to have acceptance for a
special regime for anadromous species, such as salmon, whereby the
fishing of such stocks in the high seas would normally be
prohibited. In addition, states of origin of the anadromous stocks
would be vested with exclusive management and conservation rights
throughout the full migratory cycle of these stocks.

Preservation of the Marine Environment. The coastal state has
full powers to control and limit land-based pollution of the marine
environment. The contentious question relates to jurisdiction with
respect to vessel-source pollution. Anti-pollution standards may be
established by the coastal state so long as they do not hamper
innocent passage in the territorial sea. Beyond, coastal state
regulations must conform with internationally agreed conventions
(usually negotiated under the aegis of the Intergovernmental
Maritime Consultative Organization), such as the 1973 Convention
for the Prevention of Pollution from Ships.

Arctic Waters Pollution Prevention Act. The problem for
Canada is that there is yet to be agreed upon a really effective
set of international rules to protect the marine environment in
special areas presenting a most delicate ecology and unique
navigational hazards - as in the Arctic waters. To fill in this
regulatory lacuna, the 1970 Arctic Waters Pollution Prevention Act
was unanimously adopted by Parliament, providing Canada with
extensive powers over a 100-mile environmental protection zone.
Some states, including the USA, deemed such action as interfering
with the sacrosanct freedoms of the high seas while others,
including one important maritime power, were more appreciative of
the concerns leading to the 1970 legislation.

4. PROPOSED OBJECTIVES OF ASJOC

Laws and regulations such as those mentioned above are the
legal embodiment of national policies. As such, they provide the
basis for defining operational objectives and planning programmes
for the proposed Arctic Services Joint Operations Centre (ASJOC).
For purposes of simplifying the preliminary discussion, it is
suggested that the relevant laws and regulations may be grouped
within a broader framework of national policy. On the basis of
available government publications, it may be perceived that the
national policy is designed to (1) foster economic growth,
(2) enhance the quality of life, (3) ensure a harmonious natural
environment, (4) promote social justice, (5) safeguard sovereignty
and independence, and (6) work for peace and security.

The proposed objectives of ASJOC ought to be in harmony with
laws and regulations, within the framework of national policy.
However, any concise statement of general objectives is essentially
a short "coded" version of a set of complex ideas, like a telegram.
Such a general statement may be "decoded" differently by different
people, and there always exist possibilities of misinterpretation.
In this case, the proposed Centre appears to cut across the
missions of various federal government agencies so that the
possibilities for misinterpretation may be very great and the
emotional content of the arguments may be heightened. (The authors
are well aware of the difficulties and potential conflicts of views,
but feel that time is running short and the debate must now be
entered.)

The proposed objectives of ASJOC are stated as follows, first
in a short "overall" form, and then in an expanded element-by-
element form.

Short Statement of Objectives

The main objectives of the Arctic Services Joint Operations
Centre are to provide an organization and methods for (1) central
acquisition, processing and distribution of data and information
vital to the implementation of laws and regulations in the arctic,
(2) central coordination of the allocation, assignment and dispatching of resources required for the implementation of laws and
regulations in the arctic, under delegation from responsible
government agencies, (3) provision of specialized central
information and support services, on demand, to government agencies
and private organizations, and (4) maintenance of records of
specified events and statistics related to the implementation of
laws and regulations and provision of central support services in
the arctic.

Expansion of Objectives

The main objective of ASJOC can be expanded in a number of ways. The following expansion is not unique and does not exhaust the possibilities. For the sake of perspective, the objectives are first numbered and tabulated, then described one by one. The numerical order of tabulation is not significant.

1.   Centralized Information
1.1. Surveillance information.
1.2. Environmental monitoring information.
1.3. Geographic-oceanographic data bank.
1.4. Resource availability information.
1.5. Jurisdictional information.

2.   Centralized Coordination of Resources
2.1. Policing actions.
2.2. Sea traffic management and piloting.
2.3. Air traffic control.
2.4. Search and rescue operations.
2.5. Emergency pollution control and cleanup.
2.6. Fisheries and wildlife control.

3.   Specialized Support Services on Demand
3.1. Provision of communications aids.
3.2. Provision of environmental forecasting aids.
3.3. Provision of navigation aids.
3.4. Provision of medical aids.
3.5. Provision of aids to surveys and research.

4.   Maintenance of Records
4.1. Storage of data and information.

Objective 1.1. Surveillance Information. Central acquisition, processing and distribution of data and information concerning the detection, identification, position-fixing and surveillance of (a) aircraft, (b) surface ships and vessels, (c) undersea craft, (d) fixed installations onshore and offshore, (e) landed parties onshore and on ice.

Objective 1.2. Environmental Monitoring Information. Central acquisition, processing and distribution of data and information concerning the detection, identification, position-fixing and monitoring of environmental conditions, including (a) atmosphere and weather parameters, (b) sea state and surface ice parameters, (c) land state and surface parameters such as permafrost and snow cover, (d) natural and man-made disasters (e.g., tidal waves, oil-spills, fires), (e) natural and man-made pollution of atmosphere, sea and land, (f) ecological conditions, (g) atmospheric, sea and land surface conditions affecting radio communications, radar and remote sensor operations, (h) underwater, surface and bottom conditions affecting sonar, seismic waves and underwater radio.

Objective 1.3. Geographic-Oceanographic Data Bank. Central maintenance of a system for fast acquisition of geographic and oceanographic data and information from existing data banks and archives, including (a) maps, charts and survey data of human populations, wildlife, plantlife and known ecological factors, (b) maps, charts and survey data of settlements, marine terminals, coastal and offshore installations (e.g., ice stations, oil-gas platforms), storage depots, airfields, weather stations, radio and radar stations, lighthouses, beacons, buoys, survival huts, etc., (c) maps, charts and survey data describing physical parameters - geological, glaciological, hydrographical, oceanographical, etc.

Objective 1.4. Resource Availability Information. Central acquisition and maintenance of data and information describing the sources, disposition, readiness and availability of fixed and mobile resources for allocation, assignment and dispatching on specialized tasks, including (a) national defence aircraft, ships/vessels, installations, vital equipment (e.g., decompression chambers), material, manpower, etc., (b) other government and private aircraft, ships/vessels, installations, vital equipment, material, manpower (e.g., Coast Guard vessels, ice reconnaissance aircraft of DOE, seismic survey crews of DEMR, mobile radio and radar units of DOC, commercial aircraft, oil company fire-fighting equipment, oil-spill containers, chemicals, etc.), (d) national and provincial police units, (e) mobile first aid units, medical, hospital and health care facilities, (f) communications facilities and navigational aids (modes, characteristics, precise locations), and (g) government and private research facilities and manpower.

Objective 1.5. Jurisdictional Information. Central acquisition and storage of data and information describing the missions of government agencies and roles of private organizations operating in the arctic, including (a) texts of laws and regulations relating to defence, mineral production, transportation, communications, environmental pollution, conservation, habitations and settlements, health, safety, crime, customs, etc., (b) arctic missions of government agencies, relative to bases in law, (c) roles and activities of commercial enterprises and other private organizations in the arctic, in the context of laws and regulations, and (d) organizational communications networks connecting agencies and organizations for coordination of joint tasks.

Objective 2.1. Policing Actions. Central distribution of data required to initiate policing actions by military units or national police units (RCMP), and initiation of some commands under limited authority delegated by the responsible agencies, including (a) legal warning and/or arrest of violators, (b) legal seizure of ships/ vessels, installations, equipment or material used in violation of Canadian laws, (c) enforcement of legal restraining orders (e.g., denial of passage, prevention of dumping), (d) legal blockade of territorial sea routes.

Objective 2.2. Sea Traffic Management and Piloting. Central
control of surface ship/vessel and underwater craft movements in
the Canadian arctic seas, similar to Vessel Traffic Management
(VTM) systems used in the St. Lawrence River and the Great Lakes,
including (a) granting permission for passage in designated zones,
(b) ship routing, spacing, scheduling and queuing at loading points
in designated zones, and (c) placing pilots and/or observers on
board on designated routes.

Objective 2.3 Air Traffic Control. Central control of air
traffic in the Canadian arctic, similar to Air Traffic control
(ATC) systems used throughout North America, with special provision
for control of low level movements in designated lanes (e.g., ice
reconnaissance flights along main sea routes, helicopter flights
in tending oil-gas drilling platforms, military surveillance, etc.).

Objective 2.4. Search and Rescue Operations. Central control
of air-sea search and rescue operations, including (a) rapid
response to locate aircraft and ships in distress, (b) rapid
deployment of aid to downed aircraft, vessels and fixed installations
(e.g., offshore oil-gas platforms) in distress, (c) emergency
evacuation of people to safe areas, aid stations or other vessels/
installations, and (d) emergency transport of aid supplies (e.g.,
air drop of food and clothing, medical supplies).

Objective 2.5. Emergency Pollution Control and Cleanup. Central
control of pollution under emergency conditions (e.g., tanker oil
spills), including (a) rapid location of emergency, (b) rapid
deployment of aircraft, vessels, equipment, supplies and manpower
to contain, neutralize or otherwise minimize the adverse effects
of the pollutant, (c) deployment of resources to clean up the
pollution, and (d) gathering evidence and keeping records as a
basis for levying legal penalties and preventing future occurrences.

Objective 2.6. Fisheries and Wildlife Control. Central
control of the landing of fish and taking of animals in arctic
coastal regions, including (a) advisory messages to fishing vessels
and hunting parties concerning the status of legal quotas, (b) board-
ing for inspection of catch and conditions of sanitation and health,
and (c) initiation of police action in the event of violations of
Canadian laws (see objective 2.1.).

Objective 3.1. Provision of Communications Aids. Central
provision of vital communications aids, including (a) interconnection with all existing civilian and military telecommunications and courier communications systems in the arctic, and with
standard Canadian telegraph, telephone and broadband computer
communications systems, (b) emergency message relay services on
demand, and (c) maintenance and distribution of operational status
reports on all arctic communications facilities (e.g., location,
frequency, on-off time, etc.).

Objective 3.2. Provision of Environmental Forecasting Aids.
Central provision of vital environmental forecasting aids,
including (a) communications interconnection with all existing
civilian and military environmental reporting stations in the

arctic, and with central forecasting stations in Canada (e.g., Ice
Central in Ottawa), and (b) relay and broadcast of routine
environmental forecasts on a regular basis and special forecasts
on a demand basis, dealing with weather, sea state, sea ice, land
state, snowcover, pollution, atmospheric and surface radio-radar
interference, natural disasters, etc.

Objective 3.3 Provision of Navigational Aids.   Central
provision of routine navigational reports on a regular basis and
special reports and services on demand, including (a) relay and
broadcast of status reports on all navigational aids such as radio
and radar stations, lighthouses, beacons, buoys, etc., (b) relay
and broadcast of reports on environmental conditions and general
traffic conditions related to defined air routes, sea lanes or grid
points, (d) position-fixing, (e) icebreaker or other Coast Guard
assistance, (f) towing arrangements, and (g) pilotage.

Objective 3.4 Provision of Medical Aids.   Central provision
of emergency medical aid information and services on a demand
basis, including (a) reports on location and status of first aid
stations, hospitals, health care facilities, medical and para-
medical personnel, (b) transport of medical supplies, equipment and
personnel, and (c) transport of patients from remote locations to
aid facility.

Objective 3.5 Provision of Aids to Surveys and Research.
Central provision of routine reports on a regular basis, and
special reports and services on a demand basis, including (a) communications, environmental forecasting, navigation and medical
aids, (b) emergency transport of scientific supplies, equipment and
personnel, (c) emergency connection to data banks and computing
facilities, and (d) cooperative involvement of ASJOC facilities and
personnel in scientific experiments.

Objective 4.1. Storage of Data and Information.   Central
maintenance of records and control of the storage and accessibility
of data and information related to ASJOC operations including
(a) fast-access storage at the Centre and at decentralized data
banks, and (b) slower access to decentralized archives.

5.   SOME OPERATIONAL PROGRAMME REQUIREMENTS

The objectives outlined in the preceding section are not
exhaustive. Others could be added or some could be deleted. The
list seems long and the objectives evidently cut across those of
several agencies. However, it is not our intention to propose that
the Arctic Services Joint Operations Centre replace or repeat the
functions of other agencies. On the contrary, it is proposed that
ASJOC centrally coordinate certain specialized activities of a
number of agencies.

First, ASJOC could serve as a joint "clearing-house" for vital
arctic data and information that may be chiefly acquired and
processed by other agencies before feeding into the ASJOC central

processing and storage system.

Second, ASJOC could serve as a joint "command and control" centre, with its command and control functions clearly circumscribed so as not to interfere with the missions of the cooperating agencies. It could trigger actions indirectly by messages to units of other agencies, which would then take action under their own authority in law. Alternatively, it could trigger actions directly by commanding the allocation of resources under specific and limited authority delegated to it by the cooperating agencies. We expect that ASJOC would operate in a mixed mode, indirectly for some operations and directly for others, depending on the organizational preferences of the cooperating agencies.

It would be premature to define operational programmes in detail or to design an organizational structure for ASJOC at this stage of the proposal. A good deal of discussion is necessary first to explore the objectives thoroughly; a good deal of study and analysis is required as a basis for selecting from a variety of feasible operational programmes; and a good deal of careful negotiation is required to evolve the organizational structure.

However, we can sketch a scenario in which the operational programme matches the objectives proposed in Section 4. Then we can talk about programme requirements in terms of information and resources. For convenience, let us use the same headings to describe programme elements as were used to describe objectives.

1.1. Surveillance Requirements. Both the surveillance elements and the environmental monitoring elements of the ASJOC programme would overlap to some extent in terms of information and resource allocation. Both would involve remote sensing from satellites and aircraft, including balloons and possibly remotely piloted vehicles (RPV's) in future. Both would require observations from ships at sea and various land and sea installations.

Satellites such as NOAA, LANDSAT and SEASAT are useful for environmental monitoring - NOAA to monitor clouds, sea surface ice, snowcover, etc., LANDSAT to monitor land surface and sea ice characteristics, and SEASAT to monitor sea state, sea ice, etc. (SEASAT-A is to be launched by NASA in 1978). However, only SEASAT-type satellites will be useful for surveillance, since they will carry high-resolution synthetic aperture radar (SAR), capable of monitoring ships and icebergs, in cloudy weather, day and night.

Under agreements with NASA, the Canada Centre for Remote Sensing (CCRS) presently receives NOAA and LANDSAT data directly from the satellites at its Prince Albert Satellite Station, and is presently developing a semi-mobile receiving station at Pouch Cove, Newfoundland, to receive NOAA, LANDSAT and SEASAT data. CCRS operates an effective ground data processing and image production facility to deal with LANDSAT data, and has recently proposed to develop a ground data processing facility to handle SAR data from SEASAT. The CCRS system of satellite data acquisition, data processing and information output could be coupled into the proposed ASJOC system, through the introduction of appropriate communications channels, interface equipment and operational rules.

Aircraft surveillance of maritime areas is primarily the responsibility of the Canadian Forces. However, their resources for arctic surveillance are limited. The Atmospheric Environment Service (AES) operates aircraft for ice reconnaissance and other environmental monitoring missions in the arctic during the summer navigation season. The CCRS operates an aircraft in the arctic from time to time as an experimental remote sensor platform. Other agencies of such departments as Indian and Northern Affairs, Environment, and Energy, Mines and Resources operate aircraft from time to time in support of survey and scientific missions. It is proposed that information from such non-military and military airborne operations be coordinated through ASJOC to help satisfy the surveillance objective wherever multi-purpose flights are possible.

Any joint civilian-military airborne surveillance operations would involve matters of national security. It is therefore expected that the Department of National Defence would retain primary responsibility for surveillance resource allocation and information classification, and would accordingly delimit ASJOC's responsibilities and authority to sequester surveillance resources.

Ship surveillance of maritime areas is presently carried out by both the Canadian Forces and the Canadian Coast Guard. However, resources are severely limited. It is proposed that both Canadian Forces and Coast Guard surveillance data from ships in the arctic be integrated with satellite and airborne surveillance data through ASJOC. It is also proposed that arctic coastal radar data of the Department of Transport and underwater sonar data of the Department of National Defence be integrated through ASJOC.

The effectiveness of surveillance operations could probably be increased significantly if satellite, airborne, surface vessel, fixed radar and fixed sonar were combined to optimize the locations, scales and detail of coverage provided by each system. The "integration" would be achieved by providing facilities, methods and organizational structure to interface ASJOC with the various cooperating agencies. ASJOC would deal essentially with surveillance information, rather than with direct command of surveillance resources. However, some of the information may be subject to national security classification.

1.2. Environmental Monitoring Requirements. Several agencies are involved with various aspects of environmental monitoring and protection, including the Atmospheric Environment Service (weather, sea state, sea ice and snow cover forecasting), the Environmental Protection Service (pollution standards and regulation), the Coast Guard (oil spill cleanup), and others. However, the AES is clearly in the forefront in developing comprehensive environmental monitoring systems, utilizing remotely sensed data from satellites, aircraft and ground radar stations, sampling from balloons, ice stations, etc., modelling and computer simulation of weather and other environmental phenomena, and utilization of data from other Canadian agencies such as NOAA.

The AES system of data acquisition, data processing and information output could be coupled into the proposed ASJOC system. Certain environmental monitoring and surveillance operations could be combined, as described in 1.1.

1.3. <u>Geographic-Oceanographic Data Bank Requirements</u>. Many agencies produce data and information of the kinds described in objective 1.3. Some of it is in the form of charts and publications. Some of it is stored, or suitable for storage, in computer data banks, particularly information about marine terminals, coastal and offshore installations, ice stations, drilling platforms, storage depots, airfields, radio and radar stations, beacons, buoys, etc. It is proposed that ASJOC include a computer data bank facility to store certain classes of data for fast retrieval, and communications facilities to permit somewhat slower access to data banks of various agencies and institutions (e.g., CCRS tapes of historical satellite data).

1.4. <u>Resource Availability Requirements</u>. It is necessary that the ASJOC data bank contain vital information about resources available for emergency allocation and assignment, and that such information be updated at regular intervals by resource status reports. In some cases involving military resources, it may be preferable to have communications links to military data banks, providing fast access under security rules.

1.5. <u>Jurisdictional Information</u>. Most of the information required to meet this objective is in narrative form and may be found in government libraries. The most important items, dealing with questions of jurisdiction and legal authority, could be stored in synoptic form at ASJOC facilities, indexed and available for rapid referral (e.g., in random access computer storage, or in microfiche files).

2.1. <u>Policing Action Requirements</u>. To be effective, policing actions require (a) adequate surveillance information, (b) adequate information about available resources such as aircraft, Coast Guard vessels, RCMP personnel, etc., and (c) adequate communications between ASJOC and the violators and between ASJOC and policing units. In some emergency situations, policing actions may have to be initiated by remote units under delegated authority, without clearing first through ASJOC. However, in all cases information concerning such actions should be communicated to ASJOC as soon as possible, and retransmitted by ASJOC to verifying authorities if necessary. The rules will have to be spelled out explicitly, and if necessary embodied in new legislative acts.

2.2. <u>Sea Traffic Management and Pilotage Requirements</u>. The requirements of vessel traffic management (VTM) were emphasized in a recent report by Philip A. Lapp,* particularly when arctic outbound shipping expands to meet the needs of resource extraction.

*<u>The identification of requirements critical to operations on and below ice covered waters</u>, Philip A. Lapp, Sept., 1974 (report prepared on behalf of the Ministry of State for Science and Technology under DSS Contract No. OsQ3-0532).

VTM systems are already established in various Canadian coastal regions. According to the Lapp report, "... VTM is approaching the degree of control over ships that air traffic controllers have over aircraft."

It is proposed that arctic VTM systems be integrated with the ASJOC system in the earliest planning stages, to take advantage of the opportunity to specify standard information requirements and joint resource utilization. The most important requirements are for reliable communications between the ship and the control centre and accurate position fixing.

2.3. Air Traffic Control Requirements. The arctic air traffic control requirements would be similar to those that already exist in Canada. However, arctic ATC should be coupled with ASJOC, so that the positions and status of all aircraft in the arctic are known and can be controlled to meet emergency situations (e.g., search and rescue).

2.4. Search and Rescue Requirements. The Canadian Forces have the main responsibility for search and rescue operations in Canada, and cooperate with other countries in joint operations. However, none of the four primary Rescue Coordination Centres (RCC's) are located to deal effectively with arctic operations, and the existing resources seem inadequate to cover the future arctic requirements in addition to existing requirements. It is therefore proposed that an arctic RCC be established in conjunction with ASJOC. Such an RCC would require necessary airfield facilities and communications, and would remain a responsibility of the Forces. But its operations would be coupled to those of ASJOC. Perhaps some aircraft could be utilized for multiple purposes of surveillance, policing, search and rescue, and assistance with emergency pollution control and cleanup tasks (e.g., intensive surveillance, transporting emergency supplies and personnel).

2.5. Emergency Pollution Control and Cleanup Requirements. The most important source of pollution in the arctic will probably be oilspills at sea. The Marine Transportation Administration has responsibilities related to oilspill containment and cleanup. However, the MTA resources alone will probably not be adequate to respond quickly to such emergencies over the vast areas of the arctic shipping routes. It is therefore proposed that pollution control and cleanup operations be coordinated by ASJOC, utilizing resources of the Canadian Forces, MTA, the Coast Guard, and other temporarily sequestered government and private resources. The major requirements would be a set of standing plans to deal with emergencies in various zones, and information about resource availabilities (see 1.4). Certain resource requirements for surveillance, pollution monitoring, search and rescue, and emergency pollution control are overlapping (e.g., aircraft and technical manpower). ASJOC seems to be an appropriate mechanism for coordinating multiple assignments.

2.6. Fisheries and Wildlife Control Requirements. The main requirements here are essentially those of inspection and policing,

as discussed in 2.1. It is proposed that the information and
resource requirements for such operations be coordinated through
ASJOC. Fishing vessels in arctic and subarctic waters would of
course require some of the same aids as other ships, as described
under other headings.

3.1. Communication Aids Requirements. Maintenance of
reliable communications is perhaps the most important single
requirement for almost all arctic endeavours. Present facilities
for ship-shore communication include a number of MF and HF radio
stations at key locations in the arctic, and ANIK satellite relay
stations at Coral Harbour and Resolute. An expanded system is
planned for the next five years. Direct broadcast satellites may
become operational in the 1980's. All such communications
facilities should be viewed as integral parts of the system required
by ASJOC to serve its many objectives. ASJOC will require an
advanced capability to link up the other facilities, including
conventional and satellite radio systems.

3.2. Environmental Forecasting Aids Requirements. This
programme element would require the capabilities for environmental
monitoring and communications described in 1.2 and 3.1. Special
HF facsimile transmission stations will probably be required along
the main arctic shipping routes, to transmit facsimiles of satellite
and airborne remote sensing images, ice charts, and other pictorial
aids to navigation, with minimal image degradation due to noise.
It is proposed that such facilities be linked to ASJOC for purposes
of coordination and verification of information, and for relaying
weather forecasts, ice forecasts, sea state forecasts, etc., from
AES facilities.

3.3. Navigation Aids Requirements. Accurate sea navigational
aids are required for position-fixing of vessels along shipping
routes and approaches to moorings and harbours (300 to 600 meter
accuracy), for positioning of fishing vessels (in some cases to 50
meter accuracy), for search and rescue (100 meters in bad weather),
and for surveillance, mineral exploration, surveys and scientific
purposes. Present systems of navigational aids in Canadian waters
include Loran A, Decca, Omega and Loran C, and various radiobeacons
and radar beacons. Loran A provides partial coverage in the eastern
arctic but has an accuracy of the order of 2 kilometers and is now
obsolete. Decca and Loran C do not presently cover the arctic.
Decca has an accuracy of 50 to 500 meters but a range of only 100
miles at night and 250 miles in the daytime. Loran C has an
accuracy of 100 to 500 meters but its range is 600 miles over land
and 1200 miles over water. Loran C is scheduled to replace Loran A
in the U.S. and Canada. A chain of Loran C stations will be
required in the arctic to meet the navigational requirements of the
1980's, assuming that the oil transportation scenarios are correct.
Omega will ultimately provide world-wide coverage, with an accuracy
of about 3 kilometers, but it is subject to unpredictable variations.
There are presently nine radar transponders along the arctic coast
and thirteen radio beacons operating during the navigation season.

These provide for accurate short-range position fixes in certain areas, but the system would have to be expanded to deal with the shipping forecasted for the 1980's.

It is proposed that all shipping in the arctic be tracked by ASJOC, using self-reported ship positions based on the systems described above, and independently reported positions based on surveillance systems. ASJOC could thus monitor ship movements and transmit error correcting messages or arrange for search and escort if a vessel encounters difficulties in fixing its position under adverse conditions. The Department of Communications and the Department of Transport would have primary responsibilities, but their communications and navigation systems could be linked to ASJOC for central information processing.

3.4. Medical Aids Requirements. The chief requirements for this programme element would be adequate communications and fast transportation of medical resources to people in need (or vice-versa). Aircraft and medical personnel would have to be readily available for day-night, all-weather operations.

3.5. Aids to Surveys and Research Requirements. The main requirements for this programme element would be those described in 1.3, and 3.1 to 3.4 above, and monitoring the movements and status of research parties and survey crews.

4.1. Maintenance of Records Requirements. It is expected that ASJOC will generate immense quantities of data, much of which will be useful for historical analysis purposes. Most of such data could be transmitted or transported to decentralized but accessible data banks. The main requirement will be the orderly handling of such data in formats that are compatible with modes of analysis and investigation of the agencies who will utilize the data.

The descriptions of programme elements above are rather sketchy, but they may serve to indicate the scope of a programme for ASJOC, and some of the data/information requirements and resource requirements.

6.   CONCLUSIONS ABOUT ORGANIZATION AND IMPLEMENTATION OF ASJOC

Where would the proposed Arctic Services Joint Operations Centre be located? What would it look like?

As part of an arctic development policy, the Centre should be located in the arctic. The choice of location depends to a large extent on developments that are as yet uncertain, particularly the location of oil and gas fields, terminals and shipping routes. The main facilities of the Centre could be established at a single location initially, say in the vicinity of Devon Island (on Lancaster Sound). But ultimately ASJOC could have more than one operations facility.

The Centre would require a settlement, an airfield, and a technical complex including advanced telecommunications facilities, a communications satellite station, perhaps a remote sensing

satellite receiving station, computer facilities, medical
facilities, and a variety of resources required to meet the stated
objectives. It would provide a remote posting in a hostile
environment. It would be costly. It would take time to develop
and bring up to full operational capability - perhaps five years
or more, depending on the priorities of funding and resource
allocation.

In the end, it is expected that ASJOC could provide one of the
most technically advanced and effective communications and control
centres in the world, the brain for the central nervous system of
the Canadian arctic infrastructure.

We propose that the Canadian government immediately establish
an Inter-Agency Committee for Arctic Joint Operations (IACAJO),
and that the committee embark on studies of alternative modes of
joint operations and control and the feasibility of facilities such
as the proposed Arctic Services Joint Operations Centre (ASJOC).
IACAJO would require a secretariat and program planning office.

A REVIEW OF NAVIGATION SYSTEMS

AND THEIR APPLICABILITY TO THE ARCTIC

Captain M. Walker

Canadian Forces Air Navigation School

Winnipeg, Manitoba, Canada

I have been asked to present a critical appraisal of
Navigation Systems from a user's viewpoint. With such a broad
directive, it is obviously impossible to examine each system in
detail and the reader is referred to reference 77, or other
references, for more thorough development. Our task at the
Canadian Forces Aerospace Systems Course is to train selected
officers to examine a broad spectrum of technology and apply this
to meeting operational requirements. This paper, therefore, will
be largely descriptive rather than rigorous. I will be presenting
a "state-of-the-art" overview of Navigation Systems, discussing
their operational characteristics and limitations, accuracy and
cost. I will attempt to relate these characteristics to the
operational requirements of Northern users and I propose to
suggest a methodology by which an optimum system or systems might
be chosen to best meet the requirements for a Canadian Arctic
Navigation System.

Fundamentally, all Navigation Systems are subject to error
and, though based on some deterministic principles, the final
measure of their performance is ultimately statistical in nature.
This principle, though implicit in earlier treatments of navigation,
was first clearly elucidated for the practising Navigator by
Anderson (ref 4) and error analysis forms a most important part of
the published literature. It is error analysis that allows us to
quantitatively compare Navigation Systems, and for consistency in
this paper I shall be quoting all accuracies at the 2 σ or 95%
confidence level.

Anderson also points out (ref 5) that all navigation can be
considered in terms of three basic loops - the handling loop, the
dead reckoning loop and the position finding loop (Figure 1).
Implicit in this treatment is the concept of differentiating and

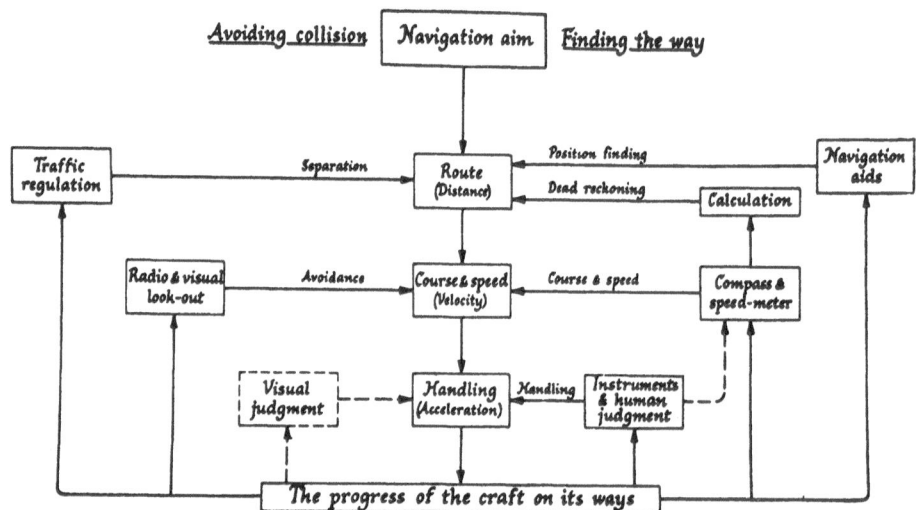

Figure 1.   A Pattern of Navigation (from Anderson, ref. 5)

integrating quantities that are themselves contaminated by error (Figure 2).

Differentiation is a "noisy" process. If I "derive" a velocity from two inaccurate fixes, then my velocity estimate is much worse and a position extrapolated or DR'd from this will have correspondingly greater error. This leads to the concept of integration as a smoothing process. A large number of randomly fluctuating zero mean errors will, by the central limit theorem, tend to integrate toward zero. The price paid for smoothing or filtering of estimates is usually time and a reduced response to higher derivatives. If the error integrated is of a bias nature or step input rather than a fluctuating zero mean error, then the output error will accumulate with time.

Any Navigation System can be thought of as a group of sensors providing position, heading, attitude, velocity or acceleration data, etc. to a processing device or computer that operates upon these raw inputs to yield output information to a display, in a format best suited to assist the Navigator in achieving his aim. These outputs will have errors which will be a function of the input errors, convolved with the transfer function of the processor. The mathematical methods used to describe the deterministic transfer functions are quite involved and are normally taught in the senior year of the associated engineering discipline (refs 100 and 131). The mathematical methods used to

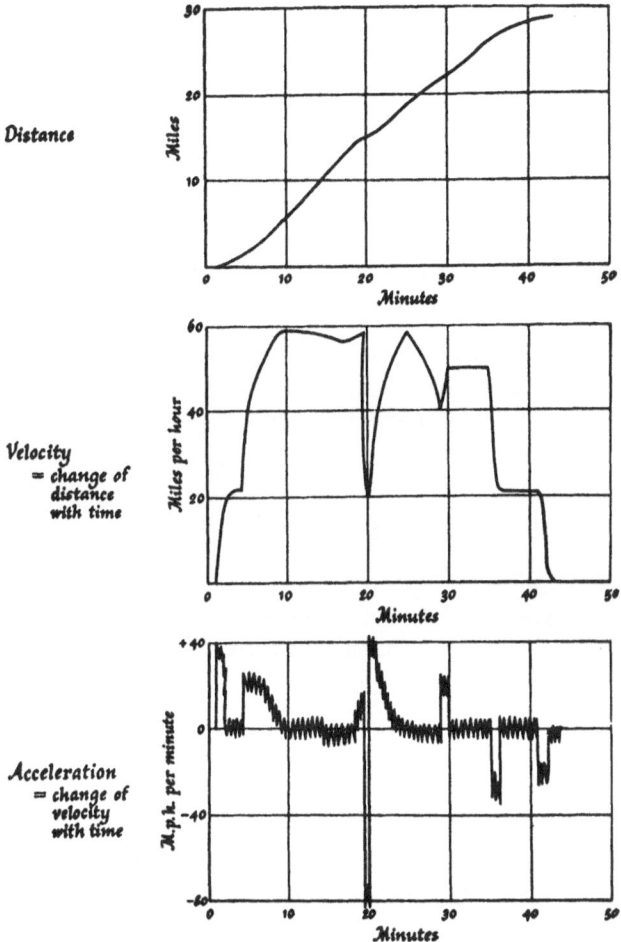

Figure 2.   Distance, Velocity, and Acceleration (from Anderson,
             ref. 4)

describe the random processes are even more involved and usually
taught at the post-graduate level (refs 132, 104 and 8). The point
is, that well established methods exist by which the accuracy of a
Navigation System can be described.

A great deal of attention has been paid in the last 15 years
to the methods of optimum filtering (ref 81), whereby inputs are
"optimally" combined to provide a "best estimate" output. The
price paid for this best estimate is computational complexity and,
occasionally, loss of dynamic response. While "filtered" outputs

are usually more accurate (or at least smoother), they still
represent only a "best estimate of" and are fundamentally limited
by the accuracy of the inputs.

As a practical sidelight to this very general philosophy of
Navigation Systems, I feel that I can safely say that since about
1970 the almost universal acceptance of digital computation has,
for all practical purposes, removed the computing element as a
source of error. Thus from a systems analysis point of view, all
errors, and for that matter all costs, in any Navigation System
are dominated by the sensor and the only costs incurred in
computation are those involved with the complexity of computation
calling for greater software and more memory.

I have dwelt on these philosophical points since we have a
very large number of systems to examine and a practical discussion
of their characteristics is possible only if we have some common
ground.

DEAD RECKONING AIDS (DR)

In this paper we will not concern ourselves with the handling
loop, though from time to time its characteristics will infringe
on our discussion. Dead reckoning, however, is the basis of all
navigation. If I know where I started and keep continuous track
of my direction and speed with respect to a given frame of

Table 1

Instruments for Manual DR

	Speed Reference	Heading Reference	Co-ordinate Frame	Computation
Air	Airspeed Indicator	Magnetic or Gyromagnetic	Maps Charts	Mental and Slide Rule (E6B)
Sea	Ship's Log Propeller Pitot or EM	Magnetic or Gyro Compass	Maps Charts Plotting Board	Mental and Slide Rule
Land	Speedometer	Magnetic or Gyro Compass	Maps Charts	Mental and Slide Rule

reference, then I can compute an estimate of my present position by integrating that velocity vector with respect to time from my time of start to the present time.

The simplest form of Dead Reckoning is a manual or mental plot using the available handling instruments of the vehicle. Because manual DR is usually taught as part of a larger professional qualification, there is a tendency to dismiss it. It should not be treated lightly since about 90% of actual navigation is done by some form of aided manual DR and any professional Navigator will maintain at least a mental plot as final check on his instruments.

The overall accuracy of manual DR is estimated as about 12½% of air distance travelled (ref 156). Marine DR is a little better but in both cases the major error source is an inadequate estimate of the motion of the air (wind) or sea (current and wind). Automated air or sea position indicators are used but require an input of wind or current vector derived from comparing the DR position with an independent fix or derived from a direct measure of ground velocity. Since the earth tends to move rather slowly, most land DR systems show error of about 1% to 2% of distance travelled (ref 86).

DOPPLER SYSTEMS

One method of measuring velocity with respect to the ground is Doppler Radar or Doppler Sonar. It can be shown (ref 49) that the Doppler shift of the energy (radar or sonar) backscattered by the surface of the earth from a minimum of three projected beams (Figure 3) can be uniquely resolved into the vehicle's velocity if the vehicle's attitude and heading can be measured (Figure 4).

Doppler radars are subject to several errors; these may be grouped generally as: a. spectrum; and b. overwater. Since the measure of the Doppler shift is a function of the beam angle and since in any practical system the Doppler beams have a finite width, then the backscattered energy shows a spread of Doppler shifts (Figure 5). The Doppler tracker, therefore, cannot make an instantaneous velocity measure since the signal rapidly fluctuates; however, by integrating and averaging many samples, a reasonably accurate velocity estimate can be made. Typical correlation times are 0.25 to 1 second, thus the Doppler cannot tract rapid accelerations.

Overwater, several errors of a bias nature are observed, the first a calibration shift error due to the difference in backscattering coefficient between land and water at various sea states, the second due to the motion of the sea surface itself under the influence of currents and wind (ref 61). These and other errors are well documented and appropriate correction techniques have been developed with a corresponding increase in complexity and price.

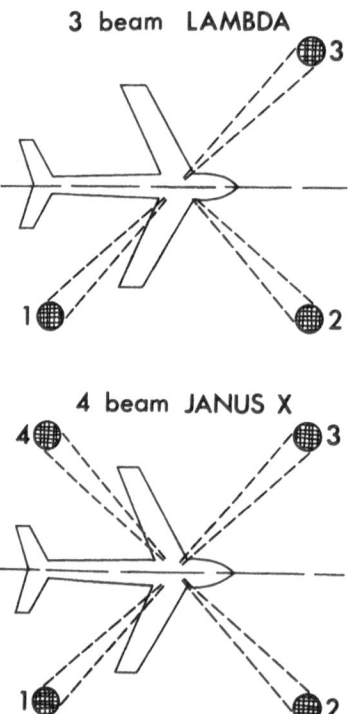

Figure 3. Doppler Configurations

Doppler Sonar for ships is entirely analogous but is further limited by: high attenuation of the beam, rendering if ineffective in water deeper than about 600 ft; backscattering from volume reverberation rather than bottom, resulting in unpredictable Doppler shifts; and quenching and acceleration problems in high sea states.

Land navigation systems, to eliminate odometer slippage or to use Air Cushion Vehicles (ACVs) with no wheels, have used very small Gunn diode Doppler radars or laser velocimeters (ref 86). It is obvious from Table 2 that the overall system error is greater than the Doppler sensor. Fried (ref 49) points out that trials of practical Doppler navigation systems have shown that the dominant source of error is the heading reference. More recent trials (ref 47) of helicopter Doppler (a popular application due to the inherent ability of Doppler to measure low and zero velocities very accurately for hover control) showed dramatic improvements by changing heading references.

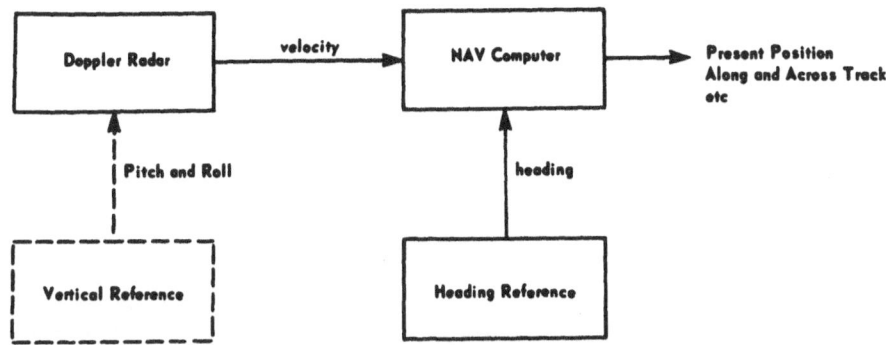

Figure 4. Doppler System

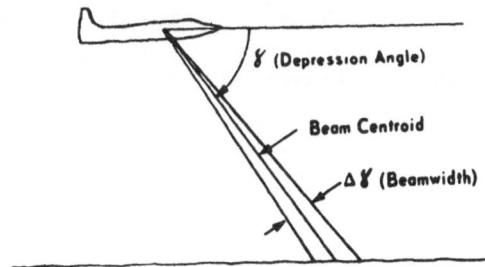

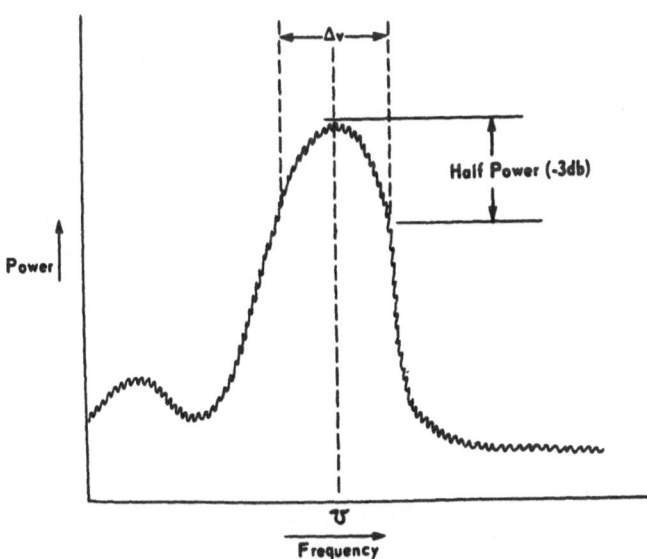

Figure 5. Doppler Spectrum (from Fried, ref. 49)

Table 2

Doppler Navigation Systems

Airborne Doppler Radar	0.2% of velocity	2% of distance travelled	$10,000 $50,000	$13,000 $75,000
Doppler Sonar	0.2% of velocity	2% of distance travelled	$40,000	Rarely used independent of satellite fixing system
Land Navigation System	Gunn diode 0.1% Laser 0.01% of velocity	2% of distance travelled	Est $1,000 for radar Unknown for laser	$5000-20,000 depending on options

HEADING AND ATTITUDE REFERENCES

The heading reference is the most important single sensor in any DR system, not only for computing present position, but for determining the pointing of the vehicle to find its destination. In terrestrial applications, the attitude reference is normally used to stabilize the other sensors but philosophically, heading and attitude references provide a portable frame of reference against which velocities can be resolved into their equivalent earth co-ordinates.

Three physical properties are exploited to yield heading and attitude reference, either singly or in combination: the earth's magnetic field, the rigidity of the gyroscope in inertial space (and its property of precession) and the earth's gravitational field.

The magnetic compass has been used for about 900 years and its properties and corrections are well understood (ref 16). For practical purposes the static errors of a bias nature due to variation (because the magnetic meridians are not the same as the geographical meridians) and deviation (caused by magnetic effects within the vehicle) can be compensated to about ±2° (depending on the compass and vehicle). The magnetic compass is also susceptible to dynamic errors caused by linear and rotational accelerations of the magnetic sensor (ref 148). Since these are large errors and unpredictable in magnitude, the pure magnetic compass is now

only used as an emergency heading and even then only when holding
the vehicle in as straight a course as possible.

Because of its high angular momentum, the gyroscope (ref 134)
has the property of rigidity in space. Unfortunately, this is a
rigidity with respect to the Universe at large or inertial space
and to make use of this rigidity compensation must be made for the
rotation of the earth with respect to inertial space (or earth
rate 15.04°/hr) and further compensation must be made for the
fact that movement around the spheroidal earth introduces another
equivalent rotation.

The simplest gyroscopic instrument is the directional gyro
indicator required by law (ref 69) for at least a backup instrument
in all IFR aircraft (Figure 6). It can be simply shown that if the
spin axis of a horizontal axis gyro is held parallel to the earth's
surface and pointed to the geographic north, then application of a
precessional torque equivalent to the earth's rotation (15.04 sin
Lat) will keep the spin axis point pointing north. Simple enough
in theory, but in practice any tilt, incorrect latitude, initial
alignment error (causing precessional tilt or topple) or motion
over the earth (causing transport wander) will result in an
apparent drift, in addition to the drift of the gyroscope itself.
Drift rates of up to 10°/hr are not uncommon with directional gyro
indicators and they must be continually reset either from the
magnetic compass or from azimuth references or astronomical
sightings when available.

GYRO-MAGNETIC COMPASS

The compensated gyro-magnetic compass using stabilized
magnetic sensor and "low drift gyro" (ref 142) is the standard
heading reference for those air operators who can afford it (and,
usually the heading reference used on most Dopplers). It has been
shown that limitations in the knowledge of the earth's magnetic
field make further improvements to the gyro-magnetic compass
unprofitable (ref 59) and accuracies of 3/4° in the slaved mode
with a 2°/hr (ref 2) drift of the free gyro represent the optimum
for this system.

STABLE PLATFORMS

The gyroscope itself, however, is capable of considerable
improvement. Inertial quality gyroscopes of 0.01°/hr are not
uncommon. The majority of directional gyro errors can be removed
by maintaining it level. Stable platforms maintained in the level
by gyro pendulous devices can result in typical heading drift
rates of about 0.4/hr. A pure stable platform, however, must be
initially aligned to some azimuth reference.

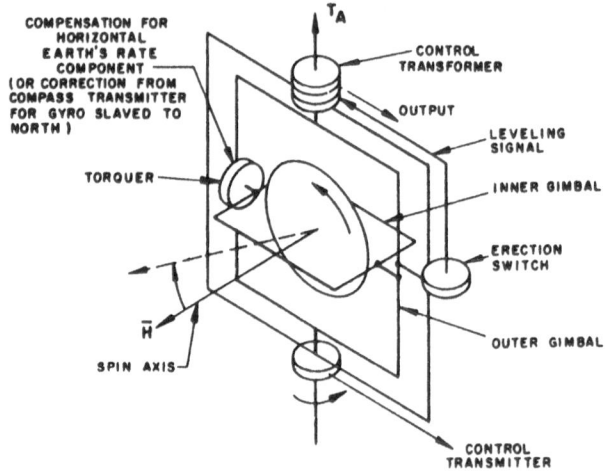

Figure 6. Directional Gyro

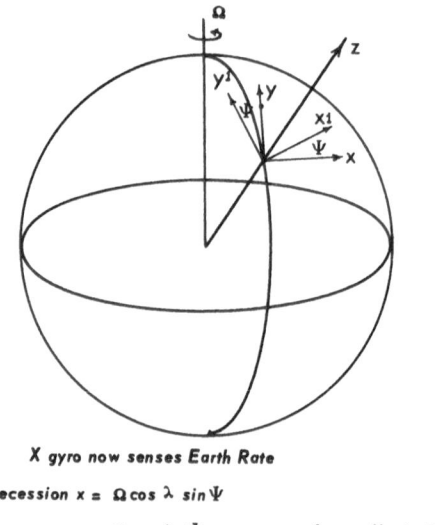

X gyro now senses Earth Rate

Precession $x = \Omega \cos \lambda \sin \Psi$

$\quad\quad\quad = \Omega \cos \lambda \cdot \Psi$ *for small misalignment*

angle Ψ

Figure 7. Gyro Compassing

INERTIAL PLATFORMS

By mounting accelerometers on a stable platform, keeping
these accelerometers level and integrating acceleration once for

velocity and twice for position, we can make an Inertial Navigator.
Inertial Navigations uses a complex series of feedback loops to
maintain level and compute position (ref 146). Gyros of better than
0.01°/hr drift and accelerometers of better than 5×10^{-5} g accuracy
are required to give position errors growing at a rate not greater
than 2 NM/hr. Inertial Platforms are gaining popularity not for
their DR position accuracy (which is modest), but for the fact
that they can provide in one self-contained unit all attitude,
azimuth, velocity and position information required to navigate.

The Inertial Navigator obtains initial azimuth alignment by
gyrocompassing (Figure 7). If the azimuth gyro is misaligned from
north, the levelling gyro will sense a component of earth rate
which will tilt the platform. This tilt will be sensed by an
accelerometer which will try to level the platform and also return
the azimuth to north (Figure 8). Two problems arise with gyrocompassing. The precession torque sensed by the levelling gyro is
a function of cos lat, hence tends to zero at the poles, and the
levelling gyro in gyrocompass mode will sense incorrectly any
velocity around the earth, as earth rate. Thus, for a gyrocompass
to operate in motion, some accurate, independent velocity must be
fed into the system.

Gyrocompasses in simpler forms have been used as the standard
heading reference in ships since about 1920. Ships' gyrocompasses
are generally accepted as accurate to about 1°. On land where zero
velocity is easy to determine, gyrotheodolites are considered
accurate to 0.001°. Several manufacturers of Inertial Navigators
have modified their IN systems for use in land surveying (refs 41
and 96). By making intermittent use of zero velocity checks, these
Position and Azimuth Determining Systems (PADS) can self-calibrate
their internal errors on survey lines of 25 miles, to better than
3 foot accuracy.

It is obvious from Tables 1, 2, and 3 that a broad spectrum
of DR systems exists, many combining the attributes of different
sensors. All DR systems have errors that grow with time. Thus,
after the error has grown beyond limits acceptable for safety or

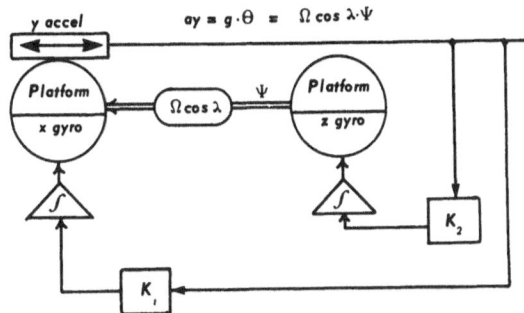

Figure 8. Dynamics of Gyrocompass Mode

operational requirements, they must be updated by some method of
positioning. If these positions are sufficiently accurate, then
methods exist (refs 82 and 159) to use this information to
calibrate out any error sources.

Positioning may be accomplished by celestial shots, visual
references or radar but the most common method of positioning is
to use radio aids. Unlike DR systems which are independent and
can be chosen by each individual to suit his own accuracy and cost
requirements, the radio aids must be provided by some agency.
Since most radio aids require very large capital outlay for ground
sites, they are usually sponsored by the nation over which the
navigation is taking place.

Table 3

Accuracy of Heading References

		LIMITATIONS	ACCURACY	COST
AIR	Compass	Subject to large static and dynamic errors	$\pm 2^\circ$ static 60° dynamic	$50-$200
	Directional Gyro Indicator	Subject to attitude errors and drifts	10°/hr	$300-$3,000
	Gyro Magnetic Compass	End of development	2°/hr free gyro $+0.75^\circ$ magnetic	$2,800-$13,000
	Stable Platform	Needs azimuth reference for initial alignment	0.2°/hr to 0.4°/hr Azimuth Drift rates	$25,000-$40,000
	Inertial	Requires independent velocity to inflight gyrocompass	0.01°/hr	$116,000
SEA	Compass	Subject to large static and dynamic errors	$\pm 2^\circ$ static 60° dynamic	$50-$600
	Gyrocompass		$\pm 1^b$	$8,000-$12,000
	Inertial	Used only on military ships	Classified	>$500,000
LAND	Compass	Must be calibrated	$\pm 5^\circ$ for hand held $\pm 0.1^\circ$ for magnetic survey	$5-$50 $540
	Gyrocompass	Must be gyro-compassed at zero velocity and used as free gyro enroute	$\pm 0.1^\circ$ gyrocompass $\pm 0.1^\circ$/hr Free Gyro	$2,000
	Gyrotheodolite	Fixed survey only	$\pm .001^\circ$	$12,000-$20,000
	PADS	Many zero velocity updates required	$\pm .01^\circ$	$300,000

RADIO AIDS

There are, or have been, in use some 200 different types of radio positioning devices (refs 71 and 138). It would be impossible to discuss all of them at this meeting, therefore we must find ways of classifying them.

All radio aids generate some form of a Line of Position (LOP). Two intersecting LOPs define a fix and three or more provide redundancy as a check on reliability or for statistical averaging. LOPs are generated either as bearings, ranges or hyperbolic lines, thus fixes can be the intersection of two bearings (bearing bearing), range and bearing (range bearing), range range or hyperbolic (Figure 9).

The accuracy of any fix depends on the accuracy to which the position line can be measured and the angle of cut of the LOPs. It is a surprisingly complex task to determine a statistically exact circle that will contain 95% of the fixes, but an acceptable approximation can be had by assuming

$$drms = \frac{\sqrt{\sigma_1^2 + \sigma_2^2}}{\sin \beta}$$

or

$$drms = \operatorname{cosec} \beta \sqrt{\sigma_1^2 + \sigma_2^2} \qquad \text{(Figure 10)},$$

where drms is the circle containing approximately 63% of the fixes, σ_1 is the standard deviation of LOP_1, σ_2 is the standard deviation of LOP_2 and β is the angle of cut (the shallower the cut the worse the fix, the best cut being 90°). The 95% confidence circle we shall use then is 1.731 x drms.

Coverage of a radio aid is of course dependent on the range at which a receiver can adequately process the signals from the ground transmitters to determine a fix. For range bearing only one need be received, for bearing bearing two must be received and for hyperbolic three must be received. But accuracy within that coverage area is strongly a function of the type of LOP. Since the most accurate practically available bearing system has an accuracy of about ±0.5°, it can be easily seen that the accuracy of a bearing LOP deteriorates to about 0.4 miles at 50 miles (tan of small angles). However, using bearing bearing it can easily be shown that the only place where β is 90° is the semicircle whose diameter is the baseline joining the two stations and that β diminishes rapidly anywhere else (in fact, so rapidly that the increase in accuracy of the bearing line closer to the station can never be used). Figure 12 illustrates.

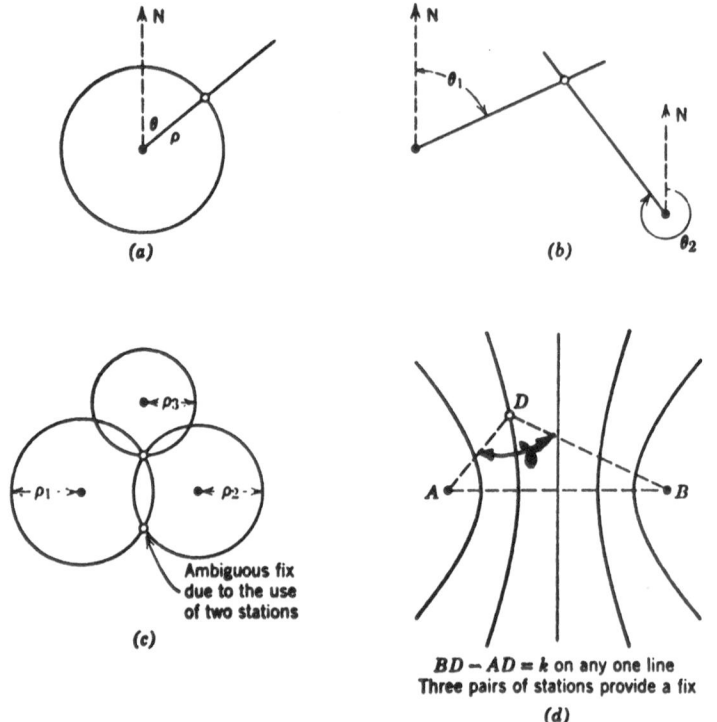

Figure 9.   Common Geometric Fixing Schemes: (a) ρ-θ, (b) θ-θ, (c)
            ρ-ρ, (d) hyperbolic (from Dodington, ref. 33)

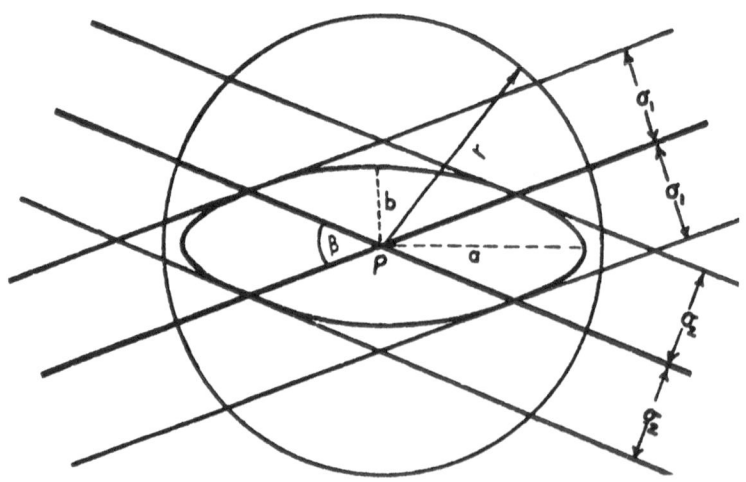

Figure 10.   Fix Accuracy

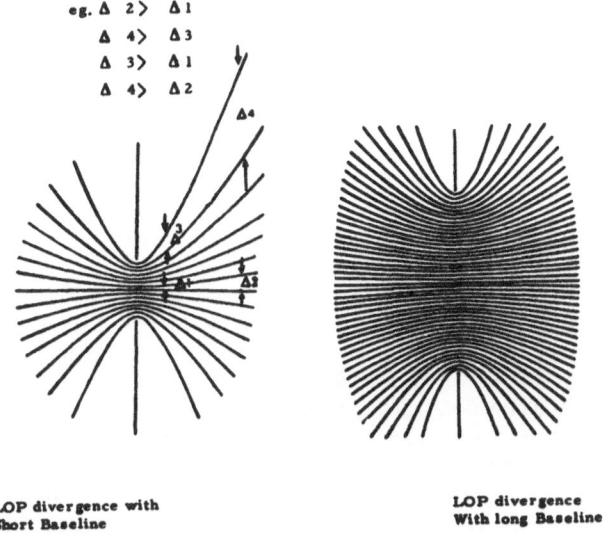

Figure 11.   Hyperbolic Divergence: since each line represents the
              locus of constant time or phase difference, between each
              line at a given point represents in the limit the effect
              that an error in measurement accuracy will have on the
              position accuracy.

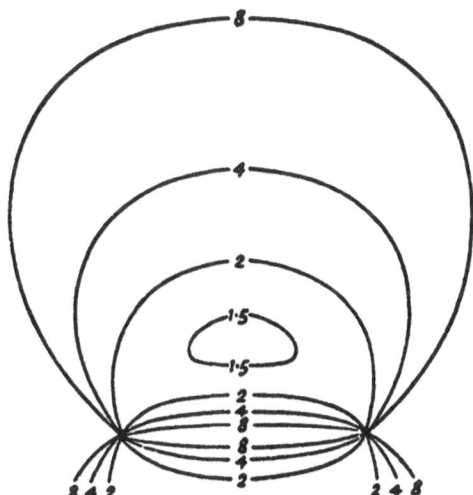

Figure 12.   Accuracy Contours for Two Bearings (95% radial errors in
              n miles assuming (i) base line = 100 n miles; (ii) 95%
              error of each bearing = 1°) (from Anderson, ref. 4).

Hyperbolic lines diverse from the baseline multiplying the effect of any error on the baseline. This divergence error is simply: error on baseline x cosec $\gamma/2$, where γ is the angle subtended at the fix by the lines joining the two stations (see Figure 9). Geometric dilution is reduced by using long baselines (if the transmitter range is sufficient)(Figure 11). Range range systems tend to have LOP errors largely independent of distance and, as illustrated by Figure 13, tend to preserve better geometry at all places except on the baseline. Range range systems are also more economical in coverage since only two stations are required for a fix. Figure 14 illustrates the accuracies obtainable by operating one given system in both hyperbolic and range range mode.

It is a simple matter then to determine coverage and accuracy contours for any given radio aid if an accurate error budget of the signal is obtainable. In general, it can be said that range range is preferable to hyperbolic, which is preferable to range bearing, which is preferable to bearing bearing. The characteristic error budget of any radio aid is a function of the propagation of the radio wave which in turn is a function of the portion of the RF spectrum occupied by that wave. By convention the RF spectrum is divided into bands (Table 4) which very broadly define the propagation mode. Again, by convention, radio waves are considered to be propagated by direct waves, surface reflected waves, surface or ground waves and ionospherically reflected, or skywaves (Figure 15). These too are gross simplifications.

Table 4

RF Spectrum

EHF	extremely high frequency	0.1 to 1 centimeter	30 to 300 GHz
SHF	super high frequency	1 to 10 centimaters	3 to 30 GHz
UHF	ultra high frequency	0.1 to 1 meter	300 to 3,000 MHz
VHF	very high frequency	1 to 10 meters	30 to 300 MHz
HF	high frequency	10 to 100 meters	3 to 30 MHz
MF	medium frequency	0.1 to 1 kilometer	300 to 3,000 KHz
LF	low frequency	1 to 10 kilometers	30 to 300 KHz
VLF	very low frequency	10 to 100 kilometers	3 to 30 KHz

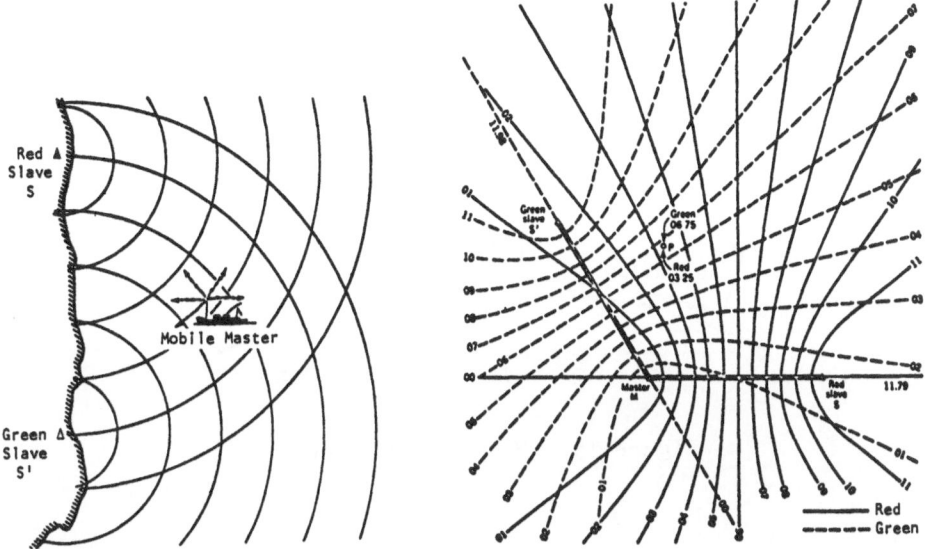

Figure 13.   Ranging (a) and Hyperbolic (b) Lines of Position (from
Bigelow, ref. 12).

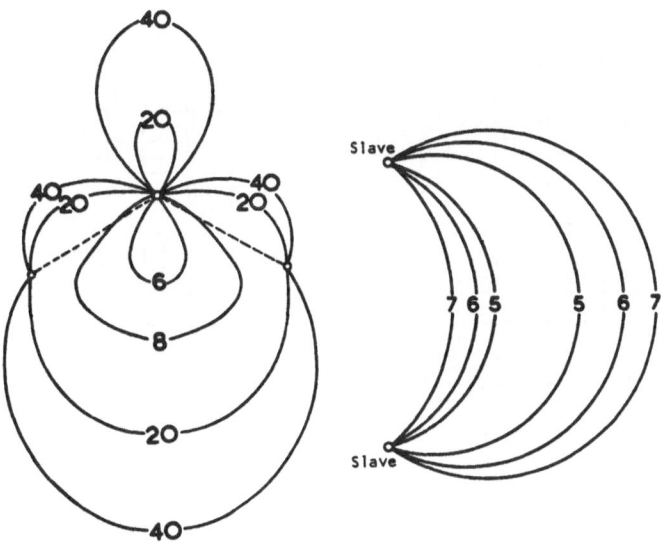

Figure 14.   Typical Relative Accuracy Contours (in feet) for Hyperbolic (a) and Two-Range (b) Hi-Fix Type A Chain (from
ref. 71).

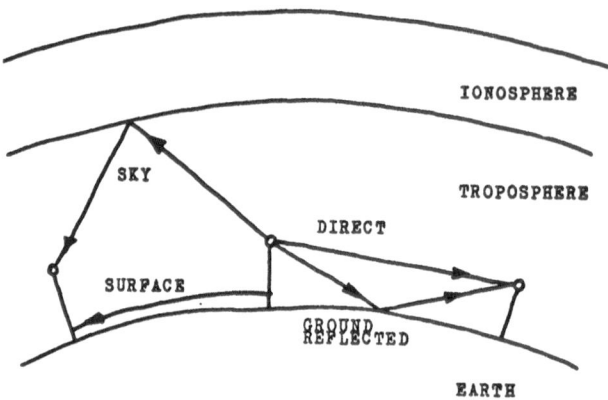

Figure 15. Propagation Modes

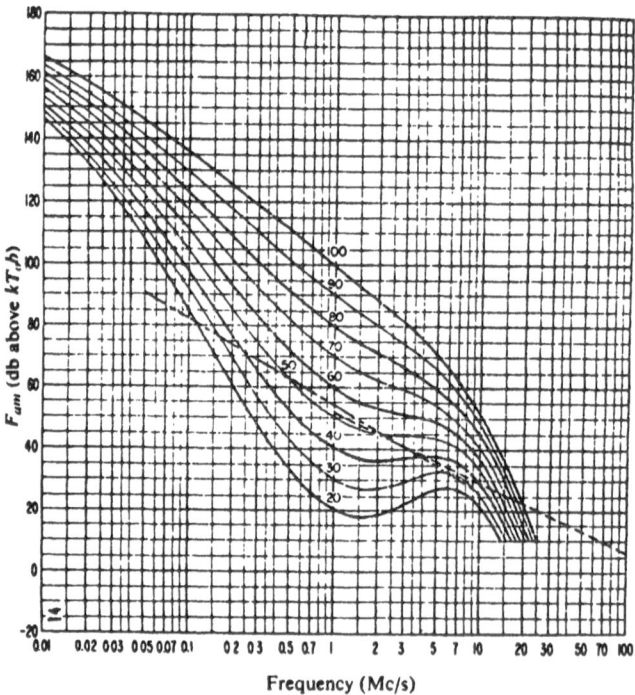

Figure 16. Variation of Radio Noise with Frequency (Summer, 0000-
            0400 h): ———— expected values of atmospheric noise;
            -•-• expected values of man-made noise at a quiet re-
            ceiving station; ---- expected values of galactic noise
            (from CCIR Report 322).

As a final set of simplifications, it can be said that antenna efficiency decreases with frequency, thus at VLF an antenna may radiate only 7% of the input power (ref 151). Propagation distance increases with a decrease in frequency, and as a result noise increases with a decrease in frequency (Figure 16). Finally, in phase measuring systems, the phase of any wave can only be measured to about 1/100 of a full cycle (without statistical filtering), thus the lower the frequency the less accurate the system.

VLF Propagation

VLF propagation has been extensively studied. Wait lists 326 references (ref 144). Burgess (ref 21) elucidates the factors involved in choosing the frequencies useable at VLF for navigation. VLF waves, because their wavelength is a significant fraction of the effective ionospheric height, can be considered to propagate in a duct mode in the spherical waveguide formed between the earth and the D layer of the ionosphere (Figure 17). Propagation is best described in terms of Transverse Magnetic Modes (refs 21 and 7) and minimum attenuation occurs at about 16 KHz (ref 70) where powers of a few kilowatts can propagate around the earth. Below 16 KHz attenuation, particularly of the higher modes, increases until mode cutoff at about 4 KHz. At about 8 KHz antenna inefficiences combined with attenuation make use of lower frequencies impractical. The suppression of higher modes leaves a single phase stable mode of propagation beyond the region of modal interference (about 300 to 600 miles) which is dominated by a diurnal phase shift due to the difference in height of the D layer from day to night (Figure 18). Several models are available to predict this variation (refs 76 and 51), the most popular being the one attributed to Swanson (ref 51).

Above 16 KHz phase instabilities due to modal interference increase until at 24 KHz the diurnal shift is difficult to determine. Experimental data by Burgess (Figures 20, 21 and 22) show this phenomenon clearly.

LF and MF Propagation

Above about 50 KHz the groundwave is relatively uncontaminated out to about 200 miles and its phase and amplitude lend themselves to quite accurate calculation (refs 73 and 145). The major factor affecting signal strength is ground conductivity which tends to retard and attenuate the groundwave. Attenuation is least over sea water (high conductivity 5 mho/meter) and greater over land (typically 0.001 mho/meter), being at its worst over Arctic tundra and icecaps (0.5 milli mho/meter and 0.025 milli mhO/meter respectively). Attenuation of the groundwave increases with frequency. If sufficient bandwidth is transmitted, then it is

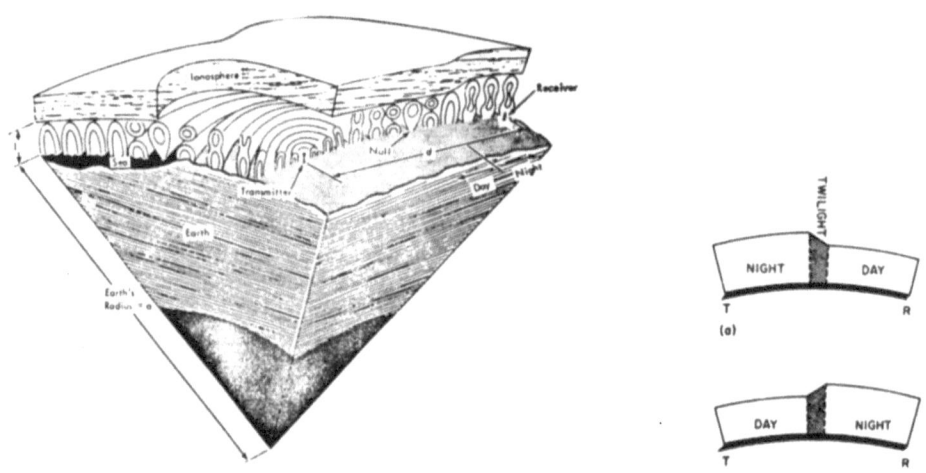

Figure 17. (Left) Approximate electric field patterns within the
earth ionosphere cavity excited by a short vertical
monopole antenna. Fields existing above the reference
reflecting height h are not shown in the figure. Note
h/a is exaggerated about 10 times (after Watt [1967].
(Right) Simplified models of sunrise/sunset transitions
in the earth ionosphere wave guide. (Both from Belrose,
ref. 7).

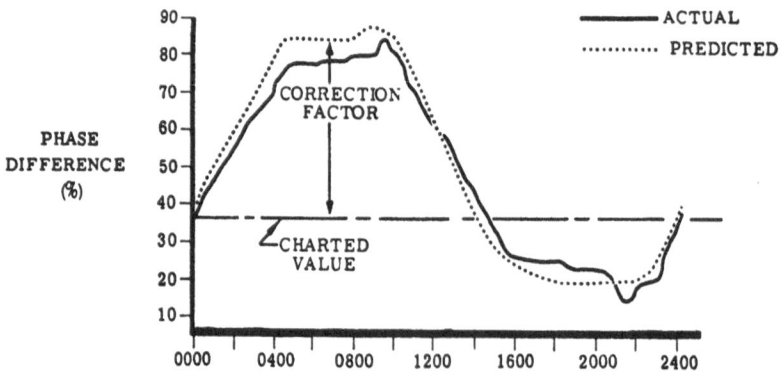

Figure 18. Typical Phase Variation (Diurnal Effect)

STATION	ON-AIR DATE/ PROJECTED ON-AIR DATE	CURRENT STATUS OF ANTENNA SYSTEM	REMARKS
A Norway	December 1975 (Reduced power - 7 KW)	Valley span installed (see Remarks)	Antenna system damaged with repair scheduled in July 1975 over a period of two weeks.
B Trinidad	February 1966 (3 KW)	Valley span installed	To be replaced by OMSTA Liberia
⚹B Liberia	December 1975	Tower being fabricated in USA	Replacing OMSTA Trinidad. Although it is expected this station will be transmitting by December 1975 in the G segment, for station tests only, it will not be accorded operational status until March 1976.
C Hawaii	January 1975	Valley span installed	
D North Dakota	October 1972	Tower installed (see Remarks)	Extensive maintenance is to be performed on the antenna system for a 45 day period beginning 1 September 1975.
⚹E La Reunion	December 1975	Tower being fabricated in USA	ONSOD electronic check-out scheduled for October/ November 1975.
⚹F Argentina	August 1975 ·	Tower erected. Ground system and electronics being installed.	ONSOD electronic checkout scheduled May 1975. This station will probably transmit in segment G for "on air" testing prior to becoming operational.
G Tasman Sea	Not applicable	Site not picked	Bilateral agreement between host country and USA has not been negotiated.
H Japan	April 1975	Tower erection completion November 1974	To be declared operational by JMSA on 30 April 1975.

⚹ It is anticipated that future station testing will be accomplished in the G segment. Station transmissions will be shifted to the proper segment when operational.

Figure 19.   Omega Navigation System Status as of April 28, 1975.

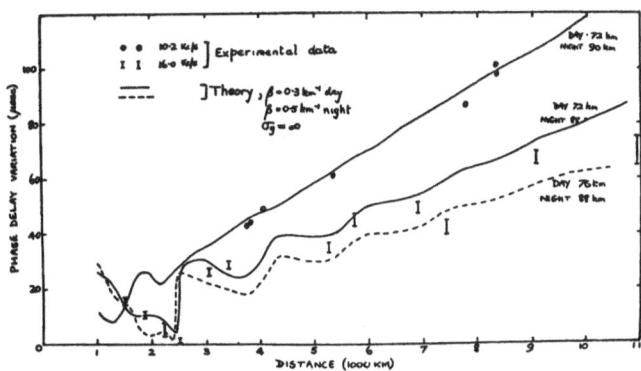

Figure 20.   Magnitude of Diurnal Pulse Delay Variation versus Distance for 10.2 kc/s and 16 kc/s.

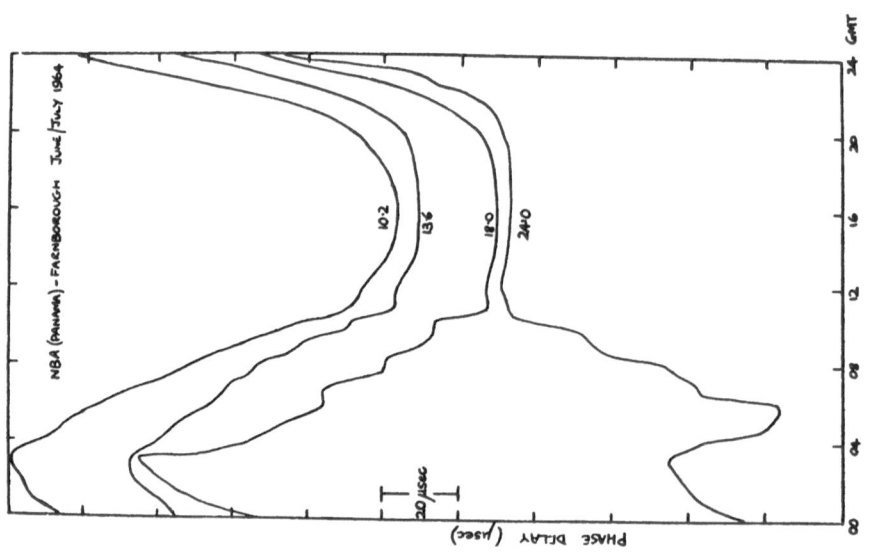

Figure 22. Diurnal Phase Delay Variation for
Various Frequencies (from Burgess, ref. 21)

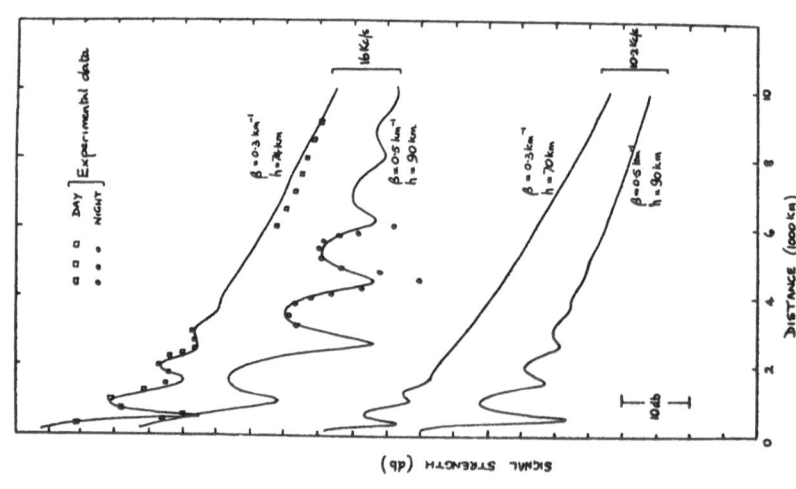

Figure 21. Field Strength vs. Distance for
10.2 kc/s and 16 kc/s (from Burgess, ref. 21)

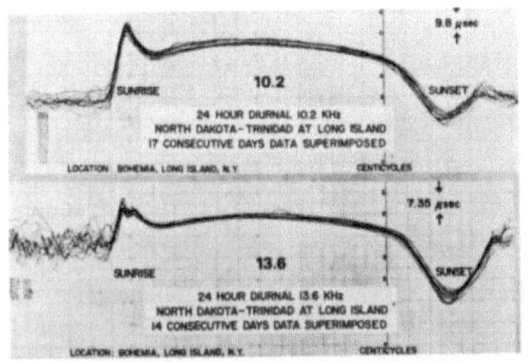

Figure 23. Omega Phase Differences, North Dakota vs. Trinidad: (a) Solar Quiet Conditions; (b) Disturbed Conditions (from Beukers, ref. 10).

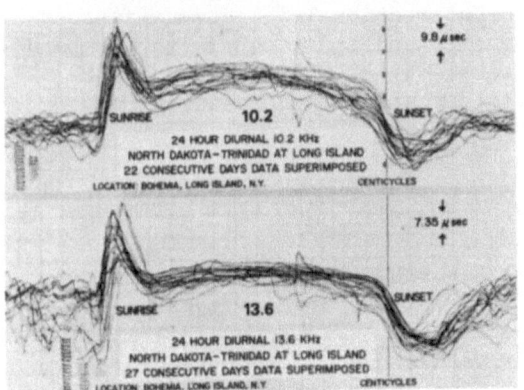

STATION	LOCATION	FREQUENCY (KHz)	RADIATED POWER (KW) Nominal	RADIATED POWER (KW) Authorized
NAA	Cutler, Maine	17.80	890	1,000
NBA	Balboa, Panama Canal Zone	24.00	150	1,000
NLK	Jim Creek, Wash. State	18.60	250	1,000
NPM	Hawaii	23 40	140 to 630	1,000
NWC	North West Cape, Australia	22.30	1,260	1,000
GBR	Rugby, Great Britain	16.00	250	300
NDT	Yosami, Japan	17.4	125	500
JXN	Helgeland, Norway	16 4	150	350
NSS	Annapolis	21 4	500 +	1,000
NAU	Puerto Rico	28.5	50	100
Ω A	Norway	12.30	1	10
Ω B	Trinidad	12.00	1	10
Ω C	Hawaii	12.20	1	10
Ω D	Lamore, N.Dakota	{12.85 / 13 10}	9 to 10	10

(frequency group bracket: 10 2 / 11.33 / 13 6)

Figure 24. Omega and Comm. Stations Available for Nav. (from Hardwick, ref. 63).

possible to separate the skywave and groundwave. Groundwave ranges of 1,200 miles are possible at 100 KHz given sufficient transmitter power. Groundwave is still useable over water up to about 2 MHz but beyond this frequency the groundwave is too attenuated for useful propagation and skywave is too unstable for navigation (Figure 26).

If attenuation, antenna efficiency, accuracy requirements and system costs could be optimized then an optimum navigation frequency could be chosen. Historically this appears to have been determined by the USAF in their early Whyn and Cytac experiments (ref 66) and by Pierce in the LF Loran experiments (ref 107); a frequency between 140 and 180 KHz or between 200 and 285 KHz with a bandwidth of about 20 KHz seemed to be the optimum choice. Hefley (ref 66) documents the failure to obtain International Telecommunications Union approval for these frequencies and as a compromise the band between 90 KHz and 110 KHz centered on 100 KHz is the only one approved for pulsed or broadband transmissions necessary to discriminate against the skywave.

VHF/UHF Propagation

Classical geometric optics describes propagation at these frequencies with attenuation following the inverse square law, though at VHF some groundwave still exists. Ground reflected waves can cause phase and amplitude fluctuations. Since the direct wave is shadowed by the earth's curvature, then either the transmitter or the receiver must be elevated, with the "horizon" being approximately $1.2 \sqrt{H}$ miles if H is in feet. Thus, the VHF UHF band is normally used for aircraft aids if the transmitter is on the ground. One way around the earth's shadow is to elevate the transmitter to satellite heights.

SHF Propagation

At these frequencies receiver self-noise becomes a limiting factor as well as a difficulty in obtaining reliable high power devices for the transmitter. Some alleviation is possible since at these frequencies the antenna is very large with respect to the wavelength. High antenna gains are possible using directional "beams".

The foregoing discussion has been treated in more detail by Johler (ref 75) and Figure 27 illustrates navaids by their spectrum position. In practice, it means the navigator has a strictly limited number of systems to choose from.

RADIO AIDS IN COMMON USE

Omega

Omega has been described in references 108, 110, 129, 137, 139, and 140. Eight 10-kilowatt stations strategically located around the globe will provide time shared signals at 10.2, 11.33 and 13.6 KHz. Since the individual stations are slaved to Universal Time UT2 by cesium beam clocks, the system can be used in either range range or hyperbolic modes. At full operational status (Figure 19), any place on the globe will receive signals from at least five stations ensuring redundancy. Also, since baselines are very long, geometry is rarely a problem.

Omega is quoted as having an accuracy of 1 mile (nautical) by day and 2 NM by night at the one sigma level. While this statement is true to a few percent, when averaged over all conditions it must be hedged very carefully. Omega fix accuracy is distinctly non-gaussian in distribution. The 1.7 NM by day and 3.5 NM by night circles will not contain 95% of the fixes. Most of the errors in Omega are due to differences in real world ionospheric height and ground conductivity from the values used in the diurnal correction model. Sudden Ionospheric Disturbances, Polar Cap Anomalies, and Modal Interference at higher frequencies all contribute errors to the tail of the Omega distribution, including some errors nearly 16 NM, though fortunately these are extremely rare (ref 18). Canadian Forces trials have shown, in a limited number of samples, fix accuracies from 2.5 to 4.4 NM at the 95% level in static tests and 1.6 NM to 2.8 NM at the 95% level in flight tests (ref 158). This data showed large bias errors indicating inaccurate ground conductivity values. Use of range range mode does not materially affect accuracy as would be expected from the geometric considerations (ref 20).

Another problem associated with Omega is fix ambiguity or lane skip. Since Omega is a phase comparison system, ambiguities occur every half wavelength in hyperbolic mode or every wavelength in range range mode, i.e., every 8 NM or 16 NM at 10.2 KHz. Although two other frequencies are provided to resolve ambiguities, the dispersion of these waves is such that under certain ionospheric conditions it is possible to resolve the wrong lane. This has led one leading manufacturer of airborne Omega sets to use only the 10.2 KHz signal and rely on a DR routine to maintain lane count.

Omega fluctuations (Figure 23) also result in poor velocity resolution (temporal decorrelation). Using Beuker's data it has been determined (ref 9) that Omega velocity resolution is about 4 knots using a one minute correlation period, which agrees well with Wright's (ref 150) ½ to ¾ knot resolution using 15 minute to one hour correlation times. Such poor velocity resolution even over long correlation times makes Omega derived velocity unsuitable for hybrid systems. In another respect, Omega derived winds have shown up to 30 knot errors in some airborne systems. The combination of

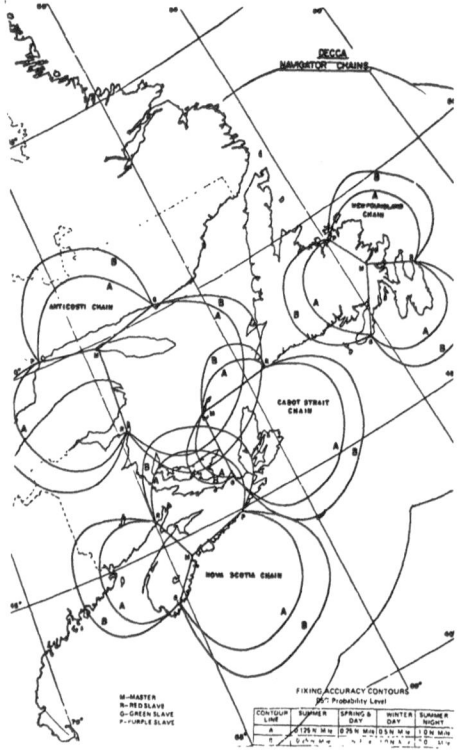

DECCA CHARACTERISTICS

1. Frequency – 70–130 KHz.

2. Range – ≈ 250 nm from master station.

3. Accuracy – better than ±100 yards (95%) near
    baseline with good angle of cut to
    ≈ ±4 nm at 250 nm.

4. Area Coverage – ground up.
    used in – Western Europe
    – East Coast of US and Canada
    – Persian Gulf
    – India
    – SW coast of   Africa.

5. A hyperbolic, phase comparison system.

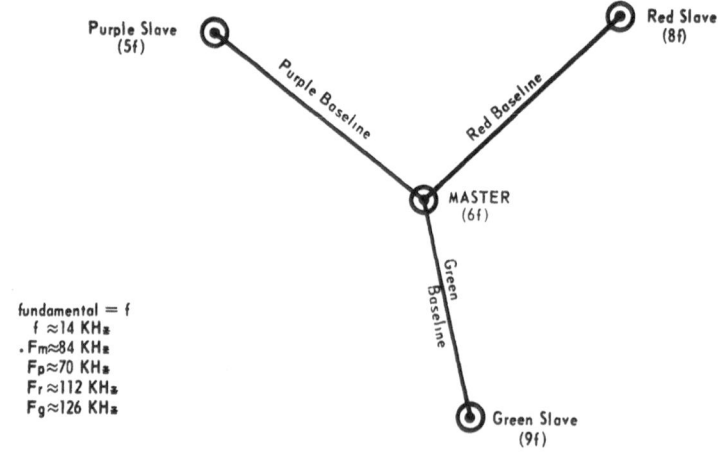

Figure 25.   (from ref. 71)

temporal and spatial decorrelation has not been measured but it is obviously this combination which causes such poor velocity resolution in airborne systems.

Differential Omega has been proposed as a Northern Navaid (refs 55 and 88) with ¼ to ½ mile accuracies 40 miles from the monitor station quoted. Beukers and Tracor data as used by Polhemus associates (ref 118) indicate that this accuracy cannot be met under night time conditions at the 95% level and that in any case real time dissemination of the data is essential because of the temporal decorrelation. Real time dissemination would require a dedicated communications link, receiver, and Omega interface, increasing complexity of the system.

Omega can be summarized as a system of great reliability and redundancy, but only modest accuracy, ideal for the enroute portion (anywhere in the world) of any system which has already been given accurate terminal aids.

Two manufacturers offer airborne Omega receivers commercially and one quotes a price of $50,000 installed. The USAF presently has a competition between six manufacturers to produce 1,000 airborne sets at a unit price of less than $15,000 (ref 27). Shipborne sets manufactured by over a dozen firms cost between $4,000 and $12,000 and at least one manufacturer has married his manual set to a low cost programmable calculator for diurnal correction and co-ordinate conversion (ref 130).

VLF

Naval communications stations (Figure 24) are phase locked to UT2 by cesium beam standards, thus they also can be used to generate LOPs. Three systems using these stations are commercially available: Global Navigation Systems "GNS 200" and "GNS 500" (ref 57), and communications Components Corporation Ontrac II (ref 26). The operation of the GNS 200 is described in reference 62. Little is known about the GNS 500 or the Ontrac II except that the Ontrac II uses a rubidium oscillator to allow range range implementation (ref 143).

The GNS 200 generates steering and range to go to a precomputed point from a known precomputed point probably using a lorhumb type algorithm as proposed by Pierce in 1948 (ref 107). This gives a very simple mechanization and an easily read display but no flexibility. This lack of positioning information and flexibility in programming alternate destination led 413 SAR Squadron to conclude that the GNS 200 was unsuitable as a fixing aid (ref 160). McLarnon of 440 Squadron Yellowknife (rev 91), however, enthusiastically supports the use of GNS 200.

Examination of Figure 24 shows that most of the VLF stations transmit at much higher power than Omega. This, coupled with uninterrupted CW transmission, permits heavier filtering. Hardwick (ref 64) reports accuracies of 1.55 NM at 95% but considers velocity

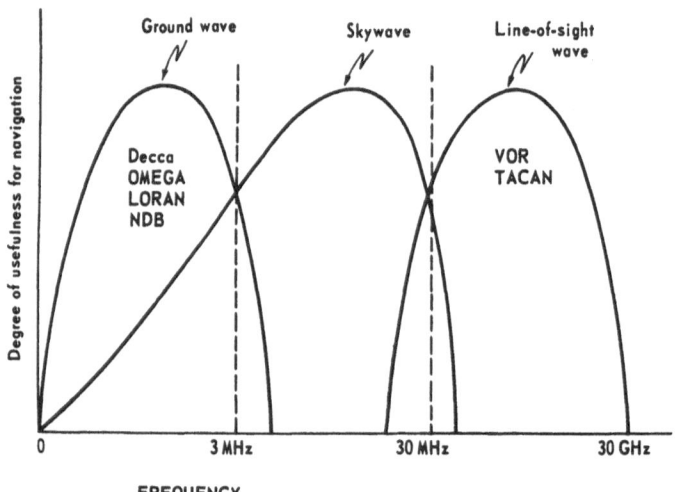

Figure 26. Skywaves are unpredictable so radio aids to navigation avoid 3 MHz - 30 MHz band (from Dodington, ref. 33).

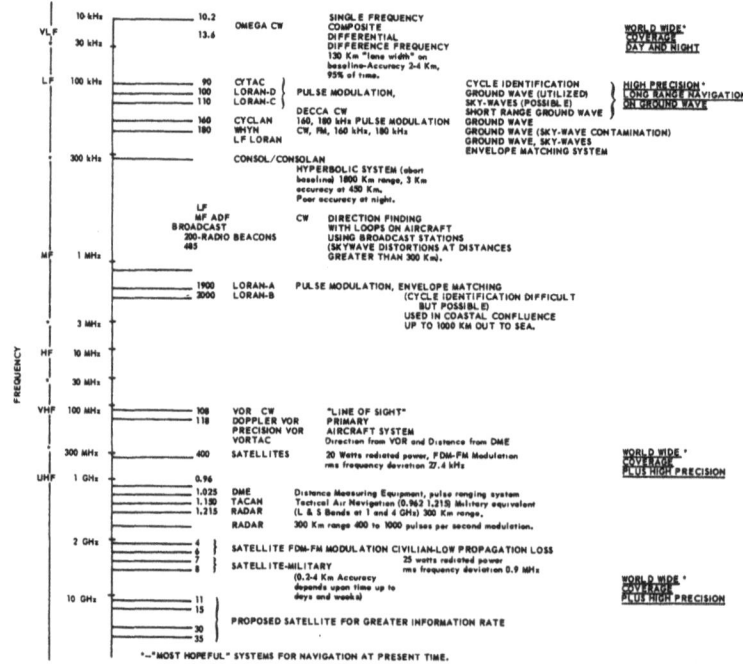

Figure 27. Navigation Systems in the Frequency Domain (from Johler, ref. 75)

resolution to be "poor" about 5 knots. Furthermore, the NAE North Star on which these trials were performed uses a rubidium oscillator to permit range range mode and heavily filtered the VLF information with Doppler (ref 63). Examination of Figure 24 also shows that all VLF comm station frequencies are above 16 KHz, thus modal instabilities can be expected. Hardwick has found them insignificant even in flights directly over Cutler, Maine, but trials by Canadian Forces have shown an along track indicator drifting which is probably attributable to modal instability.

The greatest limitation to the use of VLF communications stations for navigation is not physical but operational. The communication stations are shut down for routine maintenance (Table 5) and can also be shut down without notice for operational reasons. No formal operating policy has been published. The tragic crash of a Sabreliner (CF-BRL) on the 27th of February 1974 attributed at least in part to the unscheduled shutting down of Cutler, Maine and Jim Creek, Washington during BRL's flight (ref 24) illustrates the navigational hazards of using VLF communication stations. The GNS 200 is not approved by the Canadian Ministry of Transport as a prime navigation aid but has been granted a waiver by the United States Federal Aviation Agency for use as an R-Nav system in the Gulf of Mexico area (ref 46). Finally, it has been reported the VLF stations will adopt a minimum shift keying technique in 1975 which in turn will require modification of all VLF navigation receivers (ref 32).

Present cost of GNS 200 is about $19,000 and Ontrac II about $30,000. The only use of VLF in shipboard applications has been in research vessels and each installation is unique.

VLF can then be summarized as, Omega with restrictions.

Decca

Decca is described thoroughly in reference 71. It can be summarized as a hyperbolic CW phase comparison system operating in the LF range between 70 KHz and 90 KHz, and between 110 and 130 KHz. Because of the method of phase locking slave stations to the master, it can only be used in the hyperbolic mode. Because it is a narrow bandwidth system, skywave interference cannot be separated and this determines the useable range of about 240 NM at night under normal ionospheric conditions.

Four Decca chains operate in the Gulf of St Lawrence (Figure 25) and chains are operating in Europe, India, the Persian Gulf and Straits of Malacca. Four hundred users in Canada and 20,000 around the world attest to its popularity. Because the baselines are relatively short compared to usable range, fix accuracy is strongly a function of geometry and has been quoted as between 500 feet to 2.6 NM (ref 124). It can be calibrated to much better accuracy than this and the Decca company offers several survey versions. Decca accuracy has been monitored over the 25 years it

Table 5   Station Maintenance Schedule

DAY	STATION	GMT	EST / CDT	CST / MDT	MST / PDT	PST	YST	AST
SUN								
MON	P (NBA)	1200-1800	0700-1300	0600-1200	0500-1100	0400-1000	0300-0900	0200-0800
TUES	G (GBR)	1000-1400	0500-0900	0400-0800	0300-0700	0200-0600	0400-0500	0000-0400
	S (NWC)		1900-2200	1800-2100	1700-2000	1600-1900	1500-1800	1400-1700
WED	S (NWC)	0000-0300						
THUR	W(3)(NLK)	1700-2200	1200-1700	1100-1600	1000-1500	0900-1400	0800-1300	0700-1200
	J(2)(NDT)	2200-2400	1700-2400	1600-2400	1500-2300	1400-2200	1300-2100	1200-2000
FRI	J (NDT)	0000-0600	0000-0100					
	M (NAA)	1400-1800	0900-1300	0800-1200	0700-1100	0600-1000	0500-0900	0400-0800
	L (NPM)	1700-2300	1200-1800	1100-1700	1000-1600	0900-1500	0800-1400	0700-1300
SAT								
DAILY	O(4)(JXN)	0000-0300	1900-2200	1800-2100	1700-2000	1600-1900	1500-1800	1400-1700

(1) Omega Station T operates without scheduled interruption.
(2) J remains off an additional 2 hours on the first Thursday of each month (until 0800 GMT Friday).
(3) W off for maintenance 1st and 3rd Thursdays of each month only.
(4) O has brief interruptions occurring frequently at approximately 5 minutes before the hour.
Above information is current as of 2-15-73. To obtain weekly announcements and notice of changes in scheduled maintenance, write to: The Superintendent, Time Service Division, U.S. Naval Observatory, Washington, D.C., 20390, and request Time Service Publications Series 3, 4 and 14.

has been in service and the system error budget is very well understood. One characteristic of the statistics of Decca fixes is a leptokurtic distribution, i.e., while errors are quoted at the 95% level, the normal 63% circle in fact contains 75% of all fixes, thus Decca performance is better than a superficial reading of the literature would indicate.

Cost of a four station Decca chain is quoted as $1,868,100 with annual operating costs of $200,000 per year (in accessible temperate areas)(ref 116). Decca receivers are usually rented at about $240/month but can be purchased for about $8,000. Airborne installations normally use some form of moving map display or interface to an R-Nav system and can cost from $20,000 to $40,000.

Decca in summary is a relatively high accuracy system enjoying very great user confidence but fundamentally limited by skywave interference.

Loran-C

Loran-C has been covered previously at this meeting by Mohin. Other references can be found in refs 28 and 72. Loran-C occupies the same portion of the RF spectrum as Decca but achieves greater usable range by high power and discrimination against the skywave, using a fast rising pulse which makes the effective bandwidth of the system 20 KHz. Loran-C is slaved to UT2 by cesium clocks and can therefore be used in the range range mode but the "master" signal is identified by an extra pulse and hyperbolic mode is more common. Some receivers can also use what is called cross chain mode allowing greater flexibility in station selection.

In the hyperbolic mode long baselines reduce geometric dilution and for all except a small area around the master of Decca, Loran-C is more accurate than Decca (ref 13). The long range of Loran-C means that fewer stations are needed to cover a given area, one Loran-C chain covering the area of 10 to 25 Decca chains.

Loran-C typically has an accuracy of between 1,000 ft and 1 NM but can be calibrated to 150 ft and 1,500 ft. Recent work by Johler (ref 74) has shown that stabilities of 20 nanoseconds are achievable (consistent with the theoretical limit) and temporal correlation appears very high thus a position accuracy on any LOP of 10 ft x hyperbolic divergence may be achieved using differential monitoring. Differential corrections need not be transmitted in real time, in this case merely broadcast as a correction, and are reliable for long periods. The very high temporal correlation also means that accurate velocity resolution is possible typically 0.1 knot with 1 to 2 seconds correlation time. The effects of spatial decorrelation on moving vehicles has not been reported but work is proceeding rapidly in this direction (ref 105).

The effective range of Loran-C is dominated by three factors - ground conductivity affecting the signal strength, local noise,

and the signal processing scheme used to extract the Loran-C
information. The difference in ground conductivity between sea
water and tundra can cut the effective range typically from 1,200
NM to 600 NM for a given transmitter, noise and receiver. The
most important information to be processed is third cycle
identification. If the third cycle cannot be identified then the
absolute accuracy of Loran-C cannot be used, though repeatable
accuracy may still be good.

Loran-C transmitter site costs vary widely, but a typical
single station would cost about $4.7 million. Airborne Loran-C
receivers with co-ordinate conversion and R-Nav capability are
presently available for about $40,000 but the United States
decision to implement Loran-C in their coastal confluence (ref 43)
and the imminent decision to construct inland stations, will bring
strong market forces into play to lower this considerably.

The predicted table of costs, Table 6 (ref 114), appears quite
reasonable in the light of shipborne developments. In 1971 a
typical shipborne Loran-C receiver cost $30,000. By 1972 (under
US Coast Guard sponsorship) several receivers were available for
$3,500 and a recent advertisement in a yachting magazine indicates
that market forces have brought one manufacturer to offer a set
for "slightly over $1,000".

In summary, Loran-C is a bigger and better Decca enjoying the
sponsorship of the United States.

Table 6

Loran-C Costs

Price Goal	Receiver	Output Parameters	Flight Guidance	Computer Storage
<$2K	Acquire & Track Master + 2 sec	Course, CTE, DTG	• HSI	2 Waypoints
$5-10K	Acquire & Track Master + 2 sec	Course, CTE, DTG GS, BRG LAT-LON	• HSI • Autopilot Coupling	3 Waypoints
$18-25K	Acquire & Track Master + 3 sec	Course, CTE, DTG GS, BRG, W/V LAT-LON	• HSI • Autopilot Coupling • Approach Mode • Compatibility with ARINC 582 & 583	10 Waypoints
$40-50K	Acquire & Track Master + 3 sec Cross Chain Capability	Fundamental ARINC 582 or 583 System with Loran-C as Primary Sensor		FLT Plan Route Storage

Other LF Systems

While the band between 130 KHz and 200 KHz is assigned to
fixed or Maritime Mobile Communications Services, several powerful
transmitters use this band in Canada. Slaving these transmissions
to cesium standards and UT2 might provide useful navigation
information but only out to the skywave interference limit (unless
someone can devise a signal processing scheme whereby the broadband
signal allowed for facsimile transmissions can be used to separate
skywaves).

Around the 200 to 400 KHz region, Consol stations still
transmit (ref 71). Consol bearing require about one minute to read
and automation would be difficult. Consol accuracy is unpredictable
and only a few stations exist. Still they do provide information
and if all else fails, receivers costing less than $100 can be used
to pick them up.

Non Directional Beacons

The ITU assigns the frequency band 200 to 415 KHz and 510 to
535 KHz to "Maritime Radio Navigation" and "Aeronautical Radio
Navigation". These frequencies are now used almost exclusively
for Non Directional Beacons. These signals are processed by
direction finding loop antennae to provide a bearing relative to
the vehicle. Thus to obtain a position fix, the heading of the
vehicle must be added to the relative bearing (adding the heading
error). A general procedure for aircraft use is to "home" the
beacon and obtain a position fix "on top". Severe thunderstorms
have been known to successfully mimic NDBs, resulting in
uncomfortable (and inaccurate) "on tops". Range of a NDB is
dependent on power, ground conductivity, and local noise, with
attenuation increasing at the higher frequencies. Maximum range
is defined by ICAO as the 70 microvolt/meter contour and may be
computed using reference 145. Ranges of 150 miles for 1 Kw beacons
have been quoted in the Arctic (ref 42), but this is probably
optimistic over tundra and Phipps reports, "It is not uncommon for
the effective range of a low frequency beacon to be reduced to 20
or 30 miles" (ref 106).

Accuracy of NDBs is variable and subject to a large number of
predictable and non-predictable effects, such as quadrantal error
and conductivity inhomogeneities (e.g., coastal refraction). A 95%
figure of ±5.5° is suggested by ICAO for planning approach procedures
(ref 162) and this appears to be a reasonable value. One report
(ref 32) has proposed increasing the number of NDBs in Northern
Canada to 136, using these as an Area Navigation System. Assuming
the optimistic range of 150 NM and the best geometry possible (two
beacons on a 150 NM baseline), it is easy to show by the arguments
proposed earlier that fixing accuracy for this system can never be
better than about 25 NM 95%.

NDBs cost $15,000 to $20,000 for 1 Kw transmitters and $50,000 to $60,000 for 3 Kw transmitters. In addition, site costs run from $50,000 to $100,000 depending on location. ADF receivers for air-craft cost from $1,000 to $5,500 (ref 48) and two are required for fixing. DF receivers are mandatory on all ships over 1,600 tons and usually only one is carried with the LOP crossed with some other system for a fix.

In summary, the Non Directional Beacon is precisely that, a beacon suitable for homing and providing "on tops", but not for accurate enroute navigation. It also has the capability of being modulated to provide a voice or data uplink and may yet find its place as a differential Omega (ref 90) or differential Loran beacon.

MF Systems

The band from 535 KHz to 1,605 KHz is allocated to the local rock stations. But, in the 1,605 KHz to 2,000 KHz region lie a large number of proprietary systems such as Hi-Fix, Raydist, Toran, Rana, etc., usually used for survey work (refs 71 and 138). In general, these systems are portable. They usually operate in a CW phase locked hyperbolic mode or a phase locked range range mode, in which case only one subscriber can use the net since the survey vessel becomes master (Figure 13). Ranges are typically 90 NM to 400 NM over water.

They are accurate typically from 3 to 30 feet, but are ambiguous typically every 250 feet (on the baseline) and require skill to set up and operate.

Typically, ground stations cost about $250,000 plus another $250,000 for a year's operation, to erect, calibrate and operate. Receivers cost $10,000 to $15,000 plus any cost for options such as logger or track plotter.

Loran-A also operates in this region but since it will be phased out by 1980 (ref 43), we will not discuss it here.

In summary, survey companies keep accurate expensive information to themselves.

ILS, VOR, TACAN, DME

Between 2 MHz and 108 MHz, groundwave is attenuated with increasing severity. Skywave reflects from higher layers of the ionosphere and is unstable. In general, this band is unsuitable for navigation and is allocated to communications. The band from 108 MHz to 117.975 is allocated to "Aeronautical Radio Navigation" with 108.1 MHz to 111.9 MHz on odd decimal frequencies allocated to the Instrument Landing System (ILS) and 112.5 to 117.975 in 200 channels at 50 KHz spacing to the Very High Frequency Omnidirectional Radio Range (VOR).

ILS generates short range (15 NM) final guidance signals in azimuth to aircraft on final approach and is carefully calibrated and monitored to an accuracy of $\pm\frac{1}{4}°$. ILS channels are paired with "glideslope" signals in the 329.3 MHz to 335.0 MHz range which are monitored to $\pm1/8°$. Many shortcomings have been noted in the system, particularly "scalloping" due to constructive/destructive interference from reflected signals, and the FAA have just completed an extensive investigation of "Microwave Landing Systems" (MLS) which will be proposed to ICAO as a new international standard.

VOR is the international standard system for enroute airways guidance and about 1,200 stations are commissioned in the United States, over 70 in Canada and several hundred in Europe. Over 100,000 airborne installations are in use. VOR propagates by direct wave and therefore its range is a function of height, although a "groundwave" of sorts exists out to about 40 NM, a range that would normally be considered shadowed. This allows VOR Test (VOT). Since few aircraft fly above 40,000 ft, maximum practical range is 240 NM. Guidance signals are generated by a reference signal and a variable signal, whose phase with respect to the reference signal is a function of angle. In the aircraft, true bearings are measured by a very simple phase discriminator and generate steering signals to or from the station measured (by convention) in degrees magnetic. VORs are flight check calibrated to $\pm2\frac{1}{2}°$ but the allowable tolerance on aircraft instruments is $\pm4°$, thus the accepted criteria for approach accuracy is 4.5°.

VOR signals are horizontally polarized to reduce the effects of ground reflection, but mountains, buildings and other vertical reflectors can cause errors of up to 15°. This can be alleviated by using Doppler VOR, which uses a wider aperture antenna and different scanning technique to reduce reflections and as a bonus gives bearing accuracies to about $\frac{1}{2}°$.

TACAN. Tactical Air Navigation (TACAN) is a military system operating between 962 MHz and 1,213 MHz. Bearings are generated in a manner almost identical to VOR except that nine parasitic elements provide a "fine" bearing measurement and theoretically, TACAN should be nine times more accurate than VOR. In practice, TACAN accuracies are $\pm\frac{3}{4}°$. Use of UHF frequencies permits a reasonable aperture in a very small antenna, which is useful for portable tactical antennae or antennae mounted on ships.

DME. Distance Measuring Equipment (DME), permitting range/bearing operation, was developed along with TACAN using a transponder/responder technique. DME accuracy is ±600 ft + 0.002% of distance measured (slant range). DME proved such a useful function that it was adopted for use with VOR. DME handles many multiple subscribers by a pulse jitter technique but can become saturated, leaving distant subscribers without "lock on", and this has proven a problem in congested airspace.

Since airways are defined by VOR radials, there are confluence points at each VOR site. Furthermore, routing along airways can

result in many fuel consuming "dog legs". To alleviate this, Area
Navigation or R-Nav has been proposed, whereby computer processing
of VOR or VOR/DME gives "false VOR" waypoints which the pilot can
steer to and report over using his familiar steering instruments.

Airport ILS installations cost about $350,000 and must be
calibrated at least three times per year. VOR/DME stations cost
about $200,000 for hardward, with site costs (real estate,
construction, etc.) ranging from $50,000 to $500,000, averaging
$150,000. Doppler VOR costs from $300,000 to $350,000, with site
costs averaging $200,000, because of the large aperture antenna.
TACAN stations cost about $125,000, probably because they are
considerably smaller in size.

Aircraft installations usually use the VOR receiver for ILS
with a separate receiver for glideslope. Dual installations are
mandatory for Instrument Flight Rules (IFR) and a typical
installation (dual VOR single glideslope) would cost from $1,500
(a very minimal set of dubious reliability) to $15,000 (typical
airline TSO'd system). DME costs $1,900 to $10,000 and R-Nav can
add another $2,000 to $45,000 above and beyond the VOR/DME
necessary for inputs. Airborne TACAN sets cost $23,000.

In summary, airways enroute navigation aids have evolved
around the requirement for simple, easily read airborne equipment
which interfaces well with established air traffic control
procedures for airways. Airways aids in general provide low
accuracy positioning (though dual DME R-Nav is gaining popularity)
only to aircraft above the radio horizon.

Navy Navigation Satellite System (NNSS)

NNSS, originally called Transit, was developed by the Applied
Physics Lab of John Hopkins University to provide precise
positioning for the United States Navy (ref 136). At present, it
is the only Satellite Navigation System in existence. In mid-1967,
details of the system were released to non-military users (ref 99).
NNSS consists of four to five (depending on launch and decay) low
(650 NM) polar orbit satellites. Two stable frequencies of
approximately 150 MHz and 400 MHz are transmitted and measurement
of the Doppler shift permits an observer to calculate his position.
Length of pass, elevation of satellite, and interference between
satellites all affect the usefulness of any given pass. At the
equator only about one pass out of five will be useful due to
elevation. At latitudes over 60°N, elevation is no longer a
problem and 85% of the passes should be useful, but "bunching"
causing interference due to random rise times, reduces this to
about 60% (ref 68). Since the satellites "rise" on the average
about once every 105 minutes, it can be seen that several hours
can elapse between usable passes.

The major limitation of NNSS to marine users is the error
caused by ships' velocity affecting the Doppler count. One knot

northerly velocity can move the position ¼ NM toward the sub-
satellite point. Most marine satellite receivers are coupled to
Doppler sonar. Grant and Eaton have successfully combined range
range Loran-C for velocity determination with satellite for
calibrating the Loran-C, to give a hybrid accuracy of about 600 ft
95% (ref 58). Several other hybrid configurations have been
reported (ref 30). NNSS achieves its greatest accuracy in land
survey. If several days records (say 20 to 50 passes) are carefully
"massaged", accuracies of 30 feet or better can be obtained (ref
112). NNSS receivers have been flown in P3C aircraft, but the
velocity resolution problem is even more extreme than at sea and
the interval between fixes would make NNSS operationally unsuitable
for aircraft (ref 93).

   NNSS receivers, with the computer necessary for data reduction,
in the survey configuration cost about $60,000. Most shipborne
stand-alone sets cost in the vicinity of $100,000 and a full
Doppler/gyro-compass/NNSS integrated system will cost about $200,000
(usually used for Seismic Survey or Rig Positioning).

   In summary, NNSS is an outstanding portable survey device,
but of marginal value in short term moving situations.

   United States Defence Navigation Satellite System DNSS

   A great deal of unofficial interest has been expressed in the
DNSS (ref 103), a system which does not yet exist. Very little
firm information is available in the open literature, since the
final configuration and system parameters are not quite finalized.
McDonald (ref 84) provides a good review of the system's
considerations, but does not finalize any configuration. The
press reports (ref 98) that the system is to be called Global
Positioning System (GPS) or Navstar and will in its final
configuration (expected but not confirmed), by 1984, consist of 24
satellites, in three rings of eight satellites, in 10,400 NM orbits.
These will provide three-dimensional fixing continuously anywhere
on the globe. Each satellite will be slaved to UT2 using its own
spaceborne cesium beam clock (or possibly hydrogen maser) and will
radiate a spread spectrum ranging signal at 1.5 to 1.6 GHz. The
signal is specifically coded for jam resistance in the military
environment and maximum accuracy of the system will be available
only to those military users with a "need to know". Clear signals
of lower accuracy will be broadcast to other users.

   Position accuracy is predicted at better than 40 ft, with
velocity resolution of 0.2 fps in three dimensions for military
users (there is no theoretical reason why this cannot be achieved),
and 200 feet for clear users.

   Total cost of the system is impossible to determine from the
press articles, but $154 million is budgeted for launching and
testing the first three experimental satellites. Based on this,

it will cost at least $1 billion to launch 24 satellites with no
accounting made for the ground tracking stations. Upkeep to replace each satellite after its planned five year life will probably
cost about $200 million/year. Cost of receiver and processors is
estimated at $29,500 for military receivers and $16,300 for "clear"
receivers in the time frame predicted for operation.

In summary, Navstar with its total global coverage and outstanding accuracy, would appear to be the navigator's dream, but
civil users cannot realistically expect to use the system until the
1990 era, even if the program moves ahead on schedule, and anyone
wishing access to the coded signals would require full military
clearance and would probably be expected to pay some portion of the
development funds. Caveat Emptor.

Range Positioning Systems

In the 2 GHz and 9 GHz radar bands are a large number of line
of sight survey systems, with power limited ranges of about 20 NM.
These use transponder responder technique to generate range range
fixes. These systems are not navigation systems in the strictest
sense, but survey systems. Typical accuracies of 10 ft are
obtained. Such a system has been used for icebreaking on Lac
St-Pierre in the St. Lawrence River. The cost of this system is
unknown, but the briefings given by Shawinigan Engineering (the
system designers) indicate that it must have cost over $350,000
for one user in one small area. Range positioning systems have
also been used successfully for Arctic research. Cost of RPS type
systems ranges from $27,000 to $180,000.

RPS type systems have limited use but can be accurate and
convenient where terrain permits use of line of sight.

Radio Aid Retransmission

Navigation signals need not be processed in the user vehicle.
Any phase or time comparison signal, where a phase or time
reference is provided, can be modulated onto a carrier and
retransmitted for processing at a different location. This idea,
first proposed by Pierce, has found applications in tracking
remote objects such as buoys and weather balloons (ref 11),
reconnaissance (ref 67), Search and Rescue (refs 25 and 126), and
could be used for iceberg tracking (ref 166). Data reduced at a
home base gives the position of the tracked object. Cost of
retransmission devices has been reported as low as $25 (ref 11).

Navigation Computers

All the systems discussed generate their navigation and guidance information by a variety of techniques. In the past this has led to too strong a bias in favour of a given system. An aircraft pilot, for example, desires to know only the direction and distance to his destination, thus the very simple processing of VOR was highly desirable. In a ship with a large chart table, there is no problem plotting accurate Loran hyperbolae whose orientations are independent of track, but in a cockpit there is no space or time to do this. Even bearing information is mentally reduced to homing and beam crossing with a strong plea for DME.

The military has long recognized the value of digital computers for integrating navigation systems. Since the mid-sixties, mini-computers have been available to perform the same tasks for civil applications. Typically, an 8K, 16 bit word, moderate speed, mini-computer can handle all the major tasks - DR, co-ordinate conversion, application of corrections from tables or programmed model, and generation of steering signals. While a great amount of detail work remains, the cost and tasking of a computer is no longer a navigation problem, and from a systems standpoint, any navigation system can be considered inherently capable of processing to yield display outputs at about the same cost. In 1973 the cost of one of the most popular mini-computers used for navigation was about $20,000. In 1975 the same manufacturer offered a completely software compatible improved model at one half the weight, one half the power consumption and $1\frac{1}{2}$ times the speed (not considering the considerable improvements in microprogrammed instructions) for $5,000. Completely ruggedized computers meeting full mil spec environmental standards with similar performance are available for about $10,000. For specialized applications the "computer on a chip" or microcomputer is available for $300 or less. This price trend is continuing down under the market pressure of commercial applications.

Co-ordinate Conversion Algorithms are well documented in the literature references 50, 58, 77 and 121. Other than specific details of programming and some areas of optimal filtering, few knowledge gaps exist in navigation computers, though surprisingly enough, very few commercially available navigation computers use a best available wind to compute heading or estimated time of arrival.

USER REQUIREMENTS

Having surveyed available systems, we must consider applications. While at first glance the requirements of Land, Sea and Air users would appear to differ widely, in fact they have a large common ground. Much of the disparity in systems has resulted from each community developing its own solution to its requirements in isolation. It has been said, "Ask any user what he wants and he will reply, 'Same as what I've got now only better'".

Navigation requirements are based on two fundamental considerations - safety and economics - and though much cant is paid to the former, in reality the latter consideration usually dominates. Since economics play such a dominant part, it is worthwhile to see what common grounds exist in requirements and to consolidate these, particularly with respect to radio aids.

Air

Aircraft navigation can be divided into two phases - enroute and terminal. Enroute, air traffic control determines the safety requirements and separations. Present airways' widths are predicated on the VOR airways structure and are specified not to exceed 4 NM out to 51 NM where the lanes are allowed to expand out the ±4.5° of VOR (ref 1). Separation criteria are usually based on aircraft speed, but an uncertainty of absolute position greater than 10 NM is deemed undesirable (ref 32). Outside an airways structure, no accuracy requirements are laid down except the requirement to report (where possible) every 30 minutes to an accuracy (if possible) of 10 NM.

There is, however, a strong economic argument for accuracy. DeGroot (ref 29) argues that the 2 NM accuracy of modern IN saves $24 million per year in Boeing 747 economics, averaged over the world's fleet, by reducing the meander of 15 NM that would occur with an infrequently updated Doppler/gyromagnetic compass system. McLarnon (ref 91) reports 7% improvement in Twin Otter economics by using GNS 200 in an area where no navaids exist. Transair (one of our leading Northern carriers) confirm McLarnon's figures, reporting improvements of 5-6% in Boeing 737 economics on the Winnipeg-Resolute Bay run. At normal fuel/maintenance economics for these aircraft, this would result in a direct saving of better than $30,000 and $85,000 per year respectively.

A large class of users have stricter enroute requirements. Geophysical Survey, Forestry Spraying, Remote Sensing, and Search and Rescue aircraft all have a requirement to maintain very accurate parallel track spacing. Glicken has stated that aerial survey requirements are 500 ft or better at all altitudes (ref 56). Knutsen (ref 79) has determined that a track spacing accuracy of 1/8 NM is essential to meet Search and Rescue (SAR) requirements.

The safety aspects of SAR are obvious but the economics of maintaining accurate parallel tracks are a little more difficult to identify. The following example, however, should give some indication. The government of Quebec has installed $1.5 million worth of IN sets in their CL-215 water bombers in order to enable these aircraft to meet the stringent track spacing requirements of budworm spraying. If, say Decca coverage had been available, then installations to achieve the same accuracy would have cost about $250,000. Terminal accuracy requirements are based partly on air traffic control, but mostly on the requirements of instrument approach procedures. Stripped of the procedural details, terminal requirements can be summarized as: a Final Approach Fix (FAF) of very high reliability of 1 NM or better, which will allow an aircraft to make a standard straight in or circling approach, or to intercept the ILS in order to make an instrument approach in accordance with published procedures (Figure 28). One Nautical Mile - all else is superfluous.

Sea

Normal sea operations can be divided into three regions - open ocean, inshore or confluence where significant risk of collision with obstructions or other ships is present, and narrow channels such as harbour entrances. The safety requirements for these areas have been examined in the United States coastal/confluence regions by Blood and are shown in Table 7.

In Arctic waters, the 2 NM figure is open to question. When the most significant hazard to navigation moves ponderously, there is an increased requirement for relative accuracy between the reporting ice reconnaissance aircraft and the user ship. Further- more, the ice-recce aircraft needs some means of accurately predicting the velocity vector of the iceberg in order to extrapolate (DR) the iceberg position. Present ice-recce aircraft are reported to have a positioning accuracy of 3.3 NM to 6.3 NM 95% (ref 52). Under these circumstances, the arguments previously applied to aircraft meander also hold for a ship attempting to avoid icebergs. Although no similar reported savings can be found in the literature, a conservative estimate of 2% would result in $4,000 per year per ship saving at typical Arctic charter rates. Since about 150 ships operated in the Arctic in 1973, this translates to about $600,000 (ref 92).

Inshore, the accuracy requirement is largely predicted on the surveyed accuracy of hazards. Examination of Figure 29 and Figure 30 shows that although confluence areas have received reasonable attention, large areas are still inadequately surveyed, particularly Coronation Gulf/Spence Bay where shallow water prevails enroute to Cambridge Bay. The economic gain of accurate survey is difficult to gauge, but the risk is measured by insurance companies. The insurance on a 50,000 DWT vessel, ice strengthened to Lloyds class 1 standards, entering Milne Inlet or Pond Inlet, has been quoted as $250,000 (ref 87) above normal marine rates. How much of this is due to lack of navigation facilities or to inadequate survey is difficult to assess without consulting Lloyds, but not all the excess premium is due to ice hazards.

The cost of Hydrographic Survey is easier to measure: simply the additional costs of setting up survey nets, such as RPS, above and beyond the positioning services presently provided. For Seismic surveys: shallow water marine surveys are estimated at $800 to

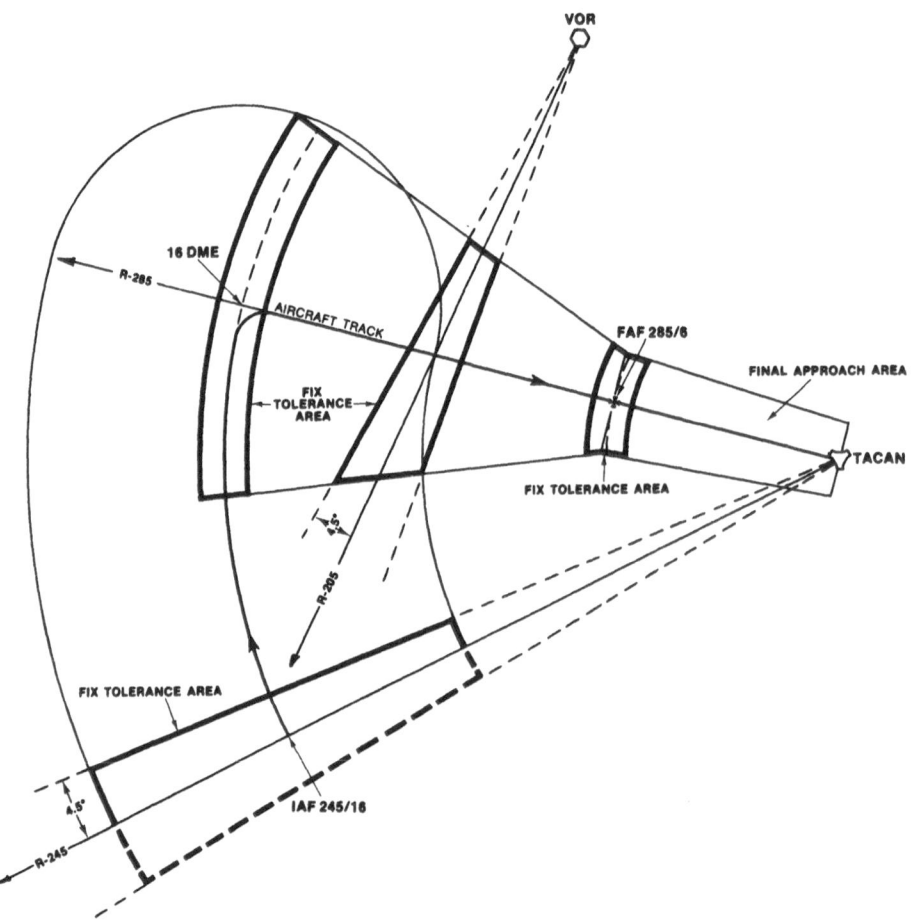

Figure 28.   Fix Tolerance Area (Approach Segments Provided) (from
             ref. 162)

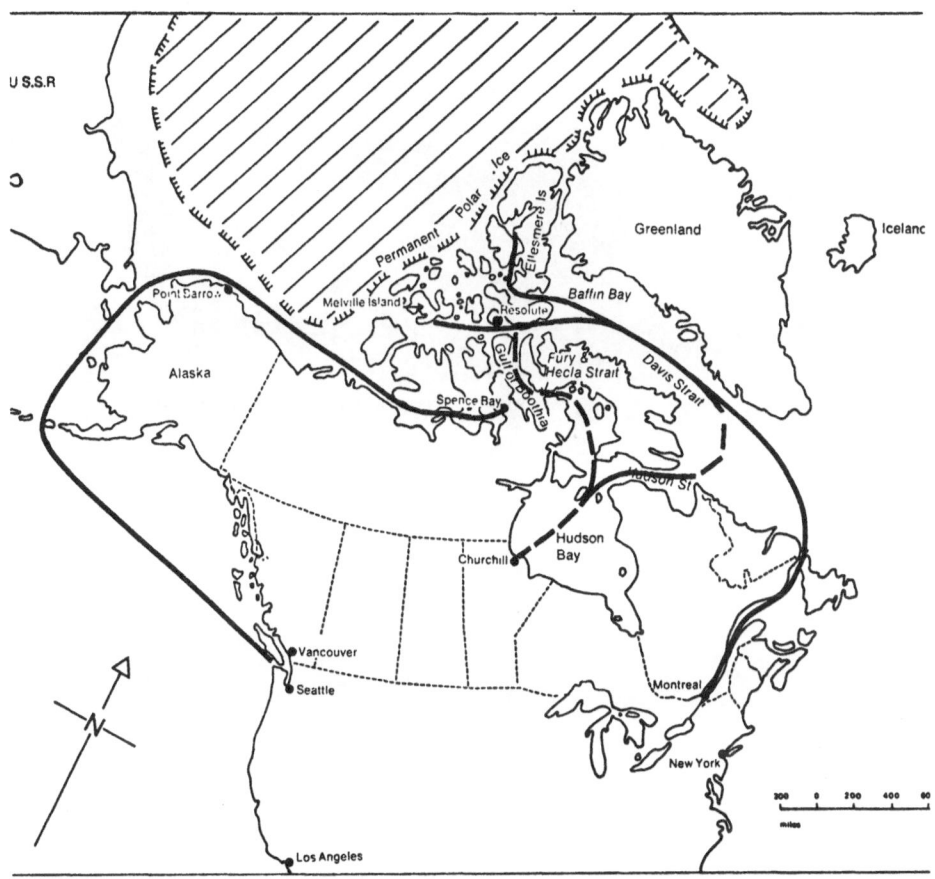

Figure 29. Arctic Islands, Marine Routes (from ref. 128)

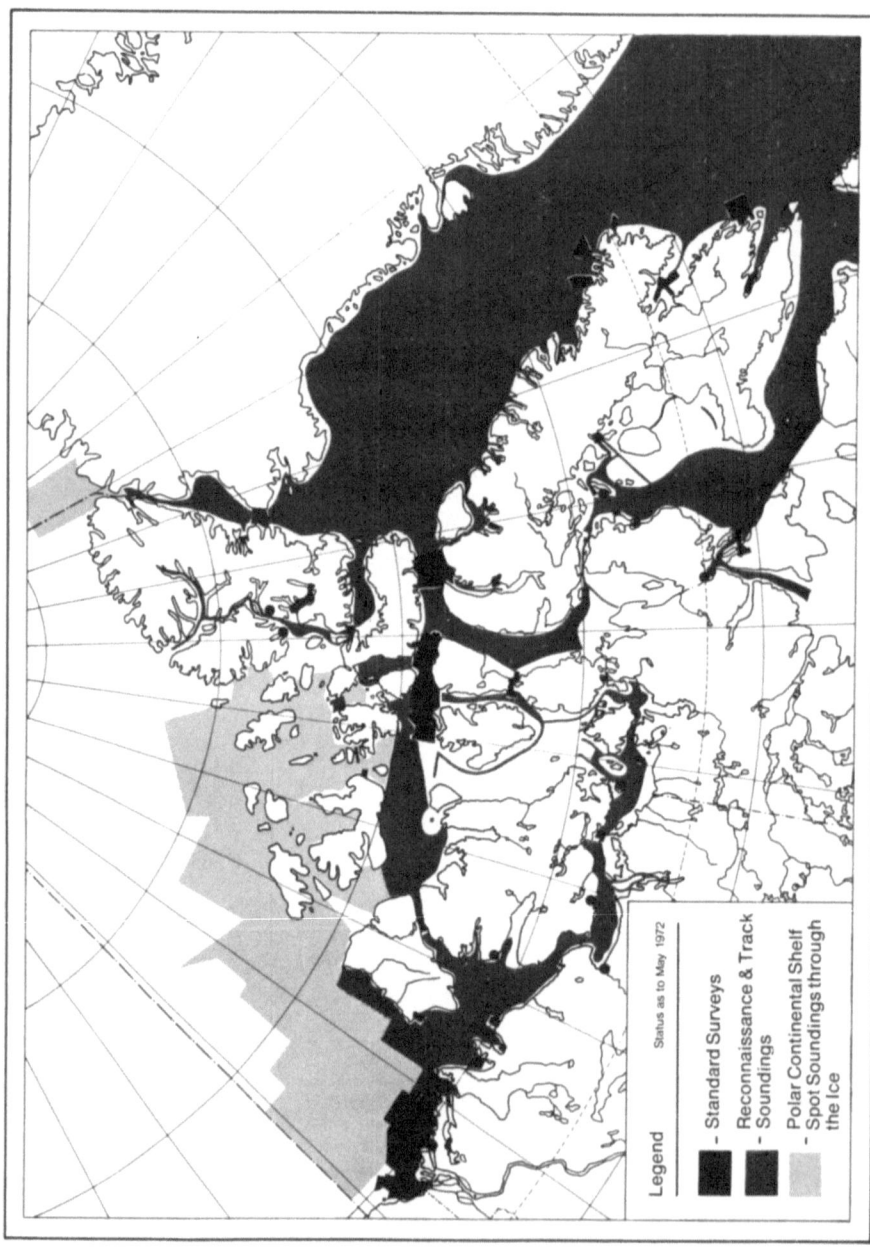

Figure 30. Hydrographic Survey, Canadian Arctic (from ref. 128)

$1,700 per line mile and deep water at $450 to $750 per line mile (ref 169). This, of course, is not all due to navigation costs, but the manning costs of setting up a survey net adds considerably to the survey and if two agencies survey the same area, not only is the cost duplicated, but the relevant data is referenced to different datums, making correlation difficult.

Table 7

Blood: Requirements for Coastal/Confluence Region

Summary of CCR Safety Related Requirements

Required Accy (n. mi.)*	General Area	Contour Locations**
0.25	Atlantic Coast	Sea Lane Inner Boundaries Cape Cod Canal Approaches
	Gulf Coast	Fairways
	Pacific Coast	Sea Lane Inner Boundaries Santa Barbara and San Pedro Channels
	Great Lakes	Harbor Entrances and Restricted Waterway Approaches
0.5	Atlantic Coast	10-Fathom Line
	Pacific Coast	10-Fathom Line
	Alaska Coast	10-Fathom Line
	Great Lakes	All Navigable Waters
1.0	Atlantic Coast	100-Fathom Line
	Pacific Coast	50 n. mi. Off Shore
	Alaska Coast	50 n. mi. Off Shore
2.0	Atlantic Coast	Outer Extremity of Gulf Stream from 28°N to 39°N Latitudes

* 95% accuracy requirement for the radio navigation aid, not the track keeping accuracy requirement of the vessel.

** Accuracy requirement applies from shore line to indicated contour location or throughout indicated area.

Land

The major difficulty in identifying land user requirements is the enormous diversity of tasks. Since nearly all land navigation is by direct visual reference, the "safety" requirement in land navigation more realistically equates to the Search and Rescue problem. It is not until a man is lost or stranded that he really needs to know more than his relative position. A recent study by the Canadian Forces Aerospace Systems Course (ref 154) identified that: a requirement for alerting within 12 hours, identification, and positioning of a SAR victim to within 15 NM could be met by a low polar orbiting satellite (CAPRICORN)(ref 80), using present Electronic Locating Transmitters (ELT) at a program cost of about $21 million. (To this author at least, the use of an orbit almost identical to Transit suggests the possibility of combining navigation and SAR functions.) This report also strongly recommended the use of navaid retransmission ELTs to achieve Omega accuracies, in conjunction with geostationary and high polar orbit satellites at a cost of $54 million. Weighed against the present $14 million per year expenditures on Search and Rescue by visual means, these expenditures seem well justified. The use of navaid retransmission for work party monitoring on frequencies other than SAR would also provide valuable checks for safety and economic reasons.

Land Survey by traditional techniques of theodolite and chain is accurate to tenths of feet and no navigation system can meet this accuracy. A large number of the tasks associated with geophysical survey, however, have no requirement for this type of accuracy. Potts and Roeber (ref 113) have identified accuracies of 50 to 100 ft as sufficiently accurate for oil exploration and geophysics. Survey lines for Seismic exploration cost $2,000 to $12,000 per line mile over land. Seismic lines are bulldozed (Figure 31 and Figure 32), causing irreparable ecological damage and considerable concern to the native population (ref 123). These latter costs are not accountable in the usual sense but must be considered in any policy decisions.

The native people themselves are desirous of some aid to navigation. They are a remarkably mobile people, thinking little of making several hundred mile journeys by skidoo to visit relatives (ref 101). Contrary to popular myth, the northern native has no "instinctive" skill at navigation. He navigates by hard learned skills, by memory of familiar landmarks and in an unfamiliar area becomes as lost as the white man. If a very low cost navigation system allowing easy reference to a map, automatic computation of direction and distance to destination with possibly an alert beacon for emergency breakdowns could be built, then it would quickly find a market in the northern communities. Economically, it might improve efficiencies on traplines for instance, upon which at least one Sachs Harbour resident earned $75,000 in one year.

Figure 31.   Survey Lines Through Tundra

Figure 32.   Survey Lines Through Bush

Ultimately, as appears to be occurring in the United States (ref 95), it may be that the secondary and tertiary land users will dominate the benefits picture. Wind measuring for meteorology, current and ice drift studies for estuary, harbour and coastline engineering may very well become major users of navaids by using retransmission techniques. Geophysicists can also make use of the accurate timing potential inherent in navaid transmissions (if they are slaved to UT2) to improve correlation of seismic shots. An accurately timed and monitored navigation net would also provide routine ionospheric sounding, and Loran-C at least has been proposed as a direct means of monitoring lapse rate and hence distant frontal passage (ref 34). The gains from these applications are impossible to assess at this time. But, because of the peculiarities of the Arctic in physics, weather and demography, they would almost certainly show far greater relative gain than equivalent applications in the south.

ARCTIC CONSIDERATIONS

Until this point in the paper, only passing reference has been made to the Arctic. This has been deliberate. Navigation problems do not change in the Arctic; they merely get worse. In some cases, much worse.

Heading Reference

To the practical navigator, the greatest single problem in Arctic navigation is lack of a good heading reference. The North Magnetic Pole wanders around the vicinity of Bathurst Island (Figure 33). Within the shaded area, the horizontal component of the earth's magnetic field is weak. Thus, a magnetic compass will not only have difficulty detecting the field but compensation made at lower latitudes may no longer be valid. In gyromagnetic compasses, the magnetic driving function may be too weak to affect the gyro. Because of magnetic meridian convergency, the driving function may force the gyro into significant errors. Bryce (ref 152) has used the work of Green and Glenny to demonstrate that dynamic errors can approach 2° in high speed (550 knot) aircraft as far south as Winnipeg.

Meridian convergency results in excess torquing signals on velocity corrected gyro compasses or excessive wander in free gyro systems. In inertial platforms, this can be computer compensated by wander azimuth techniques (the initial gyrocompassing still requires a longer time), but a free gyro must be operated in "grid". Grid gyro techniques as pioneered by Greenaway (ref 60) are a boon to Arctic navigation. However, they are not widely used outside the navigation community and for cockpit work are somewhat cumbersome, although they can be used. For instance, Bryce and Gerden

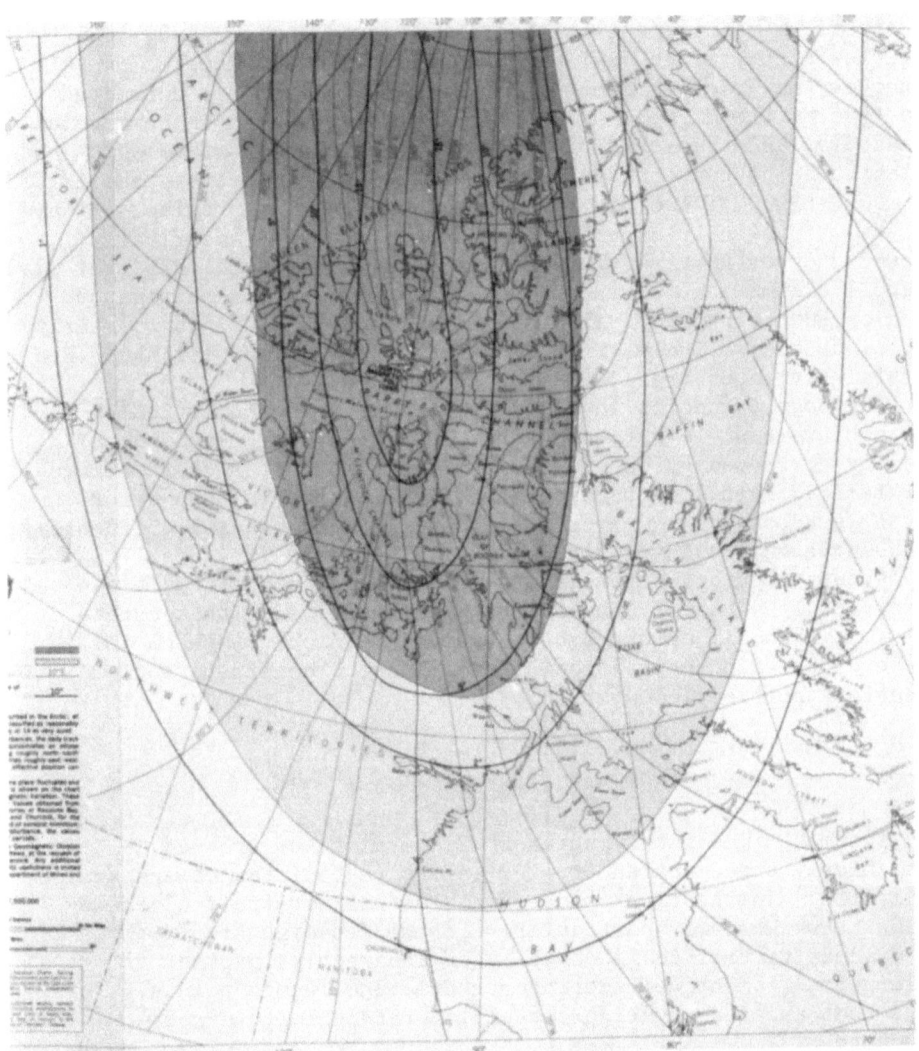

Figure 33. Magnetic Chart of the Canadian Arctic 1965.0

have worked out grid-gyro techniques for use in CF-5 tactical
aircraft (ref 153). Finally, any free gyro system requires an
external check on accuracy and although Astro techniques are used,
these have their own limitations.

A strong requirement still exists for an inexpensive,
reliable, accurate, convenient heading reference. Modern IN
technology may provide some answers. A micro-computer corrected
free gyro of the dry, tuned rotor type (now standard on commercial
IN and available separately), even with the crudest of velocity and
DR latitude inputs, can probably be built to give better than $1/10°/hr$
for about $3,000. Two new technologies which show promise, but are
not yet sufficiently mature to decide on, are the Electrically
Suspended Gyro (ESG)(ref 163) and the Ring Laser Gyro (RLG)(ref 164).
Rockwell claims that their "Micron" (ref 36) Inertial Navigator
based on ESG technology will be available for $35,000 in the early
1980s and Honeywell are offering strong competition with their RLG
Inertial Navigator (ref 78).

The cost tradeoffs involved in deciding a technological
approach are best left to industry. For practical purposes, a
velocity corrected gyrocompass DR system might offer cost savings
over full IN even though Sakran (ref 127) has shown that error
growth is unbounded under certain conditions of heading (northerly),
latitude (high), and velocity (high). Even pure IN (refs 17 and
146) shows unbounded error growth due to azimuth drift. It can be
shown (ref 159) that if any DR system is provided with accurate
position, then all errors can be bounded. Which brings us full
circle once again to positioning aids and the considerations
affecting their implementation.

Radio Aids in the Arctic

The most immediately apparent feature of the Arctic,
particularly the Canadian Arctic, is its vast distances and sparse
population. In the first place, this means that traffic density
is low. Furthermore, particularly in the Canadian Arctic, most of
the productive traffic is directly or indirectly involved in
exploration with no well defined traffic confluence points. This
means that an area coverage system is preferable to a route
structure.

The scattered population creates severe logistics penalties
in the building and maintenance of transmitter sites. Population
centers, even small ones, may not be located near a site desirable
from a navigation viewpoint. Facilities must be built for the
station operator or if automatic stations are used, maintenance
must be carried out by helicopter or aircraft. At up to $3 per
gallon of avgas, transportation of routine maintenance crews
becomes expensive and there is a strong multiplier factor in the
cost of shipping generator fuel to remote sites. Logically then,
a long range system, flexible as to the location of transmitter
sites, is preferable.

Arctic Environment

While it is rather trite to mention Arctic cold, it does have some indirect effect on navaids. Solid state devices are infinitely preferable to moving parts, either in user equipment or transmitter. While the rotating antennae of VOR and TACAN can be environmentally protected, at extra cost, they still represent the most significant causes of down time. In the Arctic, this problem can be expected to be worse.

Permafrost is a difficult material to build on. Hefly (ref 66) reported that special refrigeration units had to be used to prevent the heat of setting concrete from causing excessive settling and melting when the Low Frequency Loran tower was built at Cambridge Bay in 1948. Since that time, permafrost engineering has advanced considerably and has its own body of literature (refs 19 and 89). The LF Loran tower still stands unused, a mile off the runway at Cambridge Bay, and it is an intriguing thought that perhaps with a little refurbishing and extra top loading.... The consequence of permafrost engineering is, in any case, additional cost for all types of navaid transmitters.

Except for the cold, Arctic weather by southern standards is quite benign. Winds are generally light but can be capricious with variances of 40 knots within a few miles (strong justification for wind finding balloons). Precipitation is very light and thunder-storms very rare, except off the Newfoundland coast and possibly up the Labrador coast, where icing is a severe problem (ref 135). Icing can wreak havoc on any antenna, but is particularly hard on large well insulated structures.

A major Arctic weather problem of indirect consequence to navaids is reduced visibility due to blowing snow in winter and cloud and fog in summer. The infamous Arctic "whiteout", wherein an observer appears to be engulfed in uniform white flow, the horizon line cannot be distinguished, and judgement of distance is almost impossible, is well known. Under these conditions any surface travel is hazardous and any aircraft approach and landing is so perilous that only full Microwave Landing System capability with precision azimuth, glideslope and DME can be considered adequate (ref 161).

Under the more normal summer conditions of low cloud and fog, occurring 25%-40% of the time, which restrict takeoffs and landings 10%-20% of the time (Figures 34, 35, 36, and 37), there is a much more subtle influence on navaid selection. Airports can be fitted with MLS, but to fit each unmanned navaid transmitter with MLS would be unwarrantably expensive. Yet, if high reliability of any navaid network is to be assured, then maintenance crews must be assured of reliable access to each transmitter. To illustrate, these excerpts are quoted from a MOT report (ref 24):

"The FY (206 KHz) NDB serves as the prime enroute navigation aid at Frobisher; this beacon was off from 0010 to 0059 hrs. The operator coming on shift at 0001 performed his routine station

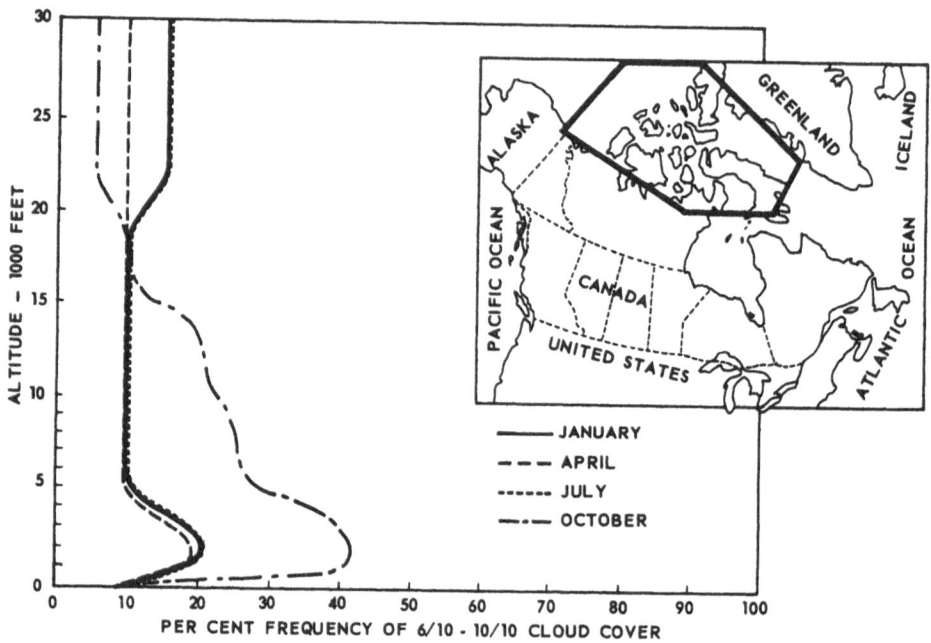

FREQUENCY OF CLOUD AMOUNTS ARCTIC

Figure 34. Frequency of Cloud Amounts in Arctic

FOG

A variety of types of fog can occur in the Arctic. They are summarized below:

a. <u>Ice Fog</u>. This fog is composed of ice crystals and is a feature of the cold winter months. It is common at inhabited areas where it is caused primarily by fuel combustion and engine exhaust.

b. <u>Arctic Sea Smoke</u>. This fog occurs over open leads during the winter and may rise quickly and drift downwind for many miles.

c. <u>Arctic Sea Fog</u>. Sea fog is predominant in the summer when pools of water on the ice pack are considerably cooler than the open water.

d. <u>Radiation Fog</u>. Radiation fog forms in the summer and autumn. It is usually very thin and does not pose much of a problem.

e. <u>Advection Fog</u>. This type of fog is formed when warm moist air moves over a cold surface, and is most prevalent in autumn when the air from the open water areas moves over the land.

AIRFIELD WEATHER LIMITS

Figures 35, 36, and 37 illustrate the percentage of time that Resolute Bay, Yellowknife, Frobisher, and Churchill are below take-off, straight-in, and circling landing limits. These percentages include only the limits imposed by ceiling and visibility and do not include the times that take-offs and landings would be precluded by crosswinds. The winds are generally light and the percentages would not be significantly increased by including crosswind limitations.

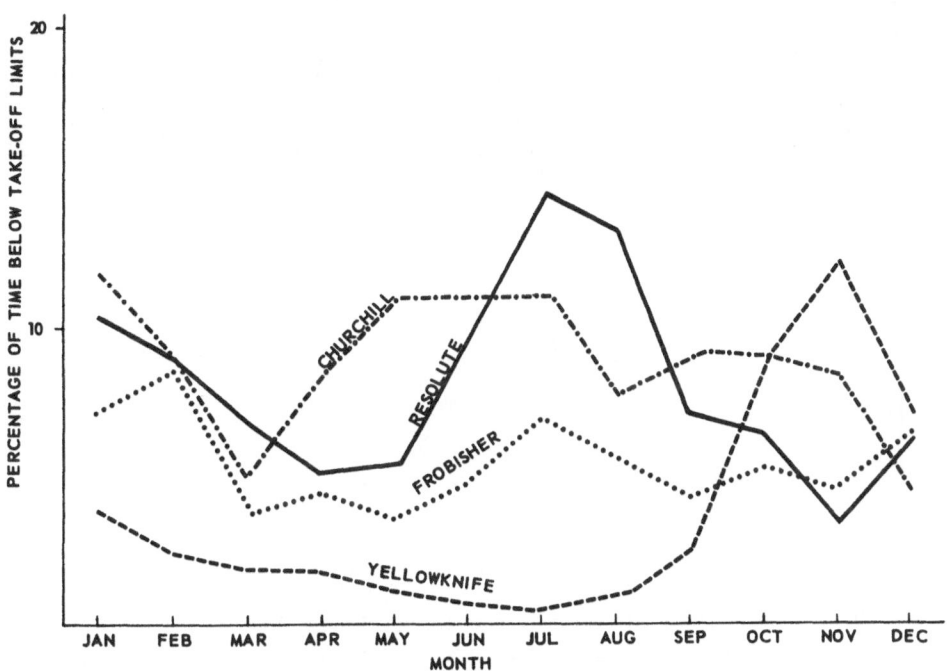

Figure 35. Percentage of Time Below Take-Off Limits

Figures 35, 36, and 37 are from data reduced from the Hourly Data Summaries for the 10 year period 1957-1966. These summaries were obtained from the Winnipeg Weather Office.

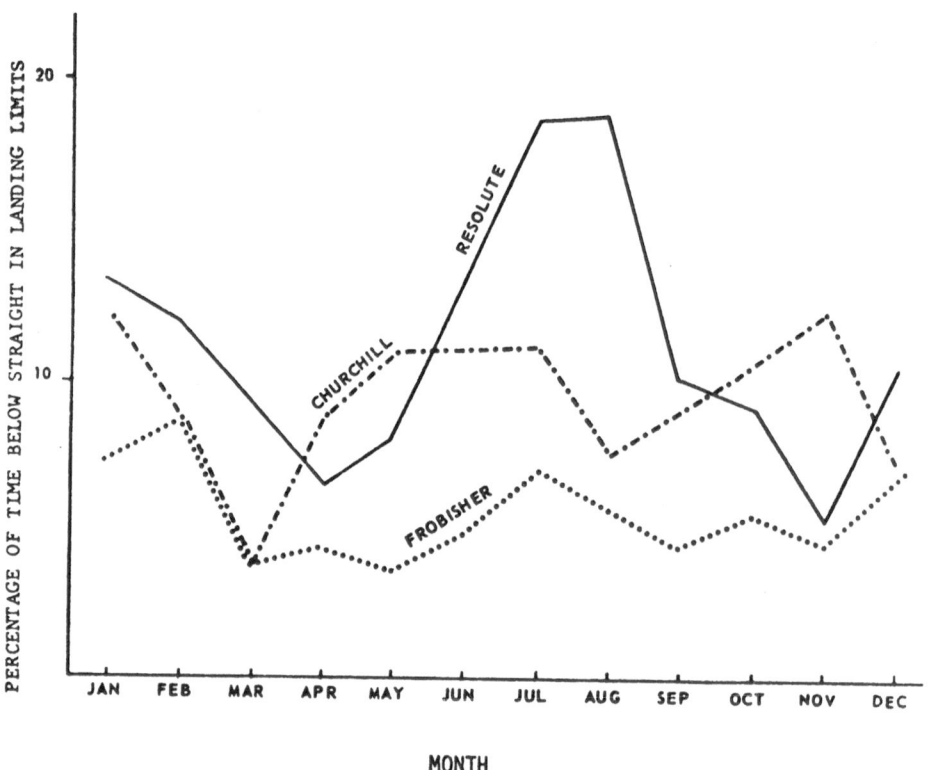

Figure 36.   Percentage of Time Below Straight-In Landing Limits
             (Note: no straight-in landings authorized at Yellowknife)

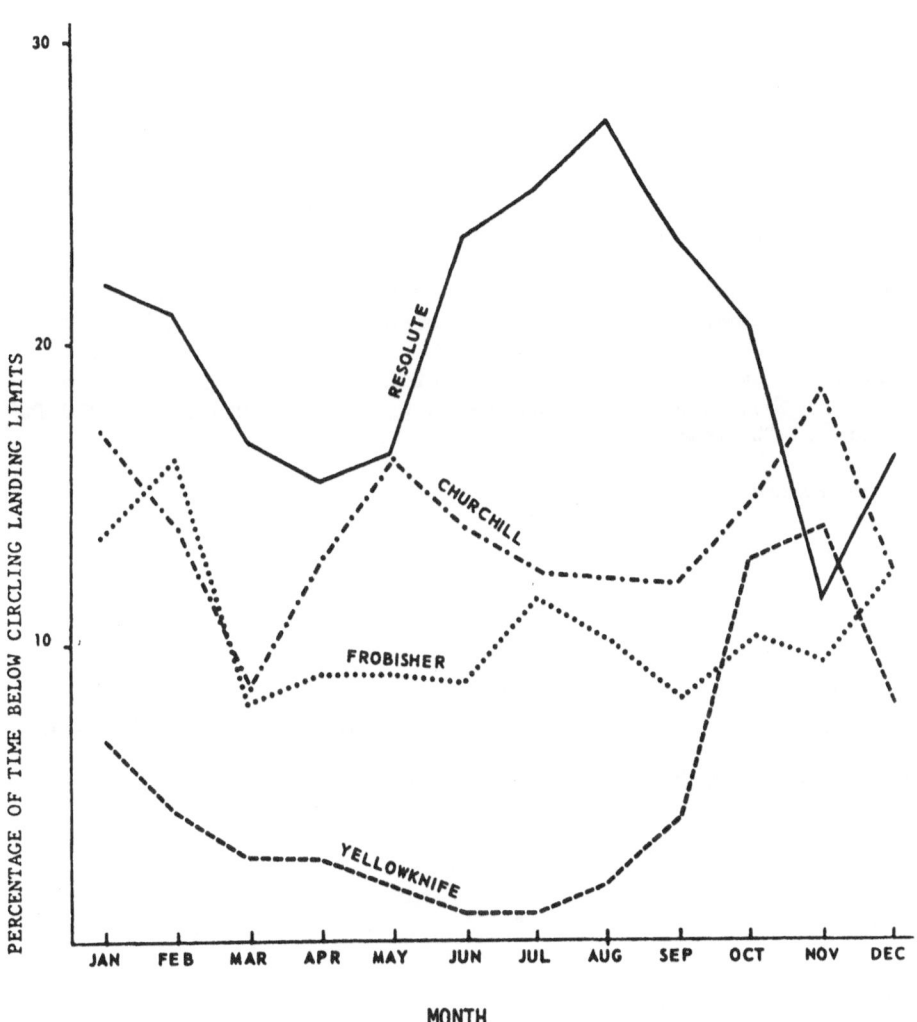

Figure 37. Percentage of Time Below Circling Landing Limits

equipment check which included verifying the operation of the FY
beacon. On the audio test, the beacon went 'in the red', a visual
indication of a transmitter outage. The operator attempted to
reset the transmitter; this was unsuccessful and attempts to have
the standby transmitter come on the line were also unsuccessful.
After a period of at least 10 minutes, the duty technician was
called 0025 to rectify the fault. The technician left his home
almost immediately and was at the transmitter site at 0042. He
discovered the tripped circuit breaker on the failed transmitter.
The technician also troubleshot the standby transmitter. He found
a cabinet door interlock switch not properly held "on", thus
preventing electrical power from reaching the transmitter unit.
After checking the operation of both transmitters, he reported the
equipment as operational at 0059...."

"There was no conclusive evidence to indicate that CF-BRL was told
of the FY NDB outage prior to commencement of descent...."

"The aircraft impacted the side of a hill (elevation 2,080 ft on a
track of 330°M). The impact scattered wreckage over an area 200 ft
wide and 650 ft long."

Northern navaids must be manned.
        The Arctic also has peculiar propagation characteristics that
affect all navaids. Belrose (ref 7) observes that, "The very low
ground conductivities in Arctic areas, and the particular ionospheric
conditions prevailing at high latitudes (due in part to solar
geometry and in part to high latitude disturbances) can lead to
rather unusual radiation and propagation conditions."
        Omega monitoring has been conducted at Frobisher Bay, Coral
Harbour, Grand Prairie, Resolute Bay and Wales, Alaska (refs 83 and
125). As yet, this data has not been incorporated into any diurnal
correction models and very little information has been published,
although reference 22 describes the manner in which the data is to
be analyzed. Trial results with airborne Omega indicate bias
errors of up to 4 miles due to the unknown effects of Arctic
propagation, and one trial reported 8 miles bias error crossing the
Greenland icecap. Absorption of VLF over the Greenland icecap is
severe, increasing to 22 db per megameter from 1.3 db per meagmeter
as the D layer comes down during the daytime. Under the effect of
solar proton events, the D layer lowers further and even greater
absorption occurs, causing virtual blackout of any VLF signal
crossing Greenland.
        At LF, the groundwave absorption and retardation are well
understood but an Arctic ground conductivity map is difficult to
find in the navigation literature. If one is available, and it
may well be at the Communications Research Centre, then it should
be made available to the navigation community. Low ionosphere may
bring in skywave earlier, but because of skywave absorption, the
skywave itself may be considerably attenuated and may not affect

CW signals such as Decca as severely as in the South. Certainly, the models to determine this theoretically are available, but no examination of this phenomenon has appeared in the navigation literature.

At MF, the low ground conductivity renders many NDBs virtually useless and by 2 MHz, it is virtually impossible to propagate a groundwave over Arctic terrain. Furthermore, at 2 MHz, phase instabilities are introduced by conductivity changes over ice covered water that render these systems useless for positioning (ref 15).

Ground conductivity affects the selection of an antenna site. For systems propagating by groundwave, the foreground of the antenna is very important. For instance, the 50 watt beacon at Berens River on Lake Winnipeg, with its radials buried into rich loan, can be received at 150 NM, while the 50 watt beacon at Cambridge Bay is often unusable at 30 NM. Again, this indicates that a long range system, flexible with respect to transmitter siting, is desirable.

Arctic propagation has been studied extensively at the Communications Research Centre. In fact, Canada's first two satellites were dedicated to ionospheric research. It is time some of this information was used in the design of Arctic navaids.

AN OPTIMUM NAVAID FOR THE ARCTIC

Having briefly examined navigation systems, user requirements, and some of the constraints imposed by the Arctic, we can now examine if it is possible to optimize navigation in the Arctic.

To adopt a laissez faire attitude is probably not optimum. Carriers, if forced by their own economics to adopt IN or to install their own NDBs, will pass these costs on to the customer. Exploration companies will continue to use their private MF location systems and pass on the cost of exploration in every tank of gas. For example, in March 1973, 162 persons were employed in "Radio Positioning Stations" in the Nova Scotia Offshore Oil Industry (ref 54). There were at least three and possibly four private chains in operation including Toran, Range Range Lambda and Accufix. It is a safe estimate to say that the exploration companies spent over $2.5 million in radio positioning services in the Grand Banks area in 1973 alone. The same positioning services could have been provided by a Loran-C station built at Goose Bay for an initial cost of $4.7 million and O&M of $300,000 per year. Goose Bay could have been paid for in two years by the exploration companies and all other users (shipping, fishing and aircraft) could have used the service free.

The economics, of course, are not that simple but models do exist that quite accurately identify the economic benefits of a government supplied common property resource. By using these techniques, McQuillan and Clough (ref 92) have identified benefits to shipping and exploration from ice reconnaissance growing from $4 million in 1975 to $147 million in 1990. This study made no examination of the navigation infrastructure that would be essential in realizing such benefits.

Benefits cannot be examined without reference to cost. If each community carries out its own benefit analyses, then many systems can be justified. Their parameters overlap and it is obviously wasteful from a systems engineering problem for the government to supply all services. The problem of choosing which system still remains.

I should like to suggest that the following philosophy be applied. Stripped of superfluities, any navigation system can be described by one parameter - accuracy. The models necessary to predict accuracy of virtually any navigation system are well documented.

Any user can be asked, under careful questioning, to define the benefits he will gain from improved accuracy. Furthermore, there is a distinct penalty to be paid for reduced accuracy or maintenance of the status quo. This penalty may appear in the ultimate price of collision, or the commercial loss involved in unwillingness to take risk, or the commercial cost of risk, i.e., underwriting. If these benefits and costs are plotted against accuracy for any given user, a curve similar to Figure 39 results. The curve may look like Figure 40 for some users, reflecting a need for greater precision under certain conditions, e.g., aircraft landing or ship entering harbour. It is most important to note that after a certain point, there is no need for greater accuracy referred to any given user, but it incurs no penalty to him at this stage.

If all users' benefits are integrated with their individual benefits weighted according to their contribution to the economy at large (for instance, reduced insurance costs due to hydrographic survey of hazards must be weighted very heavily toward the surveyor who accurately charts the hazards and far less against the shipowner who merely uses his accuracy to avoid those hazards), then a composite benefit curve can be drawn (Figure 41). The exact shape is, of course, unknown until all benefits have been assessed and integrated, but it will certainly be some form of cumulative distribution.

For any given accuracy, there is a cost penalty. This can be determined by examining dummy systems. This is a tedious but not impossible task. Although at first sight an infinite number of possibilities exist, in reality many are so quickly eliminated that the candidate systems quickly reduce to a few.

The benefit curve is now divided point by point by the cost curve and a cost benefit curve results for each candidate system. Figures 42 and 43 illustrate. The optimum system, then, is that system or combination of systems which maximizes the area under the cost benefit curve.

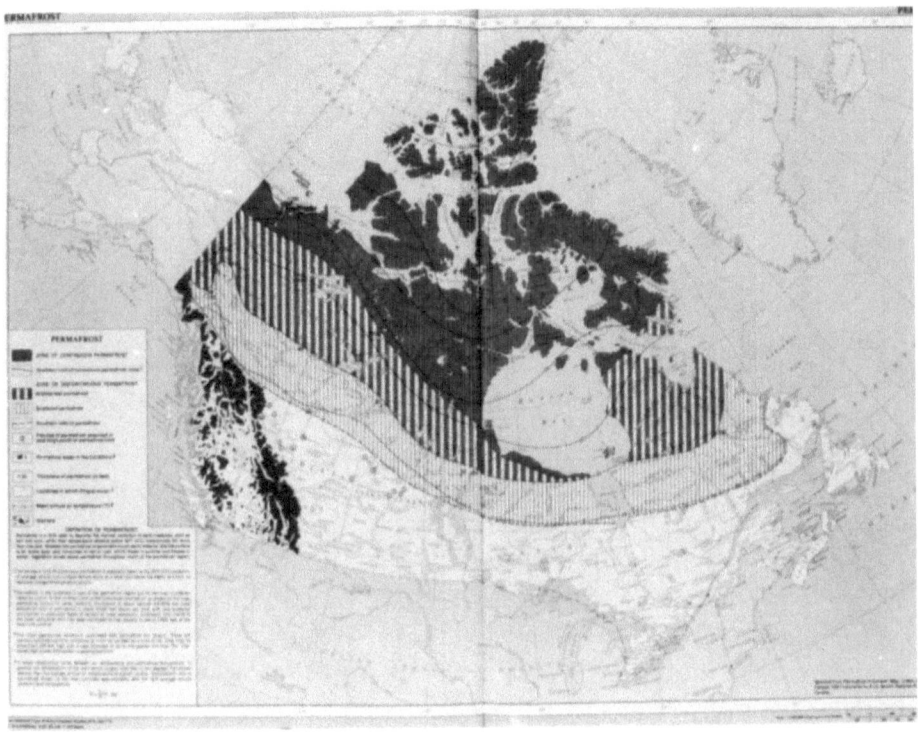

Figure 38.   Permafrost Zones in Canada (National Atlas of Canada)

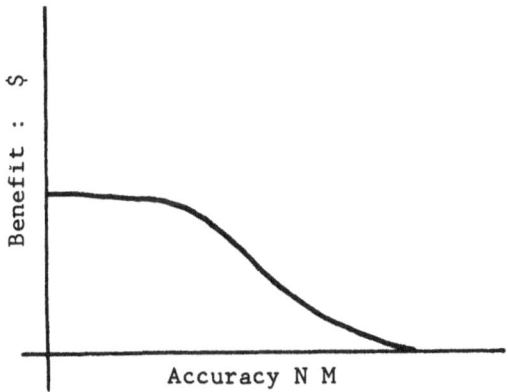

Figure 39.   Twin Otter VFR Operation

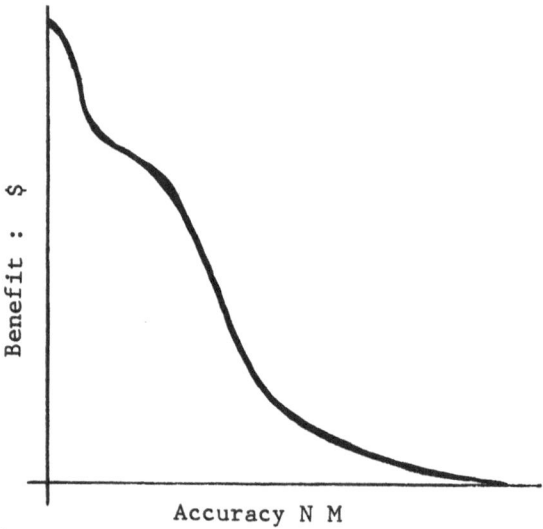

Figure 40.   Boeing 737 IFR Operation

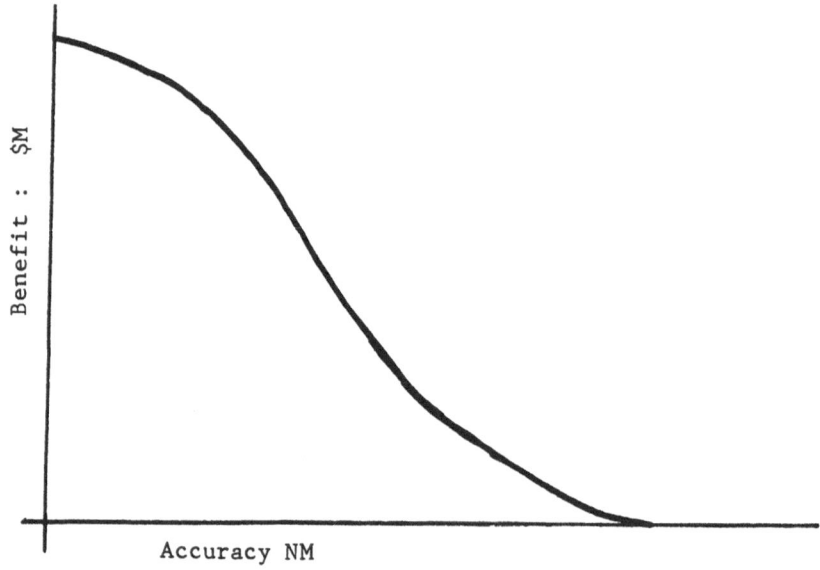

Figure 41.   All Users

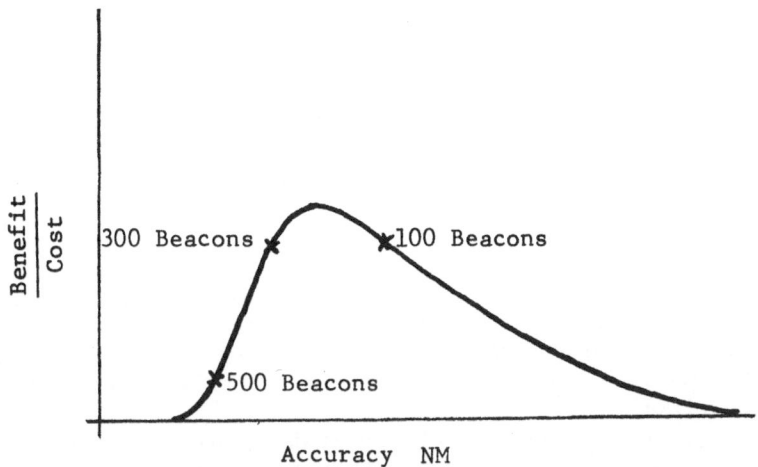

Figure 42. Differential Omega

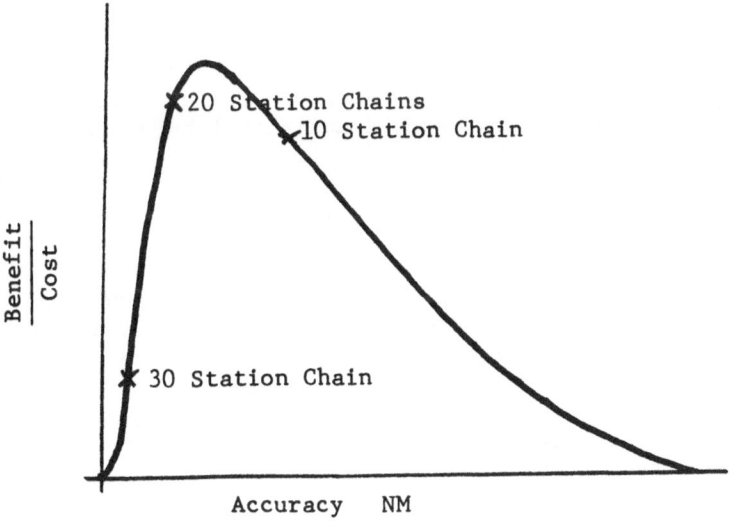

Figure 43. Loran-C

It will not be easy in practice to complete such a study;
many extraneous arguments will be advanced. But until such a study
is completed, far removed from immediate operational problems,
biases and vested interests, then each community will go its own
way and squander valuable resources.

If such a National Navigation Plan is instituted, then industry
can make firm engineering decisions in the design of navigation
systems and commonality and competition will drive the cost of user
equipment down.

What is the relevance of these arguments to the Arctic?
Examination of Figure 38 shows that the Arctic is Canada and
therefore any Canadian plan must consider those peculiarities of
the Arctic which will drastically change the parameters of such a
study. Every study of Arctic transportation (refs 6 and 102)
emphasizes the need for navigation aids in the Arctic, and at least
one Arctic mariner (CMDR OCS Robertson) agrees with this philosophy
of commonality:

"The government should not institute a program to provide a
continuous common reference navigation system to a specified
accuracy for all transportation systems, seaborne, airborne and
landborne, and that this navigation system be sufficiently accurate
to meet the continuing demand for a positioning system which will
satisfy the exploration industry in their particular investigations,
for there is no doubt but that the ultimate success of any
exploration and development program in any latitude is dependent
upon the ability of the exploration and transport vehicle to
accurately position and relocate themselves upon predetermined
points on the earth's surface.

There are today, off the shelf systems which will satisfy both
of these requirements. The provision of several such systems, all
different, would be an unwarranted additional burden on the
economic development of the area." (Ref 122).

Implementation

To formulate such a plan will require a great deal of
investigation. Similar studies have been conducted, or are still
underway, in the United States and although somewhat limited in
scope and based on different economic premises (for instance,
reference 115 examines fisheries in detail and completely dismisses
geophysical exploration, whereas these priorities are reversed in
the Arctic), they provide good starting points.

The study must identify all users, not just a sample as I have
done. By probing interview and polling, the study must identify
economic benefits, areas of operation, and benefit/accuracy
tradeoffs. The study must conduct rough engineering analyses (if
necessary under contract to a system advocate) of candidate systems,
including unusual proposals and including mixes of systems. The
study must apply rigorous cost/benefit analyses to all systems and

formulate a general policy and system specification that will allow industry to respond with appropriate hardware. There are very few facts required which are not available, with the possible exception of good ground conductivity values, which may have to be surveyed experimentally.

The study team will need the following skills:

a. economist;
b. operations research analyst;
c. radio propagation physicist;
d. radio engineer;
e. construction engineer;
f. navigator (air);
g. navigator (sea); and
h. navigator (land)(geophysicist).

These skills need not reside in separate persons or necessarily be formal qualifications. Many Northern pilots qualify admirably for the Air Navigator's job and many of the Canadian Forces Air Navigators are excellent electrical engineers.

The head of such a group requires special skills. First, he must be a diplomat to soothe losing advocates; he should be a navigator and very familiar with the Arctic and its special problems; and he must be thoroughly acquainted with all departments of government and be acceptable as dispassionate judge.

Given such talents and a _firm mandate_, it should be possible to formulate a National Plan for Navigation within one year.

Future Research in Navigation

Without a National Plan, research will be fragmented. One manufacturer will opt for IN, another for Omega, and Decca will pursue business as usual. Only two areas of research appear "system independent" at this time (except, of course, for the overall systems study itself).

We still need a good "compass" and at this point Ring Laser Gyro and Electrically Suspended Gyro technology appears very promising.

Whatever navaid is chosen, an application of propagation physics will be essential. VLF or Omega will be available anyway, and many operators will choose this system to be compatible with their global needs. There is a definite requirement for an improved "dirurnal correction" model of VLF propagation. This model should be based on sound physical principles, not just a statistical "force fit" and should model frequencies other than 10.2 KHz, if possible.

If Loran-C were to be chosen as prime navaid, then the most promising area of research left in Loran-C development is receiver signal processing. Fehlner (ref 45) shows range improvements of up to 1.8 times, using "editing" receivers (though this would not obtain against the low RF noise in the Arctic). In the Arctic, the

major limitation of Loran-C is distortion of the pulse envelope
due to low ground conductivity, and early skywave interference due
to low ionosphere, which prevents accurate third cycle identification. Many sophisticated methods have been developed in acoustic
and seismic research to deconvolve a signal from its "echo" (refs
14 and 147). These methods may very well extend the useful range
of Loran-C, changing a network of marginal coverage to one of great
redundancy. Dean (ref 28) indicates that ADF techniques work well
with Loran-C. If position is known, then it is a simple matter to
backplot by computer to obtain heading. Such processing may offer
a relatively low cost equivalent of the magnetic compass, without
the compass's limitations (ref 167).

Implicit in much of the preceding discussion has been the
assumption of a well developed precise timing technology.
Implementation of one way range range systems requires an onboard
clock of high accuracy in the user vehicle. Cesium beam oscillators
cost $23,000, rubidium about $5,000. There is a strong requirement
to develop a rugged inexpensive oscillator with stability to about
one part in 10^{11}. Full exploitation of time ordered systems
requires such a development.

The assumption of adequate computing power has also been
implicit. Standardization is a necessity: standardized diurnal
correction programs for Omega/VLF (ref 85), standardized co-ordinate
conversion algorithms and, if possible, a standardized higher level
programming language, would all contribute to more efficient
systems. The United Kingdom's "CORAL 66" and the USAF's "JOVIAL"
are examples of standardized higher languages.

The human operator cares little how his information was
obtained or processed as long as it is reliable. The interface
with the operator is still an area requiring research in human
factors engineering. Aircraft displays are, in general, highly
developed and the appropriate ARINC specifications provide good
ground rules. No such standardization is apparent in ships'
displays. Many automated bridges may be found but no well
organized body of agreed guidelines exists. Furthermore, special
problems in the Arctic may require interfacing different equipment
with the navigation displays. Much human factors engineering has
been devoted to ships' displays; more is still required,
particularly in the area of standardization.

CONCLUSIONS

Navigation systems for the most part represent mature
technology; little fundamental research is required. What is
required is sound engineering application of completed research.
Application of this research to determine sound navigation
policies will provide industry with the framework necessary to
develop the detail applications.

REFERENCES

All references to Navigation refer to the Journal of the
Institute of Navigation (United States), Suite 832, 815-15th St.,
N.W., Washington, D.C., 2005.
All references to The Journal of Navigation refer to The
Institute of Navigation (United Kingdom) at the Royal Geographical
Society, Kensington Gore, London SW7.

1 AC-90-45. "Approval of Area Navigation Systems for Use in the
   U.S. National Airspace", Federal Aviation Administration,
   Washington D.C., 18 Sep 69.

2 Ackerman, J.D. "Testing a Modern Gyromagnetic Compass", The
   Journal of Navigation, Vol. 18, No. 1, January 1965.

3 "Air Navigation Radio Aids", Transport Canada issued quarterly,
   Information Canada.

4 Anderson, E.W. The Principles of Navigation, Hollis and Carter
   Ltd., London, 1966.

5 Anderson, E.W. "A Philosophy of Navigation (Presidential
   Address)", The Journal of Navigation, Vol. 14, No. 1, January
   1961.

6 Arctic Transport, Proceedings of the Arctic Transportation
   Conference Yellowknife, NWT. December 8-9, 1970, sponsored by
   the Ministry of Transport and the Department of Indian Affairs
   and Northern Development. Information Canada, 171 Slater St.,
   Ottawa.

7 Belrose, J.S. (CRC Ottawa). "Low and Very Low Frequency Radio
   Wave Propagation", AGARD Lecture Series 29, Radio Wave
   Propagation, Technical Editing and Reproduction Ltd., July 1968.

8 Bendat, J.S. and Piersol, A.G. Random Data: Analysis and
   Measurement Procedures, Wiley-Interscience, New York, 1971.

9 Beukers, J.M. "Accuracy Limitations of the Omega Navigation
   System Employed in the Differential Mode", Navigation, Vol. 20,
   No. 1, Spring 1973.

10 Beukers, J.M. "A Review and Applications of VLF and LF
    Transmissions for Navigation and Tracking". Navigation,
    Vol. 21, No. 2, Summer 1974.

11 Beukers, J.M. "Integrated Upper Air Meteorological Sounding
    System", American Meteorological Society Second Symposium on
    Meteorological Observations and Instrumentation, March 27-30,
    1972, San Diego, California.

12 Bigelow, H.W. "Electronoc Surveying: Accuracy of Electronic
    Positioning Systems", Supplement to the International Hydro-
    graphic Review, Vol. 6, September 1965.

13 Blood, E.B. "Navigation Requirements for the U.S. Coastal
    Confluence Region", Navigation, Vol. 20 No. 1, Spring 1973.

14 Bogert, B.P., Healy, M.J. and Tukey, J.W. "The Analysis of
    Time Series for Echoes: Cepstrum, Pseudoautocovariance, Cross
    Cepstrum and Saphe Cracking", Time Series Analysis, Murray
    Rosenblatt (ed.) Wiley, New York, 1963, pp. 201-243.

15 Bourne, I.A., Ross, D.B. and Segal, B. "Phase Instability in
    Radio Waves Propagating Across Ice-Covered Seas", AGARD
    Conference Proceedings CP 33, Phase and Frequency Instabilities
    in Electromagnetic Wave Propagation. E.K. Davies (Editor)
    Technivision Services, Slough, England. July 1970.

16 Bowditch "Ch 7 Compass Error", American Practical Navigator
    United States Naval Oceanographic Office Publication HO No. 9,
    U.S. Government Printing Office, Washington, 1966.

17 Britting, K.R. Inertial Navigation Systems Analysis, Wiley-
    Interscience, New York, 1971.

18 Brogden, J.W. "Effects of Polar Cap Absorption Events on the
    Omega Navigation System", Naval Research Laboratories,
    Washington, D.C., Oct 73, DSIS No. 75-01355.

19 Brown, R.J.E. Permafrost in Canada, University of Toronto
    Press, 1970.

20 Burch, P.B. and Sakran, F.C., Jr. "Flight Tests of Two Airborne
    Omega Navigation Systems". Navigation, Vol. 21, No. 3, Fall
    1974.

21 Burgess, B. "The Influence of Propagation Conditions on the
    Design of VLF Radio Wave Long Range Navigation Aids". Paper
    presented at the conference of Deutsche Geselechaft Fur Ortung
    Und Navigation Ev Munich, 26-31 August 1965.

22 Calvo, A.B. and Bortz, J.E., Sr. "Evaluating the Accuracy of
    Omega Predicted Propagation Corrections", Navigation, Vol. 21,
    No. 2, Summer 1974.

23 Canadian Aviation "1975 Avionics Buyer's Guide", July 1975.

24 Canadian Aviation "Almost Incredible Series of Mishaps Led to
    Frobisher Sabreliner Crash", Canadian Aviation, pp. 30-38,
    December 1974.

25 Carros, J.R. "Distress Alerting and Locating System", U.S. Coast Guard Headquarters.

26 Communications Components Corp, 3000 Airway Ave, Costa Mesa, California 92626.

27 Conolly, R. "Air Force Moves Ahead on Omega", Electronics, 6 February 1975.

28 Dean, W.M. and Horowitz, S. "The Loran-C Navigation System", Navigation Systems for Aircraft and Space Vehicles, AGARDograph 55 T.G. Thorne (ed.), Pergamon Press, Oxford 1962.

29 DeGroot, L.E. "Presidential Address - Navigation and Its Relation to Society", Navigation, Vol. 20, No. 3, Fall, 1973.

30 Dennis, A.R. "A Second Generation Navy Satellite Marine Navigation System", Navigation, Vol. 21, No. 1, Spring 1974.

31 Department of Transportation National Plan for Navigation, April 1972.

32 Digital Methods Ltd. "Study to Evaluate Candidate Navigation Systems for Use in Canada North", Final Report to National Aerospace Planning Team, Ministry of Transport, Ottawa, Ontario, January 1975.

33 Dodington, S.J. "Radio Navigation", Chapter 5 Avionics Navigation Systems, John Wiley and Sons, Inc., New York, 1969.

34 Doherty, R.H. and Johler, J.R. "Meteorological Influences on Loran-C Ground Wave Propagation", U.S. Department of Commerce Office of Telecommunications Institute for Telecommunication Sciences. September 1973, Boulder, Colorado 80307.

35 Doherty, R.H. "Spatial and Temporal Electrical Properties Derived from LF Pulse Ground Wave Propagation Measurements", AGARD CP 144, Electromagnetic Wave Propagation Involving Irregular Surfaces and Inhomogeneous Media, Technical Editing and Reproduction Ltd.

36 Duncan, R.R. "Micron - A Strapdown Inertial Navigator Using Miniature Electrostatic Gyros". Paper presented to the Institute of Navigation, Washington, D.C., 13-15 March, 1973.

37 Eaton, R.M. and Grant, S.T. Rho Rho Loran-C Offshore Surveys, The Canadian Surveyor, Vol. 26, No. 2, Jun 1972.

38 Eaton, R.M. A Proposal for Loran-C Coverage of Atlantic Canada, Bedford Institute of Oceanography, Dartmouth, Nova Scotia, December 1973.

39 Eaton, R.M. "Tests of Loran-C Performance", <u>Canadian Aeronautics and Space Institute Journal</u>, Vol. 21, No. 4, April 1975.

40 Eaton, R.M. "Loran-C Compared With Other Navigation Aids in Meeting Future Canadian Needs", <u>Canadian Aeronautics and Space Institute Journal</u>, Vol. 21, No. 4, April 1975.

41 Elias, A.E. "PADS Position and Azimuth Determining System", <u>RUSI Journal of the Royal United Services Institute for Defence Studies</u>, June 1975.

42 Evans, D.E. "Navigation Aids Program for Northern Canada", <u>Canadian Aeronautics and Space Journal</u>, Vol. 21, No. 5, May 1975.

43 Federal Register, Vol. 34, No. 140, Friday, July 19, 1974.

44 Fehlner, L.F. and McCarty, T.A. "A Precision Position and Time Service for the Air Traffic of the Future", <u>The Journal of Navigation</u>, Vol. 26, No. 1, January 1973.

45 Fehlner, L.F. and McCarty, T.A. "How to Harvest the Full Potential of Loran-C", <u>Navigation</u>, Vol. 21, No. 3, Fall 1974.

46 "Filter Centre", <u>Aviation Week and Space Technology</u>, p. 51, March 1975.

47 Fisher, M.J. et al., "Evaluation of Three State-of-the-Art Doppler Navigation Systems", ECOM 4189 Commanding General U.S. Army Electronics Command, Attn AMSEL-VL-IY Fort Monmouth, N.J. 07703, January 1974, DSIS No. 75-03262.

48 Flying Annual and Pilots Buying Guide, Ziff Davis Publishing Company, 1975.

49 Fried, Walter R. "Doppler Navigation", Chapter 6 <u>Avionics Navigation Systems</u>, John Wiley & Sons Inc., New York, 1969.

50 Friedland, B. and Hutton, M.F. "New Algorithms for Converting Loran Time Differences to Position", <u>Navigation</u>, Vol. 20, No. 2, Summer 1973.

51 Gallenberg, R.J., Swanson, E.R. "Variations in Omega Propagation Parameters", Naval Electronics Laboratory Centre, San Diego, Ca 92152 NELC/TR 1773.

52 Ganong, W.F. and Hengeveld "Ice Reconnaissance and Navigation", <u>Canadian Aeronautics and Space Journal</u>, Vol. 21, No. 5, May 1975.

53 General Precision Inc. "Gyros Platforms and Accelerometers",
    Kearfott Technical Information for the Engineer No. 3, General
    Precision Inc., Little Falls, N.J., Sixth Edition, June 1963.

54 Gibbons, M. and Voyer, R. "A Technology Assessment System
    A Case Study of East Coast Offshore Petroleum Exploration",
    Science Council of Canada Background Study Number 30,
    Information Canada, March 1974.

55 Gibbs, G.L. "Omega in the Arctic and Its Rendezvous Capability
    for Logistic Supply". Paper presented at the First Canadian
    Symposium on Navigation and Resource Management, Ottawa,
    13-14 November 1974.

56 Glicken, M. "Navigation Requirements of Aerial Survey
    Operators", Navigation, Vol. 11, No. 1, Spring 1964.

57 Global Navigation Incorporated, 24701 Crenshaw Blvd, Torrance,
    California 90505.

58 Grant, S.T. "Rho Rho Loran-C Combined with Satellite
    Navigation for Offshore Surveys", International Hydrographic
    Review, Vol. 1, No. 2, Jul 1973.

59 Green, J.F. and Glenny, A.P. "Heading Definition in Commercial
    Aircraft", The Journal of Navigation, Vol. 13, No. 2, April
    1960.

60 Greenaway, K.R. Arctic Air Navigation, Arctic Research
    Defence Research Board 1951, Information Canada.

61 Grocott, D.F.H. "Doppler Correction for Sea Movement", The
    Journal of Navigation, Vol. 16, No. 1, January 1963.

62 Hardwick, C.D. and Brownley, T.R. "An Evaluation of Aircraft
    Navigation Using VLF Communications Stations with the Global
    Navigation Inc., GNS 200 VLF Navigation System in the North
    Star Aircraft", Laboratory Technical Report LTR-FR-30,
    National Research Council of Canada, Ottawa, 21 February 1973.

63 Hardwick, C.D. "VLF Development at NAE", Quarterly Bulletin
    of the Division of Mechanical Engineering and the National
    Aeronautical Establishment, Ottawa, 1 June to 31 Mar 1973.

64 Hardwick, C.D. "Aircraft Navigation Using VLF Transmission".
    Paper presented at the First Canadian Symposium on Navigation
    and Resources Management, Ottawa, Ontario, 13-14 November 1974.

65 Hastings, C.E. and Barker, A.C. "Automatic Real Time Omega
    Enhancement", Navigation, U.S., Vol. 18, No. 2, Summer 1971.

66   Hefley, G.   The Development of Loran-C Navigation and Timing,
     U.S. National Bureau of Standards Monograph 129, October 1972.

67   Higginbotham, L.D.   "Tactical Loran", AGARD Conference
     Proceedings CP 54, Hybrid Navigation Systems, 22-26 September
     1969, Technical Editing and Reproduction.

68   Huggett, W.S. and Mortimer, A.   "Observations on a Magnavox
     Satellite Navigation Receiver (702 with Automatic Acquisition
     Coupled to a HP 2115 Computer) in High Latitudes (70°N and
     49°N)", Pacific Marine Science Report 71-4.

69   "IFR Flight Instruments and Equipment Order", Air Navigation
     Orders Series 5, No. 22, Transport Canada, Information Canada,
     171 Slater Street, Ottawa.

70   Ince, A.M.   "Technical Review of EM Wave Propagation Involving
     Irregular Surfaces and Inhomogeneous Media", AGARD Conference
     Proceedings CP 144 on Electromagnetic Wave Propagation Involving
     Irregular Surfaces and Inhomogeneous Media, Technical Editing
     and Reproduction Ltd., London, 1974.

71   International Hydrographic Bureau, Radio Aids to Maritime
     Navigation and Hydrography, Special Publication No. 39,
     International Hydrographic Bureau, Monaco, 1965.

72   Jansky and Bailey Inc.   The Loran-C System of Navigation, Feb
     1962.

73   Johler, J.R., Keller, W.J. and Walters, L.C.   "Phase of the
     Low Radio Frequency Ground Wave", National Bureau of Standards
     Circular No. 573 (1956).

74   Johler, J.R. and Doherty, R.H.   "Unexploited Potentials of
     Loran-C", Proceedings of the National Radio Navigation
     Symposium, Institute of Navigation, Washington, D.C., 13-15
     November 1973.

75   Johler, J.R.   "The Impact of the Choice of Frequency and
     Modulation on Radio Navigation Systems", Navigation, Vol. 21,
     No. 3, Fall 1974.

76   Josephy, M.H. and Kaspar, J.F.   "A Polynomial Approximation
     Technique for Small Computer Skywave Correction Implementation",
     Proceedings of First Omega Symposium, The Institute of
     Navigation, Washington, D.C., 9-11 Nov 1971.

77   Kayton, M. and Fried, W.R. (eds.), Avionics Navigation
     Systems, John Wiley & Sons Inc., New York, 1969.

78 Klass, P.J. "Laser Gyro Re-Emerges as INS Contender", Aviation Week and Space Technology, 13 Jan 1975.

79 Knutsen, F.A., Capt, CAF. "Navigation Requirements and Systems for a Canadian Forces Search and Rescue Vehicle in the Airborne Search Role from 1975 to 1990", Canadian Forces Air Navigation School Report 5101, Canadian Forces Air Navigation School Winnipeg, Manitoba, 30 Jun 1975 (to be presented at the Second Canadian Symposium on Navigation, Ottawa, 18-20 Nov 1975).

80 Leigh Instruments Ltd. "Proposal for Satellite Alert and Position Determining Study for Search and Rescue", March 7, 1973, Leigh Instruments Ltd., Carleton Place, Ontario, Canada.

81 Leondes, C.T. (ed.) Theory and Applications of Kalman Filtering, AGARDograph 139, Technical Editing and Reproduction Ltd, London, February 1970.

82 Leondes, C.T. (ed.) Hybrid Navigation Systems, AGARD Conference Proceedings No. 54, Technical Editing and Reproduction Ltd, London, 1969.

83 Luken, K. "Omega Phase Shifts Due to Solar Phenomena", Proceedings First Omega Symposium, The Institute of Navigation, Washington, D.C., 13-15 November 1975.

84 MacDonald, K.D. "A Survey of Satellite Based Systems for Navigation Position Surveillance, Traffic Control and Collision Avoidance", Navigation, Vol. 20, No. 4, Winter 1973-4.

85 MacTaggart, D. "Use of Matched Airborne Omega Receivers for Rendezvous", Navigation, Vol. 20, No. 3, Fall 1973.

86 Maine, A.E. "Dead Reckoning Navigation Systems for Surface Vehicles as an Aid to Resource Exploration", Canadian Aeronautics and Space Journal, Vol. 21, No. 1, January 1975.

87 Malott, B.M. "Eastern Arctic Transportation by Sea in the 1970s", Proceedings of the Arctic Transportation Conference, Yellowknife, NWT, Dec 8-9, 1970, Information Canada.

88 McDowell, G.E. "Air Navigation Facilities", Proceedings of the Arctic Transportation Conference, Yellowknife, NWT, 8-9 December 1970, Information Canada, Ottawa.

89 McFarlane, I.C. (ed.) Muskeg Engineering Handbook, University of Toronto Press, 1969.

90 McKaughan "Analysis of a Proposed Differential Omega System", United States Naval Post Graduate School, Monterey, Calif.

91 McLarnon, R.P., Maj, CAF "Use of VLF in Air Navigation and Air
    Searches". Paper presented at the First Canadian Symposium on
    Navigation and Resources Management, Ottawa, Ontario, 13-14
    November 1974.

92 McQuillan, A.R. and Clough, D.J. "Benefits of Remote Sensing
    of Sea Ice", Research Report 73-3, December 1973, Energy, Mines
    and Resources.

93 Merkel, T.B. "Military Applications of the Transit Satellite
    Navigation System in the P3C Aircraft", Navigation, Vol. 20,
    No. 3, Fall 1973.

94 Miller, B. "Satellite Navigation Network Defined", Aviation
    Week and Space Technology, April 15, 1974.

95 Mohin, W.B., Commander USCG "Navigation Systems: Applicability
    of Loran-C to the Arctic". Paper presented at Conference on
    Arctic Systems held in St. John's, Newfoundland, 18-22 Aug 1975.

96 Moore, R.E. "The Application of Satellite Navigation and
    Inertial Guidance Systems to Surveying and Mapping". Presented
    at the First Canadian Symposium on Navigation and Resources
    Management, Ottawa, 13 Nov 1974.

97 Nard, G. "Results of Recent Experiments with Differential
    Omega", Navigation, Vol. 19, No. 2, Summer 1972.

98 "New Space Navigation Satellite Planned", Aviation Week and
    Space Technology, 15 July 1974.

99 NNSS: All the technical details of the NNSS system which were
    released on 29 July 1967 are contained in a volume called
    "Technical Documentation for Shipboard Receivers - U.S. Navy
    Satellite Navigation System", available from the:
            National Security Industrial Association
            1030-15th Street N.W.
            Washington, D.C. 20005
    The volume contains six documents:
    a. Kershner, R.B. "Status of the Navy Navigation Satellite
        System", reprinted from Practical Space Applications,
        Volume 21, Advances in the Astronautical Sciences Series,
        1967.
    b. Kershner, R.B. "Present State of Navigation by Doppler
        Measurement from Near Earth Satellites", APL Technical
        Digest, November-December, 1965 (a publication of the
        Applied Physics Laboratory of the Johns Hopkins University).

    c. Darragh, W.M. "Radio Navigation Set AN/SRN-9 (XN-5)
        Operation and Maintenance", APL Technical Memorandum
        TG-685-1, August 1965.

d.   Gutheim, G.C.   "Program Requirements for Two-Minute
     Integrated Doppler Satellite Navigation Solution", APL
     Technical Memorandum TG-819-1, April 1967.

e.   Clark, J.F.   "Near Earth Satellite Handbook Data", APL
     Technical Memorandum TG-580, June 1964.

f.   "Performance Requirements Navy Navigational Satellite
     System - Shipboard Radio Navigation Set", APL Note
     PR-T-13A, November 1967.

100 Ogata, K. Modern Control Engineering, Prentice-Hall Inc.,
     New Jersey, 1970.

101 Okpik, A.   "Address on Transportation and the Arctic
     Community", Proceedings of the Arctic Transportation
     Conference, Yellowknife, NWT, 8-9 Dec 1970, Information
     Canada.

102 Paddison, E.C. and Stone, A.M.   "Transportation in the
     Arctic". Technical Memorandum TG-1190, April 1972. The
     Johns Hopkins University Applied Physics Laboratory.

103 "Panel Discussion", First Canadian Symposium on Navigation
     and Resource Management, Canadian Aeronautics and Space
     Journal, Vol. 21, No. 5.

104 Papoulis, A. Probability Random Variables and Stochastic
     Processes, McGraw Hill, New York, 1965.

105 Pearce, D.C. and Walker, J.W.   "The Behaviour of Loran-C
     Groundwaves in Mountainous Terrain", Electromagnetic Wave
     Propagation Involving Irregular Surfaces and Inhomogeneous
     Media, AGARD Conference Proceedings CP 144, Technical
     Editing and Reproduction Ltd, London, 25-29 March 1974.

106 Phipps, W.W.   "Air Transportation in the High Arctic",
     Proceedings of the Arctic Transportation Conference,
     Yellowknife, NWT, Vol. 2, December 8-9, 1970, Information
     Canada, 171 Slater Street, Ottawa, Ontario.

107 Pierce, J.A., Mackenzie, A.A. and Woodward, R.H. et al.
     LORAN, Massachusetts Institute of Technology Radiation
     Laboratory Series, McGraw Hill, 1948.

108 Pierce, J.A.   "Omega", Transactions on Aerospace and
     Electronic Systems, Vol. AES 1, No. 3, Dec 1965, pp. 206-215
     (also published as below).

109   Pierce, J.A. "Omega". Paper presented at the conference of
       Deutsche Geseleschaft fur Ortung and Navigation Ev at Munich,
       26-31 August 1965. Conference proceedings published at
       Dusseldorf.

110   Pierce, J.A., Palmer, W., Watt, A.D. and Woodward, R.H.
       Omega, A World Wide Navigational System, Pickard and Burns
       Electronics, P & B Pub No. 886, 2nd Edition 1966. DDC Cat
       No. AD 630-900.

111   Pierce, J.A. and Woodward, R.H. "The Development of Long
       Range Hyperbolic Navigation in the United States", Navigation,
       Vol. 18, No. 1, Spring 1971.

112   Piscane, V.L., Holland, B.B. and Black, H.D. "Recent (1973)
       Improvements in the Navy Navigation Satellite System",
       Navigation, Vol. 21, No. 3, Fall 1973.

113   Potts, C.E. and Roeber, J.F. "Time/Frequency and
       Transportation", Proceedings of the IEEE Special Issue on Time
       and Frequency, May 1972.

114   Proceedings of the Loran-C workshop held in Gettysburg,
       Pennsylvania, June 5, 6 and 7, 1974.

115   Radio Aids to Navigation for the U.S. Coastal Confluence
       Region, Interim Report No. 1, User Requirements, TR-72-0354,
       March 1972, Polhemus Navigation Sciences Inc.

116   Radio Aids to Navigation for the U.S. Coastal Confluence
       Region, Interim Report No. 2, Systems Evaluation, TR-72-0661,
       June 1972, Polhemus Navigation Sciences Inc.

117   Radio Aids to Navigation for the U.S. Coastal Confluence
       Region, Supplemental Report, TR-72-0864, August 1972, Polhemus
       Navigation Sciences Inc.

118   Final Technical Report, Differntial Omega Test and Evaluation
       Program, Tracor Doc. 67-135-U, 18 Jan 1967.

119   Supplement to Final Technical Report, Tracor Doc. 67-136-U.

120   Final Report, Differential Omega Monitoring and Analysis,
       Beukers Laboratories, Report No. BLI-FR-3514, Three Volumes
       dated 19 May 1972, plus One Volume Addendum dated 27 June
       1972.

121   Razin, S. "Explicit (Non Iterative) Loran Solution",
       Navigation, Vol. 14, No. 3, Fall 1967.

122 Robertson, O.C.S., Commodore, RCN (Ret'd) "Ice Information, Marine Navigation Aids and Terminal Facilities", Proceedings of the Arctic Transportation Conference, Yellowknife, NWT, 8-9 Dec 1970.

123 Rohmer, R. The Arctic Imperative (an overview of the energy crisis), McLelland and Stewart Ltd, 25 Hollinger Rd, Toronto, Ontario.

124 Ross, D.I., Wells, D.E., Douglas, G.R. and Eaton, R.M. "Navigation at Bedford Institute", AOL Report 1970-2, Bedford Institute of Oceanography, Dartmouth, N.S., February 1970.

125 Rothmuller, I.J., Swanson, E.R., Kugel, C.P. "Omega Arctic Propagation: Synchronized Monitoring at Wales, Alaska", NELC-TE 1765 Microfiche.

126 Rupp, W.E. "Problems and Solutions in the Satellite Relay for Omega for Search and Rescue Purposes", Navigation, Vol. 21, No. 3, Fall 1974.

127 Sakran, F.C., Jr. "Long Period Error Modes of a Doppler Radar Free Directional Gyro Navigation System", Navigation, Vol. 17, No. 1, Spring 1970.

128 Science and the North, A Seminar on Guidelines for Scientific Activities in Northern Canada 1972, 15-18 Oct 1972, Mont Gabriel, Québec, Information Canada, 171 Slater St, Ottawa.

129 Scull, David C. "Omega", Radio Aids to Maritime Navigation and Hydrography, Special Publication No. 39, International Hydrographic Bureau, Monaco 1965, Reprinted 1969.

130 Sea Technology, April 1975, "Computerized Omega System".

131 Shinners, S.M. Modern Control System Theory and Application, Addison-Wesley, Reading, Massachusetts, 1972.

132 Shinners, S.M. Techniques of System Engineering, McGraw Hill, New York, 1967.

133 Smith, S.G. and Thomas, J.L. "The Long Period Errors of Doppler and Inertial Navigation Systems", Royal Aircraft Establishment Technical Report No. 66084, March 1966.

134 Sorg, H. "A Literature Survey on the Gyroscope and Its Applications", AGARD Report No. 582, Technical Editing and Reproduction Ltd, London, Dec 1970.

135 Stallbrass, J.R. "Icing of Fishing Vessels in Canadian Waters", <u>Quarterly Bulletin of the Division of Mechanical Engineering and the National Aeronautical Establishment</u>, Ottawa, 1 January to 31 March 1975.

136 Stansell, T.A., Jr. "Transit the Navy Navigation Satellite System", <u>Navigation</u>, Vol. 18, No. 1, Spring 1971.

137 Stout, C.C. "The Omega System of Navigation", <u>Advanced Navigational Techniques</u>, AGARD Conference Proceeding 28, Unwin Jan 1970.

138 "Surveying Offshore Canada Lands for Mineral Resource Development", Workshop on Offshore Surveys, December 1970, Department of Energy, Mines and Resources, Ottawa, Canada, Information Canada, 171 Slater Street, Ottawa.

139 Swanson, E.R. and Tibbals, M.L. "The Omega Navigation System", <u>Navigation</u>, Vol. 12, No. 1, Spring 1965 (also available in French), "Le Système Mondiale de Navigation Omega", <u>Navigation: Revue Technique de Navigation Maritime Aerienne et Spatiale FR</u>. Vol. 13, juillet 1965.

140 Swanson, E.R. "Omega", <u>Navigation</u>, Vol. 18, No. 2, Summer 1971.

141 "Table of Frequency Allocations", Communications Canada.

142 Treadwell, R.J. "The Perfection of the Gyromagnetic Compass", <u>The Journal of Navigation</u>, Vol. 18, No. 1, January 1965.

143 "The Pilot's Friend", <u>Canadian Flight Magazine</u>, March-April 1974, Canadian Owners and Pilots Association, P.O. Box 734, Ottawa.

144 Wait, J.R. "A Survey of Recent Research in the Propagation of VLF Radio Waves", <u>Navigation Systems for Aircraft and Space Vehicles</u>, T.G. Thorne (ed.), Pergamon Press, Oxford 1962.

145 Wait, J.R. and Howe, H.H. "Amplitude and Phase Curves for Groundwave Propagation in the Band 200 Cycles per Second to 500 Kilocycles", <u>National Bureau of Standards Circular 574</u> (1956).

146 Walker, M., Capt, CAF. "Personal Correspondence to R.M. Eaton, Bedford Institute of Oceanography", 7 Jan 1973.

148 Weems, P.V.H. "Magnetism and Compasses", Chapter 3 <u>Air Navigation</u>, Weems System of Navigation, Anapolis, Md., 1958.

149 Wells, D.E. and Ross, D.I. "Experience with Satellite Navigation During Summer 1968", Report BI 1969-6, Bedford Institute of Oceanography, Dartmouth, N.S., June 1969.

150 Wright, J. "Accuracy of Omega/VLF Range Rate Measurements", Navigation, Vol. 16, No. 1, Spring 1969.

151 Yamakoshi, Y. "Construction of the Omega Station in Japan", Proceedings of the First Omega Symposium, The Institute of Navigation, Washington, D.C., 9-11 Nov 1971.

CANADIAN FORCES AIR NAVIGATION SCHOOL

REPORTS AND PROJECTS

The Canadian Forces Air Navigation School has conducted many in-house studies in the form of student projects or staff reports. These reports usually take the form of a précis or resumé on some particular aspect of aerospace technology and are not technical papers in the usual sense. They are very useful, however, in providing brief descriptions and literature surveys. These reports are normally not disseminated outside the Canadian Armed Forces; however, copies can be made available to qualified personnel through:

Defence Scientific Information Service (DSIS)
National Defence Headquarters
Ottawa, Ontario
K1A 0K2

152 Bryce, J.C., Capt, CAF. "A Paper on the Dynamic Errors of Gyromagnetic Compass Systems", Canadian Forces Air Navigation School Guidance and Control Project, 15 Mar 71.

153 Bryce, J.C. and Gerden, V.F., Capts, CAF. "A Study on Navigation Techniques and Equipment for Tactical Aircraft in Northern Latitudes", Canadian Forces Air Navigation School Report No. 3101, 19 Jan 1973.

154 Canadian Forces Air Navigation School Report 5000, ASC 27, "A Study of Canadian Forces Airborne Equipment Required for the Search and Rescue Role from 1975-1990".

155 Canadian Forces Air Navigation School Staff Air Navigator Course Radio Aids to Navigation Précis, Canadian Forces Air Navigation School, Winnipeg, Manitoba 1974.

156 Canadian Forces Publication (CFP) 139(1), Navigator Flying Training Manual, Volume 1, Air Navigation Part One Navigation.

157 CFP 148 Manual of Instrument Flying.

158   Commander Maritime Command Final Report on Project OPVAL/AE
      397, The Operational Evaluation of the CMA-719(M) in the
      Argus Aircraft.

159   Forbes, G.J., Capt, CAF.   "Lecture Notes on Hybrid IN,
      Canadian Forces Aerospace Systems Course".

160   413 Squadron CAF.   "A Report on Operation Frozen Tusker, An
      Exercise in Northern Deployment of SAR Forces", File 0607 SAR,
      Canadian Forces Air Navigation School, Winnipeg, Manitoba,
      3 Jun 1974.

161   Gerden, V.F., Capt, CAF.   "Ground Based Aids for All Weather
      Landings", Canadian Forces Air Navigation School Guidance and
      Control Project, 21 May 71.

162   GPH 209 Manual of Criteria for Instrument Approach Procedures.
      Compiled by Surveys and Mapping Branch, Department of Energy,
      Mines and Resources, Ottawa, Ontario.

163   Lavoie, J.R.E., Capt, CAF.   "The Design Principles of the
      Electrostatic Gyro", Canadian Forces Air Navigation School
      Guidance and Control Project 0824, 5 March 1972.

164   Miller, W.H., Capt, CAF.   "The Theory and Development of Laser
      Gyro Systems", Canadian Forces Air Navigation School Guidance
      and Control Project, 8 March 1971.

165   Ross, J.A.F., Sqn Ldr, RAF.   "The GNS 200 Navigation System
      for Twin Otter", Canadian Forces Polaris, Vol. 2, No. 2, 1973.

166   Walker, M., Capt, CAF.   "A Method of Accurately Locating
      Sonobuoys", Staff Paper, Canadian Forces Air Navigation School,
      Winnipeg, Manitoba.

167   Walker, M. Capt, CAF.   "A Modest Proposal for the Loran-C
      Coverage of Arctic Canada", Canadian Forces Polaris, Vol. 3,
      No. 2, 1974.

168   White, L.R., Capt, CAF.   "Loran-C or Omega", Canadian Forces
      Aerospace Systems Course 23, Guidance and Control Project.
      Also published as a two part article in Canadian Forces
      Polaris, Vol. 1, No. 1, Vol. 1, No. 2.

169   Wolowydnyk, P., Lt, CAF.   "Notes Taken at an Arctic Systems
      Course Held in Inuvik", July 1975.

APPLICABILITY OF LORAN-C TO THE ARCTIC

Commander W. B. Mohin

U.S. Coast Guard

Washington, D.C., U.S.A.

The Loran-C Radionavigation System has been chosen to be the government sponsored navigation system for the Coastal Confluence Zone (CCZ) of the United States. This decision was published in a July 1974 Annex to the National Plan for Navigation.[1] The implementation schedule for Loran-C chains and the shutdown schedule of Loran A chains were published in the same Annex.

Figure 1 is a diagram of the planned Loran-C configuration for the coasts of the United States. The coverage limits shown have been calculated for a fix accuracy of .25 nm, 2 drms, 95%.[2] time at a -10db SNR. To make these calculations a measurement error standard deviation of .1 microseconds is assumed. Noise levels are calculated for a 20 KHz Band- with using CCIR published noise data.[3] Coverage limits were designed to meet the CCZ definition, i.e. the 100 fathom curve or 50 miles whichever was greater. Actually, all coverage limits also meet the 200 mile criteria for possible future law enforcement use if sea boundaries are changed. The coverage shown in the northeastern U.S. is an additional chain, now under consideration, obtained by tying the existing Maine, Massachusetts and Newfoundland stations together.

Figure 2 is a worldwide coverage diagram of existing Loran-C coverage. The additional coverage being added by the current expansion efforts is shown cross hatched. In the process of designing and building the new Loran-C chains, the Coast Guard has made a substantial number of improvements to ground station equipments. New timing and control equipment has allowed us to drastically reduce technician complements and all new stations are to operate in the unwatched mode.[4] This means that a technician is on the station but no live watches are required. This new equipment has also given us more control over the radiated pulse shape and other important station parameters. We are in the

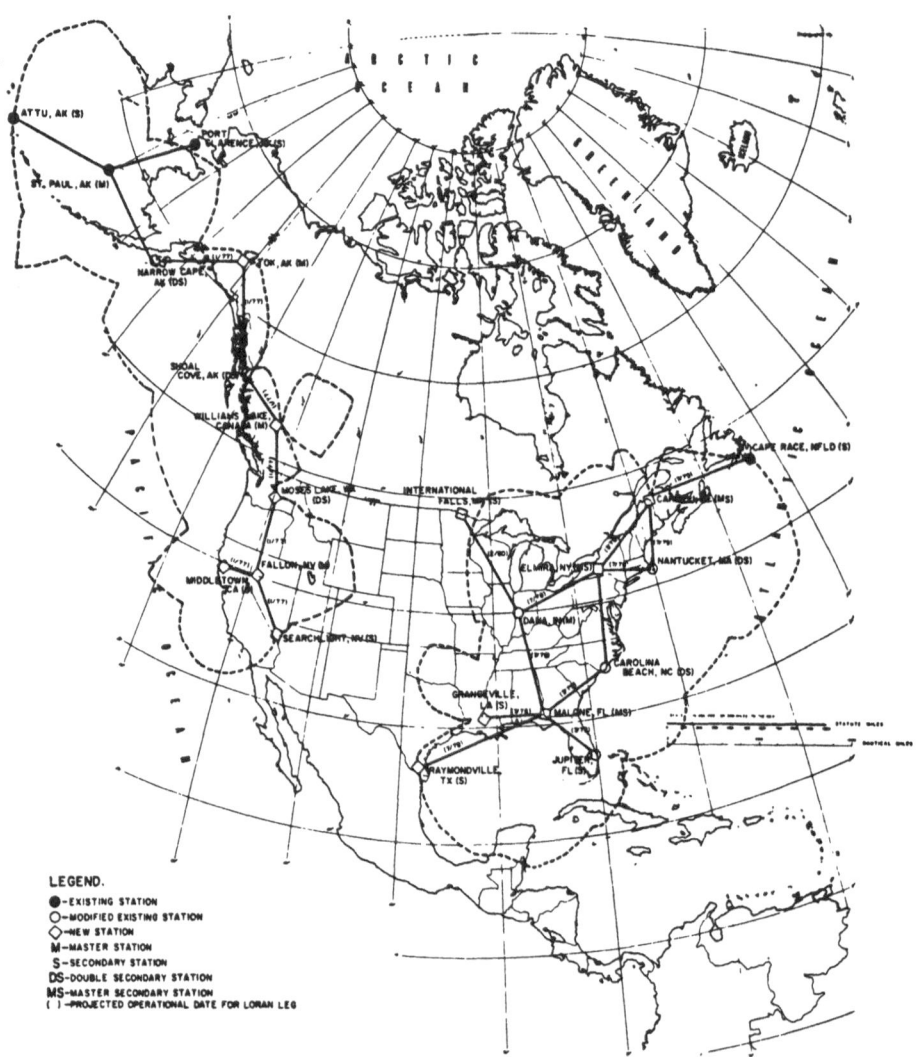

Figure 1.   Proposed Loran-C Coverage (1:3 Signal/Noise Ratio)

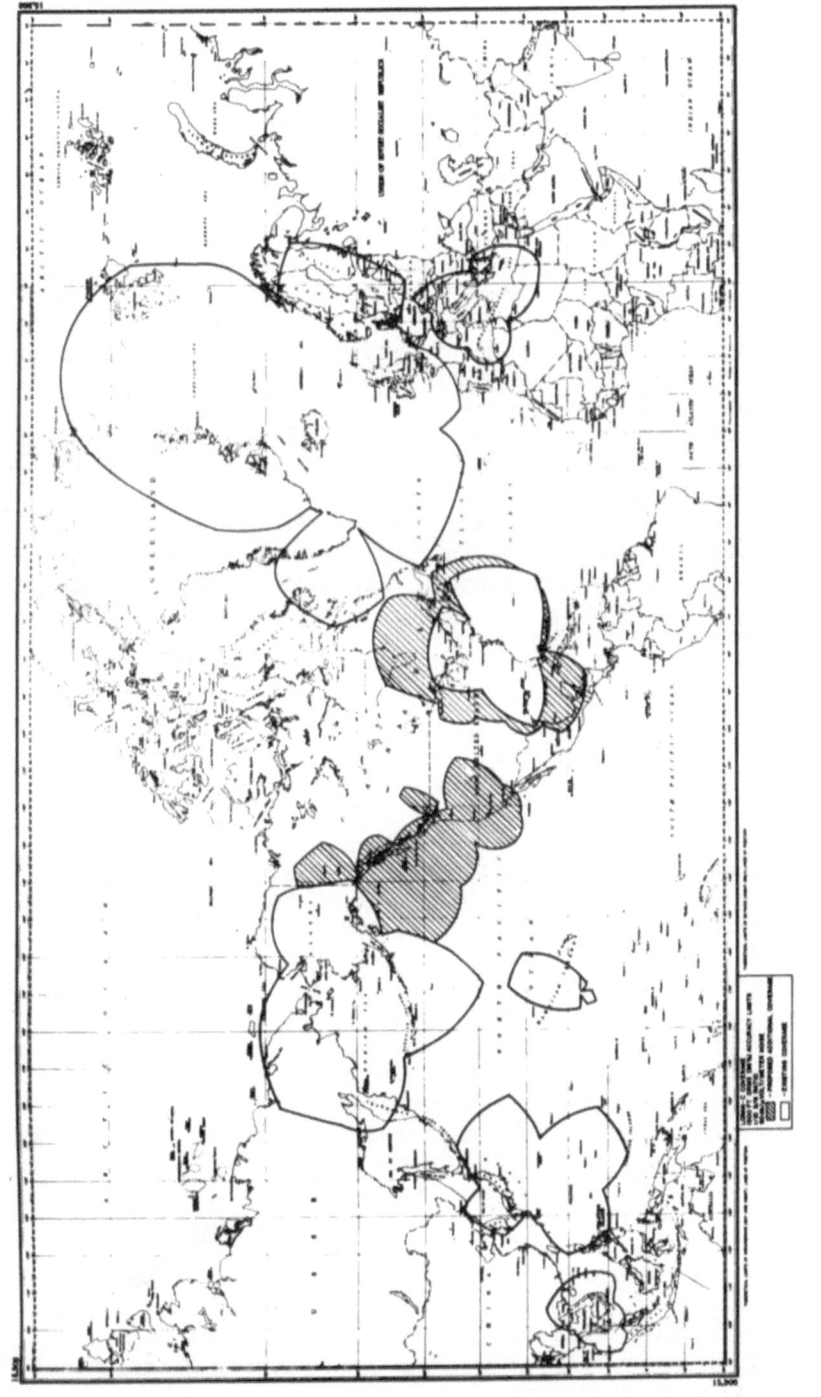

Figure 2.   Loran-C Coverage

process of developing a new solid-state transmitter that promises
to reduce maintenance even further. All of these equipments are
of modular construction and repair is effected at Central Repair
Facilities. A new unmanned chain monitor station is being designed
to allow chain control from one central point. We are also
developing automated chain control and log keeping equipments
which we hope will soon produce unmanned stations. The equipment
we are designing will require a technician within 30 minutes of
each station.

Loran-C stations are now operating in what we refer to as the
"free running" mode. This means that secondaries are not actively
synchronized to the Masters and are in effect independent stations.
Each station has three independent Cesium Standards to guarantee
the reliability of their time base which are adjusted within a
chain to 5 parts in 10^{13} of each other. We have made great
progress in simplifying complexity, reducing costs and increasing
signal stability of Loran-C stations and are continuing work in
this field to make even further gains.

There have been similar gains made in receiving equipments.
There are automatic Loran-C marine receivers available today for
about $3000. As the market expands this price will be reduced.
Receivers which track more than 2 LOP's are becoming available and
some are being developed to operate in a Secondary-Secondary Mode
which will provide increased coverage and System Redundancy. Low
cost aircraft receivers are being investigated and there appears
to be no technical reason that they should not be wholly successful. Several Loran-C guidance devices have also been built[5] and
tested successfully and other more sophisticated systems, both
marine and air, are being investigated.

The Coast Guard is presently working with the Great Lakes
Winter Navigation Board to build a small area, high accuracy
Loran-C chain to cover the St. Mary's River between Lake Superior
and Lake Huron. This system is scheduled for demonstration in
Jan/Feb 1976 to show how Loran-C can be used to guide ships
through a restricted channel when all buoyage has been removed due
to surface ice conditions. This type of test has a direct
application in the Arctic where ships must transit confined areas
with little or no surface aids to guide them.

The U.S. Department of Transportation has recently held
several meetings of government agencies to determine if there was
substantial interest in completing coverage of the entire country
by providing Loran-C coverage in the central states. Many
applications were identified where a nationwide Loran-C chain
would solve existing problems. Among these are:
(a) Possible use by U.S. Census Bureau to obtain a common
geographic data base. The Census Bureau presently does not have
adequate geographic information on rural populations and is
unable to effectively respond to many requests for geographically
based population information.

(b) Highway Safety - The States have a Federal requirement to
report highway accident positions to .1 mile in order to establish
accident statistics, aid in accident analysis and improve highway
design. At present this requirement is being solved primarily by
expensive mileage markers which provide only a lineal positioning
system along the highway. A study and demonstration of Loran-C's
ability to be used in this application is now underway in the
State of New York.
(c) The FAA is now engaged in several studies to determine how
Loran-C can be used to supplement the installed VOR/DME system for
area Navigation and non precision approach.
(d) Aerial crop survey
(e) Fire fighting
(f) Location of emergency vehicles
(g) Law enforcement
    The possibilities for surface use in both the public and
private sectors are immense. The DOT is now seriously considering
tasking the U.S. Coast Guard with design and construction of mid-
continent chains. Figure 3 is a preliminary design for interior
coverage requiring five stations and providing a great deal of
redundant LOP coverage. If these chains are funded the entire
United States could be on a common Navigation data base by 1980.
To use this capability a small low cost receiver will be necessary.
Initial studies indicate that a miniature Loran-C receiver can be
manufactured for several hundred dollars in large quantities. A
completely automatic Loran-C receiver the size of a hand held
calculator has already been developed.
    The present Loran-C system operating at 100 KHz represents a
near optimum radionavigation system when both area covered and
accuracy provided are considered. Some of the questions to be
answered for Arctic areas that have already been addressed for
other parts of the world[6] include:
(a) What is the most stringent accuracy requirement foreseen for
air, marine and land navigation?
(b) What area is to be covered - with what accuracy of navigation?
(c) Is a mix of systems desirable from the point of view of
capital costs, user costs, operational effectiveness?
(d) What is the effect of the unique Arctic problems on the
various navigation systems? For example:
1. Poor conductivity of permafrost and ice
2. Convergence of meridians
3. Magnetic compass errors of large magnitude
4. Poor weather, wilderness and a marked absence of other aids.
5. Low Ionospheric height
6. Auroral Activity
    If the Arctic is to be opened and its riches realized
Navigation will have to be provided for air, surface and marine
users.
    The air navigation picture for the Arctic was examined in a
recent study by Polhemus Navigation Science Inc.[7] under the

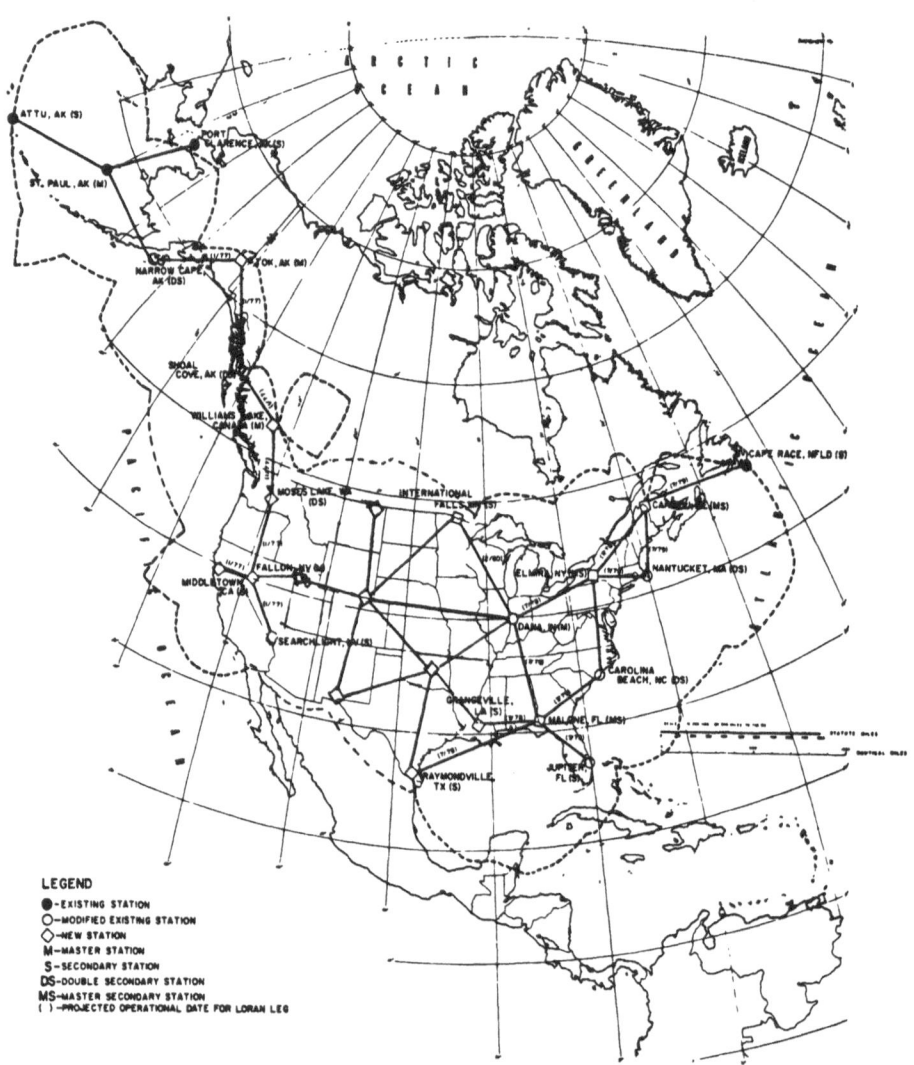

Figure 3.   Proposed Loran-C Coverage (1:3 Signal/Noise Ratio)

assumptions that:

(a) aircraft will require a fix with an accuracy of not worse than 10 nautical miles (95%) at least once each thirty minutes.

(b) The ATC surveillance requirement will continue to be satisfied by procedural rather than positive control techniques due to the high cost of land-based radar and low traffic densities.

The study suggested that a modified Non-Directional Beacon (NDB) system could be an acceptable solution for operation in the Canadian North the year 2000. This solution requires a total of 136 new NDB transmitters and modification of the current receivers to provide for retransmission of bearing and heading information in case of emergency. The cost of this alternative was placed at $3.42/cubic mile in Capital Costs. Polhemus goes on to say that "Loran-C offers the most cost effective position determination aid in terms of cubic miles of coverage. The alternate solution to the navigation and surveillance aid requirement problem is Loran-C. The cost would be of the order of $1.44/cubic mile". Another consideration in the choice of an air navigation system that must be addressed is the development of Guidance systems for both NDB and Loran-C in order to provide steering information. In the case of Loran-C these devices will provide sufficient accuracy for those aircraft navigation requirements not covered by Polhemus. These include:

(1) Search and Rescue
(2) Aerial survey
(3) Precision delivery of droppable goods
(4) Aerial law enforcement
(5) Ice reconnaissance

I believe the cost of these guidance systems will be similar for both NDB and Loran-C sensors.

Polhemus refers to the disadvantage to the user of the high cost of aircraft Loran-C receivers. At the present time a good aircraft Loran-C receiver sells for about $15K. There is no sound technical reason for this other than the fact that there has not been a requirement to develop a low cost airborne receiver. We have successfully flown $3000 marine receivers although they are not designed for aircraft requirements or environment. A task that still must be done is to define what functions a low cost Loran-C aircraft receiver should have. I believe it should at the very least track 3 LOP's and not be master dependent. A receiver of this type could easily be built given a requirement to do so.

Marine traffic will require a range of navigation accuracies. The most stringent of these are likely to be for Geophysical Survey, Vessel Navigation in restricted or dangerous waters with little or no buoyage system and fishing operations. A reasonable expectation for these applications is believed to be 50 to 150 foot repeatable accuracy. These types of accuracies are now being realized by similar users in other parts of the world. Oil survey companies have recently built and are using commercial Loran-C chains in order to obtain the range and accuracy it provides. One

of the main factors considered in choosing the requirement for .25
nm accuracy for the U.S. CCZ navigation system was the safe
navigation of super tankers carrying millions of barrels of oil.
Surely this requirement is equally, if not more, severe in the
Arctic ice. As with aircraft, guidance systems will have to be
developed for vessels in unmarked waters. We are presently working
on several of these systems to be tested in the St. Mary's River
Project early next year.

Surface vehicles in the Arctic will require some form of
navigation due to extreme weather conditions, many tracked type
behicles operating off roadways and location in the event of
emergencies. This requirement will probably be in the order of
.1 nm or better.

The area to be covered by a navigation system in any Arctic
Environment will naturally depend on the overall planning for the
area. The Canadian North will eventually be opened for all types
of exploration and traffic and hence navigation system planning
should probably address the entire area. A review of the
requirements known at this time indicates that there will be a
substantial need for precision navigation in addition to general
purpose navigation. The questions that remain concern the mix of
systems or single system that is the most cost effective for the
most users. A much more thorough future navigation requirements
study is needed for this region.

Of all the problems the Arctic Area presents to radionavigation systems the most severe is the conductivity of permafrost
and glacial ice.[8] This poor propagation is true for all electromagnetic waves having a ground wave mode of propagation. Figure 4
is a propagation chart for various values of ground conductivity
at 100 KHz. Values of conductivity vary from .025 mmhos/m for
glacial ice to 5 mhos/m for seawater. It can be readily seen that
Arctic Permafrost (.4 mmhos/m) is far from a good medium over which
to transmit an LF signal. Several unique factors have to be taken
into account therefore when discussing Arctic navigation systems.

To illustrate these problems I will go through a simplified
design of a Loran-C chain in the Arctic area. Loran-C coverage
limits are calculated by assuming a standard deviation for the
total measurement error and computing the geometrical limits of
the accuracy contour desired. In the computation a 2drms (95%)
figure is used. These coverage limits are then modified by the
range limits of the transmitters for the desired signal-to-noise
ratio. Noise figures are obtained from CCIR Report #322[2] and the
95% noise figure is used, i.e. the noise figure exceeded only 5%
of the year. These calculations are extremely conservative as
they should be in any navigation system used for safety purposes.

The first two facts that radically alter Loran-C system
design in the Arctic are the poor ground conductivities and the
low noise figure. These, to some extent, appear to offset each
other. Since almost any Loran System designed for the Arctic will
be range limited and not geometry limited, the main design efforts

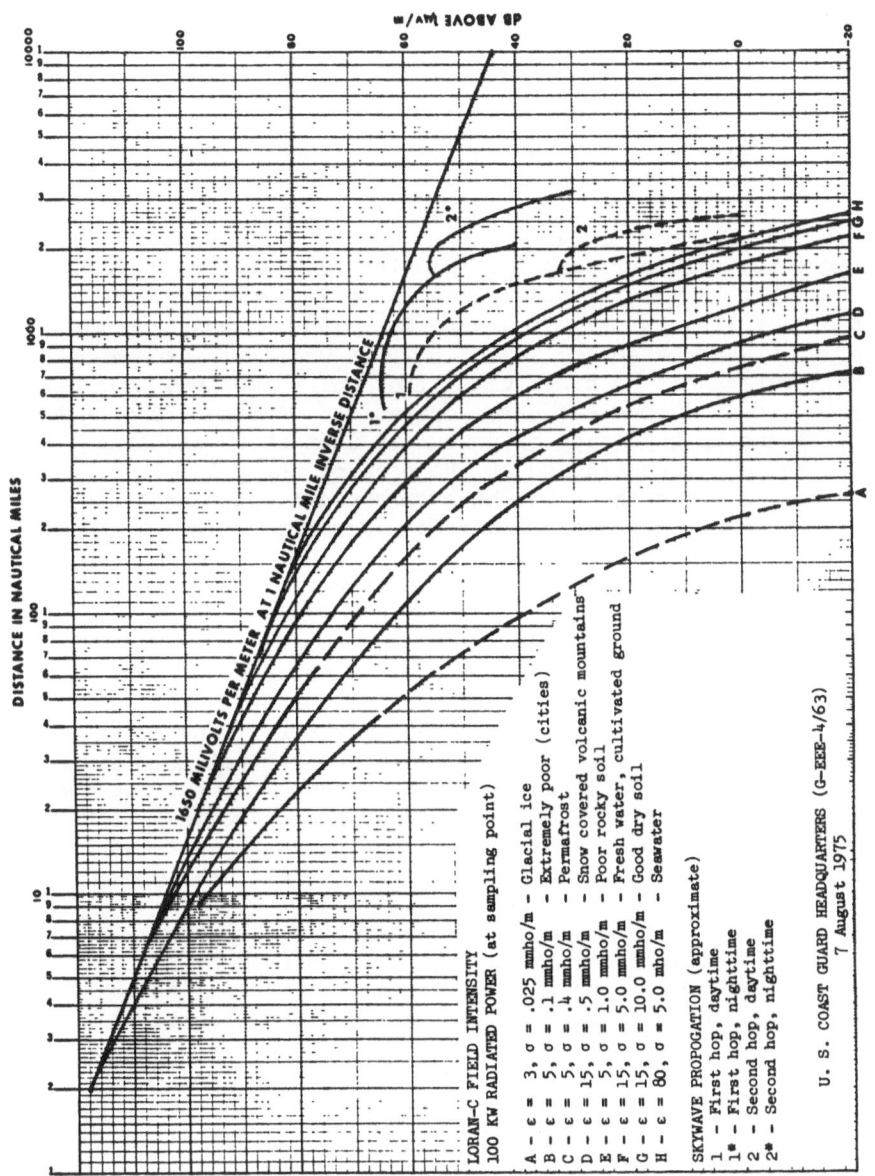

Figure 4. Loran-C Field Intensity

should concern useful range and transmitter size in order to produce an effective system.

The 95% noise figure for the Canadian Arctic is approximately 20db below that used in designing the United States CCZ system.[2,9] Referring to Figure 4 for an all permafrost path, the point where the signal strength is approximately 20db below that of a seawater path, is about 200 nm. At this point the system designed for the CCZ and an all permafrost path should have equal performance. Beyond this range the permafrost path deteriorates much more rapidly. For example a 500 KW peak power transmitter is useful to 775 nm in the CCZ system and hits -10db SNR at 410 nm over permafrost. If 50 KW stations were used over an all permafrost path, the -10db SNR effective range would be 300 nm. This leads to the conclusion that a larger number of smaller transmitters on short baselines may be a preferred solution in some arctic regions. The losses involved in permafrost and ice paths beyond several hundred miles make large amounts of power a wasteful investment.

Smaller transmitter sites will also lend themselves much more readily to unmanned remote-control operation. Another important factor influencing this type of design is that short baseline Loran-C signals can be tracked at the peak of the pulse instead of the normal 30 microsecond point because skywave interference will not be a factor at the ranges under consideration. This technique effectively increases transmitter power by 6 db.

An additional consideration that must be incorporated into Arctic system design are the complications of mixed propatation paths.[8] Glacial ice attenuates signals much more severely than permafrost and sea ice attenuates signals much less than permafrost. Initial CG studies show that attenuation of Loran-C signals over sea ice is not significantly worse than seawater. Considering that a large portion of the Canadian Arctic is an archipelago, this is a definite advantage. A Loran-C system designed for surface and marine use in the poor conductivity areas of the Arctic will be more than adequate for aircraft due to the height-gain factor. At 100 KHz the signals at 10,000 feet over permafrost are 10 db stronger than those on the ground.

All of these factors demand that any electronic navigation system being considered for the Arctic regions need be not only analyzed in detail as to user requirements and capital expenditures but must also be extremely carefully designed and planned at the same time.

Loran-C, due to its near optimal frequency for coverage and accuracy, its ability to provide ambiguous lane information and its inherent rejection of skywave interference is a strong contender for consideration in any Arctic navigation system design.

REFERENCES

1. Department of Transportation, National Plan for Navigation Annex, July 1974, Department of Transportation, Wash., D.C.

2. W. Allan Burt, et al., "Mathematical Considerations Pertaining to the Accuracy of Position Location and Navigation Systems - Part I", Stanford Research Institute, Menlo Park, Calif., April 1966 (NTIS #AD 629 609) "World Distribution and Characteristics of Atmospheric Radio Noise".

3. International Radio Consultative Committee (CCIR), Documents of the Xth Plenary Assembly, Geneva, 1963, Report 322.

4. Loran-C Replacement Equipment (An Offer You Can't Refuse), Coast Guard Engineer's Digest (CG133), Number 185, Oct-Nov-Dec 1975, Washington, D.C.

5. LCDR R. Hassard, "COGLAD", Coast Guard Engineer's Digest (CG133), Number 179, April-May-June 1974, Washington, D.C.

6. Radio Aids to Navigation For The U.S. Coastal Confluence Region, A Study prepared for U.S. Coast Guard (NTIS AD 75194), Polhemus Navigation Sciences, Inc., Burlington, Vermont, 1922.

7. Polhemus Navigation Sciences Inc. "Study to Evaluate Candidate Navigation Systems for Use in Canada North", Canadian Airspace Planning Team, MOT, Ottawa, January 1975.

8. A.D. Watt, E.L. Maxwell and E.H. Whelan, "Observations on some Low Frequency Propagation Paths in Arctic Areas", NBS Report 5574, May 20, 1958.

9. Edward Lacey, Arnold J. Kauper, Harold C. Millage, "Atmospheric Noise Measurements in the Arctic Regions", U.S. Naval Electronics Lab Report #25, 24 Nov. 1947.

OMEGA IN THE ARCTIC AND ITS RENDEZVOUS CAPABILITY

FOR LOGISTIC SUPPLY*

Graham Gibbs

Canadian Marconi Company

Montreal, Canada

INTRODUCTION

The Government of Canada and private industry is expanding exploration in the Arctic at an incredible rate. This activity is not confined to the land or permafrost areas. Exploration under the Arctic ice has been a serious activity for some time, so much so that we now read papers which quite seriously refer to submarine super-tankers for oil transportation. Navigation in the Arctic is of major concern, be it in the air, on land, on the water (ice) or under the water (ice). The aircraft is the primary form of long range transportation. I would not presume to outline the numerous problems facing the Arctic aviator. However, in an area where compass needles tend to point straight down, VOR/DME is sparse to say the least and where the weather can change with very little warning, our aircrews need to be commended. How then is Canadian Industry meeting the "Arctic Challenge"?

THE OMEGA SYSTEM

Omega Navigation represents a considerable step forward in making Arctic Exploration more efficient and safer. With accuracies of better than 2 NM (absolute) achievable the Omega System is not dependant upon moving parts which can be adversely affected by the earth's magnetic field in the Polar Regions. An interesting aspect of Omega is that it is usable in the air, on land, on and under the water.

---

*Paper was submitted for information to the participants.

Many papers have been written on Omega so yet another
description is not necessary at this time. It is appropriate,
however, to outline just some of the main features of the system
and to discuss its various forms of usability especially as applied
to Arctic exploration.

Omega is a hyperbolic navigation system which takes advantage
of the low attenuation rate and stable phase velocity of very low
frequency (10-14 KHz) signals. With ranges of 6,000 to 8,000
miles eight stations will provide redundant global coverage. The
U.S. Navy was given the original mandate to implement the Omega
network. Each station transmits three frequencies of 10.2, 11.33
and 13.6 KHz once every 10 seconds at 10 kw radiated power. The
stations have triple redundancy to ensure a very high degree of
operational availability.

A number of stations have been on the air since 1968 at
reduced power, typically one kilowatt, on an experimental basis.
These stations have provided us with sufficient coverage to prove
the practical advantages of Omega. The U.S. Omega Project Office
has now finalized international agreements with five of the six
participating nations. It is noteworthy that there will be no
user charges for Omega. The station implementation schedule is as
follows:

North Dakota: Commissioned October 1972
Norway   : Commissioned 15 December 1973
Trinidad   : Reduced power, to be replaced by a station in
               Liberia in December 1975
Hawaii   : Commissioned October 1974
Japan   : To be commissioned December 1974
Argentina   : To be commissioned March 1975
Reunion   : To be commissioned October 1975
Australia   : Negotiations in progress, scheduled full
               power February 1976

EQUIPMENT DESIGN

A number of airborne Omega Navigation Systems have been
developed. These range from totally manual receivers to fully
automatic systems. It is generally accepted now that manual or
even semi-manual systems are impractical for the airborne role.[2]
This is particularly true for Arctic and low level navigation
where a pilot cannot afford to be spending a significant amount of
time in a "head-down" position reading charts and tables. Various
degrees of sophistication in the automaticity of an airborne Omega
system are required for different applications. One degree of
sophistication, Differential Omega, which can be related to Arctic
Operation, will be discussed later in this paper.

Fully automatic Airborne Omega Navigation Systems are now in
production. One of these is the CMA-719 (ARN-115) developed by
the Canadian Marconi Company, which has been extensively flown in
the Northern regions and elsewhere.

The CMA-719 development started in 1968, as part of a shared development program with the Department of Industry, Trade and Commerce, Ottawa. The system was designed from the outset to be as compatible as was practical to the ARINC-561 INS specification. The CMA-719 comprises three units: an Orthogonal Loop Antenna (or alternate E, Field Antenna), a Receiver/Computer Unit (1 A.T.R.) and Control Indicator. Aircraft Autopilot and H.S.I. steering signals are provided together with visual navigation information at the Control Indicator. The latter comprises Greenwich Mean Time and Date (initially entered), Position Lat/Long (approximate position initially entered), nine Waypoints, Bearing and Distance, Groundspeed and Estimated Time to Waypoint, Desired Track and Track Angle Error, Actual Track Angle and Cross Track Distance, Wind Direction and Speed.

The system will synchronize to the transmitter signals automatically within 2 minutes from switch on. A feature of Omega equipments is that they can be initialized at any point in the flight envelope and also at high latitudes. Identification and tracking of the transmitted signals is automatic. Omega systems are not totally dependant upon accurate compass information. An Omega system can derive, automatically, True Track data from True, Magnetic or Grid Heading information, an obvious advantage in the Polar regions.

Since the Omega signals are effectively only transmitted once every ten seconds it is necessary to have some kind of "rate-aid" to allow the airborne system to continuously track and resolve ambiguities between transmissions. It was originally considered that this "rate-aid" must be obtained from an external sensor (TAS, Doppler, I.N.S.). This requirement, which could impose a restraint on those aircraft not having a suitable sensor, can be removed. It is practical, in an Omega System, to rate-aid the computations with the Omega Groundspeed, derived from a rate of change of position. The internal rate-aid (second order loop servo mode) is particularly practical for those aircraft whose flight envelope does not generally call for rapid manoeuvres and change of speed. Ordinarily Omega systems require a minimum of three stations to be able to calculate a position. Omega systems now in production will, in fact, use all available stations and derive all applicable LOPs to obtain a best estimate of position. In the event of a signal outage or poor Omega geometry from one of three minimum stations, the CMA-719 will revert to a 1 LOP/DR mode, that is, the system will calculate the single LOP and Dead Reckon from it. In the event of complete signal outage, automatic Omega systems will automatically revert to a DR mode.

DISCUSSION

It is important to note that Omega systems derive navigation data from fully legitimate signals intended and provided for navigation use. This is particularly important to airborne users

(Arctic operations notwithstanding) where a dedicated certifiable
system would seem to be essential. These signals are used in a
determinate way, which can be supported by analysis, and the
errors are bounded. The 10-14 KHz frequency band was set aside
for Omega only after some very extensive studies in the early 60s.
This early work is still valid and forms the basis of the Omega
Navigation System, although this aspect seems to be forgotten
occasionally in some circles. Phase propagation velocity between
10-14 KHz is consistent, fairly closely predictable and cyclic.
At even slightly higher frequencies it is irregular, unpredictable
and at 24 KHz, even slips a whole cycle during a day, due to modal
interference.[3] About the only advantages of the higher frequencies
is the higher power levels, 100 KW typical, presently deployed
worldwide (USN, VLF Communication Stations) and the current availability of these signals. Omega dedicated systems have shown that
the present Omega signal levels (10 KW full power) are quite
adequate and can be used with reasonably priced hardware. The
price differential between Omega and VLF COMM/NAV systems
currently being experienced, is not so much a function of the
receiver portion of the hardware but more a question of the higher
degree of automaticity and flexibility achievable with an Omega
System. The higher degree of automaticity is essential for safe
navigation. This applies equally to the airborne and ground
installations. The ground transmitters must be reliable and
redundant. They must also be operated within the regulations
applicable to navigation. The overwhelming disadvantage of
communications stations is obvious; NOTAMS are not issued.

ARCTIC - COVERAGE AND ANOMALIES

Redundant Omega coverage in the Arctic is primarily dependent
upon five stations situated in North Dakota, Norway, Japan, Hawaii,
Trinidad (Liberia). These stations will provide five station
redundant coverage throughout the Arctic. By the end of 1974 the
stations in Norway, North Dakota, Japan and Hawaii will all be
transmitting signals at 10 KW radiated power, the fifth station
Trinidad will continue at 1 KW until it is replaced by a station
in Liberia. Basic Omega accuracies are dependant upon several
factors: (i) Propagation prediction, (ii) signal-to-noise
considerations of received signals, (iii) geometry of usable
stations, (iv) anomalies. Propagation predictions used in Automatic
Omega software programmes are now well proven and can be written
to closely reflect the basic models published by the Defense
Mapping Agency in the U.S.A. Signal-to-noise ratios and geometry
are a function of the aircraft's position. In the Arctic excellent
geometry and station usability will prevail by the end of 1974.
No radio system is without its anomalies; the important consideration is how these anomalies are accounted for. The system must
also be designed in such a way that the display of erroneous or
ambiguous information is not permitted unless it is accompanied by

adequate warning indications. The Automatic Omega System is
flexible enough to combat the prevailing anomalies. Since Omega
is a dedicated navigation system, the applicable anomalies have
been thoroughly investigated and are monitored on a continuous
basis. In the Arctic anomalies fall into three basic categories:
(i) Ice mass attenuation of the transmitted signals (Ground
Conductivity), (ii) Sudden Phase Anomalies (S.P.As), also known as
Sudden Ionospheric Disturbances (S.I.Ds), (iii) Polar Cap
Anomalies (P.C.As), also known as Solar Proton Events (S.P.Es).

The most dramatic effect of ice mass attenuation occurs over
Greenland where the ice is several thousands of feet thick. This
effect does not occur over the Canadian Arctic since the ice is
relatively thin. Greenland causes the so-called "Greenland Ice
Shadow" of signals from the Norwegian station. This attenuation
is generally not noticeable above 60°N, due to the location of
Norway with respect to the thickest ice in the lower half of
Greenland.

Sudden Phase Anomalies (S.P.A.) are due to X-ray bursts from
solar flares. The effect is not sudden; it will build up to a
maximum in about twenty minutes, remain there for a few minutes
and take up three hours to fade away. In 1968, a year of maximum
sunspot activity, the phase dalays resulting at Omega frequencies
produced no more than an average maximum of 0.8 NM errors. Less
than 1% of all S.P.As would produce a maximum error of 4 NM.[3]
S.P.As are very infrequent even in a year of maximum activity and
are predictable.

Polar Cap Anomalies (P.C.As) are peculiar to the higher
latitudes and are similar to S.P.As but are caused by solar
protons entering the earth's atmosphere. While the effect may
last for a few days, it is not sudden and is not dramatic in
aviation terms. At Omega frequencies it would produce no more
than an average maximum of 0.8 NM error.[3] In the event such an
error is virtually within the noise of Omega signals and again are
infrequent.

Since an S.P.A. and P.C.A. will likely affect all transmission
paths to a certain extent, only a differential effect will be seen
by the navigation receiver.[3] The anomalies, discussed, were
carefully studies in the early days of Omega and in fact represent
another reason why the 10-14 KHz frequency band was chosen for the
dedicated system. At higher VLF frequencies the effects of these
anomalies can be much more dramatic and unpredictable. The
Canadian Armed Forces studies these anomalies during CMA-719 Omega
trials and deduced that Sudden Ionospheric Disturbances had little
or no effect on Omega system accuracy.[4]

ARCTIC FLIGHT TESTS

Flight testing of the CMA-719 in the Arctic started in June
1971. These flights were conducted in a Ministry of Transport

DC-3 on Ice Patrol duties, forerunner to the current Nordair
Electra Ice Patrol aircraft. The main purpose of these early
flights was to use Omega rather than experiment with it. The
routineness with which this was done was remarkable considering
the marginal signal conditions prevailing from the first
experimental Omega stations (Norway, New York, Hawaii).

Because of the nature of these trials only 22 data points
were recorded. It was, however, considered worthwhile analyzing
this data. Subsequent trials carried out by R.A.E. Farnborough
substantiated the MOT results. The 1971 Arctic trials showed that
accuracies of better than 1.1 NM (C.E.P.) were achievable under
less than ideal conditions. With the resumption of transmissions
from Hawaii and the commissioning of the station in Japan Omega
accuracy and confidence of fixes currently prevailing (August 1974)
will be substantially improved.

In September 1971 CMA-719 trials conducted by the Royal
Aircraft Establishment (RAE), Farnborough U.K., in a Comet aircraft
ventured North to Iceland. The Circular Error probable (C.E.P.)
for 54 sets of fixes backed up with photo fixes, was 1.42 NM.
For a three month period in 1973 the Aeroplane and Armament
Experimental Establishment UK conducted exhaustive trials of the
CMA-719 in a Comet aircraft. Flights were made to Greenland
although the majority were conducted in Europe, Africa and the
South Atlantic.

Nordair equipped their two Lockheed Electra aircraft with
dual CMA-719 installations during December 1972. These aircraft
are operated on regular Ice-Patrol Missions in the Arctic. The
aircraft are also equipped with inertial navigation systems, which
together with visual fixes, where possible, have been used to
assess the accuracy of Omega.

From December 1972 to the current time (August 1974) Omega
signal coverage in the Arctic has been far from ideal or in fact
normal. The Hawaiian station went off the air in February 1973
for updating to full power (due to be commissioned mid October
1974). This left Norway (1 KW), Trinidad (1 KW) and North Dakota
(10 KW). Norway went off the air during the Fall of 1973 for
updating to full power. It resumed transmissions early in 1974.
Despite some setbacks in the transition of these early experimental
stations to fully operational installations, and some aircraft
installation (noise) problems a considerable amount of flight data
was recorded by the Nordair crews. This data has been of considerable value in gaining Arctic navigation experience with Omega.
Detailed analysis of the flight data is being carried out by
Canadian Marconi at this time. During Omega operation system
accuracies are typically within 3 NM.

From November 1972 to March 1973 the Maritime Proving and
Evaluation Unit (MP&EU) of the Canadian Armed Forces, Summerside,
Prince Edward Island, conducted extensive flight tests with the
CMA-719. The trials were conducted in an Argus aircraft operated
by MP&EU. Trials routes included Summerside (Prince Edward Island),

Thule (Greenland), Keflavik (Iceland) Kinloss (Scotland), Lajes, Alert (Ellesmere Island), Comox (Vancouver Island), Bødo (Norway) and local eastern Arctic regions. The test report (4) reveals that despite the limitations of the Omega signal coverage prevailing at the time the crews very rapidly developed a high degree of confidence in the system. Six flights in the Maritime Provinces area using visual fixes were taken for local analysis purposes. The absolute accuracy in-flight varied from 1.6 NM to 2.8 NM at the 95% confidence level. Since these flights the CMA-719 Propagation prediction and Navigator software programmes have been refined and improved accuracies are now being experienced (Pacific Western Airlines B707, flight trials).

OMEGA FOR LOGISTIC SUPPLY

One of the major advantages of Omega in the Arctic is its flexibility. It is a dedicated navigation system suitable for airborne, surface and underwater navigation. The basic system can provide "absolute" accuracies of better than 2 NM. Simple derivations of the system can provide relative or absolute accuracies of better than 0.5 NM.

Frequently a "relative" capability is adequate for air-to-air or air-to-ground rendezvous. The use of matched Omega receivers can produce a rendezvous capability of better than 0.5 NM (5). Some communication link is required between the two; this could be in the form of a digital data link or voice communication. The receivers need not be of the same make, model or price class, provided a controlled match of propagation prediction characteristics is obtained. While present automatic receivers tend to use a variety of proprietary techniques for propagation correction and the simple receivers tend to use tables or pre-programmed values, all the prediction data basically comes from the same source: The Nav Oceano Propagation and Conductivity Map. It is therefore practical to make fully automatic Airborne Omega receivers covering a wide price and performance range, with matched propagation models and thus near-ideal rendezvous performance, with no sacrifice of the enroute performance. In the air-to-ground rendezvous case a standard airborne Omega receiver could be operated on the ground with a simple whip antenna. No modifications to the system would be necessary, therefore it could also be used as a back-up (spare) for the airborne installations.

A refinement of this relative capability is the Differential Omega System. Omega systems can be designed such that they will operate in the basic three frequency mode during the enroute portion of the flight and automatically switch to a Differential mode of operation when within range of VHF transmissions. A Differential Omega System is capable of providing absolute accuracies of between 0.25 and 0.5 NM and is therefore a valuable extension of the system for terminal area (supply posts) navigation. Differential Omega

takes advantage of the fact that the Omega signal propagation
characteristics at any given time are highly correlated over
relatively large areas. The system need not be complex, bulky or
expensive for optimum operation. This feature is potentially
advantageous for many Arctic operations. An airborne Omega System
and VHF transmitter could be used for the necessary ground based
components of the system (again providing airborne spares support
in the event of an airborne Omega or VHF failure). An airborne
Omega receiver coupled with an existing or dedicated VHF receiver
make up the airborne components.

The ground based and airborne VHF system are set to a
dedicated frequency. The ground based Omega System is initialized
in the normal way by inserting time (GMT), Date and Position.
The exact geographic coordinates of the ground based system are
pre-programmed into the Omega set. After synchronization, the
ground based Omega system will calculate the "Theoretical Lines of
Position" from the pre-programmed geographical coordinates, and
measure the "Actual Lines of Position" from Omega signals received
via a whip antenna. The difference between theoretical and actual
LOP's is the LOP error to be transmitted to the airborne system.
The LOP error is superimposed on the VHF signal for transmission.
This transmission is received by the airborne VHF receiver and is
used by the Omega set to correct the LOPs measured. Thus a more
accurate estimate of position is derived and displayed on the
Omega Control Indicator. VHF transmission is limited to line-of-
sight; however, this is usually adequate for terminal area use.
Many companies and organizations have already proven the techniques
of Differential Omega. Tests in France and the U.S.A. have
confirmed that accuracy improvements of 3 to 4 times can be
expected using this technique.

CONCLUSIONS

The intent of this paper was to provide an up-date on Omega
and outline some of its uses in the Arctic. It is hoped that
readers have found the paper informative and objective. Omega is
no longer a "science", it is a fact. Airborne and shipborne
production hardware is available today from several manufacturers.
By the end of 1974, Omega coverage in the Arctic will be complete.
The accuracy, reliability and flexibility of Omega is particularly
suitable to Arctic operations. The system will be playing a very
important and valuable role in the Arctic for many years to come.

REFERENCES

1. "Design and Performance of CMA-719 Computerized Omega Receiver". D. Mactaggart CMC, I.O.N. 1972.

2. "Manual Omega Receivers in the Airborne Role" CMC.

3. "Review of the Omega Navigation System". F/L D.W. Broughton R.A.E. (UK) CAW/131/3/3/AIR 14 October 1970.

4. (i) Final Report on Project OPVAL/AE397 "The Operational Evaluation of the CMA-719(M) in the Argus Aircraft" Maritime Command, Canadian Armed Forces, October 1973.

   (ii) "The Use of Omega Navigation in the Airborne-role today - Canadian Armed Forces ARN-115 Trials".

   G. Gibbs CMC NATO Meeting, Brussels, 1 March 1974.

5. "Use of Matched Airborne Omega Receivers for Rendezvous" D. Mactaggart I.O.N. Symposium November 1973.

COMMUNICATIONS IN THE ARCTIC

D. E. Weese

Telesat Canada

Ottawa, Ontario, Canada

INTRODUCTION

The purpose of this paper is to familiarize the reader with the various means of communications which are applicable to arctic regions, and to discuss the relative merits and disadvantages of each technique. The first portion of the paper will mention problem areas of importance for the implementation of any communications system in an arctic region. This will be followed by a general discussion of various communications techniques commonly used in arctic regions. The major portion of the paper will deal with satellites and their role in arctic communications. Typical satellite applications in the Canadian arctic will be presented and discussed.

COMMON CONSIDERATIONS IN IMPLEMENTING ARCTIC COMMUNICATION SYSTEMS

Before discussing the various types of communication systems, it is advantageous to consider the more significant problems associated with the installation and operation of any communication system in an arctic environment. These problems have a common basis in that all are related to the severe climate, the expensive and restricted transportation in arctic regions and the lack of local resources. Costs for implementing any system may thus become many fold the cost of implementing a similar system in a more temperate region. Some of the basic considerations are listed below and are briefly discussed.

(1) Transportation: Transportation in the arctic is more expensive and less reliable than in more temperate climates.

Typical methods are:
- Boat or barge using rivers or the ocean; this method is the cheapest for transporting heavy materials but is only possible during the summer or fall when the shipping season is open.
- Caterpillar train is sometimes used during the winter season for movement of heavy materials.
- Trucks, etc. are a cost effective method but limited to areas having roads.
- Aircraft are most valuable when timing is critical, and of course are used extensively for transporting personnel.
- Helicopters are useful where suitable airfields are not available.

(2) Prime Power: Prime power in arctic locations may be non-existent or inadequate in voltage regulation and reliability or in capacity. Frequently power generators must be constructed at communications sites to ensure reliable power for communications equipment. In such cases it is often desirable to design the communication system to have a minimum power consumption. Where power requirements are less than a Kilowatt more exotic power sources, such as wind generators, solar cells, thermo-electric convertors, fuel cells, etc. can be considered.

(3) Civil Works: In addition to the problems relating to the transportation of personnel, materials and machinery to sites, consideration must be given to special designs for civil works. Antenna and building foundations must be designed for the particular soil conditions of each site. Special consideration must be given to the temperature extremes in building construction and insulation. Depending on the local climate special de-icing may be required on antennas.

(4) Maintenance: It is desirable to design communications equipment to operate unattended to minimize operating expenses. Highly reliable equipment is therefore required, often with redundant or standby equipment which will automatically be switched on in the event of failure of the operating equipment. Equipment should also be designed to minimize the need for maintenance visits to once a year or less.

COMMUNICATIONS METHODS APPLICABLE TO THE ARCTIC

The traditional methods of providing communications in arctic regions have been:
(1) HF (high frequency) radio
(2) Troposcatter
(3) Point-to-point or Radio Relay
For some special applications such as spanning expanses of ocean, cable has been used, however, due to the expense of implementing such a system, and its vulnerability to breaks, it has not had wide

usage and will not be further considered in this paper.

The most recent, and most flexible means of providing communications in arctic regions, is through the use of communications satellites. This method of communications will be discussed in greater detail in the next section of this paper.

Each of the above techniques has its unique features, advantages and disadvantages with respect to arctic communications, and these will be briefly discussed below.

(1) HF Radio: HF radio is probably the most commonly used method of providing communications into isolated arctic locations, although in many cases it is being replaced by more reliable and higher quality radio relay or satellite systems. Normally HF radio communications is only used where 1 or 2 voice circuits are required. HF communications can be provided over large distances (up to several thousand miles) by the reflection of the transmitted radio wave by the ionosphere.

Unfortunately, the ionosphere is not a stable transmission medium. It is composed of four layers,each of which vary in height and density from day to night as well as seasonally and with the sun spot cycle. Thus the reflected signal is not stable with time, and deep prolonged fades can occur. In arctic regions, additional problems are caused by the low solar angle resulting in an irregular ionosphere and causing higher fading rates. Ionospheric storms can also disrupt HF communications for hours or even days. This effect is particularly pronounced in the arctic where auroral zone effects occur. As a result of the above propagation problems, HF communications in the arctic are often unreliable, and of poor quality. With increased usage of the HF band throughout the world, problems of interference and frequency congestion are becoming more severe, and are requiring more sophisticated means of frequency selection and assignment.

(2) Tropospheric Scatter: Tropospheric or troposcatter systems evolved in North America in the 1950's primarily to meet the communications needs of the military in arctic regions. Troposcatter systems provide much higher capacity, and also higher quality circuits than is feasible with HF communications. Whereas an HF system communicates by reflection of the radio waves off the ionosphere, a troposcatter system communicates by scattering of a small fraction of the transmitted radio wave in a common volume in the troposphere midway between the transmitting and receiving stations. Transmission distances of up to 600 miles are possible by selection of high gain antennas and very high transmitter powers, however, more common distances are 100 to 300 miles per hop. Larger distances are spanned by retransmitting the received signal to a second station. By this process of relaying the signal from one station to another, distances of several thousands of miles may be spanned, however, signal quality does deteriorate somewhat with each retransmission.

The major disadvantage of troposcatter systems is the cost associated with the large antennas (up to 120 feet square), and high power amplifiers (up to 50 Kilowatts). Troposcatter sites are nearly always manned and this also adds to the expense of operating the system.

For commercial applications a major disadvantage of a troposcatter communications system is that, like HF, it is not suitable for the transmission of television signals.

(3) <u>Point-To-Point or Radio Relay</u>: In the more populated regions of the world point-to-point radio relay using microwave frequencies is the most common means of providing medium and long distance communications over land areas. Very high capacities, including colour television transmission, is possible, and excellent quality voice and data circuits may be derived from such systems. For communications in arctic regions, however, radio relay systems have one significant disadvantage. Since radio relay systems operate on the principle of transmitting over line-of-sight propagation paths, repeaters in a system must be located, on the average, about 25 to 30 miles apart. Thus, many repeaters are required for communicating between two distant points and consequently construction and maintenance costs would be very high. In addition, radio relay systems are not capable of spanning large expanses of water, such as are often encountered in arctic regions. The normal growth of radio relay systems in Canada has been to extend into the arctic along with highway development.

A special consideration in designing radio relay systems for communications in arctic regions is the civil works associated with the construction of antenna towers which may be several hundred feet high. The cost associated with the transportation of the material, construction of the foundation and erection of the tower may be very significant for sites having difficult terrain. Another consideration in the application of radio relay systems to providing arctic communications is the security or reliability aspects of a large number of unattended radio relay sites, since the catastrophic failure of any one repeater causes a total system outage.

(4) <u>Satellite Communications</u>: In the last decade, satellite communications have been introduced, first, in the area of international communications by Intelsat and, more recently, in the area of regional and domestic communications. The rapid growth of satellite communication systems has resulted from the significant advantages they offer in providing voice, data and television service with a high quality which is independent of distance. Although earlier earth stations were large and expensive, the evolution of more powerful and sophisticated satellites has led to smaller, simpler and more economical earth stations. For example, a small transportable earth

station providing both television and voice communications is available which can be easily transported in a small aircraft of the size of a DC-3. Since communications can be directly between two terminals, with the satellite acting as a repeater, the reliability of such a communications link can be very high. Of major significance for arctic communications is the fact that intermediate repeaters are not required as in radio relay and troposcatter, and that difficult terrain or large bodies of water therefore pose no problems for satellite communication systems. Another major advantage of satellite systems is that once the satellite is available, new communication systems may be implemented quickly, particularly if transportable terminals are available. For all except the largest earth stations with complex and high power equipment, unattended operation is possible.

Since satellite communications covers a wide area, and is a new technology offering many advantages with respect to arctic communications, it will be discussed more fully in the following section.

OVERVIEW OF SATELLITE COMMUNICATIONS

In the past decade and a half satellite communications has taken dramatic strides in both technological improvements and in increased diversity and numbers of applications. This section of the paper will briefly discuss the various types of communication satellites in operation or planned for operation in the near future. Special attention will be paid to communication satellites capable of providing service in arctic regions.

Virtually all communication satellites now in use, or planned for service, are synchronous or geostationary satellites employing active repeaters. A geostationary satellite is one whose orbit is circular, and 35,790 Km above the equator. A satellite in this orbit will complete one revolution around the earth in 24 hours, and thus being synchronous with the earth's rotation it appears stationary in space to an observer on the earth. The advantage of a geostationary satellite is that it permits fixed antennas to be employed at earth stations, whereas more costly tracking antennas would be required to track non-stationary satellites. The active repeater on board the satellite receives signals from transmitting stations on the earth, amplifies the signals, changes their frequency to avoid interfering with the received signals, and re-transmits the signal towards the earth. The radiation pattern from the satellite may cover the entire earth (global beam), or only a portion of it (spot beam), depending on the design of the satellite antenna.

A satellite having higher transmit power has the advantage of being able to operate with earth stations employing smaller antennas, however, such a satellite is generally more costly to build and launch

than one having a lower transmit power. Due to improvements in
launch vehicles and in satellite technology the trend has been
towards higher power satellites, and smaller, more economical
earth stations.

CLASSES OF SATELLITE SERVICE

Communication satellites may be classified according to the
type of service they provide, each such service having different
frequency bands assigned for its use. The following are the
classes of satellite services, with a brief description of each
service:

(1) Fixed Service: Satellites in this service provide communica-
     tions between fixed earth stations, as opposed to mobile
     stations. All domestic communication satellites and Intelsat
     satellites operate in the Fixed Satellite Service.

(2) Mobile Service: Both the aeronautical and maritime mobile
     satellites fall in this category. Satellites in this service
     provide direct communications to mobile stations located on
     ships and aircraft. Examples of such satellites are the
     proposed Aerosat satellite which will provide communications
     to aircraft flying trans-Atlantic routes, and the Marisat
     satellites which will provide voice and teletype communications
     to ships located in the Atlantic and Pacific Oceans, and Marots
     satellite which will provide communications to ships in the
     Atlantic Ocean.

(3) Broadcasting Service: Satellites in the Broadcasting Service
     will have very high transmit powers enabling radio and
     television signals to be received by small community or home
     receivers. Experimental satellites having this capability are
     the ATS-F, CTS and Japanese Experimental Broadcasting Satellite.
     At this time no plans have been announced for any operational
     Broadcasting Satellite Systems.

(4) Navigation Service: This service is dedicated to satellites
     which provide navigational assistance to either ships or
     aircraft.

(5) Other Services: Other satellite services are defined for
     satellites providing such specialized services as meteorological
     measurements, earth resources exploration, and scientific
     research.

In the remainder of this paper we will concern ourselves with geostationary satellites in the Fixed Service. The next section will
discuss the limitations which such satellites have in providing
coverage of arctic regions.

COVERAGE OF ARCTIC REGIONS BY GEOSTATIONARY SATELLITES

Although geostationary satellites permit the use of fixed antennas and hence more economical earth stations, they do not provide communications visibility of the entire earth. Due to the curvature of the earth, a synchronous satellite would not be visible to earth stations located near either the north or south pole. This limitation has, however, proven to be a minor one since recent experimental tests in Canada have shown that reliable communications can be provided to earth stations located as far north as 80° N. latitude, provided some additional allowance is made for fading due to tropospheric and ionospheric propagation effects. Thus, all but the most extreme northern locations in the arctic can be reliably served by satellites in a geostationary orbit.

SOME APPLICATIONS OF FIXED SATELLITE SERVICE COMMUNICATIONS IN
THE CANADIAN ARCTIC

Both Canadian and USA domestic satellite systems are presently providing voice, data and television service in the arctic regions of Canada and Alaska. Some typical examples of services now being provided in Northern Canada are:
- TV and Radio Program distribution to some thirty locations;
- Voice and teletype service to some twenty locations;
- In addition, fifteen air transportable terminals have been procured to meet communication demands where mobility is a necessity, for example the needs arising from arctic oil exploration activities.

Current studies in Telesat have resulted in system designs suitable for the provision of the following services in the arctic:
- Small stations with stabilized antennas suitable for mounting on off-shore oil rigs, and providing voice, data and television service;
- A complete satellite communications system for the proposed Canadian Arctic Gas Pipeline which will extend from the arctic coast to the USA border. Voice communication as well as telemetry and control data would be provided through a system of some sixty earth stations.
- A data collection system using earth stations with very small antennas to transmit to a central location data collected from remote sensor platforms. Potential applications would be the collection of environmental information, water quality and level, etc.
- Extension of normal air/ground and ship/shore links in remote locations by interconnecting the ground/shore stations with co-located satellite terminals and extending the circuits to the south via a satellite link.

COMPARISON OF COMMUNICATIONS TECHNIQUES

Technique / Parameter	HF	Troposcatter	Radio Relay	Satellite
Capacity per Radio Carrier	Typically 1 Voice Circuit	Up to 200 Voice Circuits	Up to 1,000 Voice Circuits or 1 TV	Up to 1,000 Voice Circuits or 1 TV
Typical Voice Circuit Quality	Satisfactory	Good	Excellent	Excellent
Typical Availability	Poor due to unreliable propagation	Good to Excellent	Good to Excellent	Good to Excellent
System Installation Time	Dependant on Size of Station	Lengthy (1-2 years)	Lengthy (1-2 years)	As short as 1 day for transportable stations.

POSSIBLE FUTURE TRENDS IN THE COMMUNICATIONS FIELD

Rapid technological advances in the communications field in the last few decades have permitted the introduction of higher quality and more economical communication systems. It is worthwhile to reflect as to which areas of technology may provide additional advances resulting in further improvements to communications in arctic regions. Some areas where technological advances are readily foreseen are:
- Further refinement of solid state electronics resulting in lower power consumption, higher reliability and more economical hardware.
- Widespread use of micro-computers to improve control of hardware and processing of information.
- Greater use of digital communications thereby providing advantages which micro-computers can exploit.
- More sophisticated satellites, having higher power and longer life.

AREAS WHERE ADDITIONAL RESEARCH AND DEVELOPMENT WOULD FURTHER ARCTIC COMMUNICATIONS

The major obstacle to providing reliable high quality communications in the arctic is cost. Thus, improved communications can be provided if more economical solutions can be found to some of the problems facing designers of communication systems for arctic regions. Some areas which would benefit from research and development oriented towards cost reduction would be:
- Foundations for antennas, towers and buildings
- Prime power systems, particularly for low power installations (1 to 500 watts)
- More reliable and economic electronics equipment concepts
- Improved antenna designs for small earth stations to minimize size and cost and maximize performance.

SUMMARY

This paper has attempted to provide a brief summary of the type of communications systems commonly used in arctic areas, and has discussed some of the problems associated with implementing such systems in arctic regions. Some of the unique advantages which satellite communication systems offer in providing communications in arctic regions were outlined, and considered in more detail.

AN OPEN SYSTEMS APPROACH TO U.S. ICE FORECASTING:

PRESENT CAPABILITIES AND RECOMMENDATIONS

W. I. Wittmann

Naval Undersea Centre

Alexandria, Virginia, U.S.A.

A recent publication[1] defines organization theory as an emerging discipline, "evolving ... extremely complex and fraught with compounding variables...." Nothing could better describe the present state of sea ice observing and forecasting in the U.S. today whether considered in the operational or in its supporting research and development (R&D) roles. As we will see operationally, this organizational concept began with naval or military management, was greatly augmented by the National Aeronautics and Space Administration (NASA) earth sciences application's - remote sensor development programs - and finally by the private enterprise sector; the latter's most spectacular contribution being the redesign and full-scale model test of SS Manhattan as an icebreaking tanker in her arctic voyages of 1969 and 1970.[2] In R&D, too, initial developments were largely operationally supported, but derived also from publications in the international, especially Soviet, scientific literature.[3] Strictly arctic sponsorship within the Navy and earmarked especially for the studies of the broad-scale, both in time and space, behavior of sea ice was not forthcoming until the 1960's. Even then these Office of Naval Research (ONR) sponsored investigations were largely confined to the Central Arctic Basin, rather than to its marginal seas or marginal ice zone (MIZ) where shipping - and consequently ice forecasting - were applicable. It was nearly 1970 before ONR, followed shortly by the National Science Foundation (NSF), sponsored the initiation of an Arctic Ice Dynamics Joint Experiment (AIDJEX).[4] This was, and is, a first effort to understand the dynamics of the complicated air-ice-other interactions using modern technology. In the U.S., Washington and the Lamont-Doherty Observatory of Columbia University were exceptional in having had significant ONR sponsored

basic research, arctic sea ice programs prior to the AIDJEX era.
Within the Department of Defense the U.S. Army's Cold Regions
Research and Engineering Laboratories (CRREL) and the Navy's Naval
Electronics Laboratory - later to become Arctic Submarine
Laboratory, Naval Underseas Center, San Diego (ASL-NUC) were also
front runners in making useful contributions toward sea ice
behavioral knowledge. The U.S. Coast Guard (USCG) not only
participated operationally in these arctic efforts, but also
contributed throughout the program development with its ice
breakers, aerial observers and some R&D. More detailed, referenced
contributions will be disclosed below.

Canadian contributions were, and hopefully will continue to
be, vital to every forward step in the program. Although these
efforts are described by attendant colleagues and more superficially
by the author.[5]

EARLY DEVELOPMENT

To understand today's ice observing and forecasting program
in the U.S., from its earliest development to today's state-of-the-
art, it must be diagnosed in terms of modern systems theory. Since
both operational and research into the broad, large-scale behavior
of sea ice began in the U.S. in Navy it seems logical prior to
beginning our analysis of organization and management to paraphrase
from Department of the Navy's RDT and E Management Guide.[6] Any
operational system should be developed in a well-thought-out and
logical progression. First a long term of basic Research (6.1) is
directed toward increasing knowledge in those fields of physical,
environmental and other sciences relating to particular long-term
national need through providing fundamental knowledge for the
solution of an individual problem. This is followed by exploratory
development (6.2) or applied research usually differing from the
fundamental research in that it pertains to more specific hardware
or software programming and planning efforts. These efforts are
generally limited to a period of less than 3-5 years. Thereafter
follows an Advanced Development stage (6.3) which includes subjecting the hardware or software techniques ready to experiment or
test: not, however, for operational use or application. Engineering
Developments (6.4) follow. These comprise programs being
developed through the final stage or tested for operational use
but which have not yet been approved for procurement or
installation. Any support such as test ranges, construction,
maintenance requirements, etc. is included in the Management and
Support category (6.5). Finally the Operational System Development
(6.6) stage is reached. This includes all final research and
development efforts necessitated by an approved production or
service requirement.

The Arctic sea ice observational and forecasting program,
excluding the sub-arctic, internationally organized Ice Patrol

directed by the U.S. Coast Guard, did develop, however, in a
virtually reversed order. Urgent operational requirements
dictated its origin and development as a result of costly delays
and damages suffered by USN Military Sealift Command (formerly
MSTS) logistical sealift operations to the North American Arctic
in the late 40's and early 50's. From a military point of view
these operations were joint involving usually U.S. Naval, USCG and
sometimes Canadian forces. They included the resupply of Alaskan
Arctic sites along the western and northern shelves of Alaska
under the code names BARCHANGE and BAREX; they also responded to
similar shipping and logistic requirements of the U.S. Navy's
petroleum oil reserve No. 4 at Point Barrow. In the eastern Arctic
U.S.-Canadian project NANOOK was conducted jointly with Canadians
and Danish activities in the establishment of joint weather
stations in the Canadian Archipelago and northern Greenland areas
in this same period. Also, at Thule, Greenland, the U.S.A.F. had
enormous shipping requirements in the early '50's. Concurrent
auxiliary bases were established along the Baffin Bay and Labrador
east coasts; both Canadian and Greenlandic.[7] Such large-scale U.S.
DOD efforts came to an end with the establishment and resupply of
DEW line bases across the entire North American Arctic ranges
stretching from Northwestern Alaska to the Denmark straits in the
mid and late '50's.

Concurrent with these developments a simplified, sometimes
simplistic, operational ice observing (1949) and prediction program
(1953) was established with components consisting of various types
of data collection platforms, especially trained ice observers, an
Ice Forecasting Central in Washington, D.C. together with temporary
field forecast activities established at convenient, advanced
Arctic staging locals for Canadian and U.S. Naval patrol aircraft.
These transitory field activities could also be considered as a
sub-component platform at times basic for data collection. Within
the U.S. this was accomplished largely under the U.S. Navy
Hydrographic, now Naval Oceanographic Office (NAVOCEANO) direction
with close cooperation and assistance from Naval Weather Service;
support was largely, prior to 1961, operational under NAVOCEANO
direction and MSTS sponsorship. The basic user components were
the escorting Canadian, USCG and USN icebreakers together with
their convoyed ships which landed huge quantities of supplies on
unprepared beaches usually through the use of amphibious task
groups (or forces) furnished by USN forces frequently with joint
Canadian and Danish active participation.[7] In those early years,
culminating perhaps in 1960, the elements for such a closed system
were present, these are diagrammatically expressed in Figure 1.

In Figure 1 let us first consider Block A. Initially a few
dedicated individuals were selected to formulate a program with a
twofold goal. First, to avoid or minimize costly delays and damage
from ice sustained in reaching, offloading and loading at arctic
sites; many times unprepared beaches demanded amphibious landing
operations. To permit forecast-user dialogue it was quickly

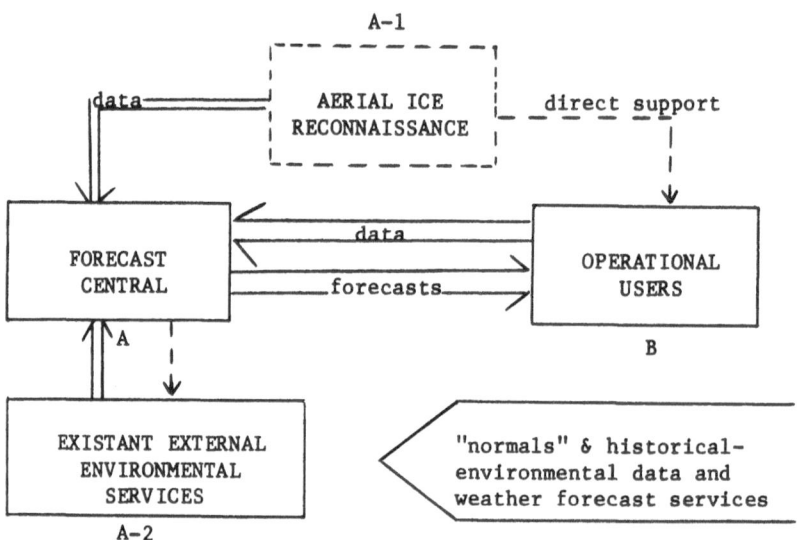

Figure 1. Description and Critique of Early Closed Systems Concept

determined to be most feasible to develop a unified ice nomenclature, standardized codes and description   for recognition of key ice features. Training of aerial ice observers with these skills, arrangement to acquire on a temporary part-time mission basis from U.S. and Canadian military patrol aircraft, and the deployment of these aircraft with the trained ice observers over the icebreakers and ships, just prior to and for the duration of fleet operations was also accomplished at the central. This sub-system or function, A-1, provided the basic data to the Forecast Central (A).

The second goal of the Forecast Central was to devise, formulate methodology for the dispatch and prediction of key ice features, both static and dynamic. Initially these "key" ice features (see Appendix I) were, edge position, stage-of-development or thickness category, floe size, degree of melt or disintegration, amount of deformation, both roughness and navigable leads of polynyas and other fractures. It also attempted to predict dynamic processes, e.g. compression or rarefaction in the ice, in terms of the icebreaker's personnel - "pressure in the ice". In other words, the forecaster would formulate advisories or forecasts. He would specify these by coordinates. In other words when articulated and transmitted by radio, radio teletype or facsimile the concept of ship routing based on optimum ice conditions, hereafter called "ice routing" was established. Two basic platforms, in addition to the aircraft, also provided basic ice information to the forecaster. First the icebreaker, together with other escorted and/or unescorted ships, was undoubtedly the earliest data platform. During our period of consideration this continued with 'breakers serving a primary ice observation and research function. Thicknesses broken enroute could readily be measured. Floe sizes, concentration of ice v. water, roughness, stage of melt or deterioration, navigable water openings - and from all this optimum routes apparent at the time of observation - were also estimated, logged and reported. Even in good weather, however, the icebreaker observation was limited to some 20 nautical miles by the range of the icebreaker-based helicopter.

But all the information discussed in the above several paragraphs yielded at best a synoptic view of the ice. Hence Block A-2 is presented to emphasize another very important function of the Forecast Central. At that time - and still probably not far from the truth - four elements were believed necessary to qualitatively project a synoptic ice situation into a forecast. These were (a) cumulative considerations of air temperature, past, present and predicted which largely controlled ice growth and/or melt; (b) surface and geostrophic wind from which ice kinematics and dynamics or ice drift could be crudely inferred (especially predicted); (c) surface ocean currents determined from published scientific studies and atlases and (d) detailed bathymetry which in combination with these earlier variables affected the offshore-onshore drift, opening or closing of leads, compaction or "pressure in the ice", influx or efflux of ice from harbors and landing sites, etc. Large-scale and predicted meteorological data was provided from NOAA (or its precedent organization) from NAVAL Weather Service, and from Canadian and sometimes Danish activities (again Block A-2). Existent bathymetric charts were improving rapidly at an accelerated rate as these early surveys were underway and at that time forecasters on rare occasion did advise ships to use overland or excessively shallow routes because the ship command and ice forecasters were working with charts of different bathymetric accuracies. The sea current data, too, suffered as most was based on mid- or end-season surveys done in past years and averaged. Most of these "historical" current determinations were used near or at the navigation season's onset. Ice atlases[8, 9, 10] and past operations reports and narratives of arctic voyages all helped, in addition, to amplify the synoptic ice picture for the forecaster; these atlases, files and data records were assembled by other activities under the direction of the ice forecaster as a further function depicted in Block A-2 of Figure 1.

Finally, in Figure 1, the function quickly grew wherein Aerial Ice Reconnaissance pilots and ice observers would provide ships in ice with "direct support". That is, they would circle over an operating ship to discuss with its command personnel the state of the ice and/or near real time ship routing problems prior to reception of the next forecast. What they learned from command personnel was also passed back to forecasters providing valuable prediction services.

In the early 1960's the significant and considerable success attained by this "closed system" of ice observation resulted in a somewhat ambivalent situation. On the one hand ice forecasters had become much more capable; on the other, this increased knowledge emphasizes the deficiencies and sometime simplistic procedures being used. Thus it also stressed the need for a structured applied research program. To add further to this need the excellent results attained by our nuclear submarines in explorator cruises beneath the ice under the direction of Dr. Waldo K. Lyon[11, 12] was on the one hand providing time space quasi-quantitative information on the one-dimensional structure of the lower 6-7 to 7/8 of the sea ice; on the other it was posing to the Ice Central questions on additional features of the sea ice whose relative importance to surface ship operational users was quite different. In shallow water the edge position and floe size distribution, even stage of development were becoming less important for such under ice operations. While transiting shallow, rough first-year or multi-year ice the depth or draft of pressure ridges or ice keels became of much greater significance. The distance between "surface openings" and likelihood of polynyas or fractures remaining stable (i.e. unlikely to close during a surfacing where men might be deployed out on the ice) were additional problems whose priority was raised.

At any rate in 1962, the Forecast Central and User Functions or System Components were augmented by a third major capability. The conceptual schematic, Figure 2, permits detailed discussion of the rather radical changes in the U.S. funding, management structure and functioning of Navy's applied or 6.2 research program directly in support of sea ice observation and forecasting from 1962 to the present.

Funding, Management and Structure of Sea Ice Forecasting In the U.S.: 1962 to 1968

Thus in 1962 two important developments occurred in the management structures of the Ice Reconnaissance and Forecasting Program. These mildly affected both the "Central", A, and the Users "B" in Figure 1, but in considering this augmented complex of functions in Figure 2, which still represents a closed system concept, it seems necessary at this time to introduce Block C, Exploratory Development. The Deputy Chief of Naval Material Development, who is also the Chief of Naval Development directly under Assistant Secretary of the Navy for R&D, ASN (R&D), controls the 6.2 monies for Oceanographer of the Navy for Ocean Science. In 1962 Navy thus budgeted funds specifically for Ice Prediction support; for the first time such applied research monies, to support Block C in Figure 2 were earmarked for improving and expanding on Navy's ice prediction capability. This modest effort was divided into two subtasks. One, and that requiring the greater

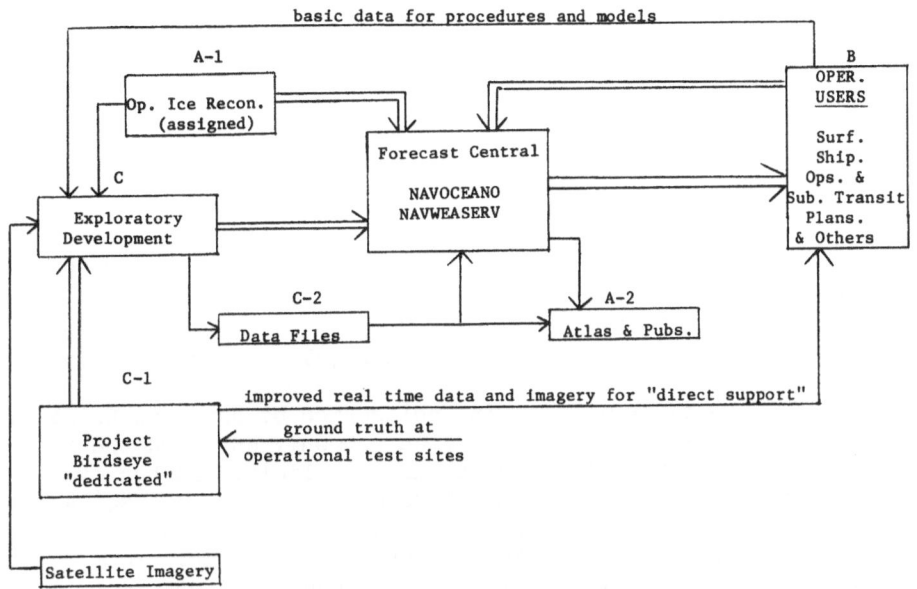

Figure 2. Growing Body of External Participants

proportionate share of the funding, was for improving ice
observational data through airborne and remote sensor - imagery
interpretation research: thus moving toward more quantitative data
and information. The second major 1962 initiative was to formulate
better procedures or models for prediction. In addition, an
aircraft was for the first time "dedicated" for full time use to
arctic ice reconnaissance and studies, PROJECT BIRDSEYE. This
meant that a platform was available to the research needs; it was
not dependent on having to deploy only during the time of resupply
of surface ship operations that could be instrumented without the
necessity of reconfigurations after every mission or series of
missions, and that it could be used in all seasons. This function
is schematically shown in Figure 2 as Block C-1. Up to 1962 it had
been most difficult to obtain military permission to go anywhere
much distant from forthcoming operations, even though each year the
only thing an ice forecaster could count on was that next year's
operations would be in a different time-space arctic location.
This arrangement through Commander, NAVOCEANO assigning technical
control of the aircraft function largely to the head of the Arctic
Exploratory effort. It also made possible the comparative testing
of various remote sensors for use in support of ice forecasting
and arctic naval operations, sometimes called "air -" versus
"ground-truth".

Some of the more important sensors and imagery initially
evaluated and interpreted included with conventional high quality

mapping cameras infrared (IR), airborne radiation thermometer
sensors, vertical mapping camera stabilized mounts and digital
controls and laser profile equipment for acquiring ice roughness
data in one dimension. The feed-back in Figure 2 from C-1 directly
to fleet users is illustrated well by the events. If an icebreaker
skipper were to ask, e.g. the height of a specific ridge, the
onboard airborne observer could much better provide this by looking
at the analogue - later digital, direct readout - even prior to
reduction - rather than guess at such a likely 10-20 foot high
feature from an altitude above 1000 ft. and at a speed of 180 knots
or greater. Thus, for the first time initial files of quantitative
data were being generated, in Block C-2, for use by modelers, Block
C and in some cases forecasters in Block A (all as indicated in
Figure 2).

In this '60-'69 period, a fine collection of data was being
generated, Block C-2 in Figure 2, which not only assisted the ice
forecaster in his art, formed the basis of a limited number of
published studies determining causal relationships, e.g. 13, 14,
15, but helped also to pin-point the areas requiring especial basic
research attention (6.1) from the scientific community. Yet,
largely because of austere funding in this modest effort, only
preparatory steps were taken to generate modern computerized data
processing and analysis facilities suitable to an adequate system
of data banks required by the complex modeling that, it was becoming
more and more obvious, would be required in making exponential
improvements in our understanding of changes induced by the complex
interaction of air-ice-ocean processes and phenomena. Defense
Advanced Research Agency through U.S. Army CRREL was later to
correct this deficiency (see below).

Before proceeding to Block D, Figure 2, let us now reexamine
the changes that were slowly occurring between the Forecast Central
Block A and the User Community Block B. Naval-DOD directions
indicated that military sea ice observing and forecasting belonged
within the "uniformed Navy". Therefore a division-of-labour was
introduced into subsystems A, B and C intended to bring this about.
An experimental forecast central was established in A with Naval
Oceanographic Office to experimentally apply ice forecasting to
surface ship operations in both the Arctic and Antarctic. "How-to-
do-it" manuals and papers would be produced primarily for complete
functional transfer of the Block A responsibility "when the
function or art was adequately developed" to Naval Weather Service
for Fleet use and application. The fleet users at this time,
Block B, remained primarily Military Sealift Command, MSC, Task
Force 43 in the Antarctic (the latter becoming more and more NSF
rather than Navy supported) and those activities concerned with
producing annual reports of ice conditions, ice atlases and other
special studies and projects. Task A under Dr. R. James at
NAVOCEANO is still functioning but with decreasing commitments.

SUGGESTED OPEN SYSTEM CONCEPT

Developments at the present are believed by the author to have set the stage for the desirability of, if not absolute requirement for, an open rather than the older closed, self-stifling concept.[16] Figure 3 is schematically intended to represent the open system that would correct deficiencies in the above suggestions. The extra-DOD and DOD management functions, not shown, would of course appear immediately below and above the D-1 and E-1 blocks, respectively; they are designated as MF.

The limited resources, including funding and manpower required in Block C, were insufficient to adequately solve the more and more detailed forecasting requirements. Especially those provided by CNM and OCEANAV to Block C (exploratory development for ice forecasting), are demanding more intense and closer associations with Block E-2 and E-3 and to a lesser extent also with extra-DOD activities depicted in Blocks D-2 and D-3. Significant help and assistance could thus be attained through separate, cooperating functions in DOD and non-DOD research activities sharing the heavy work-load burden. Foremost among the interacting groups involved in ice forecasting are the following:

ASL-NUC, San Diego: Consultative planning and participatory support in Fleet Exercises as directed by the Chief Scientist and Coordinator for Underice Cruises (Dr. Waldo K. Lyon).

U.S. Army CRREL: Bearing and other strengths of ice and other physical properties and processes more and more frequently require the scientific results produced by this Hanover, N.H. Laboratory (under Chief Scientist Dr. A. Assur and with Dr. W. Weeks' frequent direct responses and contributions).

DREO, Canada: This Defense Research Establishment Laboratory (not shown because Figure 3 is restricted to U.S. organizations) long worked independently, sometimes with U.S. activities jointly on parallel sea ice research problems. Moira Dunbar made basic contributions to all sea ice research and behavorial understanding throughout the entire program. DREO's assistance and approaches were closely watched and, in some cases, joint experiments were conducted.

Dept. of Environment, Ice Forecast Central, Halifax (later Ottawa): This parallel project to our Block A, in Figures 1, 2 and 3, was also closely monitored; data and methodology were frequently discussed jointly and, in addition, at least annual conferences with representatives of the U.S. ice forecast activity were held. Both had representatives on the WMO, WGSI (for World Meteorological Organization-Working Group on Sea Ice) in its 2-4 year periodic meetings on standardization of nomenclature, coding and data interchange and state-of-the-art.

NASA-Spacecraft Oceanography Project (SPOC): Directed first by A. Alexio, later (and now) by J. Sherman, this group interacted quickly to bring additional hard-to-get human resources, e.g. the numerical ice modeler and dynamicist, Dr. Campbell of USGS and

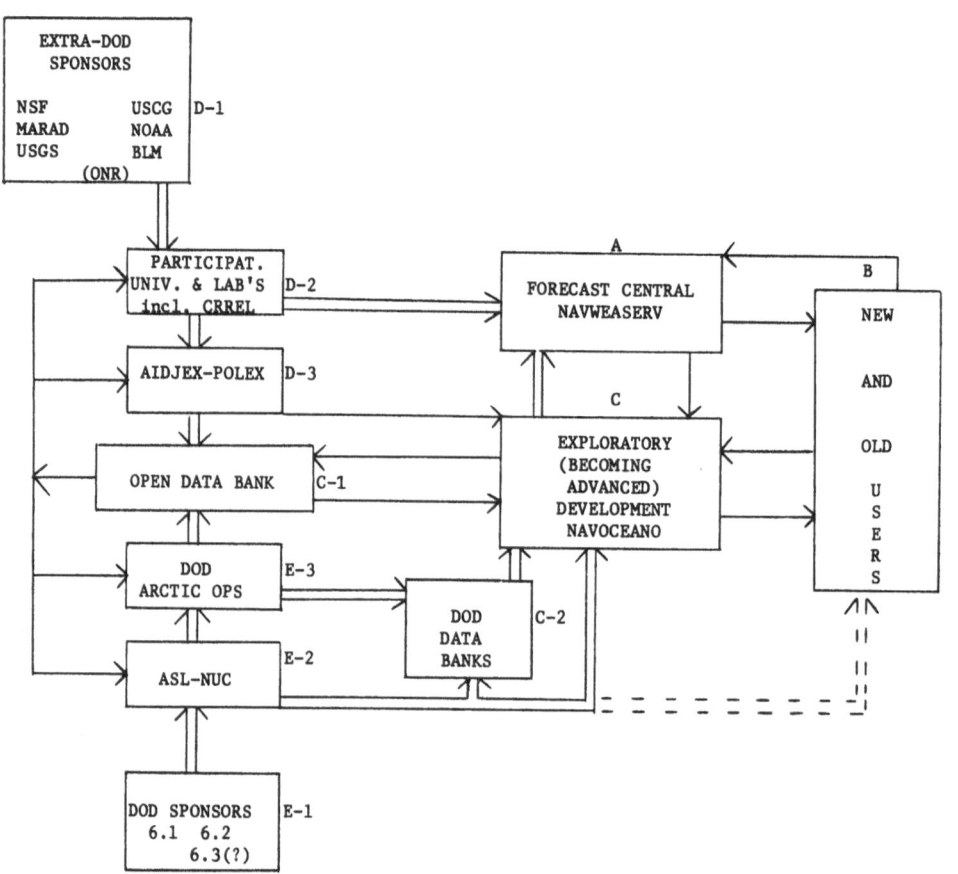

Figure 3.   Suggested Open Data System Concept

costly NASA aircraft test flights of airborne remote sensors toward
solution of the problems of acquiring, analyzing and interpreting
newer quasi-quantitative data on ice features.

NOAA-National Environmental Satellite Services (NESS): From
the launching of TIROS, the value of vidicon, later IR and scanning
radiometry had been recognized - and on an increasing basis
actually used operationally - as an aid to sea ice forecasting.
Dr. E. Paul McLain and his colleagues provided considerable
assistance in learning about sea ice forecasting needs and in
attempts to develop useful applications.[17, 18]

ONR-Sponsored Universities: University of Washington, Dept.
of Atmospheric Science and its Division of Oceanography were more
and more providing data helpful to understanding the heat and
momentum flux processes and kinetics, that were a great boon to the
forecaster. This was especially true because of the new user
requirements that were becoming expected outputs of the Navy's Ice
Forecasting System, e.g. drift of Navy Arctic Research Laboratory
Ice stations (NARLIS) and submarine support forecasts. Since both
such forecast types vary so significantly from surface ship support
type they have been - and still are - strictly coordinated in the
Exploratory Development or C Block in Figure 3.

Canadian Universities & Polar Continental Shelf Project:
Dr. E. Pounder and Dr. Max Dunbar have been both as individuals
and as producing scientists, outstanding and representative of
Canadians in providing help and assistance to the U.S. capability
attained. The former, of course, in supplementing the sea ice
physics required, the latter primarily in contributing to our
understanding of water-circulation in marginal arctic seas. Dr.
Fred Roots, in addition to scientific, provided logistical
assistance to many projects fundamental to ice prediction -
especially to those prevalent in the quasi-fast ice of sections of
the Canadian Archipelago.

AIDJEX - POLEX: Under the present direction of Dr. Untersteiner,
and with the airborne remote sensing efforts directed by Dr.
Campbell, cited above, and Dr. Wilford Weeks of CRREL, this project
was providing better data for overall understanding of the sea ice
behavior. Attention, however, was focused on the Central Arctic
viz. MIZ.

Dr. A. Collin, until recently the Dominion Hydrographer should
also not be overlooked - as is the case of many other contributants
I have surely neglected.

Other Activities: The U.S. Coast Guard icebreakers, their
captains, with daily reports on ice and post operations reports
were invaluable - all have played a key role as has the R&D
Department within that organization. Some of our best early side-
looking radar imagery was acquired through this activity.
International activities including Finland, W. Germany, Sweden,
and to a considerable extent the Danish Institute Meteorologique
who contributed their data and skills to Greenlandic Ice study and
forecasts have also provided valuable assistance to our capability.

With recent Icelandic assistance this help is especially needed
since the problems of predicting East Greenland Ice drift are far
more complex, in the author's opinion, than anywhere else in the
Arctic.

In concluding this section, the 1960-67 period then produced
greater complexity into ice forecasting, at the same time it began
to automatically degenerate as a "closed system". Somewhat ambiva-
lently in this period, when ice forecasting knowledge was
accelerating and the diversity of needs expanding, the amount of
surface shipping into the arctic and MSC requirements shrank
rather sharply and suddenly.

RECENT DEVELOPMENTS (after 1968) AFFECTING ICE FORECASTING AS A SYSTEM

In the above, from the point of view of organization and
management the implications of profound complexity are evident.
In the U.S., both DOD and a multitude of non-military governmental
activity components have been involved. Considering the looseness,
in many cases non-existence, of a management structure to integrate
and direct these widely diverse activities, it is amazing that
coordination and interaction resulting almost solely from
individual initiative and scientific acumen attained the high
degree that it did.

In the late '60's and '70's a series of new developments took
place which added even further to the complexities of the concept
schematicized in Figure 3. Basic understanding of ice behavior
increased. Perhaps, more important, the user requirements expanded
into areas wide afield from the surface ship support on which the
program had been based.

Since these developments involve highly differing emphasis on
the elements of an ice forecast, these are enumerated below and
definitions may be found in Appendix 1.

TABLE I

"KEY" ICE FEATURES

X-1   THICKNESS AND/OR STAGE OF DEVELOPMENT

      a.   Level V. Deformed

X-2   ROUGHNESS

      a.   Topside height (h) - frequency or spectra
      b.   Bottomside draft (z) - frequency or spectra
      c.   "h to z" ratio

X-3   CONCENTRATION

X-4   WATER OPENINGS

      a.   Recurrent Polynyas Fractures
         (Size-Clutter-Frequency)
      b.   Fracture Patterns (Orientation)

X-5   EDGE AND BOUNDARIES

X-6   FAST ICE

      a.   Extent
      b.   Character

X-7   FLOE SIZE DISTRIBUTION

X-8   STAGE OF MELT

X-9   SNO (DEPTH AND CHARACTER)

X-10   GLACIAL ICE

      a.   Types
      b.   Numbers
      c.   Height
      d.   Draft
      e.   Volume
      f.   Above to below water (h' to z') ratios

X-11   RECORDS OF MOTION AND DYNAMICS

      a.   Manned Drift Station Positions
      b.   Records from sequential imagery

X-12   SPECIAL DECK(S) ON PHYSICAL PROPS
        (e.g. salinity, hardness)

TABLE II

NEW USER ACTIVITIES

Y-I   AIRBORNE AND SATELLITE REMOTE SENSING INNOVATIONS MADE
         POSSIBLE THROUGH EARTH APPLICATIONS PROGRAMS E.G.: NASA
         AND NOAA'S NESS (SEE BELOW)

Y-II   MARINE RESOURCES EXPLOITATION ACTIVITIES ALONG NORTH
         ALASKAN SLOPE AND OTHER ARCTIC LOCALES
         a. OFFSHORE DRILLING
         b. PREDICTION AND ROUTING REQUIREMENTS
         c. DIVERSE PLATFORM DESIGNS
         d. TERMINAL DESIGN (FOR WITHSTANDING ICE STRESSES, ETC.)

Y-III   NEW SEA ICE-SHIP DATA FOR DESIGN OF IMPROVED ICEBREAKERS
         AND OTHER PLATFORMS
         a. USCG POLAR STAR
         b. MANY PRIVATELY PROPOSED PLATFORMS

Y-IV   CHANGING SUBMARINE UNDERICE VEHICULAR REQUIREMENTS
         (DEFENSE, EXPERIMENTAL AND COMMERCIAL)
         a. UARS

Y-V   REQUIREMENTS FOR UNDERSTANDING AND PREDICTING ACOUSTICS
         a. AMBIENT NOISE
         b. TRANSMISSION LOSS
         c. REVERBERATION

Y-VI   ARCTIC SURFACE EFFECTS VEHICLE PROGRAM OF DEFENSE
         ADVANCED RESEARCH PROJECTS AGENCY
         a. ESTABLISHMENT OF DATA BANK ON ICE TOPOGRAPHY
         b. ESTABLISHMENT OF COMPUTERIZED METHODOLOGY AND
             CAPABILITIES

Y-VII   CONSIDERATIONS OF EFFECTS OF ICE ON
         a. GLOBAL HEAT BUDGET
         b. CLIMATIC MODIFICATION
         c. AIR-ICE-OCEAN POLLUTION
         d. ECOLOGY

Y-VIII   PROJECT AIDJEX AND IMPROVED NUMERICAL PREDICTION MODELS:
         a. PRESENT EMPIRICAL MODELS OF FORMATION GROWTH AND DECAY
         b. ARCTIC BASIN MODEL (AIDJEX EXPERIMENT)
         c. ICE DYNAMICS MODELS FOR SHALLOW SEAS
         d. EFFECTS OF WAVES ON ICE FRACTURING AND EROSION (AND,
             INDIRECTLY, PRODUCTION).

TABLE III

AIRBORNE SATELLITE AND ACOUSTIC SYSTEMS AND
THEIR POTENTIAL TO MEASURE ICE FEATURES (X-i)

A. AIRBORNE

1. Multi-spectral mapping cameras
    X-1-X-11, incl.
2. Scanning cameras
    X-1-X-11, incl.
3. Laser profiler (one dimensional sampling at present)
    X-2
4. Scanning radiometers (one dimensional sampling at present)
    X-2, X-4, X-5, X-8, X-11
5. Spectrometers
    X-7 X-8
6. Radar Scatterometer
    X-2
7. Side Looking Radar
    X-4b, X-6, others in development
8. Passive microwave Imaging System
    X-1 to X-6, and X-7 to X-11 - all in development
9. Dual Frequency UHF Radiometer
    X-1

B. SATELLITE (all ice features in development except X-4 and X-5)

1. Multi Spectral and Earth Terrain Cameras (SKYLAB)
2. Multi Spectral Scanning Cameras (SKYLAB) (ERTS)
3. Television Return Beam (ERTS)
4. IR Scanning Thermal Radiometer (NOAA 2, 3 and DMS)
5. Spectrometer, Radar (SKYLAB)
6. Microwave Radiometer (SKYLAB)
        Five Freq.   (SEASAT)
7. Synthetic Aperture Coherent Radar (SEASAT)
8. Visible-Infrared Scanning Radiometer (SEASAT)
9. Compressed Pulse Precision Radar Altimeter   (SEASAT)
10. Microwave wind Scatterometer (SEASAT)

C. Topside Fathometer Component of Underice Submarine Acoustical
    Ice Suit and UARS X-1, X-2, X-4, X-10 topside profiler.

Before attempting to relate Tables I and II, it must be
emphasized that in the author's opinion, too often, the
utilization and requirement of "sea ice" data fails to take into
account its many important aspects and component features as
listed in Table I, and their time space variability. A potential
user, as a result, frequently is misled when considering that a
given satellite or aircraft, whose sensors with designated
resolutions can measure "sea ice", by not being cognizant of those
various highly different sub-categories as listed. Detail sensor
resolutions of the various airborne, satellite and acoustical
sensors required will to some degree be provided if required by
the workshop; even more effective this information - much of which
is somewhat conjectural prior to actual launch, test and imagery
interpretative work - it perhaps can be better supplied by
representatives of the Arctic Traffic and Management, Section 7.1
(surveillance and remote sensing activities) and the Arctic
Underwater Operational Systems, Section 7.5 (underwater navigation
and sensing systems). This would be in accordance with planned
procedures of the Workshop Format as stated in Section 8.2. These
section designators in the above paragraph refer to the Memorial
University "Arctic Systems Workshop" announcement of July 1974.

Table III lists the most important and/or promising Airborne
Sensors presently in use or scheduled.

Referring back to Tables I and II, a certain degree of
inexactness will always remain even after in depth cognizance is
taken of the various features (Table I) of sea ice and the diversity
of new requirements of users (Table II) and their interrelationship.
Even in support of the present surface ship operations, there is a
divergence of opinion as to what the relative import is of ice
features by various captains. Insofar as data needs, in terms of
requirement and expected resolution, Weeks and others have recently
indicated their impressions.[19] In a very timely attempt to
determine future social cost benefits, Canadian governmental
experts have, in addition, summarized their own ideas concerning
the most promising airborne remote sensors.[20] Table III, however,
represents the author's own judgment of the most valuable and
promising airborne, satellite and acoustic sensors. Table IV
represents the particular ice features and user requirements to
which these improved data would be applied. It should also be
pointed out that in the past 14 months, NSF has developed an Arctic
Offshore Program (AOP) to respond to many of those especially to
items Y-2 and Y-4 in Table II.

The Bureau of Land Reclamation recently made funds available
in addition for a similar goal and for determining environmental
impact, thus responding in Table II to item Y-7, as well; this
program is being administered by NOAA with assistance and
cooperation from the University of Alaska. Hopefully this workshop will improve these governmental agencies in effectively
responding to the difficult problem of quick widening of our

knowledge of shallow seas and coastal areas such as those bordering Alaska.

CONCLUSIONS AND IDENTIFICATION OF KNOWLEDGE GAPS

It seems obvious that the need for knowledge of ice behavior, especially in Alaskan coastal waters, will continue to expand widely and at an accelerated rate in the next decades. Although some of the progress has been accomplished within non-military agencies, notably NASA, NOAA, USGS and USCG it is suggested that:
  • A formal ice observing and forecasting system should be initiated in the U.S. within the non-DOD agencies. A lead agency should be designated perhaps initiated by the Interagency Arctic Research Coordinating Committee. The suggested, preferably, management or coordinating activity should be small but its head or staff must be entirely up-to-date in the latest engineering and scientific advances of many diverse disciplines: viz. fluid dynamics, electromagnetic physics, sea ice physics, meteorology, oceanography, glaciology, acoustics, organization theory and systems research. He must also coordinate with a parallel command activity (or, less desirable, coordinate with similar qualifications within the DOD).
  • A DOD arctic ice observing and forecasting function should be appointed by Chief of Navy Development and report directly to some scientific-operationally responsive activity such as ASL-NUC. One of the specific near-immediate requirements would be to consider initiation of an "Advanced Development Program" in Sea Ice Observing and Forecasting specifically tailored to DOD needs.
The two management or coordination activities would thus be in a much better position to prevent duplication and overlap and to assure an improved flow of information within the DOD and extra DOD activity and to prioritize and fill the knowledge gaps listed below. The severely limited available expertise thus could also be more efficiently utilized.

The urgency and timeliness of responding to these three suggested conclusions is appropriately CY 1976-7. AIDJEX will be coming to an end and the national program POLEX is believed to be too broad to maintain the direction and inertia built up in AIDJEX. Furthermore, NOAA is believed to have all the necessary components within NESS, to proceed gracefully into an expanded era of under- standing ice behavior, especially its global effects,[21] effects on pollution[22] and the environment.

Some of the knowledge gaps believed hindering advances in our knowledge of ice and its interaction with other disciplines and operational activities are as follows:

TABLE IV

MATRIX RELATING IMPORTANT ICE FEATURES, RELATIVE CAPABILITY OF PLATFORMS AND USER REQUIREMENTS

PRINCIPAL USERS USER IMPORT.	X-1 Thickness UAS	X-2 Roughness UAS	X-3 Concentration A	X-4 Open's Fracture Pattern SAU	X-5 Ice Edge ASU	X-6 Fast Ice SA	X-7 Floe Size ASU	X-8 Stage of Melt AS	X-10 Glacial Ice ASU	X-11 Ice Motion ASU
Y-2 Marine Resources Exploit	M	H	M	H	H	H	M	H	H	H
Y-3 Marine Design	H	H	H	H	L	L	M	H	H	H
Y-4 Submarine Vehicle Need	L	H	L	H	L	M	L	H	H	H
Y-5 Requirement for Acoustic	M	H	H	H	H	M	L	H	M	H
Y-6 SEV	H	H	H	H	M	H	H	H	H	H
Y-7 Broad-Scale Science Needs	H	H	H	H	H	H	M	H	H	H
Y-8b Use of Data in AIDJEX	H	H	L	H	L	L	L	M	L	H
Y-8a,c,d Use in Other Num. Models	H	H	H	H	H	H	H	H	H	H

LEGEND:
A - Airborne
S - Satellite
U - Acoustical Sensors

DEGREE OF NEED:
H - High
M - Moderate
L - Low or None

1.   In spite of the great stimulus provided by the Defense Advanced Research Project Agency (ARPA) in the 1969-74 period in their Environmental Program associated with the feasibility of an Arctic SEV,[23] and the data bank function created by AIDJEX, too little has been accomplished in the establishment of overall sea ice data banks. These are specially needed for: ice roughness, categorical ice thickness (i.e. new types, medium and thick first year, multi-year and open water), and ice motion daily from manned U.S. Soviet stations and shorter term from NAVSAT and sequential remote imagery.

2.   Numerical modelers should accelerate efforts to formulate an expression which would yield dM/dT or dV/dT over time where the time scales are weeks or months (rather than hours or minutes) and where M represents sea ice mass and V, ice volume. Crude empirical-statistical models might well suffice (see No. 5 below) rather than the shorter term model goals of present numerical endeavors. Furthermore, that widening of the forecast period would provide more realistic verification means.[24]

As a sub-component of the forecast central in Figure 3, a continual method of verification and/or quantitative evaluation must be established for most of the ice features listed in Table I.

4.   In the U.S. among all the interrelated disciplines, acousticians have seemed to work least effectively with ice problems in determining e.g. what sea ice processes contribute to ambient noise in the M12 and to what extent is reverberation affected by the ocean and changes in the physical character of the ice. The recent closer relationship of Polar Research Corp. (PRC) with ASL-NUC and the recently initiated program within Naval research (Acoustics) Laboratory should assist in correcting this deficiency.

5.   The telemetering buoy positioned and monitored by satellite must be further developed and deployed and quickly! Hopefully those buoys deployed by AIDJEX and developed by PRC should be of great assistance. Of what applicable use will a numerical model be if it requires as input, ocean current and other data, not available on a synoptic basis. Present ocean data is gathered over widely divergent areas in time and space, sometimes 20-year averages are needed, and the data are heavily biased seasonally. A quasi-synoptic automatic buoy cheap and expendable is a sine qua non for practical application if any sophisticated model of sea ice drift and deformation is to be formulated. This is especially true if realistic models are ever developed for application in the MIZ.

6.   The problems of the effects of ice by wave action (especially swells) and vertical heat transport induced by vertical heat transfer near the ice edge must be more vigorously attacked in the MIZ. This would include effects on acoustical variables induced by such ocean-ice interaction.

7.   Automated pattern recognition and imagery enhancement, methodology and techniques should be introduced as an automated ice data reduction analysis and storage system or subsystem component.

REFERENCES

1. Kast, F.E. and J.E. Rosenzweig, <u>Organization and Management: a systems approach</u>, McGraw-Hill, Inc., New York 1974.

2. Mookhoek, A.D. and W.J. Bielstein, <u>Problems associated with the design of an arctic marine transportation system</u>, Paper No. OTC 1426, Offshore Technology Conference, Houston 1971.

3. Zubov, N.N., <u>L'dy Arktiki</u> (Arctic Sea Ice). Northern Sea Route Directorate Press 1945, Translation by U.S. Naval Electronics Laboratory San Diego 1965.

4. Untersteiner, N., <u>Operations Manual for the AIDJEX main experiment</u>, Arctic Ice Dynamics Joint Experiment, Seattle 1975.

5. Wittmann, W.I., <u>The sea ice forecasting program</u>, Navigation Vol. 8, No. 4, Washington 1962.

6. U.S. Dept. of the Navy, <u>RDT&E Management Guide</u>, Publication NAVSO P-2457, Assistant Secretary of the Navy, Research and Development, Washington 1975.

7. Tait, A.J., <u>The operational concept for sea ice reconnaissance and forecasting program conducted during arctic operations</u>, in Arctic Sea Ice, NAS-NRC Pub. 598, National Academy of Science, Washington 1958.

8. U.S. Naval Oceanographic Office, <u>Oceanographic Atlas of the Polar Seas</u>, Part II Arctic, Washington, D.C. Rev. 1962.

9. --------. <u>Ice Atlas of the Northern Hemisphere</u>, Corrected Reprint, Washington, D.C. 1952.

10. German Hydrographic Institute, <u>Atlas der Eisverhaltnisse des Nordatlanticischen Oceans</u>, etc. (Ice Atlas of North Atlantic Ocean) copy held by U.S. Naval Oceanographic Office, Pub. No. 2335, Washington, D.C. 1950.

11. Lyon, W.K., <u>The submarine and the arctic ocean</u>, Polar Record 11-75, Cambridge 1963.

12. U.S. Navy Electronics Laboratory, <u>Polar Submarine and navigation of the Arctic Ocean</u>, Report No. 88, San Diego 1948.

13. Wittmann, W.I, and J.J. Schule, Jr., <u>Comments on the mass budget of Arctic pack ice</u>, UCLA Symposium, Proceedings on Arctic Heat Budget on Atmospheric Circulation, pub. by Rand Corp., Santa Monica Dec. 1966.

14. --------. Forecasting the drift path of Arlis II in 1965, Published in 2nd International Oceanographic Congress Abstracts, 29 May - 9 June, Moscow, USSR 1966.

15. Ketchum, R.D., Jr. and W.I. Wittmann, The role of drifting stations in sea ice prediction studies in Arctic drifting stations: Sponsored by ONR, AINA, Washington, D.C. 1966.

16. Ibid., Ref. 1, pp. 109-110.

17. McClain, E.P., Some new satellite measurements and their application to sea ice analysis in the Arctic and Antarctic, National Academy of Science Pub. Advanced Concepts and Techniques in the Study of Snow and Ice Resources, Washington 1974.

18. --------. Quantitative use of satellite vidicon data for delimiting sea ice conditions, Arctic 26:44-57, Montreal 1973.

19. Weeks, W. et al., The remote sensing program required for the AIDJEX model, AIDJEX Bull. No. 27, 22-44, Seattle 1974.

20. McQuillan, A.K. and D.J. Clough, Benefits of remote sensing of sea ice. Research Report 73-3, Dept. of Energy, Mines and Resources, Ottawa Dec. 1973.

21. Committee on Polar Research, National Research Council, Polar Research, a survey, National Academy of Science, Washington, D.C. 1970.

22. Campbell, W. et al. in press, Effects of an oil spill in a sea ice environment. Symposium on Remote Sensing in Glaciology, Cambridge, England Sept. 15-21, 1974.

23. Hibler, W.D. et al., Statistical aspects of sea ice ridge distributions. JGR 77-30, pp. 5954-70, 1972.

24. Sater, J.E. et al., Impingement of ice on the North coast of Alaska, in The Coast and Shelf of the Beaufort Sea, AINA, Dec. 1974.

APPENDIX I

ICE GLOSSARY

Terms arranged in Alphabetical Order

AGED RIDGE: Ridge which has undergone considerable weathering.
These ridges are best described as undulations (8.2.2.4).

ANCHOR ICE: Submerged ice attached or anchored to the bottom,
irrespective of the nature of its formation (3.3).

BARE ICE: Ice without snow cover (8.5).

BELT: A large feature of pack ice arrangement; longer than it is
wide; from 1 km to more than 100 km in width (4.4.3).

BERGY BIT: A large piece of floating glacier ice, generally
showing less than 5 m above sea level but more than 1 m and
normally about 100-300 square m in area (10.4.4).

BESET: Situation of a vessel surrounded by ice and unable to
move (12.1).

BIG FLOE: (4.3.2.3) (see floe).

BIGHT: An extensive crescent-shaped indentation in the ice edge,
formed either by wind or current (4.4.6).

BRASH ICE: Accumulations of floating ice made up of fragments not
more than 2 m across, the wreckage of other forms of ice (4.3.6).

BUMMOCK: From the point of view of the submariner, a downward
projection from the underside of the ice canopy; the counterpart
of a hummock (13.4).

CALVING: The breaking away of a mass of ice from an ice wall,
ice front, or iceberg (10.4.1).

CLOSE PACK ICE: Pack ice in which the concentration is 6/8 to
less than 7/8 (7/10 thru 8/10), composed of floes mostly in
contact (4.2.3).

COMPACTED ICE EDGE: Close, clear-cut ice edge compacted by wind
or current, usually on the windward side of an area of pack ice
(4.4.8.1).

COMPACTING: Pieces of floating ice are said to be compacting when they are subjected to a converging motion, which increases ice concentration and/or produces stresses which may result in ice deformation (5.2).

COMPACT PACK ICE: Pack ice in which the concentration is 8/8 (10/10) and no water is visible (4.2.1).

CONCENTRATION: The ratio in eighths or tenths of the sea surface actually covered by ice to the total area of sea surface, both ice covered and ice free, at a specific location or over a defined area (4.2).

CONCENTRATION BOUNDARY: A line approximating the transition between two areas of pack ice with distinctly different concentrations (4.4.9.2).

CONSOLIDATED PACK ICE: Pack ice in which the concentration is 8/8 (10/10) and the floes are frozen together (4.2.1.1).

CONSOLIDATED RIDGE: A ridge in which the base has frozen together (8.2.2.5).

CRACK: Any fracture which has not parted (7.1.1).

DARK NILAS: Nilas which is under 5 cm in thickness and is very dark in color (2.2.1).

DEFORMED ICE: A general term for ice which has been squeezed together and forced upwards in places (and downwards). Subdivisions are rafted ice, ridged ice, and hummocked ice (8.2).

DIFFICULT AREA: A general qualitative expression to indicate, in a relative manner, that the severity of ice conditions prevailing in an area are such that navigation in it is difficult (12.5).

DIFFUSE ICE EDGE: Poorly defined ice edge limiting an area of dispersed ice; usually on the leeward side of an area of pack ice (4.4.8.2).

DIVERGING: Ice fields or floes in an area are subject to diverging or dispersive motion, thus reducing ice concentration and/or relieving stresses in the ice (5.1).

DRIED ICE: Sea ice from the surface of which melt water has disappeared after the formation of cracks and thaw holes. During the period of drying, the surface whitens (9.3).

EASY AREA: A general qualitative expression to indicate, in a relative manner, that ice conditions prevailing in an area are such that navigation in it is not difficult (12.6).

FAST ICE: Sea ice which forms and remains fast along the coast, where it is attached to the shore, to an ice wall, to an ice front, between shoals or grounded icebergs. Vertical fluctuations may be observed during changes of sea level. Fast ice may be formed in situ from sea water or by freezing of pack ice of any age to the shore, and it may extend a few meters or several hundred kilometers from the coast. Fast ice more than one year old may be prefixed with the appropriate age category, old, second-year, or multi-year. If it is thicker than about 2 m above sea level it is called an ice shelf (3.1).

FAST ICE BOUNDARY: The demarcation at any given time between fast ice and pack ice or between areas of pack ice of different concentrations (4.4.9) (cf. ice edge).

FAST ICE EDGE: The demarcation at any given time between fast ice and open water (4.4.8.5).

FINGER RAFTED ICE: Type of rafted ice in which floes thrust "fingers" alternately over and under the other (8.2.1.1).

FINGER RAFTING: Type of rafting whereby interlocking thrusts are formed, each floe thrusting "fingers" alternatively over and under the other. Common in nilas and gray ice (6.4.1).

FIRN: Old snow which has recrystallized into a dense material. Unlike snow the particles are to some extent joined together; but, unlike ice, the air spaces in it still connect with each other (10.1).

FIRST-YEAR ICE: Sea ice of not more than one winter's growth, developing from young ice; thickness from 30 cm-2 m. May be sub-divided into thin first-year ice/white ice, medium first-year ice, and thick first-year ice (2.5).

FLAW: A narrow separation zone between pack ice and fast ice where the pieces of ice are in chaotic state, that forms when pack ice shears under the effect of a strong wind or current along the fast ice boundary (7.1.1.2) (cf. shearing).

FLAW LEAD: A passageway between pack ice and fast ice which is navigable by surface vessels (7.3.2).

FLAW POLYNYA: A polynya between pack ice and fast ice (7.4.2).

FLOATING ICE: Any form of ice floating in water. The principal kinds of floating ice are lake ice, river ice, sea ice, which form by the freezing of water at the surface, and glacier ice (ice of land origin) formed on land or in an ice shelf. The concept includes ice that is stranded or grounded (1).

FLOE: Any relatively flat piece of sea ice 20 m or more across. Floes are subdivided according to horizontal extent as follows (4.3.2):
     GIANT: Over 10 km across (4.3.2.1).
     VAST: 2-10 km across (4.3.2.2).
     BIG: 500-2,000 m across (4.3.2.3).
     MEDIUM: 100-500 m across (4.3.2.4).
     SMALL: 20-100 m across (4.3.2.5).

FLOEBERG: A massive piece of sea ice composed of a hummock, or a group of hummocks, frozen together and separated from any ice surroundings. It may float up to 5 m above sea level (4.3.4).

FLOODED ICE: Sea ice which has been flooded by melt water or river water and is heavily loaded by water and wet snow (9.5).

FRACTURE: Any break or rupture through very close pack ice, compact pack ice, consolidated pack ice, fast ice, or a single floe resulting from deformation processes. Fractures may contain brash ice and/or be covered with nilas and/or young ice. Length may vary from a few meters to many kilometers (7.1).

FRACTURE ZONE: An area which has a great number of fractures (7.2).

FRACTURING: Pressure process whereby ice is permanently deformed, and rupture occurs. Most commonly used to describe breaking across very close pack ice, compact ice, and consolidated ice (6.1).

FRAZIL ICE: Fine spicules or plates of ice suspended in water (2.1.1).

FRIENDLY ICE: From the point of view of the submariner, an ice canopy containing many large skylights or other features which permit a submarine to surface. There must be more than ten such features per 30 nautical miles (56 km) along the submarine's track (13.2).

FROST SMOKE: Fog like clouds due to contact of cold air with relatively warm water, which can appear over openings in the ice or leeward of the ice edge and may persist while ice is forming (11.3).

GIANT FLOE: (4.3.2.1) (see floe).

GLACIER: A mass of snow and ice continually moving from higher to lower ground or, if afloat, continually spreading. The principal forms of glaciers are: inland ice sheets, ice shelves, ice streams, ice caps, ice piedmonts, cirque glaciers, and various types of mountain (valley) glaciers (10.2.1).

GLACIER BERG: An irregularly shaped iceberg (10.4.2.1).

GLACIER ICE: Ice in or originating from a glacier, whether on land or floating on the sea as icebergs, bergy bits, or growlers (10.2).

GLACIER TONGUE: Seaward projecting extension of a glacier, usually afloat. In the Antarctic, glacier tongues may extend over many tens of kilometers (10.2.4).

GREASE ICE: A later stage of freezing than frazil ice when the crystals have coagulated to form a soupy layer on the surface. Grease ice reflects little light, giving the sea a matte appearance (2.1.2).

GRAY ICE: Young ice 10-15 cm thick. Less elastic than nilas and breaks on swell. Usually rafts under pressure (2.4.1).

GRAY-WHITE ICE: Young ice 15-30 cm thick. Under pressure more likely to ridge than to raft (2.4.2).

GROUNDED HUMMOCK: Hummocked grounded ice formation. There are single grounded hummocks and lines (or chains) of grounded hummocks (3.4.2).

GROUNDED ICE: Floating ice aground in shoal water (3.4) (cf. stranded ice).

GROWLER: Smaller piece of ice than a bergy bit or floeberg, often transparent but appearing green or almost black in color, extending less than 1 m above the sea surface and normally occupying an area of about 20 square m (10.4.5).

HOSTILE ICE: From the point of view of the submariner, an ice canopy containing no large skylights or other features which permit a submarine to surface (13.3).

HUMMOCK: A hillock of broken ice which has been forced upward by pressure. May be fresh or weathered. The submerged volume of broken ice under the hummock, forced downwards by pressure is termed a bummock (8.2.3).

HUMMOCKED ICE: Sea ice piled haphazardly one piece over another
to form an uneven surface. When weathered it has the appearance
of smooth hillocks (8.2.3.1).

HUMMOCKING: The pressure process by which sea ice is forced into
hummocks. When the floes rotate in the process it is termed
screwing (6.2).

ICEBERG: A massive piece of ice of greatly varying shape, more
than 5 m above sea level, which has broken away from a glacier,
and which may be afloat or aground. Icebergs may be described as
tabular, dome-shaped, sloping, pinnacled, weathered, or glacier
bergs (10.4.2).

ICEBERG TONGUE: A major accumulation of icebergs projecting from
the coast, held in place by grounding, and joined together by
fast ice (10.4.2.3).

ICE BLINK: A whitish glare on low clouds above an accumulation
of distant ice (11.2).

ICEBOUND: A harbor, inlet, etc., is said to be icebound when
navigation by ships is prevented by ice, except possibly with the
assistance of an icebreaker (12.2).

ICE BOUNDARY: The demarcation at any given time between fast ice
and pack ice or between areas of pack ice of different
concentrations (4.4.9) (cf. ice edge).

ICE BRECCIA: Ice pieces of different age frozen together (4.3.5).

ICE CAKE: Any relatively flat piece of sea ice less than 20 m
across (4.3.3).

ICE CANOPY: Pack ice from the point of view of the submariner
(13.1).

ICE COVER: The ratio of an area of ice of any concentration to
the total area of sea surface within some large geographic locale;
this locale may be global, hemispheric, or prescribed by a
specific oceanographic entity such as Baffin Bay or the Barents
Sea (4.1).

ICE EDGE: The demarcation at any given time between the open sea
and sea ice of any kind, whether fast or drifting. It may be
termed compacted or diffuse (4.4.8) (cf. ice boundary).

ICE FIELD: Area of pack ice consisting of any size of floes,
which is greater than 10 km across (4.4.1) (cf. ice patch).

ICEFOOT: A narrow fringe of ice attached to the coast, unmoved by
tides, and remaining after the fast ice has moved away (3.2).

ICE FREE: No sea ice present. There may be some ice of land
origin (4.2.7) (cf. open water).

ICE FRONT: The vertical cliff forming the seaward face of an ice
shelf or other floating glacier varying in height from 2-50 m or
more above sea level (10.3.1) (cf. ice wall).

ICE ISLAND: A large piece of floating ice extending about 5 m
above sea level which has broken away from an arctic ice shelf,
having a thickness of 30-50 m and an area of a few thousand square
meters to 500 or more square km, usually characterized by a
regularly undulating surface which gives it a ribbed appearance
from the air (10.4.3).

ICE JAM: An accumulation of broken river ice or sea ice caught in
a narrow channel (4.4.7).

ICE KEEL: From the point of view of the submariner, a downward-
projecting ridge on the underside of the ice canopy; the counterpart
of a ridge. Ice keels may extend as much as 50 m below sea level
(13.5).

ICE LIMIT: Climatological term referring to the extreme minimum
or extreme maximum extent of the ice edge in any given month or
period based on observations over a number of years. Term should
be preceded by "minimum" or "maximum" (r.r.8.3) (cf. mean ice
edge).

ICE MASSIF. A concentration of sea ice covering hundreds of
square kilometers found in the same region every summer (4.4.2).

ICE OF LAND ORIGIN: Ice formed on land or in an ice shelf and
floating in water. The concept includes ice that is stranded or
grounded (1.2).

ICE PATCH: An area of pack ice less than 10 km across (4.4.1.4).

ICEPORT: An embayment in an ice front, often of a temporary
nature, where ships can moor alongside and unload directly onto
the ice shelf (12.7).

ICE RIND: A brittle shiny crust of ice formed on a quiet surface
by direct freezing or from grease ice, usually in water of low
salinity. Thickness to about 5 cm. Easily broken by wind or
swell, commonly breaking into rectangular pieces (2.2.3).

ICE SHELF: A floating ice sheet of considerable thickness showing 2-50 m or more above sea level and attached to the coast. Usually of great horizontal extent and with a level or gently undulating surface. Nourished by annual snow accumulation and often also by the seaward extension of land glaciers. Limited areas may be aground. The seaward edge is termed an ice front (q.v.) (10.3).

ICE STREAM: Part of an inland ice sheet in which the ice flows more rapidly and not necessarily in the same direction as the surrounding ice. The margins are sometimes clearly marked by a change in direction of the surface slope but may be indistinct (10.2.3).

ICE UNDER PRESSURE: Ice in which deformation processes are actively occurring, hence a potential impediment or danger to shipping (12.4).

ICE WALL: An ice cliff forming the seaward margin of a glacier which is not afloat. An ice wall is aground, the rock basement being at or below sea level (10.2.2) (cf. ice front).

LAKE ICE: Ice formed on a lake, regardless of observed location (1.3).

LARGE FRACTURE: More than 500 m wide (7.1.5).

LARGE ICE FIELD: An ice field over 20 km across (4.4.1.1).

LEAD: Any fracture or passageway through sea ice which is navigable by surface vessels (7.3).

LEVEL ICE: Sea ice which is unaffected by deformation (8.1).

LIGHT NILAS: Nilas which is more than 5 cm in thickness and rather lighter in color than dark nilas (2.2.2).

MEAN ICE EDGE: Average position of the ice edge in any given month or period based on observations over a number of years. Other terms which may be used are mean maximum ice edge and mean minimum ice edge (4.4.8.4) (cf. ice limit).

MEDIUM FIRST-YEAR ICE: First-year ice 70-120 cm thick (2.5.2).

MEDIUM FLOE: (4.3.2.4) (see floe).

MEDIUM FRACTURE: 200 to 500 m wide (7.1.4).

MEDIUM ICE FIELD: An ice field 15-20 km across (4.4.1.2).

MULTI-YEAR ICE: <u>Old ice</u> up to 3 m or more thick which has survived
at least two summers' melt. <u>Hummocks</u> smoother than in <u>second-year</u>
<u>ice</u>, and the ice is almost salt-free. Color, where bare, is
usually blue. Melt pattern consists of large interconnecting
irregular <u>puddles</u> and a well-developed drainage system (2.6.2).

NEW ICE: A general term for recently formed ice which includes
<u>frazil ice</u>, <u>grease ice</u>, <u>slush</u>, and <u>shuga</u>. These types of ice are
composed of ice crystals which are only weakly frozen together (if
at all) and have a definite form only while they are afloat (2.1).

NEW RIDGE: <u>Ridge</u> newly formed with sharp peaks; slope of sides
usually 40°. Fragments are visible from the air at low altitude
(8.2.2.1).

NILAS: A thin elastic crust of ice, easily bending on waves and
swell and under pressure, thrusting in a pattern of interlocking
"fingers" (<u>finger rafting</u>). Has a matte surface and is up to 10
cm in thickness. May be subdivided into <u>dark nilas</u> and <u>light</u>
<u>nilas</u> (2.2).

NIP: Ice is said to nip when it forcibly presses against a ship.
A vessel so caught, though undamaged, is said to have been
nipped (12.3).

OLD ICE: <u>Sea ice</u> which has survived at least one summer's melt.
Most topographic features are smoother than on <u>first-year ice</u>.
May be subdivided into <u>second-year ice</u> and <u>multi-year ice</u> (2.6).

OPEN PACK ICE: <u>Pack ice</u> in which the ice <u>concentration</u> is 3/8
to less than 6/8 (4/10 thru 6/10) with many <u>leads</u> and <u>polynyas</u>,
and the <u>floes</u> are generally not in contact with one another
(4.2.4).

OPEN WATER: A large area of freely navigable water in which <u>sea</u>
<u>ice</u> is present in <u>concentrations</u> less than 1/8 (1/10). When
there is no <u>sea ice</u> present the area should be termed <u>ice free</u>,
even though <u>icebergs</u> are present (4.2.6).

PACK ICE: Term used in a wide sense to include any area of <u>sea</u>
<u>ice</u>, other than <u>fast ice</u>, no matter what form it takes or how it
is disposed (4.).

PANCAKE ICE: Predominantly circular pieces of ice from 30 cm-3 m
in diameter, and up to about 10 cm in thickness, with raised rims
due to the pieces striking against one another. It may be formed
on a slight swell from <u>grease ice</u>, <u>shuga</u>, or <u>slush</u> or as a result
of the breaking of <u>ice rind</u>, <u>nilas</u>, or, under severe conditions
of swell or waves, of <u>gray ice</u>. It also sometimes forms at some

depth, at an interface between water bodies of different physical characteristics, from where it floats to the surface; its appearance may rapidly cover wide areas of water (4.3.1).

POLYNYA: Any nonlinear-shaped opening enclosed in ice. Polynyas may contain brash ice and/or be covered with new ice, nilas, or young ice; submariners refer to these as skylights. Sometimes the polynya is limited on one side by the coast and is called a shore polynya or by fast ice and is called a flaw polynya. If it recurs in the same position every year, it is called a recurring polynya (7.4).

PUDDLE: An accumulation of melt water on ice, mainly due to melting snow but in the more advanced stages also to the melting of ice. Initial stage consists of patches of melted snow (9.1).

RAFTED ICE: Type of deformed ice formed by one piece of ice overriding another (8.2.1) (cf. finger rafting).

RAFTING: Pressure processes whereby one piece of ice overrides another. Most common in new and young ice (6.4) (cf. finger rafting).

RAM: An underwater ice projection from an ice wall, ice front, iceberg, or floe. Its formation is usually due to more intensive melting and erosion of the unsubmerged part (8.4).

RECURRING POLYNYA: A polynya which recurs in the same position every year (7.4.3).

RIDGE: A line or wall of broken ice forced up by pressure. May be fresh or weathered. The submerged volume of broken ice under a ridge forced downwards by pressure is termed an ice keel (8.2.2).

RIDGED ICE: Ice piled haphazardly one piece over another in the form of ridges or walls. Usually found in first-year ice (8.2.2.6) (cf. ridging).

RIDGED ICE ZONE: An area in which much ridged ice with similar characteristics has formed (8.2.2.6.1).

RIDGING: The pressure process by which sea ice is forced into ridges (6.3).

RIVER ICE: Ice formed on a river, regardless of observed location (1.4).

ROTTEN ICE: Sea ice which has become honeycombed and which is in an advanced state of disintegration (9.4).

SASTRUGI: Sharp, irregular ridges formed on a snow surface by wind erosion and deposition. On mobile floating ice the ridges are parallel to the direction of the prevailing wind at the time they were formed (8.6.1).

SEA ICE: Any form of ice originating from the freezing of sea water (1.1).

SECOND-YEAR ICE. Old ice which has survived only one summer's melt. Because it is thicker and less dense than first-year ice, it stands higher out of the water. In contrast to multi-year ice, summer melting produces a regular pattern of numerous small puddles. Bare patches and puddles are usually greenish-blue (2.6.1).

SHEARING: An area of pack ice is subject to shear when the ice motion varies significantly in the direction normal to the motion, subjecting the ice to rotational forces. These forces may result in phenomena similar to a flaw (q.v.) (5.3).

SHORE LEAD: A lead between pack ice and the shore or between pack ice and an ice front (7.3.1).

SHORE POLYNYA: A polynya between pack ice and the coast or between pack ice and an ice front (7.4.1).

SHUGA: An accumulation of spongy white ice lumps, a few centimeters across; formed from grease ice or slush and sometimes from anchor ice rising to the surface (2.1.4).

SKYLIGHT: From the point of view of the submariner, thin places in the ice canopy, usually less than 1 m thick and appearing from below as relatively light translucent patches in dark surroundings. The undersurface of a skylight is normally flat. Skylights are called large if big enought for a submarine to attempt to surface through them (120 m) or small if not (13.6).

SLUSH: Snow which is saturated and mixed with water on land or ice surfaces, or as a viscous floating mass in water after a heavy snowfall (2.1.3).

SMALL FLOE: (4.3.2.5) (see floe).

SMALL FRACTURE: 50 to 200 m wide (7.1.3).

SMALL ICE CAKE: An ice cake less than 2 m across (4.3.3.1).

SMALL ICE FIELD: An ice field 10-15 km across (4.4.1.3).

SNOW-COVERED ICE: Ice covered with snow (8.6).

SNOWDRIFT: An accumulation of wind-blown snow deposited in the lee of obstructions or heaped by wind eddies. A crescent-shaped snowdrift with ends pointing downwind is known as a snow barchan (8.6.2).

STANDING FLOE: A separate floe standing vertically or inclined and enclosed by rather smooth ice (8.3).

STRANDED ICE: Ice which has been floating and has been deposited on the shore by retreating high water (3.4.1).

STRIP: Long narrow area of pack ice, about 1 km or less in width, usually composed of small fragments detached from the main mass of ice, and run together under the influence of wind, swell, or current (4.4.5).

TABULAR BERG: A flat-topped iceberg. Most tabular bergs form by calving from an ice shelf and show horizontal banding (10.4.2.2) (cf. ice island).

THAW HOLES: Vertical holes in sea ice formed when surface puddles melt through to the underlying water (9.2).

THICK FIRST-YEAR ICE: First-year ice over 120 cm thick (2.5.3).

THIN FIRST-YEAR ICE/WHITE ICE: First-year ice 30-70 cm thick (2.5.1).

TIDE CRACK: Crack at the line of junction between an immovable ice foot or ice wall and fast ice, the latter subject to rise and fall of the tide (7.1.1.1).

TONGUE: A projection of the ice edge up to several kilometers in length caused by wind or current (4.4.4).

VAST FLOE. (4.3.2.2) (see floe).

VERY CLOSE PACK ICE: Pack ice in which the concentration is 7/8 to less than 8/8 (9/10 to less than 10/10) (4.2.2).

VERY OPEN PACK ICE: Pack ice in which the concentration is 1/8 to less than 3/8 (1/10 thru 3/10) and water preponderates over ice (4.2.5).

VERY SMALL FRACTURE: 0 to 50 m wide (7.1.2).

VERY WEATHERED RIDGE: Ridge with very rounded tops; slope of sides usually 20° to 30° (8.2.2.3).

WATER SKY: Dark streaks on the underside of low clouds indicating the presence of water features in the vicinity of sea ice (11.1).

WEATHERED RIDGE: Ridge with slightly rounded peaks; slope of sides usually 30° to 40°. Individual fragments are not discernible (8.2.2.2).

WEATHERING: Processes of ablation and accumulation which gradually eliminate irregularities in an ice surface (6.5).

WHITE ICE: See then first-year ice (2.5.1).

YOUNG COASTAL ICE: The initial stage of fast ice formation consisting of nilas or young ice, its width varying from a few meters to 100-200 m from the shoreline (3.1.1).

YOUNG ICE: Ice in the transition stage between nilas and first-year ice, 10-30 cm in thickness. May be subdivided into gray ice and gray-white ice (2.4).

INTERPRETATION OF SLAR IMAGERY OF ICE IN

NARES STRAIT AND THE ARCTIC OCEAN*

Moira Dunbar

Department of National Defence

Ottawa, Ontario, Canada

INTRODUCTION

Sideways-Looking Airborne Radar (SLAR) has been around for some time, and its potential for ice reconnaissance has been recognized since the early 1960s. Nevertheless, surprisingly little has been done to evaluate its characteristics in this respect and provide guidelines for interpretation of the imagery. The reasons have been mainly economic and logistic, concerning the availability of the SLAR itself and the logistic problems involved in setting up an adequate ground truth party; they do not necessarily reflect on the potential value of the technique.

There have, of course, been several studies of SLAR for ice reconnaissance, starting with an excellent presentation by Anderson (1) (2) which is of particular interest in the context of the present study because the geographical area covered was very similar. Perhaps the most thorough is the work of Johnson and Farmer (3) on data obtained during the Manhattan cruise of 1969. There is also a study of a multi-spectral SLAR experiment by Ketchum and Tooma (4) and one in the Gulf of St. Lawrence by the Canadian Atmospheric Environment Service (5) using the same instrument as the present study. Of these, however, only the Manhattan operation obtained any real ground truth, and this, although thorough, was taken for a different purpose and is

---

*A shorter version of this report was presented at the International Glaciological Society's Symposium on Remote Sensing in Glaciology in Cambridge, England, in September 1974 and will be published in the Journal of Glaciology.

described by the authors as "limited" in the context of the SLAR evaluation.

SLAR has also been successfully used for tracking ice drift, both during the Manhattan cruise (6) and by the Russians (7) who have developed what appears to be a rather sophisticated SLAR specifically for ice reconnaissance.

Ground truth is also conspicuously lacking in the present study, the background information being limited to tape-recorded visual observations (combined with a considerable familiarity with ice conditions in Nares Strait over four seasons), a small amount of photography, and infrared line-scan data, some of which can be directly compared to the SLAR imagery. Owing to the poor light conditions the photographic data, though very useful in confirming interpretation, is not good enough for reproduction and none is included in this report.

The imagery to be discussed was collected on a number of flights in a Canadian Forces Argus aircraft operated by the Maritime Proving and Evaluation Unit, Summerside, Prince Edward Island. The SLAR was a Motorola X-band AN/APS-94D, and on some flights the aircraft also carried a Reconofax 13 infrared line scanner (IRLS). Some photo coverage when light conditions allowed was obtained with a Vinten 70-mm vertical camera and a hand-held 35-mm camera.

The flights covered the whole length of Nares Strait (Fig. 1) on three occasions in 1973; January 13, March 8 and August 17. The March flight also covered two tracks in the Arctic Ocean, from Alert to the North Pole up the 62°W meridian and thence to the ice edge down the 4°E meridian. A further flight over Nares Strait was made on April 6, 1974, and additional material from other parts of the Canadian Arctic and the Gulf of St. Lawrence will be referred to.

CHARACTERISTICS OF THE AN/APS-94D SLAR

General

The APS-94D is a real-aperture SLAR which images on either or both sides of the aircraft. The blank space in the middle represents a non-imaged strip having a width of about twice the flight altitude. The ranges available are 25, 50 and 100 km a side, with corresponding scales of 1:250,000, 1:500,000 and 1:1,000,000. Of these only the first two were used.

The range resolution of the SLAR is fairly constant at about 30 m, but the azimuth resolution varies with flight altitude and deteriorates across the range of the image. At 2,000 ft (610 m) flight altitude, for instance, it is 40 m at 5 km from the track and deteriorates at the rate of 8 m per kilometre. The general image quality also seems to deteriorate with range.

The azimuth scale depends on ground speed, which is fed into the system and the scale automatically adjusted to approximate the range scale. Another automatic correction is made for aircraft drift but heading changes can cause serious distortions in the image geometry. A comparison of Figs. 7 and 8 shows considerable differences in the relationship of the track to the coasts at the south end of Kennedy Channel and even in the shape of Franklin Island. This is because the March track was not actually straight up the channel but followed the Greenland coast, entering the picture on a heading several degrees to port of the channel axis, with a correction to starboard opposite Franklin Island.

The horizontal distortions that occur on the imagery (e.g. at G on Fig. 3) are mostly due to heading changes, but a few are caused by altitude change or turbulence.

OPTIMUM RANGE/ALTITUDE COMBINATION

To establish a definitive optimum range and altitude for the APS-94D ice imagery it would be necessary to fly a series of combinations on the same day over the same ice. There are too many variables, both in the ice characteristics and the radar performance, for samples flown at different times and places to be conclusive. Nevertheless, in the course of a number of missions a considerable variety of combinations were tried, all of which point to the general conclusion that the lower the flight altitude the more detail of ice topography is obtained, and that the 25-km range gives more information than the 50-km. The very best detail was obtained on a brief sector of the run of 17 August 1973, with 500-800 ft (150-230 m) altitude at 25-km range (Fig. 5a). This was paid for, however, by a slightly sharper than usual falling-off of image quality down range. The usual combination used in Nares Strait was 2,000-3,000 ft (610-915 m) altitude and 25-km range; this proved satisfactory. At 4,000 ft (1220 m) and the same range the detail seemed to be less, and a very marked deterioration occurred in the imagery on 17 August 1973 when the aircraft went from 3,000 to 7,500 ft (915-2300 m) during one run. Detail of ice topography almost completely disappeared, and total signal return from the ice decreased sharply, leaving a very dark, low-contrast image. A section flown at 5500 ft (1675 m) was not much better, and one at 9000 ft (2750 m) had parts which show practically nothing. On the other hand quite reasonable imagery was obtained from 6000 ft (1830 m) on April 8, 1974 (Fig. 19).

For the 50-km range there is less data, as it was not used very much, but the best imagery at this range was flown at 4000 ft (1220 m) (Fig. 11). An attempt to use it at 1000 ft (300 m) was a total failure; although very good discrimination was obtained in the fairly close range, more than half the range was lost altogether. At 6000 and 7000 ft (1830 and 2150 m) fairly good imagery was obtained. A comparison of Figs. 2 and 3 shows the difference in detail between imagery at the 50-km range, flown at 7000 ft (2135 m) and the 25-km at 2000 ft (610 m), both flown over the same area on the same day.

It would seem then that for maximum ice detail the best
combination for the APS-94D is the 25-km range and an altitude
between 1000 and 4000 ft (300-1200 m), while in cases where the
extra coverage is more important than detail, the 50-km range at
a rather higher altitude is preferable.

INTERPRETATION

General

Most previous papers on interpretation of SLAR ice imagery
have included a list of ice types with descriptions of the way
they appear in SLAR imagery (3) (5). Although these appear to be
valid enough in respect to the SLAR used and the ice imaged in
each case, the practice may be fraught with pitfalls for those
trying to apply the criteria elsewhere. The problem is partly due
to an over-simplification of the characteristics of ice; all ice
of a given age category does not look alike, either to the eye or
to the SLAR. This is particularly true of multi-year ice, which
can have very different surface characteristics according to its
age and life history, but even first year ice, although in general
it is either very smooth or very rough with young ridges, can vary
considerably from place to place. To a radar it can also look
different according to whether the ice around it is older or
younger, rougher or smoother, than itself.
    Even more significant is the difference between SLAR systems,
particularly if they are in different frequency bands. Comparison
of the imagery in the various references quoted tends to show this,
quite apart from the multi-spectral study of Ketchum and Tooma (4),
which deals specifically with this question. An example may be
seen by comparing Figs. 2 - 4 with Fig. 6. The latter, which is
copied from Anderson (2) shows the same area at much the same time
of year as Fig. 4. Accompanying photographs leave little doubt
that the conditions depicted are very similar, yet the SLAR used,
an AN/APQ-56 operating in the K-band, shows the multi-year floes
as very light, and the surrounding matrix of first-year ice, much
of which is severely hummocked, hardly shows at all, giving an
impression of open pack ice. In contrast the APS-94D imagery shows
the rough first-year ice as generally brighter than the multi-year
floes.
    Therefore the statements made in this paper should be assumed
to refer only to the APS-94D, unless specifically otherwise
stated.

ICE AGE AND THICKNESS

Perhaps the most important feature for most ice observation
purposes is the age and thickness of the ice. The Nares Strait
ice covers the whole gamut of age and thickness, from the nilas

and other young stages of the North Water in Smith Sound to the
heavy multi-year floes that drift down from the Arctic Ocean.
There is therefore plenty of material to work on.

The ice in Nares Strait was shorefast in both the January and
March flights, and the same features can be recognized in great
detail in the imagery from both. The ice falls into three basic
groups: old ice floes of a great variety of sizes; very rough
deformed ice which will be referred to as first-year, though much
of it must have formed from brash ice left over from the previous
summer and is therefore of composite age; and smooth ice which may
be first-year or younger. There is very little coastal fast ice
except in the bays, and as many of these do not break up every
year, some of the ice is at least two years old. Much of it is
formed from pack ice which drifts into the bays in open seasons
and consolidates, and is therefore of various ages.

Old Ice. It has already been pointed out that ice of the same
age does not necessarily look the same, and that this is particularly
true of multi-year ice. Thus it is not surprising that the old ice
does not all present the same appearance in the SLAR imagery. In
this report the term old ice will be used, which includes second-
year as well as multi-year ice, as it was found impossible to
distinguish between them in the SLAR imagery and to try could only
be misleading.

Since X-band radar cannot penetrate the ice surface, what it
records is basically the surface roughness. If an ice floe has a
smooth surface it will appear dark, if it is very rough it will
give a bright return. Therefore it is only insofar as age can be
equated to roughness that it can be unequivocally interpreted from
the SLAR image. As there is no such direct correlation no unfailing
way of determining age exists. There are, however, various
characteristics which help, as will be shown.

In actual practice it is generally not hard to separate the
old ice from the first-year ice in Nares Strait in winter and
spring, because it is the only ice that forms discrete floes.
This, however, is due to the particular regime of the area and is
not true for all arctic channels. Old ice can be smooth or rough,
gently undulating or hummocky and therefore can vary greatly in
image quality. Examples may be found in Figs. 3 and 4 (floes A,
B and C) and Figs. 9 and 10 (floes A-D). Floe A in Figs. 3 and 4
is a particularly interesting one. It appears to be a floe within
a floe, the outer "collar" being probably younger than the inner
part. This composite quality is quite common in multi-year floes.
The smooth (dark) patches on the outer part of this floe are almost
certainly frozen melt-puddles from the previous summer. Former
puddles in fact probably account for many of the dark patches on
the old ice. This is certainly true of the shore-fast ice in the
fiord to the left of floes A and B. This fiord (Lady Franklin
Bay) did not break up in the summer of 1972 (personal observation)
but was largely covered with melt-water, which was already well
frozen over at the end of August.

That the dark areas on these floes are in fact part of the
old floe and not inclusions of thinner ice is shown in Fig. 12,
which includes an IRLS image of part of a multi-year floe in Kane
Basin, taken on March 8 (a) and an enlargement of the SLAR image
of the same floe from the January 13 flight (b). It is not usually
possible to make such comparisons, as the SLAR sees only sideways
and the IRLS straight down. In this case, however, as the ice had
not moved in any way between the two flights and the flights
tracks were not identical, it was possible to identify the IR track
from the second flight on the earlier SLAR imagery. The relationship between the two pictures can be established by the two curving
ridges at A. The dark patch (B) and the very bright rough patch
(C) in the SLAR image both have the same cold tone in the thermal
image, showing that this is not younger ice like that at point D,
which has a much warmer signature and is therefore thinner (8).
The floe shown appears to be extremely ridgy, giving a considerably
brighter SLAR signal than the surrounding deformed first-year ice
(E). The ridges in area C are in fact detectable in the original
IRLS image, though they do not appear in the reproduction.
Similarly in Fig. 13 a number of floes which appear dark on the
SLAR image (A and B for instance) are light (cold) in the IRLS,
indicating thick ice.

      Some of the tonal differences between the old floes are due
to variation in the image quality across the range, which is
characteristic of this SLAR system; floes tend to appear brighter
close to the track, as for example in Fig. 7, where floe A is
somewhat brighter than floe B. However, this is not an invariable
rule, and many differences must also be due to the characteristics
of the floes themselves. A good example is shown in Figs. 9 and
10, which were taken within two hours of each other. The different
tones of floes A and B in the two pictures are due at least in
part to their relative distance from the flight track. The right
half of floe C shows the same difference, but the light tone of
the left half in Fig. 9 is due to a flare-out effect that tends to
occur at the near edge of the imagery. Note, however, floe D,
which has much the same tone in both images although it is
appreciably nearer the track in Fig. 10. It is perhaps significant
that in both images it is on the same side of the aircraft, whereas
floes A, B and C are not. Owing to a slight tilt in the antenna
as mounted on this particular aircraft there is an unevenness of
image quality, the left side receiving rather higher backscatter
than the right. This introduces yet another variable in the image
interpretation, and may have an influence on the greater brightness
of floes A, B and C in Fig. 9.

      In the only sample we have of summer imagery (Figs. 5 and 8)
most of the floes are known from visual observation to be old ice.
They appear consistently darker than the surrounding brash ice,
and little surface detail is visible. The large floe A in Fig.
5(b) is described in the visual notes as a composite floe of
various ages, yet it presents a rather even tone. The puddles,
which were refrozen, do not show up clearly. However, all this

may be due to poor radar functioning or possibly to poor choice of
flight altitude (4000 ft (1200 m)). In the small area flown at
500 ft (150 m) (Fig. 5(a)) the surface detail on the floes is very
much clearer.

In the Arctic Ocean imagery (Fig. 11) it is much harder to
separate the old ice from the first-year, because here the pack
has not been subjected to the melting and subsequent jostling of
floes that has rounded and defined the edges of the old floes in
Nares Strait. Nor is there the same degree of visual back-up in
this area, as the light was getting poor by the time it was over-
flown. However, if the proposition that many of the small dark
patches represent refrozen melt-puddles is accepted, then most of
the ice in Fig. 11 should be second-year or older, and this is in
fact supported by the visual notes, which state that at least 7/10
of the ice visible in this area was multi-year, and most of the
rest very hummocky and of indeterminate age.

Old ice, then, seems to be characterized by a wide range of
grey-tones and may appear brighter or darker than the surrounding
younger ice. In winter conditions it may frequently be identified
by a pattern of refrozen melt-muddles which appear very dark.

First-Year Ice. First-year ice is less varied in topography
than old ice, being either undeformed and very smooth, or broken
by young pressure ridges which are sharply angular and present a
good target to the radar signal. In Nares Strait this stage is
usually carried to the extreme, the ice being reduced to a jumble
of ridges and hummocks which appears on the SLAR image as a general
bright clutter. Examples of smooth first-year ice are seen at D,
E and F in Fig. 4, at C in Figs. 14 and 15, and at A in Fig. 15.
As stated above much of this kind of hummocky ice is not strictly
speaking first-year, but is formed by freezing together of brash
ice from the previous open season. Brash ice is represented by
all the brighter areas in Figs. 5 and 8, and to judge from the
amount of it present at this date (August 17) it should account for
a fair proportion of the hummocky ice that forms in the winter.
It is very likely for instance that the ice north of Franklin
Island (at C) in Fig. 7 is of this category, with many small multi-
year floes in amongst it, such as may be seen in Fig. 8 in much
the same area.

A good sample of first-year ice appears in Fig. 20, which
shows matching SLAR and photographic images.

Young Forms of Ice. The main body of ice younger than first-
year included in the imagery under consideration is in the North
Water (Figs. 14-16). The only other areas are the refrozen leads
in the Arctic Ocean (Fig. 11) and possibly area D in Fig. 3. The
lead at D was not present at the time of an earlier flight on
December 15 (9) so it may very well have been young ice on
January 13. Areas E and F in Fig. 3 were already there in
December so must be older. Unfortunately there was a complete
undercast in this area on January 13 rendering visual observation
impossible. This points up the inability of SLAR to discriminate
ice thickness, as in all these areas the ice must have grown

considerably thicker in the eight weeks separating Fig. 3 and
Fig. 4, yet they look exactly the same in both. In March they had
all reached the first-year stage. Similarly in the Arctic Ocean
some of the leads were open, some refrozen, and some half and
half, but they all look the same in the SLAR imagery.

The North Water on January 13 was almost cloud-free, a rather
rare condition, but visibility was limited by darkness. At low
altitude, however, and from the first-rate look-out position in
the nose of the Argus, it is surprising how much can be seen in
the dark. The visual observations, made on the southbound trip
only (Fig. 14) showed an open lead about 400 m wide south of the
ice edge, followed by concentrations of about 6/10th thin ice,
consisting of nilas graduating southwards to grey ice. The leads
and open patches tended to be oriented across the channel, and
both the ice floes and leads tended to get progressively larger
towards the south.

This description is borne out by Fig. 14, or at least by the
parts of the imagery closest to the aircraft track; beyond about
5 km the young ice returned no signal at all in the north part of
Smith Sound. On the west side of the track the edge of the fast
ice shows up clearly at the extreme edge of the image, and about
the latitude of Littleton Island some returns of what look like
small ridges begin to appear. On the east side there is a very
interesting pattern of streaks of slush ice and small cakes. As
these are wind-blown features they can only form in otherwise open
water, and the grey tone, for instance between the two strings of
ice at B, is a typical sea-clutter return. The cusp-like pattern
of the streaks is curious and the cause of much speculation; its
origin is obscure.

Fig 14 was taken on the southbound trip; on the northbound
trip, exactly the same general conditions prevailed; the passage
of time, however, and also the speed of drift, is illustrated by
floe A, which appears in both northbound and southbound imagery,
and which moved south between the two passes at a speed of roughly
0.55 m s^{-1}. With a wind-speed of about 50 knots (25 m s^{-1})
recorded at our flight altitude (3000 ft (915 m)) this works out
at about the speed expected from Zubov's rule of ice drift.

Fig. 15 shows the conditions on March 8, when there were good
light conditions but an almost complete cover of frost smoke,
allowing only periodic vertical glimpses of the ice. From these
glimpses it appeared that the ice graduated from light nilas near
the head of the North Water to predominantly grey-white south of
Cape Alexander. The proportion of open water decreased from north
to south; near the head it was estimated through breaks in the
cloud to be as high as 5/10 in places.

Fig. 15 is designed to show the predominantly grey-white ice
south of Cape Alexander (b) as well as the thinner ice and more
open conditions to the north (a). It will be seen that the main
difference between them is the heavier incidence of ridges or
rafting in (b) and the presence of areas of medium grey tone in (a).

The dark pieces in (a) are undoubtedly ice, and the greyer patches must presumably be the water, no doubt with new ice forming in it. Why it should appear rougher than the ice is intriguing, as it would seem unlikely that the expanses of it are large enough for any appreciable sea chop to form, except in the area off the Greenland coast. However, as there was a strong wind blowing, there were no doubt small ripples in the open water, and it is known that small ripples can give a large amount of backscatter (S.C. Clay, personal communication).

There was only a very narrow band of open water at the head of the polynya at the point of crossing. The line of the ice edge in Fig. 14 is still precisely traceable in Fig. 15; the expanse of light grey immediately south of this line in Fig. 15 is new ice. This is confirmed by the IRLS and the vertical photography.

A further flight over the North Water was made on April 6, 1974 (Fig. 16). On this occasion visibility was good and ice conditions much the same as those just described for Fig. 15. There was, however, no open area off the coast of Greenland, a circumstance no doubt largely attributable to the winds, which were light and up-channel as opposed to strong and down-channel as on the 1973 flights. Concentrations in general were higher, especially in the narrows (north of Cape Alexander) where they varied from 6/10 to 9/10 - very little different from the area farther south, the difference being in thickness rather than concentration.

Fig. 16 shows these conditions quite clearly and is indeed easier to interpret than Fig. 15. It is, however, hard to discriminate between the various thickness categories of young ice except by a gradually increasing impression of solidity and brightness in the ridges. A line of IR imagery taken at the same time shows great variation in thickness, but is of course not directly relatable to the SLAR image.

Photographs taken with a hand-held camera, however, can be related to the SLAR and bring out some interesting points. Fig. 17 can be related to Fig. 16 by the point B which is the same in both, and it emphasizes again the lack of identity of black areas of the SLAR imagery. The area north of B is shown by the photograph to be entirely ice covered, in contrast to the open water immediately to the south, but both are black on the SLAR. Fig. 18 shows Littleton Island (below C in Fig. 16) with open water in the foreground and shorefast ice beyond. The fast ice can be compared in some detail with Fig. 16 and includes a number of floes, of which C is the largest and most easily identifiable, which are embedded in the fast ice and must therefore be older and thicker. In the photograph they appear lighter in tone, the rest of the ice having the grey look indicative of fairly young ice. In the SLAR the older floes have a fairly dark tone, the younger matrix much brighter, showing that there has been considerable deformation of the young ice. Although small in scale this ridging provides a fairly strong backscatter. The floes also have ridges on them, some of which can be seen on the original SLAR imagery (though probably not in the reproduction) but there are smooth areas between them. It may be noted that many of the small light pieces between floe C and the island (Fig. 18) can be quite clearly distinguished on the original SLAR imagery with the aid of a magnifying glass.

It will be seen that young ice forms are rather hard to interpret without ground truth or visual or other airborne back-up. There is very little doubt, however, that many of the questions raised could be answered by a thorough evaluation program with ground verification.

ICEBERGS

Icebergs show up well on the APS-94D imagery provided they are prominent enough and appear against a sufficiently uncluttered background. Examples may be seen in Fig. 15 on the extreme right just north of the Greenland coast (right of A), and in the same area in Fig. 16, where they show up very clearly in smooth first-year ice. A number may also be seen in Figs. 14 and 15 in the fast ice on the east side of the North Water, and at bottom right in Fig. 16. However, a few bergs that were noted in Kane Basin nearer to the aircraft track are lost in the general bright clutter.

CONCENTRATION

Of the flights being considered here only the August one was concerned with concentrations less than 10/10ths, except for the young ice area of the North Water. In the August imagery (Figs. 5 and 8) it is not hard to measure the total concentration; the open areas appear black and all the ice various shades of grey. The situation is not quite so simple, however, in areas and times of year when young ice may be expected to be present, or when leads refreeze, as in the Arctic Ocean (Fig. 11). Here we are once again up against the problem of smooth surfaces, which can lead to real difficulty in distinguishing between ice and water, provided both are smooth.

It is often possible to discriminate between smooth ice and water by the small roughnesses that commonly appear on the ice surface and which show as faint light lines (Figs. 3 and 4, areas D, E, F). However, very similar lines occasionally appear in water openings where small amounts of brash ice are lined up by the wind. Frequently an experienced eye can tell the difference, but there remain many cases where even an experienced eye is at a loss. Many of the small black patches north of the head of the North Water, for instance (Figs. 14-16) would be almost impossible to interpret without having seen the area. The same is true of quite a number of the smaller black areas in Figs. 3 and 4, while at the scale of Fig. 2 the whole smooth area might be open.

Two better examples are shown in Figs. 19 and 21, taken on April 8, 1974. Fig. 19 shows two areas of apparently similar ice, one in Barrow Strait (a) and the other between Coats and Mansel islands in Hudson Bay (b). They look the same but in (a) all the leads were covered with first-year ice, snow-covered and solid, whereas in (b) they were all open. Fig. 21 shows an area in Pelly Bay, where very rough old fast ice, giving a very bright return (A) contrasts sharply with newer smooth fast ice, which gives no return at all (B). In these cases positive interpretation would be virtually impossible without visual or other back-up.

TOPOGRAPHY

Surface topography shows up on the APS-94D imagery either as discrete bright signals or in the form of variations in grey tone. The difference in tone, for instance, between areas B and D in Fig. 12 (b) is probably due to minor roughnesses and undulations of the surface of B. Similarly in the same picture the overall light grey tone at E and the bright patch near C are due to the roughness of the first-year and multi-year ice respectively. In both cases the ridges and hummocks are too close together for the system to resolve. More widely spaced ridges show up as bright lines.

Without detailed ground truth it is not possible to say how reliably the ridges are picked up by the SLAR or how many are missed owing to their orientation in relation to the radar beam or other reasons, but it is certain that this occurs. Fig. 13 shows a narrow strip of relatively thin ice on the IRLS image (left of floe B) which is confirmed by photography to be smooth, but which does not appear so on the SLAR. It is oriented parallel to the radar beam about 10-15 km from the near edge of the imagery and seems to be the victim of the deteriorating azimuth resolution, which fails to distinguish it from the ridges on either side. Equally narrow strips oriented across the beam show quite clearly above floe A.

Ridges and other surface features show up better in the middle and far ranges, as would be expected, the angle of incidence of the radar beam being more oblique, but there are few cases where features that show on one of the January and March pair do not appear at all on the other. (The differences between the smooth areas in Figs. 3 and 4 are due to photographic factors; in the original imagery, though the quality is a little different, all the topography that appears in Fig. 3 can also be seen in Fig. 4). One of the very few such cases is shown at C on Figs. 14 and 15 where the topographic detail in the smooth ice in Fig. 15 is almost completely absent in Fig. 14. As there had been no change in the shape of the feature between the two flights there can have been no further ridge formation to account for the difference. Whether it is due to new snow-cover drifting against the ridges, to relative distance from

flight track, or simply to variations in functioning of the SLAR
is not clear.

Fig. 20 shows an enlarged section of SLAR imagery off the
south coast of Southampton Island (taken April 8, 1974) and a hand-
held oblique photograph of the same area. The box on 20(a)
outlines the approximate area covered by 20(b). A comparison shows
that there is a great deal of topographic detail in the SLAR image
which one might hesitate to interpret without the back-up of the
visible image. In particular the rough and smooth areas beyond
the leads in the foreground show up quite clearly. An exception
is the refrozen crack at A in 20(b), and even this is discernible
on the original, though the contrast is so low that an interpreter
would be likely to miss it altogether.

DYNAMIC FEATURES

Because of the wide coverage obtainable with SLAR, combined
with its relatively good resolution, it is unrivalled among airborne
sensors for depicting dynamic features like lead orientation, shear
zones and drift patterns. Lead orientation is well exemplified in
the Arctic Ocean imagery and is discussed in Dunbar and Lowry (10).
Figs. 2-4 show a nice example of a shear zone running from Cape
Lieber to Cape Beechey between the fast ice of Lady Franklin Bay
and the ice of the main channel. A similar line appears on the
opposite side of Hall Basin, where fast ice has formed east of a
line from Joe Island to Cape Lupton before the main channel stopped
moving. These lines are absent in Fig. 5, showing that the fast
ice must have broken up in 1973.

Drift patterns are shown in Fig. 8 in the form of eddies in
the lee of the islands, and in the North Water off the Greenland
coast (Fig. 14).

SYNTHETIC APERTURE SLAR

It is clear that the most serious drawback of the APS-94D is
its resolution, which is not adequate to resolve all necessary sea
ice features. Fig. 22 is included to show that this problem is not
the same with all SLARs. It shows a sample of imagery taken by a
Goodyear APQ-102 synthetic aperture SLAR in March 1974 in the Gulf
of St. Lawrence. The swath width is 10 miles. The resulution of
this SLAR is 50 feet (15.25 m) and both resolution and image
quality are constant across the range. The oldest ice shown is
first year. Unfortunately no APS-94D imagery of similar ice is
available to compare it with but its superior quality is obvious.
However, the smooth-surface gremlin is still present. The inlet at
top left is undoubtedly frozen and snow-covered, yet it looks
exactly the same as the open water along the coast.

CONCLUSIONS

From the evidence presented here we may conclude that SLAR has
great potential as a tool for the acquisition of sea ice data.
Although there are a number of uncertainties in interpreting the
APS-94D imagery, these are due mainly to problems of resolution,
which, as can be seen from Fig. 22, are not insoluble. Others could
be solved by a thoroughly ground-truthed test programme. The only
problem that may prove permanent is that of distinguishing between
smooth ice and water, and even this will surely diminish with
higher resolution, as with finer detail discernible the number of
cases where discrimination is not possible will decrease. However,
these points should be borne in mind when considering such programmes as direct provision of SLAR ice imagery to ships. In the
hands of an inexperienced interpreter it could be quite misleading,
as he may well tend to read it as if it were a rather fuzzy
photograph, even if he is quite accustomed to interpreting the
ship's radar. Any ship receiving such imagery should have at least
one trained interpreter on board.

As against these disadvantages there is the excellent coverage
possible with SLAR, which provides more information with less
footage of film to interpret than any other airborne sensor. This
makes it a practical tool to use for repeated surveys, whether for
ship operation, drift measurements, or other purpose, in contrast
to photography, which is not only restricted by cloud and light
conditions but requires so much flying time and produces such miles
of film as to be prohibitive except in special cases. Above all
there is the all-weather capability of radar, the only sensor that
can penetrate both darkness and cloud.

REFERENCES

1. Anderson, V.H. 1966. High altitude side-looking radar images
    of sea ice in the Arctic. Proceedings of the Fourth Symposium
    on Remote Sensing of Environment. University of Michigan,
    pp. 845-857.

2. Anderson, V.H. 1968. Radar imagery of Arctic pack ice. U.S.
    Cold Regions Research and Engineering Laboratory. Special
    Report 94.

3. Johnson, J.D., and Farmer, L.D. 1971a. Use of Side-looking
    air-borne radar for sea ice identification. Journal of
    Geophysical Research, Vol. 76, No. 9, pp. 2138-2155.

4. Ketchum, R.D. Jr., and Tooma, S.G. Jr. 1973. Analysis and
    interpretation of air-borne multi-frequency side-looking radar
    sea ice imagery. Journal of Geophysical Research, Vol. 78,
    No. 3, pp. 520-538.

5. Hengeveld, H.G. [1972]. Operation Icemap II: Side-looking
    radar trials, spring 1972. Atmospheric Environment Service,
    Environment Canada. (Internal Report).

6. Johnson, J.D., and Farmer, L.D. 1971b. Determination of sea
    ice drift using side-looking air-borne radar. Proceedings of
    the Seventh International Symposium on Remote Sensing of
    Environment. University of Michigan, pp. 2155-2168.

7. Loshchilov, V.S. and Veovodin, V.A. 1972. Determination of
    drift elements of the ice cover and ice edge movements by
    means of air-borne radar. Problems of the Arctic and Antarctic,
    No. 40 [Translated from the Russian] U.S. Dept. of Commerce,
    pp. 19-26.

8. Poulin, Ambrose O. 1973. On the thermal nature and sensing of
    snow-covered arctic terrain. U.S. Army Engineers Topographic
    Laboratories. Research Note ETL-RN-73-4.

9. Dunbar, Moira. 1974. Winter ice reconnaissance in Nares Strait
    1972-73. Defence Research Establishment Ottawa. Technical Note
    No. 73-26.

10. Dunbar, Moira and Lowry, R.T. In press. Remote Sensing of
    sea ice in Nares Strait and the Arctic Ocean, March 1973.
    Proceedings of the Second Canadian Symposium on Remote Sensing,
    Guelph 1974.

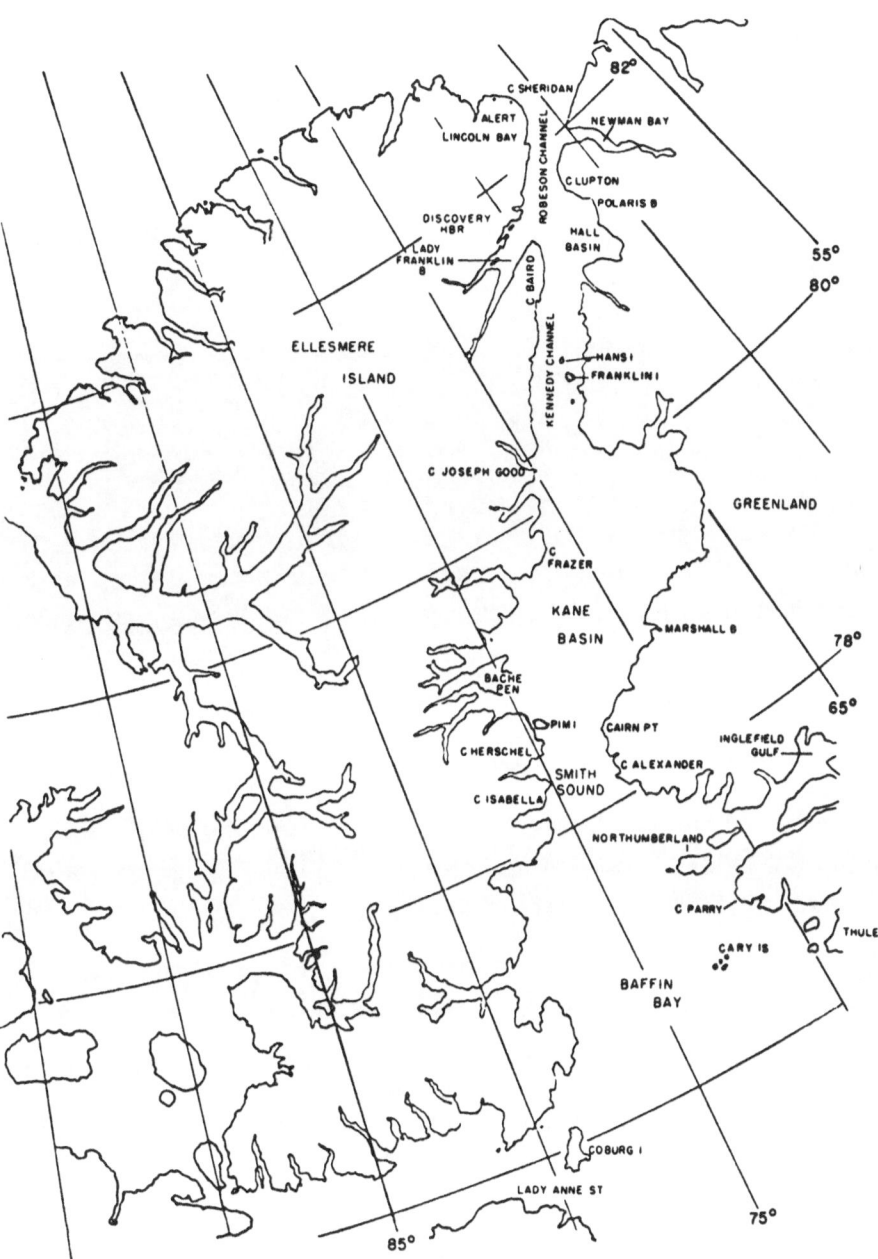

Figure 1. Location Map of Nares Strait

Figure 2.   SLAR image of Hall Basin taken January 13, 1973. Range,
             50 km; altitude, 7000 ft (2135 m). On this and other
             SLAR images the north end of the channel is at the top
             and the arrow shows flight direction.

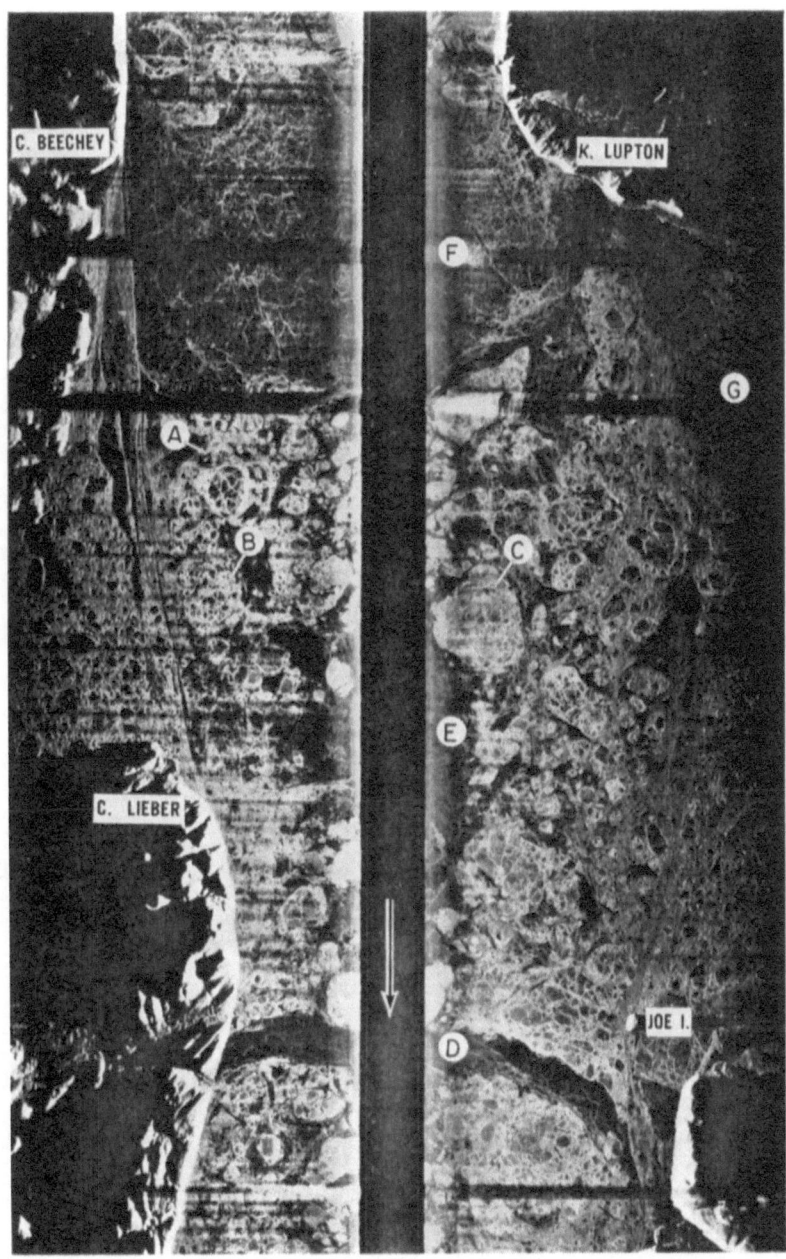

Figure 3.   Hall Basin, January 13, 1973.   Range, 25 km; altitude, 2000 ft (610 m).

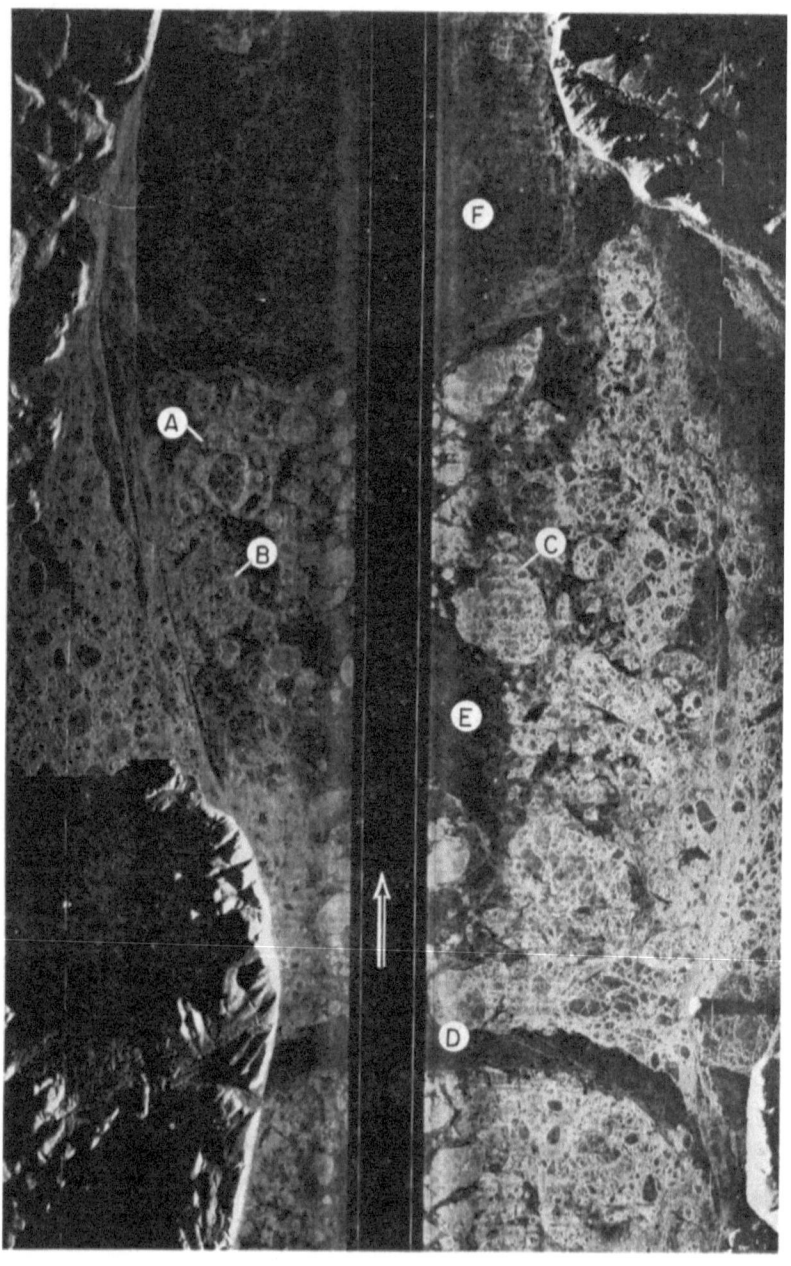

Figure 4.   Hall Basin, March 8, 1973.   Range, 25 km; altitude,
            2000 ft (610 m).

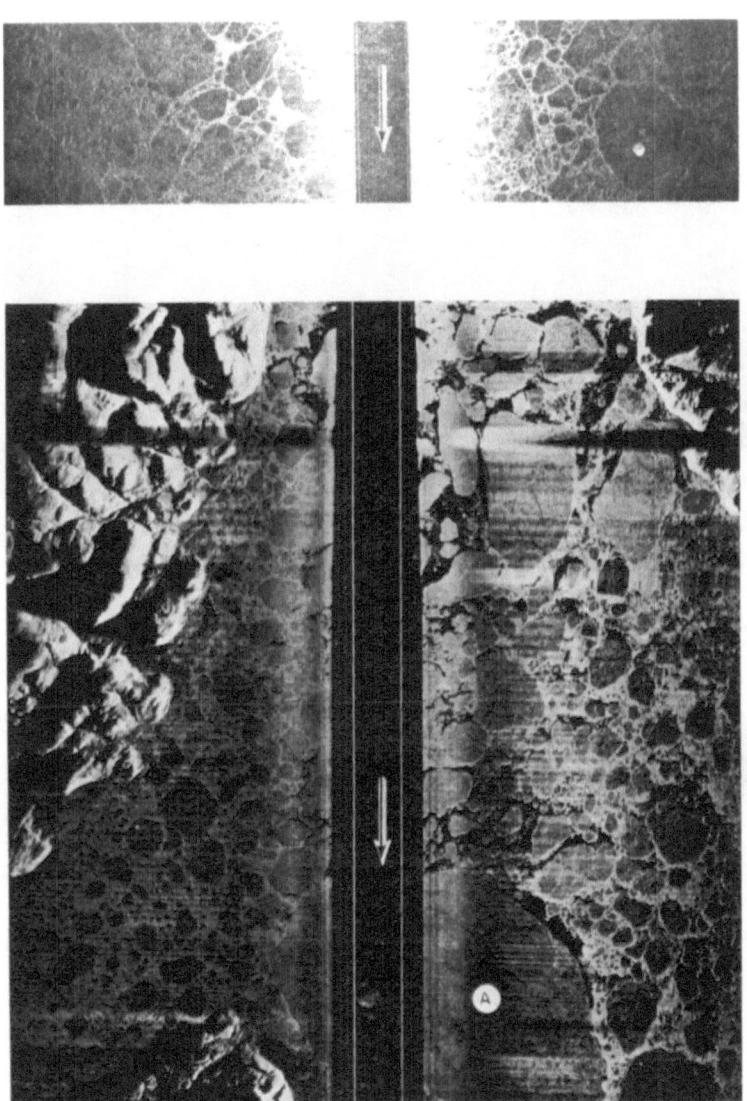

Figure 5.   North entrance to Robeson Channel (a) and Hall Basin (b)
            August 17, 1973.   Range 25 km; altitude 4000 ft (1220 m).

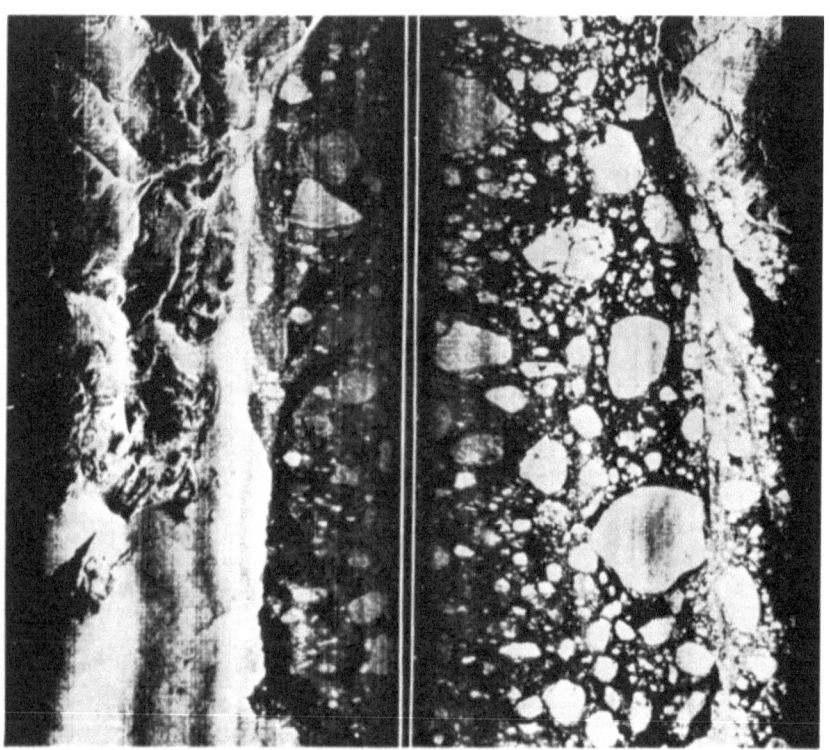

Figure 6. Northern Hall Basin, April 21, 1962, imaged by the USAF
           with an AN/APQ-56 SLAR from 60,000 ft (18,300 m).
           (From Anderson, 1968)

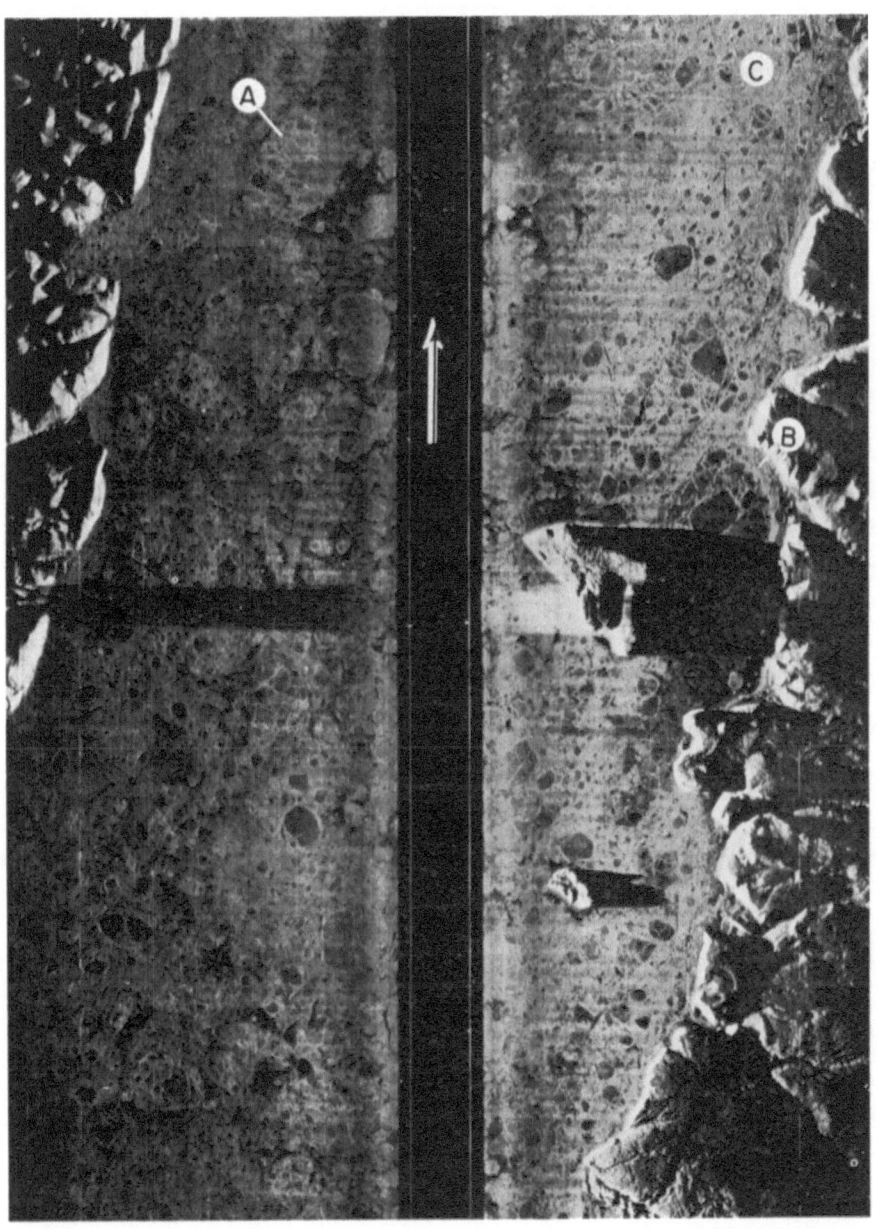

Figure 7. Kennedy Channel, March 8, 1973. Range, 25 km; altitude, 2000 ft (610 m).

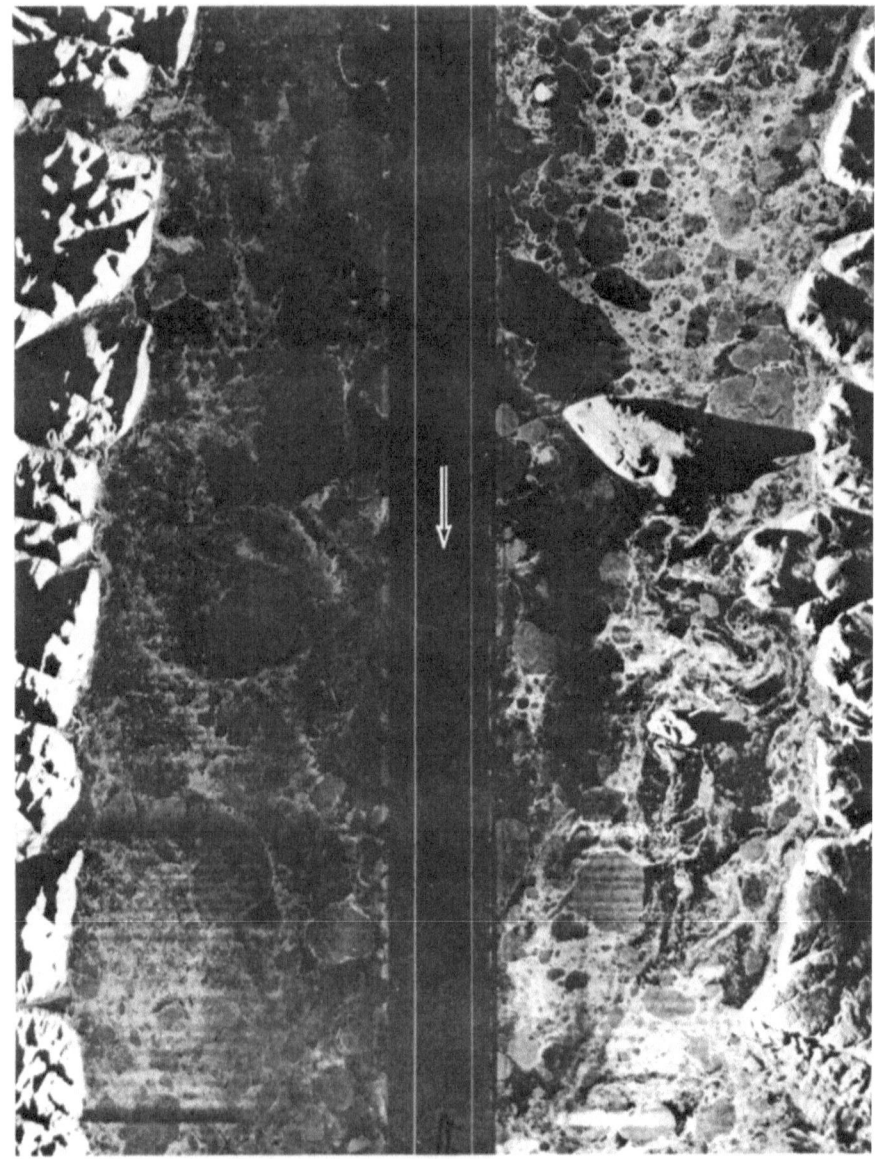

Figure 8.   Kennedy Channel, August 17, 1973.   Range 25 km; altitude, 4000 ft (1220 m).   The difference in flight altitude between this and Fig. 7 is reflected in the length of radar shadow of the islands.

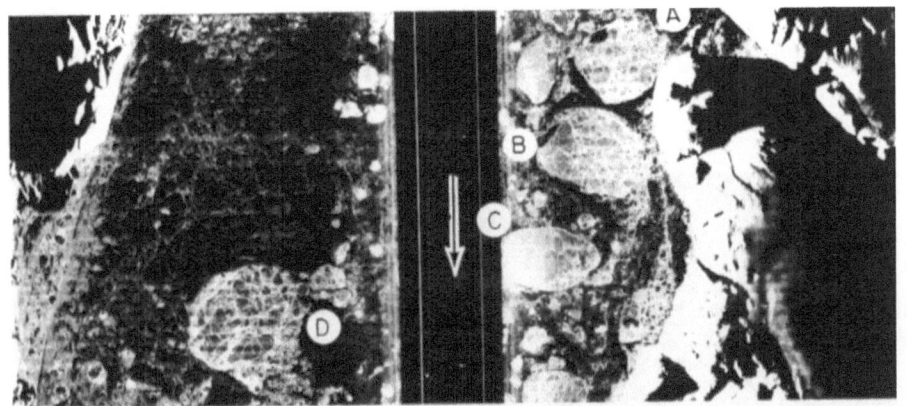

Figure 9.   Kennedy Channel, April 6, 1974.   Range, 25 km; altitude,
3000 ft (915 m);   southbound.

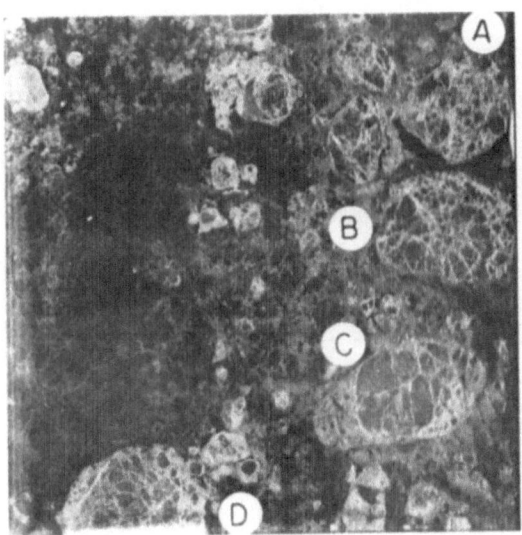

Figure 10.   The same, northbound, right side of track only.

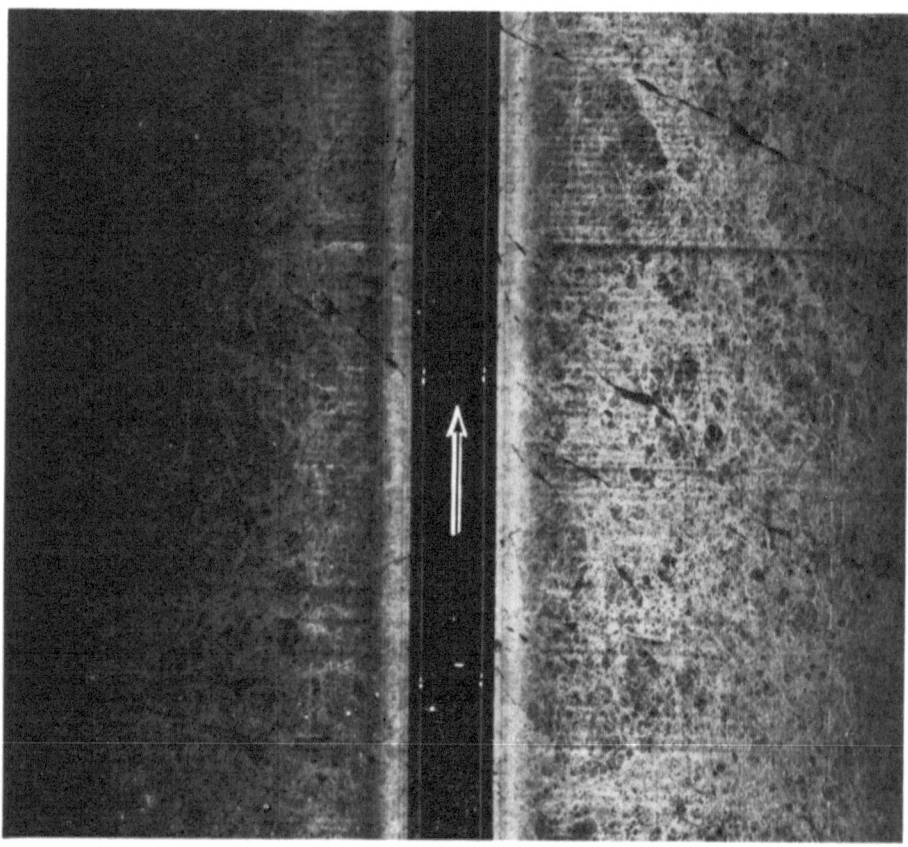

Figure 11.   The Arctic Ocean between Alert and the pole, at about
             84°30 N, 62°15 W.   Range, 25 km; altitude, 4000 ft
             (1220 m); March 8, 1973.

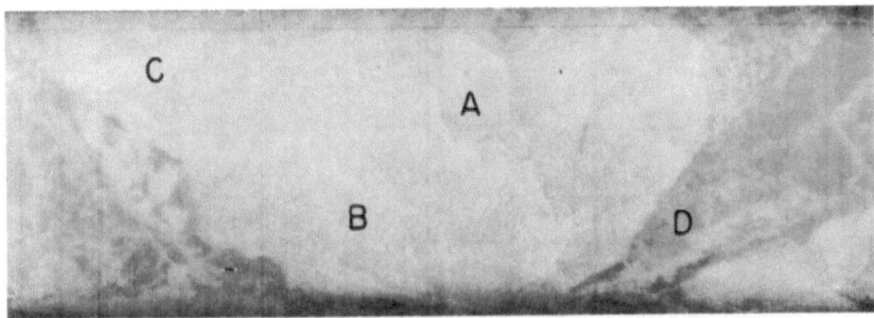

12a

12b

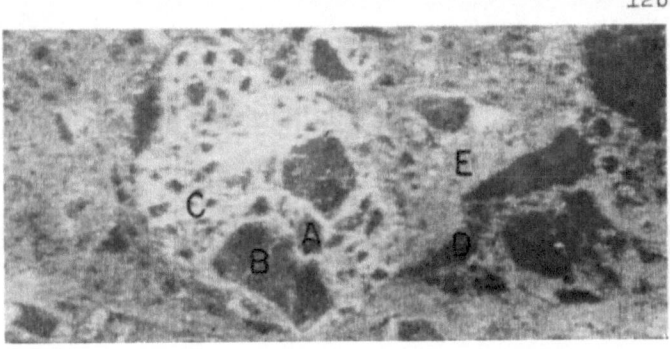

Figure 12.   (a) IRLS image of ice in Kane Basin, March 8, 1973.
Dark is warm, light is cold. Part of a multi-year
floe, with thinner first-year ice at D.
(b) Enlarged SLAR image showing the same multi-year floe,
January 13, 1973. Range, 25 km; altitude, 2000 ft
(610 m).

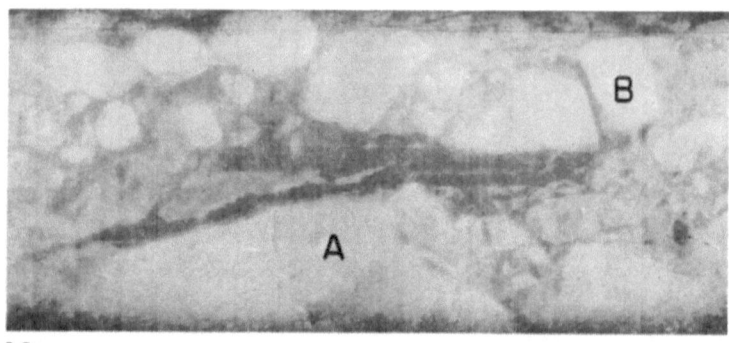

13a

13b

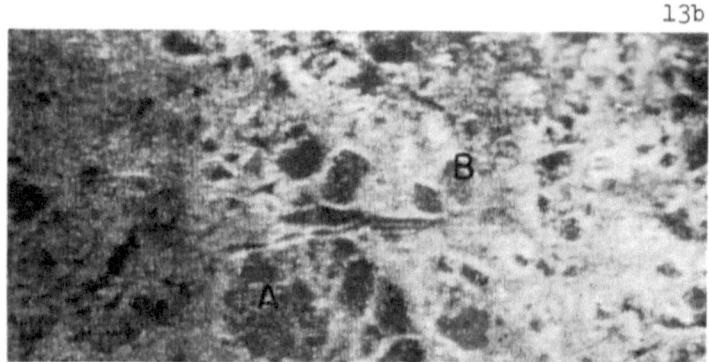

Figure 13.   (a) IRLS image of ice in Kane Basin, March 8, 1973,
              showing a patch of relatively thin ice (dark) among old
              floes.
              (b) Enlarged SLAR image of the same area, January 13,
              1973. Range, 25 km; altitude, 2000 ft (610 m).

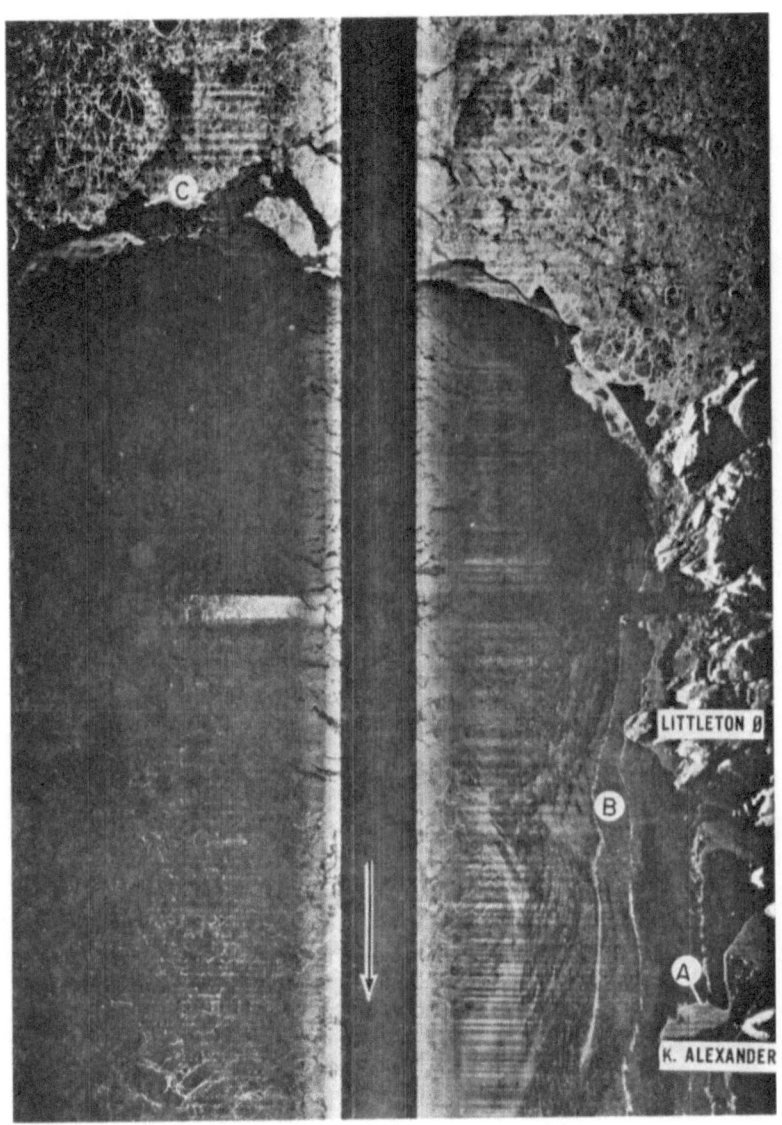

Figure 14.   Smith Sound, January 13, 1973.   Range, 25 km; altitude,
              3000 ft (915 m) showing the head of the North Water.

15a

15b

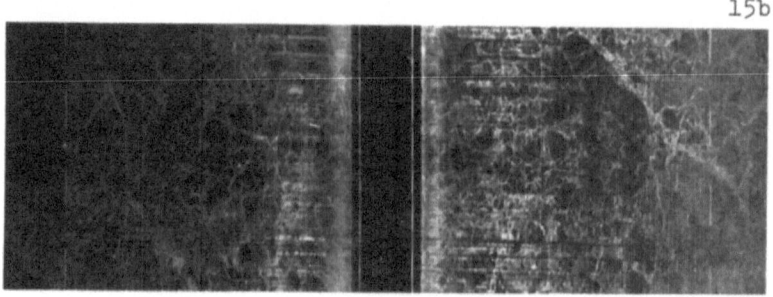

Figure 15. Smith Sound, March 8, 1973. Range, 25 km; altitude,
            2000 ft (610 m). (a) Head of North Water; (b) Area a
            few miles south of Cape Alexander.

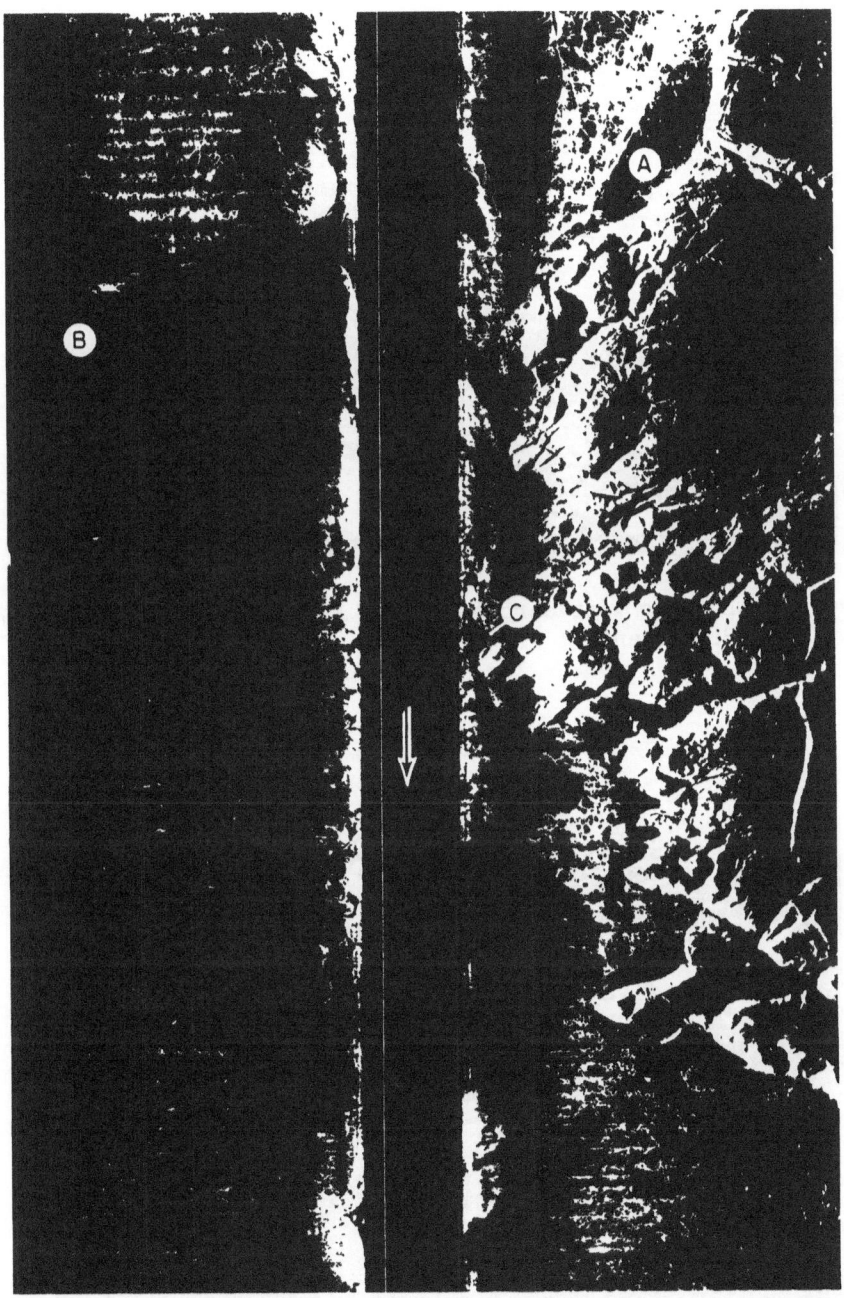

Figure 16. Smith Sound, April 6, 1974. Range, 25 km; altitude, 3000 ft (915 m).

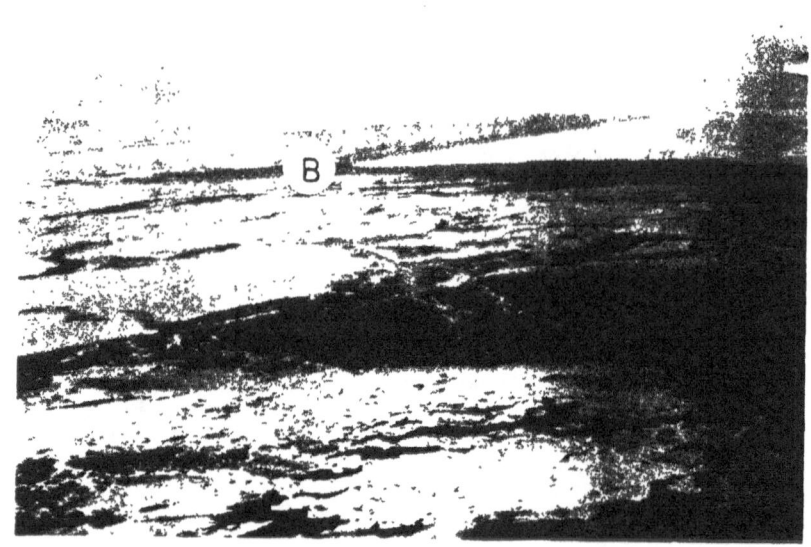

Figure 17.   Looking west across head of North Water, April 6, 1974.
             Pim Island at left rear.

Figure 18.   Littleton Island, Smith Sound, April 6, 1974.

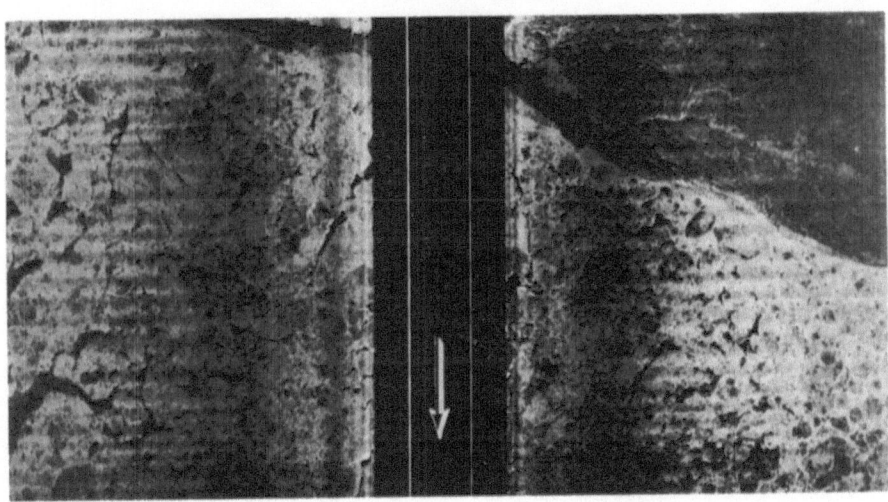

Figure 19.   (a) North part of Barrow Strait, coast of Bathurst Island
at top.   April 8, 1974; range, 25 km; altitude, 6000 ft
(1830 m); showing frozen leads.
(b) Northern Hudson Bay, S.E. coast of Coats Island at
top.   April 8, 1974; range, 25 km; altitude, 6000 ft
(1830 m); showing open leads

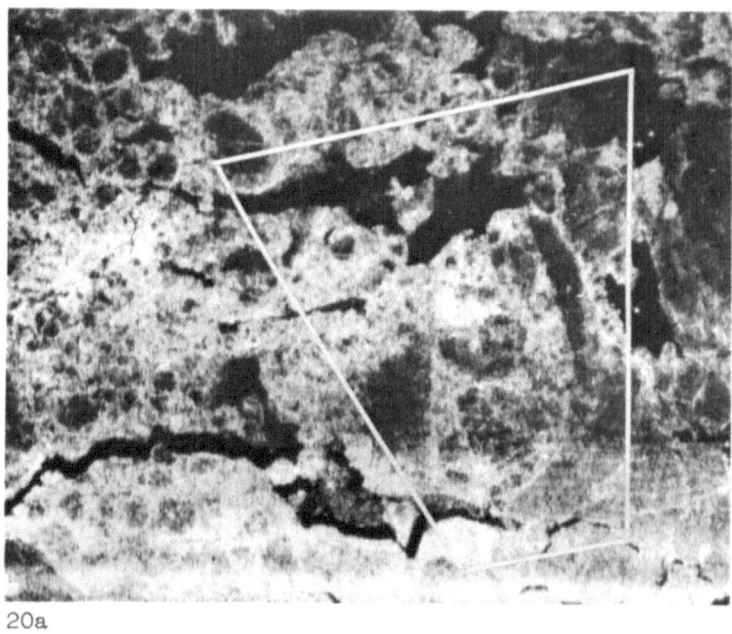

20a

20b

Figure 20.   Enlarge SLAR image (a) and photograph (b) of area off
             the south coast of Southampton Island, April 8, 1974.
             Range, 25 km; altitude, 6000 ft (1830 m). Box in (a)
             outlines coverage of (b).

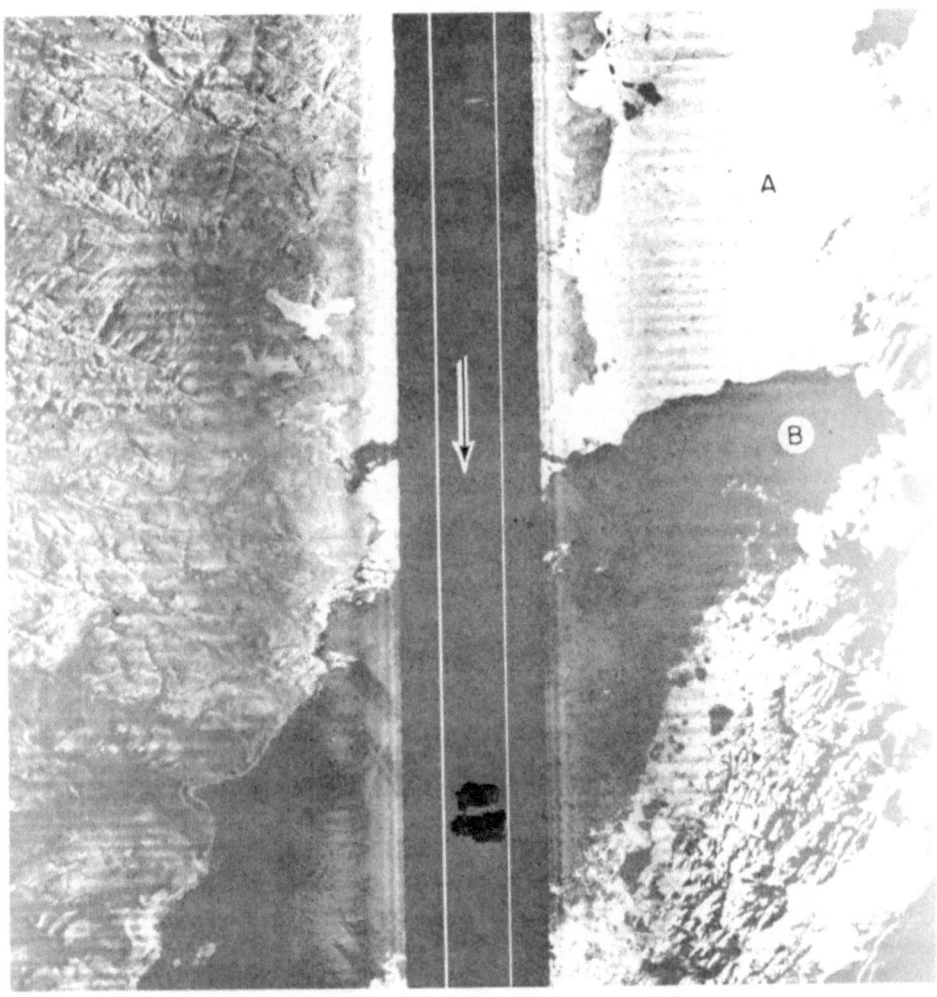

Figure 21. Pelly Bay (69° N, 90° W), April 8, 1974. Range, 25 km; altitude, 6000 ft (1830 m). Shows contrast of very rough and smooth fast ice.

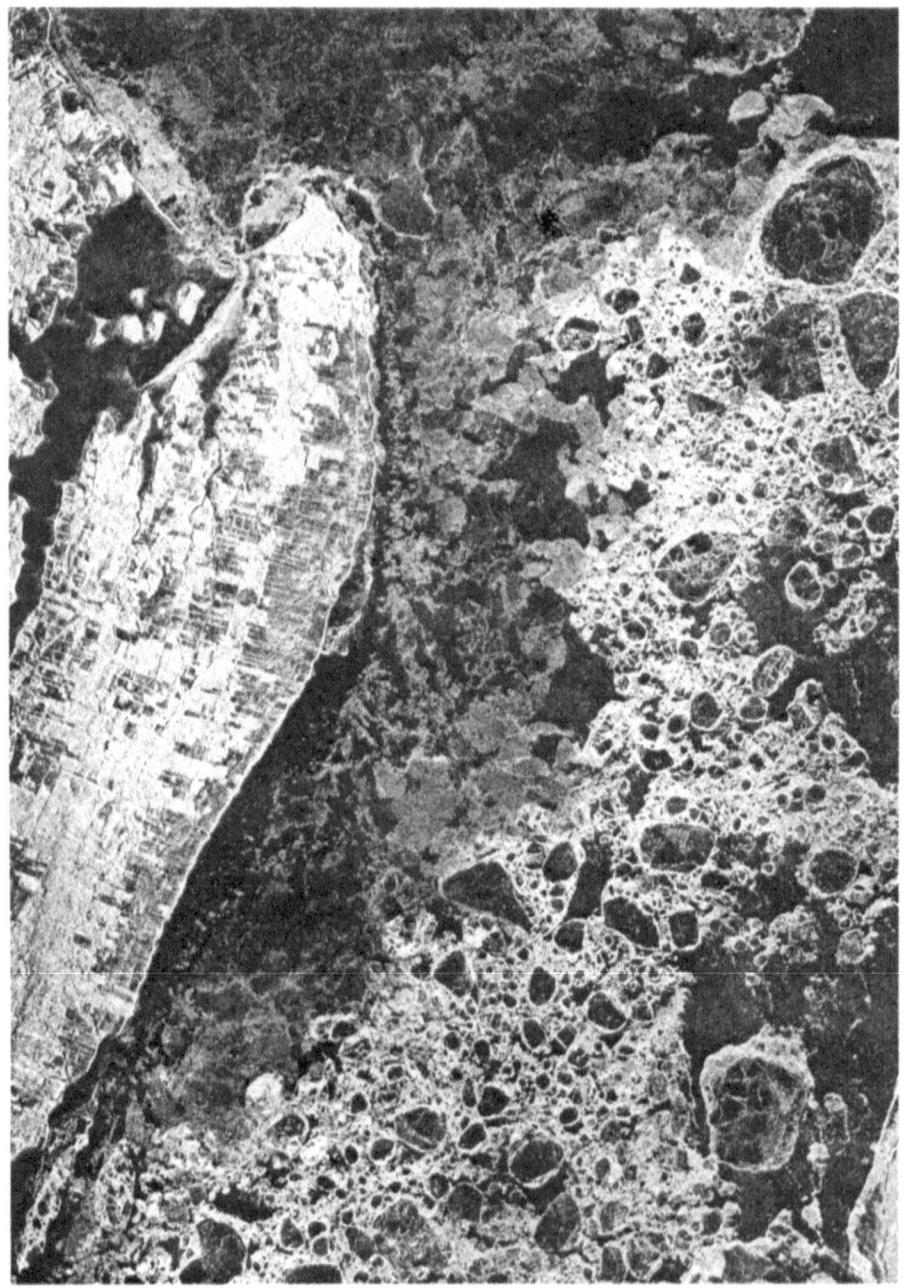

Figure 22.   Synthetic aperture SLAR image of the Gulf of St. Lawrence,
             March, 1974.   Altitude, 37,000 ft (11,399 m).   S.E. Prince
             Edward Island at left.

LABRADOR CURRENT PREDICTIVE MODEL

Ronald C. Kollmeyer

U. S. Coast Guard Academy

New London, Connecticut, U.S.A.

INTRODUCTION

Predictive numerical models of ocean current systems form an important facet to the study of the oceans. These models not only allow experimentation with oceanic physics by changing input parameters and observing the resultant effects on current patterns, but also provide a pathway towards the goal of long range oceanic predictions. Predictive meteorological models have attained a degree of success not yet realized by oceanic models. One reason for this is that input data, frequently updated, are available to the meteorologist. These real environmental input data make the difference between models formulated for parametric analysis and analytical solutions and those truly predictive models that can provide information on future changes in property distributions and current patterns. Prediction of the changes that occur in the property distributions and current patterns, generally referred to as circulation, have been long sought by oceanographers faced with interpretation of field data.

Prognostic models of ocean areas, responsive to short term baroclinic pressure changes due to property advection, mixing and local wind changes, have not been extensively explored. These models are the ocean equivalent to weather forecasting models and must incorporate the mass distribution driving of oceanic scale models and the shorter term responses to local wind patterns exhibited by embayment and coastal upwelling models. Models of this type have the requirements of synoptic initial data as well as the need for continuous information on all the open boundaries during the prediction period. It is the data acquisition problem that has caused this type of model to receive little attention.

The requirements of quasi-synoptic data, collected at great expense, limits the size of these models to regions of local interest for specialized purposes. The general model developed herein is specifically applied to exploring the possibility of making short term predictions of flow changes in the Labrador Current. The U.S. Coast Guard's International Ice Patrol is greatly interested in current prediction of the Labrador Current. Knowledge of flow pattern changes in the Labrador Current near the Grand Banks of Newfoundland would aid the Coast Guard in making predictions on future iceberg drift in the North Atlantic shipping lanes. At present, information on currents is obtained from basic geostrophic calculations using the method of dynamic heights. These steady state current determinations degrade in quality progressively after each quasi-synoptic ship survey is completed. A time dependent model using initial conditions provides hourly updated current information as local property advection, mixing and wind conditions affect the region.

MODEL FORMULATION

A flow prediction model is developed and applied to the Labrador Current as a four layer baroclinic, open boundary, computer simulation. It relies on initial temperature and salinity data gathered by hydrographic survey from a 110 x 110 km region centered at $44°22'N$, $49°36'W$, as shown in Fig. 1, and predicts advective and diffusive changes in these properties and computes the resulting current patterns with time. Wind stress is applied to the model as measured during the period of simulation. The initial survey data have been gathered from an area small enough to justify the assumption that it can be treated as synoptic information. These initial conditions are then used to compute current velocities from a diagnostic model which in turn is used in a predictive fashion as state properties are advected and mixed. The resulting flow field calculations are time dependent with transient wind fields and lateral boundary conditions updated as such information becomes available. Because only a small part of the total ocean is modeled, the vorticity techniques for determining layer velocities used by Bryan and Cox (1967), are not applicable. Instead, one assumes that the oceanic scale mean wind field and zonal insolation are imprinted on the distribution of mass, following Holland and Hirschman (1972) and relative velocities are then determined using a reference level of assumed zero motion.

A reference level at 1000m is assumed as a depth of zero water motion. This assumption is based on the findings of Scobie (1972) for the Grand Banks of Newfoundland area. Scobie found that by using the method of Defant (1961) a geostrophic level of zero motion existed between 600 and 1200m. This assumption is further

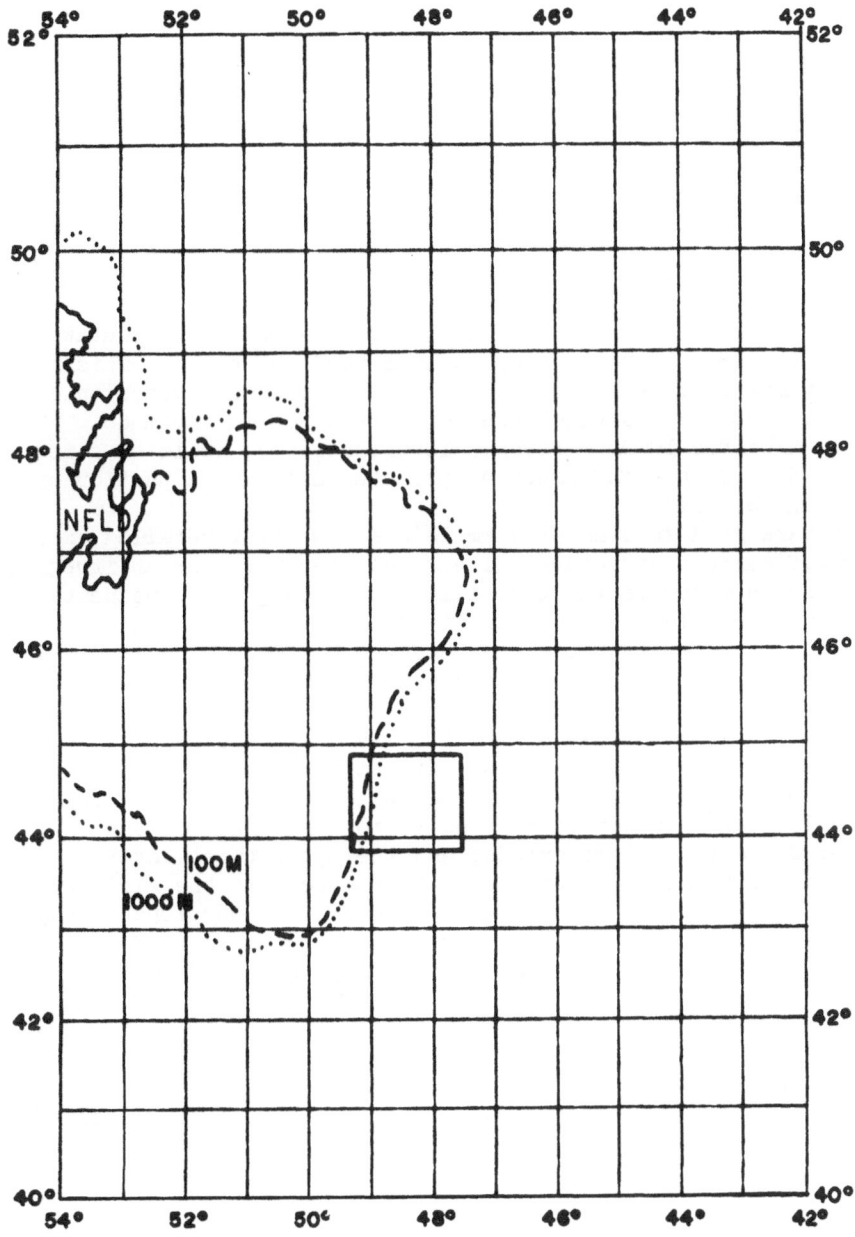

Figure 1.   Location of the area modeled. The 110 x 110 km^2 region
is located along the eastern edge of the Grand Banks
area of the continental shelf.

supported by Holland and Hirschman (1972) who found they could get
good agreement between a North Atlantic Ocean computer model and
geostrophic currents computed from a 1000m motionless depth.
Although geostrophy is not assumed in this model, one of the major
driving forces will be the horizontal pressure gradient which is
assumed to be zero at the reference depth. An abyssal layer below
the fourth layer is assumed motionless in the horizontal but
provides source water for vertical advection when it occurs due to
continuity requirements. While this motionless layer is difficult
to justify from first principles it is plausible as well as convenient for present purposes. Current measurements made in the
Labrador Current by Robe (1974) indicate near bottom oscillating
motion of tidal frequencies at depths of 900–1000m with southward
setting net currents ranging up to 10cm sec^{-1} at times. This type
of motion may be attributable to barotropic pressure waves entering
through the boundaries which are not considered in the model. The
net motion of the 1000m reference level, however, could be added
to the model if this information were available for a particular
simulation period.

A square grid spacing of 5 km is used to allow detailed examination of current structure. The model operates over an irregular
bottom and in water depths generally found along the continental
shelf. The four layers are shown in Fig. 2 with water thickness
of 70m, 130m, 300m and 500m. East-west cross sections of temperature and salinity distributions drawn from the initial survey
data obtained on Cruise I are also shown in Fig. 2. The Labrador
Current is roughly defined by the 0.0C isotherm and salinities of
less than 33.5 °/oo while the Gulf Stream is delineated by water
warmer than 5C and salinities greater than 34.5 °/oo. The upper
two model layers roughly correspond to the thickness of the
Labrador Current where it exists, and the upper three layers
encompass a substantial part of the Gulf Stream in the vicinity of
the Grand Banks area. As can be seen in Fig. 2, portions of the
model layers will have bottom incursions as the result of the
model's location along the continental slope off Newfoundland.

MODEL EQUATIONS

The model employs the following equations: the basic horizontal momentum equations,

$$\frac{Du}{Dt} - fv = -\frac{1}{\rho}\frac{\partial P}{\partial x} + \frac{1}{\rho}\frac{\partial \tau_x}{\partial z} + k_h \left(\frac{\partial^2 u}{\partial x^2} + \frac{\partial^2 u}{\partial y^2}\right) \tag{1}$$

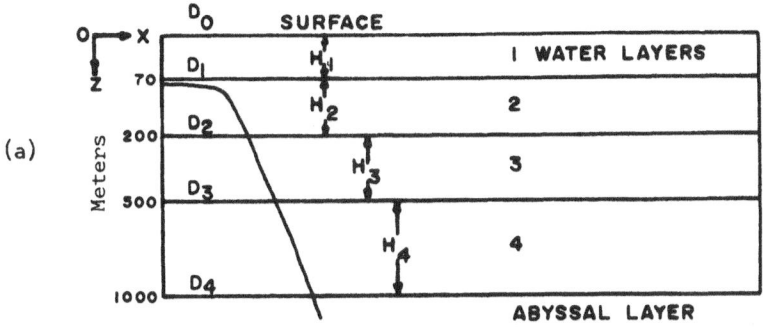

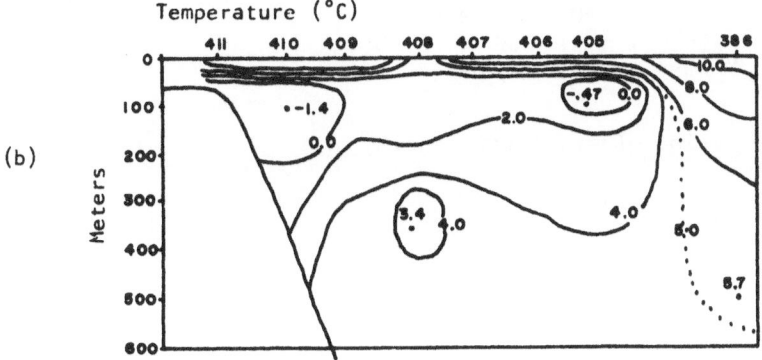

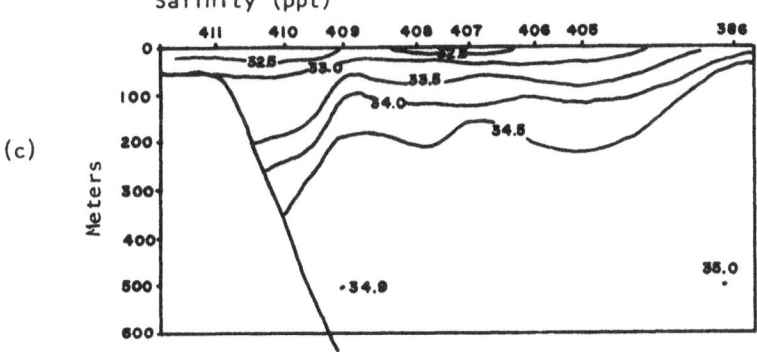

Figure 2.   Shown above as (a) is a typical layer cross section of
            the model including the incursions of bottom topography.
            (b) and (c) are east-west property sections obtained on
            Cruise I showing the general structure and location of
            the Labrador Current and the Gulf Stream.

$$\frac{Dv}{Dt} + fu = -\frac{1}{\rho}\frac{\partial P}{\partial y} + \frac{1}{\rho}\frac{\partial \tau_y}{\partial z} + k_h \left(\frac{\partial^2 v}{\partial x^2} + \frac{\partial^2 v}{\partial y^2}\right); \qquad (2)$$

the continuity equation for an incompressible fluid,

$$\frac{\partial u}{\partial x} + \frac{\partial v}{\partial y} + \frac{\partial w}{\partial z} = 0; \qquad (3)$$

the diffusion-advection equation for a state property, Θ,

$$\frac{D\Theta}{Dt} - A_h \left(\frac{\partial^2 \Theta}{\partial x^2} + \frac{\partial^2 \Theta}{\partial y^2}\right) - A_v \frac{\partial^2 \Theta}{\partial z^2} = 0; \qquad (4)$$

and the equation of state for density,

$$\rho = F(T, S, z). \qquad (5)$$

The symbols used in the above equations are:

    t = time
    u = x directed velocity
    v = y directed velocity
    w = z directed velocity
    ρ = density
    P = pressure
    f = coriolis parameter
    τ = shear stress in a vertical plane
    k_h = coefficient of horizontal momentum transfer
            (kinematic form)
    Θ = any state property
    A_h = coefficient of horizontal diffusion
            (kinematic form)
    A_v = coefficient of vertical diffusion
            (kinematic form)
    T = temperature
    S = salinity
    z = depth.

The coordinate system has its vertical origin at the surface with the positive z axis directed downward, x directed to the east and y to the north. The above equations are vertically integrated in each layer. Equations (1) and (2) are used as prediction equations for the horizontal velocities based on initial property distributions and time changing boundary conditions. Equation (3) provides for mass conservation through the determination of vertical velocities from horizontal divergence. Diffusion and advection of properties within and through the model boundaries will alter the distribution of density, which is predicted using (4), resulting in pressure gradient changes and computed current pattern variations.

The following assumptions and restrictions are used in the model formulation:

(a) Vertical averages of water properties and horizontal velocities are used within each layer;

(b) A rigid lid is imposed at the sea surface and no vertical motion of the layer boundaries is allowed;

(c) Communication of state properties between the layers derives from vertical advection. Diffusion between the layers is considered sufficiently small to neglect;

(d) Stress between the layers is proportional to the relative velocities;

(e) Hydrostatic balance is the basis for the description of the pressure field.

The vertical integration of the equations of motion is carried out as follows:

Referring to Fig. 2(a), for any layer, where i = 1, 4:

$$D_i = \sum_{i=1}^{4} H_i \, ,$$

where D is the distance from the z origin to the layer/bottom interfaces and H_i is the water layer thickness. Then for any layer when evaluated for the layer thickness H_i, the x-momentum equation integrates to:

$$\left(\frac{\partial u}{\partial t} + u \frac{\partial u}{\partial x} + v \frac{\partial u}{\partial y} - fv \right)_i =$$

$$\left(- \frac{P_x}{\rho H} - \frac{\tau x_{i-1}}{\rho H} + \frac{\tau x_{i+1}}{\rho H} + k_h \left(\frac{\partial^2 u}{\partial x^2} + \frac{\partial^2 u}{\partial y^2} \right) \right)_i \tag{6}$$

and similarly for the y-momentum equation:

$$\left(\frac{\partial v}{\partial t} + u \frac{\partial v}{\partial x} + \frac{\partial v}{\partial y} + fu\right)_i =$$

$$\left(-\frac{\mathbb{P}_y}{\rho H} - \frac{\tau_{y_{i-1}}}{\rho H} + \frac{\tau_{y_{i+1}}}{\rho H} + k_h \left(\frac{\partial^2 v}{\partial x^2} + \frac{\partial^2 v}{\partial y^2}\right)\right)_i \qquad (7)$$

where $\mathbb{P}_x$ and $\mathbb{P}_y$ are the vertically integrated pressure gradients in the x and y directions. The interlayer stress values are determined from $\tau_{x_{i-1}} = \rho\sigma(u_i - u_{i-1})$ and $\tau_{x_{i+1}} = \rho\sigma(u_{i+1} - u_i)$.

The value of σ is determined from an Ekman formulation of the form $\sigma = (.5k_v f)^{\frac{1}{2}}$, where k_v is the vertical coefficient of momentum transfer. The surface stress on the top layer τ_0 is formulated as a function of the wind velocity after Bodine (1971). In this formulation stress is determined from $\tau_0 = \rho' kW^2$, where W is the wind velocity, ρ' is the air density and $\rho' k = k_1$ for wind velocities $W<W_c$ (714 cm/sec), $\rho' k = K_1 + K_2 \left(1 - \frac{W_c}{W}\right)^1$ for wind velocities $W>W_c$, with constants $K_1 = 1.1 \times 10^{-6}$ and $K_2 = 2.5 \times 10^{-6}$.

The pressure terms $\mathbb{P}_x$, $\mathbb{P}_y$ are formulated from the vertically integrated z directed momentum equation. They are developed in such a manner that the vertically integrated horizontal pressure gradients form a driving force in each layer which is relative to the 1000m level of no motion. This requires that the horizontal pressure gradients at 1000m be zero. For convenience, a new pressure at any depth is defined by subtracting the total pressure P_T at 1000m, so that at 1000 meters the new pressure will be everywhere zero. With the assumption that air pressure on the surface is small, the new pressure at any depth z within layer i from 1 to 4 is formulated as:

$$P_i = g \sum_{k=1}^{k=i} [H_{k-1}\rho_{k-1} + (z-H_{k-1})\rho_k] - P_T ,$$

where $P_T = g \sum_{i=1}^{i=4} H_i \rho_i$.

Taking the horizontal derivative and vertically integrating over each layer, the following equations for the x direction are produced:

layer one, $\mathbb{P}_{x_1} = g \dfrac{H_1^2}{2} \dfrac{\partial \rho_1}{\partial x} - H_1 \dfrac{\partial P_T}{\partial x}$, (8)

layer two, $\mathbb{P}_{x_2} = gH_2 \left(H_1 \dfrac{\partial \rho_1}{\partial x} + \dfrac{H_2}{2} \dfrac{\partial \rho_2}{\partial x} \right) - H_2 \dfrac{\partial P_T}{\partial x}$, (9)

and layer three, $\mathbb{P}_{x_3} = gH_3 \left(H_1 \dfrac{\partial \rho_1}{\partial x} + H_2 \dfrac{\partial \rho_2}{\partial x} + \dfrac{H_3}{2} \dfrac{\partial \rho_3}{\partial x} \right) - H_3 \dfrac{\partial P_T}{\partial x}$, (10)

and layer four,

$$\mathbb{P}_{x_4} = gH_4 \left(H_i \dfrac{\partial \rho_1}{\partial x} + H_2 \dfrac{\partial \rho_2}{\partial x} + H_3 \dfrac{\partial \rho_3}{\partial x} + \dfrac{H_4}{2} \dfrac{\partial \rho_4}{\partial x} \right) - H_4 \dfrac{\partial \rho T}{\partial x} ,$$ (11)

with similar results for the y direction. In shallow water all pressure driving forces will be relative to the shelf bottom. In this manner only those pressure forces arising directly as the result of horizontal density gradients in the water column will produce motion. This technique is patterned after the pressure assumptions used for dynamic height calculations in shallow water.

Mass continuity is satisfied by allowing leakage of excess mass to occur between the layers. The sea surface in the model is a solid boundary; thus any horizontal divergence in layer one must be balanced by mass flux through its common boundary with layer two. Layer two, having its upper boundary supplying or receiving mass from layer one, may in turn exchange mass continuity through its lower boundary. Similar remarks apply to layer three. The lower boundary of layer four provides for all net mass flux requirements in the water column.

The continuity equation for incompressible flow, (3) is vertically integrated for each layer in the following manner:

$$0 = \int_{D_{i-1}}^{D_i} \left(\dfrac{\partial u}{\partial x} + \dfrac{\partial v}{\partial y} + \dfrac{\partial w}{\partial z} \right)_i dz =$$

$$\left(\cdot \left[z \left(\dfrac{\partial u}{\partial x} + \dfrac{\partial v}{\partial y} \right) + W_z \right] \int_{D_{i-1}}^{D_i} \right)_i ,$$

and using layer thickness H_i ,

$$W_{D_i} = - H_i \left(\dfrac{\partial u}{\partial x} + \dfrac{\partial v}{\partial y} \right)_i + W_{D_{i-1}}$$ (12)

where W_{D_i} is the vertical velocity at the bottom of any layer.

Problems arise when a solid bottom exists under the water column in depths of less than 1000m. The excess mass cannot then be carried off through leakage of the fourth layer.

To accomplish the reduction of the excess mass flux in those areas with a solid bottom, horizontal velocity adjustments are used to satisfy overall horizontal continuity of the water column, i.e., no horizontal divergence of the total water column is permitted. If more than one layer is involved, then vertical flow adjustments are allowed within the column to satisfy any horizontal divergence requirements of the individual layers. This is accomplished by computing the excess flow volume within any water column bounded by the sea surface and a solid bottom. These computations use the newly determined horizontal velocity values of (6) and (7). The excess flow volume is then reduced to zero by small arbitrary alterations of the horizontal velocities. The u and v velocity components share equally in the adjustment and in proportion to the layer thickness. Using this procedure, all mass flux is accounted for by velocity adjustments through the northern, eastern and abyssal boundaries of the model with the continental slope to the west forming a barrier on the bottom and western side of the water column. Mass flux between layers is then determined from (12) for an overall water column which is now nondivergent.

The flux of the state properties of temperature and salinity is controlled by (4) which is integrated as follows for all layers:

$$\int_{D_{i-1}}^{D_i} \left(\frac{\partial \Theta}{\partial t} + u \frac{\partial \Theta}{\partial y} + v \frac{\partial \Theta}{\partial y} - A_h \left[\frac{\partial^2 \Theta}{\partial x^2} + \frac{\partial^2 \Theta}{\partial y^2} \right] \right)_i dz = 0,$$

then

$$\left(\frac{\partial \Theta}{\partial t} + u \frac{\partial \Theta}{\partial x} + v \frac{\partial \Theta}{\partial y} - A_h \left[\frac{\partial^2 \Theta}{\partial x^2} + \frac{\partial^2 \Theta}{\partial y^2} \right] \right)_i = 0. \tag{13}$$

Vertical property advection occurs due to the existence of vertical velocities and large property differences between the layers. Vertical property advection is approximated over each time interval, after horizontal advection calculations have been completed as follows:

$$\Theta_i = \frac{\Delta t}{H_i} \left(\Theta_i (W_{D_i} - W_{D_{i-1}}) - W_{D_i} \left(\frac{\Theta_i + \Theta_{i+1}}{2} \right) + W_{D_{i-1}} \left(\frac{\Theta_{i-1} + \Theta_i}{2} \right) \right). \tag{14}$$

In this formulation the average property value θ_j is affected by
vertical property flux into the layer during each time period.
The amount of change is governed by the vertical velocities and
the averaged property values of the two layers.

Equation (5) is formulated after Friedrich and Levitus (1972)
and has the form:

$$\rho = 1.0 + (C_1 + C_2 T + C_3 S + C_4 T^2 + C_5 ST + C_6 T^3 + C_7 ST^2) \times 10^{-3}. \qquad (15)$$

The value of C_n provides the depth dependency for the density and
is computed from tables determined by Friedrich and Levitus (1972).
This formulation is similar to the equation developed and tabulated by Knudsen (1901) and more recently tabulated by LaFond
(1951). Bradshaw and Schleicher (1970) have reaffirmed the
validity of the original work by Knudsen. The Friedrich and
Levitus formulation is for sea water of a temperature range
between -2C to 30C and a salinity of 30 ^{o}/oo to 38 ^{o}/oo. This is
within the range encountered in the Labrador Current area. The
accuracy of this formulation is better than ±0.02 in sigma units
where sigma = $(\rho - 1)10^{-3}$.

NUMERICAL METHODS

Equations (6) through (13) are put into finite difference
form and solved for each of the four layers of the model. Centered
difference forms are used both in space and time; a double time
step is required for the horizontal velocity predictions. The
lateral friction terms of (6) and (7) are treated using the
Dufort-Frankel scheme after Richmeyer and Morton (1967) which is
unconditionally stable. The pressure equations (8) through (11)
as well as the vertical velocity equation, (12), are solved using
a centered difference scheme in space only, since no time derivatives appear.

Equation (13) is divided into diffusion and advection parts.
Diffusion is determined using a one-sided advanced time numerical
scheme and is combined with a one-sided advanced time horizontal
advection scheme patterned after Crowley (1968) and a simple
vertical advection formulation of (14). The horizontal portion
of the advection scheme uses a polynomial curve fitted to five
points in each of the x and y directions and was shown by Crowley
to produce extremely accurate property advection.

The model is initialized and spun up from rest for 100 hours
(100 time steps). Pressure is initially computed from the data
using (8) through (11) and the equation of state, (15). The
pressure gradients thus computed are applied in (6) and (7) in
increments of 10^{-2} grams sec^{-2} cm^{-2} until the full pressure
gradient values are reached. This technique prevents a large
gradient shock that can cause the velocity computations to become

unstable. Equations (6) and (7) are solved at all grid points
sequentially for each layer, through 100 time steps, after which
the velocities are fully developed. From this point on, vertical
velocities are determined, water properties are diffused and
advected, new densities and pressure gradients computed and
velocities predicted for each time step.

A time step of one hour was selected as the maximum time
period of incremental prediction in order to keep the velocity
and advection computation schemes stable. With velocities on the
order of 50 cm/sec, this one hour time step meets the stability
criterion of $\Delta t \leq \Delta s/u$ for a space centered difference scheme
where Δs is the grid spacing. This time step also satisfies the
stability criterion for the property diffusion equation of
$\Delta t \leq (\Delta s)^2/4A_h$ where A_h is the horizontal diffusivity coefficient.

INITIAL AND BOUNDARY CONDITIONS

Measured temperature and salinity distributions, averaged over
each layer thickness, as well as wind stress are used as initial
data to commence the model's operation. The boundary conditions
are specified during the operation of the model. Properties on
boundaries where current is outflowing are allowed to change as
advection from upstream dictates. Boundary regions of inflowing
current have their state property values changed in accordance
with conditions as they are known on the boundaries. In the case
of this particular model these inflowing current boundary conditions are changed in a predetermined manner based on the results
of a second survey.

The computational matrix consists of a square grid, 24 points
on each side. The outside grid points of this square form the
boundaries of the model. A second set of boundary points, outside
of the computational matrix, are required to allow the solution
of the model equations on the inner boundaries. The total storage
matrix is thus 25 x 25. The use of this fictitious row of grid
points is a relatively standard technique which allows use of the
most accurate 5 points advection scheme, Crowley (1968), right up
to the inner boundaries and then a less accurate 3 point scheme on
the inner boundaries themselves. Furthermore, it provides
velocity values just outside the inner boundaries which are
important in the advective and lateral friction terms of (6) and
(7). These outer boundaries are set equal to the inner boundaries
at each time step. Solid boundaries along the shelf represent
grid points where all velocities are zero. Property gradients at
grid points immediately adjacent to the shelf boundaries are
computed on one side only using the next seaward data points. A
flow diagram of the computational sequency scheme is shown in
Fig. 3.

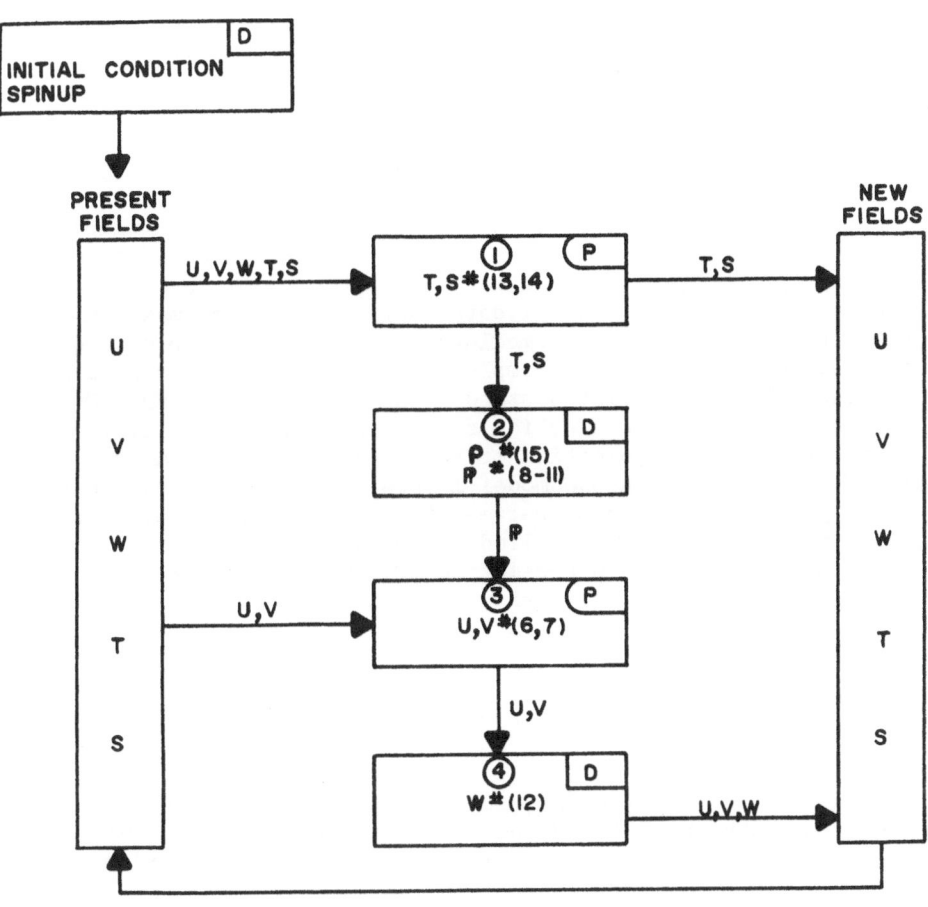

Figure 3. Flow diagram of computational sequence over each time step.

DATA COLLECTION AND PROCESSING

The data used in the model were collected during the period
26 June to 7 July 1973 by the Coast Guard Cutter EVERGREEN (WAGO
295). Two separate cruises were conducted in the survey area
shown in Fig. 1. Cruise 1 made 43 station observations in 82 hours
and Cruise II, 41 station observations in 75 hours. The interval
between cruises was 188 hours. An average one thousand meter
station cast was composed of about 2400 data levels of depth,
temperature and salinity recorded on magnetic tape. These data
were subsequently reduced in volume by using arbitrary sampling
depths. This resulted in data points every 5m from the surface
down to a depth of 75m, every 25m to a depth of 300m, every 50m
to a depth of 1000m and lastly, data every 100m to the maximum
depth sampled of 1200m. Intermediate depths were also provided
in the reduced data whenever sampled data departed from a straight
line between the above listed standard depths by more than 0.04C
for temperature and 0.06 ^{o}/oo for salinity. Model usage of the
data required the determination of vertical temperature and
salinity averages within the model layers. A short program which
vertically averaged all data between the desired model levels of
70, 200, 500 and 1000m was applied to the data.

A computer grid system consisting of a 25 x 25 point matrix
was used in the model. This grid required data at each of 625
points of the matrix. To obtain this information, the averaged
layer data for each cruise was plotted as to location on a small
area plotting sheet and hand contoured. Based on the results of
the contouring, a data point for each required grid point was then
retrieved for direct use in the model for initilization and
verification.

Weather data, specifically winds, were obtained for model use
from both measurements aboard the survey vessel and the daily
weather maps of the region as published by the U.S. Navy Fleet
Numerical Weather Facility in Norfolk, Virginia.

Bathymetry for the modeled area was obtained from Chart #802
published by the Canadian Hydrographic Service. Intersection of
the model layers with the continental shelf was determined from
the intersection of the slope depths and the top of each layer.
In this manner, the irregular shape of the western boundary of
each layer was established. The continental shelf was taken to
be flat and to form the bottom boundary of the top layer. A three
dimensional view of the model layers as well as a schematic
representation of the Labrador Current and the Gulf Stream as
they exist in the region are shown in Fig. 4.

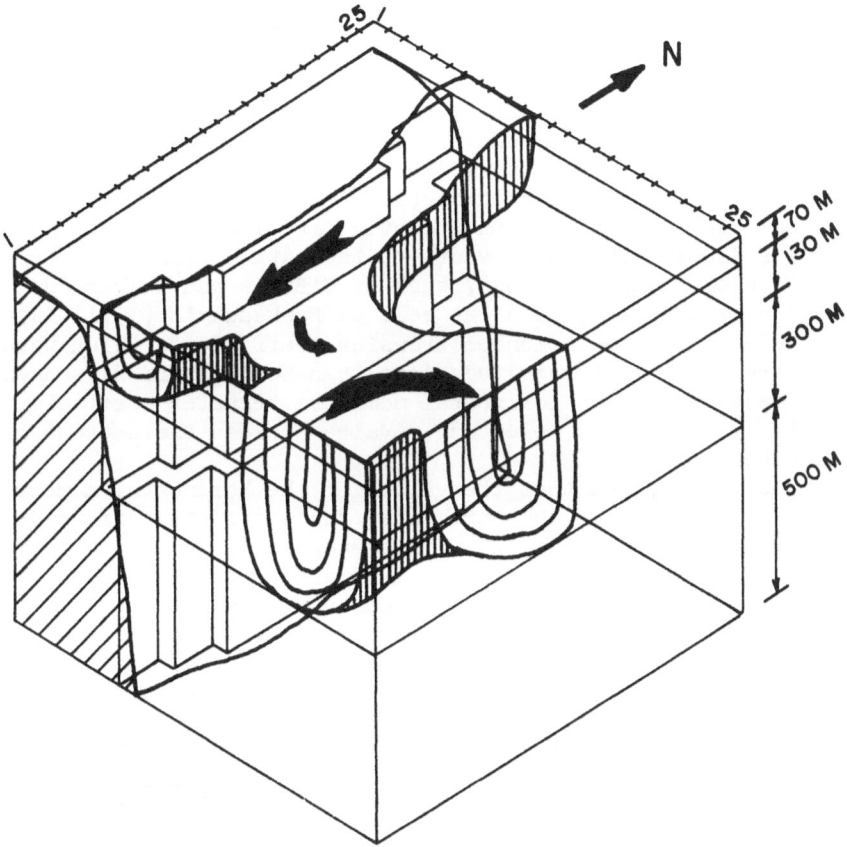

Figure 4.   A schematic representation of the modeled area showing
the four layers, the continental shelf and slope, and
the Labrador/Gulf Stream flow pattern.

TRANSFER COEFFICIENTS

Ocean model formulations for momentum and state property transfer are usually simple numerical constructions. Poor understanding of the physics of momentum and state property transfer phenomena has resulted in empirical constructions based on Fickian diffusion which are used to simulate dissipation and dispersion. These constructions require coefficients for their operation and the coefficients are difficult to estimate. Values for the coefficients of horizontal and vertical momentum transfer, k_h and k_v respectively, as used by various numerical models are shown in Table 1. Holland and Hirschman (1972) used values as high as 4×10^8 cm^2 sec^{-1} for the North Atlantic circulation. Their grid spacing was 110 km, the scale width of the entire Labrador Current model. Models of smaller areas, such as O'Brien and Hulburt's (1972) coastal upwelling model as well as Paskausky's (1971) Lake Ontario model, employed momentum transfer coefficients which were two and three orders of magnitude less than Holland and Hirschman. For the present model, a horizontal momentum transfer coefficient of 1×10^7 cm^2 sec^{-1} was selected, a value between the extremes used by other modelers.

It is often assumed that the values of the horizontal momentum transfer coefficient and the horizontal state property diffusion coefficient (A_h) are the same. This assumption is based on the concept that the transfer mechanism involves turbulent eddies whose size or mixing length is represented by the magnitude of the coefficients. Bryan and Cox (1967) used values of k_h and A_h which differed by one order of magnitude in order to obtain satisfactory results. They showed that values of A_h which approached k_h tend to limit the intensity of the flow by rapidly reducing state property gradients and thus reducing the flow dynamics. A value for A_h of 5×10^6 cm^2 sec^{-1} was used initially in the present model. This value is one-half an order of magnitude less than k_h.

Nonisotropic turbulence is generally assumed in the oceans. The values of transfer coefficients are assumed to differ greatly between the horizontal and the vertical, since stability of the water column greatly restricts the amount of vertical mixing compared with the horizontal. The formulation for the vertical friction coefficient (σ) is based on the commonly used Ekman boundary layer approximation as previously discussed. This approximation requires a value for the vertical momentum transfer coefficient k_v. A wide range of k_v values have been used in various models that employed vertical momentum transfer either between water layers or with the bottom. Table 1 lists values of k_v that have been successfully used in the past by other researchers. Holland and Hirschman (1972) find that the value of the vertical coefficient turns out to be unimportant, and that the horizontal coefficient of momentum transfer is the main

Table 1a. Values for the coefficient of horizontal momentum transfer as used by various numerical models (k_h).

Holland and Hirschman, 1972	North Atlantic Ocean	4×10^8 cm^2/sec
O'Brien and Hurlburt, 1972	Coastal Upwelling	1×10^6 cm^2/sec
Paskausky, 1971	Lake Ontario	1×10^5 cm^2/sec
Stommel, 1950 (determined from Gulf Stream data)	Gulf Stream	2.3×10^8 cm^2/sec

Table 1b. Values for the coefficient of vertical momentum transfer as used by various numerical models (k_v).

Arons and Stommel, 1956	Oceanic, bottom	1.0 cm^2/sec
Charney, 1955	Oceanic, bottom	15.0 cm^2/sec
Holland and Hirschman, 1972	North Atlantic, interlayer	1.0 cm^2/sec
Paskausky, 1971	Lake Ontario, bottom	22.5 cm^2/sec
Wert, 1970	Gulf of Mexico, interlayer bottom	$.03$ cm^2/sec 81.2 cm^2/sec

system balance term. In the present model, it was found that this
was also the case. Critical k_v values for interlayer boundaries
of 2.0 cm^2 sec^{-1} and bottom boundaries of 18.0 cm^2 sec^{-1} were used
in the present model with some variations. These values are within
the ranges given in Table 1.

RESULTS

The accuracy of the model predictions are measured by comparing
predicted property distributions with those observed eight days
after the initial data were obtained. State property distributions
are selected for model performance evaluation because they reflect
the Lagrangian history of water movement during the prediction
period. Water velocity comparisons would show only instantaneous
results at the end of the prediction period. The comparison is
accomplished by utilizing the state property distributions
obtained on Cruise I as initial conditions, and comparing predicted
results with the property distributions obtained on Cruise II. The
paths of water motion, as well as the mixing that occurs during
the prediction interval should result in a state property distribution similar to that found on Cruise II if the model is operating
usefully as a predictive tool. Fig. 5 indicates the path taken by
the Labrador Current and the Gulf Stream during a typical computer
prediction period of 188 hours. A numerical "dye" source was
placed in the Labrador Current and the Gulf Stream as they entered
through the boundaries. The "dye" was then advected by these
currents, showing the patterns of their flow during the 188 hour
prediction interval. The upper 200m are the most dynamic in terms
of velocities and definitive property gradients. These 200m would
include both layers 1 and 2. Layer 1, however, will be affected
by surface conditions such as insolation and evaporation which are
not included in the model. For this reason, layer 2 will be used
primarily for comparison purposes between predicted and surveyed
conditions. Fig. 6 shows distributions of temperature and
salinity for all layers obtained from Cruise I. Fig. 7 shows the
same for Cruise II. The incursion of the continental shelf is
indicated along the western boundary of layers 2, 3, and 4. The
Labrador Current is easily observed in the data for Cruise I for
the upper two layers. Layer 1 shows the north-south tending
current generally delineated by the 1C contour. The Gulf Stream
is seen in the southeast corner flowing to the northeast as water
greater than 3C. The salinity distribution shows the characteristic low salinity of the Labrador Current generally to the west
of the 33.0 ^{o}/oo contour and the higher salinity Gulf Stream to
the east of the 33.4 ^{o}/oo contour. In layer 2, the Labrador
Current manifests itself in negative temperature values and shows
an eastward tending filament which is also reflected in the
salinity distribution. Again, the Gulf Stream in the southeast

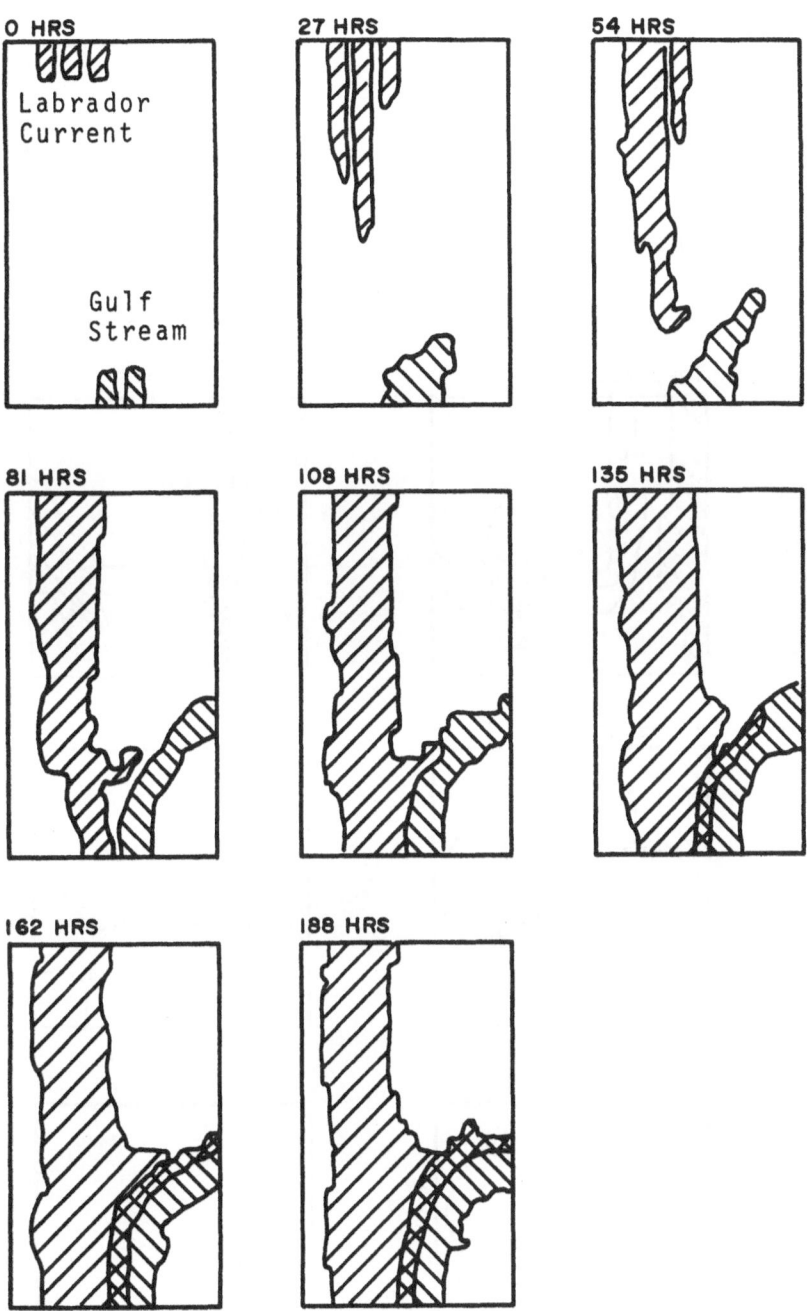

Figure 5. The paths taken by the Labrador Current and the Gulf Stream over a 188 hour operation of the model. Overlapping shaded area is a mixing region.

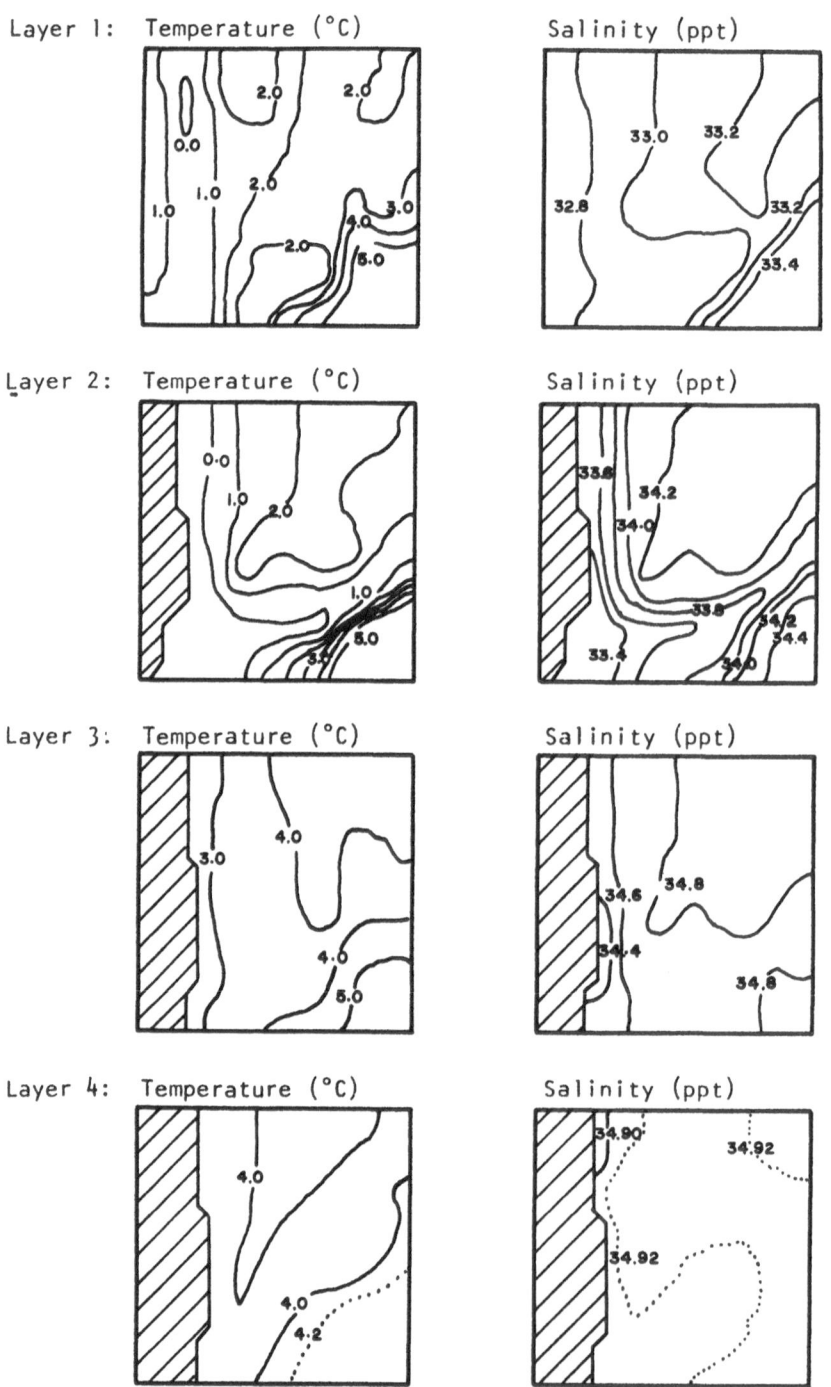

Figure 6. State property distributions for all layers obtained on
           Cruise I. These represent the initial conditions as
           measured.

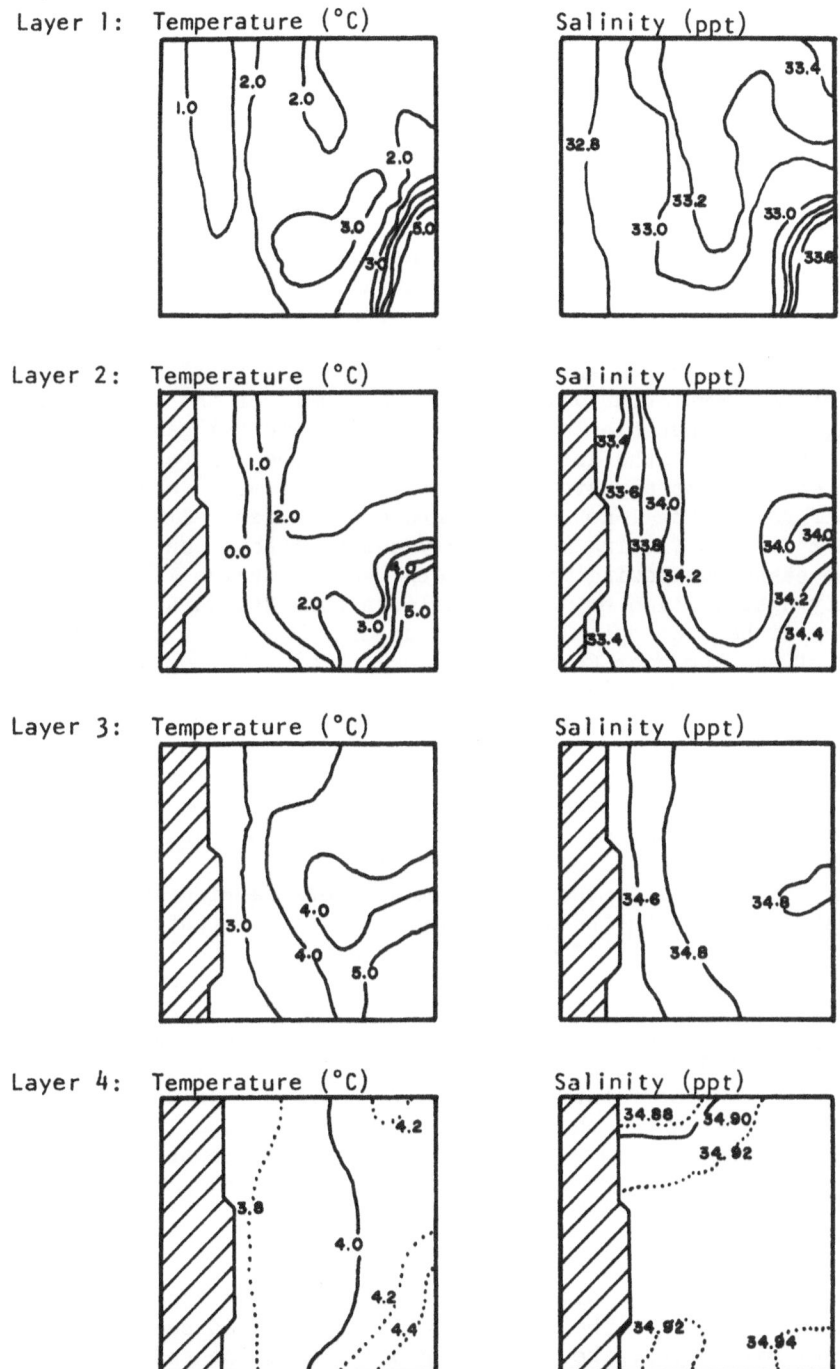

Figure 7.   State property distributions for all layers obtained on
Cruise II. Measured conditions 188 hours after initial
conditions of Fig. 6.

corner is warm, greater than 3C with a maximum temperature of 7.5C, and a salinity greater than 34.0 °/oo. Layer 3 contains warmer water along with higher salinities than found in the upper two layers. Typically, the coldest water and lowest salinities are found along the continental slope while the warmer and slightly more saline Gulf Stream can be observed in the southeast corner. Layer 4 for both cruises is characterized by a general lack of significant definition. The contour intervals of Fig. 6 and 7 for layer 4 were reduced by an order of magnitude in order to show any variation. The salinity contour interval borders on the precision of salinity determination and is therefore not interpreted here. The temperatures are almost uniformly 4.0C.

Cruise II, layer 1 shows a definite warming trend exists on the surface, particularly in the southern half of the area. This warming is attributable to solar heating due to the general lack of cloud cover during the intersurvey period. The general arrangement of current systems appears fundamentally the same as observed 8 days before. Layer 2 now shows a broader and generally colder Labrador Current flowing more nearly south without the tendency for the eastward flowing filament observed on Cruise I. The salinity distribution no longer reflects this eastward flow tendency either, except along the southern boundary. The temperatures of the Labrador Current are less than 0.0C from north to south. Layer 3 property distributions differ little from those found on Cruise I. The 4.0C isotherm in the northern half appears to have joined with its counterpart in the south thus isolating a slightly cooler water mass in the eastern central region. A similar occurrence is observed for the 34.8 °/oo isohaline. Slightly colder water appears in layer 4 along the continental shelf than was observed during Cruise I.

Many operations of the primary model were carried out using various values for the transfer and mixing coefficients in an attempt to simulate those conditions found on Cruise II. Two of these model operations, over the simulated 188 hour period, used transfer coefficients with the values of $k_h = 1.0 \times 10^7$ cm^2 sec^{-1} and $k_v = 2.0$ cm^2 sec^{-1} for both simulations; however, A_h had the value of 5×10^6 cm^2 sec^{-1} for one operation and 5×10^2 cm^2 sec^{-1} for the other. The first value used for A_h was considered to be a best estimate, while the second effectively caused little diffusion. The winds during the entire period were then constant at 765 cm sec^{-1} from 225 degrees true. The temperature and salinity conditions on the lateral boundaries were allowed to change with time linearly during the simulation period from those observed on Cruise I to those of Cruise II, 188 hours later.

The results of the simulation, described above, for layer 2 are shown in Fig. 8. The high value of A_h produced such a high degree of mixing that the property gradients at the end of the period are greatly reduced, particularly in the center of the model region, and the results compare poorly with Cruise II

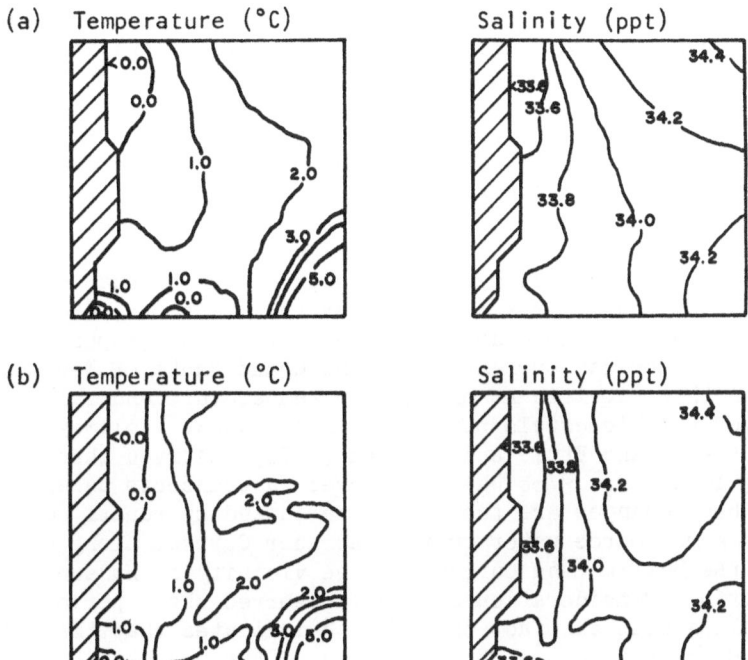

Figure 8. State property distributions for layer 2, predicted after 188 hours using the transfer coefficients $k_h = 1.0 \times 10^7$ cm^2 sec^{-1} and $k_v = 2.0$ cm^2 sec^{-1}. Boundary values varied linearly over 188 hours. A_h was varied with $A_h = 5 \times 10^6$ cm^2 sec^{-1} for (a) above and $A_h = 5 \times 10^2$ cm^2 sec^{-1} for (b).

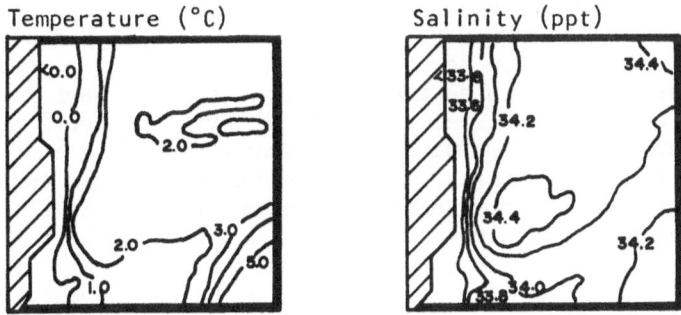

Figure 9. State property distributions for layer 2, predicted after 188 hours with all state property boundary values changing in 72 hours (heavy line) to those values found on Cruise II. The transfer coefficients used were $k_h = 1.0 \times 10^7$ cm^2 sec^{-1}, $k_v = 2.0$ cm^2 sec^{-1}, and $A_h = 5 \times 10^2$ cm^2 sec^{-1}.

(Fig. 7). The low value of A_h produced results which retained the pronounced property gradients by allowing only insignificant mixing. These results compare more favorably with those of Cruise II, particularly for the eastern half of the modeled area. The low temperature water representing the Labrador Current along the continental shelf is not shown as a contiguous body from north to south and the negative degree values entering at the northern boundary have traveled south only one-third of the length of the area.

Information on the changes that occur on the boundaries was not available, thus linear changes over the 188 hours were used for the simulations described above. Repetitive oceanographic transects of the Labrador Current along the Grand Banks of Newfoundland have shown that significant changes in temperature and salinity occurs over time intervals of less than one week. Kollmeyer et al., (1966) and Ettle and Wolford (1972) observed changes averaging 1C and 0.4 °/oo which occurred in transects taken 5 days apart. These changes are thought to be caused by runoff conditions supplying some source water to the Labrador Current along with mixing with the Baffin Land Current in the vicinity of Hudson Strait to the north. Consideration of these observed short period changes along with the fact that Fig. 8 indicates that the coldest Labrador Current water may have arrived at the northern boundary earlier in the prediction interval, suggests that the boundaries should be changed more rapidly than the 188 hours used above. To accomplish this, it was noted that a change of approximately 0.25C and 0.2 °/oo occurred in layer 2 at the boundary entrance location of the Labrador Current between Cruises I and II. These changes are less than those observed to occur over a 5 day period by Ettle and Wolford (1972). Thus, using a similar rate of change for test purposes, model boundaries were changed linearly over 72 hours, simulating a more rapid boundary change event. Figure 9 shows the results of this simulation using a low A_h value of 5×10^2 cm2 sec^{-1} to retain more definition of state property gradients. The results show the low temperature water entered the northern boundary early enough to be transported to the southern boundary during the simulation period, but the property gradients produced excessive advective velocities particularly along the eastern margins of the low temperature Labrador Current. This caused the warmer water from the northeastern boundary to move in too rapidly, pinching the Labrador Current onto the continental shelf.

Figure 10 is the result of a simulation which uses a value of 5×10^4 cm2 sec^{-1} for A_h. This is considered a compromise value between the excessive smoothing tendency and the lack of significant mixing as shown in Fig. 8. In addition, only the low temperature and low salinity of the Labrador Current entering at the western edge of the northern boundary is allowed to complete its boundary change in 72 hours. All other boundaries change linearly over the entire 188 hour simulation period. Figure 10,

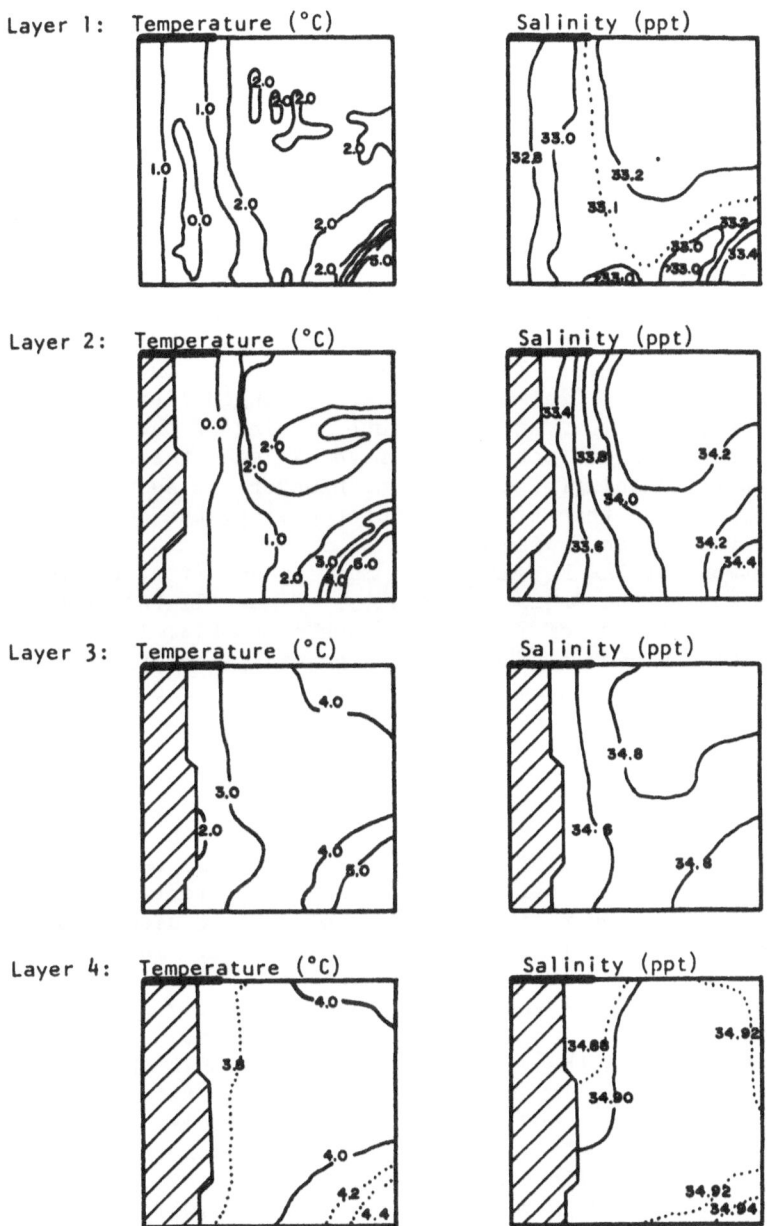

Figure 10.   State property distributions for all layers, predicted
             after 188 hours with boundary values changing in 188
             hours except at the northern boundary of the Labrador
             Current (heavy line) where change occurs in 72 hours.
             Transfer coefficients used were k_h = 1.0 x 10^7 cm^2 sec^{-1},
             k_v = 2.0 cm^2 sec^{-1}, and A_h = 5 x 10^4 cm^2 sec^{-1}.

layer 2, shows that the low temperature Labrador Current successfully transits the length of the continental shelf, and the 2C
isotherm closely simulates its position as found during Cruise II
(Fig. 7). In addition, a large area of water with temperatures
between 1 and 2C is shown to connect the north, south and east
central boundaries as found during Cruise II. The 34.0 °/oo
salinity line lies close to the position of a similar isohaline of
Cruise II while the 34.2 °/oo contour has not pushed south quite
as far. Layer 1 also compares favorably with Cruise II if the
effects of solar heating are ignored. The 0, 1 and 2C isotherm
shows reasonable agreement with that observed but the salinity is
slightly elevated above that of Cruise II. The predicted position
of the 33.1 °/oo contour lies closer to the position of the
33.0 °/oo contour of Cruise II. Layer 3 changed little from that
observed on Cruise I. The north-south joining of the 4.0C and
34.8 °/oo contours did not occur. Layer 4 lacks the contour
definition for meaningful comparison; however, it should be noted
that water of less than 3.8C is predicted along the continental
slope as observed on Cruise II.

Variations of the k_h value were made in an effort to more
closely simulate the Cruise II results. Figure 11 (a) shows
layers 1 and 2 for k_h of 7×10^6 cm^2 sec^{-1} and Fig. 11 (b) a k_h
of 2.5×10^7 cm^2 sec^{-1}. The lower value of k_h results in slightly
faster velocities. Figure 11 (a) and (b), layer 2 differ mainly
in the shape of the 1C isotherm. The lower k_h value causes the
colder water of the Labrador Current to push strongly eastward
between the 2C isotherms defining the waters of the eastern,
northern and southern quadrants. The higher k_h value produces a
better representation of the 1 degree isotherm when compared to
Cruise II, Fig. 7; however, the cold waters of the Labrador Current
along the continental shelf appear reduced in extent. Layers 1 and
2 of Fig. 11 (a) compare just as favorably with Cruise II as does
Fig. 10 where a k_h of 1×10^7 cm^2 sec^{-1} was used. The main difference between these two results is that the velocities for the
k_h of 1×10^7 cm^2 sec^{-1} are faster and more closely aligned with
those determined from the Cruise II property data.

Experiments with the value of k_v indicated that the model's
response is basically immune to this transfer coefficient. Variations between .05 cm^2 sec^{-1}, nearly in the molecular viscosity
range, up to 6 cm^2 sec^{-1} were tested as interlayer friction
coefficients with little discernible difference in the results.
This is the same finding made by Bryan and Cox (1967) and by
Holland and Hirschman (1972).

All the terms in the momentum equations, (6) and (7), were
non-dimensionalized by division with the Coriolis term. These
terms were then computed for inter-comparison throughout the model.
This was done at the end of the 188 hour prediction period corresponding to the results shown in Fig. 10. Layer 2 was used for
this analysis and the results are shown in Fig. 12. Since all of

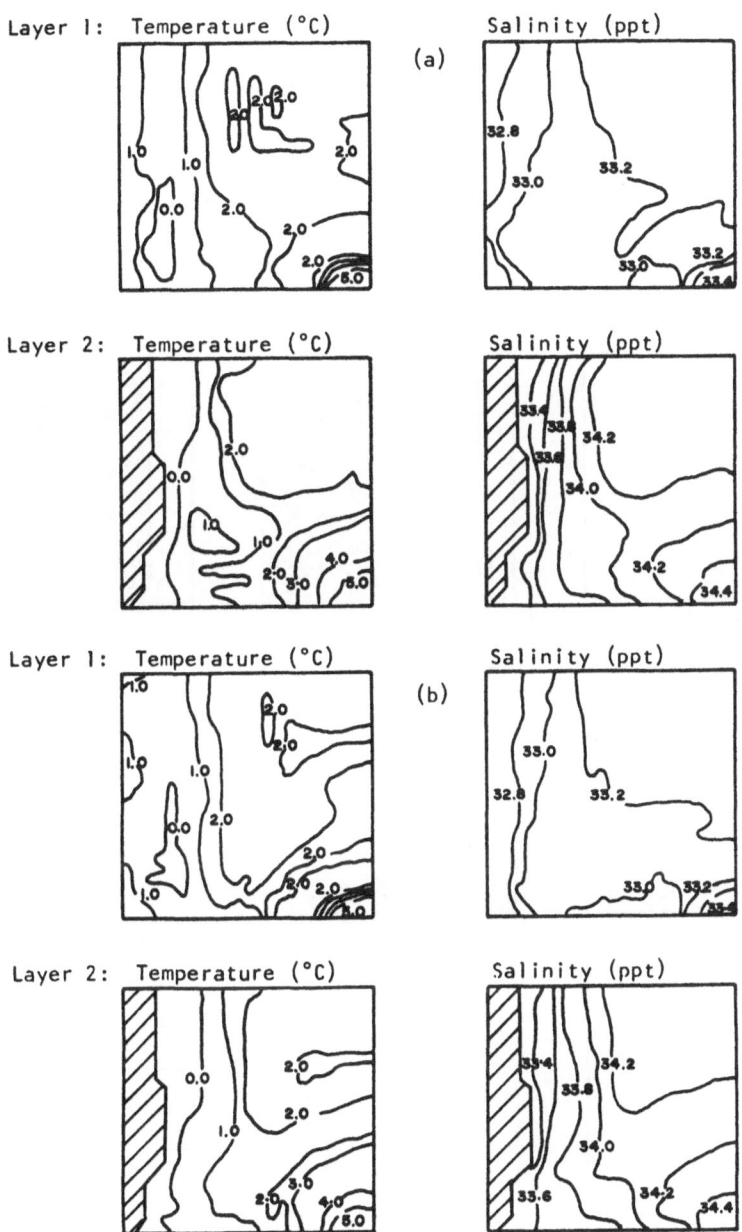

Figure 11. State property distributions for layers 1 and 2, predicted after 188 hours. These results form variations
from Fig. 10 by using a k_h of 7×10^6 cm^2 sec^{-1} shown
in group (a) and k_h of 2.5×10^7 cm^2 sec^{-1} shown in
group (b).

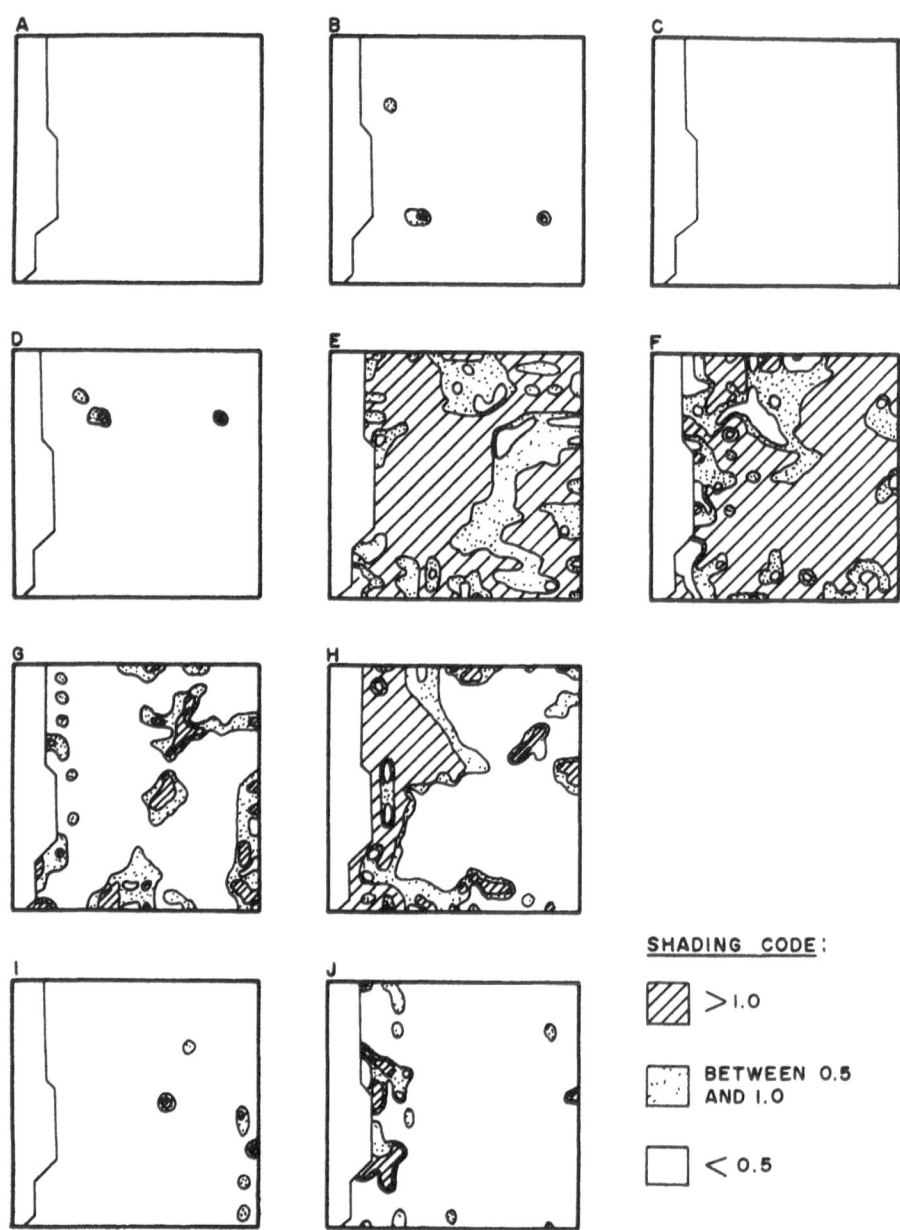

Figure 12.   Momentum equation terms for layer 2 normalized with
             Coriolis term.   Groups below are x and y directed re-
             spectively: A and B, vertical friction layer top; C
             and D, vertical friction layer bottom; E and F, pres-
             sure; G and H, lateral friction; I and J, inertial
             terms.

the terms of the momentum equation have been normalized with
Coriolis, absolute values of order 1 indicate a size which is
comparable with the magnitude of the Coriolis term. Figure 12 (A),
(B), (C) and (D) show vertical friction acting on the top and
bottom of layer 2 in the x and y directions. The vertical friction
acting on this layer in the x direction, Fig. 12 (A) and (C), is
at least one order of magnitude less than the Coriolis force. In
the stronger flowing y direction, Fig. 12 (B) and (D), a few
isolated locations tend to show some significant values. For the
most part, vertical friction appears to be of little importance.
These results support the findings of Holland and Hirschman (1972)
and others, that the coefficient of vertical momentum transfer k_V,
and therefore vertical friction, is relatively unimportant as
either a major driving force or a dissipation mechanism.

The pressure ratios shown in Fig. 12 (E) and (F) are for the
most part order 1 and greater. The x directed pressure ratios,
Fig. 12 (E) are greatest along the western half of the layer in
the vicinity of the strong density gradients of the southward
flowing Labrador Current. The y directed pressure ratios, Fig.
12 (F), appear strongest in the central and eastern region where
greater easterly flow occurs. Similarly Fig. 12 (G) and (H), the
lateral friction ratios, show the greatest x directed lateral
friction occurs in the eastern half of the layer which has mainly
east-west flow patterns while the greatest y directed lateral
friction appears next to the continental slope in the Labrador
Current region.

Figure 12 (I) and (J) show the inertial term ratios which form
a type of Rossby number. Order 1 magnitudes are observed mainly
in the y direction momentum equation. This is the direction of
the maximum flow velocities found in the Labrador Current.

The results of this analysis suggest that the model as
formulated might properly simulate the conditions found during
Cruise II by neglecting vertical friction entirely and using the
y directed inertial term only in the Labrador Current region.
Some computer time savings could then be realized at the sacrifice
of some accuracy in the solution of the model equations but the
model would then be limited as to where it can be applied.

The absolute vorticity was computed for the second layer
using the relationship:

$$\text{Absolute vorticity} = \frac{\partial v}{\partial x} - \frac{\partial u}{\partial y} + f.$$

This was accomplished for the same simulation used for the parametric study described above. Figure 13 shows the results of
this calculation. The absolute vorticity remains within 10 per
cent of its mean value along the flow patterns. The Labrador
Current and its easterly offshoot are roughly defined by the
110×10^{-6} sec^{-1} isoline while the Gulf Stream has a consistent

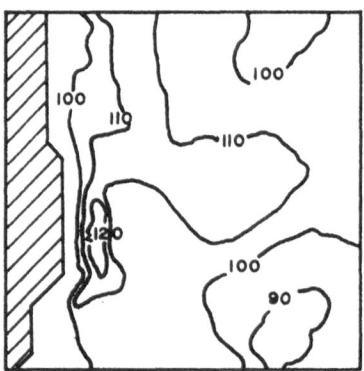

Figure 13.   Absolute vorticity (x 10^6 sec^{-1}) for layer 2 after 188
             hours of model operation.

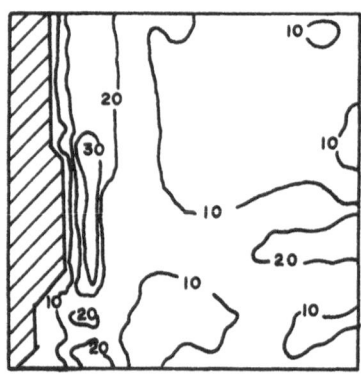   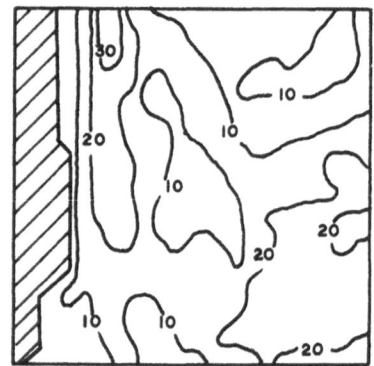

Figure 14.   Isotach (cm sec^{-1}) comparison between the results of
             188 hours of model operation, left, and computations
             based upon Cruise II property data, right.

absolute vorticity of between 90×10^{-6} sec^{-1} and 100×10^{-6} sec^{-1}.
There exists a region near the continental slope where the absolute
vorticity increases as the Labrador Current filaments stream south.
Comparing this region with layer 1 for temperature in Fig. 10,
there appears to be upwelling of cold water from layer 2. This
vertical motion into layer 1 is consistent with the requirements
of fluid column stretching in the presence of increasing absolute
vorticity.

Isotachs were also computed for layer 2 after the 188 hour
prediction period for comparison with isotachs determined from
the Cruise II current field after an initial 100 step spin up
period. Figure 14 shows that the values of the isotachs compare
favourably in the region of the Labrador Current, but are lower
in the Gulf Stream. Furthermore, the eastward flow of the Labrador
Current filament appears to be further south than predicted. These
apparent discrepancies are probably the result of the no motion
level assumption coupled with barotropic conditions not included
in the model.

SUMMARY AND REMARKS

A test of a multilayer three-dimensional, time dependent model
of a small oceanic region, using numerical calculations based on
mass distribution and wind stress was carried out. The full form
of the horizontal momentum equations was used including the
inertial terms. Initial conditions for property distribution
were used by the model along with postulated boundary changes.
A steady local wind from the southwest at 765 cm sec^{-1} was applied
to the model during the prediction period. Comparisons were made
between predicted and observed property distributions at the end
of an 8 day period. The predicted property distributions were
based on a current structure which was computed hourly relative
to an assumed motionless abyssal layer.

Fundamental flow patterns and property interactions show
encouraging results which may be further improved by addition of
more layers in subsequent models. In addition, more attention to
the details of the upper structure as affected by the wind stress
appears to be worth investigation.

The model was sensitive to the values of the horizontal
transfer coefficients. These coefficients were determined by
trial and error model operation until satisfactory comparisons
between predicted and observed property distributors were obtained.
The values which gave the best results were $k_h = 1.0 \times 10^7$ cm^2
sec^{-1} for horizontal momentum and $A_h = 5 \times 10^4$ cm^2 sec^{-1} for
horizontal diffusion. Higher values of k_h resulted in sluggish
currents while lower values produced severe cross stream velocity
gradients. Trial A_h values caused excessive mixing when of the
same order as k_h and thus lowered to produce better results. The

values used for these horizontal transfer coefficients were
consistent with those of other models of similar sized ocean
areas. In addition, the value for the vertical coefficient of
momentum transfer used was $k_V = 2.0$ cm^2 sec^{-1}. This value also
compares favorably with those commonly used, however the model
proved insensitive to its variation. Additional experimentation
with these transfer coefficients may lead to more satisfactory
results; however, the verification method of comparing property
distributions must be replaced with a more quantitative technique
before such fine tuning is attempted.

The strong horizontal density gradients of the modeled area
cause the geostrophic influence to be so great that the lateral
friction and inertial terms of the momentum equations are only
important in the strong dynamic regions of the Labrador Current
and the Gulf Stream. The lateral friction terms appear to provide
the necessary dissipation forces for the model with vertical
friction at the layer interfaces being negligible by comparison.
Lateral mixing and the advection of the property gradients appears
important in providing direction for the flow of the current
filaments.

The arbitrary method used by the model for accounting for
excess mass by direct current alteration in the horizontal,
produced acceptable results. A technique more closely aligned
with theoretical concepts like that used by Leendertise et al.
(1973) may possibly give better results; however, the small size
of the time step required by this method necessitates an excessively large amount of computer time.

Longer prediction periods coupled with information on the time
changing property conditions at the boundaries would be the next
logical step in testing the model. Verification of future models
should be accomplished in a more quantitative manner such as by
the use of drogue arrays which are configured so that they
effectively vertically integrate the currents in a water layer.
Using this method. drogue positions observed hourly could be
compared with hourly current velocity predictions. A drifting
iceberg in the modeled area would provide a satisfactory test.

It is expected that future models will incorporate a larger
area and additional layers. Furthermore, data collection for
initial conditions should be by means of expendable temperature
and salinity probes to allow for greater synopticity of the water
property information. Lastly, boundary information must be updated periodically, possibly every few days to provide information
on the changing water masses entering the model area. This information could be obtained either by ship surveys or moored data
buoys.

ACKNOWLEDGMENTS

The use of the computer at the Computer Center of the
University of Connecticut is acknowledged.
The opportunity to undertake this research was provided by
the U.S. Coast Guard's International Ice Patrol, Academy, and
Research and Development Center.
In addition, the authors are deeply indebted to the Officers
and crew of the Coast Guard Oceanographic Vessel EVERGREEN whose
diligent efforts provided the data which made this research
possible.

REFERENCES

Arons, A., and H. Stommel, 1956. A beta-plane analysis of free
periods of the second class in meridonal and zonal oceans. Deep-
Sea Res., 4, 23-31.

Bodine, B.R., 1971. Storm Surge on the Open Coast Fundamentals
and Simplified Prediction. Tech. Memo. No. 35; U.S. Army Corps
of Engineers Coastal Engineering Research Center, Maryland.

Bradshaw, A., and K.E. Schleicher, 1970. Direct measurement of
thermal expansion of sea water under pressure. Deep-Sea Res., 17,
691-705.

Bryan, K. and M.D. Cox, 1967. A numerical investigation of the
oceanic general circulation. Tellus, 19, 54-80.

Charney, J.G., 1955. The Gulf Stream as an inertial boundary
layer. Proc. Nat. Acad. Sci. U.S.A., 41, 731-740.

Crowley, W.P., 1968. Numerical Advection Experiments. Monthly
Weather Review, 96, 1-11.

Defant, A., 1961. Physical Oceanography, Vol. 1, Pergamon Press,
London, 729p.

Ettle, R.E., and T. Wolford, 1972. Oceanography of the Grand
Banks Region of Newfoundland and the Labrador Sea in 1970.
USCG Oceanographic Report 56, 280p.

Friedrich, H., and S. Levitus, 1972. An approximation of the
equation of state for sea water, suitable for numerical ocean
models. J. Phys. Oceanogr., 2, 514-517.

Holland, W.R., and A.D. Hirschman, 1972. A numerical calculation
of the circulation in the North Atlantic Ocean. J. Phys. Oceanogr.
2, 336-354.

Knudsen, M., 1901. Hydrographical Tables. CEC. Gad, Copenhagen, 63p.

Kollmeyer, R.C., T.C. Wolford and R.M. Morse, 1966. Oceanography of the Grand Banks Region of Newfoundland in 1965. USCG Oceanographic Report 11, 157p.

LaFond, E.C., 1951. Processing oceanographic data. U.S. Navy Hudrogr. Office, Publ. 614, 114p.

Leendertse, J.J., R.C. Alexander and S.K. Liu, 1973. A Three-Dimensional Model for Estuaries and Coastal Seas; Volume I, Principles of Computation. Rand Corporation, R-1417-OWRR, Calif., 55p.

O'Brien, J.J., and H.E. Hurlburt, 1972. A Numerical Model of Coastal Upwelling. J. Phys. Oceanogr., 2, 14-26.

Paskausky, D.F., 1971. Winter Circulation in Lake Ontario. Proc. 14th Conf. Great Lakes Res., 593-606.

Richtmeyer, R.D., and K.W. Morgan, 1967. Difference methods for initial-value problems. John Wiley and Sons, New York, 403p.

Robe, R.Q., 1974. Near bottom direct current measurements in the Labrador Current. EØS Trans. Am. Geophys. Un., 55, 316p.

Scobie, R.W., 1972. A comparison of current velocity determined by dynamic methods with direct current measurements in the Labrador Current. M.S. Thesis, Florida State University, 54p.

Stommel, H., 1950. Determination of the lateral diffusivity in the climatological mean Gulf Stream. Contr. WHOI No. 552, Woods Hole, Mass.

Wert, R.T., 1970. A baroclinic prognostic numerical model of the circulation of the Gulf of Mexico. Ph.D. Dissertation, Texas A&M University, 66p.

BEAUFORT SEA AUTOMATED ENVIRONMENTAL PREDICTION -

PILOT ARCTIC SYSTEM

F. B. Muller

Atmospheric Environment Service

Downsview, Ontario, Canada

1.   INTRODUCTION

The Beaufort Sea Environmental Prediction System has been designed as a result of AES participation in a major environmental impact study by the Canadian Department of the Environment in connection with proposed off-shore drilling for oil and gas. The proposed drilling area is near the mouth of the Mackenzie River off Tuktoyaktuk (Fig. 1). The AES study for the design and evaluation of the prediction system began May, 1974, and has been reported on in detail in the AES Beaufort Sea Project E-1 - Interim Report. Design-trials are taking place during this summer (1975) and the entire system was designed to permit implementation by July 1, 1976.

In view of the limited scale of operations expected in the '76 season the system actually deployed may be far short of the full design particularly in respect to the expansion of the observational system to support it. However, the main features from a system point of view will be present even if reduced in scale, so that the full system for 1977 or subsequent years will be achieved as a result of incremental growth rather than any qualitative re-design.

The system actually designed is of general interest for those concerned with management in the Arctic since its strong reliance on computerization makes it a likely fore-runner for the whole Canadian Arctic. The emphasis on computerization is not only for efficiency, but for the downstream potential to meet the anticipated requirement for a large increase in the volume, timeliness and detail of forecast information without commensurate increase in manpower and other resources.

785

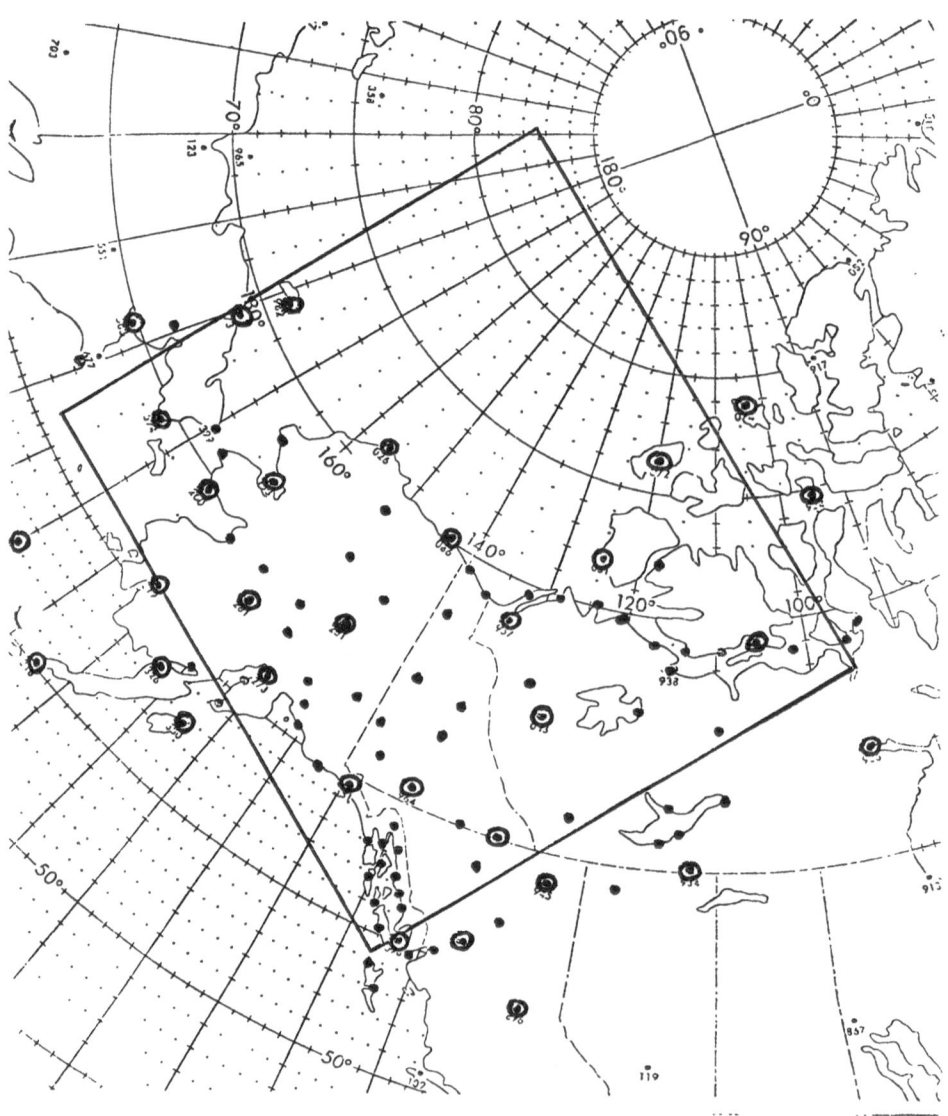

Figure 1. Beaufort Area Surface (⊙) and Radiosonde (●) Stations.

2.   THE OPERATIONAL REQUIREMENT

Drilling in 1976 is expected to be limited to two drill ships but it is to be expected that this will expand rapidly in the following years, as Canada attempts to bring the Beaufort reserves into production while waiting for the tar sands and other petroleum sources to come on-line.

Furthermore, one must plan not only for the forecast requirements for the exploration phase, but even more importantly for the extraction phase. For a production well, the economic consequences of environmental interference will be far greater than for exploratory drilling, and the scale of potential damage to the environment would be far higher.

The requirement for predictions covers all the usual aspects of the weather (cloud, precipitation, temperature, wind, etc.) and extends to ice, wind waves and water-levels. Since the season is very short for off-shore drilling, losses from unnecessary delays are very high. Thus, there is a special need for constant close-up forecast support particularly for projections out to 12, 18 or 24 hours. For forecasts this far into the future, it will be imperative to minimize the lag between the evidence for a critical environmental event, and the availability of the forecast for that event. In other words, the response time of the system to new information, i.e., new observations, must be as small as possible. The role of the computer will be to do just that, on a round-the-clock basis and as each new set of data arrives.

Beaufort Sea drilling operations are particularly sensitive to floes of multi-year ice, as well as to waves. In-shore water level may be a critical factor. Forecast support is needed

(a)   for decision-making for the start of critical multi-hour operations (e.g., running casing),

(b)   minimizing unnecessary early stages of alerts,

(c)   to guide the search for local ice that may present a threat,

(d)   and for supply operations by air and water.

The 'pay-off' from good forecast support comes in increasing drilling-time in an already short drilling-season, in improved efficiency, and, during the extraction phase, in increased production rates. In total it is considered that the system may contribute very significantly in making explorations economically viable. From a government and environmental point of view the system will permit significant reduction in the risk to the environment. From the public point of view, it will contribute considerably to early availability of Beaufort oil and gas.

3. THE BEAUFORT SEA ENVIRONMENTAL PREDICTION SYSTEM AS A WHOLE

The system as a whole (Fig. 2), specifies substantial improvements in surface and satellite observations, and provides for the linking of several operational offices. The BAB (Beaufort Advanced Base) at Tuktoyaktuk, or some such location, is the briefing office, the site of ice analysis and prediction, and the point of any final adjustment to the very short-range portion of any valid forecast. It is supported by and subordinate to the Arctic Weather Central, Beaufort Section (AWCB), 1300 miles to the south and the Ice Forecasting Centre at Ottawa, and the CMC at Montreal. The Computerized Prediction Support System is located in Edmonton under the full control of the AWCB, and uses two mini-computer processors during the first year, and three processors for the fully developed system. Each office has carefully defined roles.

The Basic Design document specifies what comes out of the system in terms of products and services, the main components of the forecast production system, how they are operated to meet the needs of the user, the plan for the Design-Trials of this summer (1976) which are now in progress, the Computerized Prediction Support System (CPSS), the Resources (including manpower, training, computer power, communications), implementation requirements, the Observational System (including surface Stations, the local ice surveillance system and satellite imagery) and complete costing.

The minimum system was costed as shown in Table 1.

Several points need noting about these costs. First, the level of the observation system required is a judgemental matter at this time and was selected to support many operations with high-quality products. Very substantial reductions may be implemented

Table 1

Cost Summary for a Minimum Beaufort Sea Prediction System

	Initial Cost	Operating Costs		
		Total in-cluding Salaries	No. of Staff	Man-Years per Season
	$ K	$ K		
Prediction Sub-System	51	240	17	10
Observational Sub-System	422	260	3	3
Total	473	500	20	13

(These costs did not include inflation or the cost of the local ice surveillance system.)

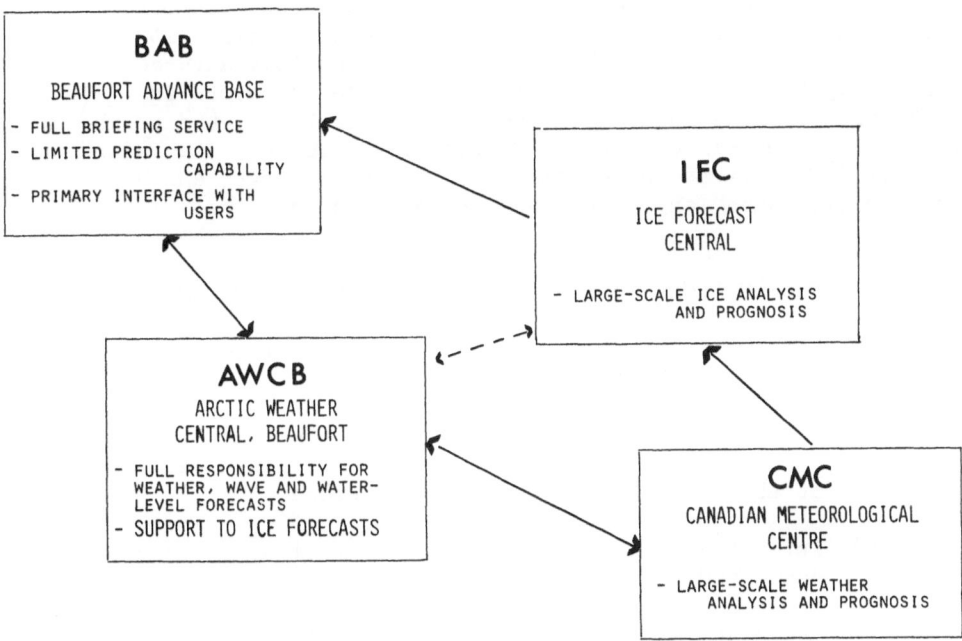

Figure 2. Beaufort Sea Environmental Prediction System.

when the amount of drilling is small and the cumulation of
environmental risks is not considered excessive. The expensive
items in the system are the deployment of automatic surface
stations on the ice (at about $75 K/yr/station plus allowance for
loss) and the purchase of special satellite reception equipment
(Very High Resolution Radiometer at about $200 K for the initial
cost).

Secondly, experience during a year or so of initial
operations with a reduced system (but including the CPSS) may serve
to demonstrate the level of service required from the point of view
of the operators themselves. One cannot be certain of this,
however, since ice conditions may happen to be very favorable.
From an environmental viewpoint, this may be acceptable for
relatively small-scale operations, and potential damage low.

Thirdly, the availability of a high-technology advanced
system may be viewed as insurance to both the operators and
environment. Like insurance, its cost should be related to the
total risk over all operations during many seasons. Its major
benefit is likely to be in seasons of marginal length and
favorability when many operations are proceeding.

4.   THE CPSS (COMPUTERIZED PREDICTION SUPPORT SYSTEM)

Computerization helps in the preparation of forecasts in the
following ways:

(a) Automated handling, filing, display and plotting of
     incoming data. Data comes in more or less steadily
     at about 200-300 characters per second.

(b) Automated watch and alert of observation and fore-
     cast products in comparison with user-defined
     thresholds for critical conditions. With many
     users and many critical environmental thresholds,
     the computer can readily scan the results of an
     hourly repetition of the basic forecast process,
     even when the output is not otherwise used, and
     thus make early warnings much more likely.

(c) Producing quickly analyses of meteorological
     charts from the latest observations.

(d) Producing quickly forecast meteorological charts
     (prognostics) 12-24 hours into the future. Large-
     scale cimcumpolar prognostics are produced
     centrally every 12 hours in Montreal and may in
     the future be updated every 6 hours. However,
     between times without a computer, the forecaster
     must subjectively assess the continuing validity of

these charts on the basis of later and more complete
data, and Regional considerations. In this the
computer can rapidly and automatically aid him. If
the meteorologist judges more improvements are
needed, these can be incorporated manually into the
computer product.

(e) Performing computations to help the forecaster
     assess the performance of the models, and diagnose
     the importance of various meteorological and
     physical processes which may contribute.

(f) Preparing 'raw' forecasts of weather elements at
     grid-points and at special sites. These 'raw'
     forecasts are partly the result of computation of
     the physical processes involved and partly based
     on sophisticated statistical analysis of past
     records. The 'raw' forecasts can then be rapidly
     scanned and edited by the forecaster as necessary.
     The elements predicted can include cloud, surface
     temperature, surface wind, precipitation (type and
     rate), visibility, etc.

(g) Ice-motion hindcasts and forecasts.

     Real-time hindcasts of ice-motion are needed to
     establish where ice previously observed is likely
     to be at the present time, given the surface
     pressures and winds that have actually been
     observed. Forecasts of future motions must also
     be made based on forecast winds, and 'tuning' of
     the model which produces the linkage between
     winds and ice-motion. Not only can observed ice
     floes be predicted in this way, but forecasts can
     be made for hypothetical ice. The latter type of
     ice forecast can be used to identify areas from
     which ice could arrive close to a drill-site to
     pose a threat, given the forecast evolution of
     the local wind pattern into the future. The
     effect of water current on ice motion in an area
     like the Beaufort Sea can be incorporated
     directly into the ice prediction module since
     both motions appear to be primarily wind-driven.

(h) Wind-generated water waves can also be forecast
     by computer, and may frequently have critical
     threshold values for certain drilling and supply
     operations. Computer-production of sea-state
     forecasts can be a big time-saver and has the
     potential of being more accurate than when done

by hand. The detailed changes of the wind for
the past 24 hours as well as the future can all
be taken into account. As with ice, hindcasts,
given observed winds and pressures, can be made to
estimate present conditions over areas from which
no reports are received.

All the eight functions listed can be repeated every hour as
surface and satellite data roll in. Experience now shows that the
complete repertoire can be done in a little over half an hour after
last data value is received, using the modest mini-computer
facilities recommended for the full system.

All products of the CPSS are subject to the control and
intervention of the forecaster at major stages in the forecast
production process. Further details on the CPSS are given in
Appendix 1. A much more complete description is in the Interim
Report referred to in the introduction.

5.   AUTOMATED ARCTIC PREDICTION

When the Design Study for the Beaufort Sea Environmental
Prediction System is completed and implemented, the basic work
will have been done to provide the foundation of a system that is
extendable from summer to year-round application, and from one
region to the whole Canadian Arctic. That is not to say by any
means that the Beaufort System as it now stands can cope with the
full range of problems or areal coverage required. However, the
necessary extension and upgrading would largely be of an add-on
nature.

Computer-power would need substantial increase. This does
not appear to present a great difficulty, even with only modest
cost increase, given the fluid state of computer evolution. A
more important item may turn out to be the scientific manpower
necessary to extend the prediction modules to cover aspects not
dealt with in the Beaufort System. Examples would be the
incorporation of water currents into ice prediction, and the
empiricism of many local effects in Arctic weather. This effort
is estimated to involve only a few man-years as compared with the
thirty or so man-years required to do the original development
work.

The important aspect is that the Computerized Prediction
Support System has the potential for continually increasing the
number of locations for which forecasts are prepared, the number
of items to be dealt with in each forecast, and the number of
items and locations for which watch and alert may be done. The
prime operating cost for such expansion is in terms of computer
rental, but not in terms of an increase of manpower in the same
proportion as the output. Manpower increases there will have to
be, to do the things the computer can't do, but with increasing
experience the CPSS will hold these to a low level.

GENERAL APPLICATION

The system and ideas outlined above have general application to the support of all major activities in the Canadian Arctic, both at their present level and as they are expected to expand. The application is thus to transportation (by land, air, ice and water), to logistics and supply operations, to all manner of weather sensitive activities (construction, research projects, pollutant control, heating, snow control, fishing, hunting, etc.) and to the everyday needs of individuals in respect to work and outdoor activities.

The first Objective of the AES is "to contribute effectively to the improvement of the national economy, the enhancement of the environment and the raising of the quality of life through full application of atmospheric and ice information". Achievement of this objective in the Arctic implies a major extension of environmental prediction services. The implication of automation as discussed is that it should be possible to provide this extension economically. Indeed the concept of the AES for 1985 as developed by its Senior Managers calls for implementation by them of a 'computerized forecast production system' for its general services. This system has been specified in outline and actions are now under way which are directed toward designing it in detail and toward organizing its achievement. The design work done for the Beaufort Sea area thus fits in with this direction of the AES and gives confidence that the AES will be able to provide the Canadian Arctic with the necessary level of forecast services.

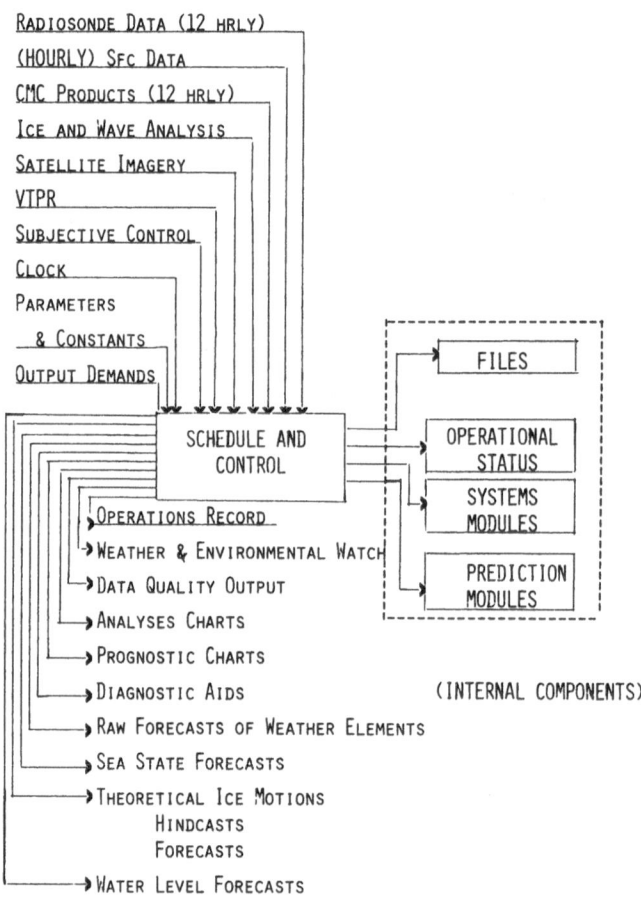

Figure 3.   Beaufort Sea Computerized Prediction Support System:
            Basic Structure.

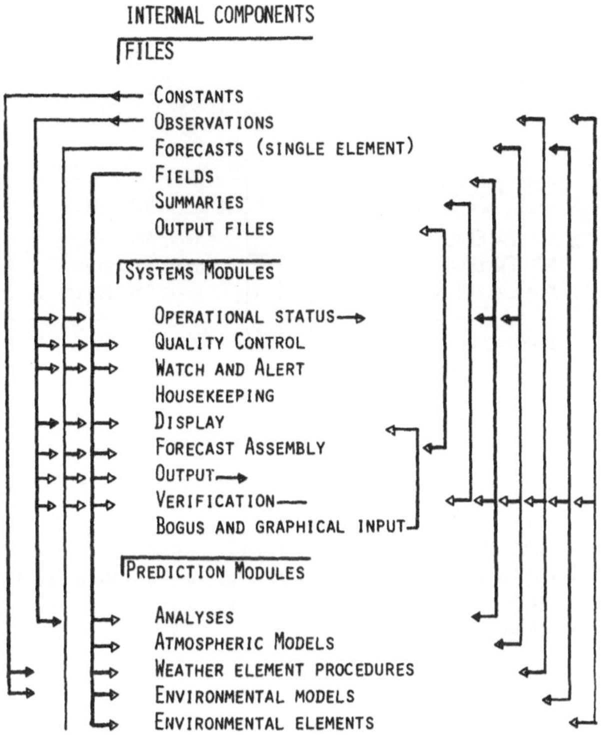

INTERNAL COMPONENTS

FILES

- Constants
- Observations
- Forecasts (single element)
- Fields
- Summaries
- Output files

Systems Modules

- Operational status
- Quality Control
- Watch and Alert
- Housekeeping
- Display
- Forecast Assembly
- Output
- Verification
- Bogus and graphical input

Prediction Modules

- Analyses
- Atmospheric Models
- Weather element procedures
- Environmental models
- Environmental elements

Figure 4. Beaufort Sea CPSS.

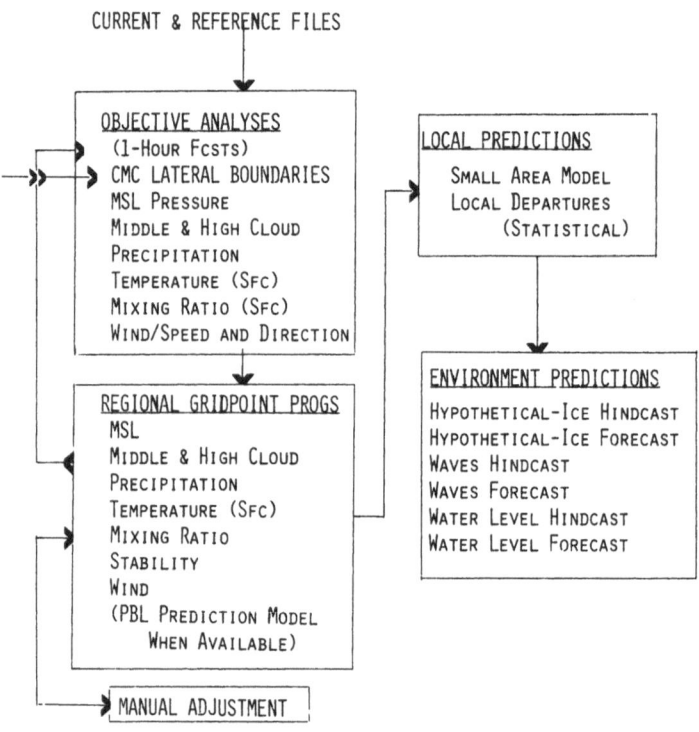

Figure 5.   Beaufort Sea Prediction Support System.

Table 1, Appendix 1

Estimated Timing for Operation of Modules of the
Computerized Prediction Support System

Hour	Function	Total (mins.)	1-3/4 hr. Total@ Response-Time		1-1/4 hr Total* Response-Time		
			A Computer	B Computer	A	B	C
h	Data extract	2	2		2		
	Weather watch	2	2		2		
	Synoptic Plot	8	8				8
	Analysis P	15	10	4	14		
	T						
	WS						
	WD						
	CLD						
	(850 MB) Stability						
	Output	2-3	3				3
	RUM II	10		10		10	
	(RUM interpolation)	2	2				
	Output (2 ref times)	1	1				1
	RUM-dependent atmospheric procedures	20		20		10	10
h+1	Data extract & wx watch	4	4				
	Environmental pred'n.	15		15		15	
	Output	1	1				
	Forecast Assembly	2		2		2	
	Forecast Watch	1	0	1		1	
	Assemble draft forecast	1		1		1	
	Messages for each location	1		1			
	Analysis for h + 1	15	12	2			
	Manual adjustment # 500 MB						
	Surface	12	12				12
	500 MB barotropic	2		2	2		
	Verification	2		2	2		
			57	60	22	39	34
			a two-hour cycle		a one-hour cycle		

Graphics input   1 min. CPU

O/A procedure on input for analysis and
         possibly for prog   2 min. CPU

                                    3 min. CPU per chart.

@ 1 hr. & 45 minutes from observation time -
         a 2-hourly cycle

* 1 hr. & 15 minutes from data. A 1-hourly cycle
         (analysis is not counted twice).

Explanation

This table gives overall timing estimates and not necessarily the sequence
of events in order to see what kind of response and cycling times might be
possible with
       (i)   Computer A (Arctic Wx Centre) plus
                Computer B (Beaufort Sea)

and   (ii)   Computer A, B plus an extra processor C.

APPENDIX 1

BEAUFORT SEA AUTOMATED ENVIRONMENTAL PREDICTION

DETAILS ON THE BEAUFORT SEA
COMPUTERIZED PREDICTION SUPPORT SYSTEM

1. General Structure

    This is shown in Figure 3.

    The CPSS consists of a computer and communications facility
plus a specially developed applications 'software' system. The
minicomputer is to have two processors, each with at least 32K
16-bit words, large discs, CRT, printer-plotter, systems software
to operate this as a distributed system, a graphics input device,
and communications links directly with CMC, BAB and IFC.
    The logical structure consists of
a) Inputs   Radiosonde data
                     (Hourly) surface data
                     Selected CMC Products (12 hrly)
                        (mainly lateral and upper
                           time-dependent boundaries)
                     Ice and Wave information
                     Satellite Imagery
                     Satellite Soundings
                     Subjective Adjustment
                     Clock-time
                     Parameters and constants
                     Output demands

b) A Schedule and Control System

c) Internal Components (Figure 4)
                     Files
                     Operational status information block
                     Systems Modules
                     Prediction Modules

d) Outputs   Operations record
                     Watch of observations and forecasts
                     Data quality analysis output
                     Data display and plots
                     Analysed charts
                     Prognostic charts
                     Raw (i.e. unedited) forecasts of
                        weather elements
                     Hindcasts and forecasts of hypothetical
                        ice, sea state, water level

2.   Prediction Modules
        Regional gridpoint models cycle back on the objective
     analyses for MSL pressure, middle/high cloud, precipitation,
     surface temperature (diurnally and local effects removed),
     surface moisture, planetary boundary layer stability and wind.
     All of these can be under the control of manual subjective
     intervention. The national Planetary Boundary Layer model
     will be integrated into the modules in advanced releases of
     the system. The PBL Model is being readied for real-time
     operation on the CYBER 76. A stripped-down version may be
     available for Regional Beaufort use.
        Local prediction procedures include
     a) the application of the 'small-area' model for deducing
         the effects of local terrain, and diabatic processes
         on surface winds,
     b) estimation of local departures from gridpoint averages.
        Environmental prediction modules carry out ice, wave and
     water-level forecast procedures.
        With input of large-scale ice-boundaries the modules
     produce estimates and forecasts of local ice and wave
     conditions. Fine-scale (20 km) ice motions are predicted for
     any floes hypothesized to be in the area. These are given
     both as forecast vectors and as trajectories of ice ending up
     within a given target area at specific times in the future.
     As spin-off to oil operators near shore and for port
     operations, water-level and surge is forecast by a special
     module.

3.   Systems Modules
        These may be considered to include the Central Scheduler
     and Control module, and the operations status module. Other
     modules provide for quality control observation and forecast
     watch, housekeeping (file-management data-transfers, saving,
     etc.), display, forecast assembly, output, graphical and bogus
     input, verification.
        Files exist for constants and parameters, observations,
     forecasts of single elements, fields, summaries, output queues,
     individual computer programs.

4.   Estimated running times for the system and its components are
     given in Table 1.
        The three-processor computed configuration for the fully
     developed system will occupy about 50% of the processing
     capacity leaving substantial capacity to the host AWCB
     computer and for enhancement.

THERMAL REQUIREMENTS OF DIVERS AND SUBMERSIBLES

IN ARCTIC WATERS

L.A. Kuehn, T.J. Smith and D.G. Bell

Defence and Civil Institute of Environmental Medicine

Downsview, Ontario, Canada

INTRODUCTION

A paramount problem in the operation of submersibles (47) and divers in cold water is the provision of adequate energy for thermal comfort for the crew or individual diver. This problem is particularly acute in the case of lockout submersibles in which several personnel are carried to and from an underwater work site under compressed gas conditions. This paper describes a general biophysical model known as Cold Diver which has been used to calculate the steady-state heat requirements for thermal comfort for all crew members of such a system in various scenarios of operation.

The human transfer capability of lockout submersibles resides in two usually spherical pressure chambers, a forward sphere and an aft sphere which may be connected by a small tunnel in some models. These submersibles may be used as transport systems for men and instruments at one atmosphere internal air pressure in a submersible mode and/or the maintenance and support of divers at an underwater work site. At present, the capability of such vehicles extends to non-lockout or submersible operations to depths greater than 600 m and lockout operations to depths of 300 m.

The forward sphere usually contains the control and life support facilities of the submersible which are operated by a crew consisting of a pilot, co-pilot and observer, all in air at a pressure of one atmosphere. The aft sphere can be used in two modes of operation, first as an extension of the submersible mode in the carrying of additional observers or equipment in air at a pressure of one atmosphere and second, in the lockout mode in the carrying of as many as two divers and a tender in compressed gas conditions. At the underwater worksite, the divers may leave the aft sphere through a lower hatch when the hydrostatic pressure equals that inside the sphere, leaving the tender to monitor their progress and to ensure their safety.

Such vehicles usually operate in association with a surface support facility, either a barge or especially equipped ship. The operational scenario usually begins with the hyperbaric compression of the divers and tender in a surface-mounted or deck decompression chamber (DDC) to the equivalent simulation of the hydrostatic pressure at the intended underwater worksite. Once compressed, these men are transferred to the lockout sphere of the submersible via a small mating tunnel for transport to the worksite. On arrival at the intended destination, the divers leave the sphere and perform their work with guidance, advice and hydraulic power tools support from the personnel in the forward sphere. After their task has been completed or a pre-determined immersion time has elapsed, the divers enter the lockout sphere for transport to the DDC where they undergo decompression or rest.

A second team of divers in the DDC could then be transported to the worksite to continue operations. Such scheduling of work and rest is very cost effective in the reduction of the ratio of decompression time to work time. Since the gas-carrying capabilities of submersibles are limited, gas requirements for compression of the lockout sphere are reduced considerably by pressurization at the surface from the surface-support gas reserves of the facility rather than the reserves of the submersible. The practice of decompression in the DDC also eliminates the need for the mixing of various breathing gases during on-board decompression. Such techniques also ensure that all crew members may be considered to be at approximately constant pressures during the extent of any submersible operation, thereby facilitating the calculation and prediction of heat requirements for thermal comfort.

At present, few submersibles have thermal insulation or any source of interior heat other than that produced by small motors and carbon dioxide scrubbers. Consequently, the internal temperature decreases quickly to the ambient temperature of the surrounding water on any cold water operation. Since almost all submersibles use unheated lead acid batteries as power sources, the cooling action of the water on such sources results in a large decline in power capability. The combination of the cold stress endured by the crew plus the reduction of power capability serves to limit the operational depth-temperature envelope for lockout submersibles. The Cold Diver program defines the heat required to enable realization of the depth-temperature envelope for which such vehicles were originally designed.

THE THERMAL SCENARIO

Figure 1 shows the three thermal environments associated with a lockout submersible. These are:

(a) Forward Sphere: Here the crew members lose heat by conduction
     to the cold metal of the inner surface of the sphere,
     respiration of cold air, radiation to the enclosing sphere
     surface and convection to circulating air. The order of
     importance of these heat transfer avenues is a function of
     ambient temperature but, in the range $0°$ to $10°C$, the order
     is: natural convection, radiation, respiration and conduction.
     The crew members are normally working and breathing at
     sedentary rates and wear work overalls in summer periods and
     insulated suits in winter periods. These remarks also apply
     to crew members in the aft sphere during a non-lockout operation.

(b) Aft sphere: On lockout operations, the aft sphere contains
     compressed gas atmospheres such as oxyhelium in which the crew
     members experience similar conduction and radiation heat losses
     to those within the forward sphere. Because of the increased
     density and thermal conductivity of the compressed gas medium,
     the respiratory and convection heat losses are increased.
     The divers and tender all wear diving clothing in transport
     to and from the worksite since there is not usually enough
     room to permit donning of the clothing. It is necessary that
     the tender also be dressed in diving clothing so that he may
     safely enter the water to aid the divers, if required. The
     clothing ensemble usually consists of a waterproof dry diving
     suit worn over as many layers of nylon pile underwear as can
     comfortably be worn and a head-enclosing diving helmet, as
     shown in Figure 2. Because of space and power limitations,
     the use of closed or free-flooding hot water suits can not be
     considered for lockout submersible operations.

(c) Lockout phase: Lockout divers experience great respiratory
     and convective heat losses during their immersion; however,
     radiative heat loss is minimal and heat loss via conduction
     is included as part of the convective heat loss (for the
     special case in which the relative water velocity equals zero).

     In this paper, the specifications and characteristics of the
Canadian Forces Submersible Diver Lockout Vehicle SDL-1 (18) are
used to provide data for the Cold Diver calculations. Manufactured
by the International Hydrodynamics Company in Vancouver, British
Columbia, it is a member of the Pisces series of submersible design.
The overall length is 6.1 m and its weight is approximately 14,000
kg in air. The two pressure chambers are made from steel of high
tensile strength, the forward sphere being 2.1 m in diameter and
the aft sphere 1.7 m in diameter. Visibility is possible from the
forward sphere through ten circular viewports, each 12.7 cm in
diameter. The existing on-board power supply consists of three
lead acid batteries of 120 volts, 28 volts and 12 volts for which
the power capacity is rated at 300 ampere-hours, based on a 6-hour
discharge.

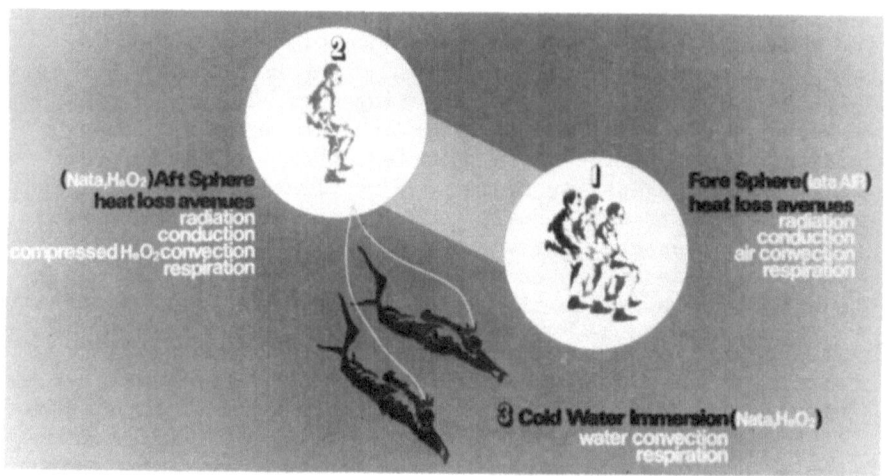

Figure　1　　Submersible diver lockout - Crew Environment

Figure　2　　Diver wearing dry suit and helmet

THE COLD DIVER MODEL

The Cold Diver program is a steady-state digital computer model of the heat requirements necessary to maintain an individual at a thermally-comfortable body core temperature or 37°C in any of the aforementioned three thermal environments. The consequences of failure to meet these requirements is a gradual lowering of the individual's body core temperature into a hypothermic state.

The Cold Diver program is applicable to ambient water temperatures in the range -2 to 37°C and to depths of 0 to 300 m of seawater. The calculations of heat losses are partitioned into natural convection, forced convection, radiation and respiration for each of the thermal environments. Transient heat losses that occur during initial deployment of the submersible in cold water are not taken into consideration.

A simple energy balance equation describes the detailed heat loss from the diver's body as follows:

$$M-W = + E \pm K \pm R \pm C \pm V \pm S \qquad (1)$$

where
  M = metabolic energy production,
  W = work output,
  E = evaporative heat loss from skin,
  K = conductive heat loss or gain,
  R = radiative heat loss or gain,
  C = convective heat loss or gain,
  V = respiratory heat loss or gain,
and S = energy storage or depletion in the body.

A number of assumptions can be made to simplify this equation. In the immersion or lockout mode, a diver is entirely surrounded by water and the breathing gas mixture contains no water vapour and usually is at ambient water temperature. Inside the submersible in the chamber-mode a crew member is surrounded by a gaseous mixture except for the parts of his body in direct contact with the chamber structure. This chamber gaseous mixture is assumed to be at ambient water temperature and to be completely saturated with water vapour due to the cooling of the submersible structure and enclosing gas from temperate conditions to temperatures below the dew point of the air initially in the submersible at the surface. Consequently there will be no evaporative heat losses from the crew members even if they were in a state of heat stress and E can be regarded as negligible. Furthermore, the conductive heat loss can be considered to be small and to be included in the convective heat loss term as the special case when relative fluid velocity equals zero.

For the chamber mode, the pertinent energy balance equation is

$$M-W \pm S = \pm R \pm C \pm V \qquad (2)$$

Further simplification occurs for the immersion mode since the radiation term can be considered to be negligible. The pertinent energy equation to lockout diving is

$$M-W \pm S = \pm C \pm V \tag{3}$$

Knowledge of metabolic energy generation and work output for various tasks is known and with calculation at the right-hand side of Equations (2) and (3), the heat required to sustain the crew member in thermal equilibrium at a body core temperature of 37°C can be calculated.

Radiative Heat Loss

Assuming that each crew member has an average clothing surface temperature, $\bar{T}_c$ and is radiating energy to his local environment which has a mean radiation temperature, T_{mrt} equal to the ambient water temperature, the net radiation exchange, R, for each individual is:

$$R = \frac{\sigma \, \varepsilon_1 \, Ar(\bar{T}_c{}^4 - T_{mrt}{}^4)}{\frac{Ar}{Ae}\left(\frac{1}{\varepsilon_2} - 1\right)} \tag{4}$$

where σ = Stefan-Boltzman constant,
       A_r = radiative area of the body, assumed to be 80% of the total body area, A, to account for interbody radiation exchange,
       A_e = surface area of enclosure,
       ε_1 = emissivity of surface clothing of crew member, taken as 0.95 for maximal estimation,
and ε_2 = emissivity of inner surface of sphere.

Other investigators (45) have assumed $\bar{T}_c$ to be equal to the mean skin temperature of 33°C, but this is a gross over-estimation and Tauber et al. (54) have suggested that $\bar{T}_c$ is normally 20°C. The Cold Diver program employs a clothing surface temperature that is calculated from the total heat flux across the suit and the convective heat losses from the clothing surface.

Total body area is calculated in square meters from the Dubois formula (23)

$$A = 71.84 \; W^{0.425} \; H^{0.725} \tag{5}$$

where W = weight of individual in kg,
and   H = height of individual in cm.
The average submersible crew member was considered to be approximately 170 cm in height and 70 kg in weight. The radiative area, A_r, was determined from the resulting value of total body area and further corrected for body position using the data of Guibert et al. (25).

Convective Heat Loss

Convective heat loss is an important avenue of heat loss from crew members during both modes of operation, either in compressed gas or in water. Natural or forced convection or a combination of both occur and are influenced in their extent by such parameters as gas or water currents and diver activity. Resulting heat losses due to both natural and forced convection are calculated by the Cold Diver program with the invocation of McAdam's Rule (39), namely that the greater of the two calculations is accepted as the total convective heat loss.

Using the basic convection formula,

$$C = U A (\bar{T}_s - T_f) \tag{6}$$

where C = convective heat loss,

A = total body area,

$\bar{T}_s$ = mean skin temperature,

T_f = ambient fluid temperature,

and   U = universal heat transfer coefficient,

and assuming that a man can be represented in convective heat loss studies by a cylinder 1.85 m long and 0.305 m in diameter, with a uniform temperature of $\bar{T}_s$ and an area of 1.8 m^2 as suggested by Tauber et al. (54), it is possible to calculate both natural and forced convective heat losses to the immediate fluid surrounding the body. In these calculations, $\bar{T}_s$ is assumed to be constant at 33°C.

The universal heat transfer coefficient, U, obeys the following relationship (54):

$$U = \frac{1}{\frac{1}{h_c} + \frac{x}{k} + \frac{1}{C_i}} \tag{7}$$

where h_c = suit-water boundary layer film heat transfer coefficient,

x = thickness of diving suit material,

k = thermal conductivity of diving suit material,

$\frac{x}{k}$ = diving suit thermal resistance,

and C_i = thermal conductance of the skin-suit interface, assumed to be 20% of the suit thermal resistance as suggested by Nevins (45).

Therefore,

$$U = \frac{1}{\frac{1}{h_c} + 1.2 \frac{x}{k}} \tag{8}$$

The validity of the cylindrical model of man in convective heat loss studies has some justification. Witherspoon et al. (61)

used a cylinder model with a diameter equal to the mean weighted diameter of 0.144 m to calculate heat transfer coefficients of immersed human subjects. A composite cylindrical model of man was advanced by Parker et al. (46) and provided good results for heat losses of man in 0.5 ata gas environments. Measurements of the boundary layer film heat transfer coefficient made in these studies correspond well with values pertaining to forced gas cross-flow over cylinders 0.127 m in diameter.

In all calculations of convective heat loss under hyperbaric conditions; i.e., for crew members in the aft sphere or locked out from it on lockout operations, the thickness of any neoprene or gas-filled diving clothing was considered to have a pressure-dependent function for thermal insulation as documented by Butler and Payne (16).

Natural Convection

The equations used in the Cold Diver model for calculation of the suit-water boundary layer film heat transfer coefficient are those given by Tauber et al. (54) and Witherspoon et al. (61). For the gas mode, the equation used was

$$h_c = 0.13 \, \frac{K}{L} \, (Gr. \, Pr)^{0.333} \tag{9}$$

and for the water mode the equation was

$$h_c = 0.53 \, \frac{K}{D} \, (Gr. \, Pr)^{0.25} \tag{10}$$

where   K = thermal conductivity of the environment,
        D = diameter of the cylinder,
        L = length of the cylinder,
        Gr= Grashoff number = $g\beta L^3 (\bar{T}_x - T_f) \, \gamma^2/\mu^2$,
        g = gravitational acceleration = 9.805 m/sec^2
        β = coefficient of volumetric expansion of fluid,
        γ = density of environmental fluid,
$\bar{T}_x - T_f$ = temperature gradient across boundary layer,
        μ = viscosity of environmental fluid,
        Pr = Prandtl number = $Cp \cdot \mu/K$,
and   Cp = specific heat of environmental fluid.

Because the mean temperature of the surface of the clothing, $\bar{T}_x$, is a factor in the calculation of h_c, the value of this temperature was obtained by iteration, matching the heat flux across the suit and across the boundary layer. If this is not done and $\bar{T}_x$ is taken as equal to the mean skin temperature of the body, then the Grashoff number and the determination of h_c are over-estimated. The iterative technique determines the actual mean clothing surface temperature for each situation. The heat flux

through the suit material is a function of the temperature gradient
across the material. For a given mean skin temperature, this is
strictly a function of the external surface temperature of the suit.
The heat flux through the boundary layer is also a function of the
mean clothing surface temperature for a given value of ambient
environmental fluid temperature. Heat flux through the suit
material must equal heat flux through the boundary layer in a
steady-state situation. By repeated estimation of the clothing
surface temperature and comparison of the heat fluxes for a given
environmental temperature, the correct mean clothing temperature
can be determined and the true amount of natural convection heat
loss calculated.

Forced Convection

The same basic assumptions as used for the natural convection
case pertain also to the case for forced convection. The equations
for suit-water boundary film heat transfer coefficients for the
gas mode and the water mode are taken from the works of Tauber
et al. (54) and Witherspoon et al. (61) respectively.
For the gas mode

$$h_c = 0.6 \; \frac{K}{D} \; Re^{0.5} \; Pr^{0.31} \qquad 50 \leq Re \leq 12000 \qquad (11)$$

$$h_c = 0.174 \; \frac{K}{D} \; Re^{0.613} \qquad 4000 \leq Re \leq 40000 \qquad (12)$$

and for the water mode

$$h_c = \frac{K}{D} \; (0.35 + 0.56 \; Re^{0.52}) \; Pr^{0.3} \qquad (13)$$

where   Re = Reynolds number = $DV\gamma/\mu$,
          D = body diameter,
and   V = fluid velocity.

Respiratory Heat Loss

The calculations in Cold Diver that pertain to respiratory
heat loss take into consideration the energy required for the
warming of inhaled gas and the addition of water vapor from the
lung surface to it. The inhaled gas is assumed to be at a
temperature equal to that of the water surrounding the submersible.
When the diver is in the immersion mode, the breathing gas is
assumed to be completely dry; the gas mode in either the aft or
forward spheres is assumed to be water-saturated. The exhaled gas
is assumed to be saturated with water vapor at an exhalation

temperature as observed by Goodman et al. (26)

$$T_e = 22 + 0.649 \ T_i \tag{14}$$

where T_e = exhaled gas temperature,

and $\quad T_i$ = inhaled gas temperature.

In the forward sphere, and aft sphere during non-lockout operations, the gas mixture is always assumed to be air but in the aft sphere on lockout operations and in the water mode, the gas mixture is assumed to be always an oxyhelium mixture with a partial pressure of oxygen of 0.3 ata, regardless of the pressure. The changes in gas properties caused by the exhaled carbon dioxide generated by the crew members were assumed to be negligible because of the great proportion of helium.

Calculations were performed for a range of respiratory minute volumes from 5 to 50 litres per minute according to the equation.

$$V = R_{mv} \rho C_p \ (T_e - T_i) + R_{mv} \rho h_{fg} \ (W_e - W_i) \tag{15}$$

where R_{mv} = respiratory minute volume,
$\quad \rho$ = density of breathing gas,
$\quad C_p$ = specific heat of breathing gas,
$\quad W_e$ = humidity ratio of gas at temperature T_e,
$\quad W_i$ = humidity ratio of gas at temperature T_i,
and $\quad h_{fg}$ = latent heat of vaporization of water.

RESULTS OF THE COLD DIVER MODEL

Figure 3 illustrates the heat loss to be expected in watts per individual for various steady-state conditions of pressure and ambient temperature. Four pairs of heat loss curves are shown, for four ambient temperatures $0°$, $10°$, $20°$ and $30°C$. One line of each pair pertains to heat loss in water while breathing oxyhelium and the other line pertains to heat loss in an oxyhelium atmosphere, both at ambient pressures and temperatures. In this figure, it is assumed that the individual has a sedentary rate of respiration equivalent to a volume rate of gas exchange of 20 litres per minute.

For each ambient temperature, the heat loss for the immersion mode is approximately twice as thermally stressful as for the chamber mode at the same pressure, for depths in excess of 10 atmospheres of pressure. This difference becomes smaller at shallower depths, the two modes being approximately equal in their thermal stress at the surface. On Figure 3, lines drawn indicating the capability of the individual to generate metabolic heat at a sedentary level of 100 watts, or at a moderate work level of 300 watts, show that neither activity condition can suffice to permit thermal equilibrium for all depth-temperature environments of interest, particularly in the ambient temperature range of 0 to

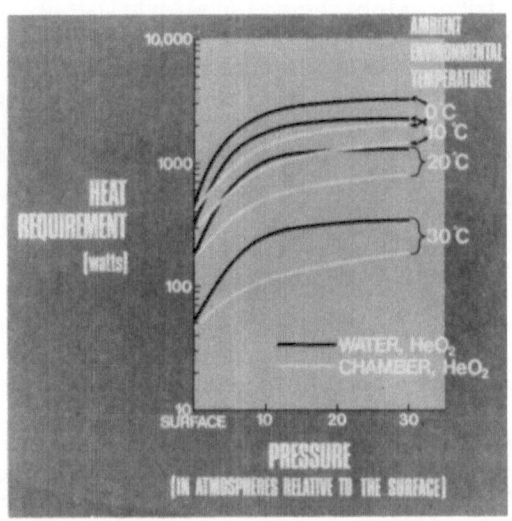

Figure 3   Heat Requirement per Individual

10°C. Thus, active heating must be supplied for thermal comfort to
the crew members in the water and compressed gas modes. This
conclusion also applies to the observers in the forward sphere in
one atmosphere of air, although their rate of onset of hypothermia
will be very slow and probably not dangerous for short missions.

The ability of the present power supply of the SDL-1 (sixty
2-volt lead acid cells with a capacity of 300 ampere hours at a
6 hour rate) to meet the thermal demand indicated by Figure 3 has
been determined. The voltage range of this battery can be expected
to vary between 2.0 volts to 1.75 volts per cell. This is a
nominal power capacity of 36.0 kw/hr at 27°C and 25.2 kw/hr at 0°C
on 6-hour rates of discharge. If all the available energy from the
battery supply could be transferred to the micro-environments of
the crew clothing suit with 100% efficiency, there would not be
sufficient energy to provide thermal comfort to the whole crew.
If one incorporates the effects of energy leaks in the system and
inefficiency of operation and the energy required for propulsion
and maintenance of the life-support systems of the submersible,
then the present energy supply is not great enough to ensure
thermal comfort for the crew for all environments of interest, even
at relatively shallow depths.

Comparison with the Literature

In 1967, Beckman (5) computed the heat required to maintain
thermal comfort in a resting diver wearing a 3/16 inch wet suit in
shallow water (presumably surface conditions) at temperatures of 0,
5, 10, and 15°C to be approximately 800, 670, 540 and 450 watts
respectively. These calculations are greater than those of the
Cold Diver model and are probably due to the difference in suits
worn and the simplified assumptions made on maintenance of skin
temperature. Beckman (6) also reported similar results in a later
review article, stating that divers wearing electrically-heated
garments under 1/4 inch neoprene wet suits require 320-340 watts
for maintenance of comfortable body temperatures in shallow water
of 5°C temperature.

In an evaluation of a free-flooding diver heat replacement
garment consisting of a 3/16 inch wet suit under which was flushed
hot water, Bondi and Tauber (10) found that the depth-compression
of the wet suit material and the permeation of the material with
helium from an hyperbaric mixture combined to reduce the insulative
value of the suit by 75%. At the surface, 895 watts were required
to provide comfort for an immersed diver and at 183 m of depth,
3100 watts were required in water temperatures of 4.4°C.

A comprehensive theoretical study by Tauber et al. (54) in
1969 showed that the heat replacement required in water of 4.4°C
temperature and at depths of 183 m may be as great as 600 to 800
watts per diver in a Personnel Transfer Capsule (PTC) and that
this requirement is dependent on the environmental temperature,

ambient gas velocity, the number of wet suits worn and time of cooling. For divers in the water at these extreme conditions, the heat replacement was calculated to range between 1500 and 3000 watts per individual, again depending on the ambient temperature, time, number of suits worn, level of activity and the fit of the suit. Later, Tauber et al. (55) determined that at least 500 watts of heat were required to maintain a diver in thermal comfort and resting conditions in cold water. These tests examined the thermal efficiency of a combination of a Wilson tubing suit and two 3/16 inch neoprene wet suits on a diver.

In 1970, Majendie (37) reported that the diver's heat requirements to 183 m at 0°C average 1.2 kw with maxima of 3.4 kw on open-circuit hot-water heating systems and 2.5 kw on closed-circuit hot-water systems.

Rawlins and Tauber (50) reported the construction of a simple mathematical model which they used to compute diver heat loss for a diver at 183 m of seawater at a temperature of 40°F (4.4°C) and breathing a 2/98 oxyhelium gas mixture. When the diver was assumed to be wearing an electrically heated suit or closed-circuit hot-water tubing suit under a 3/8 inch wet suit, the total diver heat replacement for thermal comfort in the resting case was 1500 watts. When the wet suit was only of 3/16 inch thickness under the above conditions, the calculated heat replacement was 3000 watts. Another 500 to 1000 watts were estimated to be necessary for a working diver, due to the flushing of cold water beneath the wet suit. Calculations performed pertinent to a diver wearing a 3/8 inch neoprene suit equilibrated with helium under the above conditions inside a PTC indicated that approximately 1000 watts were required for thermal comfort.

Experiments performed by Moritz and Langworthy (42) showed that a supply of heat equivalent to 480 watts for a two-hour period was insufficient to provide thermal comfort to resting divers wearing a 1/4 inch neoprene wet suit at depths greater than 3 m and in water with temperatures in the range 1.5 to 3.0°C. A superb series of experiments by Reins and Shampine (51) on the heat loss from divers wearing 3/16 inch neoprene wet suits at shallow depths showed that, in water of temperatures 12.8, 7.2 and 1.7°C, the average heat losses for 21, 19 and 9 subjects were 418, 468 and 666 watts respectively.

All the above studies agree well with the results of the Cold Diver model when allowance is made for the fact that almost all pertain to wet suit diver clothing which is not as good in thermal insulation as dry suits worn over nylon pile underwear suits (19). In shallow depths and at the surface, the Cold Diver results on diver heat loss are significantly less than literature values reported here. For deeper depths, excellent agreement exists.

EXPERIMENTAL VERIFICATION

The thermal exchanges of divers in cold water at -1.2, 4.4 and 10°C and simulated depths of 0, 10, 30.5, 61 and 91 m of seawater were measured in hyperbaric chambers. Various dry diving suits, such as Unisuit, Polar Volume, J.A.P., Piel λ 6 and Piel λ 8, were examined in conjunction with a variety of diving helmets. Three different working conditions were studied: rest, swimming and vertical exercise. The rate of heat loss and the rate of change of body heat content was also measured in operational divers at Halifax, Nova Scotia to depths of 43 m and at Resolute, Cornwallis Island, in the Canadian Arctic to depths of 76 m.

In all of these experiments, two different physiological techniques for measuring thermal exchange were applied.

The Classical Technique

In this technique, the diver's body is considered as a two-compartmental system, a central core which must be maintained at 37°C for optimum physiological efficiency and a peripheral shell which is used as a physiological thermal buffer to protect the core and consequently can be allowed to cool considerably in adverse cold environments. In this technique the core temperature and the mean skin temperature of the diver were continuously monitored to provide information on the relative thermal behaviour of these two compartments.

Core temperature, T_R, was assessed directly with thermistor rectal probes and occasionally by radio pills (8). Mean skin temperature was measured with either 12, 7 or 4 skin thermistors at various sites in the body by the following techniques.

(a) Hody Technique (32): A 12-site sensor measurement system with the sensors placed at the following locations:

T_1 = head,

T_2 = chest,

T_3 = rear calf,

T_4 = abdomen,

T_5 = lower arm,

T_6 = wrist,

T_7 = thigh,

T_8 = front calf,

T_9 = foot,

T_{10} = upper back,

T_{11} = lower back,

T_{12} = rear thigh.

The equation for mean skin temperature in this technique is

$$\bar{T}_s = 0.070T_1 + 0.085T_2 + 0.065T_3 + 0.085T_4$$
$$+ 0.140T_5 + 0.050T_6 + 0.095T_7 + 0.065T_8$$
$$+ 0.070T_9 + 0.090T_{10} + 0.090T_{11} + 0.095T_{12} \qquad (16)$$

(b) Hardy-Dubois Technique (28): A 7-site sensor measurement
system with sensors placed at the following locations as
defined earlier: T_1, T_4, T_5, T_6, T_7, T_8 T_9. The equation
for mean skin temperature in this technique is

$$\bar{T}_s = 0.070T_1 + 0.350T_4 + 0.140T_5 + 0.050T_6$$
$$+ 0.190T_7 + 0.130T_8 + 0.070T_9 \qquad (17)$$

(c) Ramanathan Technique (49): A 4-site sensor measurement system
with sensors placed at the following locations as defined
earlier: T_2, T_5, T_7 and T_8. The equation for mean skin
temperature in this technique is

$$\bar{T}_s = 0.3T_2 + 0.3T_5 + 0.2T_7 + 0.2T_8 \qquad (18)$$

Figure 4 shows the points of attachement on the body used in these
three techniques.

Because of operational limitations, it was not possible in all
cases to use one standard method for the assessment of mean skin
temperature. In all cases, the left side of the individual diver
was chosen as that for sensor attachment for all sensors not on
the bilateral axis of the body. Because of the tight neck seal on
the various suits tested, it was necessary to affix the T_1 sensor
on the left side of the neck rather than on the head.

The Hody 12-sensor technique was considered superior because
of the adequate representation given to the large muscle groups
and because the weighting coefficients were nearly uniform, thereby
reducing the error involved if one sensor read improperly or became
detached from its position. With the other methods, an error in
temperature assessment at any site possessing a large weighting
coefficient can lead to a large error in the mean skin temperature.
The fewer the sites of measurement, the greater the inaccuracy to
be expected in assessment of the mean temperature.

According to this method, the mean skin temperature and the
core temperature may be combined to produce a mean body temperature
(MBT) via the formula:

$$MBT = 0.33 \, \bar{T}_s + 0.67T_R \qquad (19)$$

Changes in MBT are attributed to changes in body heat content
according to the formula

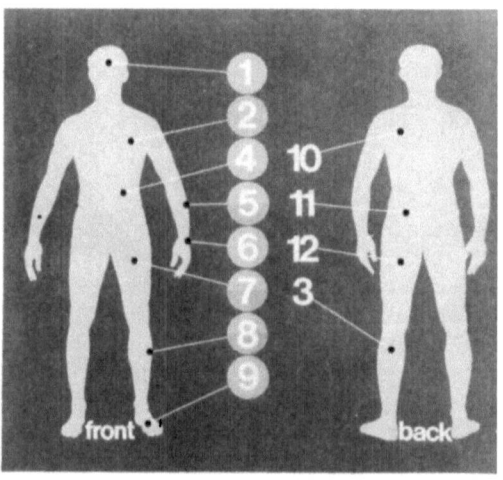

Figure 4   Site of Skin Temperature Measurement

$$S = 0.83 \ m \ \Delta \ MBT \tag{20}$$

where S = change in body heat content due to heat losses via
         conduction, convection, radiation and respiration,
      0.83 = average value of heat capacity of human flesh,
      m = mass of the diver
and ΔMBT = change in mean body temperature.

Most studies reported in the literature of heat loss from
subjects dressed in operational clothing use the classical technique
or simplified sections of it. Many workers such as Martorano (38),
Beagles and Coll (4), Moritz and Langworthy (42), Bondi and Tauber
(10), Covey (19), Veghte and Klemm (57) and McCrory (40) simply
monitored various skin temperatures with further calculation of
only the mean skin temperature. Others such as Milan (41), Reins
and Shampine (51), Boutelier et al. (11, 12) and Bell (7) rigorously
applied this technique of indirect calorimetry to determine body
heat content changes.

However, it is important to note the objections to this
technique raised by Craig (22) and Webb (58). It is a technique
of indirect calorimetry developed by Burton (14) for men in an air
environment. No evidence exists to support the use of the given
equations and weighting coefficients for human subjects in water.
The major reason to support the use of these techniques for
measurement of diver heat loss is that they have proven useful to
solve operational thermal stress problems in adverse air environments and they are readily implemented for measurements in the
water environment.

The Heat Flow Technique

The proper technique for use in direct measurement of heat
loss from human subjects is that of direct calorimetry which has
been successfully applied by only very few workers, most notably
Craig (20, 21, 22), Webb et al. (59) and Burton and Bazett (15).
These techniques usually consist of the immersion of human subjects
in insulated water calorimeters. This technique, while accurate,
is difficult to establish in laboratory experiments and of little
applicability to the measurement of heat loss in field or
operational environments.

Another technique which can be used to measure the heat loss
of divers directly and which can be used in operational evaluations
is that of heat flow discs. This method, like that of direct
calorimetry assesses the heat lost from the surface of the diver.
For complete thermal balance analysis, it is necessary to monitor
respiratory heat loss as well.

A heat flow disc is a thermopile which is an arrangement of
many small-guage thermocouples in series on opposing sides of a
flat slab or core, the resistance of which is known and is constant.
The entire assemblage is sheathed in a protective covering to form

a disc or small slab. When heat flows through the disc, a temperature difference is established across it. In the steady-state case the heat flux through the disc establishes a constant voltage generation in the thermopile and it is this voltage which can be considered as a monitor of the heat flux from a warm object when properly weighted by a geometry and surface area factor.

From early prototypes such as those developed by Hatfield and Wilkins (29) to the state-of-the-art sensors of Hagar (27) and Poppendick (48), such discs have developed from laboratory curiosities to exact measuring devices for heat flow from buildings, transportation vehicles and animals. Physiologists have been slow to apply such techniques to research on humans but recently several workers have done so in cold water studies with some success (Hong et al. (36), Nadel (44), Bell (7), Hody (33, 34)). Suitable agreement with results of indirect and direct calorimetry have been obtained by some workers but not by others. The difference in results appears to be due to the quality of sensor used.

The specification of a suitable sampling scheme for heat flow measurement from the body is most easily considered by use of the same subdivision of body area as was used for skin temperature measurement. The use of the same weighting factors and measurement sites for both temperature and heat flux transducers permits their incorporation into one unit as has been achieved by Thermonetics Corporation and described by Hody (33).

The advantage of using a sensor that measures both temperature and heat flow at the same site is that in steady-state conditions, knowledge of both these parameters permits computation of the thermal insulation afforded to a diver by his clothing and also his 'shell'. In steady-state conditions the mean heat flow from the core of a diver to his skin surface must equal the mean heat flow from the skin through the clothing to the cold water.

If the mean heat flow is $\bar{H}$, then the heat flux from the core through the shell to the skin is

$$\bar{H} = \frac{T_c - \bar{T}_s}{R_s} \tag{21}$$

where R_s = thermal resistance of the shell and the heat flux from the skin through the clothing is

$$\bar{H} = \frac{\bar{T}_s - \bar{T}_f}{R_c} \tag{22}$$

where R_c = thermal resistance of the diver's clothing.

Combining equations 21 and 22 leads to the formula

$$\frac{R_c}{R_s} = \frac{\bar{T}_s - T_f}{T_c - \bar{T}_s} \tag{23}$$

Knowledge of $\bar{H}$, $\bar{T}_s$ and T_f permits calculation of

$$R_s = \frac{T_c - \bar{T}_s}{\bar{H}} \tag{24}$$

and $R_c = \dfrac{\bar{T}_s - T_f}{\bar{H}}$

$$\tag{25}$$

the physiological and clothing insulations. This technique has
important field and testing application in the design and
evaluation of operational diver clothing and practices. Such
application has been reported for situations where steady heat loss
may be assumed (Hody (33), Bell (7)). Figure 5 shows a graph of
diver clothing insulation computed by this method for a shallow
dive in cold water.

Results of Experimental Evaluation

Application of the two techniques described here, especially
when used in conjunction with each other, has led to validation of
the Cold Diver model to depths as great as 91 m and water
temperatures as cold as -1.2°C. Such validation has been
extensively reported elsewhere (Bell (7)) but Table 1 shows a
summary of the results.

PREDICTION OF THERMAL STRESS REQUIREMENTS FOR
A TYPICAL NON-LOCKOUT MISSION

A deployment of a lockout submersible on a typical non-lockout
mission consists of a 30 minute pre-dive systems check, a launch
and two action of one hour duration, a descent time of 30 minutes
to the working site, a period of observation and/or work lasting
five hours, an ascent period of 30 minutes and a tow and recovery
operation of one hour. Figure 6 shows the thermal energy required
in watt hours for the whole of the SDL-1 crew for this 8 1/2 hour
mission for various conditions of ambient temperature and pressure.
On this figure are several lines indicating the capability of the
crew members and the SDL-1 itself to supply this thermal demand.
The levels of resting metabolism (at a rate of 100 watts per
individual) are insufficient to sustain complete thermal comfort
for use of the SDL-1 in waters of 0 to 10°C temperatures,
regardless of depth. However, it is important to realize that the

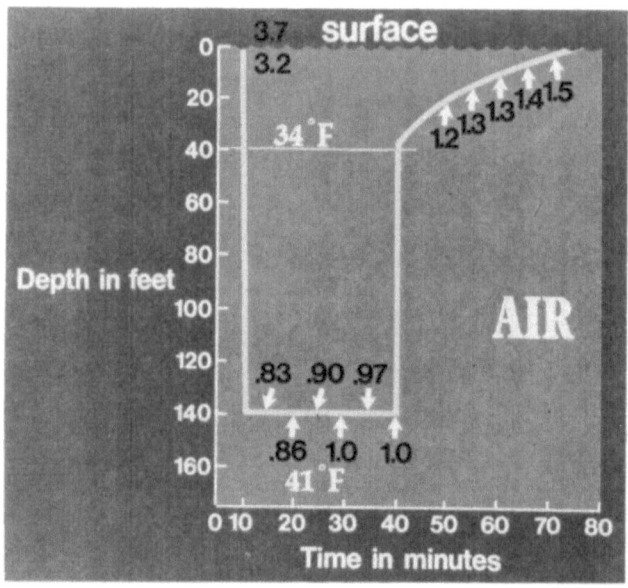

Figure 5 Suit Effectiveness during a field Dive to 138 feet on Air

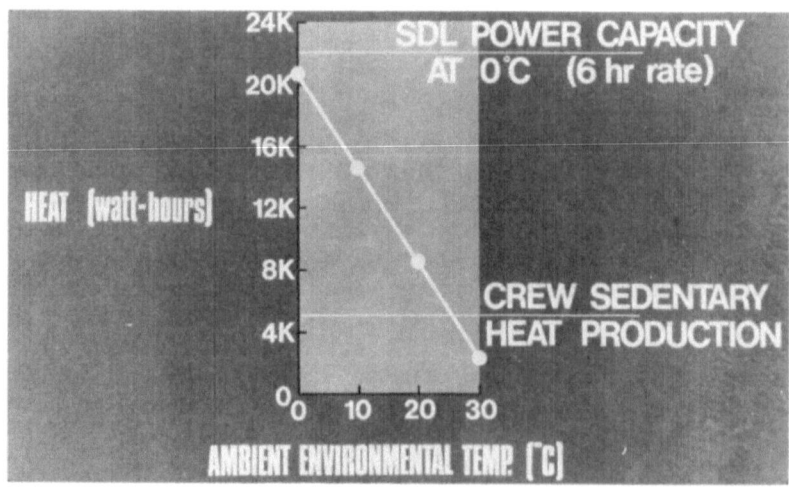

Figure 6 Heat Requirement for Typical Non-lockout Mission

Table 1   Measurements of Diver Heat Loss in Watts

Temperature °C	Pressure (atm) Relative to Surface										
	0			1			3		10	
	RV	S	WV	RV	S	WV	RV	WV	RV	WV
10°C	129	37	37	163	136	159	184	165		
4.4°C	140	88	71	226	141	208	274	210		
−1.2°C	225		155			241	329	275	475	486

Data represents averages on 2 to 7 subjects.

Task performed: RV = resting vertical position

S = swimming horizontally

WV = working vertical position

thermal demand on the crew is not excessive in these operations and although they can be regarded as losing more heat than they can physiologically sustain, the process of hypothermia will take much longer than the time of the non-lockout operation. The present energy supply of the SDL-1 contains more than enough energy to sustain the crew in thermal comfort, provided that a means of conveying the energy from the source to the men is implemented on the submersible and also provided that the thermal energy so used does not adversely reduce the operational capability of the craft as regards propulsion and systems operation.

PREDICTION OF THERMAL STRESS REQUIREMENTS FOR A TYPICAL LOCKOUT MISSION

A deployment of a lockout submersible on a typical lockout mission consists of a 30 minute pre-dive systems check, a launch and tow action of one hour duration, a descent time of 30 minutes to the working site, a lockout work duration of two hours, an ascent period of 30 minutes and a tow and recovery operation of one hour.

Figure 7 shows the thermal energy required in watt hours for the whole of the SDL-1 crew for the 5 1/2 hour mission for various conditions of ambient temperature and pressure. On this figure there are several lines indicating the capability of the crew members and the SDL-1 energy supply to satisfy this thermal demand. The levels for resting metabolism (at a rate of 100 watts per man) and working metabolism (at a rate of 300 watts per man) are clearly insufficient to sustain thermal comfort for use of the SDL-1 in waters with temperatures in the range 0 to 10°C, regardless of depth. The present SDL-1 energy supply is great enough to meet this demand provided that there was no loss in energy capacity with cooling of the batteries and that there was a method for transferring the electrical energy of the batteries to heat the microclimate of the diver's clothing. After reduction of the energy supply through cooling of the batteries and subtracting the energy required from this source for submersible propulsion and systems support, the amount of energy left for heating of the crew is such that thermal comfort can be maintained only for shallow depths in water at temperatures of 0 to 10°C.

CONSIDERATIONS OF 72 HOUR EXPOSURE FOR SDL-1 CREW

In the event of an accident in which the SDL-1 is disabled underwater in such a way that rescue is not immediately available for the entrapped crew, it is desirable that the crew be sustained in thermal comfort for a period of 72 hours or three days, during which time it is assumed that a rescue could be effected. The

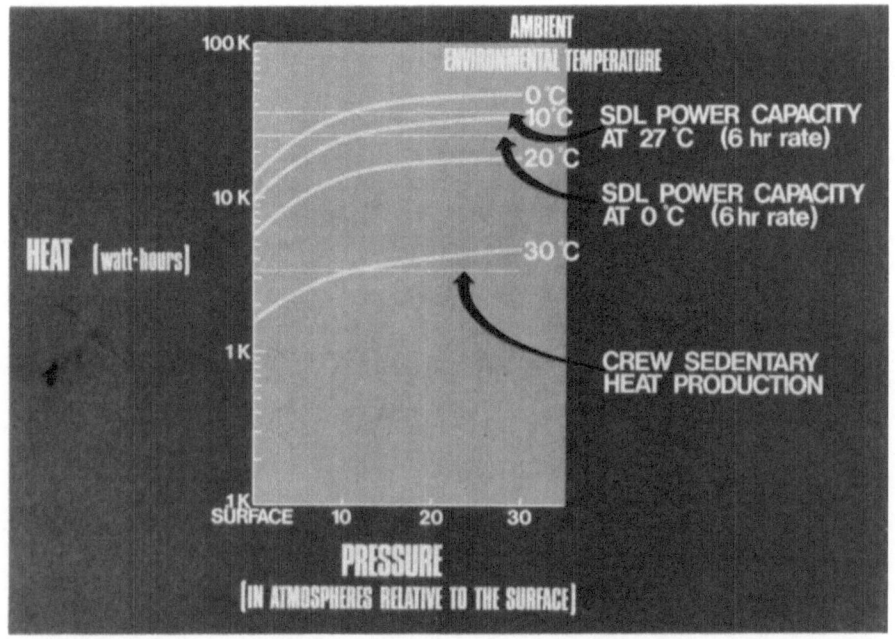

Figure 7 Heat Requirement for Typical Lockout Mission

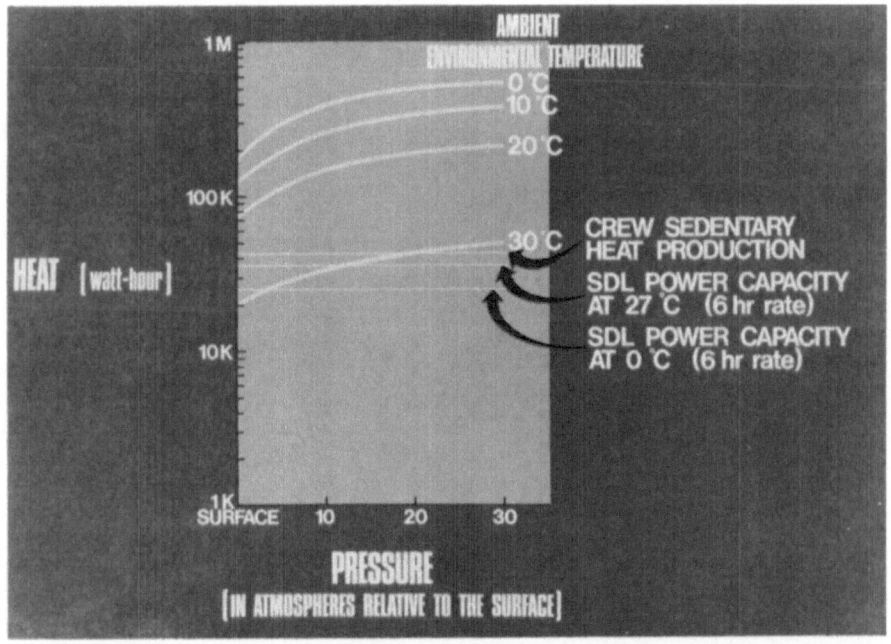

Figure 8 Heat Requirement for 72- hour Exposure

thermal requirements to achieve this objective are shown in Figure 8 for a variety of ambient temperatures and pressures. On this figure there are three horizontal lines indicating the capability of the crew's resting metabolism, the lead acid battery energy (unaffected by cold) and the combination of these two energy sources in meeting this thermal demand. It can be seen that a 72 hour exposure in water at temperatures of 0° to 10°C at any depth or aft sphere pressure would be one that would lead to hypothermia for all crew members. At shallow depths it is likely, particularly for the occupants of the forward sphere, that the crew members would be found alive but suffering from advanced hypothermia after a 72 hour exposure. For greater depths (more than 50 m) it is not likely that any crew members in the pressurized aft sphere would survive.

DISCUSSION

On the basis of 'Cold Diver' calculations, the operation of the SDL-1 with its present energy supply in cold water will be limited at great depths and for long missions by the thermal drain on crew members. Complete thermal comfort cannot be provided for the crew, even if it were possible to convey thermal energy from the power unit to the clothing microclimate of the individual members. To ensure thermal comfort for the crew to the depth and time limitations for which the SDL-1 was designed, it is necessary to implement both a new energy source and also a method to convey thermal energy from the source to the crew.

Heating of the Submersible

The Cold Diver model indicates that the heating of the entire SDL-1 (and thereby the crew) to conditions of thermal comfort would require more energy by a factor of two or three than the heating of the men individually with electrically heated or special hot water conditioned suits. Heat loss from submersibles can be greatly minimized if they were constructed according to the suggestion of Taylor (56) which is to construct each sphere as a chamber within a chamber with an insulation space between the two hulls. This space could be evacuated continuously by a vacuum pump or evacuated and sealed during the initial construction. As regards the thermal insulation of single walled pressure spheres such as those of the SDL-1, Taylor suggested internal incompressible insulation that was impermeable to gas penetration by such species as the highly thermally conductive helium. This could be achieved by protection of the insulation with a metallic foil which would also reduce the radiative heat loss to the vessel walls. Insulation mounted externally to the pressure spheres has the advantage of not reducing the internal volume of the spheres but unless the internal surface

of the sphere is silvered, there will still be a substantial
radiation leak to the sphere's internal surface. However, any good
incompressible insulation such as neoprene rubber or syntactic
foams could be mounted externally without fear of loss of thermal
conductivity by rare gas diffusion in the insulation material.

As regards the thermal cost of heating underwater habitats or
vessels, it has been reported (1) that 12,000 thermal watts are
required for heating the pressure sphere of the U.S.N. Mark II Deep
Dive System PTC. Majendie (37) in 1970 reported that the U.S.N.
Mark I Deep Dive System required a heat input of 864 watts for
every degree Celsius of temperature differential maintained between
the interior and exterior of the sphere. This heat was provided
externally to the sphere by electrical heating elements in copper
tubing embedded within one inch of syntactic foam. Such an
arrangement takes up no internal volume and the need for internal
shielding to reduce radiation leaks is lessened. This work was
amplified by Riegel et al. (52) who reported that a comfortable
temperature of 32°C inside the 80 inch diameter PTC of this system
in 40°C water is maintained by a 25 kw heat generation system when
at 800 feet of depth.

Rawlins and Tauber (50) also suggested that the preferable
situation was the heating of the entire underwater capsule and
suggested that, in the use of a PTC supplied with hot water from
the surface, 12,000 watts of thermal energy would be required and
much less if 1/4 inch of syntactic insulation was applied to the
outer surface. These conclusions were supported by earlier
calculations of Tauber et al. (54) which showed that if the PTC
was regarded as a spherical chamber, some 9600 watts would maintain
it at an internal temperature of 85°F (29.4°C) and that with
minimal insulation (1/4 inch of rubber) only 4800 watts would be
necessary. Operationally, it has been reported by Helwig and Gray
(31) of the Westinghouse Underseas Development that their heated
bell had to be covered with 1 inch of rubberized spray for thermal
comfort.

Potential Power Sources for the Submersible and its Crew

As to the type of energy sources that may be considered for
use on submersibles or on individual crew members, the choice may
be taken from such a list as that compiled by Beckman (6), namely
huge power density batteries, isotope power generators, thermionic
heaters, thermoelectric generators, catalytic fuel cells and
devices utilizing exothermic chemical reactions. Despite intensive
research, the use of batteries is still restricted by the weight of
power cells (Moritz and Langworthy (42), Frey (24) and Common and
Kettle (17)). Isotope power generators are admittedly superior in
terms of the power to weight ratio but they remain at a disadvantage
because of the high levels of harmful radiations which require
heavy shielding. Various thermionic and thermoelectric devices

have been proposed for diver heating but most are still in early stages of development and are difficult to control. Thermochemical devices (Hearst (30), Slack (53)) suffer from the same problems.

Of particular interest is the inert-cathode/magnesium anode seawater battery described by Black et al. (9) and Wilson (60) which is essentially a short-circuited seawater battery. Capable of generating 16 kw/hr of heat at a rate of 2000 watts for 8 hours, the heat is produced by placing a magnesium anode in close proximity to a cathode material such as an iron plate. The generated heat is transferred to the diver via a hot water tubing suit.

Another electrochemical system of interest is the lithium sulphur primary battery described by Holland (35) in which lithium and molten sulphur are held in porous matrices separated by a fused salt mixture containing lithium halides. These cells possess three times the energy density available from good silver-zinc batteries and have a very high power density and storage life.

Thermochemical systems have progressed considerably from the early state in which lithium salts with low melting points were enclosed within the layers of a diver's suit (30). The reaction of lithium and sulphur hexafluoride has been used in prototypes of diver heaters by Brock (13) with some success. In this reaction, lithium fluoride and lithium sulphide are produced as well as 3400 calories per gram of mixture. This heat is then used in the boiler of a Rankine power loop to convey thermal energy to water in a tubed diving suit.

One potential heat source that may be applicable to the SDL-1 would be an adaption or extension of the chemical heat engine (43) developed by the Centre des Etudes Nucleaires (CENG) at Grenoble, France. Known as a 'climataseur', this device consists of a well insulated metallic 'thermos bottle' containing molten lithium salts which is used to heat water systems which are flushed through channels in an incompressible wet suit at desired variable rates. Tests now indicate that this system may be used to heat deep divers for extended periods of time in cold water. It is suggested that a larger application of this technology may suffice for the thermal needs of the entire SDL-1 and its crew.

A second potential energy source could be an extension of the underwater automotive engine such as that which is presently being developed by the Naval Coastal Systems Laboratory in Panama City, Florida or in Japan (2, 3). This method involves the use of an internal combustion or diesel engine such as a Mazda Wankel engine so arranged that its exhaust is cleaned of contaminents and fed back into its carburetor with the addition of fuel and oxygen. Running in a closed-cycle mode, the only gas to circulate through the engine is inert nitrogen. Use of such 'clean' fuels as propane or hydrogen permits a simple process for cleaning or scrubbing of the exhaust. New storage techniques of hydrogen (62) permit its consideration as a fuel. The whole unit could be enclosed in a pressure vessel and used to supply energy requirements for propulsion, electrical power and thermal heating.

REFERENCES

1. Author Unknown (1969) "Proposals for Steady State Studies to Determine Supplementary Heat Requirements in PTC Diving Operations" U.S. Naval Medical Research Institute Environmental Stress Division.

2. Author Unknown (1974) "Diesel Engine for Sea Floor Power" Ocean Industry 9, 31-32.

3. Author Unknown (1975) "No-Exhaust Engine for Underwater" Naval Research Review, 28, 32-33.

4. Beagles, J. and Coll, E. (1966) "Diver's Body Heat Loss" U.S. Navy Electronics Laboratory Research Report 1408.

5. Beckman, E. (1967) "Thermal Protective Suits for Underwater Swimmers" Military Medicine, 132, 195-209.

6. Beckman, E. (1968) "A Review of Current Concepts and Practices Used to Control Body Heat Loss During Water Immersion" Thermal Problems in Aerospace Medicine, edited by J. Hardy, AGARDOGRAPH 111 Englhard Hanovia.

7. Bell, D.G. (1974) "Measurement of the Rate of Heat Loss in Divers at Various Water Temperatures, Depths and Working Conditions" M.Sc. Thesis, University of Waterloo.

8. Bevan, J. (1971) "The Use of a Temperature Sensitive Radio Pill for Monitoring Deep-Body Temperature of Divers in the Sea. Royal Naval Physiological Laboratory Report No. 1271, Alverstoke, England.

9. Black, S., Tucker, L., Vind, H. and Hitchcock, R. (1973) "Preliminary Development of an Electrochemical Heat Source for Military Diver Heating" U.S. Naval Civil Engineering Laboratory, Technical Note N-1315.

10. Bondi, K. and Tauber, J. (1969) "Physiological Evaluation of a Free-Flooding Diver Heat Replacement Garment" U.S. Naval Medical Research Institute Report No. 17.

11. Boutelier, C., Colin, J. and Timbal, J. (1971) "Determination of the Zone of Thermal Neutrality in Water" Revue de la Medicine Aeronautique et Spatiale, 10, 25-29.

12. Boutelier, C., Colin, J. and Timbal, J. (1971) "Determination du coefficient d'echange thermique dans l'eau en écoulement turbulent" Journal de Physiologie, 63, 207-209.

13. Brock, D. (1974) "Closed Cycle Power Sources for Underwater Systems - Thermal Systems" Journal of the Royal Naval Medical Service, 29, 130-135.

14. Burton, A. (1935) "Human Calorimetry. II. The Average Temperature of the Tissue of the Body" J. Nutrition, 9, 261-280.

15. Burton, A. and Bazett, H. (1936) "A Study of the Average Temperature of the Tissues of the Exchanges of Heat and Vasomotor Responses in Man by Means of a Bath Calorimeter" American Journal of Physiology, 117, 36-54.

16. Butler, H. and Payne, R. (1969) "Thermal Conductance of Diver Wet Suit Materials Under Hydrostatic Pressure" U.S. Naval Ship Research and Development Laboratory Interim Report 2903.

17. Common, R. and Kettle, M. (1973) "Diver Suit Heating" Marine Technology Society Journal.

18. Cox, F. (1972) "SDL-1 - Human Performance in Sustained Operations" Proceedings of Third Meeting of the Technical Co-operation Program (TTCP) Technical Panel U-2 on Human Performance and Military Capability, Defence and Civil Institute of Environmental Medicine.

19. Covey, C. (1972) "Unisuit Takes the Chill out of Diving" Undersea Technology, 13, 39-40.

20. Craig, A. and M. Dvorak (1966) "Thermal Regulation During Water Immersion" Journal of Applied Physiology, 21, 1577-1585.

21. Craig, A. and M. Dvorak (1968) "Thermal Regulation of Man Exercising during Water Immersion" Journal of Applied Physiology, 25, 28-35.

22. Craig, A. and Dvorak, M. (1972) "Head Exchange between Man and the Water Environment" Proceedings of the Fifth Symposium on Underwater Physiology.

23. Dubois, D. and Dubois, E. (1915) "Clinical Calorimetry: Fifth paper, The Measurement of Surface Area of Man" Archives of Internal Medicine, 15, 868-888.

24. Frey, H. (1966) "Electrically Heated Pressure-Compensated Wet Suits for Sealab II" Final Report on Contract No. N600 (168)63855 for U.S. Naval Medical Research Institute.

25. Guibert, A. and Taylor, C. (1952) "Radiation of the Human Body" Journal of Applied Physiology, 5, 24-37.

26. Goodman, M., Smith, N., Colston, J. and Rich, E. (1971) "Hyperbaric Respiratory Heat Loss Study" Final Report on Contract N00014-71-C-0099 to the U.S. Office of Naval Research, Washington, D.C.

27. Hagar, N. (1963) "Thin Foil Heat Meter" Review of Scientific Instruments, 36, 1564-1570.

28. Hardy, J. and Dubois, E. (1938) Journal of Nutrition, 15, 461-

29. Hatfield, H. and Wilkins, F. (1950) "A New Heat-Flow Meter" Journal of Scientific Instruments, 27, 1-3.

30. Hearst, P. (1968) "Chemical Heat Source for Wet Suits" U.S. Naval Civil Engineering Laboratory Technical Note N-998.

31. Helwig, C. and Gray, R. (1970) "Sub Arctic Diving Operations". Proceedings of Symposium on Equipment for the Working Diver, Columbus, Ohio.

32. Hody, G. (1973) "The Field Measurement of Cold Stress in the Marine Environment" Consultancy Report to the Defence and Civil Institute of Environmental Medicine, Canada.

33. Hody, G. and Kacirk, J. (1972) "Combined Skin Temperature and Direct Heat Flow Measurement in a Thermally Stressful Environment" Proceedings of Aerospace Medical Association Annual Meeting.

34. Hody, G., Kacirk, J. and Pilmanis, A. (1973) "Direct 'In Situ' Measurement of Thermal Insulation Quality Underwater" Proceedings of IEEE International Conference on Engineering in the Ocean Environment, Seattle, Washington.

35. Holland, R. (1974) "Closed Cycle Power Source for Underwater Applications - Electrochemical Systems" Journal of the Royal Naval Medical Service, 29, 136-139.

36. Hong, S., Lee, C., Kim, J., Song, S. and Rennie, D. (1969) "Peripheral Blood Flow and Heat Flux of Korean Women Divers" Federation Proceedings, 28, 1143-1148.

37. Majendie, J. (1970) "Diver Heating" Proceedings of Symposium on Equipment for the Working Diver, Columbus, Ohio, 94-117.

38. Martorano, J. (1961) "Evaluation of Diver's Dress Suits in
     Maintaining Body Temperature in Cold Water" U.S. Naval
     Medical Field Research Laboratory, Vol. XI, No. 19.

39. McAdams, W. (1954) Heat Transmission, McGraw-Hill.

40. McCrory, W. (1974) "Thermal Protection for Divers" J.
     Hydronautics, 8, 115-118.

41. Milan, F. (1965) "Cold Water Tests of USAF Anti-exposure
     Suits" U.S. Arctic Aeromedical Laboratory, Report AAL-TR-64-31.

42. Moritz, W. and Langworthy, H. (1972) "Evaluation of Marine
     Corps Battery Powered Electrically Heated Dress" U.S. Naval
     Medical Research Institute Bioengineering Laboratory Research
     Report No. 1.

43. Morocchioli, R. (1974) "Chauffage Autonome de Plongeurs sous
     Marins" Proceedings of 6'eme Coloque de l'A.S.T.E.O. sur
     l'Exploitation des Oceans, Paris, France.

44. Nadel, R., Holmer, I., Bergh, U., Astrand, P. and Stolwijk, J.
     (1974) "Energy Exchanges of Swimming Man" Journal of Applied
     Physiology, 36, 465-471.

45. Nevins, R., Holm, F., and Advani, G. (1965) "Heat-Loss
     Analysis for Deep-Diving Oceanauts" ASME Publication 65-WA/
     HT-25.

46. Parker, R., Exberg, D., Whitey, D. (1965) "Atmosphere
     Selection and Control for Manned Space Stations" Proceedings
     of International Symposium for Manned Space Stations, Munich,
     West Germany.

47. Penzies, W. and M. Goodman. (1973) Man Beneath the Sea,
     Wiley-Interscience, New York.

48. Poppendick (1969) "Why Not Measure Heat Flux Directly"
     Environmental Quarterly, 15 (1).

49. Ramanathan, N. (1964) "A New Weighting System for Mean Body
     Surface Temperature of the Human Body" Journal of Applied
     Physiology, 19, 531.

50. Rawlins, J. and Tauber, J. (1972) "Thermal Balance at Depth"
     Proceedings of the Tenth Commonwealth Defence Conference on
     Operational Clothing and Combat Equipment, Kingston, Canada.

51. Reins, D. and Shampine, J. (1972) "Evaluation of Heat Loss
     from Navy Divers' Wet Suit" U.S. Navy Clothing and Textile
     Research Unit Technical Report No. 102.

52. Riegel, P. and Glasgow, J. (1970) "Experimental Determination of Heat Requirements for the Mark I PTC" Proceedings of Symposium on Equipment for the Working Diver, Columbus, Ohio, pp. 119-130.

53. Slack, D. (1973) "A Diver Chemical Powered Wet Suit Heater" Journal of the Marine Technology Society, 7, 35-37.

54. Tauber, J., Rawlins, J. and Bondi, K. (1969) "Theoretical Thermal Requirements for the Mark II Diving System" NMRI Research Report No. 2.

55. Tauber, J., Rawlins, J. and Bondi, K (1970) "Evaluation of a Diver's Thermonuclear Swimsuit Heater System" U.S. Naval Medical Research Institute Report No. 3.

56. Taylor, L. (1966) "An Investigation of Thermal Insulating Materials for Undersea Habitats" U.S. Navy Mine Defense Laboratory, Research and Development Report 1-98.

57. Veghte, J. and Klemm, F. (1972) "Cold Water Evaluation of Environmental Diving Suits" U.S. Aerospace Medical Research Laboratory Report AMRL-TR-72-65.

58. Webb, P. (1972) "Thermal Stress in Undersea Activity" Proceedings of the Fifth Symposium on Underwater Physiology.

59. Webb, P., Annis, J. and Troutman, S. (1972) "Human Calorimetry with a Water-Cooled Garment" Journal of Applied Physiology, 32, 412-418.

60. Wilson, B. (1968) "Characteristics of an Improved Inert-Cathode/Magnesium Anode Seawater Battery" Intersociety Energy Conversion Engineering Conference Record, Vol. 1, 852-860.

61. Witherspoon, J., Goldman, R., and Breckenridge, J. (1971) "Heat Transfer Coefficients of Humans in Cold Water" Journal de Physiologie, 63, 459-462.

62. Wolf, S. (1975) "Hydrogen 'Sponge' Storage" Naval Research Review, 28, 16-22.

COLD PROBLEMS AT DIVES IN ARCTIC WATERS

Commander Bo Cassel

Royal Swedish Navy

Stockholm, Sweden

The coasts of Sweden extend from latitude north 55,5 to 66 degrees. The predominating parts of the coastline border on the Baltic Sea and a lesser part on the North Sea. The Baltic Sea has the greatest interest for Sweden both from a military point of view and for exploitation of the resources on the sea bottom.

The average temperature in the Baltic on depths of 100 meters and deeper is 2-4 centigrades the whole year. We have to face the cold problem for deep diving all the year and for shallow water dives at least eight months of the year. During severe winters the Baltic Sea is entirely ice-covered for about three months.

In former days the helmet diver was the only type of diver and the cold in the water was a minor problem. He used a lot of heavy woollen underwear and together with hard physical work this normally kept him fairly warm. Besides, these divers worked in shallow water for a couple of hours and could get warm between the dives. The real cold problems came with the mixed gases and the deep diving.

So far the deep diving in Sweden has only been performed by the Royal Swedish Navy as part of the submarine rescue program. We started in the beginning of the sixties when the new submarine rescue and deep diving vessel, the BELOS, came into service. The first heliox dives at sea were performed in 1965. All the time we have had a valuable contact with the diving colleagues in the American, British and French navies and also in some degree with the Soviet navy.

We have now under construction a new diving center in the archipelago of Stockholm. This center is designed for deep diving and submarine rescue but it is also intended for civilian deep diving and development of underwater techniques. One of the urgent aims with the diving center is to develop a saturation

diving system to 300 meters. The saturation dives will be
performed from a rescue vehicle also under construction. This
vehicle is designed to take 25 men from a sunken submarine at a
maximum depth of 300 meters and it has a compartment for saturation
diving. It has a displacement of about 50 tons and it is possible
that the vehicle will have a civilian version where the rescue
compartment instead of men can take a considerable amount of
instruments and tools.

The cold problem during deep dives is divided into two parts:
to protect the diver from <u>external</u> and from <u>internal</u> heat loss.
The body must be isolated from the cold of the surrounding water
and the breathing gas must be heated.

We use a constant volume suit with air inside called the
UNISUIT, and we have found it to work well. The underwear is
normally synthetic fur or knitted wool. Nonwashed wool seems to
be the best material but it is from an economical point of view
impossible with a new underwear for every dive. We have studied
different types of underwear with electrical heating as well as
the hot water method and it is true that such equipment can give
an agreeable temperature to the diver. A group of civilian Swedish
divers are experimenting with a method where hot air is blown
through hoses into the suit at the feet of the diver. These methods
can be useful at shallow water dives from a quay or a barge in
sheltered waters. These methods, however, require electricity,
water heater or air heater and special instruments which are
difficult or impossible to use in a bell or a compartment of a
vehicle. And deep diving is performed in the open sea and very
often under heavy sea conditions. The instruments have to be easy
to handle and nonvoluminous. It is very important that the diver
is kept warm even during long dives but it is also important that
the means to keep him warm are as simple and reliable as possible.
Wool or synthetic fur together with air in the suit is only
sufficient during short dives. Especially the hands, feet and
trunk are vital parts which are sensitive to the cold and need
additional protection. We are now testing materials of the same
type that are used by the astronauts and for car seats, and use
them as gloves, socks and trunk protection. A series of tests in
climate chamber and cold water will show to what extent this is
sufficient.

The other main problem is to give the diver a breathing gas
of a sufficiently high temperature. We have found that a deep
diver using heliox in an open circuit system can stand no more
than 10-15 minutes on 100-150 meters depth. For longer dives a
gas heater is necessary. But a gas heater is another complicated
instrument to be added to the diver's equipment. A possible
alternative is to find a semi closed or closed system as safe as
an open circuit. A Swedish company - the AGA - has for many years
worked with the Swedish navy and constructed breathing gears.
They have recently designed a semi closed breathing apparatus for
mixed gases. It can be used for nitrox and heliox and has the CO_2

absorber inside the bag. We do not know yet if this gear will be
satisfactory for all our demands. But it seems possible that the
CO_2 absorber gives the gas in the bag a sufficiently high
temperature for long deep dives. The apparatus has some new
devices which give a very low breathing resistance and it has few
complicated mechanical details. A low gas consumption is another
advantage which is important for us as for other countries without
helium resources.

The deep diver, particularly during saturation dives, must be
kept warm for at least a couple of hours in very cold water. It
is also important that he can keep warm between the dives. The
diving compartment in the rescue vehicle is designed for electrical
heating but we have not yet found a sufficient heater for diving
bells.

CONCLUSION

A deep diver has to ao his work under difficult circumstances.
Disturbing factors like cold, complicated instruments, etc. prevent
him from concentrating on his task. A good job on the sea bottom
can only be performed by a well trained diver in good condition.
Safe and simple means to keep the diver warm will facilitate the
diving.

SUBMERSIBLE ACTIVITIES UNDER ARCTIC CONDITIONS

M.D. Macdonald and A.R. Trice

International Hydrodynamics

Vancouver, British Columbia, Canada

Many of the problems encountered while operating submersibles in Arctic regions are similar to those in any remote area of the world. However, working in remote areas with severe climatic conditions becomes particularly difficult. Nevertheless, work has been done by submersibles operating in Arctic regions over the last eight years. This paper will discuss operational problems and identify areas requiring further development. Arctic areas will include Hudson Bay and the coast of Labrador.

As far as these authors know, past experience with submersibles in the Arctic is limited to seven operations. The eighth operation is currently being conducted by the submersible PISCES IV operating off the M.V. "PANDORA" in the Beaufort Sea, and the ninth by the SDL-1 operating off Labrador. These operations have been and are being conducted by three operating groups, Perry Oceanographics, Riviera Beach, Florida; the Department of Environment, Marine Sciences Directorate, Victoria, B.C., and International Hydro- dynamics Company Ltd., Vancouver, B.C. It is hoped that readers will excuse the preponderance of reporting on our own Company's operations, which is a result of the relative availability of information.

SUMMARY OF ARCTIC OPERATIONS

Year	Locations	Submersible/Operator	Purpose
1968	Arctic Islands	PISCES I/HYCO	Instrument Retrieval Biological & Geological Observations Acoustic Measurements
1970	Hudson Bay	PISCES III/HYCO	Wellhead Inspection
1971	Bering Sea	Perry Oceanographics	Biological Studies
1971	Hudson Bay	PISCES III/HYCO	Geological Survey
1972	Labrador Coast	PISCES III/HYCO	Wellhead Inspection
1973	Greenland	Perry Oceanographics	Aircraft Recovery
1974	Beaufort Sea	PISCES IV/D.O.E.	Bottom Surveys
1975	Beaufort Sea	PISCES IV/D.O.E.	Bottom Surveys
1975	Arctic Islands	SDL-1/D.N.D.-D.O.E.	Bottom Surveys

PAST ARCTIC OPERATIONS

The operation in 1968 with PISCES I was the first major operation undertaken in the Arctic with a submersible. PISCES I was air lifted to Thule, Greenland and placed on board the Canadian Coast Guard Icebreaker "LABRADOR". Working in varying conditions of Arctic ice necessitated the use of an icebreaker (Ref. 1).

In order to perform maintenace under reasonable conditions and keep the batteries warm, it was necessary to fabricate a temporary enclosure. In spite of this, problems were encountered with valves freezing and "O" Ring seals not performing adequately. This also applied to the rubber components in external electrical connectors on the submersible.

When actually diving under ice cover, a line was secured between the surface vessel and the submersible. A 3/4" diameter neutrally buoyant line worked well. If diving in open water between flows, a constant vigil had to be maintained on relative movement between submersible, surface vessel and ice.

The logistics problems in the far Arctic were always difficult and expensive. This has been true of all operations of this type. Submersible operations are usually confined to summer months. If waiting for parts to be delivered, time is lost. Invariably, because of the short operating season, this time cannot be made up. Today, a minimum HYCO sea going spares inventory amounts to some

Figure 1 PISCES—I, Polynya – McLure St. – 1968

Figure 2   Hudson Handler – 1972

Figure 3 Hudson Handler – 1971

Figure 4   PISCES IV – M.V. PANDORA II –
Department of Environment, 1974

$150,000 in value. This is particularly necessary in the far north because of the communication difficulties.

The 1970 expedition of PISCES III into Hudson Bay brought new experiences. She was carried on board the Hudson Handler. This is a relatively small 90' self-propelled barge type vessel. The lack of support facilities, i.e., surface vessels capable of supporting a submersible, resulted in the Hudson Handler being designed, built and transported by rail to Hudson Bay specifically for these operations. This operation involved the inspection of an abandoned wellhead on the ocean floor.

Although tolerable, in this instance, one dive on a wellhead took 12 hours in -2°C water. It was common to have a thin ice coating on the inside surface of the command chamber. The crew is literally inside an ice house. In the event of the submersible being stuck on the bottom, the time available for rescue before the crew succumbed to exposure would be reduced.

The danger of entrapment on the bottom, combined with the remote location and lack of readily available support facilities makes operations in these areas particularly hazardous. One can readily imagine the difficulty that would exist in duplicating the armada of surface vessels that assisted in the rescue of PISCES III off Ireland in 1973.

The summer of 1971 saw PISCES III again in Hudson Bay. During this operation, the combination of PISCES III and the Hudson Handler accomplished 3612 miles of survey track and 86 oceanographic stations during which 65 sediment and twelve hard rock core samples were recovered (Ref. 2).

Here again, in this operations, as well as the 1970 operation in Hudson Bay, logistics were a major problem. As an example, the chartered flights of computer parts and technicians were a major delay and expense.

The 1971 operation off the coast of Labrador involved new lows (highs) in operating in extreme weather conditions. This was made worse by the absence of good forecasting and the presence of icebergs. Indeed, it was the inopportune arrival of an iceberg at a drilling site that resulted in the submersible inspection being required.

The most recent excursion into the Arctic by submersibles is the operation of PISCES IV; this submersible, operated by the Department of Environment of the Canadian Federal Government, is presently in the Beaufort Sea conducting experiments. This is a repeat of last year's operation in the same area. PISCES IV operating off the M.V. "PANDORA" spent three weeks in the Beaufort Sea inspecting the bottom for ice scours, taking sediment samples and investigating the biology of the areas. Approximately eleven dives were made.

The major problem on the trip was enormous expense in terms of time and money to get to the site. A submersible onboard its mothership left Victoria on July 2nd, 1974 and the first dive was not made until August 15th, 1974. This was partly due to the

inability to predict the ice conditions in advance and the result
was a three week delay at Point Barrow waiting for ice to clear.
Also communication problems were experienced.

One other submersible company, Perry Oceanographics of
Florida, have had experience underwater with submersibles in
Arctic conditions. During the winter of 1971, work was conducted
through the ice in the Bering Sea. In this operation, surface
temperatures of -40°F. and below were experienced, causing
difficulties with O-ring seals and decreased efficiency of life
support systems. Again, communications were a problem, although,
as in other instances, were facilitated by the co-operation of ham
operators.

In 1973, a one month operation in the open sea was conducted
off Greenland. This was for the purpose of recovering an aircraft.
In this case, the problem was finding adequate shelter in the cold
weather.

SUMMARY AND RECOMMENDATIONS

The problem of working in Arctic regions can be summarized
under three headings:
1. Logistics
2. Submersible support vessels
3. Submersible technical designs.

Logistics

Logistics, while less difficult with the passing of time, are
still formidable. Ideally, to reduce the inefficiencies due to
down time and to provide for a reasonable level of safety and
emergency back up, operations in these regions should be conducted
with two submersibles. Assuming adequate support vessels are
available (heading No. 2), the increase in cost per dive would be
reduced.

The major logistical problems relate to:
1. Communications
2. Navigation
3. Transportation.
As these are not unique to submersible operations, they will not
be treated further.

Submersible Support Vessels

Here we must distinguish between transportation of men and
equipment to the area, and the support of a submersible in the
Arctic.

The normal icebreaker which is ideal for moving in ice
covered areas is far from ideal as an all weather launch and
recovery vessel, and although this problem is reduced in ice
covered areas, the task of maintaining submersible components
while the support vessel is breaking ice is both difficult and
frustrating!

To support submersible operations in the Arctic other than
just summer months and have reasonable logistics, a large submarine
support vessel capable of working under the ice would seem to be
the answer. It could be envisaged that this type of vessel would
have a range to transit between a major port and the work sites
plus an endurance of six to nine months. A system of ice snorkels
would allow crew transfer, antenna deployment, engine intake-
exhaust, etc. Underwater hatches would accommodate submersibles,
divers, drills, etc. The on-site logistics would be limited to
crew changes and mail.

This is an area to which International Hydrodynamics is
giving considerable attention, with the assistance of our Federal
Government.

Submersible Technical Design

Although all submersible operations in the Arctic have
encountered technical problems with the submersibles, this has not
been a serious deterrent. These problems have been solved, usually
on the spot. However, safety and the high cost per dive make the
normal submersible limitations enormously more significant in the
Arctic.

The single greatest need on small submersibles is power. The
present very reliable lead acid battery will have to give way to
something more efficient. This becomes ever more apparent when
operating in Arctic waters. Exactly when more power should be
available for heating and endurance, the lead acid battery is
giving less because of low temperatures.

Challenging work has been accomplished by submersibles in
the Arctic. Without attention to the factors mentioned herein, it
will be difficult to envision the present invaluable contributions
made by submersibles in other parts of the world (e.g., the North
Sea) being repeated in the Arctic. However, solutions are being
found, and with continuing research and development, submersibles
and submarines will likely be the long term solution to working
under the conditions imposed by the Arctic.

References

1. A.R. Milne. Arctic Journal of the Arctic Institute of North America, Volume 22, Number 1, 1969.

2. Offshore Technology Conference 1972, Paper No. OTC 1631.

SUBMERSIBLE OPERATIONS IN THE ARCTIC

Commander J.P.B. O'Riordan

Ministry of Defence

London, United Kingdom

INTRODUCTION

Commercial working from manned underwater units began a few hundred years ago and the concept is even older. James Sweeney's "Pictorial History of Oceanic Submersibles" provides a story that need not be repeated here (Figure 1). To the majority of people, however, submarine business has always been synonymous with waging war beneath the sea. Indeed it was naval submarine operations in the two world wars which provided the technical developments and the impetus to the commercial world.

The naval and commercial concepts are, of course, different. Few military submarines achieve a depth greater than 400 metres whilst 1000 metres is now a routine operating depth for a commercial submersible. With certain exceptions military submarines take great care to keep clear of the bottom whilst the sea floor is the work place of the submersible. These differing requirements have resulted in the commercial world largely going it alone.

During the 1960s, in anticipation of the need to support the growing activities of the offshore oil and gas industry and increasing interest in oceanographic research, a large number of submersibles of all types were built by commercial companies, mainly American, ready to capture their share of an exciting new market. By the close of the 1960s and the beginning of the 1970s that market had still to materialise to any great degree. Offshore oil and gas exploration had been spreading slowly into the relatively shallow areas of the continental shelves, and the sort of work requiring the presence of men underwater was mostly more economically carried out by 'conventional' diving teams.

With the rapid escalation of offshore activities in the last 2-3 years offshore operators have developed the necessary technology

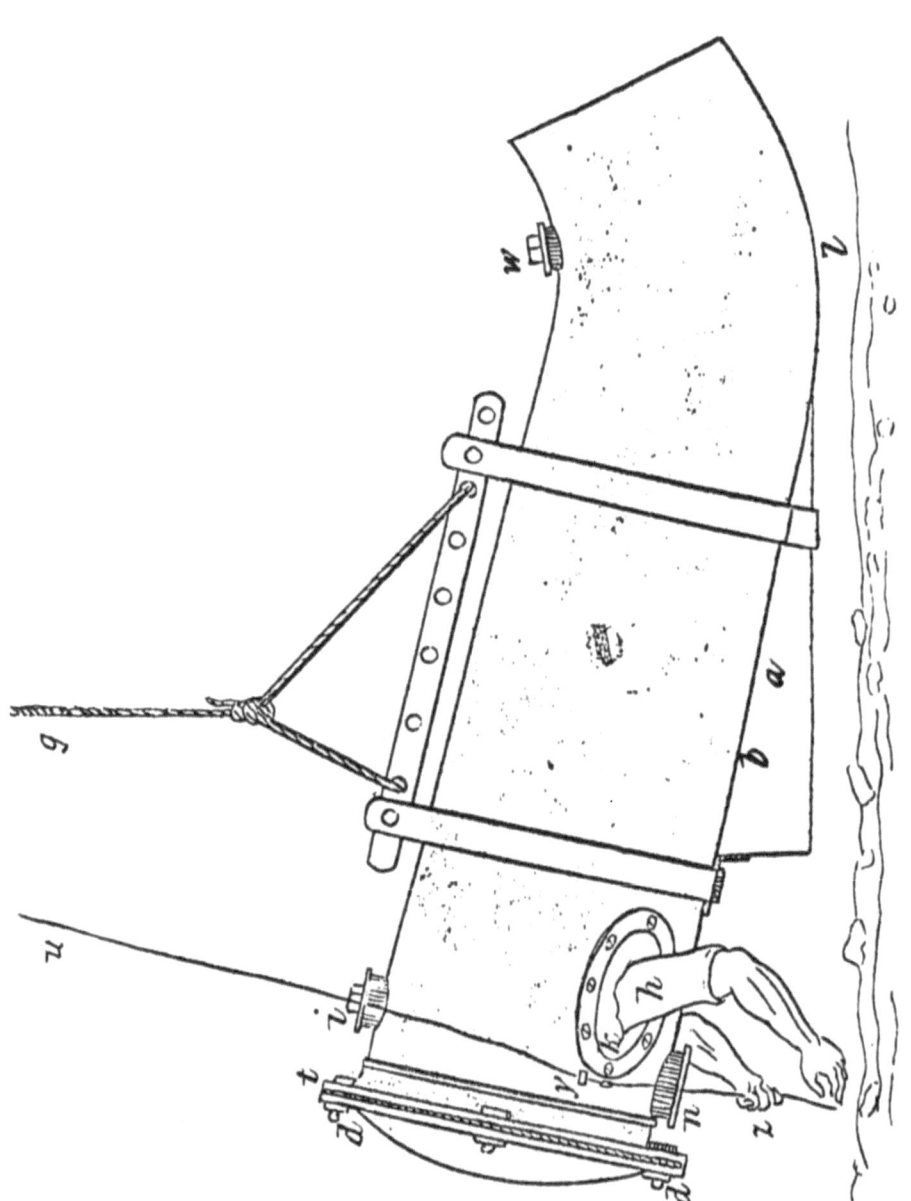

Figure 1.   Lethbridge Diving System c 1750

to explore most of the continental shelves' depths and the waters
beyond. Now the time may be approaching when there will be a
commercial need to use submersibles to exploit the vast potential
of the sea bed of the Arctic Ocean.

AIM

The aim of this working paper is to discuss and then suggest
the manner in which submersible operations could be effectively
carried out in the Arctic.

SCOPE

The paper will set the scene by highlighting the tasks for
which submersibles are needed. It will mention the type of
submersibles and their motherships which are in present day use
and will then go on to discuss how they are operated in temperate
climes. The paper will then point to the differences and
difficulties in an Arctic situation before making proposals to
overcome the difficulties.

TASKS FOR SUBMERSIBLES

The following is a list of some of the tasks which can be
performed by submersibles:
a. Underwater inspection.
b. Site and route survey.
c. Submarine rescue.
d. Unit recovery.
e. Cable and feeder line burial.
f. With tooling:
 (1) Tighten or loosen nuts and bolts.
 (2) Drill holes.
 (3) Cut cable, chains, hawsers, ropes etc.
 (4) Take small seabed rock core samples.
 (5) Cut wood with chainsaw.
 (6) Remove sediment by water jet.
 (7) Make penetrations, insert studs using Cox's gun.
 (8) Operate valve handwheels.
 (9) Paint.
 (10) Wire brush surfaces.
 (11) Welding and burning.
 In the near future the above list will be extended to include:
a. Heavy lift capability.
b. Heavy duty power supply.
c. With tooling:
 (1) Carry out explosive welding.

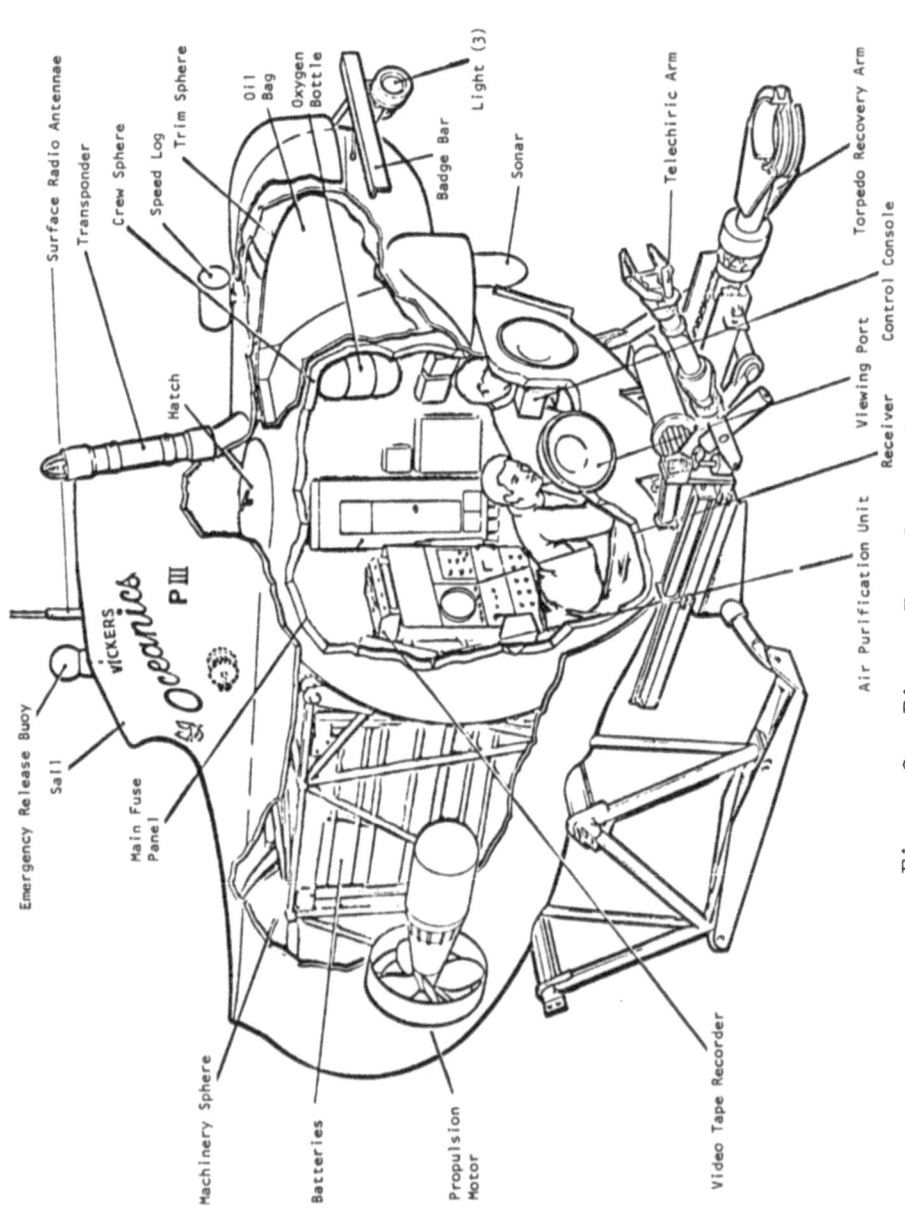

Figure 2. Pisces–Type Submersible

(2) Carry out ultra-sonic, magnetic particle, X-ray of structures, pipes etc.

TYPES OF MANNED SUBMERSIBLES

Two existing types of submersibles are chosen as being representative:

a. <u>Pisces Submersible</u> (Figure 2). The Pisces submersible developed by International Hydrodynamics of Vancouver is suitable for operating in the 600-2000 metre range. Figure 2 shows the arrangement of a Pisces-type submersible. It comprises a main sphere of 203 cms diameter and a machinery sphere of 152 cms diameter. A lead acid battery which is immersed in oil and pressure compensated is situated between the two spheres. The submersible has two propellers driven by pressure compensated oil filled electric motors (either 3 or 5 horse power). A normal dived operation is in the order of 36 hours at a speed of 2 knots. Life support systems will, however, enable a 2 to 3 man crew to remain shut down for 7 days.

b. <u>PC Submersibles</u> (Figure 3). The PC Submersible designed and built by Perry Oceanographics of Florida, is a Diver lock-out unit suitable for operating down to 400 metres. The essential features are as shown in Figure 3. It comprises a cylindrical hull with two compartments which are inter-connected by a pressure-resistant and water-tight hatch. The forward compartment is dry and carries the pilot and a diving supervisor. The after compartment carries the two divers. The divers when at work are connected to the submersible by an umbilical. The PC submersible has 2 x 10 hp motors which give it a speed of 5 knots. It carries a crew of up to 3 in addition to the two divers.

MOTHERSHIPS

To operate submersibles a support "Mothership" is required. Many are now in service and the later ones are capable of operating submersibles in sea conditions of at least sea state 4-6 (4 Metre Waves). The modern mothership has to have electrical and mechanical workshops, satellite navigation, submersible transponder tracking network and communications and a method of launching and recovering submersibles. Support ships for diver lockout submersibles need to be fitted with a saturation diving complex.

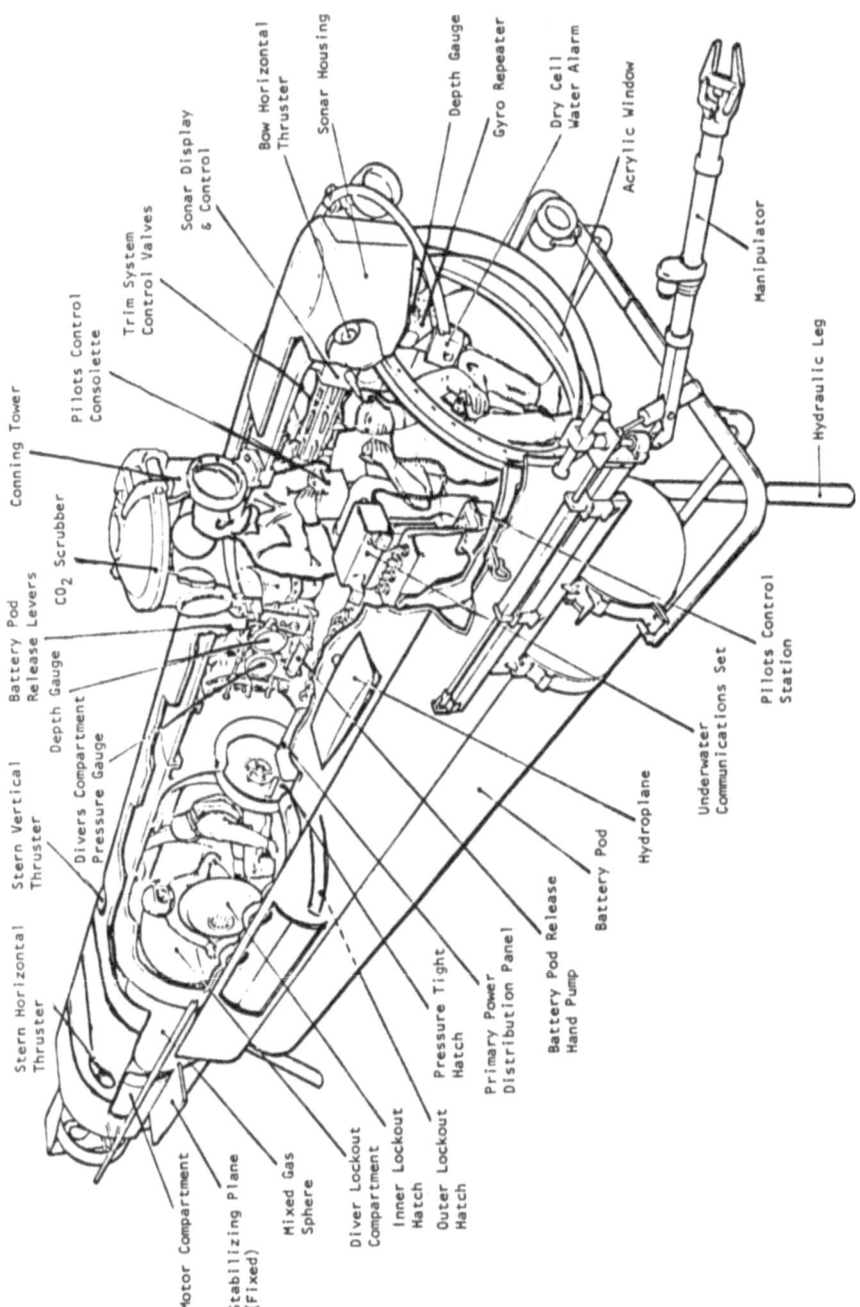

Figure 3.   PC-15 Submersible

SUBMERSIBLE OPERATIONS

Since North Sea oil and gas is close to major consumers, transport costs are reduced and more may be spent, per barrel, on exploration and extraction. Later, as prices rise, the same may well be justified further from the main consumers. The North Sea provides a nursery and proving ground for later world wide endeavours. As the experimental grows into the routine there is likely to be a place for divers, true submarines, submersibles and remotely controlled devices such as "sledges" for pipe burial.

SOME OPERATIONAL PROBLEMS

In the commercial world time is money and a Mothership with two submersibles will, given favourable weather conditions, hope to be able to keep one on task around the clock. Even a simple task such as the survey of 100 miles of pipeline will take a minimum of 50 hours for a Pisces. As has already been mentioned the latest Motherships can now operate in sea states 4-6. Even so the difficulties of launching and recovering result, and in the sea conditions in the waters around the UK has resulted in anything up to 40% "downtime" over a year. This coupled with passage time to the area can make the conduct of submersible operations long, tedious and expensive.

When on task the submersible lacks flexibility and provision of tooling has to be made well in advance of meeting the problem. Very often the underwater problems are ill-defined and two or three alternative approaches need to be worked out and tooling provided accordingly. Because of this it is foreseen that there will always be specialised tasks that divers will perform best and hence the requirement to have some submersibles which have lockout facilities.

REDUCE DOWNTIME

So far the paper has attempted to give the reader sufficient background to move from present day real life operations to look into the future. To many the prospect of the widespread use of submersibles in Arctic waters may sound far fetched or if at all possible only at a prohibitive cost. It is for consideration, however, that such operations may come about as the logical follow on from the desire of commercial companies to solve the main problem of operations today and that is to reduce "downtime" and the way in which this could be done will be discussed later.

UNDER THE ARCTIC OCEAN

As known and more accessible reserves dwindle, discoveries of
hydrocarbon and mineral resources in the far north have magnified
the importance of the Arctic. The vast potential of the sea bed
is there to be tapped. One wonders if it can be done; the area of
the Arctic Ocean is vast, its depth range likewise. Some of it is
covered by ice all the year round, some never and the remainder
only seasonally. Comprehensive studies on the Arctic stretch to
many volumes and although it can be dangerous to generalise it is
considered reasonable, at this stage, to set the scene for
submersible operations by taking a typical set of Arctic conditions.
The following gives only a brief summary of the main
characteristics of the Polar region:
a. The Polar Ice field is not a solid, complete and homegeneous
sheet of ice. The edge is not always clearly defined and
particularly in summer or after strong off-ice winds, brash ice
and small floes may be encountered as much as 150 miles clear of
the real ice edge. Predictions, even those based on satellite
information, are frequently inaccurate as under the influence of
winds, currents and seasonal variations the ice pack is constantly
shifting. The rate of drift can be anything between 0.2 and 2.6
miles per day. Ice along the coast in sheltered bays, however,
tends to remain anchored to the land.
b. Icebergs are formed when land ice or glacial ice becomes
detached into the sea. Although 7/8 of the mass of an iceberg is
below water, the draught is seldom more than 5 times the height of
the 'blockiest' bergs and may be as low as one or two times the
height of pinnacled bergs.
c. Ice Thickness. The thickness of the ice in the Polar ice cap
varies considerably. In general new ice (between 0.3 and 1.8
metres thick) forms about 8% to 10% of the area, and the rest is
old ice which has a mean thickness of some 3.6 metres. Occasionally
pressure causes ridges and hummocks on the ice and the ice keels
can be as much as 45 metres deep.
d. Leads and Polynyas are formed within the ice field by the
movement of the pack, right up to the Pole itself. They are more
frequent towards the ice edge but north south they are normally
about 25 miles apart.
e. Weather. Frequent but relatively short lived storms are a
characteristic of the Arctic winter. They are difficult to forecast
and tend to develop and subside very quickly. These gales generally
bring short high seas. When the wind blows off the ice there is
little or no swell at the ice-edge. However, with the wind in any
other direction it is common for a low swell to penetrate into the
ice field for 10 or 15 miles until the pack becomes consolidated.
f. Temperature. An air temperature at the ice edge of -37°C plus
a 20 knot wind can reduce this to an effective temperature on the
human body of -67°C. The sea water in the 45 metre surface layer
is about -2°C, in the next 75 metres it warms up to between 0 and 3°C
there on down it is about 0.7°C.

g. Icing Conditions. At -2°C salt water will freeze on exposed
surfaces and at -18°C salt spray will freeze solid on striking an
exposed surface and rapidly build up in ice blocks. Ice tends not
to form on surfaces actually washed over by the sea, and on diving
ice will clear fairly quickly once completely submerged.
h. Underwater visibility. Visibility underwater in the ice field
is generally good and horizontal ranges of 90 metres have been
reported. In summer the growth of marine organisms reduces this.
It is difficult to estimate visually the thickness of ice from
below.
i. Sonar Conditions. Sonar conditions are affected by the sound
velocity profile as well as the wind and swell. With strong on-
ice winds the noise of brash, pack, pancake and slush can make
conditions impossible. In light airs conditions may be excellent.
Generally speaking the further inside the ice edge one goes the
better the conditions.

THE SUBMARINE

As we have discussed earlier the Mothership's inability to
launch and recover her submersibles in all weathers is the chief
cause of downtime. This could be rectified by the provision of a
stable platform. The only manner in which this could be guaranteed
would be by using a submarine.

Choosing a submarine as the stable platform is a simple matter,
from this point on it is all uphill. First the type of submarine;
but in making this decision cost is going to be the decider.
Although the attractions of a nuclear powered submarine Mothership
are many and obvious it is considered that the cost of such a
vehicle would be prohibitive and its use uneconomical. Because of
this it is not considered further in this paper. The alternative
is the conventionally powered diesel electric.

THE SUBMARINE AS A MOTHERSHIP

It is often preferable to base a new concept on a well proven
system and it is therefore envisaged that the submarine mothership
would most likely be an ex-military one. Perhaps it could be an
American Guppy, a Soviet Foxtrot or a UK Porpoise. For example
the UK Porpoise has a submerged displacement of 2,400 tons, a
surface speed of about 12 knots and a good low speed endurance
whilst dived. Being 88.5 metres long it could possibly be
converted to carry up to four 9 metre long submarines. With a
dived speed of advance of between 2 and 4 knots it would probably
only have to snort, i.e. go diesel electric to charge its batteries,
for about 2 or 3 hours a day. It would be capable of a patrol
cycle in excess of 40 days. At a later date there would likely be
a requirement for a 'tailor made' mothership.

Obviously a full range of 'in depth' feasibility studies would be required but the following are some of the areas which would require investigation:

a. <u>Cost</u>. The scrap value of a 16-17 year old conventional submarine of the sort which has been mentioned would be in the region of £ 100,000. To refit the existing hull to make it operationally safe and satisfactory would cost between £2 million and £3 million, depending on the amount of conversion, and would take at least a year. The submarine would then be operational for 4-5 years, after which another, perhaps less expensive, refit would give it a further 4-5 years of useful life. To build a comparable new submarine would cost about £ 12 million, perhaps more.

b. <u>Conversion</u>. The main changes would revolve around adapting the submarine to take the submersibles. This could be done using existing forward and aft escape hatches to lock crews in and out and would be the basis for two of the docks. With two more lock out systems a total of 4, approximately 9 metre long submersibles could be carried. The top weight would necessitate re-ballasting of the submarine but should not be an insurmountable problem. Divers would be able to lock in and out either direct when one of the submersibles was away or through the PC type submersible on deck. The torpedo tubes could be blanked off or removed and the stowage compartment used as a workshop. One compartment would be needed as a re-compression chamber. Some of the sonar outfit could be removed leaving a basic set for own safety and for tracking the submersible. The radar and communications outfits would need to be retained and a satellite navigation system installed if not already fitted. Most existing submarines have some form of diver lockin/lockout down to about 150 metres.

c. <u>Refitting and Maintenance</u>. A refitting firm with submarine expertise and experience would be required and a base port with back up facilities and a dry dock would have to be chosen.

d. <u>Running Cycle</u>. 5 to 6 weeks on task plus 2 to 3 weeks for maintenance and sea trials would probably be the normal operating pattern. It is anticipated that depending on the size of crew and the level of employment of the submarine that it would cost something like £750,000 a year to operate. An estimated cost over its commercial life time is at Figure 4.

e. <u>Personnel and Training</u>. It is thought that recruitment of a trained crew would not be a serious problem, since most navies have a high throughput of submariners, leaving the service whilst still young. A crew of 40, plus the crews for the submersibles, would probably be the minimum for safety, but the interaction of power systems, life support systems, duration of dives, etc. is clearly a complex business. Once a crew had been assembled they would have to work themselves up and then go on to exercise and practice the full range of procedures for releasing, controlling, rendezvousing and recovering their submersibles. Only when they become fully · proficient should they be permitted to deploy to their operational area.

1.   TOTAL LIFE COST

ITEM	COST (£ Sterling)	TIME (Years)	AGE (of hull)
Purchase price of second hand conventional submarine	100,000	0	17
Refit and conversion	2,500,000	+ 1	
Running costs	750,000 750,000 750,000 750,000 750,000	+ 6	23
Second Refit	1,650,000		
Running costs	750,000 750,000 750,000		
Scrap	750,000	+11	28
TOTAL	11,000,000	11 years (9 operational)	

2.   OPERATIONAL CYCLE

Passage   2 weeks
Operating   3.5 weeks
Maintenance 2.5 weeks - giving a minimum time on task
                         of 23 weeks (161 days).

3.   RUNNING COST

To absorb the cost of the refitting years means that the
effective running cost per operational year is £1,220,000.
This means £7,440 a day or £310 an hour.

Figure 4. Estimated Cost of Using a Submarine
as a Mothership (does not take account of the Submersible)

f. <u>Command and Control</u>. Most maritime countries now work a system
of voluntary notification by which a firm operating submersibles
gives notice, on an in-confidence basis, to their own Ministry of
Defence. This helps overcome problems of mutual interference
between dived craft as well as providing some guarantee of assistance in the event of an accident. As submersibles and their
submarine motherships venture further afield the need would arise
for first of all some international voluntary system of
notification and then, hopefully, a statutory one.
g. <u>Maintenance and repair whilst on task</u>. With carried onboard
spares and the well tried submarine philosophy of belt and braces
for ships safety systems there should be little problem in the
Mothership being able to maintain a high operational availability.
For the submersibles it may not be so easy for only limited
internal and no external maintenance or repairs could be carried
out whilst at sea although the submersible would be able to 'plug
in' for a battery charge on its return from a mission. It is quite
likely that although a submarine would carry 4 submersibles it
would only operate 2. A possible answer, but adding yet more to
the cost, would be to have a surface depot ship looking after 2 or
more submarines and to whom they could, having chosen a window in
the weather, transfer and submersibles in need of repair.

SCENARIO

The conventional submarine is no newcomer to the Arctic but
because it has to rely on snorting to charge its battery is never
likely to venture far in under the ice. This, for the time being,
limits the sphere of operations but there is still much of the
Arctic which is ice free or on the ice edge. For instance what is
sometimes called the Marginal Ice Zone (100 miles inside and 30
miles outside the ice), allowing for seasonal variations, gives us
an area of half a million square miles in the 900 to 1500 miles
bracket from the United Kingdom.
Let us now consider how such an operation might be mounted
using a UK commercial conventional submarine which has deployed to
a forward operating base such as REYKJAVIK. Although it could have
carried its own submersibles it is more likely, to ensure that they
were in tip-top condition, that they will have gone in a ship of
trade or Mothership together with spares and a support team.
Having fuelled and topped up with stores the submarine Mothership sails with 4 submersibles (2 Pisces and 2 PC) and once in deep
water dives to head for an oil exploration area, which for the sake
of argument, is near the ice edge in the Denmark Strait. Her route
and operating areas would already have been cleared by her owners.
From now on the submarine is in her element and the transit to
her patrol area will hopefully be a routine and uneventful one.
When on location it could, if the depth of water permitted, bottom
or proceed at very slow speed to launch its submersibles. Control

of the submersibles would be by active sonar, bleepers and underwater telephone and, in the final stages of docking, visual. The
Mothership would also use her active sonar to give early warning
of ice and patrol at a safe depth of about 60 metres to keep clear.
When she requires to go shallow to communicate with 'head office'
or snort to charge her batteries she would use her upward looking
echo sounder to determine the thickness of the ice and also assist
her in looking for a 'blow hole'. It is not a problem for a
submarine, having caught a stopped trim, to lean on ice of about
0.3 metres thick, break it and then expose its snort mast. The
proximity of the submersibles to the ice would mean that some form
of 'anti roll bar' type protection would be required for them.
Should, due to the 'sea state', the Mothership be rolling at her
normal operating depth the problems of docking and undocking would
be eased by her going deep.

The working conditions for the diver may appear anti-social
but with heated suits and breathing gas he is probably a great deal
better off than his colleague digging a hole in Alaska.

SUMMARY

To summarise:

a. Methods are required to explore and tap the vast potential
of the sea bed of the Arctic Ocean.

b. Submersibles, already in existence, have the capability
of fulfilling many of the likely tasks.

c. No matter how sophisticated submersibles become there is
likely to be a place for the diver in addition to submarines,
submersibles and remotely controlled devices.

d. The environmental conditions of the Arctic should not
inhibit the dived operations of submersibles.

e. The difficulties of launching and recovering submersibles
from existing Motherships in bad weather may result in anything up
to 40% "downtime".

f. The only way in which to guarantee a stable platform and
therefore solve the problem of "downtime" is to use a submarine as
the Mothership.

g. The cost of a nuclear powered Mothership would be
prohibitive.

h. An ex-military conventional submarine could be converted
and operated commercially at an on task cost of approx. £310 an
hour.

i. The use of a submarine Mothership in the Arctic may come
about as the logical follow on to the desire of commercial firms
to reduce "downtime".

j. The conventional submarine is already no stranger to the
Arctic.

CONCLUSION

It is concluded that it is not beyond the bounds of reason for a submarine Mothership to effectively control and operate submersibles in the Arctic. Whether it will ever be cost effective depends upon how much we may have to be prepared to pay for the fuel we shall need so badly.

WORKING SESSION CHAIRMEN'S REPORTS

P.J. Amaria, A.A. Bruneau   and   P.A. Lapp

Memorial University of Newfoundland   Philip Lapp Ltd.

St. John's, Nfld., Canada   Toronto, Ont., Canada

This section contains the edited version of the original six reports submitted by the working group chairmen. The editors accept responsibility for any distortion inadvertently introduced in the summarising and editing processes.

The reports were submitted by the following working session chairmen, who were assisted by their rapporteurs:

Chairman	Rapporteur	Arctic System

Captain R. White
U.S. Coast Guard
  Academy
New London
Connecticut, U.S.A.
	Professor H. Snyder
	Memorial University
	  of Newfoundland
	St. John's, Nfld.
	Canada
		Transportation

Brigadier General
  K. Greenaway
Department of
  Northern and
  Indian Affairs
Ottawa, Ontario
Canada
	Dr. D. Dunsiger
	Memorial University
	  of Newfoundland
	St. John's, Nfld.
	Canada
		Traffic Management,
		Navigation and
		Communication

Dr. H. Schroeder-
  Lanz
University of Trier
Trier, Germany (FR)
	Dr. D. Bajzak
	Memorial University
	  of Newfoundland
	St. John's, Nfld.
	Canada
		Operational
		Information and
		Forecasting

Chairman	Rapporteur	Arctic System

Dr. R.E. Francois
University of
  Washington
Seattle, Washington
U.S.A.
	Mr. J. English
	Memorial University
	  of Newfoundland
	St. John's, Nfld.
	Canada
		Underwater
		Operational

Mr. E. Reimers
Ministry of
  Environment
Oslo, Norway
	Dr. R. Peters
	Memorial University
	  of Newfoundland
	St. John's, Nfld.
	Canada
		Environmental
		Protection

Mr. A. E. Pallister
Pallister Resource
  Management
Calgary, Alberta
Canada
	Mr. D. Grenville
	Memorial University
	  of Newfoundland
	St. John's, Nfld.
	Canada
		Social and Cultural

and

Dr. N. Orvik
Queen's University
Kingston, Ontario
Canada

ARCTIC TRANSPORTATION SYSTEMS

The issues facing Arctic transportation systems were identified as either internal or external. Internal issues are those which primarily involve the technology of transport. In summarizing these issues, no strong in-depth attempt was made to find solutions, rather to recognize and list items. Where transportation affects other systems, or vice versa, the issues were identified as external. These items identify common gaps between transportation and another area of interest.

Internal Issues

The identification and ranking of significant issues is facilitated by categorizing the commodities to be transported and applying these categories to the various modes of transportation.

The categories of commodities to be transported were identified as bulk fluid, bulk solid, supplies and people.

Bulk fluid commodities include oil, gas, and slurries. Bulk solids include ores, concentrates, and metals. Supplies includes construction, maintenance, and production materials as well as domestic items (groceries, fish, handicrafts) and general cargo. The last category is people, both in terms of transport to and from, as well as within, the Arctic.

It should be noted that energy (e.g. electricity) was identified by the group both as a commodity to be transported as well as a method, however it was decided not to include this category in this consideration.

Recognizing that most transportation is intermodal, the separate modes of transport were identified under the headings of land, sea, and air. Surface effect vehicles (air cushion vehicles) were considered separately.

With these two classes of categories, a matrix can be constructed as illustrated in Table I. This table provides a quick overview of the internal issues which relate to Arctic transportation systems. Added to the commodities category is a grouping called interface. Within each cell of the matrix the issues are identified in terms of degrees of severity. This ranking also expresses a certain experience or confidence factor, i.e. - where problems are identified as insignificant or minor, the experience or confidence factor is considered to be high. This matrix also serves to identify major gaps in knowledge so that from it a "shopping list" of research can be prepared, viz:

Pipelines - Bulk Fluid and Bulk Solid
      I   Terrain Stability
     II   Surveillance of Work
   III   Nature of Commodities
    IV   Weather

Table I

Arctic Transportation Systems
Summary of Internal Issues

TRANSPORT MODE / COMMODITY	LAND				SEA			AIR	
	PIPELINE	RAIL	TRUCK	SEV/ACV	SURFACE SHIP	SEMI SUBS	SUBS	AIR (FW)	AIR OTHER
BULK FLUID	XXX	XXX	N.A.	N.A.	XX	XXXX	XXX		N.A.
BULK SOLID	XXX	XXX	N.A.	N.A.	XX	XXXX	N.A.		N.A.
SUPPLIES	N.A.	XXX	X	N.A.	X	N.A.	N.A.		X
PEOPLE	N.A.	XXX	X		X	N.A.	N.A.		X
INTERFACE	∨	∨	∨	∨	∨	∨	∨	∨	∨

XXXX MAJOR PROBLEMS
XXX INTERMEDIATE
XX MODERATE
X MINOR

/ INSIGNIFICANT PROBLEM
∨ SYSTEM INTERFACE PROBLEMS
N.A. NOT APPLICABLE

Rail (all commodities)
        I   Terrain Stability
      II   Weather
    III   Manpower

Truck (all commodities)
        I   Terrain Stability
      II   Weather

Surface Effect Vehicles - problems are not significant

Ships - Surface (all commodities)
        I   Human Factors
      II   Ice characteristics and effect on performance
    III   Manouverability and docking
Ships - Semi-submersible (all commodities)
         (as for surface, plus)
     IV   Vertical Control
  - Submersible (all commodities)
         (as for semi-submersible, plus)
      V   Hydrographic Data
     VI   Navigation Systems
    VII   Search and Rescue

Aircraft - Fixed Wing (all commodities)
        I   Landing
  - Other (all commodities)
        I   Weather

In addition to these issues which were clearly internal and related specifically to one mode of transport, a number of general items were identified which applied to many of the commodities or modes:
- The nodes, that is where one form meets another - harbours, terminals, offshore terminals.
- System flexibility in terms of commodities carried and destination, e.g. a pipeline is extremely fixed as opposed to an aircraft or ship.
- Volume fluctuations - capacity of a mode to withstand peak loads.
- Reliability and redundancy as they blend together with damage and risk.
- Timing of implementation of a project - or the ability to transport and the time it takes to get a mode in operation.
- The group expressed a keen interest in somebody's good 'crystal ball' which would provide an ability to anticipate future technological developments. Such an ability would provide a sense of direction or a sense of push.
- Government involvement - funding as well as regulatory role.
- Habitability - of people involved with transportation system.

- Energy conservation - wise use of energy.
- Search and rescue - both in the sense of recovering personnel and property but also identification of problem areas, e.g. pipeline cracks.

External Issues

For each of the other systems under consideration, items were ranked in an order that generally reflected the group's judgement of priorities.

A.   Environmental Protection
     1.   Location and description of valuable and vulnerable areas which have to be protected.
     2.   Elaboration of permissible levels of disturbance and pollution within
          (a) different types of land areas and shallow waters;
          (b) different types of marine areas; and
          (c) specified periods.
          2.1 Guidelines for using vehicles, vessels, and surveying equipment within (a) and (c).
          2.2 Guidelines for removing natural resources within (a), (b), and (c).
          2.3 Guidelines for construction and operation activities, including baseline studies, monitoring, and inspection activities within (a) and (c).
          2.4 Guidelines for construction and operation of tailings, pipelines including outlets within (a) and (b).
          2.5 Guidelines for using different types of ships within (b) and (c) with special reference to avoiding damage to animal and plant life.
     3.   Guidelines for transportation of and contingency arrangements for contaminants within
          (a) different types of land areas and shallow waters;
          (b) different types of marine areas; and
          (c) specified periods.

B.   Social and Cultural
     1.   The need for a clear statement of Arctic region social and cultural goals, and their priorities, reflecting national, international, and Arctic region requirements.
     2.   The need for an accepted framework for assessing Arctic Transportation Systems in light of the stated social and cultural goals. This framework should include a comprehensive list of characteristics and criteria for evaluation.
     3.   The need for agreed upon procedures and institutional forms which have the required local credibility, decision making powers, and commitment in all phases of feasibility studies,

planning, implementation, operation, maintenance, and
potential decommissioning of Arctic Transportation Systems.

C. Traffic Management, Navigation and Communication
    1. Co-ordination: A major requirement is to define channels
        of effective "communication" - to state requirements and
        "deliver" data to the user. This applies equally to both
        navigation and communication systems and indeed, to
        management itself when a number of government and private
        sector components are involved. Co-ordination should
        perhaps be organized regionally, users on other lines.
        There appears to be an associated problem related to
        definition of roles. Co-ordination of documentation is
        also required.
    2. System Quality and Integration: Navigation and
        communication systems are of variable quality and may be
        committed to use by a single mode/user. It is necessary
        to maximize the ability of all such systems through
        integration which would also serve to minimize unnecessary
        duplication. These systems, along with operational
        information and forecasting, and to a degree underwater
        systems, are highly related to transport:

                         Transport

                    Co-ordination Problems

        Traffic   Operational Information   Underwater

D. Operational Information and Forecasting (OIF)

    General:
    1. Definition of desirable standards of output (level of
        services) and the accuracy in the operation of the system.
        Level of services and accuracy must be considered in
        relation to cost.
    2. Communications feedback mechanism from user (transportation)
        to OIF of actually observed data for interactive improvement
        of methods.
    3. Establishment and refinement of user interpretation systems
        (e.g. Canadian ice charts).

    Specific:
    4. Remote sensing of ice and snow: Methods do exist, however
        the step from data collection at singular points by means

of prototype scientific instrumentation to an integrated system is formidable.

5. Prediction of seismic activities (earthquakes) and wave propagation.

6. In many areas, hydrographic data are already collected by means of multispectral interpretation of satellite imagery in combination with ground truth data. The specific modification of this method for the Arctic needs developing.

E. Underwater Operational Systems

1. Updated information should be provided, both with reference to the physical environment of the Arctic underwater as well as technological developments in underwater operational systems.

2. The economic study of transportation costs in the Arctic and the possible two phase (surface and submerged) operation of the systems.

3. It should be pointed out that this system is closely related to Arctic Transportation and in many ways is a sub-group of it, consequently external issues under this heading could be considered internal.

ARCTIC TRAFFIC MANAGEMENT, NAVIGATION AND COMMUNICATION SYSTEMS

Within the framework of the group's mandate, communication and navigation systems were treated as basic elements contributing to all Arctic systems. They act as a service to, and a linkage for, resource discovery, movement, and management, contribute to the regulatory and administrative process, and serve to unite the sparse social milieu.

Internal Issues

A. Communication Systems.
Five main issues emerged from working group discussions. These were: the gaps in geographical coverage; under-utilization of existing systems; new system requirements; the inadequacy of available energy packages for communications terminals; and the social implications of systems design, development and operation.
(1) Gaps in coverage
   a. Due to the positioning of the INTELSAT satellite, communication coverage of Greenland is lacking for all but the southern portion. This gap might be partially filled by repositioning domestic satellites which provide coverage in North America (e.g. Canada's ANIK or future U.S. satellites).
   b. Coverage by geostationary satellites cannot be provided North of 80°N. This area could be most economically served by terrestrial systems which would be interconnected into southern satellite systems. (N.B.: With recent improvements in northern communications related to satellites, many arctic communities are now better served than communities in sparsely populated southern regions).
(2) Under-utilization of existing systems
Many of the sky-wave propagated communication systems do not operate at full capacity. Within the known constraints of propagation conditions and available frequency spectra, procedures can be improved and sophisticated equipment used to provide more effective utilization of channels and systems.
(3) New system requirements
   a. Small, inexpensive and portable ground transmit/receive sets are required for direct ground to satellite communication. However, there is a tradeoff between the costs of such sets and the satellites.
   b. Search and Rescue satellite systems are needed to yield a reduced search time for localization. Cost benefit analysis might demonstrate that such a system is more economical than the present mode of carrying out searches.

      c.   Systems effectiveness depends on service and
           maintenance. Better service arrangements are
           required. With appropriate modular designs, local
           people can be more readily trained to carry out
           on-the-spot service and maintenance.

(4) Energy Sources
     There is a need for low cost, reliable and stable power
     packs and sources for communication devices of all kinds
     and sizes. This deficiency in present technology is a
     general problem.

(5) Social implications
     Communication serves two purposes – the transmission of
     information related to administration of regulatory
     aspects of the north, and secondly, the transmission of
     information from a social point of view. It is
     imperative, therefore, that local people are involved in
     the operating and maintenance of these systems. It was
     felt that the standard of maintenance and operation
     would go up if local people became involved in these
     tasks.

B.   Navigation Systems
    (1) Internationally coordinated National Navigation Plans
       There is an immediate need for a commonality of integrated
       air-sea-land navigation systems. Failing this, nations
       will be faced with system duplications and escalating
       costs, and they will lose present opportunities to
       coordinate requirements between modes (air, sea and land)
       and across nations. The state of technology is such that
       acceptable solutions do exist for all modes of
       transportation. The difficulty is that no single ideal
       system exists and national and vested interests intrude
       in the selection process. Formulation of a National
       Navigation Plan will require considerable investigation
       and a formal mechanism, such as a dedicated inter-agency
       Program Planning Office which includes representative
       user groups. The study would identify all users and employ
       cost-benefit analysis of all systems. It would aim to
       produce general policy guidelines and specifications that
       will allow industry to respond with appropriate hardware
       and systems, and enable users to plan sensibly. Current
       systems which appear to go some way in meeting users'
       requirements include LORAN C and OMEGA. The concept of a
       NNP as it affects other users is discussed below under
       external issues.

    (2) Research into propagation phenomena
       A coordinated study is required of ground propagation
       phenomena as they affect low frequency electromagnetic
       waves in the Arctic, particularly over permafrost and
       glacial ice. The operation of LORAN C in this environment

also requires further study. At present insufficient
data exists which would allow for proper planning and
the determination of system configuration.
(3) Navigation charts
In certain areas a need is evident for more detailed
topographic and hydrographic surveys. Innovative methods
should be developed for producing such charts in the
Arctic.

External Issues

A. Internationally coordinated National Navigation Plan (External)
Such a plan would involve international, political, and
economic interactions. It must satisfy regulations and
regulatory bodies as well as commercial, scientific and
individual efforts and requirements.
In order to satisfy these diverse bodies considerable
investigation is required. The study must identify all users
and must apply rigorous cost-benefit analysis to all systems
proposed in an attempt to satisfy individual user needs.
The development of common policies and the taking of an
integrated approach leads to a major requirement for a
response from industry. In addition, a planning office will
be required, joining all users via representatives from the
user community.
Programs which are developed from the plan require careful
phasing to allow the users to make adjustments, both
technical and social, and to avoid undue economic strains.
Finally the plan must consider:
(a) International coordination, e.g., repercussions on
frequency allocation, and vested national and commercial
interests.
(b) Publicity to ensure acceptance of the plan.
(c) Social responsibilities for both maintenance and
operational effects.

B. Communications
Three issues emerged:
(1) Broadcasting for social, cultural and educational needs
flowing in an east-west pattern. Such programs would
aim to satisfy the needs of a northern society and would
extend beyond national boundaries.
(2) Individual Communications
Sufficient channel capacity must exist to satisfy the
inter- and intra-community traffic density. It must also
be reliable and allow for contacts to be made in a
reasonably short period.
(3) Data Collection and Dissemination
A need exists for developing common data formats

especially for data handling by computer. There is also
a need for designation of special high data rate channels
(e.g., in remote sensing) and a requirement for common
specifications of hardware.

C. Arctic Operations Systems
The original heading within the group's mandate of Traffic
Management is considered too narrow and doesn't meet with
either national or international requirements. Various
regulatory bodies are created as pressures demand and no overall concept appears to be emerging. If current methods are
allowed to continue then such bodies as ICAO, SAR and VIM will
be duplicated with regulations being written to protect
resources and the environment. In particular, (i) ocean
pollution regulations (e.g., Arctic Water Pollution Act);
(ii) fisheries and wildlife regulations; and (iii) land use
regulations must all be incorporated into a common system.

If allowed to continue, multiplicity of common elements
will grow, yielding economic strangulation. It will also
result in an insufficient response to user needs.

The following Figure shows a proposed centre to coordinate
these efforts. The centre cannot be planned within the context
of this conference but requires a Program Planning Office, not
a consultant, and would require at least two years of effort.
This Program Planning Office must operate under strict cost/
benefit savings for overall implementation over individual
units. Capital savings are envisaged for: (a) the centre
itself; (b) shared aircraft; (c) data buoy; (d) satellites
including environmental sensors; (e) communication systems;
and (f) computer utilization.

The "timeliness" of this concept should not be lost. An
immediate need exists for the development of the nerve centre
shown in the Figure - in order to make an Arctic Operations
System responsive to all users for both regulatory and
information purposes.

<u>Regulatory</u>   <u>Information</u>

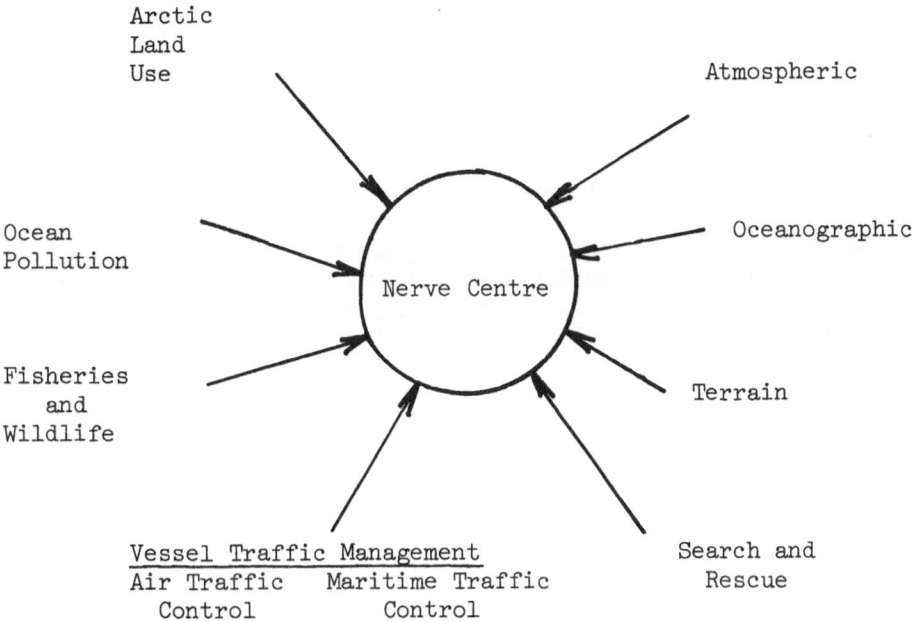

Figure 1   Arctic Operations System

ARCTIC OPERATIONAL INFORMATION AND FORECAST SYSTEMS

The keynote speaker and subsequent speakers of this working group outlined the presently applied operational and forecasting systems. These papers served as the focal point of the group's deliberations. At the present time there are three major reporting and forecasting systems, relating to weather, icebergs and sea ice respectively. The first two have a long history, the third is a relatively recent addition. With the advance of technology, existing systems can be greatly refined and new ones developed.

During the working group sessions the adequacy of present systems and the needs for further development were carefully examined. In discussion a set of internal and external issues were identified.

Internal Issues

The internal issues were recognized as being gaps and therefore as requiring immediate attention. These issues can be separated into two groups as (a) relating to the refinement of present systems and (b) relating to the development of new systems.

A. System Refinement
   1. Improvement of co-ordination and expansion of environmental information services (both national and international) including hydrographic information.

   2. Definition of the major functions of national and international observational and remote sensing systems, and general improvement of the systems.

   3. Development of archiving and retrieval systems for remote sensing systems, and general improvement of the systems.

   4. Improvement of positioning and rectification of remote sensing imagery.

B. New Systems
   1. Establishment of an advanced high-technology total environmental prediction system to provide quick response to changing conditions.

   2. Development of techniques for long-range prediction, taking into account season and secular changes (e.g. freeze-up).

   3. Development and application of seismic prediction techniques applicable to the Arctic.

External Issues

The external issues can also be divided into two groups, (a) relating to data gathering, and (b) relating to the better transmittal and use of the final product of the systems.

A. Data Gathering
   1. Information on wanted output, required standards (e.g. level of services), and accuracy in the operation and development of systems. The present interaction between user and supplier is recognized as a gap.

   2. Provisional observed input data to the operational forecasting systems. Deployment of cheap automatic sensors is required.

   3. Use of local knowledge, experience, and "observation" to contribute to local environmental data bases.

B. Product Use
   1. Establishment of systems for the transmission and reception of environmental information and prediction products in real time (including under-water reception).

   2. Establishment of user interpretation systems.
      The relations of these issues with the various other Arctic systems working groups is illustrated in the attached matrix (Table) which includes a listing of specific information required under item A.2. Since some of the issues relate to data collection, the following levels of data acquisition have to be recognized:
      (a) Scientifically gathered information in pre-planned programs by trained observers.
      (b) Itinerant data gathered from vessels, aircraft, and fixed stations of opportunity, not generally observed by trained personnel but obtained incidental to their primary job, e.g. sea surface temperatures and local weather reported from commercial ships.
      (c) Locally gathered data, mainly in the form of single point information, generally from remote areas and due to the accidental presence of the observer, e.g. iceberg sighting by commercial aircraft, seismic activity in remote places, unusual animal migration, etc.

Table I

Arctic Operational Information – External Issues Matrix

Arctic Systems External Issues	Environmental Protection	Traffic Management, Navigation and Communication	Transportation	Underwater Operational	Social and Cultural
A-1	X	X	X	X	X
A-2	X (current, wind, wave, ice, water level, tides, and surface temp.)	X (wind, iceberg, current, ice, wave, and electromagnetic propagation)	X (deployment of cheap automatic sensors)	X (ice depth, current)	X (occurrence of unusual environmental events)
A-3	-	-	-	-	X
B-1	-	X	-	-	-
B-2	X	X	X	X	X

ARCTIC UNDERWATER OPERATIONAL SYSTEMS

The committee prepared a flow chart (Figure 1) to demonstrate how it considered the problems in underwater operational systems. In order to scope the considerations of other panels the committee defined its area of operations, and following that, the why of being there. The two functions which would be carried out at these locations would be (1) observation, which is essentially the things which are included with gathering information; and (2) intervention, which is doing something in the medium or on the bottom.

The group then addressed the "how" of accomplishing these functions. There now exists a platform technology which provides essential support system. Underwater Systems must then be selected from manned vehicles, remotely piloted vehicles, and pre-programmed robots. The final questions deal with the integration problems that arise vis a vis other systems, both as Arctic Underwater Operational Systems affect them and vice versa.

Areas of Operation

Underwater operations will be required in five basic areas:
1. Marginal sea-ice zone - The zone 30 miles in from the extreme outer limits of the pack ice, or the annual ice extending out 100 miles from the ice edge.

2. Arctic atmospheric conditions with open water - Typically that area where Gulf Stream waters pass through northern latitudes, e.g. the Norwegian Sea.

3. Seasonally ice covered areas - Areas that are ice covered for reasonably predictable periods of time, for example parts of Hudson Bay.

4. Areas of permanent ice cover - Northern shore in the Arctic Ocean; the Canadian archipelago.

5. Areas affected by transient ice masses - examples are the coast of Labrador and Greenland.

Water depths considered were limited to the Continental Shelf. The goal of establishing year round operational capability in the various regions leads to basic problems as to which systems and combinations of systems can be used, based on the support-interface variations.

The unique conditions associated with the noted operational areas are logistic support, ice cover constraints on conventional techniques, intemperate surface temperatures and conditions, and sub-zero water temperatures.

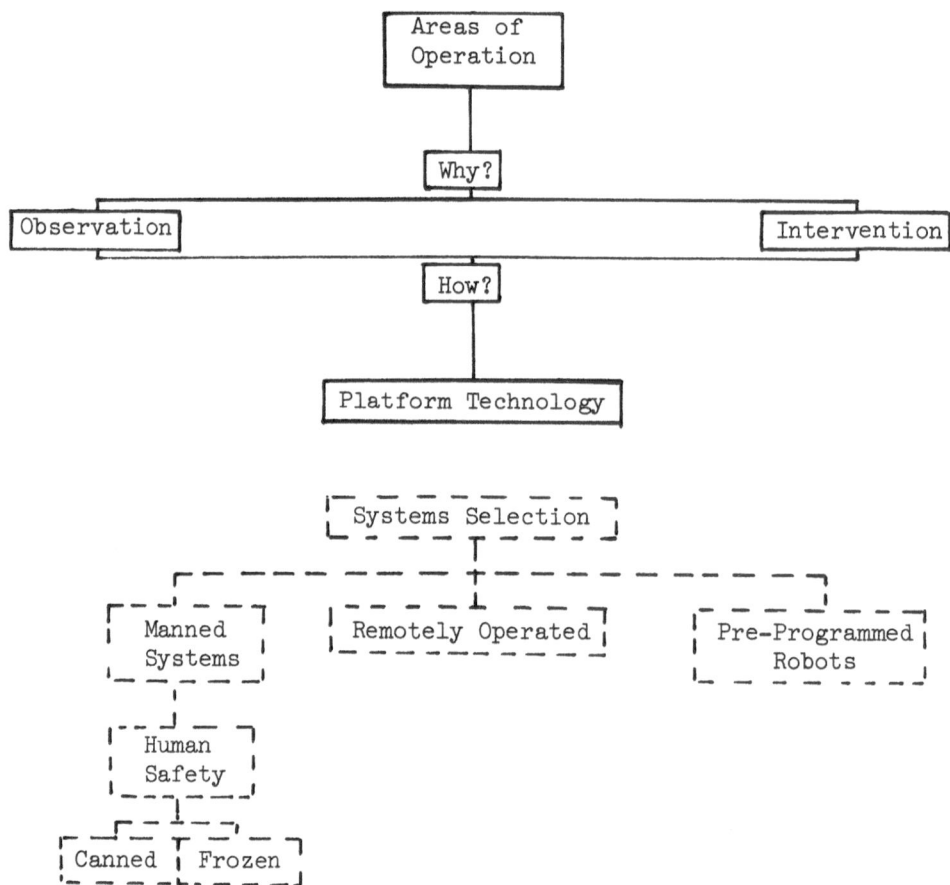

Figure 1   Arctic Underwater Operational Systems
             Problem Identification Flow Chart .

Functions (Why)

The two very general categories of functions carried out are observation and intervention.

(A) <u>Observation</u> - Observation activity includes such broad topics as inspection and data acquisition (bathymetry, hydrography, satellite ground truthing). The observational mode is generally one that provides basic background information which may lead to direct "intervention".

Both human and instrument sensors can be used to carry out these activities. The human can use both his visual and his auditory capabilities, but a far more viable set of sensors are those which combine human and "artificial" ones. Artificial sensors allow for the acquisition of data such as water column properties, acoustic information, bathymetry, current, hydrography and sub-bottom properties.

Underwater operations act as both a user and a supplier for environmental data. Technology exists to provide the information required, but this may only be on a highly localized basis. Existing broad spectrum data acquisition techniques are not always applicable to the Arctic.

(B) <u>Intervention</u> - Intervention activity includes operations, installations, repair, servicing, rescue, recovery, salvage, and manipulation.

The sub-surface technology exists to accomplish the majority of anticipated intervention tasks. Present and near-future activity is associated with resource exploration which imposes the demand of high mobility and year round capability. Existing technology gaps are in the areas of manipulator integration and local power supplies.

Platform Technology (How)

Platform technology is defined as including fixed or mobile systems for the underwater delivery and deployment of on-site observation and intervention systems. Various types of underwater systems are available to meet specific job, area, and cost requirements. Basic problems (e.g. physiology) exist in all areas, but are not necessarily peculiar to the Arctic. However, the interface conditions which affect such factors as endurance, range, mobility, energy requirements, and logistics impose critical restrictions on the freedom of choice of alternate systems.

The ability to operate in all areas and all seasons may require the use of a multiplicity of systems to accomplish a single task (e.g. manned, remotely operated and pre-programmed robot systems).

Internal Issues

A. Observation Mode.
    1. Lack of year round data on underwater environmental
        parameters.
    2. Deployment techniques are governed by horizontal mobility
        restrictions associated with the air-water interface
        conditions.
    3. There is a lack of designer-through-user co-operation and
        feedback in the area of hardware.
    4. Because of dissemination and communications gaps, there
        is a lack of accessibility to existing data and historical
        information.
    5. There is no mechanism for "piggy-back" information
        acquisition to ensure that users feed the system as well
        as draw upon it.
    6. There is a lack of real-time interpretation from monitoring
        systems over the entire season to assess the function of
        equipment during periods of inaccessibility.

B. Intervention Mode
    1. The major gap is the lack of horizontal mobility associated
        with the above-mentioned arctic constraints and the physical
        nature of underwater operational systems.
    2. Gaps are seen in the designer-through-user compatability
        and related positive feed-back (e.g. manipulator - well
        head, etc.)
    3. Timing and lead-time constraints are viewed as critical.
        For example, these constraints will seriously affect
        pollution control. A gap is seen in the ability to monitor
        in real time the integrity of pipelines, etc. and more
        seriously, in the ability to mobilize the required undersea
        repair system during winter periods of inaccessibility.

C. Platform Technology
    1. Horizontal mobility -
        a. Main logistics and transport of the chosen system to
            the general base of operations.
        b. Transport from the general base to the specific
            operational site with a high degree of repeatability
            and accuracy.
    2. Difficulties arise due to the relative motion of the ice
        surface with reference to the bottom, during the period of
        on site operation.
    3. Local energy technology gaps are considered major
        especially in areas such as:
            (i)   Base station requirements;
            (ii) Submarine and submersible requirements;
            (iii) Undersea stations;
            (iv) Instrumentation demands; and
            (v)   Diver requirements.

D. Areas for Future Consideration

The committee made the following observations:

1. Serious consideration must be given to the complete elimination of surface constraints and restrictions through the eventual use of systems such as sea-floor habitation and mobile submarine operations.

2. Local energy requirements may be met by consideration and development of thermal energy methods and other departures from the conventional.

3. Because of the unique difficulty of search and rescue operations in the arctic, co-operation and co-ordination in the development of international capability is recognized as an important function in the orderly exploitation of arctic resources.

4. Operational technology interchange should be encouraged through seminars, workshops, and technical meetings devoted to specific arctic undersea problems.

External Issues

The group had several issues placed before them in the form of questions raised by the other groups. Two general questions dealt with underwater physical and chemical capability, and underwater inspection and repair capability. As well, specific questions raised by the other groups were as follows:

1. Traffic Management, Navigation and Communication Systems group raised two points -
   (a) What is the nature of navigation requirements with regard to area coverage, accuracy, and transportability?
   (b) What are the communications requirements?

2. Transportation Systems group posed seven questions:
   (a) What is the physical environment of the underwater?
   (b) What technical developments have been achieved and what should be done?
   (c) What is the economic comparison of an underwater system versus a surface system?
   (d) What about two phase operation?
   (e) What is the impact of the other five areas?
   (f) What about navigation systems?
   (g) What about underwater iceberg profiles?

3. Operational Information and Forecast Systems group enquired as to the required standards of accuracy.

4. Social and Cultural Systems group posed four questions:
   (a) What are employment opportunities?
   (b) What are the educational and training skills required

for quality employment?
(c) What efforts could be made to provide information to
     residents regarding long-term employment and training?
(d) What plans are there to make possible more decision-
     making by residents?

The group felt there would be little or no difficulty in
providing answers to these questions, although a number of them
were too vague and non-specific to be dealt with in the time
available at the conference. Many of the questions indicated a
basic lack of knowledge and/or comprehension of the state of the
art and capabilities of underwater systems on the part of the
other groups. The working group recognized the need to correct
this communications gap.

ARCTIC ENVIRONMENTAL PROTECTION SYSTEMS

Introduction

Members of the group represented widely different education
and backgrounds and this led to considerable discussion on
environmental concepts which in turn impaired and delayed agreement
on internal and external issues.

The group felt uncomfortable with the expression, "Protection
of Arctic Environmental Systems". A better description would be
"proper use and management of Arctic Systems".

The group discovered from the internal work and from the input
of other groups that environmental protection is perceived as
something that could be but is not necessarily a major issue in
northern development. There was consensus that, while intelligent
development can take place, undesirable effects must be predicted
and minimized.

Throughout the world it is becoming increasingly apparent that
man must come to terms with the environment if civilization is to
continue to survive as we know it today. It is a misconception
that the arctic environment responds to the impact of man's
activity in much the same fashion as do other world environments.
Arctic ecosystems have unique characteristics which make them
different from systems of more temperate regions and this
necessitates a new pattern of behavior by technological man if he
is to survive as a viable entity into the future.

The expansion of human activity into the north has already
focussed attention on numerous conservation problems that were
previously unknown.

Arctic systems are characterized by a very slow rate of change.
They cycle very slowly in the cool brief summers and most life
remains dormant during the long winters. For example, this is
reflected in the longlasting effects that have been observed when
tracked vehicles have been used indiscriminately. Low productivity
is another characteristic of these ecosystems and this extends from
the vegetation to animals occupying the highest levels of the food
chain. Because arctic systems are made up of few species, the
components of the systems are very inter-dependent. This implies
that should anything happen to one species it is likely that other
species will be influenced.

State of the Art

The state of the art for environmental science could be
briefly summarized as follows:
1. We have a pretty good picture of the terrestrial macro fauna
    in the arctic but poor knowledge of the micro fauna on land
    and in freshwater.

2. In the marine arctic waters we have an incomplete picture of
    the macro fauna and we know almost nothing of marine
    invertebrates and micro fauna.

3. There is a fairly good knowledge on soil types and stability
    but movement of water in permafrost is poorly known. Only in
    the coal mines in Svalbard do we know something of this.

4. We have generally very little knowledge of short term and long
    term behavioral responses in animals exposed to the various
    kinds of human activities.

5. We have a fairly good appreciation of climate and the behavior
    of ice-cover, but limited knowledge of the marine physical and
    chemical environment.

                            Internal Issues

      The group did not agree on many issues. The following did
receive agreement as our major internal issues and these items
include virtually all the issues identified to us by other groups:
1.   Identification of the tolerance and sensitivity of arctic
environmental systems. This can be achieved by:
      a. Organized, standardized environmental information
          (physical and biological).
          This requires: general surveys and identification of
          key areas and critical periods. The vulnerability of
          arctic systems has been recognized by politicians in that
          all polar countries have established national parks and
          nature reserves. This brings us down to key areas which
          include for example, denning, feeding and spawning areas.
          The parks certainly include key areas but it is a far cry
          from what would represent sufficient protection. As a
          consequence, mapping of such areas is being undertaken as
          exemplified through IUCN polar bear feeding and denning
          areas, and DOE/DINA mapping of key wildlife areas in
          Canada (e.g. Cummingham Inlet Beluga Whales).

      b. Management of Information
          Knowledge of arctic ecosystems is international and
          circumpolar. There is an important need for an
          international environmental management and protection
          system data bank and a translation and distribution system.

      c. Interpretation and understanding of environmental systems.
          Data has no use unless it is put into the system in a
          system analysis approach which provides for valid data
          interpretation leading to an understanding of environmental
          processes.

2. Transfer and application of knowledge to environmental design
and decisions on management and development.
    There is an outstanding need for an interface between those
who provide environmental data and the developers.
3. Monitoring of the effects on the arctic environment of
operations and activities for recorrections and input into item 1.
    The above scheme is general and applies to management every-
where. What makes it different in the arctic is due to:
(1) the severe nature of the climate; (2) short period of the year
when it is convenient to obtain biological data in spite of the
fact that much data is needed year round; (3) wide fluctuations
in annual situations requiring investigations lasting several
years; and (4) this all leads to difficult and high cost
investigations.
    Those who came to this conference in the belief that this
group should provide the environmental answers to human activity
in the arctic could be frustrated at this time. There is no reason
for that. The fairly poor state of the art should be quite clear
from the way environmental issues in the north are perceived.
Input into environmental research is, generally speaking, not at
all sufficient to draw valid conclusions. We find the same lack
of environmental concern throughout the arctic.

                          External Issues

    The identification of these external issues is based on the
consensus that the current state of the art in identifying the
tolerance limits and relative sensitivities in Arctic environments
is primitive.
1. Traffic Management, Navigation and Communication
    a. Systems must be navigationally and technically safe. The
        navigation system must be capable of informing operators
        of hazardous areas, their relative location and movement.

    b. There is a need for a communication system which will give
        the operator continuous update of information.

2. Transportation
    a. Transportation modes must be designed to reduce the
        magnitude of accidents.

    b. Transportation modes should present minimum disturbance to
        the ecosystems.

    c. Transportation modes should have on board environmental
        emergency control equipment where applicable.

    d. Transportation modes should provide waste control and/or
        treatment systems.

    e. Support emergency control equipment should be stockpiled
        at key locations along transportation routes.

3. Operational Information and Forecasting
    a. These activities should provide input information to
        operational groups which, in turn, can cause environmental
        damage.

    b. The system should provide direct information on weather
        events and hazards related to management of wildlife.

    c. An information system is needed to assist in mobilizing
        various personnel, materials, and equipment in emergency
        situations. This system should be capable of providing
        information on alternative actions.

    d. An environmental data bank is needed for baseline information
        storage and retrieval.

4. Underwater
    a. Emphasis should be placed on development of technology for
        monitoring and accident detection under ice. Clean-up and
        repair for underwater operations are also required.

5. Social and Cultural
    a. Input on the nature, extent and use of natural resources
        by northern residents is needed as an integral component
        of the proposed environmental data bank.

    b. The development of communications and resource support
        systems for northern residents is needed to assist in
        environmental management activities.

    c. A mechanism for training is needed to enable northern
        residents to participate in acquisition of environmental
        data. This includes training for operation of all types
        of related equipment.

    d. Consideration of conflicts between use of renewable
        resources and non-renewable resources should be an
        integral part of the decision-making process.

ARCTIC SOCIAL AND CULTURAL SYSTEMS

Introduction

The group shared the concern of its only native northern
member that issues and problems so vital to the native peoples and
residents of the Arctic were being discussed without benefit of
their participation and a more direct representation of their views
and interests. It felt that future conferences on the Arctic
should ensure a more balanced representation between north and
south.

It should be noted, nevertheless, that the inclusion of this
social and cultural workshop in such a heavily technically-oriented
conference was to be warmly welcomed as the group felt it was able
to provide some guidance and input to the technical discussions.

Internal Issues

A framework has been outlined within which to examine the
issues. This can be summarized under seven principal headings,
selected from many, of which the first and the last - "Pace" and
"Power" - are in the group's view, the most significant.

1. Pace
The speed with which change is introduced and development
undertaken is of course a critical factor in shaping the nature and
scale of the problems which result. Social systems can usually
adapt more readily to gradual change, as the shorter the time
horizon, the greater the stress that is created. Natural resource
and infrastructure development projects are becoming even larger
with a consequent sharp increase in the complexity of the problems
that must be solved, some of which are inevitably unique. These
tend to fall into two main groups: (1) the short-term problems
associated with implementation; and (2) those which will be
encountered over a longer term during the operational period.

Continuous meaningful consultation and planning can produce
acceptable solutions given sufficient quality of effort and the
allocation of enough time and resources far enough ahead of the
decision to proceed, and subsequently when it is underway. This
ideal approach is all too often frustrated in practice by a
variety of restraints, including delays in the development of
policy due to local, regional, national and international opposition
to a particular proposal or pressures brought to bear to modify it.
A current example is the discussion of land claims settlement in
Canada.

The result is normally the one which follows any sudden
release of accumulated pressure - a surge which may overwhelm any
provisions made for handling the undertaking in an orderly manner
such as to cushion its impact on the local environment and social

system. It is therefore an important issue, in the group's
opinion, that the goals and priorities of development in the
Arctic be continually assessed with a view to correcting the
widely-accepted assumption that resource extraction takes precedence
over people.

2. Existing Systems

The social, legislative, judicial, regulatory and infrastructure
systems largely determine the manner in which new development is
introduced. These systems are evolving in the Canadian north from
systems suited to the needs of the past to systems which, it is
hoped, may be better able to handle the radically changed situation.
The systems of Alaska, Greenland and northern Scandinavia have
similar problems even though the attitudes of the governments
concerned vary widely, their land rights and planning systems are
dissimilar, and the native peoples and residents of their Arctic
regions differ in numbers and composition. For example, Denmark
until the early 1950's controlled strictly the access of outsiders
to Greenland in order to protect the native peoples, while Alaska
has been open for a century.

There are, therefore, relatively few generalizations which can
be made about the approaches to problem solving in the Arctic as a
whole. The systems peculiar to each country or region will affect
the way in which a problem is approached. But it must be emphasized
that there is an increasing community of interest amont Arctic
peoples in preserving their culture and social structure. They are
resisting the imposition of outside values and priorities and are
beginning to insist on a wider measure of participation in the
assessment of projects and in the decision-making process prior to
implementation.

These pressures are already leading to modifications of
existing systems and the changing distribution of power may result
in their further evolution.

3. Planning and Assessment

The planning process, if it is to produce acceptable results
in both the short and the long term, must be supported by adequate
data and it must involve the local population. The quality of
planning is a factor of the time available and of the resources in
both human and financial terms devoted to it. Given sufficient
resources though, it may still fail if the institutional and
organizational structures that exist to direct, support, control
and implement it are not equal to the task.

A basic criterion which must be established at the outset of
any planning operation, whether a project is in the private or in
the public sector, is the rate of return expected or required from
a proposed investment.

This economic yardstick will affect decisions on its component
parts as well as on the project as a whole. Project proponents,
using such a criterion, may argue that some facility which would

benefit local residents is not justified in economic terms.
However, in the context of regional or of national policy
objectives, rather than those of the project alone, the provision
of such a facility may well be justified. Procedures and methods
which are more sensitive to this problem need to be developed.

4. Technology
    The interface between social systems and technology is
important because of the problems it creates as well as those it
solves. The second part of this report presents interactions which
took place during the course of the conference between this group
and the other five working groups. The group felt that, while
current technological trends are probably adequate to meet future
northern needs, they are also capable of creating serious future
problems. When developing technologies for use in the North, the
needs of northern peoples should be of a higher priority than they
are at present.
    'Intermediate technology' could be usefully applied to solving
practical, everyday problems of the northern resident and in
bringing isolated people and communities closer together.
    There was also concern that more effective steps should be
taken to prepare and educate newcomers to the north in the social
and cultural systems they would be encountering.

5. Cultural and Social Values
    These are highly subjective and are expressed in the varying
definitions of what makes the north a place in which a person wants
to live. Some people are strongly in favour of new development,
while others resist it just as strongly. Dr. Kleinfeld[1] illustrated
these conflicts vividly in the context of Alaska and underlined
the effect which they had on the way in which the various existing
systems reacted to change. This study also demonstrated how
conflicts in priorities could result in technology apparently
failing to serve human needs. The underlying cause was often seen
to be the separation between those who benefited from the end
product and those who paid for it. It is essential therefore that
the various groups of people involved should maintain a dialogue.
Human values change and it is important that these changes be
recognized and understood.

6. New Approach to Arctic Development
    A social and cultural system that provides choice and which
enhances the dignity of the individual will be more enduring and
more creatively productive than one that does not. The group felt

---

[1]Kleinfeld, Judith. Economic, Demographic and Sociocultural
Effects of Development in Alaska.

that more needed to be done to open up employment opportunities
for northern residents in order for them in turn to make quality
contributions to northern development and service activities,
particularly those which were of a long-term self-sustaining
nature. This would involve more innovation and imagination than
had been shown to date in this field.

The group also expressed keen interest in the concept of a dual
approach to development in northern regions whereby certain types
of operations might be designed to be highly automated and to be
serviced from a distance. Such operations could be kept separate
and isolated from local communities if this was the preference of
the local residents. They could, however, if they wished choose
to participate where their own skills were relevant or if they
were prepared to acquire new ones. For the choice to be a real
one though, it would be mandatory for the local people to have a
high degree of involvement in the planning and development of the
project.

It was felt that this approach needed to be studied further
and worked out in detail in the context of specific projects,
particularly non-renewable extractive resource development.

7. Power
In the final analysis, what happened will be determined by
those with the power to make decisions or to influence them; which
means those with political or financial muscle and, of course, the
government machine itself. In the future, indigenous peoples must
be included among those who possess political power because they
are, even now, in a position to withhold their cooperation.

The group felt that the existing power structure resulted, all
too often, in inadequate solutions being adopted in the Arctic
even when decisions were arrived at in a responsible manner and
with the best of intentions. This may happen through failure to
include the northern viewpoint due to lack of effective input.
When industry and government pursue the same objectives together,
they make a formidable combination which cannot easily be opposed
by those who might seek to modify their approach or priorities.

Within the group there was positive support for the
aspirations of native peoples for the evolution of forms of
government in which they would exercise a much greater level of
control over cultural and educational matters and over the
development and use of natural resources in their regions. It was
felt that an increase in local authority would, over a period of
time, reduce the social and environmental cost of development and
help to create in the north a social and cultural environment
which would enhance the opportunities for future economic and
social development. The group also suggested that urgent
consideration be given to some arrangement which would facilitate
the carrying out of a review of development proposals independent
of the parties involved, taking into account the views of the
indigenous peoples together with those of industry and government.

It was felt that this would help to induce a higher quality of decision-making in respect to Arctic development.

External Issues

The pressure of the conference programme and timetable permitted only superficial communication, expressed initially in the form of broad, general questions addressed to the other groups. Further exchanges led to a number of specific suggestions being made for other groups to consider. It did not prove possible to pursue either the questions or the suggestions in any detail but they did serve to identify the interface between each group and to influence the content and direction of discussion.

The Social and Cultural Systems Group first posed four questions to the other groups:

1. What employment opportunities are presented by normal and new activities for northern residents?
   (By employment, we refer to quality participation in northern operations. By quality, we mean long-term, self-sustaining activities which utilize and develop local expertise and are adjusted to local needs.)

2. What specific education and training skills are required to ensure that those areas of quality employment now requiring southern staffing are opened to northern residents?

3. What efforts could be made to provide information to northern residents regarding long-term employment and training opportunities?

4. What procedures are, or could be, underway to direct more participation and control of decision making by northern residents?

The group subsequently offered the following further points for consideration by the other groups:

A. Environmental Protection
   1. What can be done to establish much more honest and candid communication processes so that native people will be informed about potential resource development projects when they are still in conceptual stages?

   2. What can be done to effectively include consideration of the needs and rights of native people in the early stages of decision making processes?

   3. What can be done to assist native people to understand the implication of technical reports and development projects

to their interests?

4. What can be done to utilize the unique kind of expert
    knowledge that native people have of the land and of the
    animals which they utilize?

5. How can the local people best be involved in controlling
    and protecting the areas in which they live?

6. Scientists and technologists who are involved with aspects
    of environmental protection have a unique responsibility
    to represent the interests of native society in these areas
    and to ensure that native peoples fully comprehend the
    decisions they are being asked to approve.

B. Traffic Management, Navigation and Communications
    1. What more could be done, and how, to provide east-west
        telecommunications across the North and to link communities
        in the same region with each other as well as with the
        South?

    2. What needs to be done, and by whom, to raise the level of
        operational and maintenance training for local people
        running regional radio communications systems in the North
        so as to improve their effectiveness?

    3. In what ways might radio and TV equipment and techniques
        be economically applied to improving communications within
        communities and regions in the North as well as between
        South and North?

    4. Could satellite coverage be extended to Greenland on an
        economic basis?

    5. How can the far North gain access to communications via
        satellite?

    6. Could a common distress calling frequency for the North
        be allocated and monitored?

    7. Are there navigational aids and techniques that might be
        applied to assist local travellers within the North?

C. Arctic Transportation
    1. Are there practical steps which might be taken to design
        or modify transport equipment to achieve more reliable
        operation in the North and to make it simple to maintain
        using available facilities and skill?

    2. How can an east-west transport system be developed

economically to meet the needs of northern peoples?

3. What can be done to improve local intercommunity transportation in the Arctic?

4. Social and Cultural goals need to be clearly stated so that a framework may be provided for assessing transportation requirements in the north. This would serve to identify characteristics of transport systems which are important in meeting the needs of northern peoples. Procedures and mechanics with some bite to them are required to ensure that the cost equation is not always assessed in micro-economic terms.

D. Operational Information and Forecasting
1. In what way might weather forecasting and reporting be made more responsive to the needs of local peoples and localities?

2. What are the standards, the levels, and the accuracy required for such information and how can this best be arrived at in consultation with the users?

3. Could a greater volume of observed data be usefully provided by northern residents to the forecasting system, including information on wildlife?

4. Could not more be done to provide training and employment to northern residents in such roles as meteorological technicians?

E. Underwater Operational Systems
1. What will this technology do to help ensure an acceptable life for the children and grandchildren of today's northern peoples?

2. Can new scientific knowledge gained from underwater studies allow us to: (a) better protect and manage biological systems that have their photosynthetic base in sea water; and (b) thus indirectly help to identify distributions and migrations of marine fishes and animals?

3. Are there technological components developed for underwater operational systems (e.g. small nuclear power plants) that could eventually find application in northern communities?

CONCLUSION

The questions and points listed above are representative of
those that were stimulated by the very preliminary form of
interaction and discussion that took place between the Social and
Cultural Systems Group and the other working groups. They
provide, nevertheless, food for thought and the basis for further
discussion and study.

SUMMARY OF CONFERENCE CONCLUSIONS

P. A. Lapp

Philip A. Lapp Ltd.

Toronto, Ontario, Canada

The principal accomplishments of a workshop seminar as
comprehensive as the Arctic Systems Conference can be divided into
two categories:
- the personal benefits derived by those participating
  in the conference, and

- the written proceedings which provide a permanent'
  record of what was said, and which summarize the
  collective conclusions of specialist groups immersed
  for the better part of a week to focus on very
  specific arctic-related problems.

Because those present at the conference are among the top and most
influential in their fields, by far the largest impact in the short
term will result from the personal experiences of the participants.
Each took away with him new perspectives and ideas. The
international character of such NATO Conferences assures a thorough
mixing of backgrounds and knowledge, and undoubtedly new personal
relationships were developed which, in the larger term, will help
to illuminate from a variety of directions the intensely complex
systems problems facing those responsible for the earth's arctic
regions.

Such complexity is reflected in the subject matter and inter-
relationships appearing in the reports of the working groups. The
title of each group was chosen to reflect the issues and concerns
resulting from increasing pressures to develop non-renewable
energy resources in the arctic. Thus the technologies associated
with transportation, underwater operations, navigation,
communications and environmental forecasting were selected as
central and critical to agencies involved in arctic development.

They all are highly interactive and interdependent. But all
developments now and in future must generate an acceptable
environmental impact - a requirement that carries special
significance in the arctic with its very delicate and critically-
balanced ecosystems. For this reason, special attention was
devoted to arctic environmental protection systems and their
relationships to the central technologies selected. Finally, the
conference planners recognized that affected social and cultural
systems should and would have an over-riding influence on whether
and how the arctic should be developed. A working group on such
systems focussed on issues facing northern residents as pressures
mount for energy-related development in the arctic.

Throughout their deliberations the six working groups did
identify knowledge gaps and potential directions for future
research and development. The following paragraphs summarize their
major findings and lead to some general conclusions that can be
drawn from the conference as a whole.

Arctic Transportation Systems

Transportation modes include land and ice (pipeline, rail,
track and ACV), sea (surface ship, semisubmersible and submarine)
and air (fixed and rotary wing aircraft). Commodities to be
transported are bulk fluids and solids, supplies and people.
The major knowledge gaps were identified as:
(a) Terrain Stability - for pipelines, rail, truck and airstrips.
(b) Ice characteristics - for the design of ships, semisubmersibles
     and marine terminals.
Research and development work to aid in a better understanding of
the two areas would result in safer and more economical arctic
transportation. Also identified was the need for better hydrographic data and for improved navigation, communications and
environmental forecasting which are covered by other groups.

Arctic Traffic Management, Navigation and Communication Systems

Such systems provide an essential service in the north for
resource discovery and development, transportation and
administration as well as a means to unite isolated communities to
each other and the larger centres in the south. While there are
significant gaps that need to be filled in communications and
navigation services already in place, the major knowledge gaps
were claimed to be:
(a) The effects of permafrost and glacial ice on ground wave
     propagation.
(b) Hydrographic details of the bottom in the major passages
     through the archipelago.

There is a critical need for low-cost, reliable and stable power supplies for communications and navigation devices of all kinds and sizes that will operate in the arctic environment. This requirement arose throughout the conference on many occasions, and it can be cited as a major conclusion that research and development on power sources for the arctic is a matter of urgency and high priority.

The rapidity with which communications facilities, particularly satellite systems, are being installed in the arctic, and the development of new navigation systems, such as the U.S. Global Positioning Satellite, makes it difficult to identify gaps and needs at the time of the conference that are valid for any distance into the future. However, it was observed that while communications facilities were being developed with a north-south orientation, there was a social and cultural desire as well to create east-west patterns within the arctic.

A need was identified to coordinate the large number of regulatory and information functions being performed in the north, and the group recommended the study of an Arctic Systems Joint Operations Centre. Such a multi-agency centre would coordinate the dissemination of weather, oceanographic, terrain and search and rescue information, perform regulatory services in connection with arctic land use, ocean pollution, fisheries and wildlife legislation, and operate traffic management systems associated with vessels terrain vehicles and aircraft.

Arctic Operational Information and Forecast Systems

The major environmental forecasting services required in the arctic relate to weather, sea ice and icebergs. The major knowledge gaps appeared to be:
(a) The physics of the arctic atmosphere because of
     inadequate density of measurement stations,
(b) The physical characteristics of sea ice, particularly
     the build up of pressure fields and the dynamics of pack ice.

The major needs focus on the requirement for more data through extension of weather and ice observation services. The more extensive and operational use of remote sensing is seen as one way to acquire more environmental data efficiently. Satellites, aircraft, ships and automatic stations such as data buoys are suitable platforms for the observation system. Research and development should be focussed on more efficient sensors, high-rate data handling and interpretational methods. Sophisticated forecasting systems resulting from such technology should provide for local knowledge inputs which often can be of great value in long-range forecasts particularly. Dissemination of forecast data could be through the Joint Operations Centre referred to above.

Arctic Underwater Operational Systems

Underwater operations in the Arctic entail two functions – observation such as inspection and data acquisition, and intervention which includes repair servicing, rescue, salvage, etc. Both functions can involve manned vehicles, remotely piloted vehicles and pre-programmed robots. In situations where man is included, life support systems are required; and while considerable knowledge has been gained in temperate waters, operational experience in cold waters and under the ice has been relatively limited. The major knowledge gaps are:
(a) The performance of man in cold water, and the lack of year around data on underwater environmental parameters,
(b) The factors that affect horizontal mobility including ice motion and environmental constraints.

The group identified a need to create an information system for underwater technology including an environmental data bank, and a convenient means of accessing and providing input to the system in order to foster designer – through-user cooperation and feedback. Again a requirement was stated for suitable power sources to meet base station as well as underwater vehicle and diver needs, perhaps through the use of non-conventional thermal energy methods. Research and development should be employed to close existing gaps in manipulator device technology to perform intervention functions, and in the integration of such devices with existing underwater vehicles.

Arctic Environmental Protection Systems

The proper use and management of arctic systems requires a deeper and more detailed knowledge of certain aspects of arctic biology than presently exists. The group identified the following knowledge gaps:
(a) Microfauna on land and fresh water,
(b) Micro and macrofauna in marine arctic waters,
(c) Movement of water in permafrost,
(d) Response of animals to human activities.

There is a need to organize and manage an arctic environmental data base, and provide interpretational services. Information transfer techniques must be established to keep the developers informed on environmental matters. Base line studies and environmental monitoring during development were emphasized as high priority directions for research and development.

Development-related activities such as transportation systems must be introduced with minimal and acceptable impact on the ecosystem. Research is needed in how to specify such requirements in physically meaningful and realizable terms. Again, some form of environmental data bank and information system are essential.

Social and Cultural Systems

In general, it was averred that more native people should be invited to participate in such conferences; and that northern residents should be adequately informed so that they can be involved in major decisions related to the arctic. The knowledge gaps expressed by the group revealed the vacuum within which major development decisions are being taken affecting the social and cultural fabric of the north. They include:

(a) mechanisms whereby northern residents can participate in decision processes involving land use and development.

(b) The pace at which development can occur without the destruction of existing social, cultural and value systems already present in the arctic.

(c) The means to develop and express social and cultural goals and priorities in the north.

Research and development related to "intermediate" technology would appear to be most appropriate for arctic communities. There is an urgent need to use imaginative means of delivering educational and training programs to northern residents, and to inform them of employment opportunities. Research is needed in how best to employ human resources in the arctic for environmental monitoring functions, weather stations, ice observation, etc.

It was concluded that northern residents are in a position to wield great power through withholding their cooperation, and that every effort must be made to weave them into the decision-making fabric at all levels if adequate solutions ever are to be found for the problems facing communities in the Arctic.

General Conclusions

Throughout the working group sessions there appeared to be a general concern about the need to organize, in some systematic fashion, arctic knowledge in a way that is readily accessible. While Arctic bibliographies provide such a facility for scholarly works on the Arctic in the adjudicated literature, there appears to have been little in the way of practical, up-to-date information available in any organized form. Most groups expressed an urgent need also for an arctic data bank to store historical and real-time environmental base line, weather and ice data. Such a storage system, with the necessary means of data retrieval, is entirely compatible with and could form part of, the proposed Arctic Systems Joint Operations Centres concept. The wise placement of such Centres could meet requirements of several of the arctic systems discussed and studied at the Conference. The inter-jurisdictional problems arising mainly from various government departments competing for resources and power, could be a major deterrent to the establishment of such Centres. But once

established, they could become coupling mechanisms to the
northern residents which is believed to be an important element in
successful and peaceful northern affairs.

Unquestionably the Conference was a success on all counts.
Already, a year later, some of the ideas generated either have
been implemented, or have been the subject of detailed study. The
impact will be felt for some time to come. Such results have been
derived mainly through the personal experiences of the participants.
It is hoped that the permanent record will be as valuable an
influence on the future evolution of the Arctic.

CONFERENCE DELIBERATION

Introduction

The NATO-sponsored Conference on Arctic Systems was held at
Memorial University of Newfoundland during August 18-22, 1975. In
all, 115 experts from eleven countries, both NATO and non-NATO
countries, took part. The background of the experts who attended
the conference ranged over social sciences, natural sciences,
biological sciences, political sciences, life sciences, behavioral
sciences and many branches of engineering.

The inaugural speaker at the conference was Dr. J. Rennie
Whitehead, Canadian Member NATO Science Committee, who expressed
in his address "the need for International Cooperation in the
Arctic". Dr. Whitehead's address was very well received by those
present.

The opening plenary session was held on Monday, August 18
and was chaired by Professor Donald Clough, Member of the Special
Program Panel on Systems Science. Six keynote speakers (2-Canada,
1-Germany, 2-USA, 1-Denmark) addressed the gathering, each speaking
broadly on one of the six Arctic Systems that were under
consideration.

Working sessions were held on Tuesday, Wednesday and Thursday
(August 19-21) during which twenty-nine working papers were
presented. Time was devoted to the discussion of the working papers
as well as to the identification of major issues internal to the
specific system under consideration and external issues linking it
with the five other systems.

The format used in the conference gave every person the
opportunity to participate in the group discussions, to expound his
views and to express concern with regard to the many multi-variable
and inter-related problems faced in the Arctic. In the following
section the six systems identified for the purposes of this meeting
are presented.

ARCTIC SYSTEMS UNDER CONSIDERATION

While the list was undoubtedly altered somewhat as new

suggestions were made, the following types of Arctic Systems were
proposed for detailed examination at the conference:

Arctic Environment Protection Systems

- arctic ecology
- animal and marine life protection
- oilspill prevention and clean-up systems resulting from
    accidents during drilling, from completed installations
    and pipelines, and from surface and undersea tankers
- all forms of pollution prevention, embodying detection,
    communications, command and control, and clean-up
    methodologies
- methods for prevention including standards
- system management

Arctic Traffic Management, Navigation and Communication Systems

- all forms of transportation
- mobile field survey and scientific teams
- all support service operations
- resource management
- surveillance and remote sensing activities
- navigation systems for the arctic regions
- communication systems

Arctic Transportation Systems

- alternative methods of transporting resources (oil, gas
    and minerals) to market, including ships, submarines,
    aircraft, pipelines and other types of vehicles
- transportation support systems
- ice-breaking systems, ice dynamics and engineering properties
- deployment techniques and related support systems
- system management

Arctic Operational Information and Forecasting Systems

- arctic weather forecasting, use of satellites
- ice forecasting (condition and movement), use of SLAR,
    microwave and other cloud and darkness penetrating
    systems
- organization of management systems

Arctic Underwater Operational Systems

- man in cold water
- life support systems
- underwater navigation and sensing systems
- submersibles, remote manipulator systems
- system management

Arctic Social and Cultural Systems

- social and cultural problems associated with native and
    non-native peoples operating in the north
- native peoples and arctic systems, cultural preservation
    and economic development
- housing, education and health care systems
- management of policy.

IDENTIFICATION OF MAJOR ISSUES AND KNOWLEDGE GAPS

Each of the six working groups were given the challenge to
identify the significant gaps in knowledge associated with the
system under consideration. The knowledge of those in attendance
and the materials prepared in the papers presented provided content
for the debates which ultimately resulted in the preparation of a
final report presented to the conference at large.
Through a process of communication with each of the other five
groups and in response to issues identified by other groups, each
working group in turn was required to identify the issues of
importance linking the system under consideration with each of the
other five systems. These linkages were also refined and presented
in the final report.

PLENARY AND WORKING SESSION FORMAT

The conference was unclassified and divided into six working
groups of 15-20 people, each group assigned to discuss a specific
integrated Arctic System. For each group a Chairman and a
Rapporteur were appointed. Each group also appointed five liaison
persons with responsibility to coordinate discussion and communicate
between their own group and each of the other five groups.
The progress reporting plenary sessions were held on Tuesday
and Thursday afternoons (Wednesday plenary was cancelled and
replaced by a working session) for the purpose of reporting progress
and for general briefing. During these sessions, each group
presented the framework of the issues internal to their group and

the development of external issues, linking them with the other
groups. Conclusions drawn and the identification of knowledge gaps
in the state of the art related to the Arctic Systems under
consideration were presented.

WORKSHOP SESSION FORMAT (REVISED)

Tuesday, August 19th, 1975

Workshop Session I

    Introduction of Participants   (1/2 hr.)

    Liaison Appointments

    Presentations and Discussion   (1 hr.)

    Framework of Internal Issues   (1 1/2 hr.)

Plenary I (Progress Reports)

    Chairman Presentation of Framework
    - 6 Reports @ 10 minutes plus
       discussion   (1 1/2 hr.)

Workshop Session I

    Development of External Issues   (1 hr.)

Special Session

    Meeting of Chairmen and Rapporteurs
      with the Directors of the
      Conference   (1 hr.)

Wednesday, August 20th, 1975

Workshop Session 2

    Organization and Work Plan   (1/2 hr.)

    Presentations and Discussion   (1 hr.)

    Refinement of External Issues   (1 1/2 hr.)

Liaison persons coordinating
discussion and communicating
between their own group and each
of the other five groups   (1 1/2 hr.)

Workshop Session 2 (Conclusion)

Discussion - Major Gaps and
External Issues   (1 hr.)

Thursday, August 21st, 1975

Workshop Session 3

Presentations and Discussion   (1 hr.)

Discussion - Major Gaps and
External Issues   (1/2 hr.)

Continued Discussion and Plan of
Final Group Presentations   (1 1/2 hr.)

Plenary 3 (Concluding Reports)

Chairmen Presentation of Final
Report   (1 1/2 hr.)

    - Internal: Identification
              of Gaps

    - External Issues Linking
              with Other Systems

Concluding Session   (1 hr.)

GRAPH OF GROUP INTERCONNECTIONS FOR EXTERNAL ISSUES

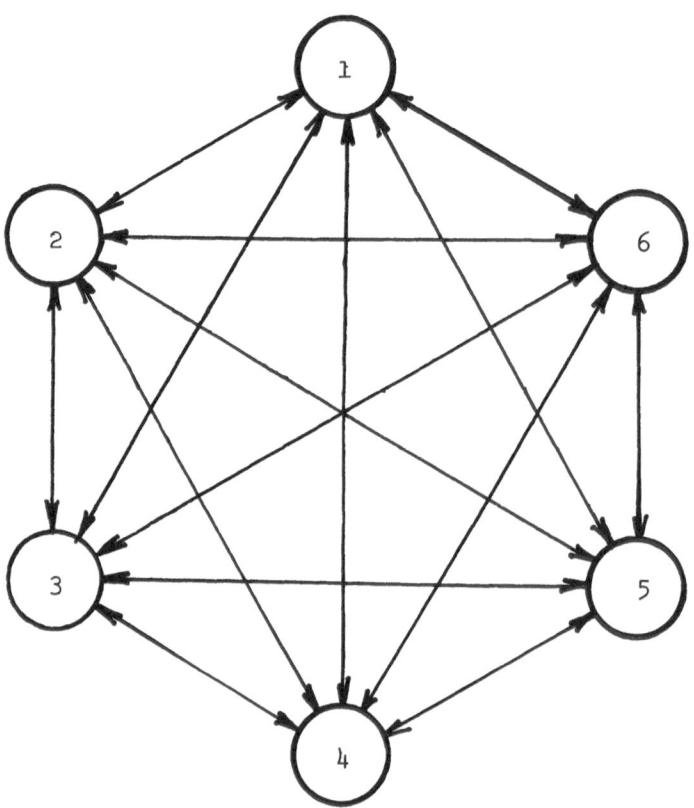

<u>6 Groups. 6 Chairmen</u>

30 one-way channels (issues)

15 two-way channels (matching issues)

- assign one liaison from each group to each other group
  total of 30 liaison people, or 15 pairs of people to
  communicate at any time during the symposium.

- each group will have 5 input channels from other groups.

PURPOSE OF USING THIS FORMAT

The objective of this conference was to attempt to deal with completely integrated arctic systems, involving large numbers of interactive elements. A finite number of such systems were considered for examination using modern systems analysis and design techniques, such that the inter-relationships within and between each of the arctic systems were systematically identified and examined in depth.

This format allowed for interaction among the six groups and created a forum of experts from various disciplines who could approach the internal and external issues of major concern in a more unified and coherent way.

APPOINTMENT OF PLENARY AND WORKING SESSION CHAIRMEN

The following were the chairmen appointed:

Plenary Sessions

Professor D. Clough
Canada

Dr. J. Keys
Canada

Working Sessions

Environment Protection

Mr. Eigil Reimers
Norway

Traffic Management, Navigation and Communication

Brigadier General K. R. Greenaway
Canada

Transportation

Captain R. M. White
U.S.A.

Operational Information

Dr. H. Schroeder-Lanz
Germany

Underwater Operational

    Dr. R. E. Francois
    U.S.A.

Social and Cultural

    Mr. A. E. Pallister
    Canada

    Dr. N. Orvik
    Canada

BRIEFING OF WORKING SESSION CHAIRMEN, RAPPORTEURS AND PARTICIPANTS

    During the Monday Opening Plenary Session, all the
participants were informed about the proposed format of the
conference. A briefing session was held with the Chairmen and
Rapporteurs on Monday evening, at which details of the conference
format and its purpose were discussed. The Chairmen in turn
explained the format to their groups during the working sessions
on Tuesday morning and adapted it to serve the particular needs of
the individual groups.

PERIODICAL DE-BRIEFING AND MEETING WITH CHAIRMEN AND RAPPORTEURS

    The Directors and plenary session Chairmen informally met with
working session Chairmen and Rapporteurs during the first day of
the working session to resolve any difficulties arising in following
the proposed format. A de-briefing meeting of Chairmen and
Rapporteurs was held on Tuesday evening at which it was decided to
cancel the Wednesday plenary session (1 1/2 hr.), replacing it
with working sessions at which the five liaison experts from each
group presented the major external issues of concern in the arctic
system with which they interacted.
    This mid-course alteration of the conference format enhanced
the degree of interaction among the six arctic systems that were
considered and maintained a high productive use of the available
experts within the imposed time constraint.

CONFERENCE ATTENDANCE

    One of the criteria established during the early stages of
the conference planning was to invite a select composite and
coherent multi-discipline group of experts with wide ranging
experience and background. This selection of the group was further
constrained by a requirement for balanced representation from NATO

and non-NATO countries and a desire to keep the number in each group small enough to facilitate active discussion by all members of each working group.

PROCEDURE FOLLOWED IN INVITING PARTICIPANTS

The procedure recommended in the NATO bulletin for selecting delegates through "national points of contact" in the NATO countries was followed. These contacts were asked to recommend experts who could make a significant contribution at the meeting. Formal letters of invitation were sent to suggested individuals explaining the objectives of the conference and the proposed format to be followed.

When inviting experts from non-NATO countries, certain protocal was followed. A number of individuals were invited from USSR, China, Argentina, the German Democratic Republic, Sweden, Poland, Finland and Japan. Individuals from the latter four countries accepted our invitation.

Invitations were sent to a number of native Greenlanders and Inuit residents living in the North. However, because of some of their own pressing problems, no Canadian Inuit could attend the meeting.

SUMMARY OF INVITATIONS AND ATTENDANCE

Name of Country	Total Invited	Invitation Declined	Invitation Accepted	Observers	Total Attended
NATO Countries					
Canada	74	15	59	27	86
U.S.A.	19	8	11	0	11
Netherlands	4	4	0	0	0
France	3	2	1	0	1
West Germany	4	2	2	0	2
U.K.	7	5	2	0	2
Denmark	12	5	7	0	7
Iceland	4	4	0	0	0
Norway	6	4	2	0	2
	133	49	84	27	111

Name of Country	Total Invited	Invitation Declined	Invitation Accepted	Observers	Total Attended
Non-NATO Countries					
Japan	2	2	0	0	0
U.S.S.R.	8	8	0	0	0
Finland	9	7	1	0	1
Sweden	3	2	1	0	1
Poland	6	4	2	0	2
Argentina	1	1	0	0	0
China	1	1	0	0	0
East Germany	2	2	0	0	0
	32	27	4	0	4
TOTAL	165	76	88	27	115

FIELD TRIPS

Churchill Falls Power Development Project - Labrador

The construction of this power development was started some twelve years ago, to harness the hydro potential of the upper Churchill River. It was built at an estimated cost of $960 million and designed with an installed capacity of 5,000 mega watts of electric power.

The power station is built underground and the cavity structure in considered to be one of the biggest in the world.

Sponsors

1.   Churchill Falls (Labrador) Corporation Limited
     Churchill Falls
     Newfoundland
     Canada
     [CF(L)Co]

2.   Iron Ore Co. of Canada
     Labrador City
     Newfoundland
     Canada

3.   Memorial University of Newfoundland
     St. John's
     Newfoundland
     Canada

Field Trip Attendance. The trip to Churchill Falls was
attended by twenty-five people. Delegates left St. John's for
Churchill Falls on August 22, and proceeded to Labrador City and
on to Montreal on August 23, 1975.

A seminar and site trip were arranged by the Management of
CF(L)Co. At Labrador City a tour of mining and mill operations
was provided by the Iron Ore Co. of Canada.

Offshore Petroleum Drilling Rig (Nfld.)

A field trip was arranged for the participants to visit
"SEDCO-J", an offshore Petroleum Drilling Rig operated by Mobil Oil
Ltd., Canada. SEDCO-J was situated about 120 nautical miles east
of St. John's, Newfoundland.

Sponsors

1. Mobil Oil Ltd. (Canada)
    Calgary
    Alberta
    Canada

2. Memorial University of Newfoundland
    St. John's
    Newfoundland
    Canada.

Special Seminar.   A two-hour seminar was arranged by Mobil
Oil Ltd. (Canada), at Memorial University, to give participants an
overview of the Company's offshore petroleum drilling activities.
This talk was very interesting and informative, and gave an idea
of the enormous undertaking which is required for any offshore
petroleum exploration.

Field Trip Attendance. Two helicopter trips were arranged,
morning and afternoon, by Mobil Oil Ltd. (Canada) to transport
twenty-three delegates from St. John's Airport to the offshore
drilling rig and return.

All expenses incurred on this field trip were borne by Mobil
Oil Ltd. (Canada).

RECORDING OF PROCEEDINGS

Plenary and Concluding Sessions

A video and audio recording of the plenary and concluding
sessions was made by the Memorial University Educational Television
Department.

Working Sessions

Audio recordings of all the six working group sessions were
made by Eastern Audio Ltd., St. John's, Newfoundland.

SOCIAL PROGRAM

Banquet

A banquet was held in the Fort William Room, Hotel Newfoundland,
from 1930 hr to 2230 hr on Wednesday, 20 August 1975.

Sponsor. The Ministry of State for Science and Technology,
Government of Canada, sponsored the banquet.

Guest Speaker. The guest speaker was the Honourable Mr.
T. Alex Hickman, Minister of Justice of the Province of Newfoundland
and Labrador, who was introduced by Dr. M. O. Morgan, President of
Memorial University of Newfoundland.

Head Table Guests.   Guests at the Head Table included
Dr. & Mrs. Alain Frecker (The Chancellor of the University and his
                          wife)
Dr. & Mrs. M. O. Morgan (The President of the University and his
                          wife)
Mr. & Mrs. J. H. Moore   (The U.S. Consul General in St. John's
                          and his wife)
Dr. & Mrs. P. Schmidt-   (The German Consul General in the Montreal
           Schlegel   Consulate and his wife)
Mrs. G. K. Sann   (Wife of the German Consul General in
                          St. John's)

Others at the Head Table included Professor and Mrs. P.J. Amaria,
Professor D. Clough, Dr. & Mrs. P.A. Lapp, Dr. John Keys, Dr. & Mrs.
T. Alex Hickman, Dr. & Mrs. A. A. Bruneau. Dr. Bruneau chaired the
banquet.

Attendance. In all, 135 persons attended the banquet.

LIMITATIONS OF THE FORMAT, CONCLUSION AND RECOMMENDATION

We have observed that in order to achieve the goals of this Conference which attempted to examine six very broad Arctic Systems, a number of criteria are essential in the design, development and conduct of the Conference format:

1. Selection of a composite and coherent multi-discipline group of experts with wide ranging experience and theoretical knowledge.

2. Clear and concise guidance to chairmen, rapporteurs and participants on the conference format to be followed, charging each chairman with specific goals and objectives to be achieved.

3. A formal schedule that allows sufficient time for interaction among working groups and individuals.

4. The establishment of a communication system involving participants from each group and allowing interaction with other groups. The mode of communication should be mainly heuristic.

5. The creation of an atmosphere before and during the conference that gives every individual the feeling that he is invited to participate in and contribute to the meeting and that the collective discussion of the group leads to the achievement of the conference objectives.

6. Logistics and physical planning should be such that an atmosphere is established for individual and group participation compatible with the conference format.

The format followed at this Conference successfully created a forum for the discussion of the pressing internal and external issues and identified knowledge gaps in each of the Arctic Systems under consideration. Productive use was made of the experts who attended the Conference within the time constraints imposed, though extra time would have been appreciated by most participants.

We recommend that such a format will be very useful for future conferences and symposia where the discussion of pressing issues involving design, formulation, analysis and synthesis of policy alternatives is of major concern.

THE CONFERENCE AS AN EXERCISE IN COMMUNICATION:

A CRITIQUE

J. G. Dawson

Memorial University of Newfoundland

St. John's, Newfoundland, Canada

General comment, appreciation and criticism of the Conference
which was solicited from the participants has already appeared in
a previous section of these Proceedings. The following remarks
are offered, at the request of the Organizing Committee, from the
point of view of an interested observer who was given the privilege
of access to all aspects of the operation and proceedings. I
should like here to express my appreciation for this most valuable
and informative experience. My remarks must necessarily be brief
and personal. Obviously a full critical study would have required
a further conference to formulate it!

Conferences are extremely expensive exercises in communication.
This is true not merely in the obvious and well recognized sense
of economic costs of travel, organization, and facilities, but
more importantly in the less obvious sense of the human time
involved. The specific characteristic of a conference, in contrast
to written or oral communication at a distance, is that it creates
for a brief and unique time a special human interface. Its success,
as a conference, must therefore be measured primarily by the
success with which this unique interface is exploited. For this
reason the mode of communication of a conference should be primarily
heuristic rather than formal. To put it another way; the success
of a conference is measured by the level and range of new questions
that the participants go away with, by enlarged horizons, rather
than by the stock of information and formal conclusions acquired.
For the latter can be acquired by standard and less costly
processes of public communication. We all know, for instance, that
nothing is more wasteful of conference time or more trying to the
patience of those attending than spending an hour listening to the
reading of a paper which by its nature could and should have been
read and digested by the participants a month previously.

The characteristically heuristic mode of conference exchanges
must be developed formally as well as informally if the conference
is to succeed and the human interface is to be fully exploited.
The former will depend very largely upon the appropriateness of
the structural procedures imposed by the organizers. The latter
is greatly influenced by the physical characteristics of the
facilities provided.

With these criteria in mind it was my impression that the
Organizing Committee achieved a very high level of success in
structuring the procedures but that the physical facilities left
much to be desired.

In the first place the procedural format adopted reduced to a
minimum the repetition of material already available, or which
should have been available, in written or schematic format. This
of course puts a heavy burden upon the office facilities of the
host institution for the massive and rapid production of typescript
copy. I think this aspect was well and efficiently handled. On
the other hand the Conference lent itself and would have benefitted
greatly from a central display of large scale maps, charts and
data presentation which would have acted as a centre of focus for
informal exchanges and an overall briefing device. Similarly the
visual presentations at general meetings were less than adequate.
We are all aware of the waste of time and frustration generated by
sitting in the dark listening to a commentary on a complex visual
presentation which can be read comfortably from the back of the
auditorium only with the aid of a pair of night glasses. Visual
materials must be in the hands of the technicians responsible in
time for them to prepare adequate transparencies and they must
operate the machines.

Secondly; the working procedures adopted within sections and
the insistence upon cross-representation between sections, the
evening de-briefing sessions leading to the general morning
presentations created and maintained a level of corporate awareness
of the problems under investigation which would never have been
achieved under less exacting conditions. Without such procedures
the Conference could easily have degenerated into at least three
or even six separate conferences which happened to coincide in
time and place, and were ineffectually linked by general sessions
which would vaguely air matters of common concern without their
being subject to any systematic scrutiny.

It is true that this imparted a hectic atmosphere to the
entire Conference which was subject to general and justified
comment. But in terms of the heuristic criterion, I think that
this atmosphere should be regarded as evidence of the success of
the formal structuring: especially since it was undoubtedly
compounded by deficiencies in the informal situation.

Thirdly the structuring was successful in maintaining a
constant interface at every level between two sorts of participants;
those whose expertise was primarily of a scientific and

technological nature, and those whose concern was for the value issues raised by ecological balance and human habitat. Without the discipline imposed by the structuring I am sure that the formal and technological aspects of the Conference would have over-shadowed the essential heuristic exchanges and detracted heavily from the unique value of the occasion. Here perhaps I might mention one matter which occasioned considerable comment, namely the absence of any formal representation from those people whose home and traditions we were discussing. The comment is understandable but on balance I think it was fortunate for the Conference and that the decision not to send formal delegates from native peoples organizations was wise. Their presence would have undoubtedly introduced a further dimension to the two modes of communication already present, namely, the adversary mode. The intrusion of this third mode into the existing structure and scope of the Conference would, I believe, have resulted only in confusion, frustration and a general lowering of the heuristic achievement.

Finally, a word about the physical facilities. Ideally a conference should always take place around one large central concourse to which participants would naturally gravitate whenever they are not occupied in formal meetings. It should be plentifully supplied with seating which can be arranged easily for informal groupings and maximum use should be made of visual displays to convey information and generate discussion. Administrative offices and seminar rooms for formal meetings should be adjacent. In short it should attract and generate the maximum amount of informal exchanges. Unfortunately such a concourse area was not available and in consequence participants tended to disperse and disappear when they became tired of standing in the relatively small area available.

ORGANIZING COMMITTEE (Scientific)

General Conference Chairmen
    Professor Donald Clough
    Professor of Management Science
    University of Waterloo
    Waterloo, Ontario, Canada

    Dr. John Keys
    Vice-President
    National Research Council of Canada
    Ottawa, Ontario, Canada

Directors
    Dr. Philip A. Lapp
    Philip A. Lapp Limited
    14A Hazelton Avenue
    Toronto, Ontario, Canada

    Dr. Angus A. Bruneau
    Vice-President
    Memorial University of Newfoundland
    St. John's, Newfoundland, Canada

Administrative Director
    Dr. Pesi J. Amaria
    Associate Professor of Engineering
    Memorial University of Newfoundland
    St. John's, Newfoundland, Canada

LIST OF DELEGATES
(Name and Affiliation)

Arctic Environment Protection Systems

Dr. John Allen
Manager
Oceans & Oceans Engineering Division
c/o James F. MacLaren Ltd.
435 McNicoll Avenue
Willodale M2H 2N8
Canada

Dr. D. S. Braden (Author)
B.P. Alaska
Anchorage
Alaska
U.S.A.

Dr. A. E. Collin (Keynote Speaker)
Assistant Deputy Minister
Ocean & Aquatic Affairs
Environment Canada
Fontaine Building
Hull
Canada

Dr. H. T. Doane (Author)
Environment Protection Service
Environment Canada
Box 2406
Halifax
Nova Scotia
Canada

Mr. L. M. Etchegary (Author)
Polar Gas Project
P.O. Box 90
Commerce Court W
Toronto M5L 1H3
Canada

Lt. J. H. Getman (Author)
c/o Rear Admiral A.B.E. Siemens
Chief Office of Research and Development
U.S. Coast Guard Headquarters
Washington
U.S.A.

Mr. W. Hindle (Author)
Vice-President
Polar Gas Project
P.O. Box 90
Commerce Court W
Toronto M5L 1H3
Canada

Mr. J. Hnatiuk (Author)
Industry Project
Manager
Gulf Oil
P.O. Box 130
Calgary
Canada

Dr. D. E. Kerfoot
Arctic Land Use Research Programme
Northern Natural Resources and
    Environmental Branch
Ottawa
Canada

Dr. John Keys (General Chairman)
Vice-President
National Research Council of Canada
Ottawa
Ontario
Canada

Mr. R. Lamoureux (Author)
Polar Gas Project
P.O. Box 90
Commerce Court W
Toronto M5L 1H3
Canada

Mr. Hans Lassen (Author)
Fishery Biologist
Danish (FISHERY)
Institute for Fishery & Marine Research
Charlottenlund Slot
DK-2900
Charlottenlund
Denmark

Mr. J. M. McBride
Arctic Co-ordinator
Atmospheric Environment Service
4905 Dufferin Street
Downsview
Ontario
Canada

Dr. G. R. Peters (Rapporteur)
Faculty of Engineering and Applied Science
Memorial University of Newfoundland
St. John's
Newfoundland
Canada

Mr. Eigil Reimers (Chairman)
Environmental Consultant
Ministry of Environment
Myntgt 2, Oslo-Dep
Oslo 1
Norway

Dr. E. F. Roots
Chief Science Advisor
Environment Canada
Ottawa
Ontario
Canada

Dr. Dan Smith (Author)
Water Pollution Control Directorate
Environment Protection Service
Ottawa
Ontario
Canada

Dr. Peter Wadhams
Scott Polar Research Institute
Cambridge
England CB2 1ER

Dr. J. Rennie Whitehead
  (Inaugural Speaker)
Advisor, Ministry of State for Science
  and Technology
Ottawa
Ontario
Canada

Mr. A. Williamson
Department of Regional Economic Expansion
Ottawa
Ontario
Canada

Arctic Traffic Management, Navigation, and Communication Systems

Professor Donald Clough
  (General Conference Chairman)
Department of Management Science
University of Waterloo
Waterloo
Ontario
Canada

Dr. D. Dunsiger (Rapporteur)
Faculty of Engineering and Applied Science
Memorial University of Newfoundland
St. John's
Newfoundland
Canada

Dr. Michael Eaton
Department of Environment
Bedford Institute of Oceanography
Halifax
Nova Scotia
Canada

Dr. M. Galipeau
Canadian Marconi
2442 Trenton Avenue
Montreal
Quebec
Canada

Brigadier General K. Greenaway (Chairman)
Department of Northern and Indian Affairs
Ottawa
Ontario
Canada

Mr. Collin Langford
Newfoundland Oceans Research and
  Development Corporation
P. O. Box 5244
St. John's
Newfoundland
Canada

Dr. Philip A. Lapp (Director)
Philip A. Lapp Ltd.
14A Hazelton Avenue
Toronto
Ontario
Canada

Commander W. B. Mohin (Author)
U.S. Coast Guard Headquarters
400 - 7th Street West
Washington, D.C. 20590
U.S.A.

Dr. L. W. Morley (Author)
Director
Canadian Centre for Remote Sensing
2464 Sheffield Road
Ottawa
Ontario
Canada

Mr. A. Simpson
Chief, Northern Communications,
  Central Region
Department of Communications
2300 - One Lombard Place
Winnipeg
Manitoba
Canada

Dr. J. Taagholt
Ionosphere Laboratory
Technical University of Denmark
Lyngby
Denmark

Mr. M. Turner (Keynote Speaker)
Chief, Policy and Co-ordination
Marine Telecommunication and Electronics
  Branch
Ministry of Transport
Ottawa
Ontario
Canada

Captain Mel Walker (Author)
Canadian Forces Air Navigation School
CFB Winnipeg
West Wind
Manitoba R2R 0T0
Canada

Mr. D. Weese (Author)
Manager, Systems Engineering Group
Telesat-Canada
333 River Road
Ottawa
Ontario
Canada

Dr. J. G. Wright
President
J. G. W. Systems Ltd.
Suite 100, 56 Sparks Street
Ottawa
Ontario
Canada

Arctic Transportation Systems

Dr. P.J. Amaria (Administrative Director)
Faculty of Engineering and Applied Science
Memorial University of Newfoundland
St. John's
Newfoundland
Canada

Mr. P. Andersen
FENCO Ltd.
St. John's
Newfoundland
Canada

Dr. P. Bergen
Transportation Industries Branch
Department of Industry, Trade and Commerce
Ottawa
Ontario
Canada

Dr. A. A. Bruneau (Director)
Vice-President
Memorial University of Newfoundland
St. John's
Newfoundland
Canada

Mr. Dan Buch
Greenland Technical Organization
Hauser Plads 20
DK-1127 Copenhagen
Denmark

Mr. E. Buchholz (Author)
Operational Research
Research and Development
C.N. Railway
Montreal
Quebec
Canada

Dr. Fen-Dow Chu (Author)
State University of New York
Maritime College
Fort Schuyler
Bronx, Ny. 10465
U.S.A.

Mr. J. L. Courtney (Author)
Policy and Economic Advisor
Transport Canada
Place de Ville
Ottawa
Ontario
Canada

Mr. Gunnar Edelmann
Naval Architect
Board of Navigation
P.O. Box 158
SF-00141 Helsinki-14
Finland

Mr. Jan Furst
President, Newfoundland Oceans Research
  and Development Corporation
P. O. Box 5244
St. John's
Newfoundland
Canada

Mr. William German (Author)
German & Milne
Suite 401
1010 St. Catherine Street West
Montreal H3B 1G2
Canada

Mr. C. W. Harry
Department of the Navy
Naval Ship Research and Development Centre
Bethesda
Maryland 20084
U.S.A.

Mr. W. B. Hunter
Vice-President, Operations
Northern Transportation Co.
9945 - 108th Street
Edmonton
Alberta
Canada

Lieutenant Commander D. J. Hussey
Directorate Maritime Engineering and
  Maintenance
National Defence Headquarters
Ottawa
Ontario
Canada

Professor T. W. Kierans (Author)
Faculty of Engineering and Applied Science
Memorial University of Newfoundland
St. John's
Newfoundland
Canada

Dr. F. Legerer (Author)
Faculty of Engineering and Applied Science
Memorial University of Newfoundland
St. John's
Newfoundland
Canada

Mr. D. C. Murdey
Marine and Ship Dynamics Laboratory
National Research Council
Ottawa
Ontario
Canada

Mr. L. Nitzki (Keynote Speaker)
Director of Research
A. G. Weser
2800 Bremen-21
Werft str. 160
Germany

Professor H. Snyder (Rapporteur)
Director, C-CORE
Faculty of Engineering and Applied Science
Memorial University of Newfoundland
St. John's
Newfoundland
Canada

Captain R. White (Chairman)
Dean, U.S. Coast Guard Academy
New London
Connecticut
U.S.A.

Arctic Operational Information and
Forecasting Systems

Dr. D. Bajzak (Rapporteur)
Faculty of Engineering and Applied Science
Memorial University of Newfoundland
St. John's
Newfoundland
Canada

Miss Moira Dunbar (Author)
Defence Research Board
Ottawa
Ontario
Canada

Mr. Jens Fabricius
Danish Meteorological Institute
Lyngbyvej 100
DK-2100
Copenhagen
Denmark

Mr. W. Ganong (Keynote Speaker)
Atmospheric Environment Service
4905 Dufferin Street
Downsview
Ontario
Canada

Professor Preben Gudmandsen
Technical University of Denmark
Build. 348
DK-2800
Lyngby
Denmark

Commander R. C. Kollmeyer (Author)
U.S. Coast Guard Academy
New London
Connecticut
U.S.A.

Mr. F. B. Muller (Author)
Meteorological Services Research Branch
Atmospheric Environment Service
4905 Dufferin Street
Downsview
Ontario
Canada

Dr. James Robar
SED Systems Ltd.
P.O. Box 1464
Saskatoon
Saskatchewan S7K 3P7
Canada

Dr. Hellmut Schroeder-Lanz (Chairman)
Department of Geography
University of Trier
D55 Trier
Schreidershof
Germany (FR)

Mr. W. Wittmann (Author)
2310 Wilkinson Place
Alexandria
Virginia 22306
U.S.A.

Arctic Underwater Operational Systems

Dr. Bo Brannstrom
SAGA Petroleum A/S. et. Co.
Maries Vei 20
Hovik
Norway

Commander Bo Cassel (Author)
Marinstaben
Sektion 3/Dyk
S-100 Stockholm 100
Sweden

Mr. J. English (Rapporteur)
Faculty of Engineering and Applied Science
Memorial University of Newfoundland
St. John's
Newfoundland
Canada

Dr. Robert E. Francois (Chairman)
Applied Physics Laboratory
University of Washington
Seattle
Washington
U.S.A.

Dr. L. Kuehn (Author)
Defense and Civil Institute of
  Environmental Medicine
Bio Sciences Unit
P. O. Box 2000
Canadian Forces Base Downsview
Downsview
Ontario
Canada

Mr. M. Macdonald (Author)
International Hydrodynamics
145 Riverside Drive
North Vancouver
British Columbia
Canada

Mr. P. Nuytten (Author)
Can-Dive Services Ltd.
Unit 3 - 250 East Esplanade
North Vancouver
British Columbia
Canada

Commander J. P. B. O'Riordan (Author)
Directorate of Naval Warfare
Ministry of Defense
Room 438, Main Building
White Hall
London SW1
England

Mr. M. Ruelokke
Manager
Underwater Visual Systems
P.O. Box 35
Mount Pearl
Newfoundland
Canada

Arctic Social and Cultural Systems

Mr. L. Auerbach
Science Advisor
Science Council of Canada
650 Kent Street
Ottawa
Ontario K1P 5P4
Canada

Dr. J. C. Beal (Author)
Queen's University
Kingston
Ontario
Canada

Mr. Claus Bornemann (Keynote Speaker)
Head of the Greenland Counsel Secretariat
c/o Ministry for Greenland
Hausergade 3
DK-1128
Copenhagen
Denmark

Mr. D. Grenville (Rapporteur)
C-CORE
Faculty of Engineering and Applied Science
Memorial University of Newfoundland
St. John's
Newfoundland
Canada

Dr. Alfred Jahn
Director of Geographic University
Polish Academy of Sciences
Wroclaw
Poland

Dr. Judith S. Kleinfeld (Author)
Institute of Social, Economic and
  Governmental Research
University of Alaska
Fairbanks
Alaska 99701
U.S.A.

Mr. Finn Lynge (Author)
Head of Greenland Radio
DK-3900
Godthaab
Denmark

Professor John G. McConnell
c/o Dr. R.T. Armstrong
Scott Polar Research Institute
Cambridge CB2 1EN
England

Dr. D. H. Morrissette
Chief, Social Research Division
Territorial & Social Development Branch
Department of Indian Affairs and
  Northern Development
Ottawa
Ontario
Canada

Dr. Ludger Muller-Wille
D-53 Bonn-Bad Godesberg
Frankenstr 9
Germany

Dr. Nils Orvik (Chairman)
Director, Center for International
  Relations
Department of Political Studies
Queen's University
Kingston
Ontario
Canada

Mr. A. E. Pallister (Chairman)
Pallister Resource Management
3rd Floor, Pacific Plaza
700 6th Avenue S.W.
Calgary
Alberta T2P 0T8
Canada

Dr. E. Peterson
Co-ordinator for MAB Sub-Program IV
Arctic and Isolated Development Program
Western Ecological Services
211-11 Fairway Drive
Edmonton
Alberta
Canada

Dr. Douglas Pimlott (Author)
University of Toronto
Ramsey Wright Building
Toronto
Ontario
Canada

Dr. Paul-Emile Victor
Director des Expeditions
Polaires Francaises
47 Avenue du Marechal Fayolle
75016 Paris
France

Observers

Dr. F.A. Aldrich
Dean, Graduate Studies
Memorial University of Newfoundland
St. John's
Newfoundland
Canada

Dr. G. H. Farmer
Memorial University of Newfoundland
St. John's
Newfoundland
Canada

Mr. W. Graham
Mobil Oil Ltd.
Halifax
Nova Scotia
Canada

Mr. P. Hood
St. John's
Newfoundland
Canada

Mr. H. Jacobs
Faculty of Engineering and Applied Science
Memorial University of Newfoundland
St. John's
Newfoundland
Canada

Mr. G. Kirby
St. John's
Newfoundland
Canada

Mr. S. T. Matthews
St. John's
Newfoundland
Canada

Mr. R. McGrath
Mobil Oil Ltd.
Halifax
Nova Scotia
Canada

Dr. J. Green
Memorial University of Newfoundland
St. John's
Newfoundland
Canada

Mr. E. Mercer
Wildlife Division
Provincial Department of Tourism
St. John's
Newfoundland
Canada

Mr. R. R. MacIsaac
St. John's
Newfoundland
Canada

Dr. L. Misra
Memorial University of Newfoundland
St. John's
Newfoundland
Canada

Dr. S. Murthi
Memorial University of Newfoundland
St. John's
Newfoundland
Canada

Mr. J. Moir
St. John's
Newfoundland
Canada

Mr. T. Northcott
Wildlife Division
Provincial Department of Tourism
St. John's
Newfoundland
Canada

Captain A. Provan
College of Fisheries, Marine Engineering,
  Navigation and Electronics
St. John's
Newfoundland
Canada

Mr. L. Proudfoot
St. John's
Newfoundland
Canada

Mr. J. Saltman
St. John's
Newfoundland
Canada

Dr. D. Rendell
Memorial University of Newfoundland
St. John's
Newfoundland
Canada

Mr. N. Riggs
St. John's
Newfoundland
Canada

Mr. D. Reddin
St. John's
Newfoundland
Canada

Mr. W. Russell
St. John's
Newfoundland
Canada

Mr. A. Roughmann
St. John's
Newfoundland
Canada

Mr. W.J. Rayan
St. John's
Newfoundland
Canada

Mr. D. Rees
Memorial University of Newfoundland
St. John's
Newfoundland
Canada

Dr. P. Smith
Memorial University of Newfoundland
St. John's
Newfoundland
Canada

INDEX